D0914239

Bridge management 4

Inspection, maintenance, assessment and repair

Edited by M. J. Ryall, G. A. R. Parke and J. E. Harding:
Department of Civil Engineering, University of Surrey, UK

 Thomas Telford

Published by Thomas Telford Publishing, 1 Heron Quay, London E14 4JD.
URL: http://www.t-telford.co.uk

Distributors for Thomas Telford books are
USA: ASCE Press, 1801 Alexander Bell Drive, Reston, VA 20191-4400, USA
Japan: Maruzen Co. Ltd, Book Department, 3–10 Nihonbashi 2-chome, Chuo-ku, Tokyo 103
Australia: DA Books and Journals, 648 Whitehorse Road, Mitcham 3132, Victoria

First published 2000

Organizing Committee: Dr M. J. Ryall, Professor J. E. Harding and Dr G. A. R. Parke

Technical Committee: Dr C. Abdunur (France), P. H. Besum (Belgium), N. H. Bettigole (USA), Dr C. Birnstiel (USA), Professor O. Buyukozturk (USA), D. Collings (UK), P. C. Das (UK), Professor M. C. Forde (UK), Dr D. M. Frangopol (USA), Professor E. Gehri (Switzerland), Dr G. Hearn (USA), N. Hewson (UK), Professor M. H. Ingvarsson (Sweden), Dr P. Jackson (UK), Professor T. Kitada (Japan), P. Mehue (France), Professor U. Meier (Switzerland), Professor C. Melbourne (UK), N. Ricketts (UK), Professor H. Sundquist (Sweden), Dr R. J. Woodward (UK), Dr B. Yanev (USA)

A catalogue record for this book is available from the British Library

ISBN: 0 7277 2854 7

Printed and bound in Great Britain by Bookcraft, Bath

Preface

The importance of sound bridge management is now accepted as a *"sine qua non"* by the bridge engineering fraternity throughout the world. Because of the huge amount of effort that has been expended in finding solutions for saving ageing and deteriorating bridge stocks, a paradigm shift seems to have taken place so that bridge engineers now consider management with every aspect of design, construction, commission and use. This has resulted in a change from a curative to a preventative mentality with regards to bridge management, and hopefully means that not only will the next generation of bridges last longer and outperform previous ones, but also the whole-life costs will be reduced, resulting in a lesser amount of money having to be set aside for maintenance.

Since the last three Bridge Management conferences, a tremendous amount of research work has gone into ways of tackling every task associated with sound management. There is still a long way to go, however, as bridge engineers work overtime and often within very tight budgets to put right failing bridges and try to find ways of ensuring that the same problems do not occur in the future. In that respect there is a move to fit new bridges with monitoring devices so that any potentially damaging mechanisms can be detected by what are essentially early warning systems. This is especially relevant for long-span bridges where the consequences of failure are likely to be more disastrous than for small-span-bridges. Such moves take us towards the age of the 'smart' or 'intelligent' bridge somewhere in the future.

The aim of the Fourth International Conference on Bridge Management was to bring together all of the very best and, more importantly, up-to-date work that has been done in the field by leading academics and practitioners. The papers are intended to provide bridge engineers with practical and economical solutions to some very pressing problems. With this in mind, the conference has been designed to be as wide ranging as possible and it is hoped that both delegates and bridge engineers at large will be able to use these proceedings to find answers to their questions and solve some of their dilemmas. There are many very good case studies and the results of new research into old problems.

Since the first conference in 1990 the 'scientific art' of bridge management has advanced by an astonishing degree and it is now well and truly out of 'short trousers' and into 'longs'. In this volume there are a total of 92 papers, covering: advanced bridge management systems; monitoring; risk and reliability analysis; rehabilitation; the use of less well known materials such as galvanised steel, FRPs; whole-life costing; strengthening; repair and assessment.

The university of Surrey, which hosts this series of conferences, has sustained an interest in bridge management for many years, as evidenced by the MSc course specialising in the area which has attracted quality students from both overseas and the UK. The Department of Civil Engineering has also been active in the field, notably in methods of determining the in-situ or 'residual' stresses in prestressed concrete bridges; the buckling of steel-plated units related to steel boxes; the use of GFRP in new bridges; the application CFRP to strengthen bridges and, latterly, research in conjunction with the Highways Agency to develop a full-reliability procedure for evaluating the load-carrying capacity of existing bridges.

Our sincere thanks go to our sponsors, who have consistently encouraged and helped us through this series of conferences, and also to our technical committee. Without the authors of the material presented, of course, there would have been no conference and so our thanks go to them, and finally thanks also go to the support staff who provide the background administration to ensure the success of the event, and the delegates for providing such a stimulating and challenging atmosphere for the exchange of ideas and information during the conference sessions.

Dr M. J. Ryall
February 2000.

Contents

Reliability based bridge management procedures

PARAG C DAS, OBE, PhD, CEng, MICE
Project Director Bridge Management, Highways Agency

INTRODUCTION

"(The Highways Agency is)... to give priority to the maintenance of trunk roads and bridges with the broad objective of minimising whole life costs .... " - A new deal for transport, Government White Paper, 1998.

"Sustainable development means living off our income, not eroding our capital base, so that we are not storing up problems for future generations. Especially important is investment in public assets" - Sustainability counts, DETR Consultation Paper, 1999.

"To take action to reduce congestion and increase the reliability of journey times"- Highways Agency Business Plan 1999-2000

"(Outcome objective) A network that is maintained in a safe and serviceable condition" Highways Agency Business Plan 1999-2000.

The above DETR and Highways Agency statements clearly set out the government's broad objectives regarding value for money, sustainability, functional needs and safety, particularly in respect of the road network.

The Agency is responsible for the maintenance and operation of the trunk road network in England which includes 16000 structures most of which are bridges. Maintenance of the structures is a major part of its activities. As such, it is particularly important to interpret the government's broad objectives in the context of bridge maintenance and adapt the procedures to satisfy those objectives as well as provide means for monitoring how well these are being met.

In order to measure and maintain the reliability of the bridge-stock in respect of the multi-objective requirements, it is necessary to adopt a management process which is capable of addressing a number of objectives in an explicit manner. The Highways Agency is currently developing such a methodology ,based on a strategic R&D programme. A number of the new procedures are now ready for trial, which will start in the near future, and others are under development. The purpose of this paper is to describe in outline the significant aspects of these procedures. A number of these procedures have been more fully described elsewhere [1,2,3,4,5,6], hence only the others are dealt with in detail in this paper.

RELIABILITY

The reliability of a system or an element of a system means its likelihood or probability of satisfying a particular design or operational objective. A system may for instance be an industrial process, a service operation such as a train service, or an infrastructure, and an element is an individual component of such a process or system. A reliability based management process is that which is particularly centred on monitoring and improving the reliability of a system.

A reliability based approach can be illustrated with the example of a train service, the main objective of which is to carry passengers to their destinations according to the timetable. Its reliability in respect of this objective can be measured in terms of the percentage of trains that arrive within 5, 10 or 15 minutes of the intended arrival time. To determine if the reliability of the service is satisfactory, a target needs to be set, such as that 95 percent of the trains are to arrive within 5 minutes of the intended arrival time. A reliability based management approach will be to gear all the component activities of the operation in a co-ordinated manner to achieve this target.

Such numerical or explicit definition of reliability makes it possible, not only to measure the efficiency of a system, but also to improve it when necessary. Furthermore, when multiple objectives need to be satisfied, for instance the train service has to be safe as well as punctual, a reliability based approach is the only rational way to manage such a system.

RELIABILITY BASED BRIDGE MANAGEMENT
From the above it can be seen that a reliability based management requires certain basic ingredients. These are - the objectives, the performance indicators and the performance targets. The procedures for planning, execution and reporting need to be focussed on these fundamental ingredients. The objectives of bridge management, which have been discussed in greater detail elsewhere [7], can be summarised as follows :-

1. *Functional*. Traffic disruptions arising from bridge works should be kept to a minimum. This means that regular maintenance and preventative actions should be adequate so that repair needs are minimised. It also means that, options for repair or strengthening procedures should be chosen in terms of the lowest traffic disruption costs. Furthermore, in prioritising work, those that will cause the greatest traffic disruptions if not carried out, should be given the highest priority.
2. *Safety*. Bridges should be safe for their intended use. They must also be maintained adequately so that defects and deterioration are not allowed to get to a stage that may cause public alarm.
3. *Aesthetics*. An acceptable level of appearance must be maintained.
4. *Sustainability*. Bridge maintenance works should be programmed in such a way that future work loads remain at a manageable level and backlogs do not accumulate.
5. *Economic*. Maintenance should be based on whole life costing.

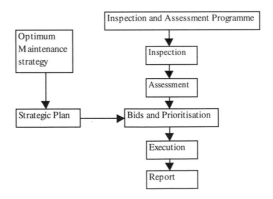

Figure 1. Component activities of a reliability based bridge management methodology

The overall bridge management procedure has a number of individual activities which are shown in Fig.1. These procedures, even those that are already widely used, such as inspection and assessment, need to be modified to suit a reliability based approach. The following are the main features of these modified or new procedures.

TARGETS AND PROGRAMME

Structural elements or other components of bridges are considered as sub-standard or unacceptable, if they are assessed to be inadequate in respect of the applied loading, or considered to be unsafe due to a critical defect or simply found to be in an unacceptable condition for other reasons. The first target, therefore, must be to repair, replace or strengthen all bridges with sub-standard parts as they are identified.

Unsafe structural elements can only be identified through inspections and assessments. The second target, therefore, is to have a comprehensive but feasible programme of inspections and assessments which will cover all the structures within stipulated periods, and to fully carry out the assessments and inspections according to programme.

In order to devise a whole life based sustainable bridge maintenance regime, it is necessary to plan a smooth level of work and expenditure for the foreseeable future. This requires the application of adequate preventative measures in good time. A long term strategic programme therefore must be developed with the optimum mix of preventative and essential work. The third target therefore will be to keep the annual preventative expenditure to the levels stipulated in the strategic plan.

The Highways Agency's current 15 year bridge rehabilitation programme, which started in 1988, is nearing its end. A new steady state successor programme, the outline of which is described below, is being gradually introduced as the supporting technical procedures become available:-

Inspections. In line with the current arrangements, all bridges will undergo a general inspection (GI) (visual inspection of all parts) every two years. In addition particular inspections (PI) of critical elements will be carried out at prescribed intervals.

Assessments. General assessments (GA) ('present state' assessments of all parts) of the whole bridge stock, will be carried out every 15 years at the rate of just over 6% per year. In addition, whole life assessments (WLA) will be carried out on elements that are either assessed to be sub-standard at the time or are likely to be sub-standard within the next 15 years. Maintenance bids will generally be based on WLA's. Particular assessments (PA) (repeat assessment of critical elements) will be carried out following each PI.

Maintenance Bids. The annual maintenance works bids will be based on the preceding yearly inspections and assessments.

Execution. All programmed maintenance work will be carried out without undue delay.

Reporting. Bridge stock condition reports containing performance indicators will be published annually.

INSPECTION

The Highways Agency is developing a new 6 part Bridge Inspection Manual, intended for trials in 2000 [2]. Apart from containing up-to-date guidance on bridge inspection, the manual is specifically aimed at reliability assessment. It is based on the 'segmental inspection' principle which requires the parts of elements important for assessment calculations to be specifically inspected. Furthermore, the time intervals of the PI's are also based on the importance of the segment for assessment purposes, determined either at the design stage or following adverse inspection findings. More critical or more deteriorated a segment is, more frequently it will be inspected.

ASSESSMENT
Present State Assessment

The currently used term 'bridge assessment' means the assessment of the structural adequacy of a bridge in its present state. Such assessments determine if a bridge element or component is sub-standard or not at the time of assessment.

In the United Kingdom, the current bridge assessment standard is BD 21 [8] with its associated standards and advice notes.

BD 21 provides the acceptable minimum safety levels in terms of load carrying capacity. However, other condition based limits in respect of appearance and public alarm, and for elements not amenable to structural analysis such as expansion joints and bearings, are also necessary. These are currently being developed.

Whole Life Assessment

The consideration of sustainability and whole life costing necessitates an assessment of the adequacy of the bridges into the future. Prediction of the future states also requires a backward estimation of their performance from new to the present time. For carrying out such long term analysis in a consistent manner, a formal procedure for whole life assessment is essential.

A new advice note entitled 'Whole life assessment of highway bridges and structures', intended for application by UK bridge authorities, is being drafted for publication in 2000. The overall procedure of whole life assessment in the advice note is schematically shown in Fig.2. It shows that the assessed current performance level is to be compared with the critical level (minimum acceptable level) appropriate for the element concerned to determine if the element is sub-standard now or is likely to be so in the foreseeable future.

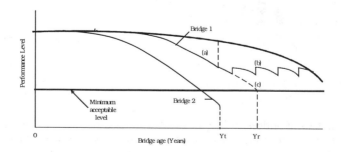

Figure 2. Options of maintenance actions

If the element is sub-standard now (at year t), the strengthening or replacement work, or any emergency measure, is to be considered as 'essential'. However, if the element is not sub-standard now, some maintenance work may still be justifiable. Such work should be considered as 'preventative'.

Once the present state performance is determined using the present state assessment mentioned above, the performance of the element is then projected into the future using a number of alternative maintenance strategies. For example Fig.2 shows that the safety (structural capacity) level of Bridge 2 at present is below the acceptable minimum level. Hence it must be strengthened/replaced, or otherwise has to be weight-restricted until the work is carried out. Bridge 1 is on the other hand at present structurally adequate, but offers a number of options. For instance, (a) it can be fully strengthened, (b) partially strengthened to maintain its current level of structural capacity, or (c) it can be left to be replaced in the near future.

For each strategy option, and for each maintenance action in the strategy, the year of the action, various costs including planning, supervision and traffic management costs and, in addition, the traffic delay costs, are submitted as bids. For instance in Fig.2, if strengthening is not carried out for Bridge 2, it will have to be weight-restricted which will result in traffic disruption. In the UK there is a standard method using a computer program QUADRO 3 [9] for calculating traffic delay costs at restrictions on the trunk roads. The whole life performance thus results in maintenance work options, including 'do minimum' options, which are entered into the bidding process described later.

OPTIMUM MAINTENANCE STRATEGIES

The necessity and options for the essential work required when a structure is assessed to be sub-standard are by and large obvious. On the other hand, when a structure is assessed to be adequate at present, preventative work is more difficult to justify since future predictions invariably contain considerable uncertainty. It is therefore necessary to develop standard recommendations regarding appropriate preventative methods for different types of structures based on optimum strategies determined using whole life costing. The following procedure which was used in the case of the Highways agency's composite bridges, is recommended for this purpose.

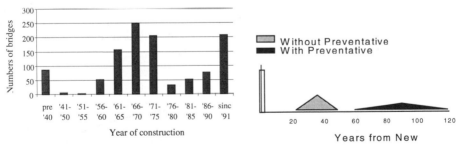

Figure 3. Composite bridges on the trunk road network in England

Figure 4. Bridge rehabilitation rates

First, the numbers of bridges in the group in the network need to be tabulated for each year of construction, as shown in Fig.3. Then the year by year rates of rehabilitation of the bridges in terms of their age, in the absence of periodic repainting, the preventative method under consideration, as well as when regular repainting is carried out, need to be determined as shown in Fig.4 (estimated by experts in this case).

The rate of application of painting in terms of bridges per year is also similarly estimated. Multiplying these rates of rehabilitation and painting with the numbers of bridges built in different years, the profiles of future numbers requiring rehabilitation with or without painting can be determined as shown in Fig.5. It has also been found that approximately 7 % of the bridges will need repainting each year.

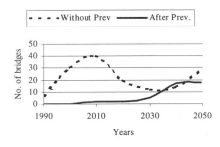

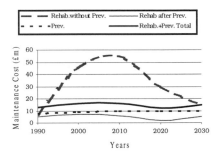

Figure 5. Predicted bridge rehabilitation numbers with and without prev-entative maintenance

Figure 6. Predicted total maintenance costs for two options

Multiplying the above annual numbers by unit costs for rehabilitation and repainting an average bridge, the cost profiles of rehabilitation with and without repainting can be determined as shown in Fig.6. It can be seen from Fig.6. that the cost of rehabilitation without using repainting as a preventative measure is considerably greater than when it is used. The recommended maintenance procedure in the case of composite bridges in the Highways Agency's network will therefore be to repaint as necessary as a preventative measure.

In the case of other preventative measures, the cost differences between the 'with or without' options may not be so obvious. In such cases, the discounted present value of each option should be determined for the choice to be clearer.

REHABILITATION RATES
It can be seen from the above methodology that rehabilitation rates with and without different preventative measures are necessary for determining bridge type specific optimum maintenance strategies. In the above example, the rehabilitation rates shown in Fig.4. are from a range covering different bridge types which were based on expert estimates. Advanced probabilistic methods can be used to determine these rates more rationally, as shown schematically in Figures 7 and 8, which are explained below.

It is known that the load carrying capacities of a group of similar bridges constructed in the same period, even when just constructed, are not all equal but follow certain statistical distribution as shown by the vertical distribution in Fig.7 for the initial load carrying capacity K_0. As the bridge group gets older, the distribution gets wider as some bridges deteriorate more rapidly than others. The age at which the group of bridges may eventually need rehabilitation, i.e. when they will reach the minimum acceptable capacity, also has a distribution as shown in the figure by the horizontal distribution. This distribution is equivalent to the rehabilitation rate shown in Fig.4, if rehabilitation is defined as the work necessary when a bridge just crosses (goes below) the minimum acceptable capacity level. Fig.7 also shows that, if preventative measures are applied to a group of similar bridges, their rehabilitation rate could be different (occurring later) from a similar group where no such measures are applied.

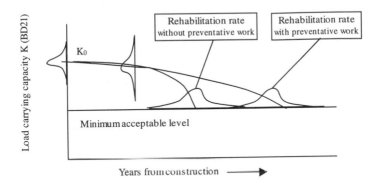

Figure 7. Probabilistic deterioration of load capacity with time

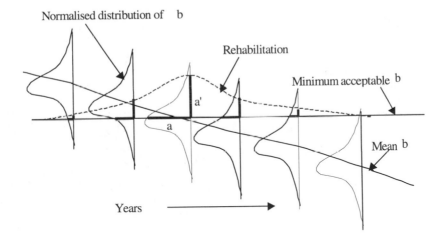

Figure 8. Probabilistic calculation of rehabilitation rate

It is unrealistic to have available sufficient statistical data from the past to obtain these distributions directly. The only possibility, apart from simple estimates, is to use probabilistic analysis as shown schematically in Fig.8 in which a number of successive vertical distributions of load carrying capacity for a group of similar bridges are shown at a period when the mean capacity is passing below the minimum acceptable capacity. The mean line shown is therefore equivalent of the ends of the mean profiles shown in Fig.7, except that the capacity is shown in terms of the reliability index β. The probability of a bridge in the group of being exactly at the minimum acceptable capacity level at different times from construction is shown by the bold horizontal frequencies in the distributions. If each of these horizontal frequencies, for example a, is now plotted vertically, for example as a', where a'=a, a horizontal distribution such as those of Fig.7 can be obtained by joining the frequencies a'. Such horizontal frequencies are the required rehabilitation rates. Thoft-Christensen, Frangopol and others [10,11,12] are now developing this methodology using time dependent reliability analysis to be applied with specific alternative preventative maintenance measures in order to determine optimum maintenance strategies for different types of bridges.

STRATEGIC PLAN

An important step in the management process is to produce a strategic plan which would provide the future expenditure profiles for different types of maintenance work covering a number of years representing an ideal mix of work in terms of logistics and funding.

A strategic plan was prepared by the Highways Agency in 1997 which is shown in Fig.9. It was based on the whole structures stock being divided into four bridge types. Optimum maintenance strategies were determined for all four types using the method described earlier for the composite bridges. The overall plan was the summation of the four optimum strategies.

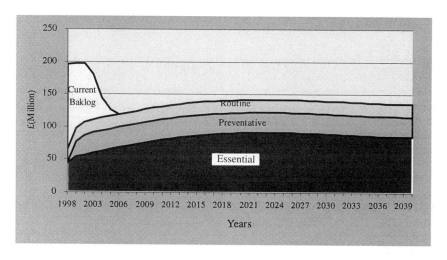

Figure 9. Highways Agency's predicted future expenditure on bridge maintenance

MAINTENANCE BIDS AND PRIORITISATION

It should be remembered that the Strategic Plan provides only an overall guidance on the extent of future annual bridge maintenance expenditure if a sustainable programme is to be maintained. The onus remains on the annual bidding and prioritisation process to actually carry out the programme. The proposed bid and bid prioritisation procedure has been described in greater detail elsewhere [4,5,6].

For the bids, a number of options are entered for each structure or element requiring some maintenance action, either preventative or essential, in the following year, according to their whole life assessments. Once the maintenance bids from all the agents are received the network level prioritisation can be carried out by prioritising first by the nature of the work, i.e. contractually or otherwise committed, essential and preventative etc., and then by work categories which are to be chosen by the authority (e.g. pier strengthening, parapet replacement, protection against scour etc.). Following this, the cost profiles are discounted to present value (PV) and the maintenance option with the lowest PV is selected for each structure or element.

Finally, all the selected bid items are totalled up to check against the strategic plan estimates. This is because the cheapest option for a scheme may not be the best option in terms of the strategic considerations of the whole stock. For instance if the cheapest options for the whole stock totalled to say 20 bridges for rehabilitation in a particular year, and the strategic plan indicates that 60 bridges are to be rehabilitated, the discrepancy needs to be investigated. In such cases there is a possibility that an unacceptable backlog of rehabilitation work may build up in the future, and hence, the project level bids may need to be adjusted to include some options which may not be the cheapest project options in order to bring some of the postponed rehabilitation work forward.

The process can be summarised as follows :

      STEP 1 - Prioritise by work type (committed, essential etc.)
      STEP 2 - Prioritise by category (pier strengthening etc.)
      STEP 3 - Select maintenance option with lowest Present Value (PV)
      STEP 4 - If the resulting cost profile for any work type is less than that in the
                    Strategic Plan, select next higher PV option

The final prioritised list should show, from the bottom of the list, decreasing funding level and increasing traffic delay cost for not providing the full funding (see Fig.10). Those involved in considering the options for different funding levels should then be able to draw a line at a given level and get an indication of the extent of traffic disruption that is likely to take place due to weight restrictions etc. which would be necessary for maintaining safety if the full bid is not funded. It should be noted that the traffic delay cost is only used here to indicate the likely level of traffic disruption.

CONCLUSIONS

The paper describes in broad outline a reliability based bridge management methodology which would be more directly aimed at satisfying the overall network objectives of functionality, safety, value for money and sustainability. The methodology comprises a number of procedures, some of which are new, and others have been available for some time but suitably modified.

Work Type	Category No	Structure/ element	Bid year cost	Cumulativ -e bid year cost	Traffic delay cost	Cumulativ -e traffic delay cost
Committed	1	***	***		***	
	5	***	***		***	
	15	***	***		***	
	33	***	***		***	
Routine	All		***		***	
Essential	3	***	***		***	
	12	***	***		***	
	17	***	***		***	
	23	***	***		***	
	50	***	***		***	
Preventative	2	***	***		***	
	16	***	***		***	
	28	***	***		***	
	41	***	***		***	
	42	***	***		***	

Figure 10. Prioritised bid list

Prediction of whole life performance is an integral part of the methodology and a new national advice note is being prepared to recommend procedures for the whole life assessment of different bridge elements and components.

A very problematic area of bridge maintenance is in determining cost-effective optimum preventative measures which can prolong bridge life without wasting scarce resources. The paper gives recommendations on how such strategies can be developed using a probabilistic approach.

ACKNOWLEDGEMENTS
This paper is being presented with the kind permission of Mr Lawrie Haynes, Chief Executive of the Highways Agency. The author is also grateful to Professor Dan M Frangopol of the University of Colorado at Boulder, Dr Toula Onoufriou of the University of Surrey and Dipl.Eng. Katja D Flaig of Messrs W S Atkins, for their very helpful comments and suggestions during the drafting of the paper.

REFERENCES
1.   Das, P.C. Development of a comprehensive structures management methodology. Management of highway structures. Thomas Telford Publications, London, 1999.
2.   Narasimhan, S. And Wallbank, W. Inspection manuals for bridges and associated structures. Management of highway structures. Thomas Telford Publications, London, 1999.
3.   Shetty, N. et al. Advanced methods of assessment of bridges. Management of highway structures. Thomas Telford Publications, London, 1999.
4.   Haneef, N and Chaplin, K. Bid assessment and prioritisation system. Management of highway structures. Thomas Telford Publications, London, 1999.

5. Das, P.C. New developments in bridge management methodology. Structural Engineering International. Vol 8 No 4., November 1998.

6. Das, P.C.. Maintenance planning for the trunk road structures in England. International bridge management conference. TRB Conference. Denver, Colorado, USA, April 1999.

7. Das, P.C. and Micic, T.V. Maintaining highway structures for safety, economy and sustainability. Eighth International conference on structural faults and repair. London, 1999.

8. BD 21. The assessment and strengthening of highway bridges and structures. Design manual for roads and bridges. The Stationery Office, London,1997.

9. QUADRO. Vol. 14 Section 1 Design manual for roads and bridges. The Stationery Office, London.

10. Frangopol, D.M. and Das, P.C. Management of bridge stocks based on future reliability and maintenance costs. International conference on current and future trends in bridge design construction and maintenance. Singapore, October 1999.

11. Thoft-Christensen, P. Estimation of bridge reliability distributions. International conference on current and future trends in bridge design construction and maintenance. Singapore, October 1999.

12. Frangopol, D.M. et al. Optimum maintenance strategies for highway bridges. International conference on current and future trends in bridge design construction and maintenance. Singapore, October 1999.

Bridge management in Europe (BRIME): overview of project and review of bridge management systems

DR R J WOODWARD and DR P R VASSIE, Transport Research Laboratory, UK

M. B GODART, Laboratoire Central des Ponts et Chaussées, France.

INTRODUCTION

Over the last 50 years the construction of the trunk road and motorway networks across Europe has required large numbers of bridges to be built. As these structures age, deterioration caused by heavy traffic and an aggressive environment becomes increasingly significant resulting in a higher frequency of repairs and possibly a reduced load carrying capacity. Deterioration is exacerbated because many modern structures are more prone to chemical degradation than their forerunners. The effects of alkali silica reaction, chloride ingress and carbonation exacerbated by low cover and poor quality materials are causing progressive deterioration of the bridge stock.

The number of bridges on the European road network that require maintenance is increasing and the direct cost of the engineering work necessary to maintain a satisfactory road network is high. There is therefore a need for rational methods for deciding how maintenance budgets should be allocated to ensure that they are used cost effectively. The most appropriate maintenance strategy for a stock of bridges is a complex subject and there are a number of issues that determine the most economic strategy. These include:

- condition of the structure
- load carrying capacity
- rate of deterioration
- maintenance treatments available and their effectiveness, lifetime and cost
- traffic management costs
- traffic flow rates and the associated delay costs
- cost of working in the future discounted to present day values
- the costs accruing from improvements such as strengthening or bridge widening
- implications for safety and traffic flow if the work is not carried out immediately.

By taking all these factors into account it should be possible to develop a programme of maintenance work optimised to achieve a standard condition at a minimum lifetime cost. This paper gives an overview of a European research project on **Bridge Management in Europe (BRIME)** which is being undertaken to develop an outline framework for a bridge management system for the European road network. It also summarises the results of a worldwide review of bridge management systems.

ORGANISATION OF THE PROJECT
The project is being co-ordinated by the Transport Research Laboratory (TRL) in the UK and the work is being done in collaboration with the Bundesanstalt fur StraBenwesen (BASt) in Germany, the Centro de Estudios y Experimentacion de Obras Publicas (CEDEX) in Spain, the Laboratoire Central des Ponts et Chaussees (LCPC) in France, the Norwegian Public Roads Administration (NPRA) and the National Building and Civil Engineering Institute (ZAG) in Slovenia.

OBJECTIVES
The objective of the project is to develop a framework for the management of bridges on the European road network, and to identify and examine the inputs required to implement such a system.

PROGRAMME
The project has been divided into eight workpackages, each of which has been further subdivided into tasks. The workpackages are listed below:

Workpackage 1: Classifying the condition of a structure.
Workpackage 2: Assessing the load carrying capacity of existing bridges, including the use of risk based methods.
Workpackage 3: Modelling of deteriorated structures and affect of deterioration on load carrying capacity.
Workpackage 4: Modelling of deterioration rates.
Workpackage 5: Deciding whether a sub-standard or deteriorated structure should be repaired, strengthened or replaced.
Workpackage 6: Prioritising bridges in terms of their need or repair, rehabilitation or improvement.
Workpackage 7: Reviewing systems for bridge management and development of a framework for a bridge management system.
Workpackage 8: Project Co-ordination

Workpackages 1 to 6 are providing the modules required to make up a bridge management system and the combination of information from these different modules is being studied in Workpackage 7 which is also examining the various systems currently available for doing this.

A more detailed description of the work carried out in Workpackages 1,2,3,4 and 6 is given in separate papers (1-4). Some of the work undertaken in Workpackage 7 is described below.

REVIEW OF SYSTEMS FOR BRIDGE MANAGEMENT
To obtain information on the BMS's currently in use a questionnaire was developed and sent to the partners in the BRIME project, and to other European countries, as well as countries outside Europe known to be well advanced in terms of bridge management (Canada, Japan, USA). The questionnaires were sent to the national highway authorities in each country and to the New York City and California State roads departments. Replies were received from everyone except Japan

and Canada.

Global description of bridge management systems
Of the eleven countries that replied eight use a computerised Bridge Management System; two do not use a BMS but are in the process of developing one and one country uses a partial one. The ages of the computerised BMS's vary from 2 to 20 years.

Documentation
For most countries the procedures for using the BMS are given in various documents such as maintenance manuals, management instructions, user manuals. In Slovenia there is no official user manual or guidelines; the available documentation is a three volume report of a research project. Two countries do not have any special documentation on their BMS.

For most countries, the BMS is used to manage bridges on the national highway network ie motorways and trunk roads. No country has its BMS linked to a road management system. However the Norwegian system has an automatic link to the road network for route number and location.

Database
All countries, except the UK and Slovenia, use commercial database software; the most popular is ORACLE although ACCESS, DELPHI and Structured Query Language are also used. Most countries use WINDOWS based systems.

The database is used for the management of both individual structures and the bridge stock in most countries. Spain uses the database mainly for the management of the bridge stock, and France uses a different database for the management of individual structures and for the management of the stock.

The BMS's are used at all the different levels that have a role in maintenance ie national, regional or county authorities and maintaining agents. They are also used by consultants in Norway and Denmark. The responsibility for maintenance is always at the national level.

Information is updated according to the type and performance of the BMS; for some it is done daily, some occasionally and for others annually or even every 2 years.

The number of datafields varies enormously; for example the Norwegian database contains 1228 fields in 147 different tables. The Finnish database contains 250 datafields. Some databases have the facility to add user-defined fields.

Bridge condition
There are 3 or 4 levels of inspection (routine, general, detailed and special). The results of general and detailed inspections are stored on the database although Norway also stores the results of measurements and investigations. In general the condition is stored for both individual elements and the whole bridge, except for Germany where the condition is stored only for the whole bridge and the UK where it is only stored for individual elements. The condition is mostly based on a 3 to 5 point rating scale.

Other information recorded on BMS
The date, type, cost and location of maintenance work are recorded in every country. For

France, it is only stored for maintenance work which costs more than 300kF and for Slovenia, type and cost are stored separately. The condition immediately before and after maintenance is not stored, except in Denmark and California.

Prediction
Most countries do not use past condition data or a deterioration model to predict future condition. The exceptions are:
- Finland which uses probabilistic Markovian models at the network level, and deterministic models at the project level.
- New York which uses past condition data and degradation of materials with time.
- California which uses past condition data.
- France and Slovenia use previous condition ratings.

Costs
For most countries maintenance, repair and, in some cases, inspection costs are stored on the BMS's; the exceptions are Germany, Norway and Slovenia.

The BMS's used in most countries do not calculate the financial consequences of traffic disruption caused by maintenance work and the associated traffic management. In UK, delay costs are calculated using either look-up tables or the computer programme QUADRO or look up tables derived from the programme.

Decisions on maintenance and repair
Most of the countries do not use the BMS to make decisions on maintenance and repair. The exceptions are Denmark that uses a prioritisation programme, Finland that uses a repair index and California that uses long term least cost optimal strategies from PONTIS. In France and Germany, decisions are based on engineering judgement. The UK uses whole life costing and cost benefit analysis. Spain makes decisions based on the cost of the repair as a percentage of replacement costs but it is not clear from their response whether this includes the traffic disruption costs.

Most countries decide when maintenance work is needed on basis of inspections and engineering judgement. Slovenia decides on the basis of increased traffic flows and the importance of the bridge to the region.   California decides on the basis of safety and an analysis of the economic benefits.   (In general it is based on technical rather than economic requirements.)

Most countries decide which is the best maintenance option to use on basis of engineering judgement. In UK it depends on the solutions available, the whole life cost appraisal, the cost of traffic management and the traffic disruption to the network.

Prioritisation
Germany, France, UK, Norway and Slovenia do not have a BMS module for generating an optimal (minimum cost) maintenance strategy subject to constraints such as a lowest acceptable level of condition. However a system is currently being developed in the UK. Such a module is used in Denmark (for repair) and Spain, Finland, New York and California.

In the optimisation process, other constraints are often applied such as cost and policy for UK, and the lowest long-term cost that prevents failure for California. Budget and bridge condition are also used as a constraint in several countries.

The BMSs used in France, Germany, UK, Norway and Slovenia, do not produce a prioritised maintenance strategy for the bridge stock when the maintenance budget is insufficient. However it does for California, Denmark, New York and Spain, although for Denmark this is only done for repair. Only Denmark and New York quantify the economic consequences of carrying out a sub-optimal maintenance strategy.

Each country uses different criteria for prioritisation. For most countries, the responsibility for prioritisation of bridges is at the national level. The exceptions are Norway and Finland where the responsibility is at the local level.

Quality control
For all countries, there is no quality control of the management of bridges, except for Finland where internal procedures are applied.

INPUTS, OUTPUTS AND ALGORITHMS FOR A FRAMEWORK FOR A BMS
As indicated above, the BRIME objective is to set up a framework for a BMS which can be used at both the project and network levels. Project level information is related to individual bridges, elements or components. It is important for specifying the maintenance requirements and retrieving data about particular bridges. Network level information relates to the entire bridge stock or to subsets of the stock such as all the bridges in a given region. Network level information is important for determining whether the average condition of bridges in the stock is improving or deteriorating and for estimating the value of the budget needed in order to maintain the condition of the network at an acceptable level. To evaluate the effectiveness of a bridge maintenance programme, it is necessary for the BMS to have in-built targets related to benchmark values for average condition of the stock, the replacement rate for bridges, the percentage of the stock with traffic restrictions and the disruption to users arising from traffic restrictions at different times. An assessment of how closely such targets are met will establish the sufficiency of the budget and the consequences associated with particular budget levels.

The first step in setting up the BMS framework is to decide the requirements. On the basis of the answers to the questionnaire and the technical literature a list of essential functions for the BRIME BMS was established; these functions are shown in Table 1 which satisfied the project objectives and was achievable. They were developed to generate a system of well defined project and network level outputs for the BMS. These are listed in Table 2.

The next phase in the development of the framework was to decide what input data and algorithms were needed in order to generate the output. Examples of inputs and algorithms for some project and network level outputs are shown in Tables, 3 and 4. The project level outputs, algorithms and inputs in Table C follow a sequence. Outputs A2 to A6 can be obtained from their respective inputs using simple database queries and reports. The outputs then act as inputs for A7 to A10 which require more complex algorithms. The outputs for A7 to A10 are then used as inputs for A11. This sequence of increasing complexity of data processing where initial outputs act as inputs for more complex algorithms is typical of management systems. It can also be seen that many of the inputs required for network level outputs are the same as the inputs for some project level outputs. This demonstrates the inherent interconnection between the project and network level systems, confirming that the BMS requires both systems in order to produce satisfactory results. When sufficient project level data has been collected it can be compiled to give correct format for the network level inputs. This indicates that the system should initially be run at project level until enough data has been collected to permit reliable operation of the network level algorithms.

Table 1   Functions of a Bridge Management System

Functions of a Bridge Management System
To provide an inventory of bridges
To record/predict the historical and future condition of elements and components
To record/predict the historical and future load carrying capacity of the bridge
To assess the rate of deterioration
To select the most cost-effective maintenance
To evaluate the cost of various maintenance options
To evaluate traffic management and delay costs
To calculate discounted costs to give a lifetime cost
Assess the implications for safety and traffic congestion of deferring maintenance work
To produce optimal and prioritised maintenance programme
To assist with budget planning

Table 2. Project Level Outputs

PRIMARY	SECONDARY
General Queries	List and count bridges meeting specified criteria.
Inspection History	List and count bridges that are overdue for inspection.
Test History	List and count bridges that are substandard.
Maintenance History	List and count bridges that are in the poorest condition
Traffic History	state.
Condition History	List and count bridges with traffic restrictions.
Load carrying capacity history	Budget needed for the optimal maintenance programme.
Posting History	The numbers of bridges with deferred maintenance due to
Optimal Maintenance Programme	a sub-optimal budget.
Prediction of variation of load carrying capacity with time.	The long term cost of deferring maintenance due to a sub-optimal budget.
Prediction of the variation of condition with time.	The prioritised maintenance programme for a given sub-optimal budget.
The effect of maintenance and/or strengthening on the future rate of change of condition and or load carrying capacity.	Prioritised maintenance programmes based on other constraints. Predictions of load carrying capacity for a specified budget.
Cost of Optimal Maintenance Programme	Prediction of the condition for a specified budget/maintenance programme.
Estimation of the cost of maintenance Based on the load carrying capacity and condition.	Routing of heavy, high, wide or long vehicles. History of different types of maintenance.
Estimation of the cost of traffic disruption due to maintenance or traffic restriction	History of occurrence of different types of defect. History of occurrence of substandard bridges. History of performance of different element types and component types. Cost rates for different maintenance options. History of performance of different maintenance methods.

Table 3 Input data and Algorithms required to satisfy project level outputs

OUTPUT	ALGORITHM	INPUTS
A1 General Queries	Database query	Inventory
A2 Inspection history condition history	Database query	Date of inspection Type of inspection Extent/Severity Occurrence of defects
A3 Maintenance History	Database query	Date of maintenance Location of maintenance Type of maintenance Area maintained Cost of maintenance Immediate effect of maintenance on load carrying Capacity/condition Duration and extent of any Traffic delays
A4 Traffic History	Database query	Date of survey Flow rate % HGVs Alternative route Additional time/distance
A5 Load carrying capacity History	Database query	Date of assessment Load carrying capacity See also inspection/condition/ Test history
A6 Posting history	Database query	Start and finish dates for any Load, height, width restrictions Duration and extent of traffic delays
A7 Predicting the variation of load carrying capacity with time	Markov Chain or Neural Network models	Outputs A2 to A5
A8 Predicting the variation of condition with time	Markov Chain or Neural Network models	Outputs A2,A3,A4 & A6
A9 Estimating the cost of Maintenance	Neural Network	Outputs A2,A3 & A5
A10 Estimating the cost of Traffic delays	QUADRO	Outputs A3 to A6
A11 Optimal maintenance programme	Neural Network Dynamic Programming Discounting	Outputs A7 to A10 Also constraints and future life required

Many outputs can be generated from the input data using the basic data processing functions of a relational database. Predictions of condition and load carrying capacity, the estimate of maintenance and traffic delay costs, and the production of optimal and prioritised maintenance programmes require more complex algorithms. These algorithms usually involve Markov

Chain transition probabilities, dynamic programming or neural networks.

Table 4. Input data and Algorithms required to satisfy the network level outputs

OUTPUT	ALGORITHM	INPUTS
B1 List/count bridges satisfying specific criteria	Relational database processing	Inventory
B2 List/count bridges overdue For inspection		Output A2 Date for next inspection
B3 List/count bridges that are Substandard		Outputs A5
B4 List/count bridges in the Poorest condition state		Outputs A2
B5 List/count bridges with Traffic restrictions		Outputs A5 & A6
B6 Budget for optimal programme		Output A11
B7 No. of bridges with deferred maintenance		Output A11 and B9
B8 Long term cost of Maintenance	Neural Network Dynamic programming	Outputs A7 to A11
B9 Prioritised maintenance Programming	Neural Network Dynamic programming	Outputs A7 to A11 Budget constraint

The final step in the development of the framework BMS is to break down the inputs into specific data fields which form the basis of the database inventory. This process for developing a bridge management system (Requirements → outputs → algorithms → inputs → data fields) ensures that the data collected has a well defined purpose and avoids the collection of unnecessary data.

ACKNOWLEDGEMENTS
The work described in this report has been part funded by the European Commission Directorate General for Transport, with balancing funds provided by the authorities responsible for the national road networks in the UK, France, Germany, Norway, Slovenia and Spain.

REFERENCES
1. Kaschner R, Haardt P, Cremona C and D W Cullington. *Bridge Management in Europe (BRIME): Structural Assessment.* Fourth International Conference on Bridge Management. Surrey, 2000.
2. Daly A. *Bridge Management in Europe (BRIME): Modelling of Deteriorated Structures.* Fourth International Conference on Bridge Management. Surrey, 2000.
3. Bevc L, Peruš I, Capuder F, Mahut B, Lau M Y and Grefstad K. *Bridge Management in Europe (BRIME): Condition assessment of Bridge Structures.* Fourth International Conference on Bridge Management. Surrey, 2000.
4. Blankvoll A, Larsen C K, Markey I, Raharinaivo A, Bevc L, Capuder M and Perus I. *Bridge Management in Europe (BRIME): Chloride ingress and bridge management.* Fourth International Conference on Bridge Management. Surrey, 2000.

Study for a bridge management system in India

JOHN C COX, Associate Director, WSP, New Delhi, India, and
STEPHEN J MATTHEWS, Director, Bridges, WSP, Basingstoke, England

INTRODUCTION

The Government of India, (GOI) through its Ministry of Surface Transport (MOST) has been undertaking the 2nd National Highways Project to improve the standards of construction and maintenance of the National Highways throughout India. The work has been supported by a loan from the World Bank.

As part of the project, the MOST required to implement a Bridge Management System. The aims of such as system are as follows:

- To have up to date knowledge of the condition and carrying capacity of all bridges on the National Highways.

- To be able to predict how the condition of bridgestock will deteriorate if left unattended

- To be able to plan a programme of maintenance, repair, and reconstruction so that the strategic highway corridors are maintained for use at a given standard of service

- To be able to predict the costs of funding these operations for the future (say 5 years ahead), and to be able to demonstrate the economic necessity for this to Central Government financial departments so that appropriate allocations can be made

In 1996, WSP (then O'Sullivan & Graham), in association with a local subconsultant in India, were awarded the contract to undertake a Pilot Study to for a Bridge Management System. (BMS)

The study comprised two phases, a preliminary phase (investigative study) of four months, followed by a design and pilot implementation phase of seven months.

In the preliminary phase the Consultant (O'Sullivan & Graham Limited*) was required to review existing organisational and financial policies in the Ministry of Surface Transport (MOST) and arrangements with Public Works Departments (PWD) in each state throughout India for managing bridge inspections and repairs. Bridge management organisations in other countries were also to be investigated to determine what lessons could be learned from the experience of others and adopted within the Indian context. The Consultant was also required to review internationally available BMS software and select the most appropriate for implementation in India.

In the second phase of the project (seven months) the Consultant was required to design a new, more effective, organisational structure for bridge inspections and repairs, and to implement the selected BMS on a pilot study basis.

The Terms of Reference (TOR) placed great emphasis on the bridge management software, referring to this as the "system".

It became clear at a very early stage in the project that the main issues were of an institutional nature. Previous experience suggested that bridge management software would serve no useful purpose without an effective bridge management organisation with well defined responsibilities, and staff who were properly trained in the operation of the Bridge Management System and rewarded accordingly.

During the preliminary phase of the project, this was discussed with the Client (MOST). The project was re-focussed to develop an appropriate organisational structure, which would:

- overcome the problems which hindered maintenance of a complete bridge inventory.
- ensure thorough and timely bridge inspections, by trained personnel.
- support timely and adequate provision of funds for inspection and maintenance works.

In 1996, funds could generally be made available for emergency repairs, but regular maintenance requirements were largely overlooked. This reflected the 'crisis management' style of bridge management.

As in many developed and developing countries, there is a tremendous backlog of rehabilitation work on the National Highway bridges in India. The Government is addressing its rapidly growing road infrastructure needs by widening existing National Highways and providing for many new Expressways. This development of the National Highway network will greatly add to the bridge management problem and underlines the need for an effective bridge management organisation to be put into place now – with, or without the support of BMS software. The new organisational structure was now referred to as the 'system'.

PHASE 1

Institutional Issues
Information was collected on the organisational structure of over twenty maintenance management organisations throughout Western Europe, USA, Asia, Australia, the Caribbean, and the Far East.

The organisations were asked to provide information on their existing staffing, systems and practices for bridge management, whether or not they made use of the private sector, and whether they utilised a computer bridge management system (BMS). Many responded directly, but only three provided details of their staffing policies.

Organograms were developed for several other organisations based on their responses and additional sources of information. The information collected provided a good overview of the typical problems faced by all bridge management organisations.

Insufficient maintenance funding was probably the most common problem, although a familiar picture developed that, throughout the world, bridge management organisations are still evolving.

Staffing structures sometimes include the private sector and sometimes restructured public sector organisations. Naturally, such changes can affect the morale of the actual staff involved, although over 15 years experience of WSP in managing similar operations in the United Kingdom has shown that with sensitive handling, operatives respond well to their new responsibilities, challenges, and rewards.

Backlog of maintenance and rehabilitation works was a common problem and that this was being tackled in a number of different ways.

Organisational and financial policies were notably weak in many countries. These policies must be mutually supportive.

Financial policy determines the way in which government allocates funds between different sectors. Organisational policy influences the utilisation of the allocated funds.

For maintenance operations, it is contradictory to have a financial policy that is designed to provide adequate funds without an organisation that enables the funds to be used in a cost-effective manner.

Conversely, there is little point in having a fully staffed and trained maintenance organisation without the adequate and timely funds to carry out the required inspections and the identified works.

As illustrated by Brooks et alia (1), Figure 1, maintenance managers are constrained by the policies that dictate legal, financial, personnel, and operational procedures. The activities of technical staff are constrained by the management decisions based on inventory, budgets, work programmes, and availability of resources.

A bridge management cycle was developed (see Figure 2.) This was intended to

- ensure consideration of the activities that comprise effective bridge management
- identify weak points in the existing system (organisational arrangements)

In India, the MOST "owns" the national highways and allocates annual maintenance funds to the PWDs in the States of India for carrying out inspections and repairs to roads and bridges.

During Phase 1 of the Study, visits were made to MOST, and to two States in India – Orissa, and Maharashtra. Staff were interviewed at all levels in MOST and State PWD's. A visit was also made to the Central Southern Railway headquarters in Mumbai to discuss and compare Bridge Management Operations on the vast Indian Railways network (which comprises mostly steel bridges, whereas the bridges on the National Highways are mostly concrete).

Information from the interviews and from MOST technical directives and other official Indian publications were used to develop the cycle of bridge management activities.

Level	Capability
Technical	• Approximate criteria planning • Materials test facilities • Effective quality control of operations • Access to research and information • Initial training and continuous professional development
Managerial	• Existence of up-to-date inventory • Works effectively planned, programmed, and monitored • Budget related to actual costs and ability to disburse • Effective cost control • Adequate plant and equipment available and effectively utilised • Availability of materials required
Institutional	• Legal power to undertake works • Rational and Functional administrative structure • Employment and training of staff of sufficient calibre • Funds to undertake works and for administration, salaries, and expenses • Financial control

The technical capabilities of a bridge maintenance operation are dependent on adequate managerial support. The managerial capabilities are dependent on adequate institutional (policy) support

Brooks Pyramid : Capabilities of a Department   figure 1

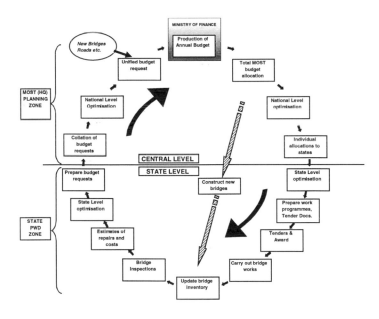

The Bridge Management Cycle (Existing organisational structure)   figure 2

The initial bridge management cycle was reviewed against the range of activities carried out by bridge management organisations in the 19 other countries that we had studied.

It was clear that this needed further items to be introduced (Figure 3) in order to include activities that would, *inter alia*, lead to optimal deployment of resources, and prioritise bridges to carry out works in order of cost-effectiveness rather than simply on a "worst-first" basis. The cycle now corresponded to the more comprehensive range of activities carried out by the more developed bridge management organisations elsewhere in the world.

The bridge management cycle was then studied in terms of the level of qualifications, experience and training needed to properly carry out each of the primary bridge management activities.

The findings were summarised in a bridge management matrix, from which it was recognised that the academic levels were placed too high for application to developing countries where, typically, there is a lack of graduate bridge engineers. This is the case in India.

MOST directives stipulate the rank of officer that should carry out inspections on different types of bridge but without stating any qualifications or experience criteria. The matrix was discussed with senior representatives from the MOST and reviewed in conjunction with a technical paper on bridge inspections produced earlier by one of the Ministry's officers (2).

A 'model' bridge management organisation was produced, which reflected the staffing structure required to carry out the activities that make up the bridge management cycle. This organisation could be private, public, or a mixture of both.

Note that the number of staff at any particular level would, in reality, be based on the size and nature of bridge population under the jurisdiction of the organisation.

A viable new organisational structure for managing the bridges on all the National Highways throughout India was then developed in full during Phase 2 of the study.

Supporting BMS software
The needs of the bridge management cycle, and the full range of activities required of an effective bridge management organisation, led to the development of a minimum specification for bridge management system software (BMSS) to support the new organisation in the execution of its responsibilities.

These included

- inventory
- inspection
- determination of optimum repair levels
- definition of minimum repair levels
- prioritisaton
- work programmes
- issue of works orders

Development of the specification included a review of the various algorithms used in BMSS - for example, would prioritisation of bridges be based on safety or condition-based criteria.

The various bridge management organisations contacted in Phase 1 were also asked to provide details of any BMSS that they use.

A review of professional publications and conference proceedings also revealed a number of companies and government departments that had produced their own BMSS. Each of these agencies was contacted and asked to provide details of their system against a checklist of items. The checklist identified minimum features required in a BMSS to meet the specification that we had developed for India.

A BMSS for India must be suitable for the very wide range of geographical and climatic conditions encountered (from mountainous terrain with severe winter conditions, to monsoon plains, deserts, and highly saline coastal regions). It proved very difficult to obtain this information within the four months of Phase 1, but enough was obtained to undertake a review of seventeen different software packages.

The software packages were first sorted into three generations (3) of development. This immediately reduced the number of candidates that would then be compared in detail against the earlier specification.

A short-list of three was made. The final selection was based on a number of factors, including:

- stage of development of the software (incorporation of appropriate up to date research)
- user-friendliness (graphical user interface, etc)
- flexibility of algorithms to include a wide range of bridge types
- capability to accommodate increasing knowledge of the deterioration regimes
- a wide range of maintenance regimes

The BRIDGIT programme from Delcan Corporation in Canada was determined to be the most appropriate BMS to support the new bridge management organisation in India.

PHASE 2

Institutional Issues
India's National Highways extend to over 33,000 km with more than 6,400 bridges. The network is expanding quickly. There is a tremendous backlog of bridge repairs.

Major rehabilitative works are already carried out by bridge design offices (called Bridge Circles) in some States, and are contracted-out to consultants in other States.

The bridge management cycle and the matrix of tasks/qualifications/experience were studied to identify the primary weaknesses in the existing organisation; viz. MOST and the State PWDs.

Frequent changes in staff assignments are a major problem. There is no separate career structure for bridge engineers. Staff could be assigned to work on public buildings or on highways, with changes occurring about every two years.

There is a clear need for well-trained and experienced bridge inspectors to identify the degree of seriousness of observed defects and thus instigate appropriate action.

Recommendations were developed during Phase 2 to introduce private sector bridge inspectors.

The selected BMSS uses 'condition states' to record the severity and extent of bridge defects. To be meaningful, criteria for reporting these needed to objective, and examined by trained personnel.

It would be necessary to develop a permanent bridge inspectors' course to provide training to new bridge inspectors in the private sector, and bridge engineers and bridge managers from the ministry and the public works departments; see Figure 4.

Training manuals were drafted for four levels of bridge management course: bridge inspector, bridge engineer, bridge manager, and BMS operator.

If a private sector company were to be expected to pay for some of its staff to be trained as bridge inspectors, and to invest in bridge access and testing equipment, it would need an incentive. The company would need assurance that a reasonable workload would be provided over which it could obtain a return on its investment.

A system of term maintenance consultancy commissions was recommended. Firms of consulting engineers would be contracted for a term, say two years, during which time they would become responsible for carrying out all inventory and inspection works on a given length of National Highway(s), and the design and supervision of repairs arising from these inspections.

Regular maintenance works could also be carried out by term maintenance contractors who could also be responsible for providing the access equipment, instead of the term consultant, as they would also need the equipment for carrying out many of the repair works. Terms of reference and tender documents were drafted for the term consultant roles.

This arrangement mirrors (in part) operations which have been used in the United Kingdom and other countries for some 15 years, which have led to cost effective maintenance for highways and bridges.

The ministry had constituted a project Steering Committee for the project. The final proposals for a new bridge management organisation, with supporting financial policy arrangements were agreed with the committee at a series of presentations and workshops.

The degree of involvement by private sector firms was left flexible, with recommended guidelines, owing to the differences in existing capabilities and staffing levels in various State PWDs.

A set of interim guidelines was also developed with the aim of improving bridge inspection practices in other states until the new organisation is implemented in those states. A draft bridge inspectors' manual was produced for use on training courses and in the field.

Supporting BMS

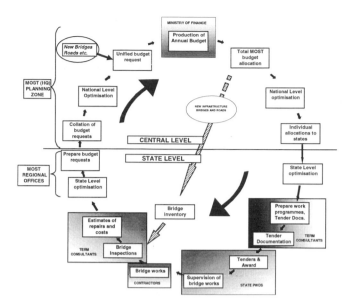

The Bridge Management Cycle – New Bridge Management Organisation - figure 3

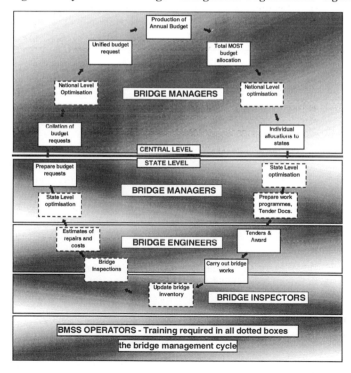

Summary of Bridge Management training requirements figure 4

During Phase 2, the user-interface of the selected BMSS software (BRIDGIT) was customised to accommodate the same inventory format as the paper system designed and used by the MOST and State PWDs.

An additional benefit of the project was the entry of <u>all</u> MOST 1989 National Highway bridge inventory. Until this time records, whilst extant, had been kept on paper. Once all of the data had been entered, considerable flexibility was immediately achieved, and the BMSS was used to analyse the data.

One of the useful features of BRIDGIT is its ability to carry out data integrity checks. A large proportion of the bridge records were found to contain errors or omissions which should be rectified once new bridge inventories are compiled at the start of the follow on implementation project. The data were then queried and sorted to produce a wide range of standard reports on the inventory and condition of the bridges.

The deterioration models and condition state descriptions within the BMS were calibrated, based largely on the knowledge of experienced Indian bridge engineers in the project team and the project Steering Committee. These models will be reviewed to update repair costs at the commencement of the trial implementation project.

IMPLEMENTATION PHASE
Since completion of the BMS Pilot Study at the end of 1997, the ministry has had to revise its proposals for implementing the new bridge management organisation. This has largely been due to problems in gaining approval from the World Bank for the terms of reference for the term consultants. It is hoped that this problem will be sorted out in the very near future.

Work is expected to proceed, with a trial implementation in three states, starting in January 2000. The "term consultants" will now become sub-consultants to WSP who will act as the main Supervisory Consultant.

Local staff will be recruited to be trained as trainers for

- bridge inventory and inspection
- BMSS operation

Training courses will be established in the MOST's own training centre, the National Institute for Transportation and Highway Engineers (NITHE). This will provide a permanent facility for training all four levels involved in the bridge management cycle: inspectors, engineers, managers, and BMSS Operators.

Trainees will include staff from the ministry, public works departments, and other local consulting firms. The success of the training facility will largely determine the success of new public-private bridge management organisation.

Towards the end of the 18 month trial implementation period there will be a review of the strengths and weaknesses of the new organisational structure. Recommendations will be made for a second stage of implementation, which is planned to include six or seven states.

Provided that sufficient funds are made available in the MOST's budget, supported by loans from international financing agencies such as the World Bank, the new organisation could be replicated throughout the country in about five to seven years. India has twenty-seven states, plus Union Territories. The number of term contracts issued every 1-½ to 2 years will have to increase significantly to achieve this target.

CONCLUSIONS

At the outset of the BMS Pilot Study, it was recognised that implementation of any "Bridge Management System" must first concentrate on the institutional issues in order to provide an effective organisation for collecting the inventory and inspection data required by BMS software.

The "system" must be people-centred. Development of a viable system must begin with identifying capabilities and available resources, then designing the system to operate within those constraints.

The development of a bridge management cycle and a study of the weaknesses in the existing organisational arrangements helped to define a new bridge management organisation. A review of skills and experience of public sector engineers led to the introduction of private sector involvement, especially for bridge inventory and inspections.

Design of repairs could subsequently be carried out either by the private sector or, in some states that already possess suitable bridge engineering skills, by the respective Public Works Departments.

An appropriate computer BMS programme was selected, aimed at supporting the new organisation; the new organisation was not designed to work around a given software package. It is considered that, with appropriate budgetary support, this system will prove viable and sustainable.

*   O'Sullivan & Graham Ltd is now incorporated in WSP

1.   Brooks, DM et alia: "Road Maintenance Overseas"; 1980 (Figure 1 is derived from the original pyramid which Brooks developed for a road maintenance organisation but the same tiered principles apply to a bridge management organisation)

2.   Sinha, NK: "Need for Improvement of Highway Bridge Inspection System"; 1990

3.   Refer to the paper by R Blakelock 'The Development and Use of Bridge Management Systems (1994) for further details of the three '"generations" of BMS software.

Decision-making processes in an advanced bridge management system

DR LUIZ CARLOS PINTO DA SILVA FILHO, MSc, PhD, Senior Lecturer, Federal University of Rio Grande do Sul (UFRGS), Porto Alegre, RS, Brazil, and PROFESSOR NIGEL SMITH, BSc, MSc, PhD, CEng, FICE, MAPM, Professor of Construction Project Management, University of Leeds, UK

INTRODUCTION

This paper describes a study carried out by the authors and discusses how the concept of the asset management of a number of individual bridges can be incorporated into a system for the structured and advanced management of a bridge stock. A new beginning was necessary and one of the first steps was to establish a "soft" or business process based diagram of the Bridge Management process to understand the constraints and information flows inherent in the system, Figure 1, (1). This guided the research into re-engineering what should constitute bridge value and investigating the effects of deterioration and maintenance work.

Bridge Management can be defined as a structured process that begins with the identification of problems (diagnosis) and progresses into the assessment of risk and the prediction of future conditions (prognosis). The examination of the present condition and the expected deterioration leads to the determination of strategies for intervention (treatment definition). The process is completed by the execution and control of repair (therapy).

The pivotal step of the process, in terms of decision-making, is the choice of which set of actions to perform for each structure during the phase of treatment definition (1). The most significant activity involved in the decision-taking is the appraisal of the effects of different intervention strategies. Due to the current perceptions on public investment and its objectives, the main question in relation to the appraisal of maintenance and repair strategies in bridges is the assessment of the ratio between the resulting increase in "value" and the cost of intervention. The authors argue that the definition of the "value" of a project that must be wide reaching if bridge management is to be justified socially.

JUSTIFYING THE ADOPTION OF THE NOTION OF VALUE

The recognition of the important role of bridges in a transport system and the imposition of more stringent regulations in terms of performance over the years has reinforced the need for having a sound process of Bridge Management. The basic conundrum of decision-making in Bridge

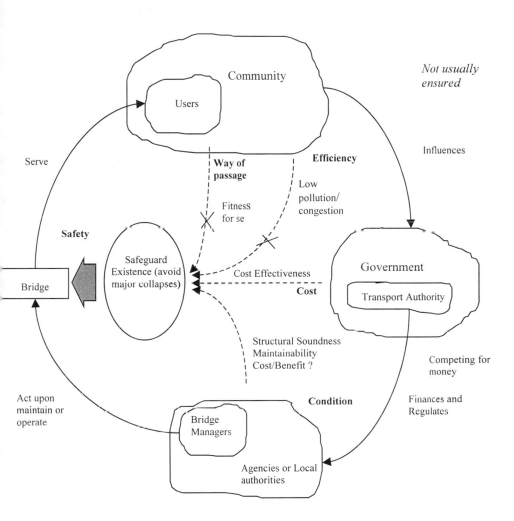

Figure 1. A soft systems view of Bridge Management

Management is to find the combination of actions over time, including the do-nothing option, which would constitute the most appropriate strategy for each element in the bridge network; or as some practitioners would admit, to identify the strategy which they would "regret least !".

In a hypothetical and idealistic situation, technical and economic considerations about each bridge in isolation would be sufficient to determine the best action to be undertaken. In the real world, particularly in the developed countries a common trend of reductions in the money available to public infrastructure owners has been observed. A study (2) shows that the total government spending on infrastructure in the U.S. has dropped sharply from 4.5% in the middle 1960's to around 0.8% per year in the late 1980's. Similar situations are found in other countries, where authorities and agencies in charge of infrastructure maintenance and renewal are increasingly being budget constrained and hence need to develop methods to determine how to spend their restricted budgets wisely.

The definition and comparison of the most suitable maintenance, repair and improvement, (MRI) strategy is not an easy task because the decision-making environment in infrastructure maintenance is usually dynamic, involves multiple objectives and is subjected to great uncertainty (3). Each bridge has its own particular characteristics and the variety and synergetic nature of defects create unique deterioration patterns that cannot be easily simulated. The decision of which projects to undertake with a limited budget will ultimately depend on the different value attributed to the effects of different levels of intervention. This approach helps establish the set (or portfolio) of projects that would represent the best value for money. An MRI programme is a combination of individual projects. The choice of the most suitable programme to execute would imply the definition of the portfolio of projects that would maximize value. From a social point of view, the conflicting interests of the parties involved (and their manifestations in terms of policy requirements) can also change over time, affecting the choice of strategy to adopt. The choice of strategy will impact the condition state and the performance of the structure. To define the best strategy for the owner it is necessary to consider the effects along the whole lifecycle of the structure (2).

APPLICATION OF THE NOTION OF VALUE TO BRIDGE APPRAISAL

Value for Money can be considered to be realised by determining the investment options that will bring the best return for the money invested (4) and has already been adopted as the standing policy in many areas of public administration, specially roads (5,6). The UK Government in particular has put a lot of emphasis into this approach and has recently instructed civil servants to examine "options for the delivery of the policy framework that minimise public expenditure cost and maximise the positive impact on the economy" (7).

To provide fair results, the concept of value must be applied consistently. Contrary to many "black-box" approaches used in other systems to deal with bridge management procedures, the research presented in this paper advocates the use of a "white-box" architecture. It adopts the stance that it is necessary to explicitly acknowledge the underlying system of values of human decision-makers to make a practical decision. It is also necessary to have a clear understanding of the various components of bridge value that the user wants to consider and to allow the use of distributional weights to express differences in the (dis)utility attributed to different impacts, such as pollution or delay.

The application of the concept of value to bridge management demands two things:

- the identification of the components of value

- the determination of the value components

IDENTIFICATION OF VALUE COMPONENTS

If Value for Money is accepted as the standing policy, the basic question becomes how to determine the value of an MRI operation. In the process of managing bridges, decision-makers must try to attend simultaneously to various requirements. The value of a MRI operation would therefore depend on the extent to which the consequences of that specific course of action would satisfy the set of requirements adopted by the authority in charge. Each possible course of action to correct or minimise a bridge problem has a positive effect because it solves some structural or functional deficiency, improving performance. Interventions however customarily also have a negative effect associated with them, because they usually adversely effect the transport function of the structure. The balance between these impacts will indicate the "value" of the MRI strategy.

The most suitable way to extract a common figure that could be used to prioritize actions would be to reduce all the utility components to a monetary basis. This would imply the definition of values for non–market commodities such as; time: for the consideration of the effects, in terms of delays, of the congestion or detours imposed by the solution being appraised. Despite the resistance of some researchers in attributing monetary values to such commodities, previous experiences have demonstrated that this can be done and that the results are reasonably accurate with the way people. The authors suggest that the use of value is positive and offers a feasible approach to making a structured analysis.

CONSIDERATIONS ABOUT BRIDGE UTILITY

Bridge utility is a complex concept to define. The examination of the classification criteria used by some authorities to define the priority of bridge activities can help shed some light on the factors that must constitute the Bridge Utility. The Sufficiency Rating Criteria of the FHWA considers the structural adequacy, the user safety, the serviceability, the functional obsolescence and the essentiality for public use. The Indiana BMS is based on the consideration of structural safety, traffic safety, community impact and investment effectiveness. As suggested by these examples, the importance given to a bridge is usually dependent on a combination of the several factors. Structural soundness is considered paramount but functional performance is gaining in importance because it is being recognised that the main role of bridges is to allow the efficient movement of people and goods.

One of the most important components of Bridge Utility is the actual value attributed by the various parties to the structure. First, there is the pure financial value of the asset, expressed by the amount of materials, time and work used in its construction. It is necessary to protect and safeguard this considerable investment and make sure that there is an adequate return from the investment made. Additionally, there is the functional value of the bridge, which depends upon the number of users and the importance of their trips. There is also the Human Value, expressed by the potential cost in lives due to accidents generated by substandard conditions in the bridge or in the case of collapse of the bridge because of excessive deterioration of its parts. Sometimes special characteristics can also raise the value of a specific bridge, such as aesthetics or historical considerations. Finally, political and legal circumstances could influence the importance given to certain structures.

Apart from the value of the bridge, the utility will have other components related to the effect of the structure on users and society. The saving of time that would otherwise be wasted using an alternative route is an example of the utility offered by the bridge caused to a user. Traffic delays [8], can cost as much as ten times more than the maintenance work. The DETR in the UK shows in a costing exercise [9], that the whole life maintenance costs of a concrete bridge of 323 m2 of deck area could be broken down as £1,163 for repairs, £2,490 for traffic management and £398,560 for traffic delays. This example clearly shows the need to investigate how these costs can be assessed and the criticality of including them in a structured way as part of the decision-making criterion

The value of an MRI intervention is closely associated with the notion of bridge utility. The relative weight given to each element will depend on the underlying system of values of the authority in charge of Bridge Management. This system of values is ultimately a reflection of the vision of public priorities held by the government and, as such, it is dynamic and might vary over time. Special circumstances can also sometimes significantly alter the existing system of values by swaying public opinion and creating more favourable conditions to push for MRI operations. For example, the 1994 earthquake in Northridge, US and the following one in Kobe, Japan in 1995 raised public sensitivity to seismic retrofitting enough to allow authorities in California to take advantage of the situation and pass a proposition committing £1.3bn to bridge repair and improvement. New legislation or changes in traffic pattern can render current maintenance policies obsolete and force the adoption of new aims and objectives. All these effects must be taken in consideration during the decision-making analysis and provisions to include them in some form in the decision models used in computer algorithms must be made. One possible way to express them is by using a utility function.

The Value for Money approach could be made by employing a subjective multi-attribute utility function to represent the effects of repairing a certain bridge structure. This approach is not aimed at "turning decision-making into a formulaic mechanical procedure" but "facilitates decision-making concerning complex issues by providing the decision-maker with an ordered structure within which to assemble and analyse a wide diversity of information" [10]. A similar treatment could be proposed to represent the individual utility of each bridge, as follows:

$$U(x_i) = UC_1(x_i) + \ldots + UC_i(x_i) + \ldots + UC_n(x_i) = \Sigma_n [UC(x_i)] \quad \text{(Eq. 1)}$$

Where $U(x_i)$ is the utility function of bridge x_i and $VC_n(x_i)$ are the various utility characteristics of the bridge x_i. What is important is the variation in utility resulting from the adoption of a certain MRI strategy, which could be represented by:

$$U(x, MS_k) = \Delta V(x,a_1)\, \delta 1_k + \ldots + \Delta V(x, a_i)\, \delta i_k + \ldots + \Delta V(x, a_n)\, \delta n_k \qquad \text{(Eq. 2)}$$

Where $U(x, MS_k)$ is the utility of MRI strategy k in bridge x, $\Delta V(a_i)$ is the variation in bridge value caused by the adoption of course of action a_i, n are the number of different courses of action considered and δi_k is a dummy variable that represents the adoption or not of the course of action a_i in the strategy k.

It would be possible to give different importance to these characteristics using a system of relative weights, as in equation 3. This is the type of utility function that will be adopted to express the various components of Bridge Utility.

$$\Delta V(x, a_1) = \alpha_1 VC_1(x, a_1) + \ldots + \alpha_i VC_i(x, a_1) + \ldots + \alpha_n VC_n(x, a_1) \quad \text{(Eq. 3)}$$

The Value of a MRI option was defined as being largely a result of the variation in utility of a bridge. The Bridges Utility is in turn a combination of various the components expressing the dimension of bridge value. Having concluded that cost is the basic dimension to which all value characteristics should be reduced, it is easy to understand why Economic Analysis has become the core of strategy appraisal in the decision-making process of contemporary BM practice.

ABMS MODEL

Since the assessment of impacts is vital in the determination of the variations in value, mechanisms to internalise the various impacts of maintenance were investigated. Special attention was given to environmental and user impacts. GIS tools were utilised to model the extent and effect of some of these impacts, Figure 2. The possibility of including second order impacts on society was discussed but it was concluded that these are still too difficult to model practicably. Work packaging was identified as a crucial factor, since the combination of works on two bridges in close proximity can contribute to an increase of value by reducing disbenefits and though the optimisation of resources.

SUMMARY

The results of the study are be used to guide the production of a prototype Advanced Bridge Management System for trial. A modular structure and open architecture is advocated, making updating, customising or upgrading the systems an easy task. Interactivity and clarity are two important design principles and the explicit use of the concept of value is considered an important step in creating a clear and understandable decision-making structure to support sound investment strategies and maximise welfare.

Figure 2. Illustration of Open Architecture

ACKNOWLEDGEMENTS

The authors would like to acknowledge the financial support of CNPq, the Brazilian agency for research and development and the School of Civil Engineering, University of Leeds.

REFERENCES

1.   Silva Filho, L.C.P., Unpulished PhD Thesis, Towards Advanced Bridge Management Systems, University of Leeds, 1999.

2.   Haas, R., 1993. Private Sector Role in Infrastructures Renewal: Challenges and Opportunities. In: McNeil, S.; Gifford, J.L. (eds.), *Infrastructure Planning and Management*. Proceedings of two parallel conferences sponsored by the Committee on Facility Management and The Committee on Urban Transportation Economics of The Urban Transportation Division, ASCE, 21-23 June 1993, Denver, Colorado, pp.41-45. (130)

3.   Scherer, W.T.; Glagola, D.M., 1994. Markovian Models for Bridge Maintenance Management. In: *ASCE Journal of Transportation Engineering*, v.120, n.1, Jan./Feb. 1994. pp.37-51.

4.   Dell'Isola, A., 1991. Every penny's worth. In: *Civil Engineering*, v.61, n.7, July 1991. pp.66-69.

5..   Bower, D.A., 1996. Evaluating the Indirect Cost of Change. In: Proceedings of the *IPMA '96 World Congress on Project Management*, Paris, June 24-26, 1996, Volume 2 – "Task". Paris: AFITEP. pp.375-382.

6.   Vassie, P.R., 1997 Whole Life Cosy Model for the Economic Evaluation of Durability . Options for Concrete Bridges, Das, P.C., Safety of Bridges, Thomas Telford, London, pp 145-150

7.   Milne, A. 1997, The Mother of All Reviews. In: *Building Homes*, Nov. pp.35.

8.   Tilly, G.P., 1997 Principles of Whole Life Cost, Das, P.C., Safety of Bridges, Thomas. Telford, London, pp138-144

9.   DOT, 1994. *Manual of Contract Documents for Highways Works*. Volume 1: Specifications for Highway Works (Original Version 1991, Amended August 1993). London: HMSO.

10.   BA 28 1992 Evaluation of Maintenance Cost in comparing alternatives for Highway Structures Part 2, design manual for roads and bridges, Volume1, section 2, London, HMSO.

A bridge management system for the South African National Roads Agency

P.A. NORDENGEN, Transportek, CSIR, Pretoria, South Africa, D. WELTHAGEN, S.A. National Roads Agency, Pretoria, South Africa, and E. DE FLEURIOT, Stewart Scott Int., Sandton, South Africa.

INTRODUCTION

Declining funds for road construction and maintenance in South Africa during the past ten years has resulted in more attention being paid to the preservation of the existing road infrastructure. The increase in the legal axle load from 8.2 to 9 tonnes in March 1996, pressure from the Southern Africa Development Community to implement a further increase to 10 tonnes and a low level of control of heavy vehicle overloading in most parts of the country do not help to alleviate the situation. Bridges and other road structures are key elements in any road network; effective management and proper maintenance of these structures is therefore essential. The economic benefits of using a systems approach to the management of structures have been proven by many authorities. Effective management requires that maintenance and rehabilitation be carried out when the greatest benefits are derived, as maintenance costs may increase substantially as serviceability levels of structures decline.

A Bridge Management System which was originally developed and implemented by the Division of Roads and Transport Technology of the CSIR for the Taiwan Area National Freeway Bureau during 1995 has more recently been modified and implemented for a number of road and rail authorities in southern Africa. The BMS was initially implemented for the city of Cape Town Municipality and Spoornet (the South African rail authority) during 1996/97, the Botswana Roads Department during 1997/98 and is currently being implemented for the South African National Roads Agency and the Western Cape Provincial Administration.

This paper describes various aspects of the implementation of the BMS for the National Roads Agency (NRA) as well as the relationship of the BMS with other systems in the Roads Agency's Integrated Transportation Information System (ITIS). A brief description of the BMS modules, their inter-relationship and the inspection rating procedure are also presented.

SYSTEM IMPLEMENTATION

The Roads Agency is responsible for all bridges on the national road network in South Africa. This comprises all major roads and freeways constructed to link the major cities such as Johannesburg, Durban, Pretoria and Cape Town. These 2 055 bridges consist primarily of road overpasses and underpasses, with a total deck area of two million m^2 and an accumulated bridge length of 140 kms. It is expected that in the future more of the urban freeways and other provincial roads will fall under the jurisdiction of the Roads Agency, which will significantly increase this total. In South Africa approximately 2.5 percent of the funds allocated to road rehabilitation projects is spent on bridge maintenance. Budgets are created with the help of the BMS. Bridge inspections are either carried out every five years under the BMS inspection programme or during road rehabilitation projects. In the latter instance bridges are inspected and repaired together with the road pavements under the same contract. The CSIR and Stewart Scott International (SSI) were awarded a contract to fulfil the following functions:

- Develop and implement a computerized bridge management system;
- Procurement of tenders for the first round of principal bridge inspections.

In order to expedite the process, the two project components were carried out simultaneously. The existing BMS's developed for the city of Cape Town and Spoornet were customized and enhanced to satisfy the requirements of the Roads Agency. The method of implementation is unique when compared with other systems.

Implementation of the bridge inspection programme

The work carried out comprised the following:

- Compilation of inventory sheets;
- Field inspections;
- Compilation of bridge inspection reports;
- Input of the inventory and inspection data into the computer using the BMS software.

For the successful implementation of this contract it was essential that persons were suitably qualified and experienced in bridge design and rehabilitation. Because of the crucial role the bridge condition survey fulfils in the BMS, and of the often-complex behaviour of bridge structures, it was felt that structural engineers with a reasonable degree of experience should be used to carry out the principal inspections. The following minimum requirements for inspectors were thus specified:

- Have a minimum of five years experience in bridge design/engineering;
- Be registered as a professional engineer or technologist in South Africa;
- Be available to carry out bridge inspections for at least 80 hours in any month;
- Be available to attend a three-day training course at own cost.

All who attended the training course were asked to submit curriculum vitae of prospective assistants. A bridge inspection team comprises the bridge inspector and an assistant, whose duty is to help the inspector during inspections. The inspector was required to provide training on bridge assessments to his assistant throughout the programme. It was indicated that assistants should be chosen from previously disadvantaged communities, as part of the Reconstruction and Development Programme (RDP) of the government of South Africa, by way of a joint venture with an associated firm.

Because of the large number of inspectors involved in the project, E-mail and the Internet provided an ideal means of communication. Through this medium all instructions were issued and all feedback received. Distribution of BMS software releases and updates was also done in this way.

It was decided at an early stage that the BMS would be operated in a paperless environment. All inventories and inspection reports are thus submitted in electronic format. Thirty-one digital cameras were purchased by the NRA and made available to the inspectors for their field photo capture. Approximately forty thousand JPEG format photographs were collected during these field inspections. As their total size precluded submission via E-mail or through the Internet, these photos were submitted either on CD or stiffy disk.

Allocations of bridges to inspectors
The road network was divided into three geographical areas, based on the travelling distance from the three regional offices of Pretoria, Cape Town and Pietermaritzburg. The following aspects were taken into consideration when allocating sections of roads to inspectors:

- The total bridge deck area allocated to each inspector was approximately the same;
- Proximity of bridges to the inspector's office location.

Control assessments of bridge inspections
Control assessments are carried out on bridge inspections by studying bridge inspection reports. The format of the written report, which includes digital photographs of all identified defects as well as standard inventory photographs, enables the client to validate bridge defect ratings. The validation is done in the form of spot checks on a small percentage of the total sample of bridges inspected. This is done as soon as the reports have been submitted. In exceptional cases, where there are significant differences between the control assessments and those of the inspector, an additional visit to the site is required by the inspector and a representative of the Roads Agency to review the ratings in question.

INTEGRATED MANAGEMENT SYSTEMS APPROACH IN THE ROADS AGENCY
The NRA is currently implementing an Integrated Transportation Information System (ITIS) to manage its assets and information. The ITIS Geographic Information System will form the kernel of ITIS. A central Oracle database will be shared by the various management systems integrated into ITIS. These include the Pavement Management System, Bridge Management System, Maintenance Management System and the Traffic Observation Management System.

Geomedia will be used as the stand-alone ITIS GIS software. GeoMedia Web Map will allow the publication of active maps on the NRA Intranet and the Internet. The software allows ordinary web browsers to be used for the extraction of and interaction with live GIS data. The required plug-in for Netscape Navigator and the ActiveX control for Internet Explorer are freely distributable to end-users. By clicking on specific features on the published active maps, viewers using the Intranet or Internet will be able to easily retrieve related information stored in the Oracle database. Depending on the purpose for which information is sought, different levels of access will be given to different viewers. Information of a general nature will be freely available to all viewers. Consultants appointed by the NRA will be given an increased level of access to the data relevant to the appointment.

The bridge management system will be used as a stand-alone system for the analysis of bridge data. Only the engineer responsible for bridge management in the NRA Head Office will be allowed to modify the BMS data stored in ITIS. Raw data collected from drawings and visual assessments of bridges will be input into the BMS and transferred to the ITIS database. When the data is required for analysis, it will be transferred back to the BMS. The analysis will be performed independently from ITIS after which the analysis results will be transferred to

ITIS. Both raw data and analysis results will be viewable by approved users using the browser technology described above.

COMPONENTS OF THE SYSTEM
The bridge management system currently consists of five modules as follows:

- Inventory
- Inspection
- Condition
- Budget
- Maintenance

The system developed for the Taiwan Area National Freeway Bureau also contains a Seismic module, as Taiwan experiences earthquakes on a fairly frequent basis. Each module contributes to a greater or lesser degree to the BMS database, and the modules are linked together as illustrated in Figure 1.

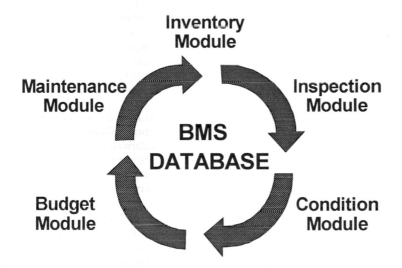

Figure 1. Modules in the BMS

Inventory module
The first step in the implementation of a BMS database is to compile the bridge inventory which consists of a record of all bridges in the network with comprehensive details of the type of bridge, construction materials, major dimensions, clearances, etc. This information is obtained from "as-built" plans and confirmed and/or measured in the field. Details such as loading and hydraulic data, where not available on drawings, are obtained from the design engineers. Depending on the availability of drawings, the collection of inventory data can be a costly exercise.

Inspection module

Each structure must be appraised at a network level with respect to its condition of serviceability and safety. The inspector completes standard inspection forms listing all the elements of a bridge structure with all the common defects normally encountered. Bridges have been subdivided into 21 items as follows:

1. Approach embankment	11. Parapet/handrail
2. Guardrail	12. Pier protection works
3. Waterway	13. Pier foundations
4. Approach embankment protection works	14. Piers & columns
	15. Bearings
5. Abutment foundations	16. Support drainage
6. Abutments	17. Expansion joints
7. Wing/retaining walls	18. Longitudinal members (deck)
8. Surfacing/ballast	19. Transverse members (deck)
9. Superstructure drainage	20. Deck slab
10. Kerbs/sidewalks	21. Miscellaneous items

The appraisal is carried out regularly for all bridges but may be required more frequently for steel bridges and bridges subject to foundation settlements or flooding. In South Africa, some bridge authorities carry out their own inspections whereas others, such as the Roads Agency, appoint one or more consulting firms to conduct the inspections. Principal inspections are carried out every three to five years, depending on the availability of funds. Monitoring inspections, to assess the deterioration of certain defects specified during principal inspections, as well as after major disasters such as floods, are carried out more frequently. The deterioration of structures is monitored by means of both principal and monitoring inspections.

Condition module

The condition module is used to prioritise the bridges in the system based on the most recent inspection data. The overall priority index is based on the priority and functional indices. The functional index gives an indication of the strategic importance of the bridge in the network and is calculated from various parameters in the inventory module. These include class of road or railway line, detour length, traffic volume, width between kerbs, type of structure and profitability of line (in the case of rail structures). Each parameter is given greater or lesser relative importance by user-defined weighting factors.

The priority index is based on the condition rating of the structure and is calculated from the D (Degree), E (Extent) and R (Relevancy) of each of the identified defects on each of the 21 predefined inspection items. More importance can be given to certain items such as deck slab, longitudinal members and piers as opposed to items such as guardrail and surfacing by means of user-defined weighting factors.

During an inspection, sub-items are inspected and rated individually, such as piers and deck spans. However, individual columns forming a single pier, or longitudinal members on one span, are considered as one sub-item.

A distinction is made between the condition index and priority index. The condition index gives an indication of the condition of the structure as a whole, taking into account each item and sub-item. For example, all nine piers (eight in good condition and one in poor condition) of a ten span bridge are included in the calculation. The priority index, on the other hand, which is used to determine the bridge ranking, only takes into account the worst rating of the sub-items of an item such as piers. Thus in the above example, only the one pier in poor condition would be used in the priority ranking calculation, and the piers in good condition are ignored.

Budget module

The main purpose of the budget module is to assist the bridge manager in allocating identified repair work into different budget years. The estimated quantities for repair that are done during inspections are used as a basis for determining budgets for the repair of each structure. During an optimisation procedure, the estimated cost of repair for each defect is compared with the relevancy of the defect to determine a benefit-cost ratio. In the case of limited budgets, maximum benefits can be achieved by first repairing items with the greatest reduction in risk to the road user and the lowest cost. In addition there is a facility whereby the bridge manager can overwrite the optimisation procedure by manually assigning selected bridges or types of repair work into a chosen budget year. The budget can then be re-optimised with the given constraints.

Maintenance module

In order to complete the cycle of the BMS, all maintenance activities that have been successfully completed are required to be entered into the system. This includes information such as actual quantity of work done, contractor, date, actual cost and any other significant comments. The system assumes that the defect no longer exists on the relevant item once the maintenance work has been indicated as complete.

INSPECTION RATING PROCEDURE

Perhaps the most important element of a bridge management system is the inspection rating or condition assessment procedure. The ability to accurately capture on paper the condition of the structure in terms of the structural integrity and the safety of the user has a major impact on the quality of the system outputs and ultimately determines the success of a BMS.

The method chosen to inspect bridges is very important in that it is the only tangible record that can be used for rating of bridges and for the repair budget predictions. Simple and more precise inspections result in more accurate analyses. The emphasis should thus be on more detailed inspections rather than superficial inspections for more accurate budget predictions.

In addition to rating identified defects, the inspector is also required to take at least one photograph of each defect, and in the case of the first inspection of a bridge, a number of standard photos of the bridge which are listed on the photographic record sheet. These include photos such as:

- Bridge from upper approach viewing along centre-line of deck (from both approaches)
- Along deck edge (both sides) - to record deck profile and deflections
- Bridge in elevation showing total deck length and full height of piers and abutments
- Underside of the deck
- Typical abutment and pier
- Upstream and downstream sides of the river from the bridge (if a river bridge)

The photographic sheet allows the inspector to write remarks about each photo and to record the direction of the photo and the camera photo number. These latter items greatly assist in the preparation of the inspection reports.

The DER rating system

The essence of a bridge inspection is to identify the defects on a bridge and their relative importance so that they may be prioritised and the available funds allocated efficiently for their repair. It is thus important to rate the degree of each defect (how bad is the defect) and the extent to which the defects exist on the respective inspection item (how common is it). However the most important purpose of the rating is to identify the consequences of the defect with regards the safety and serviceability of the bridge. This forces the inspector to not just give a visual rating of the defect but to look at the defect from a global point of view and to try and understand its influence on the structural integrity of the bridge. Because of the complexity of a bridge this last rating is very important; two defects that look the same may have significantly different influences on the bridge when one considers the safety of the motorist.

The rating system which has been used in the approach to condition assessments is referred to as a DER rating system and has the following components:

- D   represents the **degree** or severity of the defect
- E   is the **extent** of the defect on the item under consideration
- R   is the **relevancy** of the defect. This rating considers the consequences of the current status of the defect with regard to the serviceability of the bridge and the safety of the user (pedestrian, cyclist, motorist, passenger).

In addition to the above three ratings, the inspector is also required to rate the **urgency**, U, to carry out the remedial work to repair the defect. This rating considers possible future events that could adversely affect the defect, and provides a procedure for applying time limits on the repair requirements. Together with the urgency rating, the inspector is required to identify the remedial work activity (and estimated quantity) that must be carried out to repair the defect. The repair activity is selected from a standard list that is different for each bridge item. Activities include, for example, repair spalled concrete (all concrete items), backfill erosion/scour damage (approach embankment), remove sand, debris and vegetation (surfacing) and reinstate expansion gap between deck and abutment (abutments). Each of the repair activities has a unit rate that is used in the budget module to determine an *estimated* budget for the repair of the structure.

The rating is essentially a four point system (1 to 4), with the value of zero providing a way of identifying alternative meanings. The rating system is summarised in Table 1.

Table 1. Details of the four point rating system

Category	0	1	2	3	4
Degree (D)	Not applicable	None	Fair	Poor	Critical
Extent (E)	Unable to inspect	Local			General
Relevancy (R)	Uncertain	Minimum	Minor	Major	Maximum
Urgency (U)	Monitor only	Routine	Within 5 yrs	within 2 yrs	A.S.A.P.

It is possible to use one overall condition rating by combining the above three ratings but it is more difficult to be consistent. By considering each of the above ratings separately the inspector is able to concentrate on each aspect without confusing one for the other, and consequently obtain a more accurate rating of defects. It also simplifies the rating procedure and provides a more precise picture of the actual condition of the bridge to the bridge owner. With this method one can also produce more accurate budget predictions and maintenance, repair and rehabilitation actions to be used for preliminary work schedules used to carry out the work. In essence the bridge owner has a clearer and more accurate picture of the condition of the bridges in the network.

CONCLUSIONS

The BMS that has been developed and implemented for the South African National Roads Agency and various other road and rail authorities in southern Africa and in Taiwan has some unique characteristics:

- The inspection procedure focuses on defects only, making inspections simpler;
- A relevancy rating of each defect is required, which forces the inspector to evaluate the consequences of defects;
- Elements in good condition are not rated, thus reducing computer input;
- Selected defects can be monitored only, and can be excluded from the budget calculations.

During the implementation of the system a number of lessons have been learned:

- There is little to gain by using inexperienced personnel for carrying out principal inspections, as these are particularly important and play a key role in the BMS. Monitoring inspections may be carried out by less qualified personnel because they can use previously completed inspection sheets to compare ratings and in so doing learn from the experience of others. Furthermore, when using inspectors with adequate experience in bridge design, they are able to provide valuable advice to the client on recommended repair procedures.
- Adequate descriptions of photographs taken at the time of inspections greatly facilitates the compilation of bridge reports when returning to the office, particularly when a number of bridges are inspected during one trip. It is recommended that more photographs be taken than are actually required as these can be invaluable when discussing various defects with colleagues or the client, and may save an additional visit to the bridge. This is especially applicable to the use of digital photography where the additional cost of additional photos is negligible.
- Following a systematic approach during inspections ensures that all defects are noted and rated. An important lesson learnt is that inspectors need to pay attention to detail, as it is often the apparently minor defects that provide the solution to the cause of other major defects such as settlement and rotation.
- Special equipment for the inspection of bridges at network level is very rarely required. A good pair of binoculars is more than adequate for most bridges. A good quality camera with a flash and zoom lens was found to be essential. The use of digital photography was found to be highly efficient with regard to the subsequent incorporation of photos into the database.

ACKNOWLEDGEMENTS

This paper is based on projects carried out by the Division of Roads and Transport Technology, CSIR, in conjunction with Stewart Scott International and is published with the permission of the Directors both organisations. The permission of the South African National Roads Agency to publish information specific to the implementation and operation of their BMS and other systems is also gratefully acknowledged.

Testing moveable bridge operation

DR CHARLES BIRNSTIEL
Consulting Engineer, Forest Hills, New York, USA

INTRODUCTION

Movable bridges are machines and the span drive and stabilizing machinery of these bridges are subjected to faults as are other electro-mechanical-hydraulic devices. The faults may be caused by improper installation, wear, deterioration due to environmental conditions, structural modifications, foundation settlement, inadequate maintenance, and misoperation. Some faults may be diagnosed from visual observations of movable span motion and wear of machinery parts, and others, aurally.

However, quantitative measurements are helpful in diagnosing faults, especially on bridges equipped with sophisticated speed control devices. Such performance tests may aid in detecting; excessive imbalance of bascule and vertical lift spans, ineffective torque equalizers, defective braking, incorrect limit switch settings, or unfavorable adjustments of speed controllers for electric motors. In addition, improper operation of air buffers, trunnion bearings, and span guides can be detected from such tests.

Some of the parameters that are measured and recorded versus time and movable span position during performance tests are: torsional shear strain in shafts; voltage, current, and power drawn by electric motors; and fluid pressures.

The equipment required to perform such tests is not elaborate in terms of contemporary technology. Shear strains can be measured using foil-type electrical resistance strain gages in conjunction with signal conditioners/amplifiers. Voltage, current, and electric power measurements require only commercially available transducers. Draw-wire and capacitance-based transducers are available for measuring linear and angular displacements. A wide variety of transducers are manufactured for converting fluid pressure to a voltage signal. The analog signals may be recorded directly on a paper strip chart recorder or digitized and stored in a "personal" electronic digital computer for later data processing. Some tests we conducted since the early 1980's required simultaneous recording of 10 channels of data; strain gage, electric, fluid pressure, and displacement.

With increasing frequency bridge Owners are specifying that transducers and data collection equipment be included as part of new permanent machinery and control installations. The objective is to improve bridge management efficiency.

Two case studies of bridge operating tests are cursorily described subsequently. They are; 1) measurement of the imbalance of a bascule leaf by the dynamic strain gage procedure, and 2) an electric performance test of a span drive vertical lift bridge.

BASCULE LEAF IMBALANCE MEASUREMENT

In the dynamic strain gage procedure for measuring bascule leaf imbalance, strain gages are affixed to shafts of the drive train in a manner that will indicate the torsional strain on the shaft surface during operation. In effect, custom transducers are manufactured in place. Strains during leaf opening and closing (when wind velocity is low) are recorded on a strip chart and then numerically analyzed to obtain curves of leaf imbalance moment (M) versus leaf opening angle (theta). It has become conventional to compute the coordinates of the center of gravity (R, alpha) based on an assumed total leaf weight. Notation is defined in Figure 1.

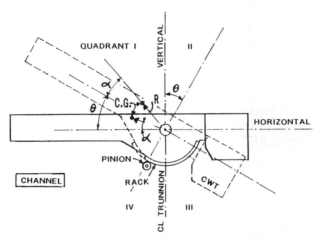

Figure 1. Notation for location of center of gravity.

The bridge tested is a 50-year-old simple trunnion bascule with a forward leaf length of about 70 ft (21m), that is powered by an electro-mechanical drive as shown in Figure 2. The total leaf weight was taken as 1050 kips (477 tonnes).

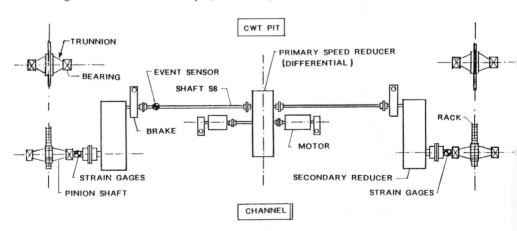

Figure 2. Arrangement of span drive machinery.

Two strain gage rosettes were attached to each pinion shaft, back-to-back (spaced 180 degrees apart, circumferentially) in order to minimize the shaft bending effect. Each rosette comprised two foil uniaxial grids that were oriented 90 degrees to each other and bonded to a stainless steel shim. The shims were spot welded to the shafts so that the angle between a foil grid and the shaft axis was 45 degrees. The two rosettes on each shaft were wired in the four-arm Wheatstone Bridge configuration and connected to one channel of a signal conditioner/amplifier. The output from the signal conditioner/amplifier was wired to a Linseis strip chart recorder which produced a trace of torsional shear strain versus time.

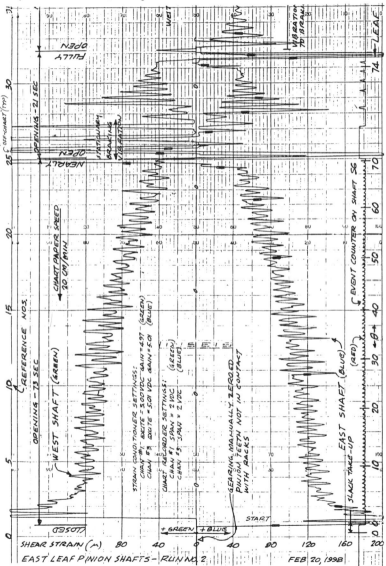

Figure 3. Strip chart of shear strain recorded while raising (opening) leaf.

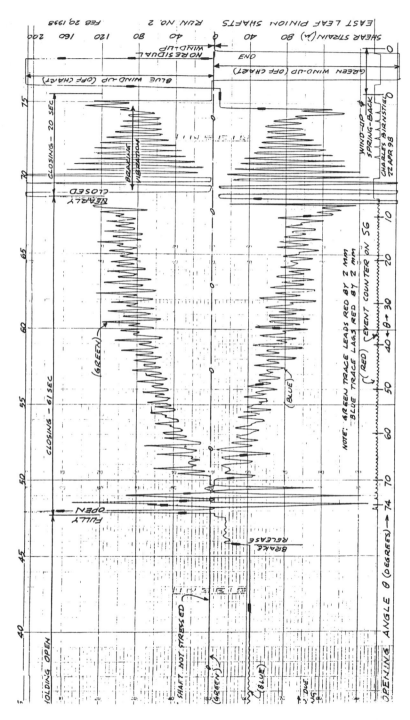

Figure 4. Strip chart of shear strain recorded while lowering (closing) leaf.

In order to record leaf position corresponding to the shaft shear strains, an event sensor was mounted on Shaft S6 and the output fed to the third channel of the strip chart recorder. One magnet was attached to the shaft. The event marker pen (red) made a trace on the lower edge of the chart paper. In this way all three pen traces, the two for shaft strains and that for leaf position, are related because they a simultaneously recorded on the same strip chart.

Figure 3 is a photocopy of the strip chart produced during an opening of the leaf and Figure 4 is that made during the closing of that cycle. Note that the positive signs for the two traces are in opposite directions so as to minimize superposition of traces. The three pens are offset (in the longitudinal direction of the chart) to avoid physical interference during recording. When reading the strip chart, the pen offset (See Figure 4) should be taken into account.

The shear strain traces of Figure 3 in the range 3 < theta < 68 degrees and of Figure 4 in the range 68 > theta > 10 degrees reveal the following:

o   Gearing has significant backlash (likely due, in part, to wear) as indicated by the small fluctuations of the traces having a short period (1 sec).

o   The differential in the primary speed reducer is functioning as intended because the pinion shaft torques are essentially alike at any angle.

o   The leaf is span heavy (and substantially so) within the range because positive torque is required to raise the leaf and positive torque is required to hold the leaf back during lowering.

o   Acceleration time is too short at all speed changes; resulting in unnecessarily high stresses in shafting.

o   Stops at Nearly Open and Nearly Closed positions are sudden, creating high strains due to braking and causing leaf to vibrate severely.

o   Only one shaft is holding leaf in the open position.

Three imbalance measurement runs were recorded for each leaf. The shaft shear strain data shown on the strip charts was reduced to digital form by dividing the area under the strain traces in the constant (almost) velocity region into segments and determining the average ordinates to the traces for each segmental area. The torques in the West (green) and the East (blue) pinion shafts were computed from these strain ordinates using basic structural mechanics. These torques were added together at common opening angles and multiplied by the gear ratio of the racks to the pinion shafts (which is 9.16 to 1) giving the moment (M) acting on the leaf. These values were plotted versus the opening angle in Figure 5. A positive moment indicates moment necessary to open the leaf during opening or a retarding torque during closing.

A curve representing M = WRcosine(theta+alpha) was computer-fitted through the opening and closing data. W = total weight of leaf plus counterweight and the other terms are defined in Figure 1. The difference in the ordinates between the opening and closing curves represents, at any angle, the sum of the opening and closing friction due to the rack and pinion engagement and the friction in the trunnion and pinion shaft bearings.

The friction during closing is not necessarily equal to that during opening at the same leaf angle. Furthermore, the friction may vary with the opening angle. For this bridge the friction is very uniform with respect to the opening angle, indicating that the trunnion bearings function well and there is little, if any, binding in the rack/pinion engagement.

On the assumption that the friction during opening equals that during closing at any opening angle, average points were computed from the initial cosine fits. Through these points a cosine curve was computed-fitted. This curve (thick curve) gives the probable imbalance moment of the leaf referred to the trunnion axis, at any opening angle. It shows that the leaf is span-heavy for all angles of opening. The values of R and alpha, sufficient to locate the center of gravity of the moving mass in one plane, are given in the figure. The means of R and alpha from the three runs were averaged to give the leaf imbalance.

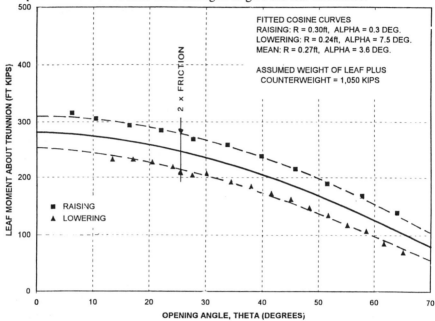

Figure 5. Imbalance curve, East Leaf, Run No. 2.

ELECTRICAL PERFORMANCE TEST OF A VERTICAL LIFT BRIDGE
A 60-year-old double track railroad bridge of the span drive vertical lift type, as shown in Figure 6, was experiencing electrical control problems shortly after replacement of the original electrical system. The new normal span drive is powered by two 250 HP AC wound

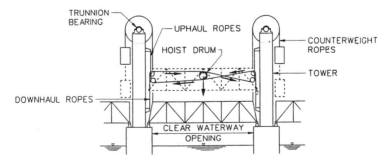

Figure 6. Span drive vertical lift bridge.

rotor motors with a synchronous speed of 900 RPM. Permanent resistance is connected into in series with the secondary windings to reduce the full load speed to 675 RPM. In this way the operating characteristics of the motor are improved for this application and the rated full load torque of the motor is maintained. Speed control is accomplished by thyristor controlled primary voltage drives utilizing feedback control.

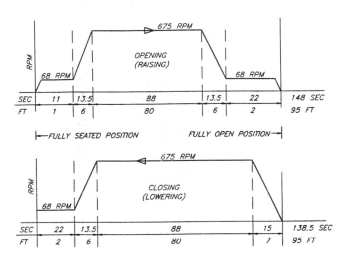

Figure 7. Design bridge operating profile.

The bridge structure is of substantial construction with a lift span of 332 ft (101m) and the total weight of the moving mass is 7200 kips (3270 tonnes). However, although the structure is rigid, the wire rope span drive is not, and the "softness" of the mechanical system causes difficulties for the feedback control. It was reported that bridge operation was frequently aborted by the over-speed switch. It was also observed that the motor currents were high and fluctuated widely.

A test program was implemented that eventually required 18 span lifts. During each run the following parameters were simultaneously recorded on a strip chart; motor primary current, primary power, secondary current, tachometer-generator output voltage (the feedback), and revolutions of the transverse shaft. The transverse shaft drives all four wire rope hoist drums and its rotation is a measure of lift span position (after accounting for rope stretch).

Figure 8 is the first part of the strip chart that was produced while raising the lift span in Run No. 5. Note the large fluctuations in the feedback signal (which measures creep speed), primary current, and power. As marked, there was improper adjustment of the brake mechanism which resulted in early application of brake torque. With the object of improving performance, successive adjustments to the feedback signal lead time and the system gain were made on the speed regulator card. The adjustments were actually more involved than describe herein. Suffice it to say that the adjustments were successful in obtaining smoother operation and reducing the swings of current. However, because the basic soft mechanical system could not be altered, there was a practical limitation to the improvements.

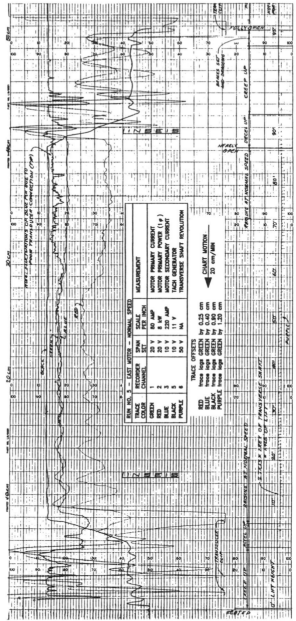

Figure 8.   Strip chart - Raising span with East Motor only driving at normal speed.

SUMMARY

Measurement of selected parameters and recording them with respect to time can aid in diagnosing movable bridge operating faults and planning corrective action. They can be made with comparatively inexpensive electrical components that are commercially available. Bridge management may be improved by utilizing this technology.

Full-scale model testing to investigate the sheer capacity of concrete slabs with inadequately anchored reinforcement

DENTON, S.R., PB Kennedy and Donkin Limited, Bristol, UK, IBELL, T.J., Bath University, Bath, UK, POSNER, C.D., PB Kennedy and Donkin Limited, Hatfield, UK.

INTRODUCTION

The current programme of bridge assessments in the UK has identified that many existing concrete bridges do not satisfy the current design requirements for reinforcement anchorage at supports, in some instances resulting in an apparent shortfall in shear capacity. This particular problem has been addressed, in part, in the assessment standard for concrete highway bridges, BD44/95[1], which permits an effective bar size to be used in assessing shear capacity based on the actual anchorage length present. Research has shown, however, that the requirements of this code can be conservative[2,3].

The assessment of a substantial number of similar reinforced concrete approach-spans to overbridges on the M4 has highlighted that they all appear to have insufficient shear capacity due to a particular reinforcement detail that appears not to provide sufficient anchorage at a support. Unfortunately, the anchorage present in these structures, nominally 1.25 bar diameters beyond the bearing centreline, is shorter than that for which test results have been reported, so the findings of this previous work could not be extended to their assessment with confidence. However, there were a number of reasons to suggest that their actual shear strength would exceed the capacity determined in accordance with BD44/95.

A programme of full-scale laboratory-based model tests has therefore been undertaken to establish the shear strength of these slabs. An innovative testing method has been developed, combining proof load tests and failure tests, to overcome the particular problem of accounting for the variation in shear strength with load position (or shear span).

OBJECTIVE

The objective of the testing programme was to establish, with confidence, whether the shear capacity of the approach-spans exceeded that necessary to sustain the most onerous effects due to live loading. Whilst it was recognised that the resulting test data might assist in reviewing the assessment requirements for anchorage at supports, this was not the principal focus of the project.

TESTING PHILOSOPHY

A detailed analysis of different options for loading regime and specimen geometry was undertaken to establish the most advantageous and cost effective configuration, and particularly so that:

(i)   The specimens behaved principally in plane strain;

(ii)   The behaviour of transverse reinforcement was realistically modelled; and,
(iii)   The lateral distribution of shear force in the specimens reflected that experienced in
        the approach spans.

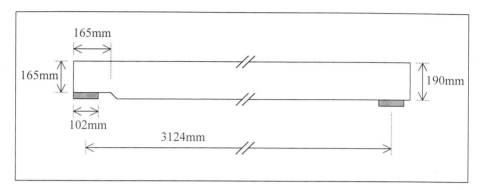

Figure 1 : Specimen Geometry

It was concluded that full-scale specimens should be tested, matching the geometry and material properties of the existing slabs as closely as possible, but with a width of 0.508m (20"). The specimen geometry is shown in Figure 1. Furthermore, it was concluded that a discrete load should be applied close to the critical end of the specimen, and across its full width.

One disadvantage of using a discrete load is that the shear strength of reinforced concrete slabs or beams is known to be sensitive to the position of the applied load, as identified by Kani[4]. This effect results, in part, from changes in the shear failure mechanism as the load is moved away from the support (see Clark[5]).

Level	Loading
Load Level 1	BD44 strength assessment load (unfactored) with no enhancement
Load Level 2	Intermediate Load Level, between levels 1 and 3.
Load Level 3	$\gamma_{fl} \times \gamma_{f3} \times$ Assessment Live Load effect
Load Level 4 - Target Level	$\gamma_m \times \gamma_{fl} \times \gamma_{f3} \times$ Assessment Live Load effect

Table 1 : Proving Load Levels

It was essential therefore that the testing procedure accounted for this variation in capacity, particularly as it was thought that the critical load position might differ from that found in previous shear tests because of the variation in slab depth close to the support (see Figure 1) and because of the extremely short anchorage. In order to overcome this problem and derive the greatest amount of data from each test (thereby increasing the cost-effectiveness of the testing programme and limiting the number of specimens required) a testing scheme was developed combining both proving load tests and a failure test for each specimen.

The purpose of the proving load tests was to establish that the slabs had sufficient shear capacity to carry a particular loading applied over the full range of potentially critical load positions. The proving load tests were undertaken at four different load levels, as described in Table 1.

Load Position	Location	Load Level							
		1 (kN)		2 (kN)		3 (kN)		4 (kN)	
1	1.0d	40		58		77		97	
2	1.5d	41		60		77		97	
3	2.0d	42		61		76		96	
4	2.5d	43		63		76		95	
5	3.0d	45		65		75		95	

Table 2 : Applied Proving Loads

Experience indicates that the critical load position is highly likely to lie within the range d to 3d, where d is the effective depth of the section, equal to 130mm. The procedure for undertaking the proving load tests was, therefore, to apply each load level at load positions giving a clear separation of 3d, 2.5d, 2d, 1.5d and d between the load and support bearing in turn before progressing to the next load level. The fourth load level was the target level corresponding to the required capacity to sustain Assessment Live Loading in accordance with BD21/97[6]. The load levels used in the tests are summarised in Table 2.

SPECIMENS
The properties of the test specimens were matched as closely as possible to the existing approach-slabs. The main longitudinal reinforcement in the bottom face of the test specimens comprised five 20mm deformed bars at 102mm spacing. Three 12mm deformed bars were placed longitudinally in the top face of the specimens. Transversely, 12mm bars were placed in the top and bottom faces of the specimens at spacings of 204mm and 102mm respectively. The cover to the reinforcement was 25mm.

On the as-built drawings the bearing at the critical end of approach-spans were shown to be a continuous rubber strip 4" wide and approximately ½" in depth. Similar bearings were used in the tests.

The as-built drawings showed a nominal anchorage of the main longitudinal reinforcement beyond the centreline of the bearing equal to 25mm. From considerations of contemporary tolerances, those specified on the as-built drawings and statistical analysis of anchorages measured following the exposure of some bars on one structure a "worst credible" anchorage beyond the centreline of the bearing was determined as 6mm.

Eight specimens were tested, four having an anchorage beyond the centreline of the bearing of 25mm and four having an anchorage of 6mm. The tests were undertaken in two phases with four tests in each phase, as summarised in Table 3.

The as-built drawings specify a concrete grade of 6000/¾, corresponding to a characteristic cube strength of 41MPa and an aggregate size of approximately 20mm. Results of tests on cores taken from existing slabs indicated strengths somewhat higher than 41MPa, as would be expected. The slabs were tested between 31 and 42 days after casting. The concrete used in the first and second phases of testing had cube strengths of 33MPa and 50MPa when the tests were undertaken. These were intentionally at the lower end of strengths likely to be encountered in the whole population of approach-spans.

Test	Phase	Anchorage	Concrete cube Strength	Failure Test Load Position	Notes
C1	1	6mm	33 MPa	3d	Worst Credible Anchorage Test
C2	1	6mm	33 Mpa	2.5d	Repeat 1 of Test C1
C3	2	6mm	50 MPa	2d	Repeat 2 of Test C1
C4	2	6mm	50 MPa	1.5d	Repeat 3 of Test C1
C5	1	25mm	33 MPa	3d	Nominal Anchorage Test
C6	1	25mm	33 MPa	2.5d	Repeat 1 of Test C5
C7	2	25mm	50 MPa	2d	Repeat 2 of Test C5
C8	2	25mm	50 MPa	1.5d	Repeat 3 of Test C5

Table 3 : Test Specimens

TEST PROCEDURE

The test rig is shown diagrammatically in Figure 2. The slabs were loaded using a hydraulic jack (under displacement control), with the load applied through a load cell and a loading beam to distribute the load laterally across the specimen. PTFE sheets were placed between the load cell and loading beam to minimise any lateral load applied to the specimen. Furthermore, a rubber pad of known (and low) shear stiffness was also placed between the load cell and loading beam. The relative lateral displacement of the top and bottom of this pad was measured during the tests using electronic displacement transducers to enable any lateral load applied to the specimen to be measured. A rubber strip 4" wide and ½" thick was placed between the loading beam and specimen.

Rather than moving the hydraulic loading jack, it was decided to place the test specimen on a rigid frame which was itself moved between each proving load test to enable the load to be applied at the required position.

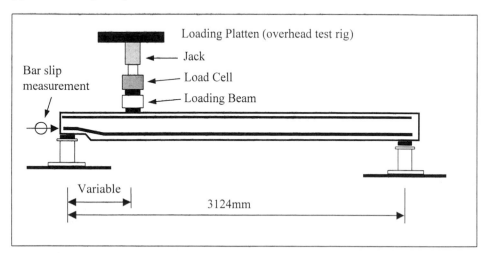

Figure 2 : Test Rig

Electronic displacement transducers were used to measure the deflection of the slab. These were connected to an electronic data logger, which simultaneously recorded the load applied to the specimen as the displacement of the slab increased. Displacement transducers were also used to measure the horizontal displacement of the specimen.

During casting, a small cavity was formed in the end face of the slab using a short length of plastic tube so that the slip of one reinforcing bar could be measured using a displacement transducer. The displacement measured by this transducer was also recorded using the electronic data logger.

RESULTS

Proving Load Tests
All eight slabs carried the proving loads in all positions. No shear cracks were observed. Flexural cracking occurred under the load and between the load and the further support (i.e. outside the shear span). The crack spacings were reasonably constant and approximately equal to 100mm. The flexural cracking was greatest with the load 3d from the support and diminished as the load approached the support.

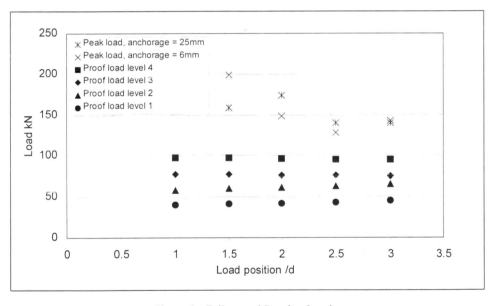

Figure 3 : Failure and Proving Loads

Failure Load Tests
The load positions for the failure tests and the failure loads are summarised in Table 4. The load positions refer to the clear spacing between the loading and the support bearing and are given as a multiple of the effective depth of the slab at the support, which is equal to 130mm.

The variation in Failure Load and Proving Load with load position are plotted in Figure 3.

Specimen	Anchorage	Load position	Failure Load
C1	6mm	3d	143 kN
C2	6mm	2.5d	128 kN
C3	6mm	2d	149 kN
C4	6mm	1.5d	199 kN
C5	25mm	3d	140 kN
C6	25mm	2.5d	140 kN
C7	25mm	2d	174 kN
C8	25mm	1.5d	159 kN

Table 4 : Failure Load Test Results

Load deflection plots for the Phase 1 and Phase 2 slab tests are shown in Figures 4 and 5 respectively. Similar failure sequences were observed in all cases, as follows:

(i)   A small crack initiated from the change in section depth 52mm from the support between a load level of 100 and 120 kN.

(ii)   At failure a complete shear crack developed from the edge of the bearing towards the edge of the loading pad. The reinforcement slip measurements indicated that the formation of this shear crack and initiation of significant slip of the reinforcing bars occurred simultaneously.

(iii)   With the formation of this shear crack, the load carried by the specimen immediately reduced but remained greater than 80 kN in all cases.

(iv)   As further displacement was imposed on the slab, the load was found to increase (with the exception of slab C4). At the same time, the initial shear crack widened, further cracking occurred and the slip of the main longitudinal reinforcement increased.

(v)   As the displacement was further increased, the applied load decreased.

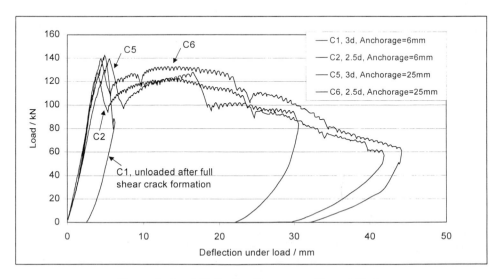

Figure 4 : Load-Deflection Response – Phase 1 Tests

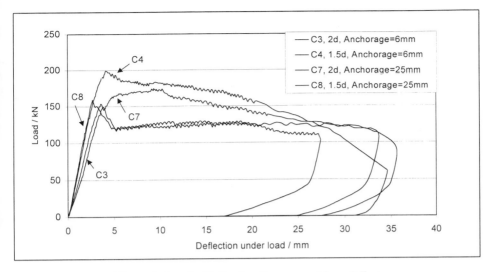

Figure 5 : Load – Deflection Response : Phase 2 Tests

DISCUSSION

The test results indicate that the shear capacity of the slabs is very much greater than that predicted by BD44/95. It appears, therefore, that the bond developed by the reinforcing bars over the support bearing is far greater than predicted. This is not entirely unexpected. Firstly, the assessment code only considers reinforcement anchorage beyond the centreline of a flexible bearing and secondly the concrete over the support bearing where the bars are anchored is under considerable compression.

The "bond" behaviour of deformed bars is characterised by the formation of inclined radial compression struts equilibrated through the development of hoop tension in the concrete around the reinforcement (see e.g. Reynolds[7]) . Bond failure is thus generally observed to occur through splitting failure of the concrete on a plane passing through the reinforcement. The width of the test specimens and the presence of transverse reinforcement tended to prevent the formation of vertical "splitting" cracks and the compression from the bearing acted against the formation of horizontal "splitting" cracks. Both these effects will have increased the effectiveness of "bond" in the anchorage region.

An important, and somewhat unexpected, aspect of these tests was the ductile response recorded for all specimens. In all cases, the ultimate capacity was reached in a controlled failure, with much energy clearly being dissipated by the constituent materials at more-or-less constant rate, usually at a load above the target load level. The shear cracks started to form at an average of about 2/3 of the final collapse load, in line with previous research on concrete bridge decks[8].

From the measurements of the longitudinal shear displacement of the rubber pad below the loadcell, it was possible to estimate the horizontal membrane force present in each specimen during loading. In all cases this horizontal force was found to be very low just prior to the maximum load being applied. Although the horizontal force increased with increasing displacement after the formation of a complete shear crack, it remained low and never exceeded about 10% of the ultimate applied load. This result provides confidence that horizontal forces developed had a negligible effect on the ultimate capacity, although it is conceivable that they did contribute in part to the ductility of the response at very large displacements.

From Table 4, the influence of the load position on the failure load is not immediately obvious. This is because specimens C1, C2, C5 and C6 were of lower concrete strength than the others. However, it does appear that a load position in the region of 2.5d is critical. This matches well with the results of Kani[4], even with a very short support anchorage and a stepped support region. It can also be seen from Table 4 that the longer anchorage length did not always give a greater peak load, although this was the case for what appears to be the critical load position.

CONCLUSIONS
The test results indicate that the approach-spans have sufficient capacity to carry Assessment Live Loading, and therefore, the objective of the research has been met. Using a novel laboratory testing procedure, it has been possible to establish this result with confidence across the entire critical shear span range (from d to 3d) using only eight test specimens.

The research also suggests that in some instances, particularly when slabs are supported on continuous bearings and are reinforced with deformed bars, the effect of very low support anchorage on shear capacity is much less than that predicted by the current UK bridge assessment code.

ACKNOWLEDGEMENTS
The authors thank the Highways Agency for permission to publish this paper. The views expressed in this paper are not necessary those of the Highways Agency. Particular thanks are expressed to Mr B Barton of the Highways Agency and Mr J Shave of PB Kennedy and Donkin Ltd.

REFERENCES
1.   The Highways Agency, *The assessment of concrete highway bridges and structures,* BD44/95, 1995.
2.   Clark, L.A., Baldwin, M.I. and Guo, M. 'Assessment of Concrete Bridges with Inadequately Anchored Reinforcement', Bridge Management 3, Ed. J.E. Harding, G.A.R. Parke and M.J. Ryall. E&FN Spon, London, pp223-232. 1996.
3.   Cullington, D.W., Daly, A.F. and Hill, M.E. 'Assessment of Concrete Bridges: Collapse Tests on Thurloxton Underpass', Bridge Management 3, Ed. J.E. Harding, G.A.R. Parke and M.J. Ryall. E&FN Spon, London, pp667-674. 1996.
4.   Kani, G.N.J., 'The riddle of shear failure and its solution', ACI Journal, Proceedings, Vol 61, No. 4, April 1964, pp 441-467.
5.   Clark, L.A. Concrete Bridge Design to BS5400, Construction Press, London, 1981.
6.   The Highways Agency, *The assessment of highway bridges and structures*, BD21/97, 1997.
7.   Reynolds, G.C. Bond strength of deformed bars in tension. C&CA, Technical Report 42.548, May 1982, pp23.
8.   Ibell, T.J., Morley C.T and Middleton C.R. *An upper-bound plastic analysis for shear*. Magazine of Concrete Research, Vol.50, No.1, March 1998, pp67-74.

Three-dimensional centrifuge test of Pontypridd Bridge

C. SICILIA, University of Wales Postgraduate Research Student, T.G. HUGHES, Reader, Cardiff University and G.N.PANDE Professor, University of Wales, Swansea

ABSTRACT
In 1756 William Edwards finished Pontypridd Bridge, one of the most significant historical bridges in Wales. At a span of 44m, Pontypridd Bridge remained the longest single span UK bridge for several decades. This paper describes the building and testing of a 1/55th linear scale stone model of Pontypridd Bridge tested at the equivalent 55 gravities in a geotechnical centrifuge scale. The paper contains details of the construction of the model, including the cylindrical voids so necessary to the stability of the structure, the model instrumentation and the testing of the model to destruction. Conclusions are drawn on the behaviour of 3-D masonry arch bridges.

INTRODUCTION
There has been a significant recent effort concentrating on the experimental behaviour of masonry arch bridges. Series of interesting simplified test models had previously been performed[1] that have been very useful for the early stages in calibrating numerical simulations. More recently a series of ten tests on redundant structures have been undertaken by the TRL and are reported in a state of the art review[2]. These tests have been very useful from a mechanistic point of view. However, they have been of a limited use in the calibration of numerical models, because of the lack of geometrical and material information, inherent to the tests on redundant structures. Series of small full-scale laboratory tests have been performed at different institutions[3,4] and these have proved more useful in calibration studies, although the stress states within these structures were limited by their small dimensions.

More recently the Masonry Research Group in Cardiff University has started to use centrifuge modelling to produce masonry arch bridges test results. Initially the centrifuge-modelling technique was validated and subsequently a series of parametric studies on two-dimensional models has been initiated. The work presented here is part of a broad research program to investigate the 3D behaviour of masonry arch bridges[5]. The test presented in this paper is the first three-dimensional model equivalent of a masonry arch of significant dimensions tested in a laboratory.

The test detailed has the advantage over tests on redundant structures, in that the proportions and dimensions are real and the geometry, material properties and boundary conditions are known. Working with small structures also greatly reduces costs, time and equipment. The only disadvantage is the relative geometrical peculiarity of the particular structure chosen. This bridge was selected as it represents a significant challenge and is a historic structure of considerable interest.

The main objectives of the work are to obtain mechanistic information, to help in understanding the behaviour of masonry arch bridges, particularly related to the behaviour of the spandrel walls. The test also provides quantitative information, i.e. a complete set of deformations. In addition a detailed post mortem following the test has been undertaken.

MODEL SCALE AND CENTRIFUGE MODELLING
In masonry arch bridges, the importance of gravity conditions when working at small scale has two main aspects:
- Self-weight is the main stabilising force on masonry arch bridges and is therefore essential for modelling to properly include gravity. This is not the case with beam structures where self-weight is a destabilising action and small in comparison with the live loads.
- Soil and masonry behaviour is stress dependent and so, to have the same behaviour in the scale model and on the full-scale version, identical stresses are required.

A simple scale analysis shows that for these two effects to be respected, a straightforward solution is the use of materials n times heavier (where 1/n is the scale of the model) and of identical strengths and deformational properties. The centrifugal acceleration is used to provide enhanced gravity conditions, which effectively make the materials n times heavier.

The work presented in the current paper is part of series of tests at 1/55 scale. The big jump on scales between this and the scales used in the past required us to first study the repeatability of the tests. This was undertaken with two identical two-dimensional models. Satisfactory results allowed us to move onto more complex 3-D structures, from which the current result is presented.

A more detailed description of centrifuge modelling, scale laws, of this preliminary work and on Cardiff Geotechnical Centrifuge can be found elsewhere[6,7].

PONTYPRIDD BRIDGE
The bridge selected for the test was Pontypridd Bridge[5] over the river Taff valley and is shown in Figure 1. The bridge was completed, after several attempts, in 1756 by William Edwards. The first attempt was a three or four span bridge that was washed away by the river Taff. The following attempt, a single span bridge, failed on decentring due to the excessive weight on the haunches, opening at the crown. After that, and according to some chroniclers even another failed attempt, the bridge, as we know it, was finished. The features of that very Welsh structure are the result of this long history. The 44m long span, which remained the longest in Britain for several decades, protected it from the river. The six cylindrical holes on the haunches, the possible use of charcoal as backfill, the hump at the crown, as well as the stepped backfill profile to obtain an optimum distribution of weight, to be resisted by a dramatically slender arch (1/56) were the consequence of the economical problems faced by the builder.

The final structure seems now to have passed the exam of history as, near its 150 anniversary, it remains intact and in use for the former coal mine community of Pontypridd.

DESCRIPTION OF THE MODEL
Model and scales: The model contains the arch, backfill material, spandrel walls (inc. parapet walls) and rotational restriction of the wing walls. The global dimensions are scaled at 1/55, however for the masonry constituents a different scale is used. The homogenisation technique[8] assumes that, provided the unit is of a small scale in relation to the structure, it is only necessary to ensure that the relative size of the unit and mortar joints is correct. Using

this theory and due to the impossible difficulties that working with 1/55 scale block and mortar joints would produce, the masonry on the arch is of 1/10 scale in the circumferential directions and 1/55 in the radial (ring thickness) direction. Similarly the spandrel walls are at 1/20 scale in their front elevation but are 1/55 in their thickness.

Figure 1. Pontypridd Bridge

<u>Materials</u>   The masonry fabric is made out of Pennant sandstone blocks and lime mortar. The latter was selected, primarily because it was used on Pontypridd, but also because the bond strength of the mortar is important in the modelling of arch bridges. Previous workers have detected a considerable variability on model tests and have associated it with the variability of either the shear bond strength of masonry[9] or the masonry tensile strength[10]. The use of a very weak mortar reduces this dependency and good repeatability has been achieved on previous tests of the series. It is important to note that the structure has a single ring, which excludes ring separation and allows us to concentrate on the global concepts.

The backfill material was not chosen to match the one used in the real structure, mainly for the lack of consistent information. However as the primary goal of the test was to produce sensible results to be used on calibration, this was not seen as a limitation. A well-graded local limestone was chosen and compacted over the structure at its optimum moisture content (7%). A comprehensive materials test program was carried out and the results are presented in Figure 2. This includes constituents and composites properties as well as interfaces properties.

<u>Construction:</u> The construction process of a model structure of such a scale is complex. A general view of the model is contained in Figure 3. Construction was naturally undertaken at unit gravity. The main problem was building the spandrel walls; these needed to be constructed prior to compaction of the fill, as otherwise the backfill material would dry and harden. As a result, the spandrel walls were required to be sufficiently strong to absorb the stresses of the compaction. Following the completion of the arch ring, laid against an aluminium former, the two walls are build against fixed vertical faces and include the cylindrical holes. After 28 days the fill is laid and compacted with the wall supports still in place. To generate the cylindrical holes within the fill, PTFE tubes were put in the appropriate position and pulled once the filling and compaction was complete. Additional fill was used on the side remote from the live loading so as to reduce the effect of the rigid support provided by the strong box and is shown in Figure 4. After this, the two lateral wall supports were removed. The arch ring former was removed at this stage.

The model includes some rotational restriction of the wing walls. A 5 mm thick neoprene sheet was used between the fill and the strongbox. This had the advantage of reducing the space required and to provide a well-defined boundary conditions easy to model numerically.

PONTYPRIDD BRIDGE		Scale: 1/55th
GEOMETRY (Prototype)		
Profile		Segmental arch
Radius (intrados)		27.08 m
Span (intrados)		42.73 m
Rise at centre span (intrados)		10.44 m
Arch thickness		0.76 m
External width		4.95 m
Spandrel and parapet walls thickness		0.47 m
Slope surface		20 %
Cylindrical holes (diameter)		2.7, 1.7 and 1.0 M
MATERIALS (model)		
Stone		Pennant sandstone
Dimensions arch blocks		12.5 x 14 x 18.1 mm
Dimensions spandrel blocks		7.5 x 8.5 x 19
Specific weight		2.61 T/m^3
Compression strength		77 MPa
Poisson's ratio		0.15
Young's modulus (Sec 30%)		18235 MPa
Mortar		*(3:1 sand:lime in volume)*
Compression strength (25 mm cube)		1.38 MPa
Young's modulus (Sec 30%)		1970 MPa
Poisson's ratio		0.181
Tensile strength (flexion test)		0.58 MPa
Soil		Crushed limestone
Angle of friction		45 °
Cohesion		0.172 MPa
Max dry density		2.374 T/m^3
Maximum particle size		5 Mm
Arch masonry (wallettes)		3 by 5
Compression strength		33.80 MPa
Young's modulus (Sec 30%)		3104 MPa
Interfaces		
Masonry-soil on shear	Angle of fric.	35 °
	Cohesion	0.024 MPa
Masonry on shear	Angle of fric.	32 °
	Cohesion	0.063 MPa
Bond wrench test		0.017 MPa

Figure 2. Geometric and material properties

Instrumentation   Fifteen free armature linear variable displacement transducers (LVDT) were used to monitor the arch deflection. The transducers used were unidirectional and so the movements of the arch obtained do not represent absolute movements, but the projection of the absolute movements in the radial direction. Six 6 mm sub miniature diaphragm pressure sensors were used to monitor the fill-arch extrados interface pressures. To analyse the three-dimensionality of the structure's behaviour, both instruments were distributed in two lines of gauges, one on the centre line of the structure and another under the one spandrel wall. The instrumentation was completed with an LVDT measuring the loading beam movements and a load cell. Video and digital camera images were recorded during the test.

SELF-WEIGHT AND LIVE LOAD TEST
The first part of the test was the application of the dimensionally correct self-weight of the structure by spinning it up to 55g. During the application of the self-weight, two trends were observed. On the one side, the whole structure moved down, as the normal stresses along the arch increased. At the same time, there was a slight sway towards the live loaded side, this was thought to be caused by the larger amount of fill over the abutment remote from the live load end. This effect was not considered to affect the results as the movements detected were small and this same phenomenon could be reproduced during calibration of the numerical model. During this phase, central and spandrel wall transducers showed identical variations. The pressure transducers demonstrated some confusing results, although the g level increments could be traced on the results. The data provided during this phase of the test is very useful. Standard tests do not provide these results, as the application of the self-weight is a continuous process as part of construction. It is however something that all numerical methods require and therefore represents useful calibration information. It provides information on the stiffness of the undamaged structure.

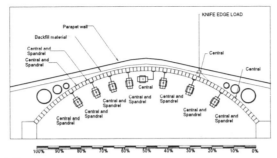

Figure 3. Instrumentation

Figure 4. Bridge model

During the time at constant 55g, prior to loading, the displacements stabilise. After this, whilst still at 55g, a knife-edge load was applied with a 15 mm-width loading beam, positioned at 26% of the span. The load was applied with deflection control, at 0.075 mm/min, this allows us to obtain the post-peak behaviour. Two loading-unloading cycles, to test the elasticity of the structure, were applied up to 15% and 30% of the predicted failure load. The behaviour shown in the cycles was very plastic: Even at these low percentages of the failure load, hardly 50% of the movements induced were elastic. This is undoubtedly due to the weak and plastic mortar used and to the geometry of the structure.

Following the proof loads, the structure was monotonically loaded up to failure. The structure clearly swayed away from the line load as the line load increased, it also moved downwards under the live load and slightly upwards on the remote side. Up to 250 N, the pressure gauges show a general increase in pressure on both ends (expect for the spandrel gauges). At higher loads the gauge positioned at 10% of the span starts to show a decrease in pressure. This was probably showing that the failure mechanism had initiated and that the arch at this point was rotating downwards around a hinge that was closer to the abutment than the gauge. This phenomenon was also detected on previous 2-D tests undertaken as part of this and on previous full-scale work[9]. For loads above 250N, the extra-strength achieved was probably due only to the backfill and the spandrel walls weight restricting the movement.

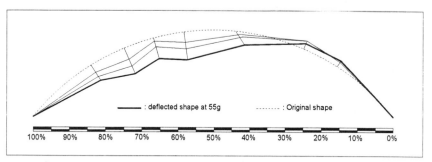

Figure 5. Deflected shape under self-weight at 15g, 30g, and 55g magnified 100 times

At around 400 N, a loss in the structure's stiffness was detected, see Figure 7, and the central pressure gauges at 78% started to show the pressure increasing at a faster rate. The phenomenon was not detected at the gauge at 81%. This seems to be produced by the tangential spandrel-arch separation. It is difficult to check the relation between the two phenomena, as only detailed post mortem visual information is available on the detection of cracks. However it seems quite probable, as the spandrel-arch crack is clearly visible on the video images at least from 500 N. An increase in the pressure recorded in the gauge at 10% was from then on also detected, probably due to a slight change in the hinge position, due to the spandrel-arch separation, or to a stress transfer from the backfill-spandrel interface to the backfill-arch. At 570 N the load started to plateau, reaching a maximum value of 588 N before dropping off slowly. During this period, the arch displacements increased rapidly. The pressure gauges showed no comparable increase, indicating that the maximum structure strength had been achieved. The arch then moved freely as a mechanism and the soil-spandrel walls ensemble had reached it maximum contribution. A drop on the central pressure gauge at 81% was finally detected, but not at 78%, probably indicating that the fourth hinge was located in between the two gauges.

The structure was not loaded to catastrophic failure; once the structure showed that no more strength could be mobilised the test was stopped. The stabilisation or drop off in the extrados pressures, as the movements kept on increasing showed that the structure had reached its maximum and that more movements into the fill would not provide the extra pressures that would produce extra strength. At the end of the test just three visible hinges were located. However, it was considered that the structure failed in a four-hinge mechanism, as a three-hinge snap-through would have shown a much more sudden and catastrophic failure corresponding with the peak load.

A considerable amount of cracking was found on both spandrel walls. The spandrel-arch separation was clear, running from 16% to 34% of the span, with a crack width of 1 mm at its maximum, close to the line load. A crack in the fill at the crown was visible as well, aligned with the position of the third hinge. Figure 8 presents a summary of the main results.

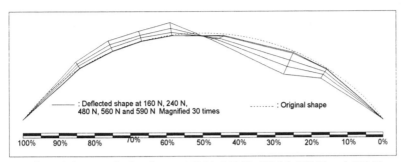

Figure 6. Deflected shape under loading at 160 N, 240 N, 480 N, 560 N and 588 N magnified 30 times (just the active values are magnified)

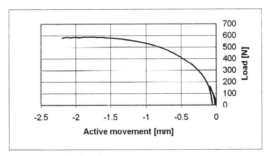

Figure 7. Load-deflection under the load

DISCUSSION
The displacement transducers on the centre line and on the spandrel walls line showed identical results throughout the entire test, indicating that the structure behaved in a two-dimensional way. However, this cannot be a general conclusion for arch bridges as this structure is quite narrow with just five blocks spanning the arch width. Previous work has shown both two and three-dimensional behaviour[9,10]. Further work on the three-dimensionality of the behaviour is to be undertaken especially under asymmetrical loading.

Load position	26%
Failure load (N)	588
Mvt. At 28% at 55g	0.125 (1/6215)[1]
Mvt. At 28% at 95 % of FL (560 N) (mm)	1.21 (1/640)
Stiffness between 0-10% of FL (kN/mm)	2.99
Stiffness between 60-80% of FL (kN/mm)	0.345
Hinge positions	10%, 29%, 52%, 80%(?)

Figure 8. Results summary

---

[1] In brackets is the ratio movement to span

From the results of pressure transducers, it is difficult to draw general conclusions and it is inappropriate to look at absolute values. Trends are important and even they need a certain selection. It is clear that all the pressure transducers worked, as they reproduced the g level and the load-unloading cycles. However, the small size of the transducers compared to the larger size of the aggregate particles aggravated the results. Previous workers[9] have encountered similar problems, even at full-scale tests, and they suggested that it would be much easier to measure pressures in the soil close to the extrados.

In the case of the gauges under the spandrel wall, the situation was significantly worse, as we were trying to pick up the contact pressures between two rigid bodies with just two transducers. In general, it is considered more and larger transducers need to be used, so that they average stresses over a larger area and the global behaviour can be better captured.

From the visual information at the end of the test, a great deal of understanding of the behaviour of the spandrel walls was obtained. An idealised crack pattern of how the spandrel walls broke, in order to allow the arch to move is presented in Figure 9. The test showed that in order for the arch to form its failure mechanism, the spandrel walls needs to break, which does not happen so clearly with the backfill material, which is better able to deform. The pattern should be confirmed in future tests but is considered useful information, especially for mechanism analysis that includes the spandrel walls.

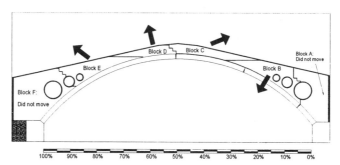

Figure 9. Spandrel wall crack pattern

CONCLUSIONS
A comprehensive set of results has been obtained, although with limited soil/masonry interface pressures, and some understanding on the behaviour of 3-D masonry arch bridges has been proposed. Specifically, on the way arches deform and the mechanism of failure involving a mechanism of failure on the spandrel walls.

The complete material data, the known dimensions of the structure and the laboratory controlled test conditions make of the test suitable for calibrating numerical simulations. The results of the structure under self-weight represent a significant advantage over full scale tests.

The repeatability of the results and success of the construction technique, as well as the cost and time involved in the test preparation makes this kind of test a very attractive alternative for future work. The work shows that 3-D as well as 2-D masonry structures can be modelled on the centrifuge.

AKNOWLEDGEMENT
The authors are grateful for funding from the University of Wales through a grant award from the 1997-98 Academic Support Fund and EPSRC initially under grant number GR/K/04262 and currently under grant number GR/L/63310. Their support is fully acknowledged.

REFERENCES

[1]- PIPPARD, A., ASHBY, R. (1939). An experimental study of voussoir arch. *Proc. ICE Vol. 10 p383-404.*

[2]- PAGE, J. (1987). Load testing to collapse of masonry arch bridges. *Structural assessment: The use of full and large scale testing. Garas, F.K., Clarke, J.L., Armer, G.S.T. pp.265-271 Butterworths.*

[3]- HARVEY,W., VARDY, A., CRAIG, R., SMITH, F. (1989). Load test on a full scale model four metre span masonry arch bridge. *Dpt of Trans TRRL CR 155. TRL, Crowthorne.*

[4]- MELBOURNE, C. (1989). Load test to collapse on a full scale six metre span brick arch bridge. *Dpt of T. TRRL CR 189. TRL, Crowthorne.*

[5]- HUGHES, T., PANDE, G.,SICILIA, C (1999). William Edwards Bridge, Pontypridd. *Proc. of the SWIE.*

[6]- DAVIES, M.C.R., HUGHES, T.G., TAUNTON, P.R. (1995). Considerations in the small scale modelling of masonry arch bridges. *Proc. 1st Int. Conf. on Arch Bridges. Bolton. pp 365-373. Ed. Melbourne.*

[7]- TAUNTON, P.R. (1997). Centrifuge modelling of soil/masonry structure interaction. *PhD Thesis, Cardiff Univ.*

[8]- PANDE, G.N., LIANG, J.X., MIDDLETON, J. (1989). Equivalent elastic moduli for brick masonry. *Comput. and Geotech. 8, 243-265.*

[9]- GILBERT, M. (1993). The behaviour of masonry arch bridges containing defects. *PhD Thesis, BIHE*

[10]- NG, K.H. (1999). Analysis of masonry arch bridges. *PhD thesis. Napier University.*

[11]- HARVEY, W.J. (1995). Loaded ribs or complex systems- a personal view of our ability to model arch bridge behaviour. *Proc. 1st Int. Conf. on Arch Bridges. Bolton. pp 29-36. Ed. Melbourne.*

Evaluation of measured bridge responses due to an instrumented truck and free traffic

DIPL.-ING. STEFAN LUTZENBERGER, DR.-ING. WERNER BAUMGÄRTNER,
Lehrstuhl für Baumechanik, TU München, Germany

INTRODUCTION

Dynamic measurements at bridges are usually carried out to study the realistic static and dynamic response of special members, such as deck plates or webs caused by transient loads as traffic or flurries. Especially to get detailed information of the parameters of the corresponding vehicles it needs a lot of additional effort, as stopping of traffic is necessary. As partners in the European research project WAVE (Weighing-in-motion (WIM) of Axles and Vehicles for Europe, O'Brien et al. 99) we took part in an experiment at a bridge in France where measured surface roughness data and measured dynamic wheel loads were provided by the partners. The understanding of the responses of different locations is the basis to establish procedures to derive truck and wheel loads from measured strains (WIM).

A numerical simulation method is used at our institute for several years based on a commercial FE program (Baumgärtner, Penka 98). Particularly with respect to measurements at a bridge and a truck, this method can provide time-dependent strain, force and acceleration curves which can be compared to measurement records. the second objective and to reveal the complex dependencies between truck impact and strains at different locations of a spatial bridge structure (Lutzenberger, Baumgärtner 99). The FE model of the bridge was updated in correspondence with the eigenfrequencies derived from measured strains and accelerations.

Other fields where permanent strain measurements are essential are the monitoring of structural behaviour and the evaluation of the service life of bridges (Baumgärtner et al.95).

This paper will focus on the discussion of the measurement records taken from specific locations of the bridge in relation to a truck with known wheel loads and trucks out of the free traffic.

STRUCTURES: BRIDGE AND TRUCK

BRIDGE

The Belleville bridge is situated along the autoroute A31 between Metz and Nancy in France. Research work was done with the two span access bridge with spans of 54.9 m and 51.7 m . The structure consists of a steel box, a steel girder and a concrete plate (Figure 1) and has a height of 2 m and a width of 15 m.

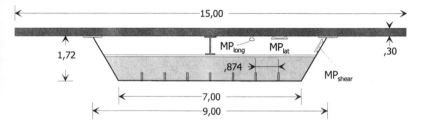

Figure 1: Cross section of Belleville bridge with typical locations of sensors (MP)

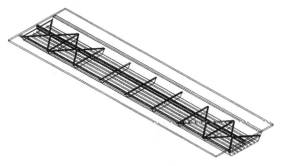

Figure 2: Structure of Belleville bridge, first span

Dynamic measurements of strains and accelerations were performed under free traffic. The ANPSD (averaged normalised Power spectral density) was computed by normalising and averaging the PSD of several measurement points. The Eigenfrequencies and the corresponding Eigenmodes were identified (Fig. 3, Lutzenberger 96). A modal analysis with a FE model (Fig. 2) was performed for a better understanding of the structural behavior.

TRUCK

The investigations were performed with a special truck of the technical research centre of Finland (VTT). It is a three axle lorry with axle spacings of 4.20 m between the first an second axle and 1.2 m between the second and third axle. The second and the third axle coact due to a mechanical connection. Strain gauges and accelerometers are installed on the axles and allow to measure the dynamic wheel loads. The dynamic behaviour of the vehicle was studied with the help of a FE model (Fig. 4, Eckl 99) and by analysing measurement records. An experimental mode shape identification was performed and the FE model was adapted to the measured Eigenfrequencies. The axle loads were 58.8 KN for the first axle, 86.2 KN for the second axle and 72.2 KN for the third axle. The Eigenfrequencies of the truck are 2.2 Hz (body, front deflection), 3.1 Hz (body, rear deflection), 8.3 Hz (interaction of the rear axles).

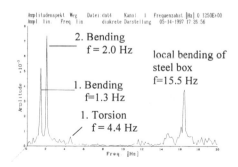

Figure 3: ANPSD of measurement record

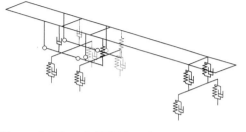

Figure 4: FE model VTT truck

SURFACE ROUGHNESS

As the roughness of the road is the major influence factor for the dynamic excitation of the truck and therefore for the dynamic forces applied to the structure, the roughness profile is essential to interpret the measured data. High dynamic wheel loads cause strong vibrations of the bridge and influence the accuracy of truck parameter detection. The use of a stochastic surface description of the road profile is not realistic to study the influences to measurement records. Local characteristics have to be regarded. The road profile at the Belleville bridge for both wheel tracks was provided by the LCPC/Paris (Fig. 5a).

The road roughness in front of the bridge is much higher than on the bridge and there is a relatively high jump at the joint at the beginning of the bridge. As the relevant part of the dynamic excitation is the frequency range of 1.5 – 15 Hz (eigenfrequencies of the truck), the road profile was high pass filtered (Fig. 5b). Most of the dynamic excitation is caused by the unevenness of the road in front of the bridge and by the joint at the beginning of the bridge. Compared to this the surface on the bridge is comparatively even. Therefore the dynamic excitation of the vehicle is quite high before it reaches the bridge. This results in very high dynamic axle loads at the beginning of the bridge and lower ones at the end of the bridge.

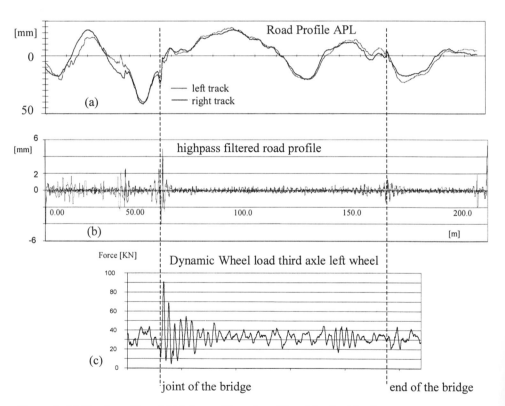

Figure 5: (a) Road profiles, (b) highpass filtered profiles, (c) wheel load

MEASUREMENT ANALYSIS

Measurement equipment

At Belleville bridge, strains were measured using displacement sensors, mounted at the end of a steel tube with a length of 0.5 m (Fig. 6). The average strains over this length therefore can be calculated according to $\varepsilon = \dfrac{\Delta l}{0.5\,m}$. The signals were anti aliasing filtered, amplified, digitalized with a sample frequency of 100 Hz and stored in a computer. The sensor locations are shown in Figure 7.

Figure 6: Measurement sensors

Discussion of sensor signals

Measurement records at bridges show two major effects: mainly global and mainly local reactions of the bridge. If a sensor provides significant values during the whole time the truck passes the bridge, this is called a global reaction. If significant values are available only when the truck is close to the sensor we call it a local reaction. The steel box as a main structural element gives a global response. Construction elements as the concrete plate which are close to the neutral line of global bending are directly excited by the wheel loads and therefore show a local reaction.

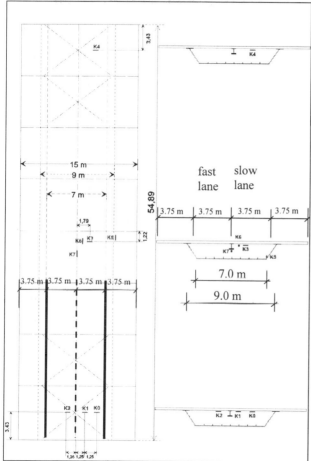

Figure 7: Sensor locations

The main components of sensor signals recorded during the crossing of a truck are shown in Figure 8. Whereas each signal contains all components, a lot of sensors are dominated by two or three of them.

Examples are given in the Figures 9 to 11. In this example a four axle truck with two single axles and a tandem axle enters the bridge at 3.7 s, hits the middle support at 4.9 s and leaves the bridge at 7.3 s

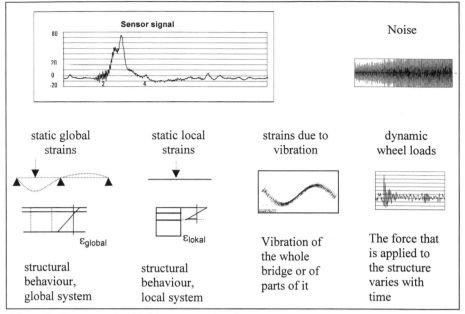

Figure 8: Components of bridge sensor signals

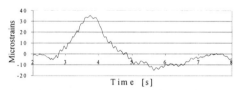

Figure 9: Measurement record Channel 5

Sensor location: Bottom of the steel box. Mainly global bending and bridge vibration. The measurement record is similar to the influence line of the moment of a two span beam. The information about the axles of the vehicle disappears.

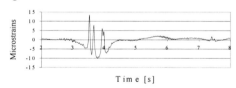

Figure 10: Measurement record Channel 6

Sensor location: Bottom of the concrete plate close to neutral axis. Sensor orientation: longitudinal direction. Mainly local bending. The record is similar to the superposition of four m_y moment influence lines of a plate according to four axles.

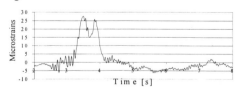

Figure 11: Measurement record Channel 7

Sensor location: Steel girder (which supports concrete plate in longitudinal direction)
The measurement record shows a superposition of global and a local reactions.

Parameter identification

VTT truck

Lateral Position: The cross section shows that in case of lateral bending the concrete plate can be regarded as a two-span plate (Fig. 12). The bending moment in the transverse direction caused by a truck, has an opposite sign in the left span. Therefor the sensor on the left and on the right lane can be used to identify the lane of the vehicle passing the bridge (Fig. 13). The flat influence line in the longitudinal direction and the vehicle dynamics prevents axle identification.

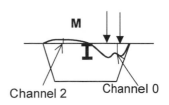

Channel 2   Channel 0

Figure 12: lateral bending of the concrete plate

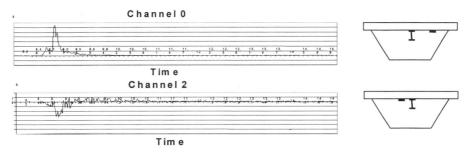

Figure 13: Different signs in signals of channel 0 and channel 2

Speed: Two local sensors at different longitudinal positions can be used to calculate the speed of vehicles according to: $v = \dfrac{\Delta s}{\Delta t}$. To achieve a good accuracy the sample frequency and the distance of both sensors must be adapted to the problem. The distance between both sensors was 48.03 m. The resulting speed is 22.24 m/sec.

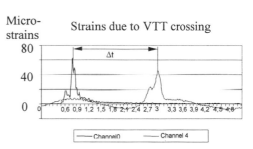

Figure 14: Δt for speed calculation

Dynamic amplification factor: Dynamic amplification factors for locations with global characteristics can be calculated by comparing the maximum strains in the original record to the maximum strains in a generated static record where dynamics are filtered out with the use of a a low pass filter. For measurement points with local characteristics this procedure is not possible. By low pass filtering the sharp peaks in the static part of the record, corresponding to axles, would also disappear.

Free traffic

Axle identification, axle distances: The measurement record for the longitudinal bending of the concrete plate shows all axles. The reason is that the influence surface (see Figure 15) has a sharp peak around the sensor location.

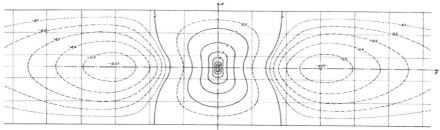

Figure 15: Influence surface m_Y of a plate supported along two edges (Pucher 1973)

As the axle identification is the most critical point, a verification on the basis of a video tape, that shows the traffic during the measurements, was made. The following figure (Fig. 16) shows that even a 5 axle truck with a tridem axle and the axles of a car (in a local zoom) can be identified. Using the vehicle's velocity the axle distances can be calculated. For this measurement record the calculated axle spacings are 3.65 m between the first and the second axle, 5.22 m between the second and the third axle and two times 1.31 m for the tridem axles. An inaccuracy results from the sample frequency that was found to be too low for this problem. With the chosen 100 Hz and a speed of 26.1 m/sec, the maximum error can be up to 0.261 m. For this task a higher sample frequency has to be chosen.

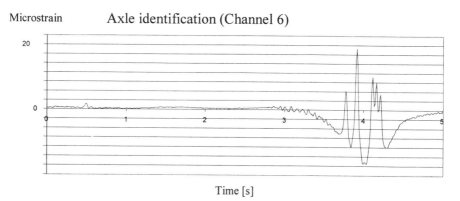

Figure 16: Axle identification of a car (two axles) and a five axle truck

Lane identification: As vehicles on the left lane yield positive strains and vehicles on the right lane negative strains in the signal of channel 2, this location can be used to identify the lane of the vehicle /Fig. 17).

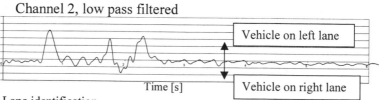

Figure 17: Lane identification

CONCLUSIONS

Compared to the approach to measure the deflections in the middle of a bridge, which represent an integrated reaction of the bridge, the measurement of strains give detailed information about structural behaviour. This is essential to be able to derive the truck parameters.

Our investigations with the measured strain records lead to clear results of the static behaviour of the concrete deck plate, the steel girder and steel box due to crossing trucks. In contrary to strains at the bottom of the steel box which have relative wide influence lines the strains recorded at the deck plate in longitudinal direction with quite narrow influence lines are well suited to separate trucks and even axles in a clear manner.

For a study of the dynamic reactions of the bridge it was helpful that the wheel loads of the VTT truck led to very high dynamic excitations caused by the surface roughness and the jump at the beginning of the bridge. The time-varying dynamic amplification of the strains dependent on the structural member and direction of measurement were presented.

The records of selected locations on the deck plate were used to derive the velocity of the truck of concern. Sensors in longitudinal direction proved to give a clear separation of the axles of a truck and in combination with the derived velocity the distance of the axles can be evaluated.

Further research work is necessary for a reliable prediction of the precise transverse position of a crossing truck. Another task in the sense of WIM will be to establish a method to use local and global responses simultaneously to evaluate sufficiently precise truck parameters and wheel loads.

References:

Baumgärtner W., Geissler K., Waubke H., " Updated service life evaluation of bridges using measurements", IABSE report 73, Extending the lifespan of structures, IABSE Zürich 1995.

Baumgärtner W., Penka E., " FE Simulation der Interaktion einer Brücke mit darüberfahrendem Fahrzeug", In: Finite Elemente in der Baupraxis, Ed. Wriggers P. et al., Ernst und Sohn, Berlin 1998.

Eckl Michael, "Modellierung des VTT-Trucks in MSC-Nastran, unter Berücksichtigung der Nichtlinearität von Blattfedern", Diplomarbeit, Lehrstuhl für Baumechanik, TU München 1999.

Lutzenberger Stefan, "Analyse modaler Größen großer Strukturen unter ambienter Anregung", Diplomarbeit, Lehrstuhl für Baumechanik, TU München, 1996.

Lutzenberger S., Baumgärtner W. 99, "Interaction of an instrumented truck crossing Belleville bridge", In: Weigh-in-motion of Road Vehicles, ed. B. Jacob, Hermès Science Publications, Paris, 1999.

O'Brien E., Dempsy A., Znidaric A., Baumgärtner W. 99, "High-Accuracy Bridge-WIM Systems and future Applications", ", In: Weigh-in-motion of Road Vehicles, ed. B. Jacob, Hermès Sciences Publications, Paris, 1999.

Pucher Adolf, "Influence Surfaces of Elastic Plates", Springer Verlag Wien, New York, 4. Auflage, 1973.

trials on a variety of bridges [11]. The AE system detects and locates active defects, such as crack initiation, anywhere in a structure using externally mounted sensors, positioned metres away from defects. This method allows non-invasive 100% volumetric assessment of structures and is therefore particularly effective for box girders, which has led to the development of the BOXMAP™ technique.

STRATEGIC CONSIDERATIONS FOR AE BRIDGE MONITORING

Before carrying out an Acoustic Emission test there are strategic considerations to be made. Three main factors, structural detail, the acoustical behaviour of the structure and the monitoring strategy, namely Global, Semi-Global and Local, will dictate the number and location of sensors.

Box girders may have some of the following design features; diaphragms (with and without access holes), diagonal-bracing, stiffeners, and external access holes. All of these features influence the way that the elastic stress wave travels, decays and is located by the software. It is therefore necessary to review technical drawings to make use of structural details as wave-guides to ensure the full extent of the structure is monitored.

The location and number of the sensors to be used is a function of signal attenuation, which is affected by structural geometry, plate thickness, design, paint characteristics, and background noise. In every case there will be unwanted acoustic emission "noise" from external sources such as traffic activity, expansion joint impact, bearing movement and airborne electromagnetic interference (EMI). It is essential to filter any extraneous emissions from collected data so that emission from defect growth is recognisable. Background noise is overcome using a range of methods; guard sensors are used to shield sensor arrays from sources of noise e.g. expansion joints; pattern recognition programmes filter out identified noise sources that are incorporated into the cracking data. These methods have been developed using the experience gained during 6 years of bridge monitoring work with Federal Highways in the USA and during numerous bridge trials in the UK, and fatigue testing at Cardiff University.

Structures can be continuously or intermittently monitored with permanently installed sensors from the site or remotely via a modem. Technological advances and increased accessibility to the Internet allow "remote from site" real time structural monitoring. Permanent mounting of cables and sensors with pulse and self-test capability, eliminates the need for repeated access, giving rapid payback compared with the cost of aerial lifts and aerial walkways. Short duration on-site monitoring can provide qualitative and quantitative information that can effectively assess the structure over a period of days using temporarily mounted sensors.

Three common monitoring methods are used to obtain differing levels of information;
- Global
- Semi-Global
- Local Area

The radical "Global" integrated inspection and evaluation monitoring system discussed by Carter and Holford [12] has not yet been completed. However, "Global" monitoring using large sensor spacing still offers maintenance engineers the ability to identify damaged structures and focus further monitoring or inspection on areas sustaining damage. Trials have shown that an entire box girder can be monitored for one day and reveal a number of active areas. Physical Acoustics and Cardiff University are currently developing techniques to reduce the number of sensors required to globally monitor a structure.

"Semi-Global" monitoring provides precise source location, but requires an increased number of externally mounted sensors to give 100% volumetric monitoring, i.e. capable of locating damage anywhere in the structure, including shear studs, diagonal-bracing, welds, stiffeners and diaphragms.

"Local" area monitoring is used to assess active sources and known defects, using a small array of sensors around the area of interest. Crack orientation, length, activity, and position can be determined with accuracy over a short monitoring period, whilst continuous or intermittent monitoring at this level can identify the direction of crack growth and the rate of propagation. This information is essential for the bridge engineer to determine the status of the known defect and schedule maintenance or repair.

Acoustic Emission monitoring strategies all offer different levels of information about structural integrity and defects. AE monitoring using all of the strategies in conjunction may provide priority based assessment and maintenance methods where structures need to be fully assessed. "Global" monitoring is initially used to grade the structure according to its condition. Structures with active defects can then be "Semi-Globally" monitored to identify and locate significant sources, which in turn can be "Locally" monitored, or assessed using traditional NDT methods, where access is possible to acquire detailed information.

FIELD TRIALS

One of the earliest acoustic emission tests of a steel bridge was carried out by Pollock and Smith [13] in 1972 on a military steel bridge. Since then there have been many more tests on a wide variety of structures, both steel and concrete. Huge experience was gained by Physical Acoustic Corporation during local area monitoring of identified defects and repairs for Federal Highways (F.H.W.A.). This work prompted Physical Acoustics to develop a Local Area Monitoring system, L.A.M., a remote AE system weighing less than 12 kg. It possesses 8 AE channels, 4 high-speed parametrics (used for strain, vibration, and deflection gauges) and 6 slow-speed parametrics, with a power consumption of only 12 watts.

Large area monitoring of steel structures, encompassing both "Global" and "Semi-Global" monitoring procedures have been developed by Cardiff University in conjunction with Physical Acoustics Limited. The first trial on Trecelyn Viaduct in 1996 sought to establish the level of background noise typically encountered in bridge structures and to study the attenuation of signals in large structures. Emissions from expansion joint impact were found to "ring" throughout the structure, which required digital filtering. Bolted connections were found to significantly reduce the signal strength, yet artificial sources were adequately picked up at distances in excess of 20 metres.

Extensive trials were carried out in 1997 on Saltings Viaduct, a 26 year old composite box girder on the A465 near Neath, South Wales (Figure 1).

They investigated the monitoring potential of three different strategies:

- Large scale structural acoustics, which is fundamental in the functionality of "Global" monitoring
- Large area "Semi-Global" monitoring
- Local area monitoring of sites identified by "Global" monitoring

Figure 1: Saltings Viaduct

This study showed that active areas were identifiable at 10-metre sensor spacing, many of which coincided with internal diaphragms. Local monitoring of the most active diaphragm identified a strong source from the access hole cut into the plate. Recent Finite Element analysis [14] showed this to be an area of particularly high stress, which has led to on going fatigue tests on box girder sections with a variety of diaphragm designs.

MOTORWAY TRIAL I - PRELIMINARY INVESTIGATION
Recent motorway bridge trials were carried out in December 1998 and January 1999. During preliminary investigations a composite box girder bridge was monitored during the morning rush hour. Prior to monitoring, an acoustic investigation was conducted using an artificial source to confirm location accuracy for a variety of details within the structure, including stiffeners, diaphragms and the internal welds. The bridge was monitored using a Semi-Global, externally mounted array. This test revealed that the bridge was in a very good condition with a very small amount of data collected, with no signs of any active damage mechanisms. This preliminary investigation enabled an optimum test set-up to be determined for the following major bridge trial.

MOTORWAY TRIAL II - MAJOR TRIAL TEST DETAILS
A 40m span composite box girder was studied over a two week period, during which it was semi-globally monitored for a one-week period between 5am and 10pm. The most significant AE sources, identified during investigations in the first week, where locally monitored during the second week. The girder lies underneath the cheveroned area between the 3 lanes south bound and the merging area of a two-lane slip road, Figure 2. This girder was chosen since

transverse deformation across the deck induce high stresses in the diagonal bracing members, shown in Figure 3.

In order to identify any sources in the girder and attachments, the beam was monitored using a semi-global array that gave 100% volumetric monitoring. Two-dimensional location was achieved by "un-wrapping" the structure, with damage locations shown on an "opened-up" diagram. Integrity of the cross braces was assessed using linear source location between sensors at either end of each cross brace. An initial AE study identified the expansion joint as a large source of extraneous noise. To filter this noise from the test data, guard sensors were placed to the ends of the girder.

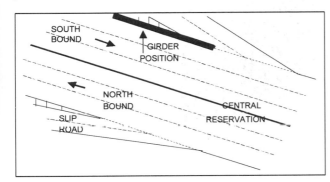 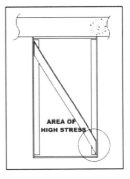

Figure 2: Position of box girder.

Figure 3: Box girder cross section.

One day trial monitoring in the first week identified 62 sources. Local arrays were set up around the two most active sources.

RESULTS OF THE TRIAL

From the 62 identified sources, 18 were identified as significant and analysed in detail. These were all found to be active on a daily basis. All but one of the 18 significant sources were found to be located at the end of the diagonal bracing. The main source "Λ", identified by "Semi-Global" monitoring during the first day and subsequently monitored by a local array, is crack like and approximately 250 mm long, as illustrated in Figure 4. It was located below the end of a diagonal bracing connection at the bottom of the flange. It is difficult to state whether the source is in the internal seal weld or bottom flange/web weld. The absence of emission from the central region of the suspected crack suggests that there is no further growth in this area.

AE activity showed a strong correlation with traffic flow. During periods of peak traffic flow, the average traffic speed dropped and at times traffic was stationary. These periods saw a reduced amount of emission. Most activity occurred when there was a high percentage of lorries travelling at a high speed, ~55-65mph. Figure 5 shows the correlation between strain due to free flowing traffic and AE events detected from the most active source, source "A".

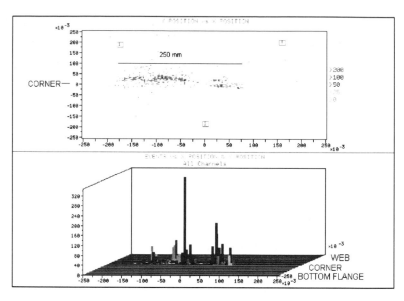

Figure 4: Local monitoring location graphs for Source "A"; top graph shows activity in 2D, lower graph shows AE activity along the length of the crack like source.

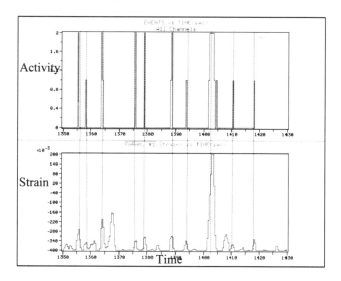

Figure 5: Correlation between events from Source "A", and strain, over the same time period during peak but free flowing traffic conditions.

CONCLUSIONS

Acoustic Emission monitoring trials on motorway bridges using "Semi-Global" and "Local" sensor arrays have demonstrated that AE is a very effective method for detecting damage in bridges and steel structures, and for continuous monitoring of defect growth.

A major advantage is that there is no need for internal access or paint removal. In addition it is 100% volumetric testing, and damage location is achieved with relatively few sensors. Global monitoring has been developed as a method for identifying damaged structures; further research is being carried out by Cardiff University and Physical Acoustics Limited to develop this method to its full potential.

REFERENCES

1. Bridle, R., "The Economics of Maintenance, Repair and Strengthening." TRRL Document 1991.
2. DTp., "Bridge Inspection Guide", HMSO – Department of Transport, 1984.
3. Russel, H., "Fears for Flyover as Cracks Appear", New Civil Engineer, pp 6, 30th July 1992.
4. D.Fowler, "Bridge Cracks Paralyse M50", New Civil Engineer pp 5, 13th February 1992.
5. Fisher,J.W., "Hundreds of Bridges - Thousands of Cracks"Civil Engineering, Volume 55, No. 4, April 1985.
6. Prine, D.W. and Hopwood, T "Improved Structural Monitoring with Acoustic Emission Pattern Recognition", Proc. of the 14th Symposium on Non-Destructive Evaluation, San Antonia, Texas 1985.
7. Physical Acoustics Corporation, "Acoustic Emission For Bridge Inspection" Report No. FHWA-RD-94- prepared for FHWA and U.S. Department of Transportation, June 1995.
8. Davies, A.W., Holford, K.H., Sammarco, A., "Analysis of Fatigue Crack Growth in Structural Steels by Classification of Acoustic Emission Signals", Proc 1994 Engineering Systems Design and Analysis Cong, PD Vol. 64.8.2,Volume 8, Part B, ASME, USA pp 349-354 ISBN 07918-1280-4, 1994.
9. Davies, A.W., Holford, K.H., "Health Monitoring of Steel Bridges Using Acoustic Emission", Structural Assessment: The Role of Large and Full Scale Testing, Joint Institute of Structural Engineers / City University, E and FN Spon. pp 54.1-54.8. ISBN 0-419-224-904, 1-3 July 1996.
10. Yan, T., Holford, K., "Acoustic Emission Analysis Using Pattern Recognition Technique", 23rd European Conference on Acoustic Emission Testing, Vienna, Austria, pp 225-229, 1998.
11. Holford, K.M., Cole.P.T., Carter, D.C., Davies, A.W., "The Non-Destructive Testing of Steel Girder Bridges by Acoustic Emission". 14th World Conference on Non-Destructive Testing, Vol.4, pp 2509-2512, ISBN 8120411269, 1996.
12. Carter,D., Holford, K.M., "I.M.A.G.IN.E: Letting bridges do the talking" INSIGHT Vol 38, No 11 pp 775-779, November 1996.
13. Pollock, A.A., Smith, B., "Acoustic Emission Monitoring of a Military Bridge", Non-destructive Testing, Vol 5, No 6, pp 164-186, 1972.
14. Pullin, R., Carter, D.C, Holford, K.M., Davies, A.W. "Bridge Integrity Assessment By Acoustic Emission - Local Monitoring", Proc. 2nd International Conference on Identification on Engineering Systems, Swansea pp 401-409, ISBN 0860761584, 1999.

Localization and identification of cracking mechanisms in reinforced concrete using acoustic emission analysis

S. KÖPPEL,
Research associate, Institute of Structural Engineering, ETH Zurich, Switzerland
T. VOGEL,
Professor, Institute of Structural Engineering, ETH Zurich, Switzerland

ABSTRACT

In the context of a research project that aims at improving the capabilities of acoustic emission (AE) analysis in assessing internal damage in reinforced concrete, laboratory tests have been carried out. Qualitative and quantitative acoustic emission techniques were applied to locate and distinguish different cracking mechanisms.

During preliminary tests, specimens and loading arrangements with defined damage locations and processes were chosen. In order to study the interaction of reinforcement and concrete several pull-out tests were performed. Double punch tests were conducted to monitor AE occurring due to tension cracks. In a second phase reinforced concrete beams were subjected to static and repeated loading. The AE parameters and waveforms of signals obtained by eight broadband piezoelectric transducers were processed and stored by a measuring device. In a subsequent analysis AE sources were located. Furthermore, a classification of different fracture mechanisms on the basis of the similarity of some waveforms was attempted.

INTRODUCTION

Non-destructive and integral assessment of reinforced concrete structures is largely limited to visual methods. Defects that do not appear on the surface can hardly be detected. Whereas the non-destructive evaluation (NDE) of metal and composite materials is well developed, concrete poses difficult problems. The distinction between normally occurring heterogeneities and damage that affects the safety of the structure is ambiguous. In addition, concrete structures are usually unique and universal failure criteria do not exist.

AE is defined as the spontaneous release of localized strain energy in a stressed material for example resulting from microcracking and can be recorded by sensors on the surface. Thus AE-analysis potentially allows one to monitor processes occurring within a structure. In order to be able to evaluate the condition of a structure it is required not only to know about the existence and location but also about the nature of a detected process.

In the present work AE, emitted during the loading of various reinforced concrete specimens, is investigated. Since the location and nature of fracture mechanisms leading to failure are known for these specimens, the results of an AE analysis can be verified or calibrated.

SPECIMENS AND TEST SETUP

Three types of tests with concrete specimens were conducted in order to observe acoustic emissions due to different failure mechanisms. Specimen geometry and loading arrangement are given in Figure 1.

First pull-out tests were carried out to examine signals caused by the interaction of steel and concrete in the bond zone. Four concrete cubes with side lengths of 200 mm and a centred rebar Ø 14 mm were used. The bond length was chosen as 42 mm which equals three rebar-diameters or five rib-spacings. Signals are expected to be caused by the loss of adherence at the steel-concrete interface first. The bond force will then be transmitted via interlock of steel ribs to the surrounding concrete and signals will occur through shearing action. Microcracks might also form in the tension ring around the rebar that balances inclined compressive forces.

To study the AE behaviour of concrete under tension, a double punch test was conducted. As the determination of the direct tensile strength was not of primary interest, this indirect test was suitable because of its simple and clearly defined loading conditions. Concrete cubes with side lengths of 200 mm, similar to the ones used for pull-out tests, were used. Using cubes instead of cylinders allowed one to place sensors on all surfaces of the specimens. Signals were expected to be emitted in the radial tensile cracks. The location of these cracks was more or less predetermined by the specimen geometry.

Pull-out, A1-A4   Double-punch, S1   Three point bending, B1-B2

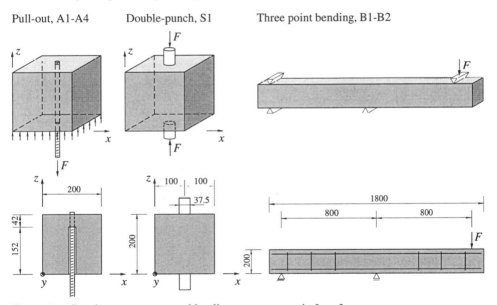

Figure 1.   Specimen geometry and loading arrangement in [mm]

A third series of tests was carried out investigating signals occurring in a beam subjected to bending. Two beams with a length of 1.80 m and a rectangular cross-section, 200 mm high and 150 mm wide were subjected to three point bending. The longitudinal reinforcement consisted of two rebars Ø 14 mm in the tension zone and two rebars Ø 10 mm in the compression zone. Stirrups were only used to hold the longitudinal reinforcement in place during casting. To reduce possible sources for acoustic emission and keep the conditions simple, no stirrups were placed in the region above the middle support. Signals were only monitored in this region where stresses due to bending were highest. Acoustic emissions were expected to be caused by the formation and growth of bending cracks and at higher loading rates by the loss of bond and yielding in the longitudinal reinforcement.

For all specimens normal density concrete with 330 kg of Portland cement per m^3 and a water-cement ratio of 0.47 was used. The mix proportion was 0.35 : 0.3 : 0.35 for aggregate sizes 0-4 : 4-8 : 8-16 mm. Reinforcement consisted of hot rolled steel Topar S500 with an elastic modulus of 202 GPa, a yield strength of 502 MPa and an ultimate strength of 610 MPa.

Loading
During the tests, the load was increased in steps up to failure. After every step the specimens were partially stress-relieved and only reloaded after a certain period of time. For the pull out tests the load was decreased by 20 to 30 % over a period of three minutes. In figure 2(a) the load-time history of a pull-out test is given as an example. As the main interest in all preliminary tests was collecting signals to be quantitatively examined, not too much emphasis was put on a testing procedure with exactly defined loading rate and load steps. Since for pull-out and bending tests loading was applied using hydraulic jacks and manual pressure regulation, this would not have been feasible anyway.

Load cells were used to monitor applied forces continuously and displacement transducers with a sensitivity of 1 μm measured deflection and slip for bending and pull out tests respectively.

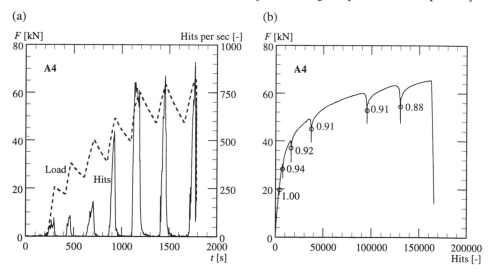

Figure 2.   (a) Development of hit-rate and load during pull-out test A4; (b) load against cumulative hits and Felicity ratios during pull-out test A4

AE measuring system
The AE measuring system AMSY4 by Vallen GmbH was used to acquire AE data during the tests. Eight channels with transducers, preamplifiers and A/D converters were employed simultaneously. Piezoelectric transducers (PZT) with a broadbanded and calibrated frequency response between 50 and 250 kHz (± 3 dB) and a diameter of 13 mm were applied.

Preamplification was set to 40 dB and every channel triggered independently with a threshold of 34 dB (sensor output). A band-pass filter with a range of 20 to 500 kHz was employed. AE parameters and transient waveforms were stored in two different files. Because of the system's limited acquisition capacity (around 100 waveforms per second) the amount of acquired data had to be reduced. Only wave-forms of signals with a peak amplitude of over 45 dB were stored, by means of a front-end filter, set accordingly. A waveform consisted of 2048 data points with a digitizing rate of 10 MHz equal to a stored time of 205 µs with a pretrigger time set to 50 µs. Data points were measured with a resolution of 16 bit.

RESULTS
During the pull-out tests, AE signals were recorded throughout the whole test with a sharp increase in the hit rate after exceeding a load level of 50 to 60 % of maximum load. The specimen subjected to double punch loading emitted no AE signals up to a load level of 75 % of ultimate load. The hit rate increased first slowly, than sharply just before the brittle failure of the specimen. In the bending tests only few AE signals were recorded before cracking. Then most signals were due to the growing and widening of existing cracks. After exceeding a certain width, cracks no longer acted as AE sources but rather as a shield, preventing signals from propagating across it.

Qualitative analysis
The set of AE parameters extracted and stored for every signal includes peak amplitude, duration, rise time, counts and energy. The analysis of these parameters can give a good overview of the development of acoustic emissions during the test. As an example in figure 2(b) the applied load is plotted versus the cumulative hits measured during the pull-out test A4. During the first reloading, acoustic emission activity only set in after reaching a load equal to the previous maximum load (validity of the Kaiser effect). With increasing load level, AE-activity started earlier during reloading phases at 94 % down to 88 % of the previous maximum load. The decrease of this ratio (Felicity ratio) is related to the damage accumulation in the bond region. While the development of this ratio was similar for all pull-out tests, bending tests displayed a completely different behaviour. Influences due to specimen geometry or loading and measuring arrangement can hardly be considered and thus statements on the basis of this qualitative analysis are not transferable. Analysing AE parameters only, does not allow distinguishing between noise and signals or even a classification of different fracture types.

Localization
The knowledge of the location of a source is looked upon as an absolute requirement for quantitative analysis of AE data. Thus very much emphasis was put in a 3D-localization as exactly as possible. Signals that hit all eight channels within a short period of time (depending on the sensor arrangement) were assumed to originate from the same source and were combined to one event. The location of the source was then calculated using the program HypoAE. This program is a derivate of Hypo71, designed for the calculation of earthquake hypocenters. It was further developed and adapted for use with AE Signals by Drs. Lani Oncescu and Christian Grosse [3]. HypoAE uses the arrival times of the longitudinal waves (p-waves) at all channels. As the problem is over-determined with eight (instead of four) channels an iteration algorithm can be applied. The mean residual resulting from this algorithm is a value for the accuracy of the localization. Transient waveforms stored by the acquisition unit had to be transformed into a format readable by HypoAE.

The first p-wave arrival times were determined manually interpreting the signals in the time domain. For the localization a p-wave velocity of 4.8 m/ms was assumed. This value was determined in various tests using artificial sources. It has been found to be equal for all specimens with an accuracy of ±0.1 m/ms regardless of the existence or direction of reinforcement. For propagation lengths of more than about 200 mm, attenuation became to strong and the arrival time of the first p-wave could no longer be extracted from the waveform with certainty. A localization of AE sources on the basis of shear or longitudinal components, which have higher amplitudes but propagate at lower velocities, would require a knowledge of the corresponding arrival times at all channels. During evaluation these arrival times were not considered to be detectable unambiguously and thus this alternative procedure not feasible by simple means. The percentage of signals detectable in the described way is summarized in Table 1 for all test series.

Table 1.   Average number of AE sources located during the different test series in relation to the total number of signals registered

Test	Hits	Waveforms	Events	Events, localized accurately
Pull-out (A1-A4)	100 %	38 %	8.8 %	7.5 %
Double-punch (S1)	100 %	53 %	4.5 %	1.9 %
Bending (B1-B2)	100 %	45 %	4.4 %	3.1 %

During evaluation up to over 1000 sources per test could be located. To maintain clarity, AE sources were represented for each loading stage separately. As an example, localization results of the pull-out test A3 and the double punch test S1 are plotted in figures 3(a) and 3(b), respectively. Both diagrams only show sources accurately (with residuals smaller than 6 mm) calculated during a selected loading stage.

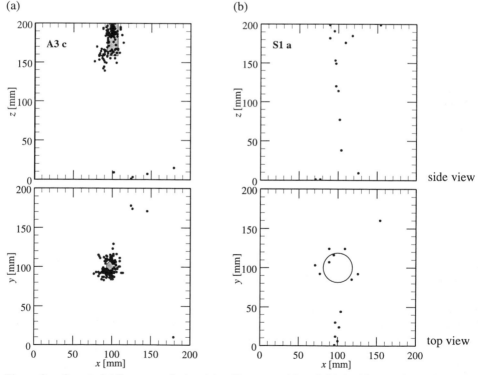

Figure 3.   Located AE sources during (a) pull-out test A3 and (b) double-punch test S1

In the pull-out tests AE sources were detected in the bond area as expected. According to the asymmetric profiling of the rebar, most sources were concentrated at the ribs, aligned in the y-direction (see figure 3(a), top view). A second group of AE sources was located below the bond area, probably along compression trajectories. Finally some AE sources were located at the bottom surface of the test specimen where it was supported.

During the double-punch test the large part of the events was recorded just before or during failure. The location of AE sources indicated the crucial crack oriented in a vertical (y-z) plane from the loading cylinders to one side of the specimen. AE-sources directly under the loading cylinders were situated on the surface of a cone. Even if very few events were recorded during this test, their location gives an idea of how failure occurred.

Basic waveform analysis
The waveforms of localised events were analysed applying the Fast Fourier Transform (FFT) with a rectangular time window over 50 μs before and 100 μs after the first threshold crossing. Resulting frequencies ranged mainly between 40 and 300 kHz, while frequencies over 150 kHz were rare. Most signals showed one or two peaks between 100 and 150 kHz corresponding to wavelengths λ around 40 mm. Using the far field condition defined in [1] as $r \gg \lambda/2\pi$ a minimum distance r between AE source and sensor of several centimetres is required.

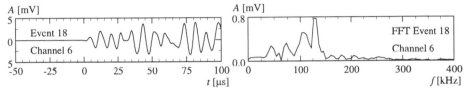

Figure 4.   Typical waveform recorded during bending test B2 and corresponding FFT

Looking for similarities in the waveform, a number of signals recorded during the first two loading stages of bending test B2 were compared visually. Sensors were mounted on three sides of the beam distributed over a length of 150 mm at the middle support. Events located with a sufficient accuracy originated from the two cracks on both sides of the group of sensors and are plotted in figure 5.

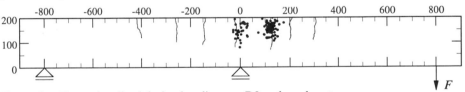

Figure 5.   Events localized during bending test B2 and crack pattern

A visual inspection revealed, that two events with high similarity in all channels always originate from the same region only a few millimetres apart. As one would expect similar fracture mechanisms to be distributed more or less uniformly over the crack plane, the wave propagation path and potential reflections on the specimen surface must have a strong influence on the waveform. To avoid any influence of the propagation path, AE signals of sources with small relative distances (clustered sources) but emitted at different loading stages were compared. In figure 6 the waveforms (FFT) of four different events received by channels 5 and 6 are plotted as an example. While all four events were located within about 10 mm relative distance near the upper surface just over the middle support, events 13 and 18 were registered during or just after crack formation. By the time the events 85 and 104 occurred, the crack responsible for the acoustic emission had reached a width of 0.1 mm near the source location and a depth of 120 mm (see figure 5). Similarities within the two groups are obvious. While the frequency distribution of signals emitted at an early loading stage exhibit different peaks between 40 and 150 kHz and a slight one at 260 kHz, the later signals, probably due to friction in the existing crack tend to be more monofrequent with 110 kHz.

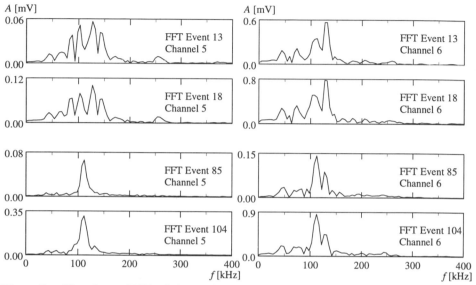

Figure 6. Waveforms (FFT) of clustered events received by channels 5 and 6; events 13 and 18 (above) during the crack formation and events 85 and 104 during the crack opening phase of the bending test B2

In order to quantify the similarity of two particular waveforms, a simple procedure was applied. The two FFT-diagrams are normalized to an area of 1 under the curve to eliminate the influence of the absolute amplitude and then subtracted. The remaining area, plotted grey in figure 7, is a value inversely proportional to the similarity of two waveforms. The application of this procedure yielded values from 0.1 for almost identical signals up to 1 for signals with only few common frequencies. In addition to this comparison, a more sophisticated procedure, estimating the coherence spectra of two events channel by channel, was applied. While the similarity of two events could be calculated easily, the association with a certain fracture mechanism is uncertain and requires further evaluation of the data, for example using moment tensor inversion.

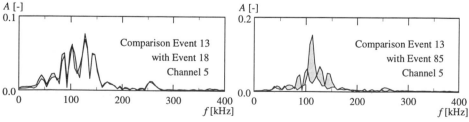

Figure 7. Comparison of two pairs of waveforms (FFT) measured by channel 5 during bending test B2

CONCLUSIONS

During pull-out, double-punch and bending tests with reinforced concrete specimens AE signals were measured by eight transducers (PZT), recorded in an acquisition unit and subsequently analysed. For every signal a set of parameters was extracted on-line. Simultaneously, signals exceeding a defined peak amplitude were digitized and stored as transient waveforms.

The absolute number and amplitude of AE signals occurring during the tests depended strongly on various circumstances, for example the specimen geometry, sensor arrangement and loading rate. Considering every test by itself, the development of AE activity exhibited a good relation to damage accumulation in the specimen (figure 2). This purely qualitative analysis, however, proved to yield no statements applicable to changed conditions.

AE signals recorded by all eight channels were localized using a program on the basis of the different p-wave arrival times. Considering only signals stored as transient waveforms and detected by all eight channels reduced the amount of data by about 90 % to 95 % (table 1). The remaining data contained only little noise and could be located generally with a high accuracy. For all tests the location of AE sources corresponded well with visually observed cracks. During the bending tests the following restrictions were encountered: With increasing distance (about 200 mm for the given sensor sensitivity and preamplification) between the source and the sensor, attenuation became too strong, the p-wave arrival time could no longer be detected and therefore localization failed. Existing cracks posed an obstacle to wave propagation and limited the area where AE sources could be localized (figure 5).

The waveforms of located signals were transformed into the frequency domain using FFT. A visual comparison allowed to form different groups of events with sets of quasi identical waveforms originating from one cluster (figure 6). Influences due to different propagation paths being avoided, different waveforms were associated with different source mechanisms.

ACKNOWLEDGEMENTS
The authors gratefully acknowledge the financial support of the Swiss National Science Foundation.

LITERATURE
[1] Aki, K.; Richard, P.G.; *Quantitative Seismology, Theory and Methods;* W.H. Freeman and Company; New York 1980; 932 pp.

[2] Berthelot, J.-M., Ben Souda, M., Robert, J.L.; *Identification of Signals in the Context of Acoustic Emission in Concrete;* Journal of Nondestructive Evaluation, Vol. 13, No. 2; New York 1994; pp. 63-73.

[3] Grosse, C. U.; *Quantitative zerstörungsfreie Prüfung von Baustoffen mittels Schallemissionsanalyse und Ultraschall;* Dissertation University of Stuttgart; Stuttgart 1996; 168 pp.

[4] Grosse, C.U., Reinhardt, H., Dahm, T.; *Localization and Classification of Fracture Types in Concrete with Acoustic Emission Measurement Techniques;* International Symposium Non-Destructive Testing in Civil Engineering (NDT-CE); Proceedings Vol. 1; Berlin 1995; pp. 605-612.

[5] Landis, E. N.; *A quantitative acoustic emission investigation of microfracture in cement based materials;* Northwestern University; Evanston 1993; 185 pp.

The CORROSCOPE system for the diagnosis, detection and repair of reinforcement corrosion in concrete

DR K N TUSCH, Colebrand Ltd, Warwick Street, London, UK
PROFESSOR V MAZOURIK, Colebrand Ltd and Russian Academy of Sciences
DR P R VASSIE, Transport Research Laboratory, Crowthorne, UK

INTRODUCTION

Many reinforced concrete structures suffer from corrosion of the reinforcing steel at some point during their life. Corrosion is normally caused by the presence of chloride salts or carbon dioxide in the structures environment. It results in damage to the concrete such as cracking, spalling and delamination as well as a reduction in the cross section of the steel. A loss in strength or hazards due to falling concrete can be consequences of the damage caused by reinforcement corrosion.

During the last decade a number of tests have been developed to investigate various aspects of the corrosion of steel in concrete [1]. The general purpose of these tests is to recommend the extent, type and priority for remedial action. Each test delivers specific information and it is essential to carry out a range of tests providing different sorts of information if a complete appraisal of the corrosion process, sufficient to recommend remedial action, is to be achieved.

There is a conceptual problem concerning the interpretation of these tests because they are mainly based on electrochemical techniques, reflecting the mechanism of the corrosion process. The units of measurement often involve amps, volts and ohms hence it is not surprising that civil engineers sometimes find it difficult to interpret the results of these tests in engineering terms. This paper describes a recently developed software package called CORROSCOPE which enables the results of corrosion tests on concrete structures to be interpreted in engineering terms, enabling the remedial action to be correctly specified.

THE CORROSCOPE PHILOSPHY

CORROSCOPE employs a range of frequently encountered corrosion tests and visual observations as listed below:

- Chloride depth profiles
- half cell potential
- concrete resistivity
- delamination soundings
- cover depth

- carbonation depth
- potential gradient
- corrosion current
- location of cracks and spalls
- areas vulnerable to corrosion

Bridge Management 4, Thomas Telford, London, 2000.

The tests are carried out over all or part of the structure or structural element. Simple non-destructive tests such as half cell potential, potential gradient, concrete resistivity, delamination soundings, cover depth and visual observations are usually made on a closely spaced grid. Semi-destructive tests such as chloride depth profiling and carbonation depth and complex non-destructive tests such as corrosion current are carried out at a lower density. The results of the tests are interpreted in such a way that they provide answers to the questions listed in Table 1.

These answers provide sufficient information to make a rational specification of an effective remedial treatment. The specification includes the best type of remedial action, the extent of repairs and some indication of their urgency or priority. The latter is obviously only based on the condition of the structure and it is appreciated that prioritising maintenance work depends on a range of diverse operational factors of which condition is just one. The answers should also provide the engineer with a clear picture of the condition of the structure thereby enabling correct decisions to be made.

Table 1. Questions answered by corrosion tests on concrete structures

Question	Possible Answers
What is the current or likely future cause of corrosion.	Chlorides, carbonation or both.
When did corrosion start or when will corrosion start.	Estimate of year of corrosion initiation.
Where is corrosion currently taking place.	Map of structure indicating areas where corrosion is likely to be occurring.
What type of corrosion is occurring.	Localised or general or both.
What is the rate of corrosion.	High or medium or low.
Where has the concrete been damaged by corrosion.	Map of structure indicating the damaged areas.
Which areas are vulnerable to chloride induced corrosion.	Map of structure showing areas vulnerable to chloride induced corrosion.
Which areas have low concrete cover depth.	Map of structure showing areas with a low cover depth.
What is the most appropriate maintenance option.	The type of maintenance and specific method is specified.
What is the extent of concrete repairs needed for a durable repair.	Map of structure showing areas requiring repair.
What is the current priority for repair work.	Low, medium or high based on the rate of deterioration.

CORROSION TESTS FOR CONCRÈTE STRUCTURES

Despite much research a single test which determines the location of corrosion sites and how much steel is remaining has not been found. Thus a range of tests is employed each of which supplies a piece of the jigsaw puzzle and together they provide a nearly complete picture of the corrosion. Another reason for using a range of tests is that their interpretation is sometimes difficult leading to apparent contradictions [2]. The use of a suite of test methods allows inconsistent data to be identified and rejected. The primary and secondary function of each test is given in Table 2 and it can be easily seen how the information from each test relates to the questions posed in Table 1.

Table 2. Functions of Corrosion Tests for Concrete Structures

Test	Primary Function	Secondary Function
Carbonation depth.	Cause of corrosion.	Location of corrosion.
Chloride depth profile.	Cause of corrosion.	Location of corrosion.
Half cell potential.	Location of corrosion.	Initiation time.
Concrete resistivity.	Rough estimate of rate of corrosion rate when corrosion has initiated.	Probability of high corrosion rates during life of the structure.
Potential gradient.	Type of corrosion.	Rough estimate of the rate of corrosion.
Corrosion current.	Rate of corrosion.	Location of corrosion.
Cracks coincident with reinforcement.	Indication of location of general corrosion.	Indication of possible site of falling concrete.
Non coincident cracks.	Possible site of future corrosion.	Indication of other deterioration mechanisms.
Spalls.	Site of general corrosion.	Site of loss of bond and falling concrete.
Delaminations.	Site of significant loss of bond and possible large concrete falls.	Site of general corrosion
Rust stains.	Evidence of general corrosion although location of stain does not necessarily correspond with the corrosion site.	
Cover depth	To interpret carbonation and chloride measurements.	To locate areas vulnerable to corrosion.

CORROSION TESTS AND MAINTENANCE MANAGEMENT

The management of particular structures to satisfy some target such as a required life or a minimum lifetime cost is usually called project level management. The policies involved in the management of a stock of structures are usually referred to as network level management. The corrosion tests under discussion in this paper are primarily associated with project level maintenance management although the data accumulated can also contribute to network level management. Maintenance is used in its most general sense and encompasses routine maintenance, preventative maintenance, repairs, rehabilitation and full or partial replacement. The main questions that need to be answered in the maintenance management of a structure are:

- when should maintenance be carried out
- what type of maintenance is appropriate
- how much maintenance is needed
- for how long will maintenance work last before further work is needed.

It is important to know what is causing the corrosion since this has a major influence on the most appropriate type of corrosion. The location of corrosion as defined by half cell potential, iso-potential contour maps helps to decide how much repair is needed. For corrosion caused by carbonation the extent of repairs corresponds closely within the area of damaged concrete, although it is normally somewhat larger. For corrosion caused by chlorides the area of concrete repairs needed to produce a durable repair is usually much greater than the area of damaged concrete because

- the localised corrosion commonly associated with chloride induced corrosion often does not produce visible concrete damage
- Localised anodes provide a degree of cathodic protection to the surrounding areas which may, therefore, have a high chloride content without causing corrosion. These chloride contaminated but non-corroding areas must be replaced in order to avoid the development of incipient anodes in the future.

The required extent of concrete repairs may be so great that other repair techniques like cathodic protection may be more appropriate. The extent of repairs indicated by the half cell method can be imprecise but visual observations, carbonation, chloride and corrosion current tests in combination with the half cell method enables the extent of repairs to be specified effectively.

The type of corrosion is important because it influences the type of maintenance, the extent of repairs, and indicates the location of possible sites of major losses in reinforcing steel cross section.

The rate of corrosion helps to decide the priority of repair. It may be possible to defer maintenance for a period on a structure which is corroding only slowly without incurring significant further deterioration whereas for a structure with a high corrosion rate delaying maintenance could shorten the structures life or necessitate more costly and complex remedial work in future. Corrosion current and potential gradient measurements give a reasonable estimate of rate whereas concrete resistivity measurements only indicate the degree to which the concrete will conduct corrosion currents. If the resistance is very large then the corrosion rate could be negligible when corrosion initiates. In this sense resistivity measurements can be used to screen structures to find those that are unlikely to have significant corrosion problems during their life.

Cover depth measurements can be used to identify in advance areas on the structure which are vulnerable to corrosion thereby enabling additional preventative maintenance or monitoring to be adopted.

Visual observations of damaged concrete are valuable for identifying areas of a structure that could present a hazard due to falling lumps of concrete so that loose concrete can be removed in a planned way safeguarding the public. Visual observations are also used to define areas of low steel-concrete bond.

USING THE CORROSCOPE ANALYSIS PACKAGE

The data from each test method is stored on a data logger on site and subsequently down loaded to a computer for analysis. CORROSCOPE can handle elements of irregular shape and nodes of the grid where measurement was not possible. CORROSCOPE plots the data on a plan of the structure. Data recorded over a regular grid can be plotted as contour lines or as shaded colour zones. Non gridded data such as visual observations are plotted to show areas of spalling and delamination. Cracks are represented as straight lines on the plan. CORROSCOPE permits three types of analysis which can be used independently although the benefits are maximised by the complementary application of all the analysis methods. The methods are:

- co-ordinated cross-wires on multiple screens
- user defined criteria
- automated analysis
- expert module

The co-ordinated cross-wire method enables the data for a number of measurement techniques to be plotted on identical plans of the structure or structural element. There is one plan for each technique and these can be viewed simultaneously in window tiling mode. Each window has cross-wires to define the position on the plan and the cross-wires on all the plans are co-ordinated allowing the values for different measurement techniques, at a user defined position, to be easily compared on a single monitor screen. The measurement values at the cross-wire position are displayed on each window. The position of the cross-wires is controlled using the mouse. This method is particularly useful for those who are experienced at interpreting corrosion survey data.

The user defined criteria method enables those areas which satisfy the criteria to be shaded on a plan of the structure. The criteria may be as complex as desired and involve a number of different measurement techniques. A simple and commonly used example is to find all the areas on the structure where the total chloride content of the concrete exceeds 0.3% by weight of cement **and** the half cell potential is less than $-350mV$ vs $Cu/CuSO_4$ reference cell. A small variation on this criteria would be to replace the **and** operator with the **or** operator. A full set of logical operators is available for defining criteria. These criteria are often used on bridges to define the area requiring concrete repairs [3]. This technique requires less experience than the cross-wire method although some experience is needed in order to set up relevant criteria.

The automated analysis method is better suited to users who have less experience in the interpretation of corrosion survey data.   All the available data sets are analysed simultaneously to provide answers to the questions listed in Table 1. The operator can decide whether CORROSCOPE analyses the data for an entire element of a structure or alternatively to analyse in detail a specified part of an element which may be structurally critical. In addition to answering these questions the automated analysis module of CORROSCOPE carries out the following functions:

- Recommends the techniques and the measurement density to use on a structures based on a set of parameters describing the general features and environment of the structure.
- Checks for inconsistencies either within or between data sets of all measurement techniques; this is helpful because it draws attention to possible errors in data collection or to exceptional conditions in the concrete or its environment.
- Warns if insufficient measurement techniques or data points have been used to reliably answer the questions in Table 1.

The expert module is particularly useful for analysing unusual situations. The expert module is implemented as a Microsoft Access Database and has three main components: Expert Table, Implications Table and Actions Table. In the expert table the data for each type of measurement is assessed over a user specified area and allocated to one of three categories based on well established criteria. The categories are:

- data indicates corrosion is taking place (Yes)
- data indicates that corrosion is not occurring (No)
- data is indeterminate about corrosion (Uncertain)

The last category is necessary because for some techniques like half-cell potential there is a range of measurements over which the interpretation is uncertain. A single record in the data table consists of an allocation of yes, no, or uncertain categories to each type of measurement carried out as shown in the example below:

Half cell potential	Yes
Corrosion current	Yes
Carbonation	No
Chloride	Yes
Potential gradient	Yes
Resistivity	Uncertain
Visual indications	Yes

This can be summarised by the vector [Y Y N Y Y U Y]. Each record in the expert table is linked to an implication and action in their respective tables. For example the above record has the following implication and recommended action:

Implication: Corrosion due to chlorides is occurring. It probably consists of both localised (high potential gradient) and general (visual indications) corrosion.

Action: The damaged concrete must be repaired using the concrete repair technique. Future corrosion must be stopped by preventing further ingress of chlorides and applying cathodic protection or chloride extraction.

DATA REPRESENTATION
Data can be represented as contour lines, shades of colour or numeric labels or any combination of these. These representations can be exported to other software packages such as word processors or desktop publishers. CORROSCOPE is a windows 95 application. Three examples of data representation and analysis are shown in Figures 1 to 3.

Figure 1 shows a set of half-cell potential data represented as a probability density function and cumulative probability. The percentage of data less than a chosen half cell value is reported. This provides guidance on the extent of corrosion on an element or structure.

Figure 2 shows plots that are used to indicate the prognosis of carbonation on a structural element. The upper graph shows how the carbonation depth is expected to vary with time at a particular location. The lower plot incorporates the cover depth information to indicate the area of the element that is at risk of corrosion and how this will vary with time.

Figure 3 indicates as a shaded region the area of an element that satisfies the chosen criteria, superimposed on a cover depth contour plot.

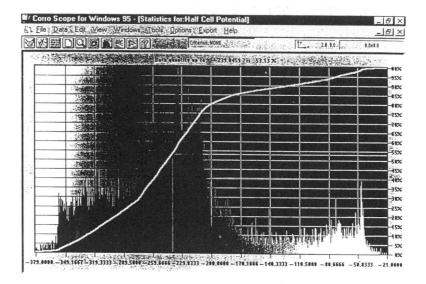

STATISTICAL DISTRIBUTION OF INSTRUMENT DATA

Figure 1   Statistical Analysis of Data

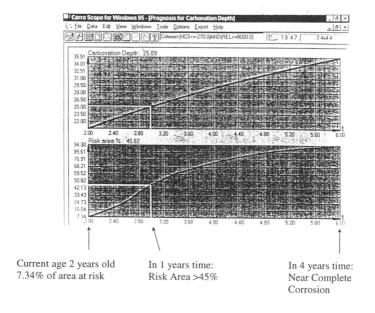

Current age 2 years old   In 1 years time: In 4 years time:
7.34% of area at risk   Risk Area >45%   Near Complete
                                                      Corrosion

Figure 2   Prognosis for Carbonation Data

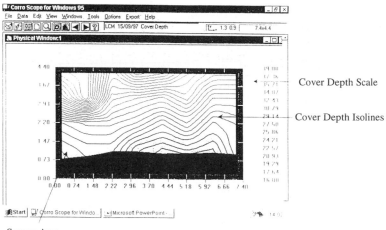

Survey Area

Criteria Zone: Blue Shaded Area
Half-Cell < -350 mV & Resistivity < 8000 ohm cm
Therefore > 90% Probability of Corrosion

Figure 3   **Area of Element Satisfying user Specified Criteria**

REFERENCES

1.   Broomfield, JP. Corrosion of Steel in Concrete, E & F N Spon London (1997).

2.   Vassie, PR. Measurement Techniques for the Diagnosis, Detection and Rate Estimation of Corrosion in Concrete Structures, pp 215-229. Controlling Concrete Degradation Ed. RK Dhir. Thomas Telford (1999)

3.   Department of Transport. Inspection and Repair of Concrete Highway Structures. BA35/90 (1990).

Site testing and monitoring of pier crossheads suffering from chloride induced reinforcement corrosion

G HUGHES (WSP), J S C SMITH (WSP), C P WARREN (WSP)
Basingstoke, United Kingdom

ABSTRACT

This paper summarises the results of some of the site testing and site monitoring carried out on the pier crossheads of the M4 Elevated Viaduct in West London. Overall, the results are consistent with, for instance, high levels of rebar corrosion being associated with high levels of chlorides, high levels of half-cells, high corrosion currents and low concrete cover etc. It is however clear that no single measurement is a good indicator of rebar loss of section, other than direct measurement having extracted the rebar. Rather, all the site testing results should be considered together to give an estimate, and the estimate should then be treated with caution.

The results show significant rebar loss of section for half-cells greater than 250mV. This compares with the UK guideline to repair areas where the half-cells are greater than 350mV. A reason for the difference might be that half-cells measured on site can vary about ± 100 mV due to seasonal effects.

INTRODUCTION

The M4 Elevated was built in 1965 and is a 2.9km long viaduct in West London. The deck joints over the piers have probably leaked since the structure was built. The result is that water containing de-icing salts has leaked down the deck joints and caused extensive chloride induced reinforcement corrosion in the pier crossheads.

In 1991, WSP took over responsibility for monitoring and maintaining the West London road network on behalf of the UK Highways Agency. Shortly afterwards, inspections of the M4 Elevated brought to light problems that were to concern the investigation and assessment team for the next eight years. The deck joints over the piers were leaking, and probably had been since the structure was built. The result was that water containing de-icing salts had leaked down the joints and caused chloride induced reinforcement corrosion in the pier crossheads. The corrosion products had caused delamination and spalling of the concrete, and there were large areas of loose concrete and rust stains (Figure 1).

Immediate action was taken to remove the loose concrete which might have posed a danger to users of the A4 which runs under the crossheads, and investigations and assessment were commenced with a view to maintaining public safety and rehabilitating the structure.

This paper summarises some of the results of the site investigations on Crosshead number 45 which was one of the worst condition crossheads and which was subjected to a barrage of tests. It also summarises some of the results from other crossheads carried out over a period of eight years to try to obtain information on the rate of deterioration.

THE STRUCTURE
The structure was designed by Sir Alexander Gibb and Partners and construction was completed in 1965. The decks consist generally of 17m simply supported spans of precast prestressed beams with a concrete slab. They are supported on 129 reinforced concrete piers, each pier consisting of a central column with a crosshead. The crossheads in section are of inverted T shape with the nibs at the sides supporting the decks. The crossheads and nibs vary in depth, deeper in the middle and tapering towards the ends, making the nibs at the ends of the cantilevers the critical locations with regard to load carrying capacity (Figure 2).

There are 25mm gaps between the crossheads and deck ends, and up to 60mm gaps between the tops of the nibs and the soffits of the decks. These narrow gaps make the faces between the crossheads and decks inaccessible, essentially hidden faces. These hidden faces have reinforcement which is critical to tie the nibs which support the decks back into the bulk of the crosshead (Figure 2).

SITE INVESTIGATIONS
In order to quantify the problems, site investigations were carried out. One of the first exercises was to rank the crossheads on the basis of delamination on the exposed faces. Then, a sample of crossheads were monitored for half-cells and delamination starting in 1992, initially at 3 month intervals but now on an annual basis, to try to obtain data on rate of deterioration. These initial investigations were all carried out on the exposed crosshead faces out on site. As assessment progressed, it became clear that there was a need to obtain information about the condition of the hidden faces of the crossheads. This information was crucial because of concerns about the possible corrosion of the reinforcement for tying back the nibs into the bulk of the crosshead. Attempts were made to take photographs and measurements within the narrow joint gaps, but this proved very difficult and the results were questionable.

Therefore, the decision was taken to remove one of the crossheads in order to test its hidden faces. This was a major undertaking given the location of the structure and, the decision to go ahead was only made by the Highways Agency after careful and detailed consideration of all options. Crosshead 45 was chosen because it was one of the ones in worst condition. The removal of Crosshead 45 and its replacement with a new one has been described elsewhere (Concrete Engineering International, Vol 3, No 1, Jan/Feb 1999). This paper summarises the testing subsequently carried out on the original Crosshead 45.

TESTING OF CROSSHEAD 45
The 200 tonne original Crosshead 45 was cut up into three transportable sections, transported to the Highways Agency maintenance depot at Heston (Figure 3), supported on stillages, and testing commenced. The testing carried out covered all surfaces of the crosshead, with testing nodes on a 500mm grid. The tests comprised:

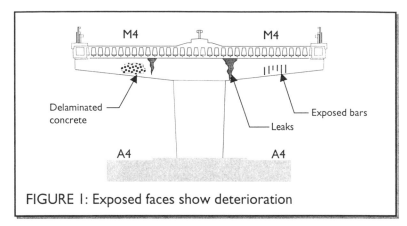

FIGURE 1: Exposed faces show deterioration

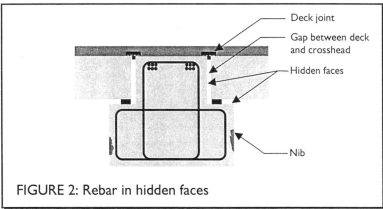

FIGURE 2: Rebar in hidden faces

FIGURE 3: Part of original crosshead transported away

a)   Visual, crack mapping etc
b)   Delamination of the concrete surface, by hammer sounding [Delam]
c)   Concrete cover, using a cover meter [Cover]
d)   Chloride ion content as percentage of weight of cement, by obtaining dust samples at
     5-25, 25-45, 45-65mm depth, so that chloride level at the reinforcement level could be
     interpolated [Chlorides]
e)   Half-cell potentials, using a copper/copper sulphate electrode [Half-cells]
f)   Corrosion currents, using a linear polarisation resistance LPR technique [Corrosion
     current]
g)   Loss of cross sectional area of rebar, by exposure of rebar using water jetting, removal
     of rebar, and direct measurement using callipers [Rebar Loss]

The testing was carried out using state of the art equipment, and by highly experienced
personnel, over a period of approximately 1 year.

RESULTS FOR CROSSHEAD 45

First view of the results indicated that there were three basic types of surfaces:

a)   Solid surfaces, with the concrete appearing dense and sounding solid when tapped
     with a light hammer
b)   Delaminated surfaces, as identified by hollow sounds when tapped with a light
     hammer
c)   Spalled surfaces, where the concrete cover had already fallen off and the rebar was
     exposed

The spalled surfaces were generally located at corners, in particular at the front edge of the
nibs.   The rebar at these spalled surfaces had been directly exposed to leaks and the
atmosphere for an unknown length of time and was badly corroded, more so than for the other
areas where there was concrete cover remaining.   The results from these different types of
surfaces were separated out in the analysis.   The test results are illustrated in Figures 4A to
4F.   By reference to the figures:

a)   Figure 4A :   Rebar loss plotted against type of surface
     Average rebar loss for the solid, delaminated and spalled surfaces was 3.1%, 12.2%
     and 16.9% respectively, an increasing trend as expected.   The results for the solid
     surfaces are of most interest, because the other areas were already delaminated or
     spalled and would certainly need repair.   For the solid surfaces, the rebar loss was
     generally low, even zero, but the results show that there can be up to say 50% rebar
     loss by pitting corrosion without any delamination as a warning sign.

b)   Figure 4B :   Rebar loss plotted against chloride content
     The chloride content varied in the range 0% to 5%, with average 0.71%.   This
     compares with the critical value of 0.3% in UK Highways Agency Advice Note BA
     35/90.   Basically, the crosshead was extremely contaminated with enough chlorides to
     allow corrosion to start anywhere.   The test results from the crosshead are of no use in
     testing the 0.3% value, but the results are not inconsistent with this value.   It is noted
     that once a critical threshold of say 0.3% is passed, the results indicate that rebar loss
     is not a function of chloride level.

c)   Figure 4C :   Rebar loss plotted against half-cells
        The half-cells vary up to about 700mV with average 318mV. It is seen that there is
        significant rebar loss above say 250mV. (This compares with the guideline given in
        UK Highways Agency Advice Note BA 35/90 to repair areas greater than 350mV.)
        The results show that half-cells less than 250mV indicate relatively insignificant
        corrosion, but half-cells above 250mV show anything from zero to say 50% rebar loss.

d)   Figure 4D :   Rebar loss plotted against corrosion current
        The results show a large number of results with zero or small rebar losses associated
        with low corrosion currents. However, it is seen that there is a large amount of scatter,
        with quite a lot of high rebar losses associated with low corrosion currents. Better
        results had been expected from this new high technology tool. It may be that the
        active corrosion zones move around over the years to produce the scatter in the results.

e)   Figure 4E :   Rebar loss plotted against cover
        The results are generally consistent with more rebar loss for low cover compared with
        high cover, as expected. It is interesting that corrosion can still occur with cover of
        almost 80mm. Also, close inspection of the results reveals proportionately more
        delamination and spalling for low cover compared with high cover, again as expected.

f)   Figure 4F :   Rebar loss as function of cover and half-cells
        This figure plots only those areas where there was 10% or more rebar loss ie. the
        worst areas. The results indicate that signifcant rebar loss occurs with half-cells
        greater than 250mV, and that delamination and spalling occur for covers less than
        30mm.

There is clearly a large amount of scatter but the results are as expected, generally consistent
with:

Cause	Symptom	Result
High chlorides	High half-cells	Deterioration
Low Cover	High corrosion currents	Rebar loss
		Delamination and spalling

A full statistical analysis of the above data and other data is ongoing with a view to more
exactitude in estimating the level of rebar loss from non-destructive testing. Best fits,
confidence limits, probabilities etc will be developed as part of the exercise. However some
tentative observations can be made. The likelihood of finding significant rebar loss on the
crossheads appears most likely at (in order):

a)   Delaminated and spalled areas
        Delamination and spalling is caused by corrosion, so they are very good signs of
        corrosion. Spalling often occurs at corners such as the nib front corner of the
        crosshead.
b)   Half-Cells >250mV
        Corrosion is an electrochemical process and high half-cells indicate the reaction is on-
        going, or could start.

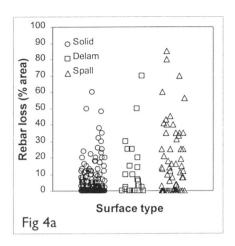

Fig 4a

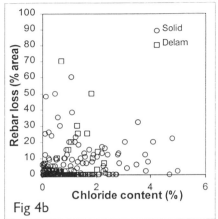

Fig 4b

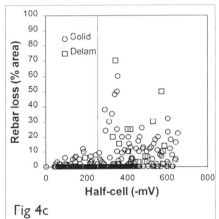

Fig 4c

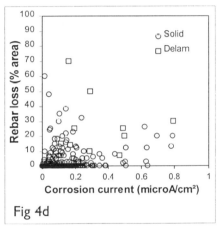

Fig 4d

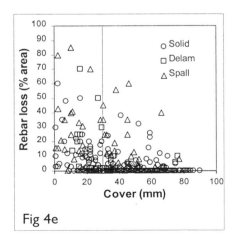

Fig 4e

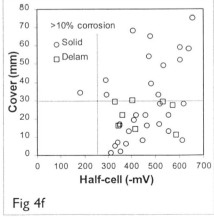

Fig 4f

FIGURE 4: Test results

c)   Cover <30mm, say
     Low cover means the rebar has little protection from salts and corrosion can lead to delamination and spalling.
d)   Corrosion current > 0.3 microamp/cm²
     Corrosion is on electrochemical process so more current indicates more active corrosion.
e)   Chlorides >0.3%
     Chlorides break down the passive protection of the rebar. For the crossheads, it seems there are enough chlorides everywhere to initiate corrosion and so it is not effective as a criterion for finding corroded rebar as regards the crosshead.

OTHER RESULTS

Given the space constraints for this paper, it is only possible to quickly summarise some of the other testing results for the M4 Elevated:

a)   Expected accuracy of chloride readings at 0.3% is about ± 0.1%, from sending samples with known chloride contents to various laboratories.
b)   Half-cell readings vary with season, moisture and temperature. Expected accuracy at 350mV is about ± 100mV, from measuring half-cells on the crossheads for eight years.
c)   All the crossheads have been subjected to leaks. Crossheads with the more extensive leaks are in the worst condition as expected.
d)   Half-cells on the exposed faces of one of the crossheads were measured before and after installation of a new deck joint to stop the leaks. The half-cells 'cooled' about 50mV over 9 months as the concrete dried out.
e)   Half-cell readings on a representative sample of the piers have remained essentially constant over the last eight years, indicating little if any deterioration. (This has been helped by an ongoing programme of deck joint replacement to stop the leaks).
f)   The average half-cell reading on the crossheads is about 300mV. The structure is about 35 years old, so if the crossheads have deteriorated at a constant rate, the rate is: 300/35 = 8mV per annum.
g)   Average half-cells on individual crosshead have remained fairly constant over the last eight years, but with some of the "hot spots" moving from one location to another.
h)   Average loss of section of the rebar due to corrosion is now about 4%, so average corrosion rate over the last 35 years is about 0.1% loss of section per annum. However, individual bars have corroded up to 50% say and show severe pitting.
i)   Another result, and the main result for those involved in the investigation and assessment of the M4 Elevated, is that the condition of the hidden faces of crosshead 45 was essentially the same as the exposed faces below. Removing the 200 tonne crosshead had given the hoped for result.

DISCUSSION

The results as described above are relevant to the M4 Elevated and its exposure conditions. It is noted that a full statistical analysis of the data is at present ongoing, so final conclusions have not yet been reached. Nevertheless, the following observations are tentatively put forward:

a)   Site testing involves inaccuracies and scatter of results.

b)   The existing guidelines in the UK are that areas with chlorides greater than 0.3% and half-cells greater than 350mV should be repaired. The results show significant rebar loss for half-cells greater than 250mV. A reason for the difference in half-cell value might be that half-cells can vary about ± 100mV due to seasonal effects.

c)   Determining areas of delamination with a light hammer is surprisingly accurate, and a full delamination survey using a light hammer and marking the delaminated areas with a paint marker or similar, should be an essential part of any condition survey. This low technology method is often overlooked, but gives good useful results.

d)   On the basis of the test results, it would appear that the likelihood of finding significant rebar loss appears most likely at (in order):

   •   Delaminated and spalled areas
   •   Half-cells >250mV
   •   Cover < 30mm say
   •   Corrosion current > 0.3 microamp/cm^2
   •   Wet areas
   •   Chlorides > 0.3%

It is noted that the M4 crossheads have wet areas and high chloride levels almost everywhere, so not unsurprisingly they appear to be relatively poor indicators of corrosion areas. Therefore, the order as above is valid only for the M4 Elevated crossheads and other wet and very chloride contaminated structures. More generally, wet areas and chlorides should be nearer the top of the order.

e)   No single measurement (other than direct measurement after extracting the rebar) is a reliable indicator of level of corrosion. Rather, the results should be viewed together to give an estimate, and then that estimate treated with caution.

f)   A fundamental problem could be that the active corrosion zone moves around over the years. Therefore, high half-cells or high corrosion currents might be indicative of active corrosion zones, but not necessarily the worst corroded zones.

g)   It is very difficult to determine the exact existing condition of a structure without resorting to extensive destructive testing, and more difficult to gain accurate estimates of the rate of deterioration. This presents significant difficulties in formulating bridge management strategies without erring on the very conservative and safe side with commensurate increased costs.

ACKNOWLEDGEMENT

The authors have submitted this paper with the kind permission of the UK Highways Agency. Any opinions expressed are those of the authors.

Continuous acoustic monitoring of steel tendons and cables in bridges

D W CULLINGTON and T BRADBURY,
Transport Research Laboratory, Crowthorne, Berkshire, RG45 6AU
P O PAULSON,
Pure Technologics Ltd, Calgary, Canada, T2R 1L5

ABSTRACT

This paper describes laboratory trials and site installations of an acoustic monitoring system, developed in Canada, for detecting the fracture of stressed high-tensile steel wires in structures. The fracture of a stressed wire releases energy that can be detected as an acoustic event by surface mounted sensors. Characteristics of the event provide information on the position of the event and its cause. Trials at TRL have demonstrated that the system can reliably detect wire fractures in grouted post-tensioned structures and hanger cables, successfully rejecting other acoustic events. Installations on the Railway Viaduct in Huntingdon in the UK and the Bronx-Whitestone suspension bridge in the USA have shown the system to work in practice. Wire fractures can occur for a number of reasons. Acoustic monitoring is useful because the wires are often inaccessible for visual examination and fractures cannot generally be detected by non-destructive inspection techniques

1. INTRODUCTION

High tensile steel wires have many structural applications as individual wires, or in the form of tendons, cables or ropes. Typical structures include pre-tensioned and post-tensioned concrete bridges and buildings, pipelines, cable-supported bridges and anchors in ground engineering. All of these structures have a common feature. Generally, the wires are difficult to inspect because they are inaccessible. Post-tensioned tendons are located deep within a bridge deck, and suspension cables are difficult to inspect non-destructively below the surface layers. The problems are even greater in ground engineering applications. Wire failures can occur in service due to corrosion, stress corrosion and fatigue. Very often the loss of a few wires is not critical but eventually, for reasons of safety, repairs or replacement of the cable or structure may be required. For the effective and economic management of such structures, information is needed about the number and position of wires that have fractured.

The Transport Research Laboratory has carried out evaluation trials of the SoundPrint® acoustic monitoring system for detecting wire fractures in steel tendons and cables. The system works by continuously listening for the characteristic acoustic events that accompany the fracture. Most of the trials have been undertaken for the Highways Agency, as part of a programme of research on post-tensioned bridges.

2. THE SOUNDPRINT® ACOUSTIC MONITORING SYSTEM.

The SoundPrint® system was developed by the Canadian company Pure Technologies originally to detect wire-breaks in unbonded (ungrouted) tendons in the floor slabs of office buildings. Potentially it can be applied to any structure in which the integrity of steel cables is difficult to appraise visually. Research has indicated that NDT is generally not able to detect broken wires and monitoring is therefore a practical alternative.

The system has been described elsewhere (Halsall et al, 1996; Paulson, 1999). Briefly, when highly stressed steel wires fracture there is a release of energy, which is transmitted through the structure. Sensors (accelerometers) attached to the external surface of the structure are used to detect these acoustic events. Each sensor is connected using coaxial cable to an on-site data acquisition unit, such as the one shown in Figure 1. The system monitors continuously but collects no data until triggered by an acoustic event lying within pre-set limits. Software filters are then applied to reject events of no further interest. Events that successfully pass these tests may be wire fractures. They are sent via the Internet to Canada for the events to be classified and their position calculated.

Figure 1. 16-channel SoundPrint® acquisition unit.

Other events that might generate acoustic responses in a bridge include vehicles going over discontinuities in the road surface, small objects such as stone chips striking the concrete and expansion joint defects responding to trafficking.

Unless a monitoring system has previously been proved to work in a comparable location, evaluation trials are desirable to:

- confirm that wire breaks can be detected in that particular application
- demonstrate they can be distinguished from the non-break events expected on the site
- show that they can be captured reliably in the presence of the ambient noise.

Pure Technologies initially proved that wire breaks could be detected in office buildings with unbonded tendons. The operation of the system on a bridge with grouted tendons was believed to be more difficult than in buildings for two reasons. Grouted tendons would probably release less energy on fracture than unbonded tendons, and non-break events and background noise were likely to be more dominant.

3. LABORATORY TRIALS

TRL has undertaken a number of evaluation trials of the SoundPrint® system. Laboratory trials are an integral part of the evaluation process because, compared with site, the environment can be better controlled and wire breaks created for test purposes without damaging a structure in service.

3.1   Trials on post-tensioned tendons

Corrosion is known to have caused the fracture of wires in post-tensioned concrete bridges, and non-destructive testing has proved to be of limited value in detecting the presence of broken wires (Cullington et al, 1996). Invasive inspection is overwhelmingly the method adopted but this can be carried out at only a small number of locations for reasons of cost and damage to the bridge.

For the first trials at TRL, the SoundPrint® system was used to detect trial wire breaks in the tendons of two post-tensioned concrete members. One, the 30m long Bank Lane Unit, contains fully-grouted ducts. The other, a 10m long beam, was specially constructed at TRL and contains a combination of well-grouted and poorly-grouted ducts. Acoustic events were created by cutting, grinding and corroding the wires to which access was gained by coring or drilling into the ducts. Accelerated corrosion was obtained by means of anodic dissolution using an electrode positioned close to the corrosion site and a small current imposed though a saline electrolyte. In addition, other types of acoustic events were created by dropping, throwing or rocking objects, dragging chains etc. This was done to test the system's ability to reject non-break events.

Two types of trial were carried out: open trials in which Pure Technologies were informed of the tests being carried out, and blind trials in which details of the events were not revealed to Pure Technologies until after their report was received.

The initial trials demonstrated that the SoundPrint® system could detect and locate ungrouted and partially grouted wire breaks as well as some fully grouted breaks. It was also capable of rejecting most non-wire break events. However, areas for improvement were identified in the detection of low-energy, fully-grouted wire fractures and the rejection of small sharp impact events, some of which were occasionally incorrectly classified as wire breaks.

With data from the problem events supplied by TRL, Pure Technologies reconfigured the system using new sensors, new software and a higher sensor density. It then reliably detected fully-grouted wire breaks and rejected small sharp impact events. Following this success, and a recommendation from Thorburn Colquhoun, the Highways Agency decided to install a system on a bridge as described in Section 4 of this paper.

A further step on the Bank Lane unit was the design and development of an external wire-break rig. This rig is fixed to the concrete surface and, as the name implies, can be used to create wire breaks in a stressed strand held in the rig. The results of the final series of trails using this rig are given in Table 1.

3.2 Trials on a hanger cable

Hanger cables are subjected to dynamic loading from wind and traffic. As a result, they may undergo a combination of corrosion and fatigue that eventually leads to wire fractures. Detection of fractured wires is difficult, as only the wires on the outside of the cable bundle are visible, and even this may require the removal of wrapping at the cable ends where fractures are most likely to occur. Acoustic monitoring has the potential to detect fractures anywhere in the cable, including within the sockets.

The trials used a 2.8m length of spiral-strand hanger cable 36mm in diameter. This was installed in a hanger-cable fatigue rig at TRL. The rig has two servo controlled hydraulic actuators, one to provide tension in the cable and the other to produce dynamic transverse oscillations about the cable's midpoint, as shown in Figure 2.

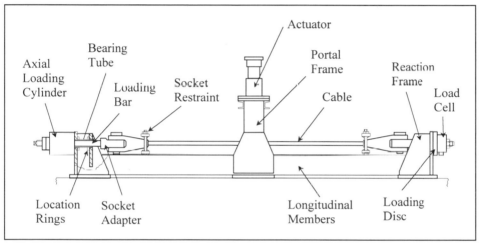

Figure 2. Diagram of hanger cable test rig.

A SoundPrint® system was set up with two sensors mounted on the cable, one on each socket. Only two sensors were required, because the location of acoustic events was needed in one dimension - along the length of the cable. The test rig generates significant background noise within which the monitoring system has to operate. Initially, therefore, a DAT sound recording of the rig in operation, picked up by the two sensors, was sent to Pure Technologies for examination.

To commission the system, the hardware filters were tuned by creating small impacts on the cable at the ends and in the middle. A dxf geometry file was created and loaded onto the system and the software filters were supplied by Pure Technologies. Calibration of the system was straightforward and took only a matter of hours.

White rings were painted on the cable near the neck of the sockets and at the centre on both sides of the central clamp. These were an aid in checking for breaks in the outside wires. A fracture results in the wire moving due to the release of strain and displacing the white line.

Values for axial strain and oscillation amplitude were based on those experienced by cable hangers in service and from previous fatigue trials at TRL. The axial load was initially set to 36 tonnes and the amplitude of the oscillations to ± 20mm with a frequency of 1Hz. Figure 3 shows the completed set up.

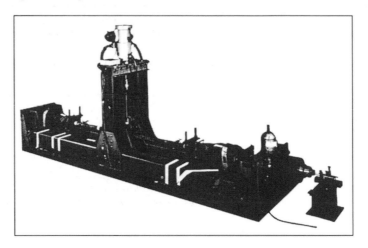

Figure 3. Photograph of hanger cable test rig

After 83000 cycles, the SoundPrint® system had registered 15 wire breaks. The experiment was stopped and the cable was removed from the sockets, cut and dismantled to discover how many wire-breaks were present and their location. These data were then compared with the SoundPrint® report. Figure 4 shows a time domain plot of a wire break event captured during the experiment. A TRL researcher heard this event (and later one other) as it occurred, thus corroborating the fact that the system had correctly identified a break.

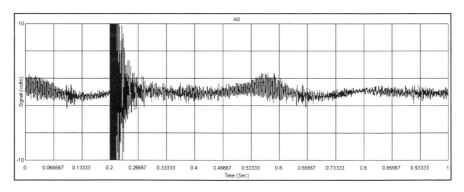

Figure 4. A one-second recording showing a hanger cable wire fracture and background noise.

On comparing the results from the destructive cable inspection with the events reported by Pure Technologies, it was concluded that the SoundPrint® system had located the wire fractures to within 300mm and generally within 100mm. A complete audit of the number of wire breaks created and detected was not possible, because some breaks were present in the cable from an earlier trial. However, the results as given in Table.1 indicate that none were missed. It was also noted that the frequency and time domain plots were similar to those for wire breaks in unbonded tendons in post-tensioned bridges.

3.3   On going Trials

Research is in progress at TRL into the measurement of concrete strain that occurs as wires fracture. A post-tensioned test beam is being monitored using Vibrating Wire strain gauges and a four-channel SoundPrint® system. A 20mm diameter hole was drilled into the specimen to gain access to the tendon and a corrosion cell was set up to corrode the wires. When a wire fractured, an operator who fortuitously was present heard it. At that moment SoundPrint® also recorded an event. The plot of this event and the data for the VW gauge nearest to the site of corrosion are given in Figures 5 and 6. It can be seen that the VW gauge shows a change in strain when the wire break occurred. Data evaluation is still in progress.

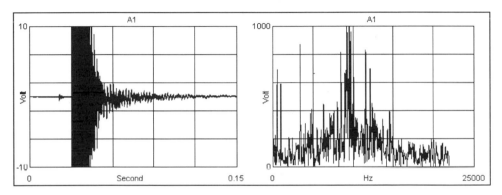

Figure 5. Time and frequency domain plots of a wire break heard by an operator

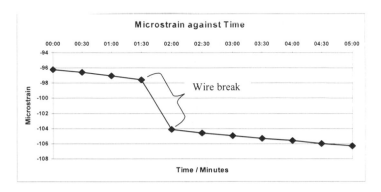

Figure 6. Graph from laboratory trials showing the change in microstrain due to the wire break as recorded by the VW gauge nearest to the site of corrosion.

4. SITE INSTALLATIONS AND TRIALS

Two site installations of the system are described in this section, one in England carried out under the supervision of TRL and another in the USA carried out under the supervision of Weidlinger Associates Inc of New York.

4.1   Railway Viaduct Huntingdon England

The post-tensioned A14 Railway Viaduct at Huntingdon was suitable for the first UK installation of a monitoring system for several reasons. The structure had been the subject of a Special Inspection that had indicated the presence of voids, water and chlorides in the tendon ducts, but no significant corrosion of the strands. Further structural investigations were in progress that would be supplemented by a clear indication of the presence or absence of actively fracturing wires. The structural form contained features that lent themselves to monitoring, in particular half-joints, which are difficult to inspect.

The fact that the prestressing system in the Railway Viaduct is apparently in a good condition indicates that the structure has a long potential life and will require economic management for the foreseeable future. The high volume of traffic using the route makes it essential to maintain the structure in service with minimum interruptions and appropriate regard for safety.

Following a commission from Thorburn Colquhoun, who are responsible for the Railway Viaduct as part of the Area 8 maintenance contract for the Highways Agency, a SoundPrint® acoustic monitoring system, was installed on the viaduct in mid 1998 (Cullington et al 1999). It comprises 32 channels positioned over a 48m length. Since the commissioning, numerous trials have been carried out on site. These mainly comprised facsimile wire break events created by a Schmidt hammer or spring impactor applied to the soffit, and external wire breaks as mentioned in section 3.1, created on the internal web of a cell of the structure. Table 1 contains a summary of the results. Continuous uninterrupted operation is important for this type of monitoring. After an initial settling in period, the system uptime has been over 99%. No naturally-occurring wire breaks have been detected.

4.2   Bronx-Whitestone Suspension Bridge USA.

A SoundPrint® system was installed on the main cable of this bridge in northern USA. Acoustic sensors were attached to six consecutive cable bands on the main cable monitoring a length of approximately 84m. The SoundPrint® data acquisition unit was located at deck level. The cable itself was undergoing maintenance and the outside wrapping had been removed revealing the high tensile steel wires. This enabled blind trials of the system to be carried out by cutting six wires, the results of which are summarised in Table 1.

Location	Nature of trials	Outcome
Bank Lane Unit	External wire breaks	25 breaks, five blind, all detected, 22 located within 0.2m and 3 within 0.5m
Hanger cable rig	Fatigue test fractures	End A1, breaks count exact (3 new); End A0, 6 old + 6 new breaks exceeds actual by 2.
Strain gauged beam	Corrosion fracture of internal wires	Breaks detected at TRL; data analysis in progress.
Railway Viaduct	External wire breaks and facsimiles	44 external wire breaks and facsimiles, 18 blind; 43 detected 18 within 0.2m and rest within 0.5m, 41 events correctly classified.
Bronx Whitestone main cable	Wires cut mechanically	6 wires cut blind, 5 detected within 0.22m and one within 0.7m

Table 1 . Results of trials at various locations

5. CONCLUDING COMMENTS

The SoundPrint® system has proved to be an effective way of monitoring the fracture of stressed high tensile steel wire in structures. There are many possible applications of this technology to the safe and efficient management of structures containing these elements. The system is already fully functional but remains responsive to new challenges. TRL is actively pursuing the application of SoundPrint® technology to other areas not yet evaluated elsewhere. Pure Technologies are co-operating with this work.

6. ACKNOWLEDGEMENTS

The paper has been written with the permission of TRL and the Department of Transport Environment and the Regions. The views expressed are those of the authors and not necessarily those of the sponsors. The authors wish to thank the many individuals and organisations who helped with the gathering of these data, including Donald MacNeil of TRL David Ball of HA, Tony Wakeman of Thorburn Colquhoun and John St Ledger of WS Atkins.

7. REFERENCES

Cullington, D W, M E Hill, R J Woodward and D B Storrar. *Special inspections on post-tensioned bridges in England: Report on progress.* FIP Symposium 1996: Post-tensioned concrete structures. The Concrete Society, 1996, pp 482-491.

Cullington, D.W., MacNeil, D., Paulson, P. and Elliot, J., (1999), *Continuos Acoustic Monitoring of Grouted Post-Tensioned Concrete Bridges.* Paper Presented at the 8th International Structural Faults & Repair Conference, London, UK.

Halsall A.P, W E Welch and S M Trepanier. *Acoustic monitoring technology for post-tensioned structures.* FIP Symposium 1996: Post-tensioned concrete structures. The Concrete Society, 1996, pp 521-527.

Paulson, P.O. (1999*), Practical Continuous Acoustic Monitoring of Suspension Bridge Cables.* Transportation Research Board 78th Annual Meeting January 10-14 1999. Washington DC. TRB, 2101 Constitution avenue NW Washington 20418.

Continuous acoustic monitoring of bridges

J.P. FUZIER and J.P. DOMAGE,
Freyssinet International & Compagnie, Velizy - Villacoublay Cedex, France
J.F. ELLIOTT and D.G. YOUDAN,
Pure Technologies Ltd., Calgary, Canada

ABSTRACT
This paper describes the application of acoustic monitoring systems on a number of projects, particular the Bronx Whitestone Bridge in New York. Reference is made to systems on the Alex Fraser Bridge in Vancouver, The George Washington Bridge in New York and a structure at Huntingdon in the UK. Extended acoustic monitoring systems are presented, which provide Damage Surveillance, show Seismic Response and allow interrogation of data on a dedicated Structure Internet Site.

Background details of the development are given.

INTRODUCTION
The use of high-strength steel wire has contributed greatly to advances in bridge design and structural performance over the last hundred years of so. Unfortunately, this increase in strength has not been accompanied by a corresponding increase in durability. In addition to conventional dissolution corrosion (or rusting), these high-strength steels are susceptible to failure through brittle fracture caused by stress corrosion, hydrogen embrittlement and fatigue. These corrosion mechanisms cause a significant loss of ductility in the steel, and failure can occur without a gradual loss of cross-section in the wire (Figure 1). Loss of ductility is not always accompanied by a reduction in ultimate strength. Studies have shown that galvanized wire is more prone to embrittlement than non-galvanized wire when the galvanizing is damaged or locally depleted. It is important to note that, although chlorides and other contaminants can accelerate embrittlement, water and oxygen are the only ingredients necessary to initiate the process. For this reason, embrittlement has been found in post-tensioning strands in completely enclosed, climate controlled high-rise buildings where rainwater has entered the tendon system during construction.

The presence of corrosion in high-strength steel wire in bridges can have serious consequences. The Ynys-y-Gwas Bridge in the United Kingdom collapsed in 1985 as a result of corrosion of the post-tensioning system. A major program of inspection of the grouted post-tensioned bridge stock in the U.K. was then initiated. At least four other Bidges were decommissioned and replaced because of serious corrosion problems. In 1992, The Highways Agency in UK banned the use of grouted post-tensioning in their bridges pending the development of improved detailing, grouting procedures and quality control measures. Corrosion has also been found in post-tensioned bridges in France, Germany, Italy, Denmark, Japan and the United States. Severe corrosion and extensive wire failure has been discovered in the main cables and in the hanger systems of a number of suspension bridges in North America and Europe. This has resulted in expensive rehabilitation projects in many instances and replacement of the main cables of the Tancarville Bridge in France and the General Grant Bridge in the U.S. Complete hanger systems have had to be replaced on the First Severn Crossing and, more recently, on the Forth Road Bridge. Corrosion-induced failures of stay cable wires have been documented in China, Argentina, Venezuela and Germany.

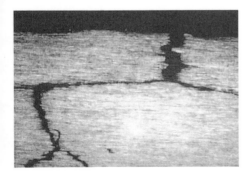

Figure 1. Cross-section through post-tensioning wire showing brittle failure mechanism.

Figure 2. Standard sensor for buildings and parking structures.

In most cases, corrosion of high-strength steel wire in bridges is not visually evident. Grouted post-tensioned tendons are obviously not visible. The circumferential wire wrapping on suspension cables and the steel or polymeric protective sheathing on stay cables preclude non-intrusive visual inspection. Even for the hangers (suspender ropes) of suspension bridges, where corrosion is most likely to occur at the deck-level socket connections, the absence of visible wire breaks in the outer layer of wires cannot eliminate the possibility of internal breaks.

Engineers have had to rely on intrusive investigations or on the use of available non-destructive evaluation techniques to provide some information about the condition of these components because of their inaccessibility. These methods have serious disadvantages in terms of sampling reliability, accuracy, cost, and disruption.

INVESTIGATIVE TECHNIQUES
Most of the methods used to investigate the condition of these bridges involve a representative sampling of the overall structure. (A summary of the techniques used to investigate many forms of bridge construction in included in the paper by Elliott). Hence, there is a possibility that corrosion and wire failure in localized areas may not be detected. A comprehensive health monitoring system that can provide information on deterioration for the entire structure would be useful for determining where to investigate and would also provide information about future performance once an investigation or repair has been completed. A continuous acoustic health monitoring system developed for this purpose has been applied to all three of the bridge types described above.

DEVELOPMENT OF CONTINUOUS ACOUSTIC MONITORING
The principle of examining acoustic emissions to identify change in the condition of the structural elements is not new. However, until recently, continuous, unattended, remote

monitoring of large structures was not practical or cost-effective. The availability of low-cost data acquisition and computing hardware, combined with powerful analytical and data management software, resulted in the development of a continuous acoustic monitoring system called SoundPrint®, which has been successfully applied to bonded and unbonded post-tensioned structures and bridges in Europe and North America since 1994.

The system has been developed from the observation that failures of prestressing wires generated an audible acoustic response. Pure Technologies reasoned that if these events had frequency or energy characteristics sufficiently different from ambient acoustic activity in a structure, it would be possible to identify the events, as well as their location and time of occurrence, with an appropriate instrumentation, data acquisition and data analysis arrangement. This would permit the non-destructive identification of broken strands so that these strands could be replaced periodically as part of a long-term, cost-effective structural maintenance program.

A prototype monitoring system was installed in the 6,000 square-meter ground floor of a building in Calgary, Canada in February 1994. The system consisted of an array of sensors (Figure 2) connected to an acquisition system with coaxial communication cable. The sensors were broadband piezo-electric accelerometers, which were glued to the underside of the concrete slab with cyano-acrylate adhesive. Sensor locations were chosen so that an event occurring anywhere on a slab could be detected by at least four sensors. The structure was divided into three acoustic zones delineated by expansion joints. A total of 60 sensors were used resulting in a density of one sensor per 100 square meters of slab area. A spatial multiplexing technique was employed to acquire data from the sensors using only 32 acquisition channels. For unbonded structures, sensor density varies between one per 35 to 100 square meters, depending on the geometry of the structure.

The goal of continuous automated monitoring combined with low-cost, centralized data processing was central to the development of the technology. Original software consisted of a commercially available data acquisition package located at the site computer, and a proprietary data analysis and report generation package located at the processing facility.

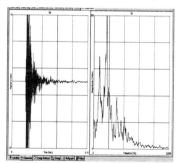

Figure 3. Time domain and frequency spectrum plots of wire break detected by sensor 10.0 m. from event.

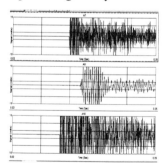

Figure 4. Time domain plot showing relative arrival time of signal at different sensors

The data acquisition software was later replaced with more suitable proprietary software. As a partial strand replacement project was being undertaken coincident with the installation of the system, it was possible to acquire data from many wire breaks. This information was used to train the data processing software to "recognize" wire breaks. When events possessed all the known properties of a wire break, they were classified as "probable wire breaks". Events possessing some of these properties were classified as "possible wire breaks". All other events were classified as "non-wire break events". By analyzing the time taken by the energy wave caused by the break as it traveled through the concrete to arrive at different sensors, the software was able to calculate the location of the wire break, usually to within 300   600 mm of the actual location. From the beginning, the capability of the system to accurately identify and locate wire break events was remarkable. Independent testing showed the system to be 100% correct when spontaneous events classified as "probable wire breaks" were investigated. Figure 3 shows a typical acoustic response to an unbonded wire break at a sensor 10.0 m (32.8 ft.) from the break location. Figures 4 and 5 illustrate how the system locates events.

Initially, data transfer from the site to the processing centre was accomplished through the use of a direct dial-up modem connection. However, within nine months of system commissioning, automated transmission of data using Internet protocols was achieved. This was a major advance towards the goal of automated, cost-effective monitoring of multiple sites as no human intervention was needed for routine system operation and long-distance call charges were more or less eliminated.

Presently, over 300,000 square meters of unbonded post-tensioned slab in twenty structures are being simultaneously monitored.-The analytical software is capable of automatically generating reports summarizing the time and location of wire breaks and other significant events. The operating efficiency of the system over the monitoring period is also recorded on the reports.

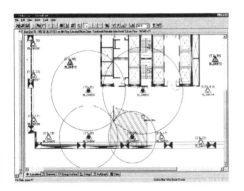

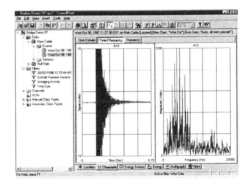

Figure 5. Analytical software generates graphical representation of event location algorithm for high-rise building slab.

Figure 6. Time domain and frequency spectrum plots of wire break detected by sensor 5.0 m from event.

MONITORING OF GROUTED POST-TENSIONED BRIDGES

In February 1997, the Highways Agency in the U.K. appointed British engineering consultant Transport Research Laboratory (TRL) to evaluate the acoustic monitoring system for use on grouted post-tensioned bridges. Initial tests were carried out on a free-standing bridge beam at the Laboratory's Crowthorne facility.

The work by TRL is only briefly summarised here, as more comprehensive details are included in a paper by Cullington et al. reported at this conference. The test protocol consisted of causing accelerated corrosion of grouted wires; as well as external wire breaks using a test rig designed to simulate fully-grouted and partially-grouted wire failures. Other events caused by impacts were also generated. The beam was monitored from Pure Technologies' processing center in Calgary, Canada, where reports were generated summarizing the event classifications and locations. In a combination of open and blind testing, the system correctly identified all 25 wire breaks generated.

To test the system under normal highway operating conditions, instrumentation was installed on a section of a highway bridge at Huntingdon, 90 km north of London. The bridge has cast-in-place 32-meter main spans with cantilever spans that support precast beams. The cast-in place-spans are of grouted post-tensioned box girder construction. Sensor density on the cantilever slab is approximately one per 25 square meters. Additional sensors were placed on the main span to provide information about signal attenuation.

Sophisticated software filters were designed to eliminate most spurious acoustic activity at the site-based acquisition system and only relevant events are transmitted to the Calgary processing center. The system correctly identified 41 out of 44 test events generated. The system will remain in place on this structure for the foreseeable future. The success of the testing at Crowthorne and Huntingdon is likely to lead to the system being included in the Highways Agency's list of approved monitoring methods.

MONITORING OF SUSPENSION BRIDGES

Main Cable Monitoring - In October 1997, the monitoring system was tested on the Bronx Whitestone Bridge in New York City. This bridge, with a main span of 701 meters, was opened to traffic in 1939 and is owned and operated by MTA Bridges and Tunnels, an agency of the Metropolitan Transportation Authority of New York. The monitoring system was installed during a rehabilitation of the main cables. This work involved removal of the circumferential wire wrapping, repair of broken wires and the application of corrosion-inhibiting oil to the wires. Consequently, it was possible to cut wires in the cable to test the system's recognition and location capabilities. Single sensors were attached to six cable bands, each 12.2 m apart. An array of three additional sensors was placed around two of the cable bands to evaluate radial location capabilities.

A portable acquisition system was set up at deck level and the testing was done while construction work was in progress. Six wires were cut within the monitored section in a blind test. The system correctly classified the events and located them longitudinally with errors ranging from 0.0 m to 0.7 m. Radial location using all four sensors on a cable band was accurate to within 7.5°. Acoustic events caused by steel chisels being driven between the wires were easily identified and filtered. Figures 6 and 7 show plots of the response of sensors to a wire break and construction activity respectively.

Analysis of the data generated during the test showed that sensors mounted on alternate cable bands would be able to provide information of sufficient quality to permit reliable identification and location of wire breaks. A complete system based on this configuration is presently being installed.

The acquisition unit will be located in the Bronx anchorage house, and data will be transmitted from the sensors to the acquisition system through a coaxial trunk line attached to the existing messenger cable. Durability issues and ease of installation and maintenance were major factors in the design of the hardware. The sensor mounting brackets are designed to permit installation without modification to the cable band assembly and without damaging the paint system.

Data will be transmitted over the Internet to the Calgary processing center, where it will be analyzed and archived. The data will also be routed from the Calgary center to a second data processing computer located at the bridge administration building, and placed on a secure Internet site available to the owner. This will permit bridge operations personnel to review the data and reports using the same proprietary processing software in use at the processing center.

Suspender Rope Monitoring - Testing is currently in progress at TRL and on the George Washington Bridge in New York City to determine the effectiveness of the monitoring system for long-term damage monitoring of suspender ropes. The TRL study is also included in the paper by Cullington et al. within this conference. Results indicate that the system is capable of detecting and locating fatigue failure of wires in a locked-coil rope. A system was installed on the George Washington Bridge in November 1998 to monitor three sets of suspender ropes. The purpose of the installation was to evaluate the durability of the system over a complete climatic cycle. To date, the operating efficiency of the system has been over 99%. Later in 1999, the suspender ropes will be removed for routine forensic examination. During the removal process, it will be possible to cause failure of wire by accelerated corrosion and other methods so that the "soundprint" of these types of events can be acquired by the monitoring system. This information will be used to develop processing filters for a potential permanent installation.

MONITORING OF CABLE-STAYED BRIDGES

The monitoring system is particularly suitable for use on cable-stayed bridges. Because of the fact that, after construction is completed, visual inspection of cable components is not possible, the system can provide reassurance about the long-term integrity of the wires. In order to evaluate the performance of the system in this type of application, a prototype system was installed on the Alex Fraser Bridge in Vancouver, British Columbia. This bridge has a main span of 450 metres and has a total of 192 stay-cables consisting of parallel wires protected by HDPE sheathing. The wires are anchored in zinc castings. The longest stay-cable is approximately 250 m.

Acoustic sensors were installed on the deck and pylon anchorage assemblies of forty-eight stay cables on the southeast section of the bridge (Figure 8). The sensors were connected by coaxial cable to the acquisition unit located inside the southeast pylon leg.

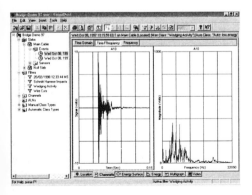

Figure 7. Sensor response to steel chisel impact 6.0 m distant.

Figure 8. Sensor Detail at Deck Anchorage

It was not possible to cut wires in the cables to calibrate the system. Therefore a rebound hammer was used to verify that events with similar frequency and energy characteristics to wire breaks occurring at one end of a stay-cable could be detected and identified at the other end. This capability was confirmed for all lengths of stay-cables on the bridge. No wire breaks have been detected during the period of operation of the system.

A continuous vibration monitoring system has recently been installed on three of the cables. The same acquisition system is used for both systems.

REPORT GENERATION AND PRESENTATION

Because the data acquisition and processing software permits continuous real-time data transmission and analysis, it is possible to make on-demand reports available to authorised users through a web page interface. Automatic e-mail notification alerts the user to the occurrence of a significant event.

Using password protection, the user can enter sites for which they have clearance and generate reports to their own specifications. User-defined parameters include reporting period, event classifications, event locations and temporal distribution of events. An example of a typical user-defined query is shown in Figure 10.

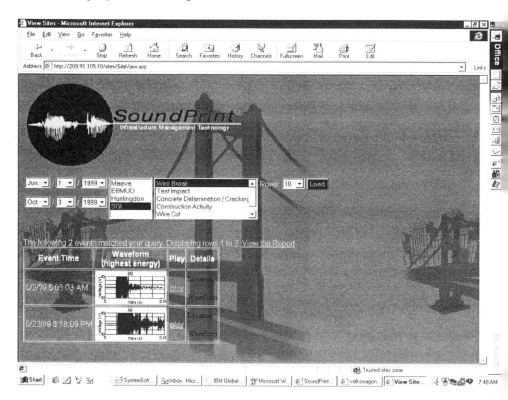

Figure 10. Typical user-defined query using Web interface.

Once the parameters have been defined, presentation-quality reports can be generated in either Word or PDF format.

These capabilities provide users with immediate access to information about the performance of their structures or infrastructure. This addresses one of the main concerns with instrumentation systems, i.e. the collection of large amounts of data, which requires intensive manual post-processing to provide useful information.

NEW APPLICATIONS

The continuous acoustic monitoring principle has recently been adapted for damage surveillance of bridges and other structures. Using the acoustic system as a "trigger", video or other media records of events of interest can be acquired. The use of the acoustic trigger eliminates the need for continuous media recording and review.

This system can be used to detect and record instances of damage to bridges caused by over-height highway vehicles or ships.   Once an event has been acquired, the data file can immediately be transmitted to a designated destination for review and appropriate response. Systems have recently been installed on two bridges in Alberta, Canada to detect and record truck impacts. This allows the owner to improve his safety management of the structure and to recover the cost of repair.

This approach can also be used for seismic damage surveillance, where agencies are faced with the challenge of rapidly determining infrastructure damage, possibly over a wide area, to a seismic event. The availability of remote surveillance information can help to prioritize the allocation of limited resources. The first such system was installed on a prestressed concrete water tank in California in June 1999. This system detected a wire break in the tank 5 hours after an earthquake occurred on August 17th, 1999. The earthquake had a magnitude of 5.0 (Richter) with an epicentre approximately 40 kms from the tank location.

In addition to acquiring video and acoustic records, the system can be configured to sample structural properties, such as natural frequency response, before and after an event. This kind of information could provide evidence of changes in the structural performance of the element or structure. Additional structural instrumentation can be integrated with the monitoring system for this purpose.

The application of continuous remote acoustic monitoring to the detection of fatigue crack development in steel bridges is presently being investigated. The frequencies of interest for this type of deterioration are much higher (>150 kHz) than those for prestressing wire breaks and for concrete cracking (4 to 25 kHz). Because these higher frequencies attenuate very rapidly through the acoustic medium, global monitoring of bridges would probably not be cost effective with conventional piezo-electric technology because of the sensor spacing required. However, localized monitoring of high-risk areas would be practical.

SUMMARY

The development of fast, inexpensive computing and data acquisition hardware, combined with the availability of low-cost Internet data transmission has led to development of a dependable remote continuous acoustic health monitoring system for bridges and other structures. The ability of the system to identify and locate events of interest in noisy environments has been verified for unbonded and grouted structures as well as for suspension bridges.

The information provided by the system can be used to accurately identify localized areas of deterioration in very large structures using widely distributed sensors. Systems installed prior to intrusive inspection of the structure can help to determine where to focus the inspection. Systems installed after inspection or repair can ensure that the long-term durability performance of the entire structure can be quantified. The ability of the system to determine the time and location of significant events permits confident statistical modeling of deterioration.

The adaptation of the system to provide information from other instrumentation or recording media, using the acoustic data as an acquisition "trigger", has resulted in its use as a continuous surveillance system for impact and seismic damage.

Real-time, user-defined, on-demand generation of reports using a web interface allows users immediate access to useful information from the instrumentation system.

REFERENCES
Elliott, J.F., "Continuous Acoustic Monitoring of Bridges", International Bridge Conference, Pittsburgh, Pennsylvania, July 13 – 16, 1999.

Cullington, D.W., Bradbury, T., Paulson, P.O., Continuous Acoustic Monitoring of Steel Tendons and Cables in Bridges, Bridge Engineering Conference, Guildford, Surrey, UK, April 16 – 19, 2000.

Optimised bridge management with permanent monitoring systems

M.E. ANDERSEN, Ph.D., A. KNUDSEN, M.Sc., P. GOLTERMANN, Ph.D., F.M. JENSEN, Ph.D.: RAMBOLL, Virum, Denmark, and F. THØGERSEN, M.Sc.: The Danish Road Institute, Roskilde, Denmark

ABSTRACT
Selecting the optimal maintenance and rehabilitation strategy (M&R-strategy) within the actual budget is a key point in bridge management for which an accurate assessment of performance and deterioration rate is necessary. For this assessment, the use of a permanent monitoring system has several advantages compared to the traditional approach of scattered visual inspections combined with occasional on-site testing with portable equipment and laboratory testing of collected samples.

This paper identifies the main bridge owner requirements to permanent monitoring systems and outlines how permanent monitoring systems may be used for performance and deterioration rate assessment to establish a better basis for selecting the optimal M&R-strategy.

The presented Case History is the Skovdiget Western Bridge, selected for field testing of the integrated monitoring system developed during the Brite/Euram project "Integrated Monitoring Systems for Durability Assessment of Concrete Structures" (SMART STRUCTURES). The project started in September 1998 and aims at reducing the costs of inspection, maintenance and rehabilitation of existing concrete bridges and minimising the corresponding traffic regulations. The objective and plan for monitoring the selected bridge are described based on the results of an initial inspection combining visual inspection, on-site testing with portable equipment and laboratory testing of collected samples.

INTRODUCTION
An ageing and deteriorating bridge stock presents the bridge owners with the growing challenge of maintaining the structures at a satisfactory level of safety, performance and aesthetic appearance within the allocated budgets. This task calls for optimised bridge management based on efficient methods of selecting technical and economical optimal M&R-strategies. One of the crucial points is the assessment of the current condition and future development of deterioration and performance.

Traditionally durability and performance assessment of concrete bridges has been based on scattered visual inspections combined with occasional on-site testing with portable equipment and laboratory testing of collected samples. This traditional approach has some apparent drawbacks such as:
- Traffic interference during inspection and testing
- High costs of providing access to the structural elements

- Scattered data makes prediction of the time for initiation of deterioration and the future damage growth less accurate
- Infrequent inspection and testing may allow deterioration to progress between inspections to an extent that makes efficient and cheap preventive M&R-strategies impossible

The use of permanent monitoring systems has several advantages once the system is installed:
- Traffic interference is reduced
- The costs of access to the structure and resources for inspection and testing are reduced
- Structural elements with difficult access are easily monitored
- Frequent collection of data enables more reliable trends for the development of deterioration and performance, e.g. the progress towards initiation of deterioration and future damage growth.

BRIDGE OWNER REQUIREMENTS
The bridge owners' main interest in permanent monitoring systems for durability and performance assessment of concrete bridges is to reduce the inspection, maintenance and rehabilitation costs as well as the traffic interference and yet still maintain a satisfactory level of safety, performance and aesthetic appearance.

A permanent monitoring system should assist the bridge owner in selecting the optimal M&R-strategy by providing:
- Input to the M&R-strategies in order to select the optimal preventive and remedial actions
- Timely warnings of initiation of durability and structural problems making preventive actions possible
- Timely warnings to ensure the safety of the structure, in particular for structures that can collapse without any prior sign of deterioration
- Input to future inspections of the structure including the time, extent and frequency of inspection, structural elements and deterioration mechanisms to focus on and additional on-site or laboratory testing (if any)
- Improved knowledge of the performance to validate the design assumptions
- Improved knowledge of the individual deterioration mechanisms and their interaction

Permanent monitoring systems should be able to monitor the relevant deterioration mechanisms such as corrosion of reinforcement initiated by chloride ingress or carbonation, freeze-thaw damage and alkali-silica reaction damage. Ideally, one or more measurable key parameters that can describe the progress towards initiation and subsequent growth of damage should be identified for each deterioration mechanism as well as simple deterioration models allowing for prediction of the initiation and growth of damage. Furthermore, the structural performance and the effects of e.g. mechanical damage due to deterioration, vehicle impact, overloading or loss of prestressing should be covered by the monitoring system.

A number of additional requirements to the sensors, data acquisition system and software for data presentation and analysis are essential to the bridge owner:
- Sensors should have a long service life, be easy to install and have a limited need for maintenance and calibration
- The data acquisition system should allow for adjustable measuring frequency (preferably on-line control), two-way communication between site and office, and send out warnings if parts of the system have stopped working
- Software should be an integrated part of the monitoring system, be compatible with standard programs and give simple presentations and analyses of the measured data

- Simple models for predicting the time for initiation and future growth of damage should be incorporated in the software. Warnings should be sent out in due time before initiation of damage and before damage reaches critical levels. It should be possible to calibrate the models by adjusting the critical levels.

DURABILITY AND PERFORMANCE ASSESSMENT

Assessing the durability and performance of concrete bridges as a part of selecting M&R-strategies involves several complicated decisions. A permanent monitoring system should assist the bridge owner in handling these questions and decisions to ensure optimal planning of the future management of the bridge including inspection, maintenance and rehabilitation. The use of a permanent monitoring system for durability and performance assessment can be described based on Figure 1.

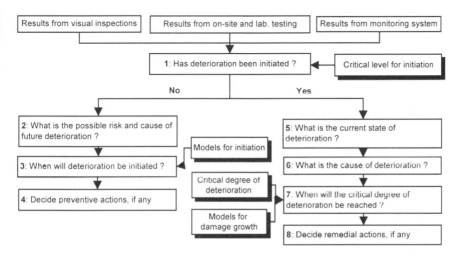

Figure 1 Durability and performance assessment

The first step (1) is to determine whether deterioration has started or not. Usually this question is answered based on visual inspection, on-site or laboratory testing. However, with an identification of critical levels for initiation of deterioration of the key parameters measured with the monitoring system, the results from the monitoring system can be used to determine whether deterioration has started or not.

If deterioration <u>has not</u> started, the bridge owner is interested in identifying the possible risk and cause of future deterioration (2) and in estimating when deterioration will be initiated (3) in order to make timely and cost efficient decisions regarding preventive actions (4). This requires an identification of the level for initiation of deterioration of the measured key parameters (see Figure 2 to the left) as well as models for predicting the future development of the key parameters for initiation of deterioration.

If deterioration <u>has</u> started, the bridge owner is interested in knowing the current state of deterioration (5) and the cause of deterioration (6). It must be determined whether the results from the monitoring system alone are sufficient to answer these questions or whether monitoring has to be supplemented by visual inspection, on-site or laboratory testing.

The next step (7) is to estimate when the critical degree of deterioration has been reached in order to determine cost efficient remedial actions (8). This requires an identification of the critical degree of deterioration (see also Figure 2 to the right), i.e. the degree of deterioration above which the condition of the structure is unacceptable for safety, performance or aesthetic reasons. Additionally, the critical level of the measured key parameters for development of deterioration (damage growth) corresponding to the critical degree of deterioration must be defined (see also Figure 2 in the middle) as well as models for predicting the future development of the key parameters for damage growth. When the future damage growth has been estimated, it can be determined when the critical degree of deterioration will be reached.

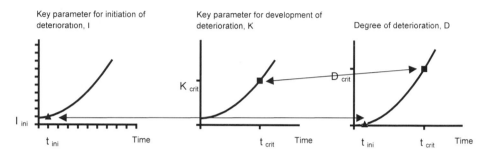

Figure 2 Left: Key parameter for initiation of deterioration; Middle: Key parameter for development of deterioration; Right: Degree of deterioration

SMART STRUCTURES PROJECT
The Brite/Euram project SMART STRUCTURES aims at reducing inspection, maintenance and rehabilitation costs and traffic delays as well as increasing the knowledge of deterioration mechanisms in practise by developing an integrated monitoring system for existing concrete bridges including a number of new sensors. The research consortium, with RAMBOLL as project coordinator, consists of 8 partners, representing bridge owners, sensor producers, consultants and testing institutes.

The development of the integrated monitoring system will be combined with enhanced deterioration models incorporating laboratory test results, on-site-measurements and data from the monitoring system. A number of new sensors for monitoring existing structures will be developed covering the following deterioration mechanisms:
- Chloride initiated corrosion
- Carbonation initiated corrosion
- Freeze-thaw damage
- Alkali-silica reaction
- Mechanical damage (structural performance)

These deterioration mechanisms will be monitored through key material parameters (temperature, moisture, chloride content, pH, corrosion risk/initiation time) and key mechanical parameters (strain, deflection, vibration).

The Skovdiget Western Bridge has been selected for field testing of the monitoring system.

CASE HISTORY – THE SKOVDIGET WESTERN BRIDGE
History of the Skovdiget Bridges

The twin motorway Skovdiget Bridges (Eastern and Western respectively) are located near Copenhagen and were constructed in 1965-67. The Skovdiget Bridges take a motorway across two local roads, a dual railway line and a parking area, see Figure 3.

Figure 3 Left: The Skovdiget Western Bridge; Right: An overview before a load testing.

Each bridge is approx. 220 m long and 22 m wide. The concrete superstructure consists of two main girders joined with transverse ribs at 2 m intervals and cantilever wings outside the girders. The average length of the 12 spans is approx. 20 m except at the ends. Each bridge was cast in five sections and post-tensioned longitudinally and transversely.

In the late 1970'ies, substantial damage was registered on both bridges. Repair work was initiated on the Eastern Bridge including replacement of the old waterproofing and surfacing, substantial partial concrete replacement and an overall improvement of the drainage system. However, as the repair work progressed, a number of additional damages were discovered. Surface water, which had penetrated through the waterproofing, was retained in the boxes. Alkali-silica reactions were discovered. The concrete was completely disintegrated in some areas (especially the upper surface of the bridge deck near the gulley and drains). Many of the prestressing cables were corroded and in some cases cable ducts were uninjected. In addition to the originally planned repairs, additional repairs of the prestressing, bearings and expansion joints had to be carried out.

Inspections of the Western Bridge in the late 1970'ies and early 1980'ies revealed a condition similar to the Eastern Bridge before repair. Due to the costly repair of the Eastern Bridge, it was decided not to repair the Western Bridge and consequently monitor the deterioration rate closely to ensure structural safety. Since the 1980'ies, the Western Bridge has been monitored with visual inspections four times a year, load testing in 1984, -88 and -93, and measurements of the movement of selected points on different structural elements once a year.

In 1998, a probabilistic-based management plan was developed for the Western Bridge. The probabilistic analysis showed that the safety was above the codified required safety and that the bridge would fulfil the safety requirements for further 8-10 years, if deterioration was allowed to continue unchanged. The management plan resulted in a detailed inspection to determine the condition of selected reinforcement groups and cables. Based on this updated information it was decided to carry out repair work to improve the drainage system of the bridge including new surfacing, new drains and gulley, and new expansion joints.

In 1999, it was decided to use the Skovdiget Western Bridge for field testing of the integrated monitoring system developed in the SMART STRUCTURES project. The installation of sensors in the bridge will be initiated in autumn 1999. The repair work will thus create valuable changes in the environmental exposure of some structural elements and corresponding changes in deterioration rate to be registered by the monitoring system.

Objective of Monitoring
In the SMART STRUCTURES project, the primary objective of monitoring is to test and document the monitoring system including the new sensors and the monitoring plan therefore focuses on providing the best possible documentation for the individual sensors, the complete monitoring system and the deterioration models. It is only a secondary objective to monitor the condition of the bridge itself.

The field testing of the monitoring system should include monitoring of chloride penetration, carbonation propagation, corrosion risk and progress, freeze/thaw deterioration, alkali-silica reactions, temperature and moisture changes as well as structural performance. In the Skovdiget Western Bridge most of these deterioration mechanisms may be found in different degrees of severity.

As a basis for the monitoring plan, an initial inspection of the Skovdiget Western Bridge was made to determine which structural elements to monitor and to provide background data for the calibration of the sensors.

Initial Inspection
A superficial visual inspection of the entire bridge was carried out to identify a number of structural elements to focus on in a more detailed inspection. It was decided to focus monitoring on two comparable spans that seemed to be in different state of deterioration (span no. 3 and 6) and an end support. The initial inspection comprised the following structural elements:
- four columns exposed to deicing salt in span no. 3
- two ribs in span no. 6 (both in good condition) and two in span no. 3 (one in poor condition and one in good condition)
- a part of the Western main girder in span no. 3 and 6 (outer side of the Western web)
- a part of the Eastern and Western edge beam in span no.6
- the Southern end support

The detailed inspection included visual inspection, hammer survey, HCP measurements, concrete cover measurements, break-ups to reinforcement and collection of cores for laboratory determination of chloride and moisture profiles, carbonation depths as well as petrographic analyses.

The results of the initial inspection are summarised in Table 1 identifying damaged areas. For all structural parts (except end supports), it is also possible to identify undamaged areas for monitoring.

Plan for Monitoring
The plan for monitoring is summarised in Table 2. In total, more than 100 sensors will be installed in the Skovdiget Western Bridge, including sensors for measuring temperature, humidity, resistivity, chloride content, pH, corrosion risk/initiation time, deformations and vibrations.

CONCLUSIONS
This paper identifies the main bridge owner requirements to permanent monitoring systems and outlines how permanent monitoring systems may be used for performance and deterioration rate assessment to establish a better basis for selecting the optimal M&R-strategy.

The Skovdiget Western Bridge is presented as a Case History. The bridge has been selected for field testing of a permanent monitoring system. Based on an initial inspection, a monitoring plan is presented for the bridge. Installation of sensors is scheduled to be initiated in the autumn 1999.

ACKNOWLEDGEMENT
The authors would like to acknowledge the European Communities, Brite/Euram project "Integrated Monitoring Systems for Durability Assessment of Concrete Structures", BRPR-CT98-0751 and the other project partners FORCE Institute, Bundesanstalt für Materialforschung und -prüfung, S+R Sensortec GmbH, Osmos Deha-Com SA, Autostrade S.p.A. and Deutsches Zentrum für Luft- und Raumfarth, Institut für Flugführung.

Mechanism	Description
Chloride penetration	Relatively high chloride contents (also in large depth in the concrete) were measured in the end support, the Western main girder, one of the inspected ribs and the Eastern edge beam. In all these locations, the concrete was severely cracked. Chloride ingress was also present in the four columns exposed to deicing salt, but to a smaller extent.
Carbonation	The general carbonation depth was negligible (2-3 mm) except in local areas of poor concrete quality and along cracks.
Corrosion	In most locations, no corrosion or light surface corrosion was observed. However, in some locations local corrosion (pitting) or surface corrosion was recorded, usually in generally cracked areas and near coarse cracks. The concrete cover varied between 20 and 45 mm in general. However, the cover varied between 4 and 110 mm in the columns.
Freeze/thaw deterioration	It was not possible to identify areas with freeze/thaw damage alone. Freeze/thaw action on cracked areas with alkali silica reactions may, however, have contributed to the damage development.
Alkali-silica reactions	Several structural elements with severe cracking caused by alkali-silica reactions were identified including the end support, a part of the Western main girder, one of the inspected ribs and a part of the Eastern edge beam.
Moisture and temperature changes	Moisture content was measured in several structural elements and a number of structural elements with relatively high moisture content and with very non-uniform moisture content was identified. High moisture content was measured in the end support, in a part of the Eastern edge beam and in a part of the Western main girder. High moisture content and non-uniform distribution was also recorded in one of the inspected ribs. In all these locations, the concrete was severely cracked. The temperature in the concrete was not measured in the initial inspection.
Structural performance	The superstructure was in a bit worse condition in span no. 3 than in span no. 6. The Western main girder was in worse condition than the Eastern. There are ribs in very poor and very good condition.

Table 1 Overview of the results of the initial inspection of the Skovdiget Western Bridge

Mechanism	Plan for monitoring
Chloride penetration and corrosion	Sensors: Temperature, humidity, chloride, resistivity, corrosion risk/initiation time Main objectives: - Verify the chloride sensor by calibration against results from laboratory testing - Calibrate the model for chloride ingress by comparing chloride content measurements and chloride ingress predictions - Test the corrosion risk probes by verifying the prediction of initiation of corrosion by break-ups to the reinforcement - Check the critical chloride level for corrosion initiation by comparing the measurements of the chloride sensor and the corrosion risk probes Structural elements to focus on: Columns, end support, ribs and main girders (areas where corrosion is expected to be initiated within a limited time period)
Carbonation and corrosion	Sensors: Temperature, humidity, pH, resistivity, corrosion risk/initiation time Main objectives: - Verify the pH sensor by calibration against results from laboratory testing - Calibrate the model for carbonation by comparing carbonation depth and pH measurements with carbonation predictions - Test the corrosion risk probes by verifying the prediction of initiation of corrosion by break-ups to the reinforcement Structural elements to focus on: Columns, main girders and ribs (areas where carbonation is present, i.e. areas of poor concrete quality)
Freeze/thaw deterioration	Sensors: Temperature, humidity, optical-fibre sensor for deformation measurements Main objectives: - Temperature and humidity will be measured to check critical levels defined for initiation of freeze/thaw damage - The deformations across horizontal cracks will be measured to check whether the cracks widths change due to freeze/thaw action Structural elements to focus on: The Western main girder, ribs, end support and Eastern edge beam
Alkali-silica reactions	Sensors: Temperature, humidity, pH, optical-fibre sensor for deformation measurements Main objectives: - Temperature, humidity and pH will be measured to check critical levels defined for initiation of alkali-silica reaction damage - The deformations across horizontal cracks will be measured to check whether the cracks widths change due to further alkali silica reactions Structural elements to focus on: The Western main girder, ribs, end support and Eastern edge beam
Temperature and moisture changes	Sensors: Temperature, humidity and resistivity (water content) Main objectives: - Verify the humidity and resistivity sensors by calibration against results from laboratory testing - Monitor the drying out of selected structural elements caused by the new wearing course, improved drainage system or rehabilitation of expansion joints - Monitor the temperature and humidity difference between the two edge beams Structural elements to focus on: End support, ribs and edge beams
Structural performance	Sensors: Optical-fibre sensors for deformation and vibration measurements Main objectives: - Check the time development of deformations and vibrations - Check the difference in deformations and vibrations between undamaged and damaged structural elements Structural elements to focus on: Main girders and ribs

Table 2 Overview of the monitoring plan for the Skovdiget Western Bridge

Monitoring to prolong service life

DR S. MEHRKAR-ASL, Associate, Gifford and Partners Consulting Engineers, Southampton, SO40 7HT, United Kingdom

INTRODUCTION

Prolonging life of existing structures is an important aspect of the work of structural engineers. These structures covers a variety of types including bridges and buildings. In bridges the adverse environmental conditions, damage due to earthquakes or floods, human intervention such as the use of de-icing salts and increasingly heavy traffic loads require bridge inspection on a regular basis and detailed assessment of their safety for use when deemed necessary.

In terms of buildings, deterioration with time, damage due to earthquakes or tornadoes, change of use and/or even re-arrangement of the load bearing components due to architectural needs also requires their inspection and assessment when necessary.

The assessment of many of these structures results in some restriction of use or otherwise relaxation of the safety margins. For bridges limitation of available funds and perhaps prioritisation of the strengthening or replacement programmes often means they need to remain in use for a number of years. In buildings similar restrictions can apply, however, in the case of historical structures, there is no choice but to come up with alternative solutions to provide longevity.

One of the methods that is becoming more popular amongst engineers is the use of monitoring techniques to establish the real behaviour of the structures and to determine the on going rate of deterioration in order to implement repair works at the right time.

This paper describes the use of monitoring techniques as an approach to prolong life of structures through describing the general strategy, available instruments and techniques. In addition, case studies are presented for projects on which monitoring is being used.

GENERAL STRATEGY

The flowchart in Figure 1 shows the basic strategy for deciding whether a monitoring system is required. Obviously, the reliable functionality of a structure has to be in doubt in the first place which could be as a result of a number of factors such as change of use, increased loading, deterioration and damage due to fire, blast, subsidence, earthquake etc. This would normally require a load capacity assessment of the structure which starts by the collection of all the relevant information such as drawings, geotechnical investigations, previous inspections and records. This basic process has been used in the assessment of bridges in the United Kingdom using BD21[1] for the Highways Agency. A similar procedure is operated for buildings using "Appraisal of Existing Structures" for the assessment of buildings[2]. The assessment would normally start from a simple and conservative approach based on the design parameters. However, it could become more complex making use of the existing material properties, sophisticated analytical methods and supplementary load testing[3-4].

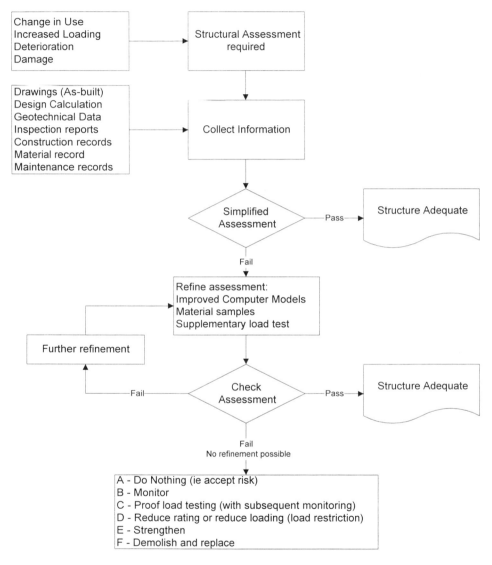

Figure 1. Flowchart showing strategy for assessment leading to monitoring

If all the attempts to pass the structure fail, the options available to the owner start from "Do Nothing" which is also often implemented while further investigations are carried out. However, this approach could have significant legal implications from Health and Safety point of view. The next strategy level is to monitor the structure. Provided monitoring is viable, it has a number of advantages over the rest of the options in Figure 1, which are as follows:

- Steps have been taken to improve the safety of the public or the users.
- Monitoring is providing a warning system.
- Rate of deterioration, if any, can be determined.
- Resources and funds can be released for more urgent projects (or developments).
- Future prioritisation for strengthening, etc. will be based on a performance based criteria.

MONITORING SYSTEM

The viability for the implementation of a monitoring system has to be established first. The following questions have to be answered.

- Does the deficiency to be monitored have a sudden failure mechanism? If yes, then the monitoring may not be the right solution or it has to be much more comprehensive.
- What components need to be monitored and to what accuracy? This affects the choice of instruments.
- What is the frequency of data sampling and is there any access limitation? Electronic devices and a data logger(s) may be needed.
- What is the design life for the system? The system has to be designed for the effects of weather and vandalism.
- How is the data to be accessed? A modem link and telephone may be required.

Other consideration which may indirectly affect the choice of the instrumentation are as follows:

- Time available for installation of instrumentation
- Requirements for traffic management
- Requirements to remove non-structural components to reach the main structural component to be monitored
- Purpose of the monitoring (ensure safety of users, improve engineering understanding, legal arbitration, determine cause and effect, or simply find out about trend of an event)
- Cost (sometimes this is the first thing to be considered but not always)

Monitoring Instruments

There are a number of different monitoring devices. The choice of which to be used, is dependent on a number of factors which have been referred to in the previous section. In Table 1, the details of a range of suitable monitoring instruments are given. This list is not exhaustive and manufacturers should be contacted as new devices with improved performance and resolution are being developed.

Data Acquisition

This is an important aspect of any monitoring system. In the circumstances where access is difficult or dangerous or very expensive or frequency of data sampling is very high, it is usually essential to have a datalogger and to use devices, which can be read electronically. The collected data can then be downloaded to a portable computer for subsequent data interrogation. It could be cost effective to use a telephone modem link to collect the data remotely. With the reliability of mobile phones, it has become possible to collect the data remotely from most inaccessible areas.

Table 1.   Monitoring Instruments

Device	Measuring	Accuracy	Suitable for	Sampling
Tell-tale	Movement	0.1mm	Masonry etc	Manual
Calliper	Movement	0.05mm	General use	Manual
Vernier	Movement	0.01mm	General use	Manual
Dial gauge	Movement	0.02mm	General use	Manual
DEMEC	Movement /Strain	0.001mm /10×10^{-6}	Steel and concrete strains and small crack movements	Manual
LVDT	Movement	0.01mm	Steel, concrete and masonry	Electronic
VW Crack	Movement	0.001mm	Steel, concrete and masonry	Electronic
VW Strain	Strain	5×10^{-6}	Steel, concrete and masonry	Electronic
Tilt	Rotation	1 - 5 Seconds	Global rotation	Electronic
Inclinometers	Rotation	10 Seconds	Global rotation	Electronic
Accelerometer	Acceleration	1 to 5% range	Vibration	Electronic
ERS	Strain	1×10^{-6}	Steel strains	Electronic
Thermocouples	Temperature	0.1deg C	General purpose	Electronic
Thermistor	Temperature	0.1deg C	General purpose	Electronic
PR	Temperature	0.01deg C	General purpose	Electronic
Laser	Movement	0.1 to 1mm	Access limitation	Electronic
Precise levelling	Movement	0.1mm	General purpose	Manual
Photogrammetry	Movement	1 to 5mm	Large areas	Manual
Anemometers	Wind speed	0.3m/s	General purpose	Electronic
Load cells	Load	1% of max	Bearings and cables	Electronic
LPR	Current	10^{-9} Amp	Rebar corrosion in concrete	Electronic
Corrosion cells	Potential	10^{-3} Volt	Rebar corrosion in concrete	Electronic
Corrosion current	Current	10^{-9} Amp	Rebar corrosion in concrete	Electronic
Resistivity	Resistance	$0.5k\Omega.cm$	Rebar corrosion in concrete	Electronic
ER Probe	Resistance	1% thickness	Loss of steel in concrete	Electronic

DEMEC = DEmountable MEChanical strain gauge	PR = Platinum Resistance
LVDT = Linear Variable Displacement Transformer	
VW = Vibrating Wire (Crack or Strain) gauges	For further information on corrosion monitoring refer
ERS = Electric Resistant Strain gauges	to BRE Digest 434, Corrosion of reinforcement in
LPR = Linear Polarisation Resistance	concrete: electrochemical monitoring. November 1998
ER Probe = Electrical Resistance Probe	

Data Management and Presentation

The data from manual monitoring systems or those collected by a datalogger have to be processed systematically to identify spurious readings and ignore those that are erroneous. However, no data should ever be thrown away. The normal method of presentation of data is by the use of a spreadsheet software, where key readings could be presented in a tabular format while the bulk of the data is converted into graphs. The facilities available in today's spreadsheet software packages have made them an integral part of almost all monitoring systems. The spreadsheet can be programmed to produce simple reports of the results of the monitoring, carry out statistical analysis and give warning if certain limits are reached.

Purposely developed software packages are also available which can be used to collect and handle data. In addition, these software packages can be used to interrogate the results and produce reports. Such a system was used on the Dee Estuary Bridge[5], north of Wales, where a multitude of instruments were installed in a pilot scheme to implement remote bridge management in the UK.

Trigger Levels
The main objective of any monitoring is to decide whether there is a need for an intervention when set of readings exceed certain predetermined levels. These are normally referred to as "trigger levels" and their determination is a matter of engineering judgement, previous records, performance of similar structures in similar circumstances and occasionally political and social considerations. For major structures, the trigger levels should be determined by a panel of experts.

CASE STUDIES
Several different case studies are detailed below which give the reasons for adopting certain types of instrumentation, their arrangement, method of data acquisition and the type of results that can be obtained.

Langstone Harbour Bridge
The Hayling Island bridge was built in 1955 to replace an old timber bridge, which connected the island to the mainland at Langstone. The present bridge is prestressed concrete and consists of 29 simply supported spans of 9.75m and additionally four spans of 2.14m located at roughly the fifth points. The decks are supported at each end by five reinforced concrete piles and a reinforced concrete capping beam cast in situ. The 9.75m spans consist of 16 rectangular prestressed concrete beams 457mm deep, placed side by side, jointed with a dry packed mortar and transversely stressed with twelve 5mm diameter Freyssinet wires. Each beam was post-tensioned with two 28.6mm diameter Lee-McCall bars laid to a parabolic profile. A full description of the construction procedure was published by Melrose and Eyre in 1957[6].

The first span from the mainland side of the bridge was load tested by the Cement and Concrete Association in 1957[7]. In addition, two precast beams were load tested to destruction, one at an age of 3 months and one at an age of 2 years. These tests indicated prestress losses of 3% and 22% respectively.

In 1989 further work was carried out by Gifford and Partners. In situ concrete stresses were measured by taking 3 instrumented stress-relief cores[8-9] and 3 instrumented slot-cuts[10] in spans 1,4 and 14 from the north end of the bridge. These spans were chosen after obtaining a finger print of the performance of all the spans under a 45 tonne axle similar to HB which traversed the bridge over night to reduce disruption to traffic. Each span was instrumented with 12 vibrating wire gauges on the soffit at midspan. Strains were measured in the longitudinal and transverse directions of the deck and across the longitudinal joints between the beams.

The estimated prestress loss in the longitudinal direction ranged between 5% to 33% which was comparable to those of 1957[6]. The loss in transverse prestressing was about 44% with a remaining compressive stress of $1.7N/mm^2$ across the longitudinal joints between the beams.

Spans 4 and 14 have each been monitored with 9 VW strain gauges since 1989. The strains in the longitudinal and transverse directions are measured at three points in each span. Also at these locations, the transverse strain across the longitudinal joints between the beam is monitored. The main intention of monitoring is to pick up any changes in the transverse concrete stresses, which could be lost as a result of loss of prestress in this direction. The transverse prestressing, due to this form of construction, is vulnerable to corrosion. It has been found that strains in the bridge have not changed significantly since the start of the monitoring.

A3/A31 Guildford Flyover
The A3/A31 Flyover, comprises a two span single cell pre-cast segmental externally post-tensioned concrete box deck supported on a reinforced concrete pier. The main span is 50m over the A3 and the side span is 20m. The structure was built in the mid 1970's and since then has had series of problems. Corrosion of some of the tendons resulting in some cases in their severance, necessitated a structural assessment, including in situ stress determination and an upward load testing to estimate the remaining level of prestress. This was carried out by first conducting a significant amount of initial analysis to identify the critical areas and levels of prestress loss required to cause the initial cracking of the concrete.

Initially the bridge was instrumented and monitored for a period to identify the diurnal temperature fluctuation effects on strains. The instrumentation was extended to 16 vibrating wire gauges to cover the concrete segments and the corresponding joints between the segments at midspan and over the central pier. Midspan deflection was also measured with the use of LVDT's. Upward load was applied in increments up to 100tonne at mid main span. The maximum upperbound of prestress loss was determined in the absence of any crack detection in the joints. In addition, the relative stiffness modulus of the segment joints to that of the precast sections was determined to be 59%.

The bridge has been strengthened in the mid 1990's as the condition of the external prestressing was considered not to be predictable. However, the monitoring and the load testing conducted allowed time to complete the design of the strengthening without the need to close the bridge. More details of the bridge and its historical problems which have lead to its strengthening are included in a paper by Brooman and Robson[11].

A Motorway Bridge in England
A major motorway bridge in England which carries a carriageway of a motorway over another motorway has been monitored over the past 7 years. It is a four span in situ concrete bridge with two suspended spans over the motorway below. The monitoring has been as a result of cracks at the throat of the concrete Mesnager hinges supporting the suspended spans which were detected during the inspection undertaken for the assessment of the bridge. The initial investigation measured dead load stresses in the reinforcements across the hinge. Then the crack widths and the strains in the reinforcements were monitored. This monitoring utilised 8 LVDT's and 6 VW strain gauges and the system was connected to a datalogger which was manually downloaded approximately every six months.

Recent inspections have indicated that crack widths in certain parts of the hinges have increased. These were carried out using go/no-go purpose made stainless steel (filler gauge) rods to measure crack widths. During the inspection material sampling was taken from the throat of concrete hinges and precise levelling was also carried out. As the system in place was only limited to the centre-line of the bridge, a monitoring system was devised by the author to extend the exiting system by another 26 LCDT's, 2 thermocouples, a datalogger and a modem mobile phone link. The present arrangement is required in order to provide an easily accessible and frequent monitoring system. The condition of the bridge is constantly under review and the monitoring is to be further expanded by the undertaking of an endoscope examination of the hinges (both non-destructively and by drilling holes) and using the X-ray technique to look at the condition of the reinforcement.

Given the stage the investigation has reached presently it is not possible to report any results. However, it is possible to state that the daily movements of the bridge at the hinges and the behaviour of the reinforcement and concrete around the cracked throats have become a subject that may require a programme of research to determine the likely effects from corrosion, fatigue and the magnitude of the applied live and permanent loads.

The above monitoring regime ensures that the safety of the structure is not in any doubt, by monitoring any changes in the condition of the bridge.

A Historical Building in London

A major grade one listed masonry building in London is being monitored presently using crack mapping, precise levelling, manual crack width measurements, 12 VW crack and 6 tilt gauges. The system was set up by the author as cracks have been widening since a recent modification to the structure. The monitoring was recommended to identify the need for any underpinning. In addition, due to the size of the project the renovation of this building is carried out in phases. Therefore it is important to keep a record of the events for any possible future discussion and investigation as to the cause of the movements and the corresponding consequences. The VW crack and tilt gauges are being monitored via a telephone modem link to a datalogger. It is hoped to be able to release some results in due course.

CONCLUSIONS

Monitoring is a tool that can be used by engineers to:

- Base their decision for the safety of the structure on facts rather than assumptions.
- Determine the priorities based on the severity of performance criteria.
- Distribute the efforts, resources and funds in a more effective way.
- Limit the expenditure when it is not really necessary.

In addition, the information derived from the monitoring of the structures have provided engineers with:

- Improved understanding of structural performance.
- Refinement in analytical methods to estimate the theoretical performance and effects of strengthening options.
- Verification of new designs and their structural behaviours to improve design methods.
- Collection of information to apportion cause and effect in disputes.

REFERENCES
1.   BD21 "The Assessment of Highway Bridges and Structures". Highways Agency, UK.
2.   Institution of Structural Engineers "Appraisal of Existing Structures". London, IStructE 1996.
2.   BA54/94 "Load Testing for Bridge Assessment. Highway Agency". Highway Agency, UK.
4.   "Guidelines For The: Supplementary Load Testing of Bridges" The Institution of Civil Engineers National Steering Group for the Load Testing of Bridges, Thomas Telford, 1998.
5.   Curran, PN. "Dee Estuary Bridge - The Design of the Post-tensioned Asymmetric Cable-stayed Crossing". Proceedings of FIP Symposium, London, UK, 25-27 September 1996, "Post-tensioned Concrete Structures", pp 94-101.
6.   Melrose, JW and Eyre, WA. "Langstone Bridge, Hampshire". Paper No.6227. Institute of Civil Engineers, London 1957.
7.   Rowe, RE. "Loading tests on Langstone Bridge, Hayling Island". C & CA Technical Report, TRA/289, January 1958.
8.   Mehrkar-Asl, S. "Direct Measurement of Stresses in Concrete Structures". PhD Thesis, University of Surrey, 1988.
9.   Mehrkar-Asl, S. "Concrete Stress-relief Coring: Theory and Application". Proceeding of FIP Symposium 1996, London, UK, 25-27 September 1996, "Post-tensioned Concrete Structures", pp 569-576.
10. Forder, S. "Calibration of Saw Cutting Technique for In Situ Stress Determination". MEng Project Dissertation, University of Surrey, 1992.
11. Brooman, H. and Robson, A. "A3/A31 Flyover - Case History of an Externally Post-tensioned Bridge". In the proceeding of the 3rd Bridge Management - Inspection, Maintenance, Assessment and Repair, edited by Harding, Park and Ryall, published by E & FN Spons, 1996, pp 473-479.

Fatigue prediction by monitoring and parallel testing

J. PEIL and M. MEHDIANPOUR
Steel Structure Institute, Technical University Braunschweig, Germany

ABSTRACT
The prediction of a realistic life cycle of steel bridges is an important task for the owners. The usually used model show only small reliability. The model is chain of a load-model, a system-transfer-model and a damage model. The results of these coupled models are usually unreliable, especially the influence of the unsafe load- and damage models controls the reliability of the result. It is shown that monitoring at critical points and a time reduced parallel testing leads to very sufficient results.

INTRODUCTION

Due to the enormous amount of steel bridges the costs for bridge rehabilitation are increasing enormously. Thus, it will be necessary to take advantage of the real service life of structures which is generally longer than estimated by usual theoretical models. On the other hand, a lot of structures show unexpected damages, due to increase of actions, higher strengthening of the structure, corrosion by environmental conditions, carelessness during construction and supervision [18]

Life cycle prediction of steel structures under dynamic random loading is usually subjected to considerable uncertainties. The reason are in general the unreliable models used:

Actions are known only approximately. To consider random loads as traffic, wind or wave loads, statistic parameters of the process in the past and the future should be available. Often, the parameters are not known or they can be estimated only roughly.

The local stresses or strains at critical constructional details are determined by means of static or dynamic (i.e. mechanical) models. The designer often uses assumptions in order to simplify the calculation model. Beside these systematic errors, random errors due to the deviation of geometrical data and material properties occur.

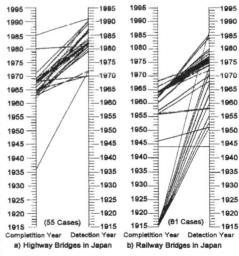

Fig. 1: Year of Completion and Damage[18]

Repeated loads cause low or high cycle fatigue problems at critical points of the structure. Damage models as e.g. Palmgren-Miners Cumulative Damage Model or models based on fracture mechanics, often lead to considerable systematic deviations e.g. [1]. Further more, the material behaviour is not deterministic, too. Using a conventional proof of the fatigue of a detail, the fuzzy assessment of the correct detail category (notch class) is an important problem in practice.

Investigations based on fracture mechanics are unreliable as well, since the results depend on the definition of an (unknown) initial crack.

The three models are interacting as a chain, the result of one model serves as an input for the next one, the likelihood of all results must be multiplied. Therefore the reliability of the final result is only small.

The uncertainties of the three models can almost be avoided or minimized using monitored data of the random strains at the critical points and additional tests in the laboratory. The method can be used for existing and for new structures. The reliability of the result is high.

OVERVIEW
In a first step, the random strains at critical constructional detail are monitored. In a first approach the load histories could now be classified by means of the Rainflow-Method. The resulting stress collectives could than be used for common Cumulative linear or nonlinear Damage rules [2, 3, 4, 5 8, 10, 11, 12, 14]. The result of such life cycle assessment will be always be uncertain [13]. To achieve a more reliable prediction, it is essential to avoid the damage model as well.

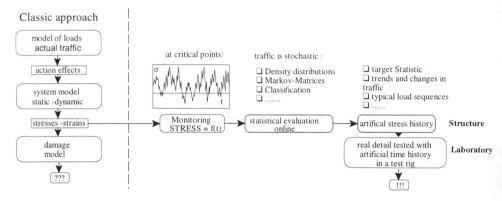

Fig. 2: Schematic procedure

The strain history is then be statistically evaluated. Using these statistical parameters, an artificial strain history taking into account the past and the estimated future of the loading situation is generated. The artificial strain history will take into account e.g. clustering of trucks, cyclic traffic jams. Due to the monitored strains, a load model and a system model is not required anymore, thus the uncertainties of both models are avoided.

The artificial generated time history, based on the monitored statistical data, is now used as an input for a digital controlled test rig, in which a sample of the actual constructional detail (or a part of it is tested. To shorten the length of the tests, the test frequency is increased, small i.e. non-damaging stress cycles and breaks of action are eliminated. The updating of the statistical parameters at any time interval allows the identification of long term changes of the loads. Concerning the prediction method, two different procedures can be used:

a) The whole life cycle is tested in one step, the specimen will be destroyed after testing. After the identification of a change in the load characteristic compared with the expected ones, new tests with artificial strain histories based on the new statistic situation are performed, see Figure 3. Expected trends could be implemented

b) Only a few years of the future are simulated in the test. The specimen will not be destroyed. After testing, the specimen will be stored (close to the structure to assure the same environmental impacts). The statistical parameters of the random process is monitored meanwhile. After having reached the end of the chosen time interval, the stored specimen will be tested again for the next interval using the new part of the artificial strain history. If there is a difference between the fatigue states of the specimen and the constructional detail, the fatigue state must be adjusted before starting the next testing interval.

This method approaches the fatigue life of a structure step by step with higher accuracy. Short tests and requirement of only one specimen for all tests are further advantages.

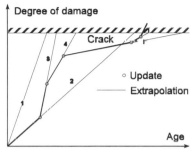

Fig. 3. Linear extrapolation

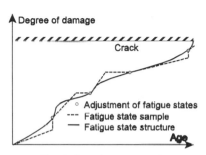

Fig. 4. Step by step testing

If a structure is monitored from the beginning, the procedure is straight forward. To predict the remaining fatigue life of an *existing* structure, load histories from the past would be required. This should not be a problem, because e.g. traffic changes in the past are known and an approximative determination is possible. The mechanical model of the structure must be known, to determine the local strains at the critical points. The model can be validated by means of the knowledge obtained by the measurements, thus they are pretty precisely.

Due to the consideration of random loads, the proposed method is more complex than methods have been applied some years ago in air craft engineering. Load histories describing landings and take offs of special air crafts are used to test the overall gear construction, e.g. [1].

An important point is the assessment of critical constructional details with notches and / or high local stresses, which should be monitored. If the structure shows typical weak points - this usually appears with older structures - the assessment procedure using classical models is straight forward. If the safety of all constructional details is well-adjusted - that usually happens with modern structures - the critical details must be identified in a probabilistic way. Different possible failure paths leading to different ultimate limit states must be investigated, taking into account the scattering of the basic parameters. The critical details are then determined by means of the dominating contribution to the probability of failure of the structure. If the structure already exists, the scattering can be reduced strongly, because the interesting basic parameters of the structure can be measured.

If a large number of critical details must be monitored, the measurement effort may be high. To reduce this effort, transfer functions are determined to obtain the local strain history from measured strain histories of neighbouring details. This should be done by means of test loadings to reduce errors of the mechanical model.

However a validation of the described method is usually impossible because the remaining life cycle of real buildings is pretty high (much longer than any research project). To avoid this problem, so called building substitutes (BS) with different weak points are tested under different load histories in the laboratory.

The load histories cover a broad field of real loading processes i.e. narrow and broad band processes under variable amplitudes (low cycle (LCF) and high cycle fatigue (HCF)). Due to the small dependence of the life cycle on the strain rate if HCF is investigated, the speed of the loading process can be increased and so the real life cycle of the substitute structure in the laboratory can be dramatically reduced. Monitoring of strains at critical constructional details and parallel testing of small specimens in the test rig allows a validation of the presented method.

To enhance the experience with real buildings under real environmental conditions, a highway and railway bridge are monitored since some month. The evaluation of the statistical parameters of the strain histories is determined in real time on site.

BUILDING SUBSTITUTES (BS)

The BS shows details with dimensions which are similar to the real details. They are provided with a number of different details as weldings, holes and other notches. Two types of building substitute are investigated up to now, see Figure 5, 6. The BS in Figure 5 shows large quadratic holes near the support, high local stresses at the hole corners will arise. The BS of Figure 6 has smaller corners and additionally a butt weld at the middle of the tension flange and a short stiffener to carry to local load of the actuator. The notches are designed such that (following the classical Miners-rule) at all critical details, the number of load cycles for a crack initiation under a sinosoidal process will be the same. Two different load histories are used at the moment:
- sinosoidal load history for calibration
- broad band random load process.

The amplitudes of both processes are adjusted such that LCF and HCF phenomena will occur.

As discussed above, specimens are needed to perform parallel tests in a digitally controlled test rig. The shape of the stiffener specimen and the welding specimen is equivalent to the real details. The specimen describing the notches at the (rounded) corners of the quadratic holes is designed such that the field of the main stresses at the corners of the real structure is mostly identical with the main stresses of the specimen. To ensure this, Finite-Element calculations of the real BS and different shapes of specimen are performed to determine the final shape of the specimen. To speed up the testing time of the specimen two notches are designed in a specimen, so that two test can be performed with one specimen.

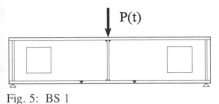

P(t)

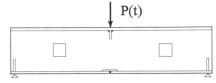

P(t)

Fig. 5: BS 1   Fig. 6: BS 2

gure 7 shows for example the calculated main stress field near the corner in a global view. Figure
shows the global main stress field of the specimen. Figure 9 and 10 compare the local main stress
eld around the corner of the BS1 and the specimen. One confirms that the stress field is mostly
entical. The gradient of main stresses, which controls the grow of a crack is almost the same in the
S and the specimen.

   The investigations are focused on the determination of the crack initiation. If the crack is initia-
d and growing on, the comparison of stresses in the BS and the specimen is no more valid, becau-
 of the different internal redistribution of forces.

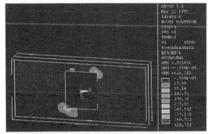

Fig. 7: Principal stresses (BS1)

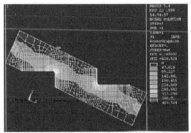

Fig. 8: Principal stresses in specimen

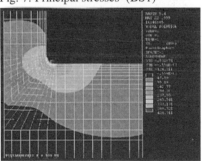

Fig. 9: Principal stresses (detail)

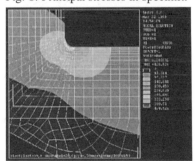

Fig. 10. Principal stresses in specimen

gure 11 shows the test assembly of the BS2 and of the specimen in the test rig. Two girders are
ted in series (time parallel) in order to reduce the fatigue testing time. The specimen are tested in
digitally controlled test rig (with the measured time series, just as the BS in series.

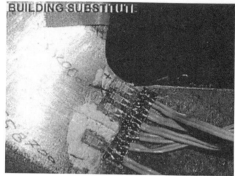

Fig.11: Building Substitute and corner with a crack

Fig. 12: Specimen tested in series with zoomed corner crack

One can see that the FE-Prediction of the direction of the crack is almost exact, the strain gages po
sitioned in this direction are situated directly over the crack in the BS and the specimen as well. .

Figure 13 shows first results of the proposed method. On the x-axis the number of cycles in the real building (substitute) is plotted, the y-axis shows the number of cycles in the specimen. Due to the wide range of cycle numbers both numbers are plotted in a logarithmic scale. The circle indicates the mean value of the cycle numbers where the crack occurs. The bars in show the rms of the tests. It can be seen that the method gives very good predictions of the real life until the crack occurs.

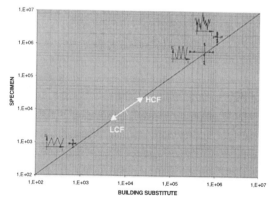

Fig. 13 Comparison of results

The ultimate state used up to now is the occurrence of the first crack, because this event usually needs repair measures. It may happen that the crack or a couple of cracks shift the position of the weak point. In this case the monitored strains are no more convincing. To take into account these effects the overall system is investigated adding the predicted cracks into the structure, observing the development of the crack using the fracture mechanics theory , predicting the life cycle of the actual crack and looking for changes in the behaviour of the structure. Fig. 14 shows a FE-model of the BS where the crack at the left corner edge is added.

Fig. 14: Overall system with added crack

MONITORING OF EXISTENT BRIDGES

In order to test the method and the measurement equipment under real environmental conditions, a highway bridge with large traffic was instrumented first. Figure 15 describes the situation. Up to recent investigations this structure does not show any severe fatigue problem, but it shows several advantages for the first tests: short distance to the Institute, large traffic, hollow cross sections of the main girders where the measurement facilities can be housed safely.

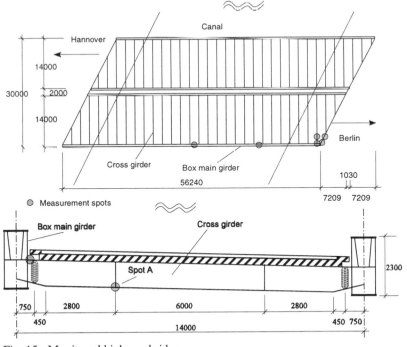

Fig. 15: Monitored highway bridge

The critical constructional details are determined in advance by means of the existing static calculations estimating the effect of notches [15, 16, 17]. As already mentioned the investigated highway bridge does not show any severe fatigue hazard. Thus for comparisons, the measured strains are amplified by a factor 6. A usual fatigue investigation based on the EC 3 using the monitored but amplified stress-collectives results in a life cycle of about 5 years. The measured life cycles of the specimens are more than 10 times higher than the calculated life cycle!

This result should not be generalized. Other investigations [6] with longitudinal stiffeners of orthotropic steel decks show that the experimental results show shorter real life cycles than the classically calculated ones.

At the moment, measurements on a new railway truss bridge are starting, where fatigue problems govern the design.

ACKNOWLEDGEMENTS
The financial support of the Deutsche Forschungsgemeinschaft DFG (German Research Council) within the framework of the Collaborative Research Field SFB 477 "Monitoring of Structures" is gratefully acknowledged.

REFERENCES
 1 Haibach,E.: Betriebsfestigkeit - Verfahren und Daten zur Bauteilberechnung. Düsseldorf, VDI-Verlag 1989.
 2 Hahin,C., J.M. South, J.Mohammadi, R.K.Polepeddi: Accurate and Rapid Determination of Fatigue Damage in Steel Bridges. Journ. Struct. Eng., Vol 119, 1992, No.1, 150-168.
 3 Jacob,B.: Definition of Load Spektra. Bericht IVBH-Colloq. Copenhagen, 1993, 37-44.
 4 Sokolik,A.: Experimental Investigation of Traffic Load on Highway Bridges. Bericht IVBH-Colloq. Copenhagen, 1993, 85-92
 5 Fujiwara, M. et.al.: Stress Histogramme and Fatigue Life Evaluation of Highway Bridges. Bericht IVBH-Colloq. Copenhagen, 1993, 301-308.
 6 Clormann, U.H. , Seeger T.: Rainflow - HCM, Ein Zählverfahren für Betriebs- festigkeitsnachweise auf werkstoffmechanischer Grundlage, Stahlbau 3, 186
 7 Mielentz,F., J.Knapp, E.Thebis: Anpassung von Dehnungsmeßstreifen an batteriegespeiste Daten-sammler. Messen, Prüfen, Automatisieren, 1993, 14-17.
 8 Peil,U., H.Nölle: On Fatigue of Guyed Masts due to Wind Load. Structural Safety&Reliability (Ed.Schueller&Yao). Rotterdam, Baalkema 1994
 9 Peil,U.: Baudynamik. In: Stahlbau Handbuch I. Stahlbau Verlags GmbH, Köln, 1993.
10 Peil,U., R.Egner: Zur Ermüdung von leichten Metallfassaden im Wind. Bauingenieur 109 (1994), 109-115.
11 Peil,U., M.Mehdianpour: Ermittlung der Restlebensdauer ermüdungsbeanspruchter Tragwerke durch Monitoring. In: Vorträge Stahlbautag 1996 Bremen. Stahlbau-Verlag GmbH, Köln 1996
12 Reppermund,K.: Probabilistischer Betriebs-festigkeitsnachweis unter Berücksichtigung eines progressiven Dauerfestigkeitsabfalls mit zuneh-mender Schädigung. Dissertation Hochschule der Bundeswehr, Neubiberg 1984.

13 Schütz,W.: Fatigue life prediction by calculation: Facts and fantasies. Structural Safety&Reliability (Ed.Schueller&Yao). Rotterdam, Balkema 1994, 1125-1131.

14 Andkjær Nielsen, H. Agerskov & T. Vejrum: Fatigue damage accumulation in steel bridges under highway random loading. Proc. 1st European Conf. On Steel Struct. Eurosteel '95. Athens 1995

15 Geißler, K.: Restlebensdauerberechnung von Stahlbrücken unter Nutzung detailierter Beanspruchungsverläufe. Stahlbau 64 (1995), Heft 3, 79-88.

16 Kunz,P., M.A.Hirt: Reliability Analysis of Steel Railway Bridges under Fatigue Loading. Bericht IVBH-Colloq. Copenhagen, 1993, 53-60.

17 Waubke,H., W.Baumgärtner: Traffic Load Estimation by Long-Term Strain Measurements. Bericht IVBH-Colloq. Copenhagen, 1993, 427-434.

18 Mikami, J ,Sakano, M., Shibata, H.: Database of damaged steel bridges. Technology Reports of Kansai-Univ. No.35 (1993) 185-196.

Possible shortcomings of certain non-destructive monitoring methods for bridge cables

PROF. M. RAOOF, and T. J. DAVIES M. Eng., Civil and Building Engineering Department, Loughborough University, Loughborough, Leicestershire, UK.

INTRODUCTION

The orthotropic sheet theory has previously been reported by the first author and his associates [1, 2] for obtaining reliable estimates of the coupled axial/torsional stiffnesses for axially preloaded spiral strands and wire ropes, which have been found to vary between the two limiting values of full-slip and no-slip, as a function of the external load perturbations. The axial and torsional stiffnesses for small load changes have been shown to be significantly larger than for large load changes, because small load disturbances do not induce interwire slippage. In the presence of interwire friction, and for sufficiently small external load disturbances, the wires stick together, and the cable will effectively behave as a solid rod (with allowance being made for the presence of gaps between the individual wires): these conditions are known as the no-slip regime. When large variations in the external load take place, with its associated large changes in the interwire contact forces within the various layers of helical wires, the tangential force changes between the round wires in line-contact will be large enough to overcome interwire friction and induce sliding movements on the interwire line-contact patches: these conditions are, on the other hand, known as the full-slip regime. Cable manufacturers have traditionally provided large numbers of axial stiffness results based on their shop measurements, however, such results invariably relate to the full-slip axial stiffness in the present terminology.

In the present paper, previously reported analytical (closed-form) solutions will be used for identifying various responses of multi-layered and large diameter spiral strands to different types of impact loading with a detailed analysis of the coupled extensional/torsional wave propagations (based on the full-slip and/or no-slip constitutive relations) along the cable. The present numerical results will clarify the controlling (i.e. first order) effect of the lay angle on the axial/torsional full-slip and no-slip stiffnesses and, hence, wave propagation characteristics of axially preloaded helical cables. Such results, are believed to have significant practical implications in terms of the non-destructive methods for in-situ detection of individual wire fractures under service conditions in, say, offshore floating platform and bridging applications to be discussed later.

THEORY
Constitutive Relations for Helical Cables

This topic has been covered at some length by the first author and his associates elsewhere [1, 2] where, for the extreme cases of either full-slip and/or no-slip for the constitutive equations relating the cable (strand and/or wire rope) tension, F, and torque, M, to the cable deformations, it has been postulated that

$$\frac{F}{E_s} = A_1\varepsilon + A_2\Gamma \qquad (1a)$$

$$\frac{M}{E_s} = A_3\varepsilon + A_4\Gamma \qquad (1b)$$

where, A_1-A_4 are the constitutive constants dependent upon both the cable material and construction. In eqns. (1), ε = axial strain = $\partial u/\partial x$, Γ = twist per unit length = $\partial\theta/\partial x$, and E_s = Young's modulus for steel.

The full-slip case of the postulated linear form of the constitutive equations has been experimentally verified. It has been shown that, within experimental accuracy, $A_2 \approx A_3$, which is compatible with the Maxwell-Betti reciprocal theorem for linear elastic structures. Refs. [3, 4] provide a simple means of obtaining the no-slip and full-slip estimates of A_1-A_4 for axially preloaded spiral strands, while Ref. [2] gives a detailed account of theoretical formulations for predicting the no-slip and full-slip values of A_1-A_4 for axially preloaded wire ropes with an independent wire rope core. It is, perhaps, worth mentioning that unlike the full-slip stiffness coefficients which (for all practical purposes) are independent of the level of mean axial load on the cable (spiral strand and/or wire rope) and are a sole function of the construction details, the no-slip values of A_1-A_4, for a given cable construction, have been found to be a function of the mean axial load on the cable.

Dynamic Analysis
Jiang et al. [5] considered a coupled system such as a cable with one end fixed at X = 0 and subjected to sinusoidal forms of the excitation functions for axial force, $F_0(t)$, and torque, $M_0(t)$, at the other end X = h, while at time t = 0 there was assumed to be no motion. Raoof and his associates [6, 7] extended the work of Ref. [5] to cases when a helical cable is subjected to various forms of impact loading at one end, with the other end fixed against any movement. The three impact loading functions considered in Refs. [6, 7] were all of the general form

$$F_0(t) = F_0 g(t) \qquad (2a)$$

$$M_0(t) = M_0 g(t) \qquad (2b)$$

where, F_0 and M_0 are the amplitudes of the external load disturbances. Three distinctly different forms of g(t) were used corresponding to unit-step, triangular and half-sine loading functions [6, 7]. Full derivations of the theoretical formulations for all of these three types of impact loading functions are reported elsewhere [6, 7], and there is little point in repeating such complex formulations here, particularly in view of extreme space limitations. It is, however, worth mentioning that due to typographical errors, some of the equations in Ref. [6] are not exactly correct, and some of the charts have been incorrectly labelled. These minor errors have been identified and corrected in Ref. [7] and, as a result, the interested reader should read the two papers in conjunction with each other in order to gain a complete picture.

RESULTS AND DISCUSSION
Numerical results have been obtained for three different 10 m long, 127 mm outside diameter axially preloaded multi-layered spiral strands with nominally the same geometrical

parameters apart from their lay angles which were 12°, 18° and 24° (with their construction details given elsewhere [8]), experiencing a mean axial strain of 0.002867 which roughly corresponds to one third of their ultimate tensile strength, and assuming Young's modulus for steel E_s = 200 kN/mm², the corresponding Poisson's ratio v = 0.28, with ρ = 7850 kg/m², and the other parameters such as m (the mass per unit length) and I (the mass moment about the cable axis per unit length of the strand in the unloaded configuration) calculated using the relationships in Ref. [6]. Based on the so-called exact formulations (i.e. not the simplified version) of the orthotropic sheet theory, the full-slip and no-slip constitutive constants for these strands are given in Table 1. It should be noted that particularly the no-slip values of A_2 and A_3 in Table 1, are (at first sight) not close to each other: this is due to the rather small values of these constants in the nominally torsionally balanced spiral strands in which, although the no-slip A_2 and A_3 constants for individual layers were, indeed, found to be fairly similar, the accumulation of small errors in the course of algebraically adding up the contributions of the counter-laid layers in order to predict the overall values for the whole strand has led to such apparent (although not practically significant) anomalies.

Table 1. Values of the full-slip and no-slip constitutive constants for the three 127 mm diameter strands as calculated using the orthotropic sheet theory.

	Lay Angle (degrees)	A_1 (mm²)	A_2 (mm³)	A_3 (mm³)	A_4 (mm⁴)
Full-Slip	12	8836.60	-2576.17	-3011.91	928782
	18	6860.21	-3602.98	-4325.86	1838043
	24	4520.34	-4889.42	-5193.44	3104408
No-Slip	12	9388.95	-1324.69	-783.62	3447433
	18	8373.84	-1769.99	-428.91	3878215
	24	7491.03	-2879.24	-826.59	4693255

Figs. 1(a-c) show the variations of the axial displacements, at time t = 0.001163 sec, along the length of the 127 mm (α = 12°) diameter cable for both the full-slip and no-slip regimes, with the end of the cable, at position X = 0, fixed and the other end of the cable at X = 10 m, subjected to unit-step, triangular and half-sine impact loading functions, respectively, with the duration of the impact load A = 0.00052 sec, F_0 = 50 kN, and M_0 = 0. Figs. 1 (d-f) show the corresponding rotational displacements along the length of this same cable for the three loading functions, respectively, at time t = 0.001163 sec, based on the full-slip and no-slip regimes as a function of the distance, X, along the cable. Figs. 2 (a-c) compare the variations of the axial displacements, as a function of time, at the centre (X = 5 m) of the 127 mm (α = 12°) diameter cable for the unit-step, triangular and half-sine impact loading functions, for both the full-slip and no-slip regimes. Figs. 2 (d-f) show the corresponding rotational displacements, as a function of time, at the centre of this same cable for the three impact loading functions, respectively, based on the full-slip and no-slip regimes.

Figs. 3 (a-f) and Figs. 4 (a-f) show the variations of the axial and rotational displacements, as a function of the distance along the cable, and as a function of time at the centre of the cable, respectively, for the 127 (α = 18°) diameter cable. Similarly, Figs. 5 (a-f) and Figs. 6 (a-f) show the variations of the axial and rotational displacements, as a function of the distance along the cable, and as a function of time at the centre of the cable, respectively, for the 127

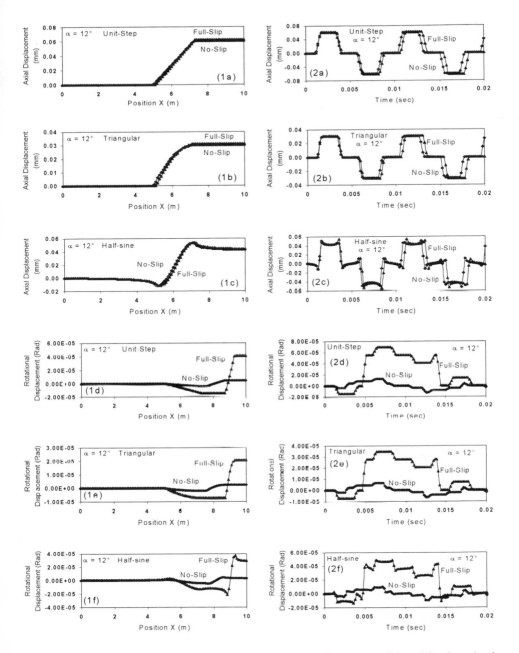

Figs. 1 (a-f). Comparison of the axial and rotational displacements along the cable at time t = 0.001163 (sec), subjected to an impact load of duration A = 0.00052 (sec) for the no-slip and full-slip conditions: α = 12 degrees.

Figs. 2 (a-f). Comparison of the axial and rotational displacements at the middle point of the cable X = 5 m, as a function of time, subjected to an impact load of duration A = 0.00052 (sec) for the no-slip and full-slip conditions: α = 12 degrees.

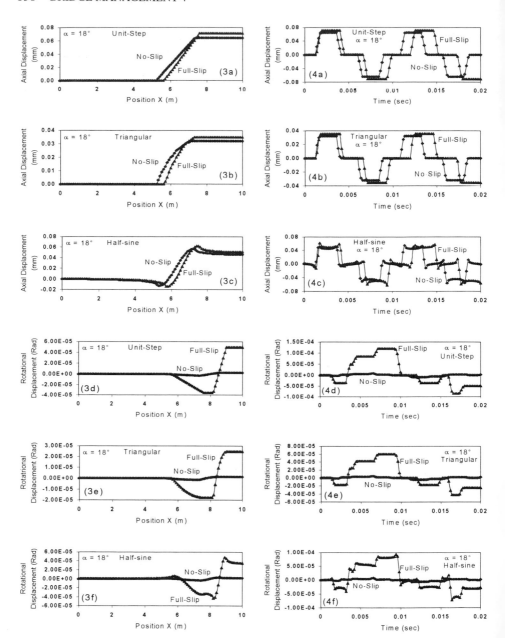

Figs. 3 (a-f). Comparison of the axial and rotational displacements along the cable at time t = 0.001163 (sec), subjected to an impact load of duration A = 0.00052 (sec) for the no-slip and full-slip conditions: α = 18 degrees.

Figs. 4 (a-f). Comparison of the axial and rotational displacements at the middle point of the cable X = 5 m, as a function of time, subjected to an impact load of duration A = 0.00052 (sec) for the no-slip and full-slip conditions: α = 18 degrees.

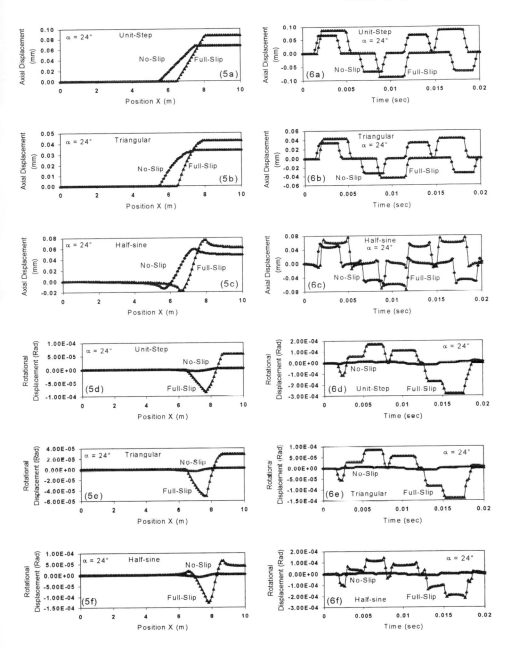

Figs. 5 (a-f). Comparison of the axial and rotational displacements along the cable at time t = 0.001163 (sec), subjected to an impact load of duration A = 0.00052 (sec) for the no-slip and full-slip conditions: α = 24 degrees.

Figs. 6 (a-f). Comparison of the axial and rotational displacements at the middle point of the cable X = 5 m, as a function of time, subjected to an impact load of duration A = 0.00052 (sec) for the no-slip and full-slip conditions: α = 24 degrees.

(α = 24°) diameter cable. In all the plots in Figs. 1 – 6 (a-f), the same values of F_0 = 50 kN, M_0 = 0, A = 0.00052 sec, and t = 0.001163 sec have been assumed.

From the graphical results it is evident that some rather significant differences exist between the full-slip and no-slip wave propagation characteristics. An important observation is that as the lay angle increases, within the practical limits, then the differences between the two bounding solutions of various full-slip and no-slip wave propagation characteristics become very important. Finally, Table 2 shows the rather significant extent by which certain wave characteristics (such as amplitudes and speeds) differ for 12° $\leq \alpha \leq$ 24° depending on whether the full-slip or no-slip solution is adopted with the parameters in Table 2 being independent of the type of impact loading function [6, 7]. These findings are believed to have practically significant implications in connection with the non-destructive methods for in-situ detection of individual wire fractures under, say, axial fatigue loading, whereby the fracture of an individual wire sends a small but measurable shock wave(s) along the cable which is picked up by the electronic boxes. Most importantly, the present results throw considerable doubt on the validity of the traditional methods for calibrating such so-called black boxes. Very briefly, instrumentation experts calibrate their devices by picking up what they call significant effects which are (under laboratory conditions) often simulated by deliberately fracturing a wire in a newly manufactured and axially loaded cable at the end of which the black box signals (waves) are picked up. However, in old and fully bedded-in cables in practice, the cable structure is compacted in such a way that (with an individual wire carrying a small fraction of the total axial load on the cable) the amplitudes and speeds of the axial and torsional waves released by the fracture of an individual wire are governed (because of the small magnitudes of perturbation forces involved) by the no-slip stiffnesses, which are significantly different from the full-slip stiffnesses which govern the behaviour of newly manufactured cables [9] originally used for calibrating the black boxes. It is, therefore, suggested that caution should be exercised in interpreting the data obtained from such devices under service conditions, using the traditional methods of calibrations based on the full-slip behaviour of newly manufactured specimens.

Table 2. Estimates of axial and torsional natural frequencies ω_1 and ω_2, axial and torsional wave speeds C_1 and C_2, respectively, and the ratios of torsional to extensional oscillations R_1 and R_2 [6], for both the full-slip and the no-slip regimes.

	127 mm outside diameter spiral strand					
	α = 12 degrees		α = 18 degrees		α = 24 degrees	
	Full-Slip	No-Slip	Full-Slip	No-Slip	Full-Slip	No-Slip
C_1 (m/sec)	4197.18	4324.20	3742.83	4134.65	3023.36	3889.41
C_2 (m/sec)	1080.25	2082.23	1567.88	2279.13	1988.52	2448.77
R_1	-0.230	-0.069	-0.502	-0.048	1.283	-0.109
R_2	3202.91	5439.06	1569.83	3293.49	524.02	1570.37
ω_1	0.002383	0.002313	0.002672	0.002419	0.003308	0.002571
ω_2	0.009257	0.004803	0.006378	0.004388	0.005029	0.004084

It is also, perhaps, worth mentioning that the exact form of the impact loading function, which relates to the sudden fracture of a wire inside the cable, is unpredictable and probably impossible to determine experimentally. However, it is thought that the widely varying forms of impact loading functions adopted in the present paper should reasonably cover the range of possibilities. The final results, based on all of the three types of impact loading functions, have shown and supported the view that the full-slip wave propagation characteristics are significantly different from the corresponding no-slip wave characteristics, and for a given strand, this difference is increased as the lay angle increases. Bearing this in mind, it is, therefore, concluded that such electronic devices should be calibrated using old and fully bedded-in (in preference to newly manufactured but prestretched) cables which have seen service conditions for a number of years.

CONCLUSIONS

The present paper extends the previously reported work of the first author and his associates who developed closed-form solutions for predicting the various characteristics of coupled extensional-torsional waves induced by various forms of impact loading at one end of steel helical cables (spiral strands and wire ropes) with the other end fixed against any movement. In previous publications, the theoretical analysis centred on a 39 mm outside diameter spiral strand, while this paper presents detailed results based on three different 127 mm outside diameter spiral strands with widely varying lay angles (within current manufacturing limits) to enable the effect of variations in the lay angles on various wave propagation characteristics to be appreciated. It is argued that, due to the presence of interwire friction in axially preloaded helical cables, for sufficiently small levels of load perturbations (due to fracture of an individual wire) applied to fully bedded-in (old) cables, one should use the no slip version of the constitutive relations. Significant differences have been found between a number of axial/torsional wave characteristics induced in cables subjected to unit-step, triangular and half-sine forms of impact loading functions, depending as to whether the no-slip or full-slip version of the constitutive relations are used in the analysis. It is demonstrated that the use of the no-slip version of the constitutive relations is even more critical as the lay angle increases for a given strand construction. The present findings may have significant practical implications in relation to currently adopted techniques by industry for calibrating the electronic boxes, which are subsequently used for in-situ detection of individual wire fractures under, say, fatigue loading associated with cable supported structures.

REFERENCES

1. M. Raoof, and R. E. Hobbs, "Analysis of Multi-layered Structural Strands", J. Engng Mech., ASCE, vol.114, July, 1988, 1166-1182.
2. M. Raoof, and I. Kraincanic, "Analysis of Large Diameter Steel Ropes", J. Engng Mech., ASCE, vol. 121, no.6, 1995, 667-675.
3. M. Raoof, and I. Kraincanic, "Simple Derivation of the Stiffness Matrix for Axial/Torsional Coupling of Spiral Strands", Computers and Structures, vol.55, no.4, 1995, 589-600.
4. M. Raoof, "Methods for Analysing Large Spiral Strands", J. Strain Analysis, vol.26, no.3, 1991, 165-174.
5. W. Jiang, T. L .Wang, and W. K. Jones, "Forced Vibration of a Coupled Extensional-Torsional System", J. Engng Mech., ASCE, vol.117, 1990, 1171-1190.
6. M. Raoof, Y. P. Huang, and K. D. Pithia, "Response of Axially Preloaded Spiral Strands to Impact Loading", Computers and Structures, vol.51, no.2, 1994, 125-135.
7. M. Raoof, T. J. Davies, and C. F. Scott, "Coupled Axial/Torsional Response to Impact Loading of Helical Cables", Proc. Third Int. Symposium on Cable Dynamics, Trondheim, Norway, August, 1999, 25-30.
8. M. Raoof, "Effect of Lay Angle on Various Characteristics of Spiral Strands", Int. J. of Offshore and Polar Engineering, Vol. 7, No. 1, March, 1997, 54-62.
9. M. Raoof, "Comparison Between the Performance of Newly Manufactured and Well Used Spiral Strands", Proc. Instn Civ. Engrs, Part II, vol.89, Mar.,1990, 103-120.

Long term measurements for fatigue loading of prestressed concrete bridges

K. ZILCH and E. PENKA
Lehrstuhl für Massivbau, Technische Universität München, München, Germany

Abstract

The experimental set-up and the results of long-term measurements on prestressed concrete bridges are presented. The measurement system records bridge data such as cross-section temperatures, concrete surface strains and changes of crack widths. Together with the results of an analytic calculation this data can be used to evaluate the bearing behaviour of cross-sections in the areas of the coupling joints under the combined loading of temperatures and dynamic traffic.

1 Introduction

In Germany, thermal stresses were neglected in the design of prestressed concrete bridges before 1970. This only has a small influence on cross-sections with high permanent loading. However sections with low loading are often underdimensioned according to modern design methods. Especially the points of zero moment are effected, where the tendon-coupling assemblies are often positioned. The lower fatigue resistance of these structural members may lead to fatigue cracking and as a consequence later on to the failure of the cross-section.

Analytic calculations had been carried out to estimate the risk of a fatigue failure in these areas. However, variability of dead loads, loss of prestress and temperature loading sometimes lead to results, which differed from the actual appearance. Therefore long term measurements were performed, attaining a more realistic bearing behaviour of the cross-section. The two strategies and their combined results are presented for the following example of a bridge, where changes of crack width up to 0.35 mm were observed in the preliminary stages.

2 Structure

The examined motorway bridge crossing a river is a prestressed concrete bridge which was constructed in the years 1961-1964. A flanged beam with 2 webs was chosen as the cross-section and over the columns, bottom slabs were added to strengthen the compression zones. The bridge consists of 7 spans (span lengths: 33 m + 43 m + 57 m + 104 m + 51.5 m + 40 m + 30.5 m = 359 m), which were set up step-by-step with falsework. The columns have a height of approximately 15 m. The described measurement is performed in the 57 m wide span in the area of the coupling joint.

3   Long Term Measurement

3.1   Measurement System

The measuring equipment which works automatically with 32 analog inputs is a special product of the company Schuehle in Ravensburg (Germany). This system incorporates the features of intelligent data loggers, multi-channel signal analysers, transient recorders, programmable controllers and state-of-the-art signalling systems.

3.2   Data Telecommunication

To transfer data from the data logger to the personal computer in Munich a GSM mobile phone system is used. About 16000 values can be transmitted per minute with this connection. The measurement can be controlled at any time and furthermore major parameters can be changed. Thus an undetected breakdown of the measuring system is prevented and frequent visits to the bridge are unnecessary.

3.3   Measuring Set-up

The measurement system was installed at the beginning of October 1998. At the moment 26 sensors are in use. With the 6 remaining analog units further sensors can be included in the set-up of the installation.

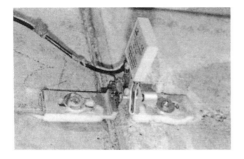
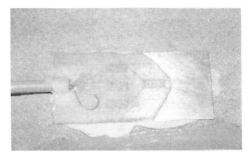

Figure 1. Displacement Transducer   Figure 2. Strain Gauge

Presently 6 displacement transducers (Figure 1, Figure 3: R1-R6) are used to specify the changes of the crack widths over the height of the web. When choosing the sensors, special attention was paid to long term accuracy. This was best achieved with sensors based on strain gauges. To detect the differences between the crack and the flanking uncracked cross-sections 6 temperature compensated strain gauges (Figure 2, Figure 3: D1-D6) are placed on the webs. With this measurement set-up it is possible to determine which percentage of the cross-section is reacting in a non-linear way during the crossing of a truck. 4 more strain sensors are located in the main span of the bridge.

In addition, 10 measurement points (Figure 3: T1-T10) are distributed via the cross-section to record the temperature loading of the bridge. The selection of the measurement points was based on a preliminary computational investigation and the results of Knabenschuh [1]. The measuring set-up for this bridge is shown in Figure 3.

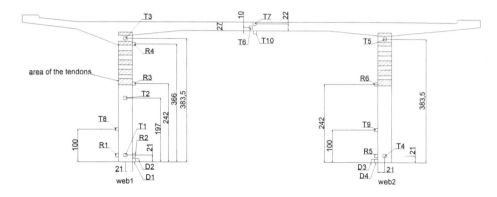

Figure 3. Measuring Set-up

3.4 Evaluation of the Results

For a good representation of dynamic changes the measurements are performed with a rate of 100 Hz [2] except for the temperature measurements. The amount of resulting data is partially analysed on site and thus minimised. Afterwards this data is automatically transmitted with the GSM modem. This is used as a basis to investigate the relations between temperature changes, dynamic traffic loads and development of the crack widths.

3.4.1 Thermal Loading

The aim of this evaluation is to get the vertical temperature gradient ΔT_v , the horizontal temperature gradient ΔT_h and the superstructure temperature T by using the mean values of the 10 temperature measurement points (see 3.3). Therefore a post-processor was developed, which makes it possible to convert the temperatures of the measurement points into cross-section temperature values.

The cross-section was divided into a net with 4 nod elements. To obtain the temperature distribution of the cross-section the following assumptions had to be made:

- the temperature of the deck is constant at the same height. For this reason the measurement points T6 and T7 are used.
- The bottom side of the deck has the temperature T10.
- The sunlit surface of web 1 has the temperature T8.
- The residual surfaces of the webs have the temperature T9.
- For the inside of the webs the temperatures T1 – T5 and a linear interpolation are used.

With these values the processor calculates the temperature gradients ΔT_v and ΔT_h and the superstructure temperature T by using the following relationships:

$$\Delta T_v = \frac{\Sigma(A_i \cdot T_i \cdot y_i) \cdot h}{I_y} \tag{1}$$

$$T = \frac{\Sigma(T_i \cdot A_i)}{A_{ges}} \tag{2}$$

where:

A_i = element area
T_i = element temperature
h = height of superstructure
y_i = distance of element to centroid
I_y = moment of inertia
A_{ges} = area of cross-sections

Figure 4 shows the frequency of the vertical temperature gradient ΔT_v, which was calculated from the temperatures of the 10 measurement points. The temperature data used for this figure includes the period from October 1998 to early September 1999. So a whole cycle of the year is not yet represented. But the period from April to August, which contains the highest vertical temperature gradients, is included. Therefore it could be expected, that the frequency of high values will decrease a little for the whole cycle of the year.

The maximum value is calculated with 8.3 K, the minimum value with –4.7 K and the mean value with 1.0 K. That means, that the thermal loading behaves as was expected when one compares this with other flanged beam bridges. Only the maximum values are about 1 K lower than on related bridges [3]. The reason for this is the great height of the webs of this bridge.

This data can now be used for analytic calculations as a realistic thermal loading of the bridge concerned.

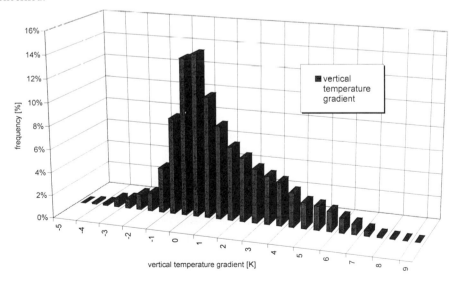

Figure 4. Frequency of vertical temperature gradient ΔT_v

3.4.2 *Traffic Loading*

The weight of each lorry crossing the bridge can only be determined with an extensive and thus unrealistic measurement. But it may be assumed that every 5 minutes at least one heavy lorry crosses the bridge, the weight of which is approximately constant. Furthermore stress variations of the tendons or of the reinforcement steel cannot be measured in a direct way. If the concrete is removed to measure steel strains, the bond conditions will be changed and the results will be different because of redistributions in the cross-section.

Therefore the changes of the crack width are measured at the coupling joint and the maximum, minimum and mean values are extracted in 5 minute intervals. The dependence of the amplitudes, resulting from this, on thermal loading was examined. But predictions about the stresses in the tendons could only be made together with an analytic calculation, which is described in the following. The evaluation is shown in Figure 5 for the amplitudes of all 6 displacement transducers and the vertical temperature gradient ΔT_v.

It can be seen, that the crack width is related to the vertical temperature gradient ΔT_v. Especially the values of the displacement transducers on web 1 show strong changes of their mean values. The differences between displacement transducer R1 and R2 can be attributed to the splitting of the crack on the interior of the web. These two measurement points are positioned over the crack in the height of the lowest layer of the reinforcement steel. From this it can be determined, that the cross-section behaves in a non-linear way for vertical temperature gradients ΔT_v greater than –1 K. The transducer R3 is positioned at the height of the lowest tendon. It can be seen, that the crack opens in this area at a vertical gradient of about +1 K. With a temperature gradient of +4 K the crack opens over the whole height of web 1 (R4).

Web 2 shows a different behaviour. The crack opens there in the region of the tendons (R5) at a vertical temperature gradient of about +5 K but then opens very fast along the height of the web (R6).

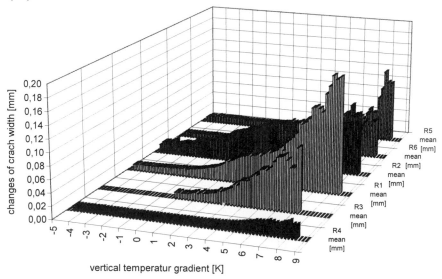

Figure 5. Relation: Amplitude of crack width – vertical temperature gradient ΔT_v

4   Analytic Calculation

A realistic model of the lorry traffic (40 to) is given in Eurocode 1 T3 [4] with the lorry type 3 of the loading model 4. Therefore the variations of the tendon stresses were calculated with this model. This lorry type causes variations of stresses which are relevant for fatigue. The variation of stresses are calculated with a distinct cracked condition, as is postulated in [5]. The loading can be divided equally to both webs, because of a restriction of the traffic. The calculation was performed for a vertical temperature gradient of +7 K and the additive safety element ΔM, which was developed in the Standard DIN 4227 [6] for coupling joints. This element takes the variations in dead load and redistributions into account. To consider variations of the prestressing force, the statically indeterminate prestressing is set to 100% and 90% of the value $t = \infty$ for all calculations. The bonding conditions for tendons and reinforcing steel are assumed to be equal, because both types of steel are smooth round bars.

The following figure shows the relation between the variations of stresses in the lowest tendon layer $\Delta\sigma_S$ and the moment M. It can be seen, that these variations strongly depend on the magnitude of the moments, caused by the temperature gradient $M_{\Delta T}$ and the safety element ΔM. It can also be realised, that the variation of stresses are below the tolerable stress variations of $\Delta\sigma_Z = 110$ N/mm². However differences are expected between the actual loading and the loading used for the calculation, because of the observed high changes of the crack width. For this reason the calculations should be based on a measurement, to get more realistic values of the stresses.

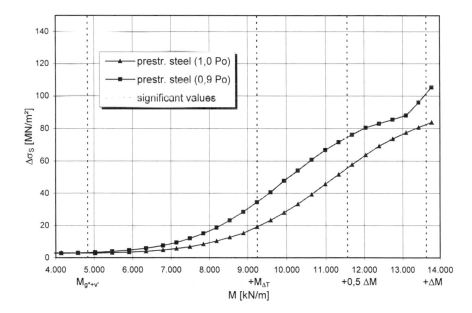

Figure 6. Relation: Moment - oscillation width of tendon

5   Combination of the Two Strategies

For the combination of the two strategies a few adaptations have to be made. To get only the lorries, which are similar to the lorry type 3 of loading model 4 (see Chapter 4), and not the heavy trucks, which are subject to approval, single extreme values of the crack width are taken out for every class of temperature gradient. The relation between the measured crack width and the temperature gradient has to be calibrated on the relation between the crack width and the moment. Therefore the points of decompression for R1 and R3 and the calculated decompression points are used. With this step the x–axis is defined and it is now possible to get the actual moment ΔM. This value can be used for a new analytic calculation to estimate the life span of the bridge in a more realistic way.

In Figure 7 the similarities between the amplitudes of the crack width and the variations of stresses can be seen. The deviations between the two curves in the cracked condition can be attributed to the deranged bond conditions. But the point of non-linearity is clearly visible, which allows the actual moment ΔM to be detected. The reason for the deviations of the curves beyond a moment of 10.000 kN/m is the small amount of crack width values, because temperature gradients over +6 K only appeared with a percentage of 3% for the whole measuring period.

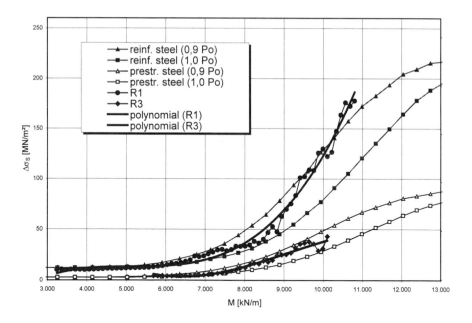

Figure 7. Comparison: Amplitudes of crack width and variation of stresses

6   Conclusion

The cross-section has now been observed for nearly the cycle of a whole year and the strong influences of the thermal loading on the crack width can be seen.

The combination of the two strategies, analytic calculations and long term measurements, provides sufficient possibilities to obtain the actual loading of a cross-section by the determination of the actual moment ΔM. The next aim is to determine time dependant changes in the bearing behaviour of the cross-section and the loading of the bridge. Furthermore, the influences of the parameters like traffic loading, bond conditions, transverse redistribution and self-equilibrating stresses have to be checked.

7   References

1. Knabenschuh, H., *Temperaturunterschiede an Betonbrücken*, Bundesanstalt für Straßenwesen, Bergisch Gladbach, 1993.
2. Baumgärtner, W. & H. Waubke, *Dynamische Messungen an der BAB Brücke Fischerdorf*, München, 1997.
3. Zilch K. et al., *Entwicklung der Kombinationsbeiwerte für den Grenzzustand der Ermüdung*, Bundesanstalt für Straßenwesen, 1999
4. ENV 1991-3: Eurocode 1 Teil 3 – Grundlagen der Tragwerksplanung und Einwirkungen auf Tragwerke, Verkehrslasten auf Brücken, 1995.
5. Handlungsanweisung zur Beurteilung der Dauerhaftigkeit der vorgespannten Bewehrung von älteren Spannbetonbrücken, Bundesanstalt für Straßenwesen, Abteilung Brücken- und Ingenieurbau, 1998
6. DIN 4227 Teil 1, Spannbeton; Bauteile aus Normalbeton mit beschränkter oder voller Vorspannung, 1988
7. Frenzel, B. et al., *Bestimmung von Kombinationsbeiwerten und –regeln für Einwirkungen auf Brücken*, Heft 715, Forschung Straßenbau und Straßenverkehrstechnik, Bonn - Bad Godesberg, 1996.
8. Rücker, W.F. et al., Ergebnisse der automatischen Dauerüberwachung an der Westendbrücke in Berlin, BAM-MAN, Berlin, 1996.
9. Zilch, K. & E. Penka, Versuchsbericht, Projekt Nr. 2542, München (unpublished), 1998.

Monitoring the environment in the box section of externally post-tensioned concrete bridges

Dr R J WOODWARD, Transport Research Laboratory, UK and Mr D MILNE, Highways Agency, UK.

INTRODUCTION

There has been increasing interest in recent years in the design and construction of post-tensioned concrete bridges with external unbonded tendons. This was given a significant boost in 1992 with the moratorium on the construction of post-tensioned bridges with internally bonded tendons as a result of problems associated with grouting of the post-tensioning ducts. It was envisaged that external post-tensioning would facilitate the inspection and replacement of the prestressing system.

To assist in the design of these structures the Highways Agency issued a Standard, BD 58 (DRMB 1.3.10), and an accompanying Advice Note, BA 58 (DRMB 1.3.9) in 1994 (1,2). These documents were based on a review of existing knowledge that included recommendations on where further research was required.

TRL have undertaken a major project for the Highways Agency to investigate all aspects of the design and construction of post-tensioned concrete bridges with externally unbonded tendons. This included the design, construction and testing of a quarter scale model of an externally post-tensioned bridge (3,4) and an investigation of the corrosivity of the environment inside the concrete box section in which tendons are housed. A knowledge of the environment inside the box section is required to help determine the level of protection required to prevent corrosion.

To provide this information, stressed steel samples and steel coupons were placed inside the box sections of two bridges: Botley Flyover in Oxfordshire and River Camel Viaduct near Wadebridge in Cornwall. In addition temperatures and relative humidities both inside and outside the box sections are being monitored.

Monitoring started at the end of 1995 and this paper describes the findings during the first three years exposure.

STRUCTURES BEING MONITORED

Botley Flyover

The bridge is situated at the interchange of the A34 with the A420 on the west side of Oxford. There are two structures and they share a similar form of construction. Both are three span continuous overbridges with spans of 12.7m, 29m and 10.75m. The south bridge was selected to host the test samples and monitoring equipment.

The bridges, which were built in 1972 and designed by Gifford and Partners, are a combination of in-situ and precast elements subsequently stressed together. A total of eight external unbonded tendons consisting of 19 no. 18mm diameter strands run the full length of the three spans between anchorages positioned at the deck ends. The tendons are located within the deck voids and are deflected by deviators within the cross beams, diaphragms and intermediate web stiffeners.

Strands within each tendon are individually sheathed except at the anchorages where the tendons are enclosed in plastic ducting. Tendon deflection at the deviators is achieved by saddles cut from lengths of rolled circular hollow section lined with PVC pipe. Corrosion protection is provided by a combination of grout and grease. At the anchorages the void between the duct and the cable is filled with grout. At the web stiffeners the voids are filled with grease. At the pier cross beams structural grade concrete was used. Drawings indicate that at the diaphragms grout was used to fill voids on the north bridge whereas grease was used on the south bridge.

Entry into each of the cells and precast beams is through 600mm diameter access holes positioned in the soffit or the side of the beam. There is no physical access between cells. The access points also act as vents and have lockable steel grills. All access points except those for the end spans are positioned over the carriageway and lane closures are required in order to gain access.

Because of the access restrictions only two of the end span cells have been inspected. There was no evidence of water leakage into either of the cells inspected. The surface of several intermediate web stiffeners had been painted with a bituminous coating. Only stiffeners on the outer edges of the structures seemed to have been treated in such a manner. The reason for this is not known. A grease treated wrapping tape has been extensively used at the ends of the anchorage ducting, presumably to seal the duct for grouting. The wrapping has also been used to seal sheathing where individual strands have been inspected.

River Camel Viaduct

The River Camel Viaduct is a single box beam that carries the A39 over the River Camel to the north of Wadebridge. It comprises nine spans, the outer spans being 37.5m and 42.5m, respectively and the inner ones 54m. It was designed by Gifford and Partners and opened to traffic in July 1993 (5).

The bridge is of in-situ construction and was built span by span, the construction being from the approximate fifth point in one span to the fifth point in the following span. Each span was cast in ten pours with the anchor block being cast first.

The deck is post-tensioned with a VSL system, each tendon comprising 23-25 no. 15.7mm diameter strands inside 140mm diameter HDPE ducts with a wall thickness of 8mm. The ducts are joined by heat shrink connections and were temporarily supported during construction to the approximate cable profile by suspension from the top slab. There are sixteen tendons over the piers and fourteen along the remainder of each span with ten in the end spans. The tendons were stressed ten days after the last pour.

The strands, which were designed to be replaceable, protrude approximately 1.5m at the live end anchorages to enable the tendons to be de-stressed. The protruding tendons are covered with an anchorage cap which is filled with a wax void filler. This is also used to protect the tendons where they pass through the anchorages and anchor block. The concentric tubes through the anchorage block enable the cable and anchorage assembly to be removed after the cables have been de-stressed.

Both structures are in good condition and have fully functional waterproofing systems. The boxes in both bridges were clean and dry and there was no evidence of water leakage from either the carriageway or the deck drainage system. However some condensation was observed, particularly in the end spans of the River Camel Viaduct (see below).

DESCRIPTION OF MONITORING SYSTEM
The temperature and humidity both inside and outside the two structures are being recorded every hour. In addition, stressed strand samples, wires from free lengths of strand and weight loss coupons, are exposed within the box sections and weight loss coupons have been placed outside the structures.

The stressed strand samples were loaded to 70% of their ultimate tensile strength and some of these samples are instrumented with load cells that record the load in the strand. Deflectors, with a radius of curvature of 2.5m, similar to the deviators found in typical bridges, are incorporated into some of the stressing rigs. HDPE pipe has been placed between the strand and the deviator to simulate the actual interaction of strand and duct.

The wires from lengths of strand and the weight loss coupons are being used to give a measure of corrosion rates. Corrosion is being monitored visually and the coupons are being used to provide data on corrosion rates. Unfortunately the weight loss coupons placed outside the River Camel Viaduct have been stolen.

Within each structure, the air temperature and humidity are being recorded local to the samples. The surface temperature of the concrete inside the concrete boxes, and the external air temperature and humidity are also being recorded.

The samples in the River Camel Viaduct were positioned towards the centre of the bridge. However following the observation of condensation in one of the end spans it was decided to place another set of coupons and unstressed wires in the end span at the west end of the bridge during August 1998. This was to determine whether corrosions rates would be higher in areas where condensation was likely to occur.

RESULTS

Strand samples and weight loss coupons
The loads in the strands on both structures have remained stable after an initial settling in period. The small fluctuations now occurring are probably caused by the components of the test rigs expanding and contracting due to the temperature variation.

Some of the steel coupons have been returned to TRL, cleaned by immersion in Clarke's solution (concentrated hydrochloric acid inhibited with stannous chloride and antimony trioxide to prevent attack on the underlying steel) and weighed. The weight losses recorded are given in Table 1 and are expressed as an equivalent loss in thickness.

Botley Flyover

Spots of corrosion started to develop on both the samples of wire and the coupons exposed in the boxes fairly soon after installation, although it was much less pronounced than on the samples exposed outside. The horizontal coupons are more severely corroded than the vertical ones and corrosion is more pronounced on the top surface than underneath. After a year the top of the horizontal coupons had a light coating of rust speckled over the surface whereas the other coupons only had a few spots of light surface rust on them. The amount of rust has continued to increase over the second and third years.

Table 1. Weight loss measurements on steel coupons

Bridge	Location	Time Exposed (months)	Average annual loss in thickness (μm)	
			Horizontal	Vertical
Botley	Inside	3.5	Not measurable	Not measurable
		10.5	0.6	0.2
		22	0.5	0.3
		34.5	0.5	0.4
	Outside	10.5	10.0	5.4
		22	6.5	5.7
		34.5	6.3	5.8
River Camel	Inside	5	0.7	Not measurable
		13	0.7	0.3
		24	0.7	0.2
		36	0.7	0.2

After three years the stressed strands are still generally clean with a few small specks of surface rust which cover about 10 to 15% of their surface area. The rust is occasionally more developed in crevices between the wires. Condensation had been observed on the strands in the test rigs.

River Camel Viaduct

Observations made at the River Camel suggest a similar pattern of corrosion is developing. Initially it appeared to be occurring at a slightly faster rate than at Botley but after three years there was little to choose between the two sites.

Small areas of localised rust staining are evident on the wires of the stressed strands and, as at Botley, it is occasionally more developed in the crevices between the wires. This may be due to moisture from condensation collecting in the troughs between the wires. There is no evidence of pitting corrosion.

During the most recent visit to Wadebridge, the end span at the east end of the structure was found to be damp with condensation on the soffit of the top slab, the walls of the box and the ducting. The visit was made during a period of heavy rain and there was considerable leakage of water through the expansion joints into the anchorage chamber which created a very damp environment. The condensation reduced away from the end of the structure and the sections where the monitoring system and coupons had been installed and the end span at the west end of the bridge were dry.

Temperature and humidity

The temperatures and humidities recorded at the two structures show that the environment within the box structures is more stable than that outside. Temperature and humidity inside the box sections follow the general trend of the ambient atmospheric conditions although values are less extreme. It is usually warmer inside the boxes than outside, particularly during the summer.

The external daily temperature cycles and fluctuations in relative humidity are generally more extreme at the River Camel Viaduct than at Botley possibly because it is in a more exposed environment.

The temperatures and humidities measured at the two locations being monitored at the River Camel are virtually identical although the humidities measured in the end span are generally, but not always, higher than those measured nearer the centre of the bridge.

DISCUSSION

Over recent years it has been found that corrosion rates on composite steel bridges are considerably reduced if the steel beams are enclosed (6). Table 2 shows the corrosion rates measured both inside and outside the enclosure on a number of bridges. It can be seen that even though corrosion rates inside the enclosure are considerably less than those outside they are still higher than those measured inside the concrete boxes at Botley and River Camel. The results obtained indicate that the corrosivity of the environments at the positions being monitored inside the two box sections is two to ten times lower than in an enclosed bridge.

Table 2. Typical corrosion rates in various environments.

Bridge	Location of Coupons	Period of exposure (months)	Annual corrosion rate (μm)
Conon	Enclosure	17	2 - 3
	Exterior	17	18
Tees	Enclosure	17	2-3
	Exterior	17	49
Second Severn Crossing approach roads	Enclosure	12	1 − 6
	Exterior	12	35 − 90
County bridges	Enclosure	12	3 - 8
	Enclosure	12	3 - 9
	Enclosure	12	3 - 7

The corrosion rate measured outside at Botley is also comparatively low. This is probably because the coupons were placed on the bearing shelf (to provide some protection from vandals) and thus they are not fully exposed to the wind and rain.

While these results show that the corrosivity inside a concrete box girder is significantly lower than inside a bridge enclosure, corrosion rates tend to change with time so short term testing may not be reliable and longer term measurements are required to confirm these findings.

Although corrosion rates are low, the top surface of the horizontal coupons was almost completely covered with light surface corrosion after two years and there were also spots of corrosion on the wires and strands. The increased severity of the corrosion on the top surface of the coupons was probably due to dust particles settling on the surface of the steel. The amount of corrosion observed has continued to increase during the three years of exposure.

The observation of condensation in the end spans in the River Camel Viaduct when the inner spans were dry requires further investigation. It indicates that the corrosivity of the environment can vary along the length of the structure and the monitoring has been extended to include measurement of corrosivity, temperature and relative humidity in one of the end spans in the structure.

CORROSION PROTECTION

While the results obtained to-date show that the corrosivity inside a concrete box is relatively low, the spots of rust observed on unprotected steel strands indicate that some form of corrosion protection is necessary and that bare wires are not an option. However, initial indications are that simple corrosion protection measures are all that is required. The only prerequisites are that the tendons are inspectable and the box is designed to prevent the ingress of contaminants.

The most common methods of corrosion protection at present are to house the tendons in wax or cement filled ducts. This has the disadvantage that tendons cannot be inspected directly. For grouted systems it is necessary to locally remove the sheathing and expose the tendons. For wax filled ducts strands can be withdrawn and examined external to the duct. However the low corrosivity found in the concrete box sections being monitored shows that as long as there is no evidence of leakage of deleterious materials into the box then the frequency and detail of tendon inspection can be kept to a minimum.

Coating systems that provide corrosion protection without requiring the tendons to be housed in a duct would appear to offer the ideal solution. The strands can be examined along most of their length for corrosion and wire fractures, while the coating provides protection against the limited corrosion that takes place in the box. At locations where strands are not visible ie where they pass through deviators and at anchorages, the unwinding of fractured wires would provide visual evidence of corrosion. Coating systems available include:

- galvanising
- paint
- epoxy coating.

All these systems have their limitations. Galvanising has the advantage that it is not easily damaged in handling and installation, and BA 58 states that "Unducted galvanised tendons have the advantage of being easily inspectable throughout their length". However it has not been used in bridges in the UK and it is believed that there are several reasons for this:

- it is a sacrificial coating and it consumable with time
- damage to the coating where it is anchored
- concern among engineers about the risk of hydrogen embrittlement
- cost; a whole life cost study undertaken at TRL showed that of four options considered for

protecting external tendons (galvanising, tendons in wax filled ducts, tendons in grout filled ducts with wax at the anchorages, and individually sheathed strands in wax filled ducts), galvanising was the most expensive.

Paint systems would be difficult to apply after installation as it would only be possible to coat the outer faces of the wires. Any subsequent relative movement of individual wires would lead to a risk of ingress of moisture between wires and corrosion of unprotected steel. Epoxy coated strands have been used in the USA but there is concern about pinholes and local damage to the coating and problems such as poor adhesion that cast doubt on its use (7).

An alternative solution is to use individually sheathed strands in grease filled sleeves. This has the advantage that the strands are protected against corrosion yet fractures of individual wires can be detected by looking for deformations in the sheath caused by movement of the wires when they fail. This has been done previously on structures where wire failures have occurred by shining a torch focused to a pencil beam along the surface of each strand to cast shadows behind any irregularity in the sheathing. The limitation of the method is that it is not possible to detect fractures within the anchor block and at deviators.

The first eight externally post-tensioned bridges built in the UK all used individually sheathed strands and wire fractures have been found in two of them (8,9). In one instance the strands were protected by Bridon's Metal A - an aluminium resin - and individually sheathed within a 38mm diameter PVC sleeve. Failures were attributed to the presence of aluminium in the resin although the exact cause was never unambiguously determined. In the second instance failures were attributed to problems during construction. On both bridges the strands were in PVC sleeves and this was also considered to be a contributory factor although this was not conclusively proven. No problems have occurred in the remaining bridges, one of which used PVC sleeves and the remainder polypropylene.

Strands in grease filled polypropylene sleeves are currently commercially available. One system uses a multi-layer protection whereby the sleeved strands are themselves housed in wax or grout filled HDPE ducts (10). The use of grout or wax filled ducts increases the cost, and makes the strands difficult to inspect. Given the low corrosivity of the environment it is questionable whether this additional protection is really necessary.

Another advantage of individually sleeved strands is that the corrosion protection is applied under factory controlled conditions, before delivery to site. Thus site operations are limited to ensuring that the sleeves are properly stored before construction and not damaged during installation. However care is required where strands are protected at anchorages and where they pass through deviators. The use of plastic lined deviators should reduce the risk of damage during installation and stressing.

CONCLUSIONS

- Conditions inside the two bridges were more stable and less extreme than the outside environment.
- The corrosion on samples inside the structures was much less than outside.
- The corrosivity of the environment inside concrete box girder bridges is very low, less than that inside a bridge enclosure.

The findings indicate that there is little risk of corrosion of tendons in grouted or wax filled ducts as long as the concrete box sections remain dry and free of contamination. Future inspections should therefore concentrate on checking that there is no leakage of deleterious materials into the box. Where this is the case direct inspection of the tendons can be limited to a small number of spot checks.

Of the corrosion protection systems available, all would appear to offer good corrosion protection.

ACKNOWLEDGEMENTS
The work described in this report was carried out in the Civil Engineering Resource Centre at the Transport Research Laboratory. The authors would like to acknowledge all the staff who were involved in the project, in particular Mr M Mckenzie, Mr N Anderson and Mr A Frost.

REFERENCES
1.   Design manual for roads and bridges. *The design of concrete highway bridges and structures with external and unbonded prestressing.* BD58. The Stationery Office, London. 1994 (DRMB 1.3.9).

2. Design manual for roads and bridges. *The design of concrete highway bridges and structures with external and unbonded prestressing.* BA58. The Stationery Office, London. 1994. (DRMB 1.3.10).

3. Daly A F and P Jackson. *Design of bridge with external prestressing: Design example.* TRL Report TRL 391. Transport Research Laboratory, Crowthorne 1999.

4. Woodward R J and A F Daly. *Design of bridge with external prestressing: Construction and testing of a model bridge.* TRL Report TRL 392. Transport Research Laboratory, Crowthorne 1999.

5.   Hollinghurst E. *River Camel Viaduct, Wadebridge.* The Structural Engineer, Vol 73. No. 3/4. April 1995. pp99-102.

6. Podolny W. *Corrosion of prestressing steewls and its mitigation.* PCI Journal. September-October 1992. pp34-55.

7. Porter M G. *Repair of post-tensioned concrete structures.* Concrete Bridges Investigation, maintenance and repair. Proceedings of a one-day symposium, London 25 September 1985.

8. McKenzie M. *Corrosion protection: the environment created by bridge enclosure.* TRL Research Report 293. Transport Research Laboratory, Crowthorne 1991.

9. Brooman H and A Robson. *A3/A31 Flyover - Case history of an externally post-tensioned bridge.* International Conference on Bridge Management. University of Surrey. April 1996.

10. External Prestressing. 1990.   Service D'Etudes Techniques des Routes et Autoroutes. Paris. February 1990.

The benefits of staged investigations of concrete highway structures

Dr DONALD PEARSON-KIRK
PB Kennedy & Donkin Ltd
Calyx House, South Road, Taunton, TA1 3DU, UK

Abstract
Interactions between singular causes of deterioration of concrete structures are numerous and can greatly accelerate the overall rate of deterioration. Causes may relate to inadequacies in design and/or construction or may result from agents acting on structures. The paper will discuss the need to improve the cost-effectiveness of investigations of structures to determine their condition. This may be achieved by the application of staged investigations where in turn the presence of a problem is identified, the cause or causes of the problem determined and the most appropriate solution to the problem selected. Case studies are presented that demonstrate the benefits of staged approaches to investigations.

1   Introduction

Infrastructure is expensive, and so should be appropriate, economic, and lead to improved quality of life and safety, whilst resulting in the minimum of adverse environmental impacts. Problems resulting from inappropriate design and poor construction practices have become apparent in many counties. Even in countries where design procedures, construction practices, and maintenance of structures have been apparently well carried out, unpredicted deterioration of structures has occurred.

Many factors need to be considered in order to predict durability and life of a structure. These include experience, results of assessments and tests and the effect of the actions of agents on different components of the structure [1].

The maintenance, repair and replacement of structures or parts of structures provides substantial experience on durability, but little data is collected systematically in a form that can be used to predict durability with any certainty. Life cycle costing techniques are being developed in order to aid decision making relating to the maintenance, repair or replacement of a structure.

The predicted service life of a structure may be assessed in one or more of the following ways [1]:-

- By reference to experience with a similar structure in similar circumstances.

- By measuring the rate of deterioration over a short period and estimating when the limit of durability will be reached.

Bridge Management 4, Thomas Telford, London, 2000.

- By interpolation from accelerated tests that will shorten the response time to the action of the agent(s) causing deterioration.

Service life prediction and management strategies for structures benefit significantly from good quality advice on the existing and future condition of those structures.

2   Causes of deterioration

Interactions between singular causes of deterioration of concrete structures are numerous and can greatly accelerate the deterioration.   Causes may relate to inadequacies in design and/or construction or may result from agents acting on structures. Various organisations have reported on causes of deterioration to concrete structures, for example the Organisation for Economic Co-operation and Development (OECD), which reported in 1988 [2] that singular causes of deterioration to 800,000 concrete highway structures in decreasing order of importance were as follows:-

Chloride contamination;
Sulfate attack;
Non-conformance to the project design;
Thermal effects;
Inadequate design;
Insufficient reinforcements/or insufficient size of reinforcement bars;
Insufficient quality of concrete;
Insufficient concrete cover;
Insufficient protection against rain water;
Lack of maintenance;
Alkali-silica reactions;

Over 320,000 of the structures surveyed were classified as being structurally deficient.

3   The need for testing

In-situ testing and sampling of structural materials can be time-consuming and expensive, but is of benefit if it is closely specified and controlled, properly organised and supervised at the appropriate level. Valuable information can be obtained for use in the assessment of the load carrying capacity of a structure, in the assessment of the state of deterioration of the structure and in formulating possible repair, strengthening and/or replacement strategies [3].

The use of material strength testing is often undertaken to determine whether or not worst credible strength as determined by testing will result in capacity assessments higher then those achieved by using characteristic (default) strengths. The use of worst credible strength may permit reduced material factors and these may give a two-fold benefit in possibly improving the assessment of a structure.

The condition assessment of a concrete structure in terms of durability can be better determined when condition testing is undertaken, with emphasis principally being directed towards the condition of the concrete and the reinforcement.

That condition assessment can be improved by testing will be demonstrated by the author later in this paper, and this was shown to be the case in a nationwide study in the United Kingdom reported in 1989 [4]. In 1986 the Department of Transport had commissioned this condition survey of 200 randomly selected concrete bridges, with the condition of the bridges [4] being classified as:

Classification	Visual Inspection	Visual Inspection with Condition Testing
GOOD	25	60
FAIR	114	99
POOR	61	41

With condition testing the assessment of the overall condition of the 200 bridges improved from over 75 per cent of the bridges being in fair and poor condition to almost 80 per cent being in good and fair condition.

One general conclusion from studies has been that service life prediction and the selection of management strategies can benefit significantly from:-

- Well targeted testing regimes specified by experienced engineers
- Proper execution of site investigations by appropriately trained engineering technical staff
- Appropriate interpretation of site and laboratory testing results by engineers experienced in testing and monitoring concrete structures.

Better quality advice can then be given to managers of structures in order that their structures may be managed more cost-effectively.

4.   Assessing the durability of concrete structures

Assessing the durability of concrete structures is essential for the cost-effective management of those structures, with assessments being improved by consideration of the results of testing and/or monitoring of the structures. In the last 15-20 years there has been a rapid growth of tests and testing and in 1997 the Concrete Bridge Development Group (CBDG) formed a task group to prepare a technical guide on 'Testing and Monitoring the Durability of Concrete Structures' under the chairmanship of Mr J J Darby. The task group addressed the need for improvements to the specification, in carrying out of testing and in the interpretation of results. The guide [5] is now being prepared for publication.

The structure maintenance process is improved by undertaking testing and/or monitoring in three phases [5], these being:-

- Condition monitoring phase in which irregularities are detected
- Diagnosis phase in which causes of those irregularities are determined
- Solution development phase in which the best course of action is selected

The technical guide [5] stresses the need for a desk study of available documents and for a preliminary site inspection prior to planning the testing and/or monitoring programmes. Detailed information is provided in the guide on site and laboratory tests both for determining physical structure and response and for determining corrosion activity and probability of corrosion.

The benefits of a staged approach to investigations are demonstrated in the case studies undertaken by PB Kennedy & Donkin Limited and outlined below.

5.   Post-tensioned bridge

Leakage from the ends of the cellular deck of a post-tensioned bridge was noted during an inspection. Evidence suggested that leakage had been occurring for a considerable time with drain holes in the structure becoming blocked and ponding of water occurring at diaphragms. Ducted post-tensioning cables in cell webs and in bottom slabs could have been adversely affected by water and water borne-salts. A staged programme of non-destructive testing was carried out to determine the condition of the superstructure and the substructure.

This single span bridge carries a farm access road over a motorway. The deck comprises four hollow sections of post-tensioned, pre-stressed concrete, with pre-cast concrete deck slabs. Each trough section has internal bracing in the form of web stiffeners at approximately 4.9m centres. The deck is supported on reinforced concrete abutments. Chamber 1 carries a single 50mm PVC water supply and 2 No. PVC communication ducts, whilst chambers 2, 3 and 4 carry brine mains of 24", 20" and 16" diameter, respectively, these mains being supported at each stiffener.

Stage 1 of the investigations comprised a Principal Inspection with very limited testing to abutments and to the deck soffit adjacent to the deck ends. Results of the inspection and testing suggested that there was significant cause for concern, and it was recommended that physical examination of reinforcement should be considered together with further corrosion and materials testing.

More extensive corrosion sampling and testing was undertaken to the soffit of the deck (Stage 2 Investigations) and subsequently to the abutments (Stage 3 Investigations). Testing was generally carried out on a 500mm grid basis, reducing to a 100mm grid in areas of highly negative half-cell potentials. Half-cell potential measurements and concrete resistivity measurements were considered in order to locate 'Hot Spots' where corrosion of reinforcement would be most likely. At these locations concrete cover was removed to determine the extent and severity of any corrosion to reinforcement. Concrete dust samples were obtained by percussion drilling at incremental depths and were analysed for chloride ion concentration.

For the superstructure, the sections of the test areas with half-cell potentials more negative than -350mV (CSE) were confined to the deck soffit beneath Chambers 2, 3 and 4 - the chambers carrying the brine mains. Resistivity measurements were taken in the small grid test areas with the majority of the measurements being less than 10 kohm cm, indicating that an active state of corrosion could be expected.

Chloride ion concentrations were at very high levels throughout the bottom slabs to Chambers 2, 3 and 4 'upslope' of certain of the diaphragms. Chloride contamination of the soffit probably resulted from road traffic spray but the high concentrations, up to 20 times the accepted critical value, towards the top of the bottom slabs probably resulted from the ingress of waterborne salts from the chambers. Exposure of reinforcement showed general and pitting corrosion to be present.

The areas of the abutments with highly negative half-cell potentials were for the most part beneath the ends of Chambers 2, 3 and 4 - that is the chambers carrying the brine mains. Resistivity measurements undertaken in the small grid test areas were all less than 10 kohm cm and thus there was a high probability of active corrosion of reinforcement. Chloride ion concentrations at the level of reinforcement were up to 12 times the critical value and thus chloride induced pitting corrosion was to be expected. Exposure of reinforcement showed general and pitting corrosion to be present, with up to 25 per cent loss of cross-sectional area of reinforcing bars.

Reports on the Stage 2 and Stage 3 Investigations recommended that the condition of the post-tensioning systems be determined at particular locations.

The investigation of the condition of the post-tensioning system in the deck (Stage 4 Investigations) showed slight/medium surface deterioration of post-tensioning ducts. No evidence of significant deterioration to prestressing tendons was observed in the 53 exposures undertaken, however in two of these exposures small voids were found in ducts.

Chemical analysis of grout samples showed that some 25 per cent of the samples had chloride ion concentrations above the level at which it is considered that chloride induced corrosion may be initiated. All the locations that had high chloride ion concentrations were in the slabs below Chambers 2 and 3, the chambers considered to be carrying the mains most actively used for the transportation of brine. The reinforcement in the top of the slab at certain locations was found to be suffering from pitting corrosion, with a loss of cross sectional area of up to 40 per cent. Concrete towards the top of the slabs beneath Chambers 2 and 3 contained pitting corrosion products and the concrete was often very damp/wet. The petrographic examination of cores suggested that brine had been present over a long period of time.

Concrete samples from cores tested for alkali content showed that the mean alkali content was 7.5 kg/m^3 , which is in excess of the currently recommended upper limit of alkali content (3 kg/m^3) for the avoidance of alkali silica reaction. Given the presence of moisture and potentially reactive fine and coarse aggregates, alkali silica reaction could be possible in the future.

It was considered that refurbishment of the bridge probably would be feasible but it would remain a sensitive structure, requiring careful maintenance and regular inspection, with the risk that the prestressing tendons would corrode in a potentially aggressive environment.

It was recommended that if the bridge deck was to be retained then:-

(A) the brine in the mains in Chambers 2 and 3 had to be diverted;

(B) the decommissioned mains needed to be removed, the top surfaces of the bottom slabs should be subjected to corrosion testing and sampling and the condition of reinforcement and post-tensioning ducts should be determined; and

(C) if the results of the investigations in (B) above indicated that refurbishment would be feasible then repair options needed to be considered.

6. Reservoir spillway bridge

Thruscross reservoir in North Yorkshire is impounded behind a large mass concrete gravity dam which carries a five span reinforced concrete road bridge across the spillway crest. The dam was constructed in 1966 and is the newest of several dams in the Washburn Valley which supplies water for Leeds and the surrounding areas. The dam is 182m long at crest level and the spillway bridge is 36m above the valley floor.

During a regular inspection by Yorkshire Water staff it was noticed that pieces of concrete were spalling from the underside of the deck and rolling down the spillway. PB Kennedy & Donkin was appointed by Yorkshire Water to investigate the problem and to assess the capacity of the bridge.

An inspection of the underside of the bridge deck with limited corrosion sampling and testing was undertaken (Stage I Investigation) that revealed that the main longitudinal beams showed evidence of extensive deterioration [6].

The cause of the deterioration appeared to be due to two main factors:

• The exposed position of the bridge.
• The nature of the deck construction.

The deck was designed and built as five simply supported spans with many elements being precast and assembled on site. In addition there was no waterproofing layer. As a result of these factors many leakage paths had developed in the deck through which road salts could penetrate and run down the faces of the main beams.

It was also found that there was deterioration to the pier supports and abutment faces and it was obvious that some further investigation of both the pier and the deck would be required to determine the extent and severity of the deterioration (Stage 2 Investigations).

Results of the Stage 2 corrosion testing and laboratory testing of concrete samples indicated extensive corrosion of reinforcement in 17 of the 25 beams, with chloride ion concentrations being at very high levels throughout the greater part of the beam cross sections. Exposures of reinforcement confirmed extensive pitting corrosion with considerable loss of cross-sectional area of some bars. Diaphragms between beams and deck slabs were similarly affected. Continuing deterioration would lead to a progressive reduction in structural capacity of the beams.

Testing of the substructure showed there to be significant concrete and reinforcement deterioration due to the effect of water and de-icing salts in winter.

Petrographic examination and chemical testing of cores taken from the substructure (Stage 3 Investigations) confirmed considerable external contamination of concrete to great depths. Alkali silica reaction (ASR) was found to be already occurring within the concrete and alkali contents in the concrete were high. It was considered that electrochemical desalination or cathodic protection, both of which were under consideration as remedial measures, could increase the risk of further ASR in the concrete.

The main conclusions and recommendations based on the investigations carried out included:

- The extent and severity of deterioration to concrete and the reinforcement in the deck were of such magnitude that it would not be feasible to repair and/or strengthen the deck elements.
- Further deterioration of the sub-structure could be expected with the continuing ingress of water and water-borne salts in winter.
- Chloride/sulfate contaminated concrete in the substructure should be removed, corroded reinforcement cleaned or replaced, concrete removed made good, and bearing shelves should be waterproofed.
- An impervious layer should be provided between the roadway surfacing and the top surface of the body of the dam to prevent the further ingress of water and water-borne salts.
- Areas of the upstream face of the dam damaged by freeze-thaw action/or contaminated by chlorides/sulfates should be repaired.

7.   Conclusions

Particular attention has to be paid to testing and monitoring the durability of concrete structures in order that those structures may be managed in the most cost-effective manner. Staged investigations enable irregularities to be detected, causes of deterioration to be diagnosed and the solution to problems to be determined.

It is essential that investigations be well targeted, be carried out by well trained engineering staff and that results of investigations be interpreted by experienced engineers.

References

[1]   British Standards Institution. (1992) Guide to Durability of Buildings and Building Elements, Products and Components. BSI, London.

[2]   Organisation for Economic Co-operation and Development (1988). Durability of Concrete Road Bridges. Road Transport Research Programme. Paris.

[3]   Frostick, I. (1996) The Use of Testing in the Determination of Material Strengths for Assessment and the Prediction of Durability. Proceedings of the Concrete Bridge Development Group Annual Conference. Dunchurch, UK March.

[4]   Wallbank, E J. (1989) The Performance of Concrete Bridges. A survey of 200 Highway Bridges. Report prepared by G Maunsell and Partners for the Department of Transport, HMSO, London.

[5]   Darby, J.J., Capeling, G., George C.R., Pearson-Kirk, D., Dill, M.J. and Hammersley, G.P. (1999) Testing and Monitoring the Durability of Concrete Structures. Concrete Bridge Development Group. Technical Guide No.2 (being prepared for publication).

[6]   Craddy, M.F. and Pearson-Kirk, D. (1996) The Restoration of Thruscross Reservoir Spillway Bridge. Journal of Concrete Repair, London. April.

Tay Bridge – the real benefits of in-situ strain and pressure measurement

DUNCAN SOOMAN
BSc CEng MICE, Senior Project Manager, Railtrack Scotland

RON WARNER
BSc CEng MICE, Technical Director, Pell Frischmann Consultants Ltd

The Tay Bridge carries the main Edinburgh to Aberdeen line over the River Tay estuary at Dundee. (A general arrangement is shown at Appendix A with cross section at Appendix B).

The bridge is a Class A listed structure and the bridge 'proper' comprises 84 spans between Wormit and the former Dundee Esplanade Station. There is some debate as to whether or not the rising approach viaduct of some 40 brick arches at the Dundee end is also included. The bridge is famour for the disaster in 1879 when the collapse of the original structure during a severe storm led to the loss of 75 lives and to this day remains the worst rail bridge disaster in UK history.

The history of the event is well documented elsewhere but the significance to the civil engineering profession is that the wind loading used prior to was $10lb/ft^2$ and hither to was $56lb/ft^2$ which, in effect, similar to the wind loadings predicted from modern design codes today.

After the disaster, the bridge was reconstructed using salvaged trusses from the original bridge as well as new material. The 'metal spans' are wrought iron trusses with steel trough deck. This is believed to be the first significant use of structural steel in bridge construction in the UK.

The structure can best be described as "light" and "lively" under load and although it had been maintained reasonably by British Rail since nationalisation of the rail industry in 1947 it passed to Railtrack in 1994 in a rather tired state.

The upshot of this was that HSE(HMRI) issued a notice to Railtrack to:-

- Undertake a condition survey
- Undertake a structural assessment

Railtrack engaged Pell Frischmann who had represented us in a similar situation on the Forth Bridge and as a consequence of the initial assessment work, freight trains were not permitted on the structure.

Feasibility Studies and Value Management exercises suggested that singling of the track over the bridge might best facilitate the return of freight traffic but preliminary structural and signalling design concluded that this was not an option to be progressed.

Bridge Management 4, Thomas Telford, London, 2000.

As a consequence, Railtrack engaged Pell Frischmann to carry out a robust costed option appraisal for discreet solutions which ranged from essential repairs only (maintaining the freight exclusion) to strengthening the bridge to carry full RA10 loading.

As this work progressed, the full RA10 (and RA8) options were obviously far too expensive and two schemes were chosen for full development:-

1.   Repair bridge for structural and H&S aspects to current traffic levels (ie excluding freight).
2.   As above but include freight at 20mph comprising loco plus 22½ tonne axle wagons.

Even with these load options, the manipulation of various bridge codes was suggesting considerable volume of steel plate strengthening were required and it was considered prudent to understand how the structure actually behaved under load to assist in the design of repairs and strengthening.

To this end, Pell Frischmann engaged Strainstall to install some 70 gauges which were in place for a 4 month period during which wind speeds of up to 80mph were experienced, all types of traffic and a test train to simulate freight.

The results of the tests were very encouraging in that actual forces in the members correlated well to those predicted by mathematical modelling although not as high thus reducing the volume of steelwork to be applied and hence dead weight.

The cost of the gauging was c £70k x c £30k for analysis and project management costs which I believe had led to a saving of approx 100 tonnes of plating which in this environment is estimated to save £1 million and 4 programme months.

For a bridge of this length, the payback for this exercise is clearly exceptionally high.

Bridge and other structural design codes are by necessity conservative in approach, since they are developed in order to be applied to a wide range of structural forms, with materials of varying grades and properties.

Several areas lead to conservatism in the design and assessment of present day structures and in cost terms, the level of conservatism might not be considered unreasonable. For example, when considering the differences in mass per unit length of various structural sections it is noted that typically there are bandings of between 10 and 20% between consecutive serial weights. The effect of an increase in cost of a new structure due to this conservatism is generally not significant since much of the cost of a new structure is related to preliminary costs and the labour content, rather than raw material content.

When considering older structures which may require strengthening as a result of the findings of an assessment, it is important to focus on the areas where conservatism may arise and lead to the apparent need for strengthening which may not be required if a more precise assessment is carried out.

In the case of the Tay Bridge assessment several areas of concern were noted which could have been jeopardised by a lack of precision.

Additionally, during inspection of the bridge it was noted that many members oscillated at even low wind speeds and that some members displayed a tendency to buckle during the passage of trains. In all of these cases, the members concerned are flat sided members probably considered as tension only members by the original designers.

Calculating the windloading in accordance with the Highway Agency Bridge Directive, BD37/88, and treating the flat sided members as simply supported, it was possible to calculate the stresses in the members using the simple Engineers Theory of Bending. However, this calculation suggested that under full wind loading the members would fail by a large margin.

Two options were available. The first was to develop a strengthening proposal based on the assumptions made and noted above. However such a repair would need to have been applied to the end two diagonal, flat sided members at each end of each affected truss, that is 16 members per span with potentially some 61 spans affected.

The second option was to undertake some research into the behaviour of the members which would be supported by in-situ load testing. It was considered that the assumptions made in the preliminary analysis were extremely conservative but in the absence of in-situ testing, it was not possible to reliably predict the actual mode by which the members would resist the applied wind load.

Pell Frischmann Consultants Ltd (PFCL) advised Railtrack that there would be some benefit in carrying out in-situ testing. However the recommendation carried the caveat that whilst the results may indicate that the members would probably be satisfactory, there was still a possibility that strengthening may be required to resist wind loading.

Having considered the above, Railtrack instructed PFCL to undertake some preliminary calculations based on theories previously postulated by PFCL regarding the possible behaviour of the members, followed by in-situ testing which could be used to confirm or otherwise the method by which the members resisted wind loads.

The following theories regarding wind loading were postulated:-

1) The wind loading was calculated in accordance with BD37/88 was excessively conservative.
2) No allowance is made in the calculation for direction effects.
3) The allowance in the code for shielding is over conservative.
4) The flat sided members do not act as simply supported but as built-in members.
5) By virtue of the large lateral displacements, much of the wind loading is carried by membrane action.

It was considered that the truth rests somewhere between all of these theories.

As an example of how these theories affect the stresses within the members; the maximum stresses calculated using the Engineers Theory of Bending are of the order of $200N/mm^2$ for a simply supported flat sided member. The deflection at midspan is calculated to be around 200mm using the same theory. If the member is assumed to be encastre at each end, the stresses at midspan are approximately one third of those for an assumed simply supported member.

If adequate tension restraint can be provided at the ends of the member and a lateral deflection of around 50mm occurs, then the tensile (membrane stress) is of the order of one tenth of the stress for an assumed simply supported member. Based on these scoping calculations it was decided that sufficient measuring devices should be installed to enable the following to be determined.

1)   Surface pressure for given wind speeds and direction on the windward member.

        Equipment required.   Pressure cells
                                             Anemometers
                                             Wind direction meters

2)   Surface pressure for internal members for given wind speeds and direction.

        Equipment required.   As above.

3)   Stresses and strains in the members for estimating the proportion of load carried by membrane action and bending.

        Equipment required.   Strain gauges at ends and midspan on both windward and leeward faces.

The diagram at Appendix C shows the location of the various sensors.

By recording the strain guage data from the leeward and windward faces, it was possible to determine the stress gradient across the section caused by bending effects, and the mean stress in the section caused by membrane action.

By recording strain gauge data from adjacent to the ends of the members and at mid span, it was possible to again differentiate membrane and bending effects and by comparing the bending effects at the ends and at midspan, determine the degree of end fixity.

From these figures we were able to determine how the wind pressure at low speeds generated mainly bending in the member, but as the speed increased and therefore the deflection increased, a significant proportion of the wind loading was resisted by membrane action.

What was also apparent from this exercise was that for individual members the proportions resisted by bending and membrane action were different and that the wind speed at which membrane action dominated also varied. It was concluded after trial calculations in which various amounts of lack of fit were tried, that the transition point between bending and membrane action was a function of the member lack of fit.

Whilst the strain guaging was set up it was recognised that a large proportion of the cost of this investigation was associated with "preliminaries" and other items not connected with the site activities and therefore for a small increase in expenditure, a significant number of additional guages could be installed.

Additional gauges were therefore installed, which could be used to measure and monitor strains in key members, due to the passage of trains. If the axle loads and configurations were known and the position of the trains relative to the positions of the members, comparisons could be made of the calculated strains to the measured strains.

The diagram at Appendix A also shows the location of these gauges.

The description given previously highlights that a number of spans were lifted as single spans and made continuous with adjacent spans after being positioned, thus dead loads were supported by single spans simply supported whereas subsequent loads, eg superimposed dead loads and live loads were supported by two span continuous structures. Any lack of fit of the members providing continuity would affect the behaviour of the structure. Similarly, due to the level of redundancy in the trusses, changes in the section properties of the main members would affect the distribution of loads throughout the entire truss.

Of particular concern in the analysis was the contribution of the ballasted trough decking to the stiffness of the troughs. For the purpose of the analysis, the troughing was modelled as a thin finite element deck on top of a space frame representing the 4 trusses and the cross bracing. Changes in the stiffness of the elements representing the troughing brought about changes in the apparent stiffness of the entire system.

Similarly because of the depth of the deep, low girders (16'-0" ie approximately 5.0m) even the continuity of light troughing and the rails across expansion joints coupled with bearing friction, could provide a limited fixed end moment.

An initial comparison between the computer model and the actual strain guage outputs using a real train of known weight and axle spacing, suggested that the computer model was significantly overestimating the forces in the members where their primary function was to resist bending, however, where the function of the member was primarily to resist shear, the agreement was much better. The conclusion that must be reached from this observation was that the boundary conditions used in the computer model did not accurately reflect the actual conditions.

The boundary conditions adopted in the computer analysis assumed frictionless bearings at the free ends of the trusses, under the bottom chords and no restraint at the upper levels of the top chords, deck and track.

By introducing spring restraints in the longitudinal direction at bottom and top chord levels and varying the stiffness of the springs (ie the K-value) it was possible with only a low k-value (expressed in kN/mm), to provide boundary conditions in the computer model which resulted in a very close. The values used would not have been unreasonable given the condition of the bearings etc.

Conclusions

The use of real time monitoring of wind speed, pressure and direction coupled with real time strain gauge data collection enabled informed decisions to be made regarding the extent of any strengthening which might have been required to resist local wind effects. The exercise has demonstrated that for this particular structure type, location and orientation conventional codified methods for calculating wind pressure and shielding effects are conservative.

Additionally the use of strain gauging for determining actual strains, which could then, using the results of material testing, be converted into stresses and forces in members, was an extremely valuable way of giving confidence to the computer analysis and a suitable method for enabling boundary conditions such as bearing stiffness etc to be evaluated.

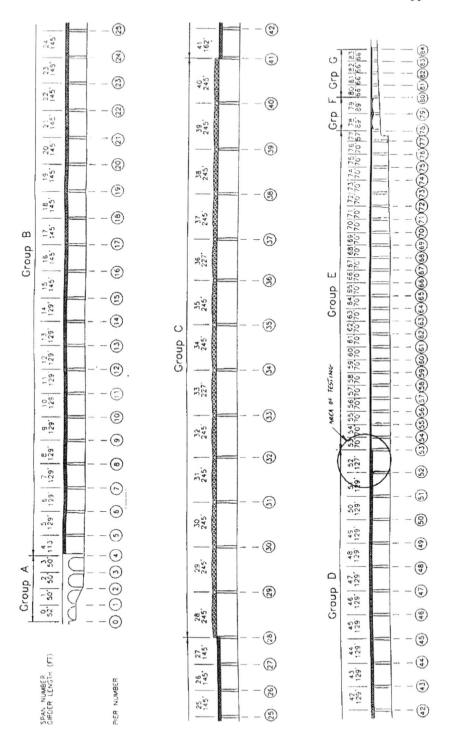

Appendix B

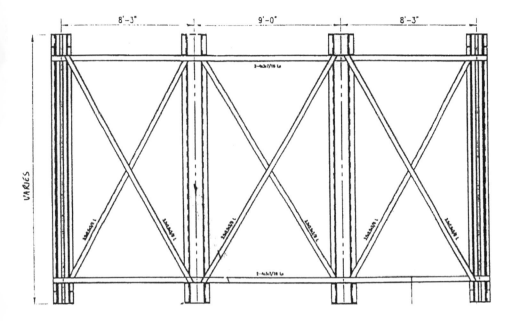

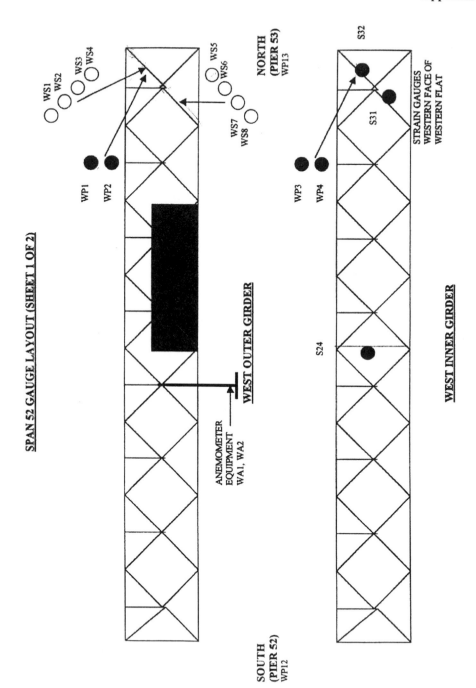

BRIME – chloride ingress and bridge management

A. BLANKVOLL, C. K. LARSEN , I. MARKEY, F. FLUGE
Norwegian Public Roads Administration (NPRA)
A. RAHARINAIVO, Laboratoire Central des Ponts et Chaussées (LCPC)
L. BEVC, M. CAPUDER, I. PERUŠ
Slovenian National Building and Civil Engineering Institute (ZAG)

INTRODUCTION

This paper describes one of eight work packages of the European funded project BRIME (Bridge Management in Europe). The project will develop a framework for a bridge management system for the European road network that will enable the bridge stock to be managed on a rational basis. The paper presents the results of the work package dealing with modelling chloride ingress and monitoring corrosion within a bridge management system. Further details concerning the project BRIME are given elsewhere in these proceedings and on the TRL web-site (http://www.trl.co.uk/brime/index.htm).

Prediction of deterioration is an important aspect of bridge management for estimation of remaining service life and planning future maintenance tasks. The objective of the work package is to consolidate and improve existing knowledge concerning the modelling and surveillance of chloride penetration in concrete. Chloride ions are considered the primary cause of corrosion in concrete bridges. The results of this work package will help public authorities establish investigative procedures to monitor the danger of and predict corrosion of their concrete structures. As such it will be an important tool in:

-   increasing the durability of new concrete structures by allowing the identification of and ranking in order of importance the predominate factors affecting corrosion;
-   deciding the optimal time to carry out preventative maintenance or repair;
-   assisting in long term budget planning.

The work package has the following three main tasks:

1. Creation of a databank of condition parameters for several concrete bridges including local exposure conditions (micro-climates). Extra investigations are performed to confirm/confute the predictions made by the models selected in Task 2.

2. Selection, use and assessment of several chloride ingress models. The models investigated are: Fick's 2nd law (used as reference model); Selmer - Poulsen model (LightCon model) with improvements by Mejlbro (Hetek model). The use of neural network models is also evaluated.

3. Investigate the requirements of a bridge management system that incorporates prediction models, condition surveys and monitoring. Assessment of residual service life and a probabilistic approach is also addressed.

The project finishes at the end of 1999 and the work presented here is fully described in [1]. As the project is not fully completed, the results presented here reflect the state of the work at the time of writing and not the final conclusions of the project.

CONDITION SURVEY DATA - DATABANK

Data from Slovenia, Norway and France were collected and analysed. Even though chloride analysis has been performed on a large proportion of all concrete bridges, only a limited number of structures could be retained for further analysis. This was done to limit the number of calculations but also because there is only a limited number of bridges where there is a complete set of data. The project was also primarily interested in structures with good chloride profiles (accurate and measurements at several depths), taken from several locations on the structure and taken at several ages at the same locations.

The Slovenian bridges are (affected by de-icing salts): Ivanje Selo, Slatina, Škedenj2, Preloge and Šepina bridge. The Norwegian bridges are coastal bridges: Gimsøystraumen (bridge chosen for the method of inverse cores), Hadsel and Sandhornøya bridge. The French bridge is: A11 PS12-10 (affected by de-icing salts).

CHLORIDE INGRESS MODELS

Chloride penetration is mainly due to a combination of chemical and physical processes. The most important processes are:

–   Diffusion; a process due to a gradient of chloride concentration in the concrete. Gradient means that the chloride concentration is higher at the concrete surface than in its core. As chloride is dissolved in water, diffusion process occurs only in pore solution inside concrete,
–   Capillary suction of chloride contaminated water; a process that takes place in empty or partly filled concrete pores. It means that water (moisture) content and concrete porosity are the main parameters that influence capillary suction.

Some chloride binding reactions occur between cement components (chloroaluminates, etc.) and chloride ions. These reactions are either physical adsorption, chemical reaction, or a combination of the two. Chloride binding is strongly influenced by climatic conditions. All the prediction models, which were applied in BRIME project, are based on the diffusion process but they include several supplementary assumptions.

–   Fick's law describes a pure diffusion process. Any diffusion law is valid only in concrete which is permanently saturated with water. It means that it is not valid in the concrete surface layer which sometimes can be dry. So, climatic conditions and concrete porosity determine how thick this concrete layer is where the diffusion law cannot apply.
It should be noticed that the diffusion coefficient is determined for a given substance (chloride ion, etc.) entering a given material. If this material changes, for example, after ageing, this coefficient also changes.

- LightCon and Hetek models are based on the diffusion process. However, boundary conditions that are constant in Fick's law may be time dependent, e.g. chloride content on concrete surface. Concrete porosity and cement type are important parameters in this model.
- Vesikari model is based on a feature of diffusion law, which states that a relationship exists between times t and depths L, for which chloride content has a given value $(t = K.L^2)$. According to this model, factor K depends on concrete water-cement ratio and on the environment. This model can be used in the design phase for concrete bridge decks, but not with condition survey data.
- CAE-HNN. In this model, determination of the whole chloride profile at a certain location is based on a set of measured data with similar features using neural networks. The ingress of the chlorides is based on diffusion. For this reason a substitute diffusion coefficient is calculated between the measured points. As there is currently not enough data available of chloride profiles at the same locations at different times, Fick's law is used to make time prediction.

Extensive investigations on chloride ingress in concrete form the basis for the research on chloride induced rebar corrosion. In this project, observations are taken from real structures, from field exposed test specimens and from laboratory tests. The records comprise readings on structures and test specimens made of different concrete grades, and subjected to different environmental conditions and exposure times.

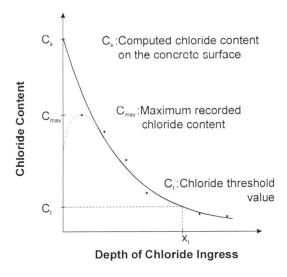

Figure 1. Simplified model for chloride ingress in concrete

The chloride profile, figure 1, is a simple illustration of some of the essential parameters when modelling chloride ingress in concrete and hence, also the service life of the structure. As a simplification the service life is taken as the initiation period and the propagation period is conservatively neglected.

Fick's second law is adequately described in the literature and is not presented here. The LightCon [15] and Hetek [14] models are similar and only the Hetek model and the neural networks are briefly presented here.

In order to estimate the future chloride ingress into concrete the Hetek model assumes that at least the following information is available:
- The age t_{in} of the concrete structure at the time of inspection
- The age t_{ex} of the concrete structure at the time of first chloride exposure
- The composition of the concrete, i.e. the type of binder, and the w/c-ratio
- The environment of the concrete, i.e. atmospheric, splash or submerged (ATM, SPL and SUB),
- The chloride profile of the concrete obtained at an inspection time $t_{in} \gg t_{ex}$, e.g. by the chloride diffusion coefficient and the chloride content of the concrete surface.

When it is not possible to obtain reliable information from the specification of the concrete structure, the inspection must be supplied with a thin section analysis of the concrete in question. Testing the concrete by NT Build 443 [17] and/or by the »method of inverse cores« [1,2] may strengthen the estimation of the future chloride ingress into the concrete.

The HETEK-model for future chloride ingress into concrete is based on the following assumptions:
- Chloride C in concrete is defined as the »total, acid soluble chloride«
- Transport of chloride in concrete takes place by diffusion. There is an equilibrium of the mass of ingress of (free) chloride into each element of the concrete, the accumulation of (free and bound) chloride in the element and an ongoing diffusion of (free) chloride in the element towards a neighbour element, and so on
- The flow of chloride F is proportional to the gradient of chloride $\dfrac{\partial C}{\partial x}$. The factor of proportionality is the achieved chloride diffusion coefficient D_a
- The achieved chloride diffusion coefficient D_a depends on time, the composition and environment of the concrete
- The boundary condition C_s is time-dependent of time t, and the composition and environment of the concrete
- The initial chloride content of the concrete C_i (per unit element of the concrete) is uniformly distributed at time t_{ex}
- The relations used for the determinative parameters with respect to the environment (ATM, SPL and SUB), the time and the composition of the concrete are documented at the Träslövsläge Marine Exposure Station on the west coast of Sweden (south of Gothenburg), cf. [14].

A hybrid neural network-like approach (CAE - HNN) was developed by ZAG and involves an empirical treatment of the phenomena. This is very suitable for problems where models are based on the experimental data. It was shown elsewhere [7], that such an approach corresponds to the use of the intelligent systems.

We assume, that the complete phenomenon, in our case in-depth chloride ion penetration, is characterised by a sample of the measurements on N testing specimens that are described by a finite set of so called model vectors:

$$\{\mathbf{X}_1, \mathbf{X}_2, ..., \mathbf{X}_N\} \qquad \text{... /Eq. 1/.}$$

Such a finite set of model vectors will be called a database in the subsequent text.

In formulating the modeler of the phenomenon $Cl^- = Cl^-(x, h, o, wt, c)$ we further assume that one particular observation of a phenomenon can be described by a number of variables, which are treated as components of a vector:

$$\mathbf{X} = \{x, h, o, wt, c, Cl^-\} \qquad \text{... /Eq. 2/,}$$

where x is depth, h height above sea level, o orientation, wt wetting, c variable which describes concrete cracks and Cl^- chloride ion concentration at depth x.

Vector $\mathbf{X}$ can be composed of two truncated vectors:

$$\mathbf{P} = \{x, h, o, wt, c; \#\} \text{ and } \mathbf{R} = \{\#; Cl^-\} \qquad \text{... /Eq. 3/,}$$

where $\#$ denotes the missing portion. Vector $\mathbf{P}$ is complementary to vector $\mathbf{R}$ and therefore their concatenation yields the complete data vector $\mathbf{X}$. The problem now is how an unknown complementary vector $\mathbf{R}$ can be estimated from a given truncated vector $\mathbf{P}$ and sample vectors $\{\mathbf{X}_1, \mathbf{X}_2, ..., \mathbf{X}_N\}$. By using the conditional probability function the optimal estimator for the given problem can be expressed as [7-9]:

$$r_k = \sum_{n=1}^{N} A_k \cdot r_{nk} \qquad \text{... /Eq. 4/}$$

where

$$A_k = \frac{a_n}{\sum_{j=1}^{N} a_j} \quad \text{and} \quad a_n = \exp\left[\frac{-\sum_{i=1}^{L}(p_i - p_{ni})^2}{2w^2}\right] \qquad \text{... /Eq. 5/.}$$

r_k is the k-th output variable (e.g. Cl^-; k is equal to 1 in a given problem), r_{nk} is the same output variable corresponding to the n-th model vector in the data base, N is the number of model vectors in the data base, p_{ni} is the i-th input variable of the n-th model vector in the data base (e.g. x, h, o, wt, c), p_i is the i-th input variable corresponding to the model vector under consideration, and L is the number of input variables. w describes the average distance between the specimens in the sample space and is called smoothing parameter.

A general application of the method does not include any prior information about the phenomenon. Because in some cases there is still lack of data, *a priori* information is needed to better fit a particular phenomenon. By a relatively simple improvement [10], the method can be effectively used for the modelling of many problems in civil engineering. Furthermore, CAE(conditional average estimator) stems from a probabilistic approach and phenomena are not treated just deterministic.

For the application of the CAE-HNN a database is needed. It consists of model vectors, what can be presented in general case in matrix form as:

$\mathbf{mv}_1$ =	p_{11}	p_{12}	...	p_{1L}	r_1
$\mathbf{mv}_2$ =	p_{21}	p_{22}	...	p_{2L}	r_2
...	...				...
...	...				...
$\mathbf{mv}_N$ =	p_{N1}	p_{N2}	...	p_{NL}	r_N

... /Eq. 6/.

The main task in the first step is therefore to represent the measured data and, if necessary, *a priori* knowledge about the phenomenon in vector or matrix form. Finally, in the second step the choice of appropriate value of smoothing parameter is needed. The parametric study has shown that the appropriate value for modelling chloride ion penetration into concrete is $w = 0.15$. Due to the lack of experimental data on time dependence of chloride ion penetration, Fick's 2nd law is used for time prediction.

RESULTS

The models are to calculate the time to corrosion initiation for three different cover depths and three different threshold values (critical chloride content). These are: cover depths: 25, 30 and 50 mm and threshold values: 0.4, 0.7 and 1.0 % of cement weight. The results for one location on one bridge are given on the next page. It should be noted that the Hetek model is the only model of the four that can take into account two or more chloride profiles. All the other models have based their calculations of the last profile (7.5 years). For the nine bridges, a total of 76 locations will be calculated with all the models. It is too early to say which models give the best answers but the time, data required and complexity involved in using the models will also be taken into account in their evaluation.

PROBABILISTIC APPROACH TO SERVICE LIFE

Predictions of chloride ingress at one point of a structure are of little value and a more global approach is needed. The approach described here uses data obtained during the OFU Gimsøystraumen Bridge Repair project[3,4] and the Durable Concrete Structures project [6].

The chloride load for Norwegian bridges is mainly a function of the height above sea level. [6] shows maximum measured chloride content in the concrete, representing 1200 chloride profiles sampled from 30 bridges, all more than 15 years old. The recordings are obtained at different heights above the sea level, from all sides of the cross-sections and from bridges

Sandhornøya, NP A3, East

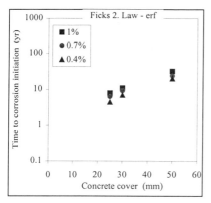

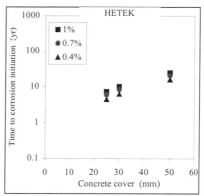

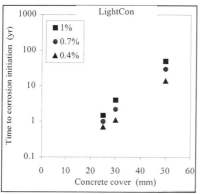

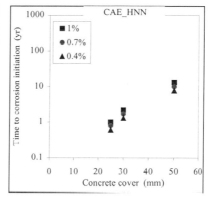

Measured chloride profile

Depth (mm)	7.5	22.5	37.5	52.5	67.5			
Cl (% conc.)	0.43	0.032	0.004	0	0.006			3.5 years
Depth (mm)	5	15	22.5	37.5	52.5			
Cl (% conc.)	0.462	0.512	0.106	0.011	0			7.5 years

Cover depth (mm)	Cl_crit (% cement)	Time to corrosion initiation (years)			
		Ficks 2.	HETEK	LightCon	CAE HNN
25	0.4	4.5	4.5	0.7	0.6
25	0.7	6.4	6	1	0.8
25	1	8	7.4	1.5	1
30	0.4	7	6.3	1.1	1.3
30	0.7	9.3	8.3	2.2	1.7
30	1	11	10.2	4	2.2
50	0.4	20	15.9	14	7.8
50	0.7	24	20.5	30	10.1
50	1	32	24.8	50	13.4
Calculated D (mm²/yr)		17	22.9	10	-
Calculated C_s (% concr.)		1.22	1.15	1.22	-

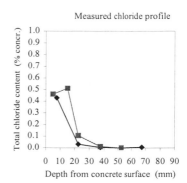

Measured chloride profile

exposed to different environmental conditions. On the basis of these findings the exposure conditions, represented by the maximum measured chloride content near the concrete surface, have been classified in four exposure zones, mainly governed by the height above sea level: 0-3m; 3-12m; 12-24m and above 24m. This kind of classification based on in-situ data is a very important element in future durability design standards.

The inspections on Gimsøystraumen bridge, performed after 12 years in service also consisted of measurements of concrete cover, chloride analyses and visual observations in more than 200 locations distributed over the 21 investigated cross-sections along one section - a 126 meter long post-tensioned box-girder. The height above sea level of the bridge deck in this section varies from 10.4 to 18.4 meters.

The design concrete strength was grade C40 for the superstructure. Cement content was 375 kg/m³ OPC and no silica fume was used. Obtained strength varied between 36.5 and 54.0 MPa with a mean of 43.2 MPa. Concrete of grade C40 normally corresponds to a bulk diffusion coefficient of $12\text{-}15\cdot10^{-12}$ m²/s when tested according to NT Build 443 after 28 days hardening.

The concrete cover was specified at minimum 30 mm. The average concrete cover was determined at 29 mm with a standard deviation of 5.5 mm. This implies that approximately 50% of the rebars have concrete cover less than the specified 30 mm and 10% less than 22 mm. The statistical distribution of the concrete cover, based on more than 3500 independent readings, is shown in figure 3.

The in-situ diffusion coefficient for the bridge section was computed based on chloride analysis of samples of concrete powder drilled from 4 holes at each location. The average in situ diffusion coefficient after 12 years of exposure was determined to $1.1\cdot10^{-12}$ m²/s with a standard deviation of $0.25\cdot10^{-12}$ m²/s.

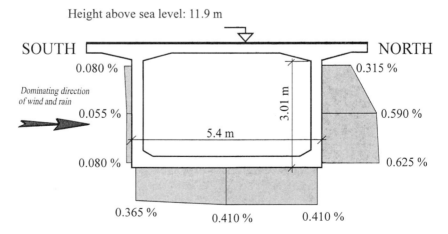

Fig. 2.   Chloride content on the concrete surface C_s, computed from measurements on the cross-section 11.9 m above sea level, Gimsøystraumen bridge.

The bulk diffusion coefficient for a 12 year old "chloride free" concrete sample, drilled from the middle of the superstructure and tested according to NT Build 443, was determined to 7.0 $\cdot 10^{-12}$ m^2/s. Taking the ages of concrete hardening into consideration this value corresponds roughly to a bulk diffusion coefficient of approximately 14 $\cdot 10^{-12}$ m^2/s after 28 days of hardening.

The maximum chloride content obtained on the cross-section located 11.9 meters above sea level varied between 0.07% Cl$^-$ and 0.38% Cl$^-$ of concrete mass on the windward and the leeward side respectively. Curve fitting of the measured values gave a maximum computed chloride content C_s on the leeward concrete surface of 0.625% Cl$^-$ of concrete mass, figure 2, and an in-situ diffusion coefficient of 1.4 $\cdot 10^{-12}$ m^2/s .

On the basis of the investigations referred to and summarised below, the chloride ingress and critical depth in the concrete after 10-12 years exposure, i.e. the age of the bridge when inspected, has been computed using the following values.

Chloride load:
- Leeward side C_s = 0.625 % Cl$^-$
- Windward side C_s = 0.010 % Cl$^-$

Material resistance:
- In situ diffusion coefficient after 12 years exposure
$$D = 1.1 \cdot 10^{-12} \text{ m}^2/\text{s}$$
- standard deviation $s = 0.25 \cdot 10^{-12}$ m^2/s

The critical depth was computed on the basis of a threshold value of 0.07% Cl$^-$ of concrete mass.

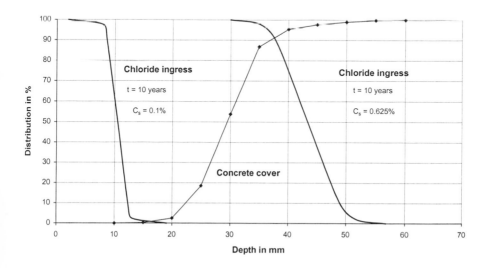

Fig. 3.   Statistical distribution of concrete cover and critical depth

The most probable critical depths after 10 years of exposure are shown in figure 3 for the windward and the leeward side of the structure respectively. Figure 3 shows clearly that the probability for rebar corrosion on the windward side is negligible. However, on the leeward side, the probability for depassivation and rebar corrosion exceeds 90 percent. These results concur with the visual observations of no signs of corrosion on the windward side and active corrosion on the leeward side.

MONITORING AND BRIDGE MANAGEMENT

Today's chloride ingress models are too precarious to automatically initiate a maintenance repair. An understanding of the corrosion process, the limitations of the models and the uncertainty surrounding the measured data are all necessary before any reliable decision can be made. As such, only experienced engineers or corrosion experts should be allowed to act upon the results generated by a chloride ingress model. In addition, engineering judgement will still play an important role in assessing the extent of the damage, the associated maintenance/repair cost and in combining different maintenance tasks from different elements/bridges in order to optimise the limited resources available.

It is important to understand the limitations and possibilities a chloride ingress model can have in a bridge management system (BMS). The model cannot predict how much reinforcement will corrode every year nor can it predict with any certainty when corrosion will initiate. However, it can predict when there will be a certain danger of corrosion initiation. As such, chloride ingress models should be used for assessing possible future maintenance, but not for assessing structural capacity or deterioration.

To fully exploit the possibilities of chloride ingress models, inspection routines should be modified. This will allow reference zones to be established, improve the measured data and increase the reliability of the models. In addition, data collection from all bridges needs to be systematised to facilitate future exploitation. This will greatly benefit neural network models but also allow for new models to be developed.

As a final note, durability surveillance must be based upon and compliment the existing inspection programme of the bridge stock.

REFERENCES

[1]   Deliverable D8: Bridge management and condition monitoring. Project: BRIME (Bridge Management in Europe) RO-97-SC.2220. European Commission under the Transport RTD Programme of the 4th Framework Programme. 1999 (to be published).
[2]   Poulsen E., Frederiksen J. M. The method of inverse cores. Private communication 1998. (to be published).
[3]   Proceedings. Int. Conf. Repair of concrete structures: From theory to practice in a marine environment. Ed. A. Blankvoll. Svolvær, Norway, May 1997.
[4]   OFU Gimsøystraumen bru. "Climatic loads and condition assessment, final report" (in Norwegian). Publication no. 85, Norwegian Public Roads Administration (Statens vegvesen), Oslo, Norway. 248 p. 1998.
[5]   Bestandige betongkonstruksjoner (Durable Concrete Structures). Instrumentert tilstandsovervåking - Ett skritt videre (Project report "Condition monitoring - one step further" (in Norwegian). NBI, Oslo, Report no. 4.1. 1999.

[6] Something Concrete about Durability. Fluge F, Jakobsen B. 5th International Symposium on Utilization of High Strength / High Performance Concrete, Norway, 20 – 24 June 1999.

[7] I. Grabec and W. Sachse, Synergetics of Measurement, Prediction and Control, Springer-Verlag, Heidelberg, 1997.

[8] I. Grabec, "Self-organization of neurons described by the maximum entropy principle", Biol. cybern. 63 (1990) 403-409.

[9] I. Peruš, P. Fajfar and I. Grabec, "Prediction of the seismic capacity of RC structural walls by non-parametric multidimensional regression", Earthquake Engng & Struct. Dyn. 23 (1994) 1139-1155.

[10] Fajfar, P. & Peruš, I., A non-parametric approach to attenuation relations, Journal of Earthquake Engineering, 1 (1997) 319-340.

[11] Raharinaivo A., Grimaldi G. "Methodology for monitoring and forecasting the condition of a reinforced concrete structure, under corrosion". IABSE Symposium Expanding the lifespan of structures. San Francisco (USA) August 23 - 25, 1995.

[12] Baroghel-Bouny (V.), Chaussadent (T.), Raharinaivo (A.) "Experimental investigations upon binding of chloride and combined effects of moisture and chloride in cementitioud materials". RILEM International Workshop on chloride penetration into concrete, Saint-Rémy-lés-Chevreuse, Oct. 15 - 18 1995.

[13] Francy O., Bonnet S., Francois R., Perrin B., "Modeling of chloride ingress into cement-based materials due to capillary suction". Proceedings 10th International Congress of the Chemistry of Cement, H. Justnes, ed., Gothenburg, Sweden, June 1997, Vol. 4, Paper 4iv078, 8pp.

[14] J.M. Frederiksen (EDT.), L.-O. Nilsson, P. Sandberg, E. Poulsen, L. Tang, A. Andersen: »HETEK. A System for Estimation of Chloride Ingress into Concrete, Theoretical background«. Danish Road Directorate Report No. 83. 1997 Denmark.

[15] Maage M., Helland S., Carlsen J:E:: »Chloride penetration into concrete with light weight aggregates«. Report FoU Lightcon 3.6, STF22 A98755 SINTEF. Trondheim, Norge 1999.

[16] Mejlbro L.: »The complete solution to Fick's second law of diffusion with time-dependent diffusion coefficient and surface concentration«. Proceedings of CEMENTA's Workshop on Durability of Concrete in Saline Environment. Danderyd, 1996 Sweden.

[17] NT Build 443 (1995-11) Nordtest method: Concrete hardened; accelerated chloride penetration. Published by Nordtest, ISSN 0283-7153.

Time dependent stress variation of a composite two-I-girder bridge – Chidorinosawagawa Bridge

DR. M. NAGAI, Professor, Nagaoka University of Technology, Niigata, JAPAN
DR. Y.OKUI, Assoc. Professor, Saitama University, Saitama, JAPAN
T. OHTA and H.NAKAMURA, Japan Highway Public Corporation, Hokkaido, JAPAN
M.INOMOTO, K.NISHIO, K.OHGAKI, and A.YAMAMOTO, Kawasaki Heavy Industries, Ltd., Chiba, JAPAN

ABSTRACT

This paper presents an outline and innovative web stiffening design of Chidorinosawagawa Bridge, which is a four span continuous composite two-I-girder bridge, constructed on a highway route in Hokkaido, Japan. In this bridge project, jack-up and down procedures were employed at intermediate supports during erection in order to avoid cracking in the concrete slab under dead load and, as a result, to attain higher durability of the bridge. Field measurement was carried out to confirm whether the expected prestress in the concrete slab is introduced or not. It was confirmed that the expected prestress was introduced, and that both measured and calculated values show good agreement.

INTRODUCTION

In Japan, since 1980, the number of the construction of continuous composite girder bridges had been limited. The main reason to avoid this type of bridge comes from the fact that thin concrete slabs of some of the bridges, designed from 1960 to 1970, suffered severe damage due to mainly heavy truck loading. However, the main cause of the damage of the concrete slab has been clarified through extensive researches carried out by using moving wheel load testing machine [1], and countermeasures to attain high durability of concrete slabs were also recommended. One of the countermeasures is to employ the prestressed (in the transverse direction) concrete slab. In addition, the reduction of the construction cost is a key point in order to start the project and it has been reported, among many bridge types with a span length from 40 to 70 meters, that the composite two-I-girder bridge is the most economical solution [2].

From these recent backgrounds, Chidorinosawagawa Bridge was designed as a four span continuous composite two-I-girder bridge with a prestressed concrete (PC) slab, and is the first application of this bridge type to highway bridges in Japan.

When designing this kind of bridge system, we have to pay attention to time-dependent stress variation. In this bridge project, since the completion of the concrete slab, the stress in the concrete and girder have been monitored. Concretely speaking, the stress variation of the concrete slab and steel girders due to creep and shrinkage are being monitored. Furthermore, an analytical approach is applied to simulate time-dependent behavior and the results are compared with those obtained from measured data.

In this paper, first, the structural characteristics of Chidorinosawagawa Bridge incruding an innovative web stiffening design method are introduced. Second, construction sequence and employed monitoring system for time-dependent behavior are presented. Finally, analytical results accounting for time-dependent effects due to creep and shrinkage in the concrete slab are compared with the field data.

OUTLINE OF CHIDORINOSAWAGAWA BRIDGE

Chidorinosawagawa Bridge (see Photo 1) was constructed in Hokkaido Island, northern part of Japan. It is a four span continuous composite two-I-girder bridge and has a total bridge length of 194 meters (46.5+53.0+53.0+40.4 m) as shown in Fig.1(a). Figure 1(b) shows a cross section of the bridge. A total width of the bridge is 11.4 meters. The prestressed concrete slab with a thickness of 320 millimeters is supported by two I-girders only. These two girders have a depth of 2.9 meters, and are connected with small-sized cross beams arranged at a distance of 8.5 meters in the longitudinal direction.

Photo 1 Chidorinosawagawa Bridges

In this bridge design, an innovative stiffening design method for a web was employed. Figure 2 illustrates modifications of a stiffening design of the web. Figure 2(a) shows the originally designed web plate based on the conventional stiffening design used in Japan, which aims at the reduction of the steel weight. In order to reduce the fabrication cost, less number of horizontal stiffeners is preferable, and the stiffening design as shown in Fig. 2(b) is now becoming popular in Japan. In Chidorinosawasgawa Bridge, an innovative stiffening design as shown in Fig. 2(c) is employed. In Japanese Specification for Highway Bridges (JSHB) [3], the aspect ratio has been stipulated to be kept less than 1.5. However, this bridge was designed to have an aspect ratio of around 3.0. In addition, a depth-to-thickness ratio designed was 158, which is a higher value compared with the value of 118 stipulated in JSHB.

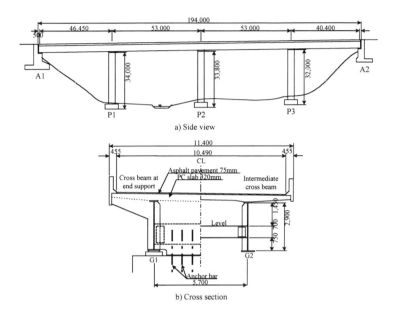

a) Side view

b) Cross section

Figure 1. Side view and cross section of Chidorinosawagawa Bridge

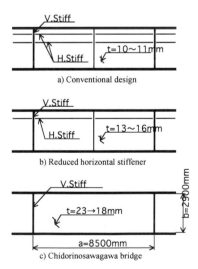

a) Conventional design

b) Reduced horizontal stiffener

c) Chidorinosawagawa bridge

Figure 2. Stiffening design of web

In general, the ultimate bending strength of plate girders are affected by interaction between the strengths of compressive flange and web. Since the compressive flange is connected with a concrete slab, the thickness of the slab is considered to be reduced. Safety against bending and shear buckling instabilities of the plate girder was examined and confirmed based on the analytical (elasto-plastic finite displacement analysis) and experimental studies [4,5].

In the design of composite girder bridges, we have to pay attention to the stress transfer from a concrete slab to steel girders due to creep and shrinkage effects. A creep coefficient of 2.0 was employed according to JSHB [3]. Since the expansive concrete was used, an ultimate shrinkage of 150 μ was employed, which is smaller than 200 μ specified in JSHB for conventional composite girder bridges.

ERECTION

A launching erection method was employed for the construction of the steel girder, then the concrete slab was cast. Figure 3 shows the casting sequence of the concrete slab. The numbers in Fig. 3(a) are the order of concrete casting. This procedure is sometimes called "piano method". In this procedure, in order to avoid concrete cracking, jack-up and down was carried out at intermediate supports. Amounts of jack-up and down and sequence are also presented in Fig. 3(b).

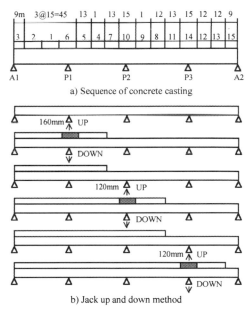

a) Sequence of concrete casting

b) Jack up and down method

Figure 3. Sequence of segmental concrete casting and jack up and down procedure

FIELD MEASUREMENT

As explained in the previous section, in order to avoid concrete cracking under dead load, prestress (compressive stress) was introduced into the concrete slab with the jack-up and down procedures during construction. An important point is to confirm whether or not the expected prestress is introduced, and identification of the time-dependent stress variation is also an important issue.

Figure 4 shows arrangement of sensors at the intermediate support P1. At both concrete and steel girder surfaces, strain gauges and thermometers were glued, and the monitoring were started just after the completion of the concrete work.

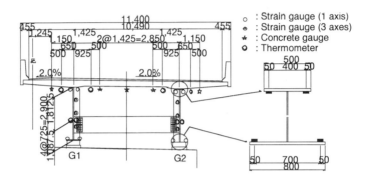

Figure 4. Arrangement of monitoring sensors

ANALYTICAL STUDY

Figure 5 shows an analytical model, in which concrete slabs, cross beams and steel girders are modeled by beam elements [6]. The slab element is connected with girder elements through rigid bar, thus the effects of creep and shrinkage of the concrete on global behavior can be taken into account.

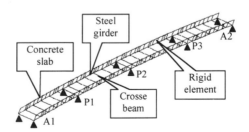

Figure 5. Structural model for time-dependent analysis due to creep and shrinkage

Table 1 shows a classification of analyses. An open circle means the effect is considered and a notation cross means it is not taken into account. For example, Analysis-A takes into account of all effects such as a casting sequence (piano method), creep and shrinkage effects under construction, and those after completion.

Table 1. Classification of analyses

	Analysis-A	Analysis-B	Analysis-C
Segmental concrete casting	◯	◯	✕
Creep and shrinkage during segmental concrete casting	◯	✕	✕
Creep and shrinkage after concrete casting	◯	◯	◯

Figure 6 shows the predicted normal stress at the upper surface of the concrete slab (Analysis-A). In the figure, "0-year" means the completion of the concrete work. In ten years, as is predicted, the compressive stress decreases, while the tensile stress increases. Figure 7 shows a comparison of the stress increments due to creep and shrinkage evaluated with different analyses. Between "Analysis-A and B", a good agreement is obtained. In the actual design stage, the erection sequence was taken into account, but creep and shrinkage effects during segmental concrete casting are neglected. From this result, validity of the assumption employed in designing was confirmed. However, since the stress distribution obtained from "Analysis-C" is largely different from that of "Analysis-A and B", a difference of the produced stress is prominent. This is mainly due to the creep effects.

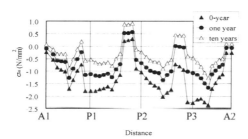

Figure 6. Distribution of stress at upper surface of concrete slab and its time variation

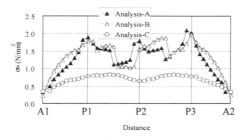

Figure 7. Comparison of distributions of stress increments with different analytical models

Figure 8 shows predicted time-dependent variation of the prestresses (compressive normal stresses) introduced with jack-up and down procedures at the intermediate supports P1, P2 and P3. Due to the creep and shrinkage effects, they gradually decrease. However, converged values are obtained in about ten years, which are still compressive stresses.

Figure 9 shows a comparison of measured and calculated values at P1. Just after casting concrete, both calculated and measured values show a good agreement with each other. From this result, it is confirmed the expected prestress was introduced. In Fig. 9, a measured value shows a variation in one year. However, after 360 days, a good agreement is obtained.

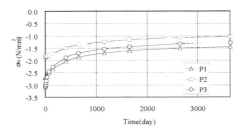

Figure 8. Time history of stresses in concrete slab at intermediate supports

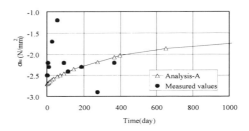

Figure 9. Comparison of measured compressive stress in concrete with analytical result

CONCLUDING REMARKS

An outline of Chidorinosawagawa Bridge including an innovative web stiffening design method was introduced. We emphasized that time-dependent stress variation in the concrete has an important factor in the design of composite bridges. The followings are results obtained from this study.

(1) It is confirmed that a designed value of prestress is introduced in the concrete.

(2) In one year monitoring, the reduction of compressive stress obtained from measured and calculated values shows a good agreement with each other.

(3) From the above result, a predicted reduction of prestress obtained from calculation will be expected.

REFERENCES

[1] S. Matsui: Technology development for bridge decks - innovations on durability and construction -, Bridge and Foundation Engineering, Vol.31, No.8, pp.84-94, 1997 (in Japanese)

[2] M.Virlogeux (translated into Japanese by Dr.Kasuga): Composite bridges, from classical to innovative designs, Bridge and Foundation Engineering, Vol.31, No.8, pp.30-47, 1997 (in Japanese)

[3] Japan Road Association: Specifications for Highway Bridges in Japan, 1997 (in Japanese)

[4] K.Ohgaki, J.Yabe, Y.Kawaguchi. T.Ohta. K.Kawashiri and M.Nagai : Experimental study on ultimate strength of plate girders with large aspect ratio and web width-to-thickness ratio, Proc. of 2nd European Conf. on Steel Structures, Praha, Czech Republic, Vol.1, pp.141-144, 1999

[5] T.Ohta, K.Kawashiri. M.Nagai, K.Ohgakji, A.Isoe and K.Sakugawa : Analytical study on stability of erection stage of composite two-I-girder bridges with reduced stiffening members, Jour. of Structural Engineering, JSCE, Vol.45A, pp.1263-1272, 1999 (in Japanese)

[6] DIANA Nonlinear Analysis User's Manual, Release 6.1

Risk based external corrosion management

MARK DUNHAM, Duffy & McGovern EXCOR Ltd., Logic House, Harfreys Industrial Estate, Great Yarmouth, Norfolk NR31 OLS.

Owners and operators are continually seeking more efficient ways of managing asset maintenance costs. In particular, costs associated with corrosion protection of assets can be high, and that investment needs to be focused where and when it's most needed. Yogi Berra once said, "If you don't know where you're going, you'll wind up somewhere else."
Our 30 years experience of inspecting and protecting onshore and offshore assets using a considerable range of coatings, insulation, and fireproofing, which has given us unique knowledge of this process and the characteristics and performance of most systems.   We recognised that the industry needed a better way to control and target fabric maintenance expenditure. We have therefor developed a risk-based computerised system (***EXCOR*** is the working title) which provides a rational approach to planning for fabric maintenance.

It is important to note that this system will work with and alongside other integrity management tools which may also deal with *internal* corrosion / erosion.

Why has the System been developed?

Coatings are normally applied to steel structures simply to prevent corrosion. Historically, the need for coatings maintenance was gauged by the extent of coating breakdown, and the <u>actual</u> requirement for corrosion protection remained largely un-quantified, lost in the mists of the original design brief. This is especially true of more mature structures. Painting is expensive, and the logic behind spending significant amounts of money on coating structures is not clear to budget managers. Thus, savings on fabric maintenance have generally been achieved by eliminating or deferring painting programmes, often subjectively without informed risk assessment. Current inspection and asset integrity analysis is either not available or not used to inform the decision making process.

Our brief was to design a relatively simple, rational, and auditable system for optimum-cost maintenance planning over the life of the asset.   To do this, it needed to establish risk from corrosion, sift and prioritise asset items requiring maintenance before the end of required life on the basis of their criticality and predicted condition. By this methodology we could decide what maintenance is required, *when it is required,* and whether it can be deferred or completely eliminated.

Bridge Management 4, Thomas Telford, London, 2000.

What is it?

We defined external corrosion management as monitoring, control and mitigation of corrosion - externally focused. When designing the system we considered significant factors affecting the following:
– The asset purpose and life
– The asset elements, their significance, location, purpose and condition
– The current protection (coating) system and condition.
– Performance of coating protection systems within various environments.
– The corrosive severity of various environments
– The management of inspection, survey, and work planning.

With current condition information provided (or agreed default values entered), we designed the system to estimate the probability of coating breakdown and consequent metal loss over time for each element of the asset.   Taking account of the criticality assigned to each item, the location, current protection and condition, it will then predict time to next maintenance, and thereafter time to (theoretical) failure of the item from corrosion.

Planning is then done on the basis of grouping items by criticality, location (Work Area), and other factors into "workable" projects, with estimated costs.

How Does it Work?

There are five inter-related modules:

- Work Site Information and Definitions
- Risk Assessment and Criticality Determination.
- Condition Survey Assessment.
- Standing Data.
- Fabric Maintenance Planning Module.

Each is considered briefly in turn.

Work Site Information

This provides written procedures for the collection of data for developing the computer application and database.   The following table describes briefly the sets of data required for the Work Site.

Data set	Description:
Compan y	Organisation name, address and contacts
Budget	Financial group, budget holders and administrators, budgeted amount.

Area	
Work site	Usually an individual asset within the budget area, such as a bridge or offshore installation. Additional information for the Work Site will include: • Required life • Resource constraints for use in Work program planning • Loss of Appearance consequence rating (for use where appearance is a major factor).
Work area	Division of Work Site into Work areas (logical to efficient maintenance work). Corrosion rate zoning (establishes the likely corrosion rate for that area)
Survey Area	Division of Work Areas into efficient *Survey Areas* (logical to efficient inspection / survey) Collection of data for Condition
Item, Item Part, Item Assembly	*Item part* is the smallest piece of the asset to be individually maintained, e.g. a pipe or girder. Sometimes these are part of a larger *Item*, which in turn comprises part of a complete *Item Assembly*. This facility provides for Item Parts to be grouped into functional areas. Information for Item Parts includes: • Name and Identifier • Criticality type (e.g. Channel, pipe etc.) • Size, specification and surface area • Required life (defaults to required asset life) • Last known condition and metal loss (a default is available).

Most of the required data can be obtained from asset registers, drawings and engineering/inspection records. Pro forma hard copy collection sheets can be provided for the client and/or others to capture the necessary information, or it can be directly entered to the system. An estimation module is included which can use historical data to predict current condition, thus getting more value from previous inspections. We felt it important to minimise the need for an initial condition survey to populate the database.

The emphasis is on minimising the set-up time and cost, so the more client-provided information there is the better. However, a range of default values can be used where specific values are not available, and the set up can be as general or detailed as required.

Risk Assessment and Criticality Determination

The Criticality Rating of the subject provides a mechanism to prioritise maintenance planning according to the amount of risk that can be tolerated. Items will be scheduled for maintenance according to their Criticality and condition, and only if required before the end of life. This allows the operator to:

- Focus on critical items to tailor maintenance planning.
- Review current condition by Criticality.

The module provides for a semi-automatic assessment of risk, dependent upon the probability of failure together with the consequences of failure.

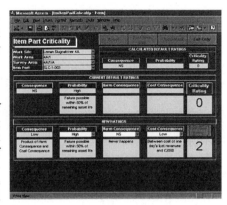

Substrate Failure Probability
The system assesses the probability of substrate failure relative to the remaining life required, based upon the Item Part service, details, condition (inclusive of coatings) and environment. It is pre-loaded with calculators for Pressure Pipework, Vessels and Structure. Based upon this calculation it sets the following ratings:

HIGH   Failure possible within 50% of remaining asset life
LOW   Failure possible between 50 and 150% of remaining asset life
NS (Not significant)   Failure unlikely within 150% of remaining asset life

Failure is defined as the point when consequences occur.

Substrate Failure Consequences (relative to service)
Substrate failure consequence ratings follow a set of defaults that relate to the item service. For instance, consequences for structural items are considered from loss of structural integrity, and for containment piping, for loss of containment. The default rating assignments follow the standard definition of consequence criteria, together with an option to include cost consequences.

Harm. Loss of life or serious injury (one that impairs an individual's quality of life or restricts future employment)
HIGH if probable (where it will happen most times when failure occurs)
LOW if possible (where it will happen rarely when failure occurs)
NS if it never happens (where it will not happen when failure occurs).

Cost. Cost at three levels
HIGH is one or more days lost operation or equivalent cost
LOW is between cost of one day's lost revenues and £2,000
NS when cost is less than £2,000.

These Rules are brought together in our Matrix below:

CONSEQUENCE OF FAILURE

		HIGH	LOW	NS
PROBABILITY	HIGH	4	2	0
OF FAILURE	LOW	3	1	0
	NS	1	0	0

The matrix is weighted in favour of the Consequence rating.
A simple override facility is included for clients who wish to use other criteria, either in general or in specific instances.

Condition Survey

We recognised that users wish to get more value out of their inspections. This module provides for capture, storage and reporting, of condition information for the components of the Work Site. Survey areas are defined within work areas and identified by means of key sketches and component listings. Written procedures for data collection minimise ambiguity.

The module contains printable blank survey forms to capture:

- Coating type and current condition, assessed to BS 3900 : Part H3 : 1983, ISO 4628/3 – 1982
- Current metal loss (pit depth etc.)
- Current metal thickness (e.g. flange thickness)
- General substrate condition
- Difficulty rating and zone

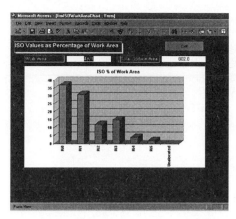

Default condition information can be used where general condition is known and a specific detailed survey is thought unnecessary.

The module provides the operator with a means of recording, reviewing and using condition data to aid planning. Written and graphic reports can be produced using a number of different report choices to give, for instance, an overview of the condition of the entire Work Site, or to isolate particular areas of concern by Item Parts.

The information held in this module is used in conjunction with Criticality and Standing Data within the Planning Module.

Standing Data

This module provides a library of 'stand alone' data for the following:

- Coating performance
- Zones, including an Environment Severity Rating and Corrosion Rate
- Substrate Condition rating, including Work rate factor.
- Difficulty rating based upon substrate arrangement and access

Coating performance

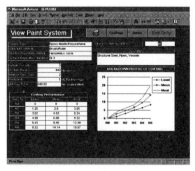

Substantial historical evidence from Duffy and McGovern archives and other publications has enabled a library of coating types to be compiled. Least, most and mean times to breakdown (using BS 3900 Ri Scale) are plotted and can be displayed as a graphic. These are matched to the Item Part, adjusted for Criticality Rating and used together with other Standing Data to predict optimum maintenance and theoretical failure. Choice of coatings is usually a trade off between cost, time/resources to apply, and life. Unit costs and productivity for each coating system are used together with other standing data to provide estimates for maintenance, and for comparison of coating and maintenance choices.

Zones, including Environment Severity Rating and Corrosion Rate

The performance of a coating system is dependent upon a number of factors, key to which are:

- The environment (i.e. the location and nature of substrate)
- The condition of the substrate immediately prior to coating

These are used to modify the Coating System performance.

Zones are used to define areas within which the environment is similar, and the deterioration of the Coating System and subsequent Corrosion Rate will also be similar. These are used to modify the Coating performance, and subsequent time to theoretical failure.

Unit Cost

Cost (excluding access) is also dependent on a number of factors. Primary factors included within standing data relate to:

- Difficulty of area
- Surface condition of area

Both of these factors are used to modify the estimated cost of the maintenance activity.

Standing Data is not a 'fixed' module and we recognised when building the system that clients may have their own norms and historical data, which could be used to good effect. Hence, the tables within this module can be updated to take account of past and future experience to build a 'learning data base'.

Planning Module

This module uses all relevant Standing Data, Work Site Information, Condition Data (or estimated condition where up to date Condition Data is not available) and Criticality Rating to provide:

- Next maintenance date
- Theoretical failure date

Based upon this information it allows the user to 'try out' different groupings and scenarios to select the optimum solution, both in terms of timing and coating choice, to suit their needs and constraints. It also shows the user where certain items may be safely deferred or even eliminated. A range of different reports, by criticality, Ri% breakdown and others are available to help planning.

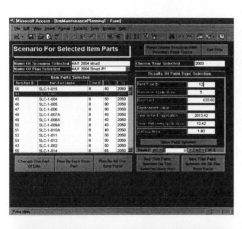

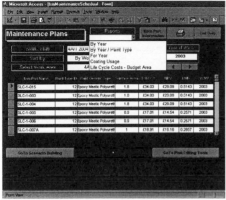

We built in the ability to look forward and predict estimated breakdown and failure in the future, and then plan and budget for maintenance. The client is able to plan efficient maintenance of whole areas when the majority of Items require it, and still identify holding activities required for specific Items in the interim.

The system allows the client to build scenarios using different groupings, years and Coating Systems, enabling comparison of costs and scheduling, building into Optimum Life Cycle maintenance plans, together with costs at Net Present and Net Future Value. This will also allow the cost of replacement to be compared with the cost of maintenance. Work Plans within Budget Area and Work Site can be produced for current and future years to build in to overall Asset Planning systems.

What are the benefits?

In summary, we have built a system that empowers the client through the following benefits:

- Auditable risk based approach (complimentary to established Safety Management Systems)
- Focuses maintenance where its needed
- Reduces maintenance cost through more efficient planning
- Develops cost estimates for Asset Life Cycle
- Enhances the value from condition inspection
- Provides work packages based upon substantial coatings experience
- Facilitates flexibility and informed decision making
- Predicts maintenance and failure

Conclusion

Owners and operators are only too aware of the continuing struggle to maintain and increase margins through focused cost reduction and Life Cycle cost management. Whilst there are plenty of tools available dealing with internal corrosion and painting planning, we believe that the area of external corrosion *management* has been neglected for too long. Those involved with the corrosion protection of assets will be aware that a significant part of their budget is allocated to external corrosion inspection and mitigation, and that unforeseen failures are expensive.

We have designed a system to hand control of external corrosion management back to the client. Relatively simple and transparent but powerful, it will squeeze more value from inspection and survey results, facilitate a risk based rational approach to planning maintenance, and provide the flexibility to design and update an optimum life cycle maintenance package for the Asset. Should circumstances change, the client is able to modify the asset maintenance program with full information, and quickly determine what must be done from what could be done. We believe this approach balances both the needs for prudence and safety, and economic efficiency.

Acknowledgements:

Mark Wilson BSc, CEng, MICE.
Michael.J.Pursell MSc, CEng, FIChemE, FICorr.
Robin Burden
Steven Marsh

Probabilistic-based bridge management implemented at Skovdiget West Bridge

F.M. JENSEN, A. KNUDSEN, I. ENEVOLDSEN, RAMBØLL, Denmark
Bredevej 2, DK-2830 Virum, Denmark. fnj@ramboll.dk, www.ramboll.dk, phone: +45-45986088
E. STOLTZNER, The Danish Road Directorate, Denmark.
Niels Juels Gade 13, DK-1059 København K, Denmark. est@vd.dk, www.vd.dk, phone: +45-33933338

ABSTRACT
This paper describes the implementation of a probabilistic-based bridge management plan for the Skovdiget West Bridge, Denmark in 1998-99. Skovdiget West bridge is a 30 year old 220 m long post-tensioned concrete box-girder bridge with serious deterioration of both concrete, reinforcement and cables. By use of deterministic load carrying calculations combined with traditional lifetime estimates, the bridge would either require major rehabilitation or replacement. The application of a probabilistic-based management plan postponed major rehabilitation, repair and replacement, and gave a saving of more than EUR 10 million compared to use of a traditional deterministic analysis and lifetime estimation.

KEYWORDS: probabilistic-based, management plan, decision analyses, rehabilitation, deterioration.

INTRODUCTION
Bridge owners today face the combination of an old deteriorating bridge stock and limited budgets. Additionally, many of these ageing bridges are vital for the infrastructural system, consequently rehabilitation projects are associated with large road user inconvenience costs. Probabilistic-based safety management and assessment tools were recently developed as an attractive approach for extension of the service lifetime, thereby giving bridge owners the possibility of reducing or postponing costly rehabilitation projects.

Today, on selection of a repair and rehabilitation strategy for a deteriorating structure, the estimate of the remaining lifetime will often be based on experience. In order to avoid serious bridge failures a conservative estimate has to be used for prediction of the remaining lifetime. The use of a conservative estimate is necessary even for the most experienced bridge inspection engineer. Additionally, the effect on the extension of lifetime as a result of various repair and rehabilitation options is associated with some uncertainty due to the lack of an efficient tool to properly evaluate the effect. This uncertainty with respect to the remaining lifetime – and the effect of repair and rehabilitation options – means that there is a lack of information when determining the optimal bridge rehabilitation strategy, which may often lead to a non-optimal spending of budgets.

By application of a probabilistic-based approach, the probability of failure P_f of the deteriorating bridge is calculated. By use of a model for the deterioration, P_f as a function of time may be determined. The lifetime is then determined as a 'criteria time', that is the time till the probability of failure passes a user-selected criteria. This criterion is typically a requirement to the maximum allowable probability of failure.

This way the lifetime can be directly associated with the safety of the bridge. In the probabilistic approach you still need to estimate the deterioration of concrete and reinforcement, but the uncertainties are included in a rational manner. Additionally, the probabilistic approach can be used to compare the various repair and rehabilitation options considered. The probabilistic approach also provides information about the importance of individual parameters that contribute to the safety. This supplies the bridge owner with a much better tool for inspection planning by focusing on important parts of the structure and for comparing repair and rehabilitation options. This level of information ensures optimal inspection planning and optimal lifetime spending of budgets.

Recently, bridge owners have changed from the traditional approach for lifetime estimation and selection of repair strategies to a more advanced and individual approach in order to optimise budgets.

The Danish Road Directorate has taken a lead in using the individual approach, and in 1998 the Danish Road Directorate in co-operation with RAMBØLL initiated work on the development of a probabilistic-based management plan for the Skovdiget West Bridge.

By using a probabilistic-based management plan – combined with a cost-optimal plan based on probabilistic-based assessment and decision analysis of several rehabilitation options – the Danish Road Directorate was able to reduce and postpone a costly rehabilitation or replacement project due for the severely deteriorated Skovdiget West Bridge.

HISTORY and STRUCTURAL SYSTEM

The Skovdiget identical twin bridges are located Northwest of Copenhagen and were constructed in 1965-67. Each carries a 3-lane motorway across two local roads, a railroad and a parking area. The bridges carry a major approach road to Copenhagen. The average daily traffic is 53,000 vehicles.

Each bridge is 220 m long, 20 m wide and has 12 spans. The bridges are concrete post-tensioned, combined box-girder and beam-slab bridges.

Figure 1. The West bridge seen from Southwest, 1999 (right). Below the midpoint of the West bridge looking North, 1999. Main box girder 3 is on the right and main box girder 4 on the left. The East bridge is seen on the right on the picture (left).

Due to poor workmanship and unfortunate design, both bridges started to deteriorate shortly after construction. In the late 70'ies substantial damage was registered on both bridges. The damage was related to un-injected or poorly injected post-tensioned cable ducts, drainage was insufficient and poorly made, the area around gulley poorly made, a bad waterproofing on whole bridge deck and an uneven bridge deck. The gulley is located over main girder 4. The water on the bridge deck around the gulley caused corrosion damage to reinforcement and cables and frost damage to the concrete.

East Bridge

Major repair was done on the East bridge in 1978 including partial concrete replacement of bridge deck (especially around the gulley), repair of post-tensioned cables (using vacuum-injection of partly injected cable ducts), new waterproofing, new base and wearing course and repair of the bearings and expansion joints. Since the repair in 1978, the East bridge has been inspected according to the normal guidelines for bridge inspections in Denmark. The East bridge was last inspected in 1998 and it was found in general to be in a good condition.

West Bridge

The cost of rehabilitation of the East bridge was EUR 3 million in 1978. A new bridge would cost EUR 8 mill. in 1978. Since the repair on the East bridge proved so costly, it was decided to leave the West bridge without repair. Instead it was decided to monitor the West bridge closely. The development of deterioration was to be monitored in order to determine when the safety was no longer acceptable. This has been done by close visual inspections 4 times a year. These inspections include a thorough visual inspection and measurements of crack widths and deformations. In addition to the frequent visual inspections and deformation measurements of selected structural parts, a test loading of the West bridge was carried out in 1984, 1988 and 1993. During the test loading the deformation of the main and secondary spans was measured while the bridge was loaded with 90 and 60 tonnes trucks. The test loading showed small deformations compared to the deformations when close to failure.

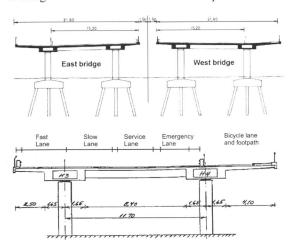 

Figure 2. Overview of the two Skovdiget bridges, location of lanes on West bridge (left). Picture from the last test loading performed on the West bridge in 1993 (right).

The last special inspection was performed on the West bridge in 1995. The inspections were focused on the main girder no. 4 and especially the area around the gulley. It was found that the concrete in the area around the gulley was severely deteriorated, there was reinforcement with pitting corrosion, un-injected cables and cables with loose strands indicating total loss of load bearing capacity. The serious deterioration was primarily related to main girder number 4. Inspections in main girder 3 and the western wing showed only minor and local deterioration.

PROBABILISTIC-BASED MANAGEMENT PLAN

A new test loading for Skovdiget West was due in 1998 (following a 4-5 year interval between tests). A test loading is costly and does not give sufficient information to be used for lifetime predictions, evaluation of the present and future load carrying capacity or information to decide between various rehabilitation options. Based on this fact combined with the results from the last special inspection in 1995, which locally showed severe deterioration, it was decided to implement a management plan based on probabilistic methods. By doing this the Danish Road Directorate obtains a better basis for decision, for minimising maintenance budgets and for extending the lifetime of the bridge - all with the prerequisite that the requirements to the overall safety are observed. Additionally, a model is obtained which may be used to financially and technically evaluate different repair and rehabilitation initiatives in the future.

In the following the actual implementation of the probabilistic-based management plan for Skovdiget West is described.

Formulation of problem
The development of an optimal management plan requires information on expected future maintenance budgets, the value and the importance of the bridge and the importance of the bridge for road users. This in connection with the safety requirements for the bridge is decided in close co-operation with the bridge owner.

Safety requirements for the bridge
The legal basis, either national or international authorities, establishes the fundamental requirement for the application of a probabilistic-based assessment. For Skovdiget the background documentation behind the codes in the Nordic countries *'Recommendation for Loading and Safety Regulations for Structural Design'*, see [1], describes in detail how a probabilistic-based assessment can be performed in accordance with the requirements for the safety level in the Nordic countries. Furthermore, [1] specifies the principles of modelling of uncertainties including model uncertainties. The requirement in the ultimate limit state for the structural safety is in [1] specified with reference to failure types and failure consequences, i.e. safety class with requirements for the formal yearly probability of failure P_f. The formal probability of failure can, using the standard normal distribution function Φ, be related directly to the reliability index β by the relation $\beta = -\Phi^{-1}(P_f)$. Based on analysis of critical failure modes for Skovdiget West, the safety class 'high' and limit state type 'failure with remaining bearing capacity' is appropriate. This equals the limits on the probability of failure and corresponding reliability index to $P_f \leq 10^{-5}$ and $\beta \geq 4.26$.

Identification of critical failure modes
The deterioration of Skovdiget is related to an overall deterioration of the whole bridge and a more concentrated deterioration around the gulley due to bad drainage. Initially, a deterministic analysis was made to identify the critical points and critical failure modes. A critical point is determined by the combination of load carrying capacity utilisation combined with deterioration at a given point for a given failure mode.

These analyses identified main girder cross section at column supports and at construction joints in the long span on main girder 4 as primary critical points. The critical failure mode was combined bending and shear, with shear failure as the dominant failure mode. It was found that transverse girders, transverse ribs and wings on the main girders were less critical (secondary critical points).

Probabilistic-based assessment model
The probabilistic-based safety assessment model is the core in the management model and must therefore be developed with a specified set of specifications. The model must be able to:

- evaluate the safety now and in the remaining lifetime of the bridge,
- analyse realistic failures and set of failures,
- identify critical areas for monitoring,
- include detailed information on deterioration, loading, failure mechanism, output from test loading, etc.
- take into account the uncertainties related to strengths, loads, traffic, models, etc.
- identify important parameters for the safety, sensitivity analysis, and
- update the safety of the bridge based on new or updated information, e.g. results from a previous or a future test loading or inspections.

For this project an influence model for the global bridge structure was developed based on elastic analysis. For the main girder cross section a capacity model was developed with the characteristics that:

- the capacity of the cross section is maximised,
- each reinforcement group and cable group may be modelled independently, and
- the model is made operational for automatic inclusion into a general-purpose program for reliability analysis.

The model developed is capable of realistic modelling of the variation of deterioration and by performance of an optimisation for load bearing capacity the extension of lifetime can be maximised. For Skovdiget West a model for main girder cross section capacity with separate modelling of 28 reinforcement groups was implemented, see Figure 3 (left). Since the concrete, reinforcement and cables below the gulley are more deteriorated than other parts of the cross section, a model for optimal force distribution was made, thereby ensuring that all remaining reinforcement is taken into account, see Figure 3 (right).

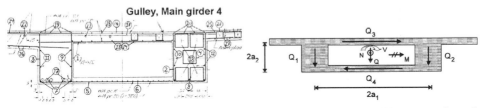

Figure 3. For both main girders the reinforcement and cables were divided into a number of reinforcement groups each modelled individually (left). A model for optimal load carrying capacity of a main girder was made (right).

Deterioration modelling
Based on all available information from previous inspections performed on the West and East bridge and from the major repair of the East bridge in 1978, combined with experience in rate of deterioration for typical damages, the remaining reinforcement area for each individual reinforcement or cable group was estimated. This was done for three critical structural points identified earlier and for the years 1978 (major repair of East bridge), 1998 (now) and estimated for 2008. Each estimate was associated with a coefficient of variation depending on the information level available when making the estimate.

Stochastic Modelling & Traffic Modelling
The uncertain variables related to traffic, load, resistance and modelling are modelled as stochastic variables with corresponding statistical distributions including parameters specific to Skovdiget West and the level of knowledge concerning materials, loads and mechanical modelling.

For this project a separate stochastic modelling of the traffic load was made. The developed traffic model is specific to Skovdiget West and is therefore not as conservative as a general modelling of traffic loads, which has to be valid for a larger group of bridges. By development of an individual modelling of traffic loads a higher safety can be obtained. The Skovdiget model is based on combinations of thinned Poisson processes with extreme loads in two lanes, since extreme loading in three lanes was found to be unlikely based on the actual appearance of vehicles in lanes. Statistics of heavy vehicles just north of the bridge were used to develop the traffic load model specific to Skovdiget West.

Calculation of safety taking deterioration into account
Using the probabilistic-based assessment model, the deterioration modelling and the stochastic modelling of traffic, load, resistance and modelling, the reliability index as a function of time may be estimated. In Figure 4 (left) the reliability as a function of time is shown for the three critical points:

- main girder 4, column support, south of long span, shear with moment failure check
- main girder 4, construction joint south of long span, shear with moment failure check
- main girder 3, column support south of long span, shear with moment failure check

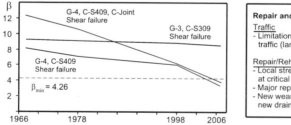 

Figure 4. Reliability index for critical points as a function of time. Repair and rehabilitation options considered the management plan for Skovdiget West (right).

The minimum required safety – or reliability index – is indicated in Figure 4 (left). The result of the analyses showed, with the given assumptions regarding the state of deterioration, that at present (1998) the safety of bridge is above the critical safety level. Using the estimates of future deterioration the 'criteria time' – or remaining lifetime – is found to be 7-8 years.

In addition the sensitivity analysis identified the critical parameters in the modelling and the critical reinforcement groups. This was later used to determine which groups needed to be inspected closely and which reinforcement groups were less important for the safety.
Table 1 shows the approximate increase in reliability index when the standard deviation is halved. It is seen that a substantial increase in reliability index may be obtained if the uncertainty modelling can be improved. This means that detailed inspections of selected reinforcement groups are advantageous since they may lower the standard deviation - or even increase the estimated value - and thereby increase the safety of the bridge.

Table 1 Approximate increase in reliability index when reducing the standard deviation to 50% of the value used.

Stochastic Variable	G4,S409,1998 Column support	G4,S409, 2008 Column support	G4,S409,1998 Construction Joint	G-4,S409, 2008 Construction Joint
Self-weight	0.32	0.06	0.06	0.05
Traffic load	0.55	0.12	0.09	0.10
Yield stress, shear reinforcement	0.55	0.37	0.18	0.15
Concrete compressive strength	0.75	0	0	0
Shear reinforcement	1.00	1.20	0.30	0.03
Post-tensioned cables	0.15	0.05	2.6	1.29

Based on the result of this analysis several repair, rehabilitation and other initiatives were analysed to extend the lifetime of the bridge.

Repair and rehabilitation options
In order to choose the correct rehabilitation plan for extending the lifetime, a number of possible options were analysed. These may be divided into 3 categories related to: traffic, repair & rehabilitation and information updating, see Figure 4 (right).
From the list of possible combinations of options the following were chosen for further analysis based on initial investigation of all combinations of options:

    0) No initiative
    1) Safety updating inspections
        1-2) Safety updating inspections & test loading
        1-3) Safety updating inspections & 'survival' wearing course
        1-4) Safety updating inspections & strengthening against shear failure
    2) Test loading
    3) 'Survival' wearing course and new drainage system
    4) Strengthening against shear failure at critical points

To choose the best – that is best technical and cost-optimal – rehabilitation option, the change in either information level or change in the rate of deterioration was estimated for each of the 7 selected combinations shown above. For this change in level of information or estimated change in the rate of deterioration the change in reliability – $\Delta\beta$ – was calculated using the developed probabilistic-based assessment model. This change in reliability is equivalent to a change in criteria time or change in remaining lifetime.

Cost-optimal probabilistic-based management plan
For each of the options a net-present value calculation was performed including both direct cost and indirect cost (road user inconvenience cost). A net-present value analysis of postponed options was also made. This gave a list of corresponding remaining lifetimes and costs to be used as input when selecting the optimal management plan. The cost optimal management plan was 1-3) Safety updating inspections & 'survival' wearing course. The estimated criteria time after implementation of 1-3) was estimated to 15 years – up from 7-8 years – with an estimated cost of EUR 0.9 million. In addition to the proposed option the following was done in order to verify the model made and to update the management plan.

- Safety controlling inspections to verify deterioration modelling of critical reinforcement groups.
- Continuous periodical visual inspections and special inspections
- Update of management plan based on 'safety updating' and 'safety controlling' inspections.

IMPLEMENTATION OF COST-OPTIMAL REHABILITATION PLAN
The management plan was implemented in 1999. The management plan included detailed inspections to verify the assumptions and to update the management plan. Secondly, if inspections did not show deterioration not accounted for in the analysis, a new wearing course and a new drainage system was to be implemented at Skovdiget West bridge.

Inspections in 1999
In 1999 two types of inspections were made. A number of 'safety controlling' inspections to ensure that the overall assumptions about deterioration on the bridge were correct, and 'safety updating' inspections at critical points identified during the analysis. The safety controlling inspections verified the overall assumptions regarding areas of critical deterioration. The safety updating inspections focused on critical areas and critical reinforcement groups identified during the previous reliability analysis, and extended the service life by 7-10 years due to more detailed deterioration information about the critical reinforcement groups.

Rehabilitation in 1999
The second part of the management plan was initiated in 1999 after completion of the inspections and verification of the management plan. In the period July-September 1999 a new wearing course was put on the West bridge. The drainage system was altered by adding two new gulleys two metres on both sides from the old gulley. By doing this water was moved from the area around the old gulley where the deterioration was most severe.

CONCLUSIONS
In the present work a probabilistic-based management plan was implemented for Skovdiget West. Probabilistic-based safety management is an attractive approach for extension of the service lifetime, thereby giving bridge owners the possibility of reducing or postponing costly rehabilitation projects not otherwise possible using a traditional deterministic analysis.

The applied approach included stochastic modelling of all the uncertainties of importance for the safety problem, a new load carrying capacity model taking variation in deterioration into account and a new traffic model specific to the bridge. The selection of the best technical and cost-optimal repair and rehabilitation option was done by using the probabilistic-based model in connection with net-present value estimates.

The result was the implementation in 1998-99 of a cost-optimal plan based on probabilistic-based assessment and decision analyses of several rehabilitation options, giving a saving of more than EUR 10 million compared to traditional deterministic analysis.

REFERENCES
[1] Nordic Committee for Building Structures (NKB) *"Recommendation for Loading and Safety Regulations for Structural Design"* NKB report no. 35, 1978 & NKB report no. 55, 1987.

[2] Danish Road Directorate (Vejdirektoratet) *"Rules for Determination of the Load Carrying Capacity of Existing Bridges"*, Danish Road Directorate, 1996.
Written in Danish: *"Beregningsregler for beregning af eksisterende broers bæreevne"*, Vejdirektoratet, Vejregeludvalget, April 1996.

[3] Mohr, G.: *"Traffic on motorways"*, February 1990.

[4] PROBAN, General Purpose Probabilistic Analysis Program, Veritas Sesam Systems, 1992.

[5] H. O. Madsen, S. Krenk & N. C. Lind : *"Methods of Structural Safety"*, Prentice-Hall, 1986.

[6] O. Ditlevsen & H. O. Madsen : *"Structural Reliability Methods"* John Wiley, 1996.

Examination of uncertainty in bridge management

Parag C Das OBE, BTech., PhD, CEng, MICE
Project Director Bridge Management
Highways Agency, UK

Toula Onoufriou BSc, PhD, DIC, CEng, MICE
Reader and Director of Research
Civil Engineering Department, University of
Surrey, UK

KEYWORDS

Bridges, management, probabilistic methods, uncertainties, research, reliability

ABSTRACT

There are many uncertainties associated with the process of bridge management, from inspection and assessment to fund allocation and execution of the works. These uncertainties should be identified and reduced as far as possible if the predictably increasing needs for bridge maintenance in future years are to be met in a cost-effective manner.

Research in bridge engineering so far has mostly concentrated on the engineering aspects of the activities. Considerable research and development work is being carried out in many countries around the world for better materials, innovative forms of construction, improved inspection and monitoring techniques and state-of-the-art evaluation procedures and criteria. It is an appropriate time now to undertake co-ordinated research into the uncertainties in the process of bridge management. For this, it is necessary to form an overview as to how future research work could be focused to address the deficiencies. Clearly, the level of resources employed in individual areas of research should reflect the accuracy that is warranted by the end use.

The paper first presents an overview of the bridge management process and develops flow charts which identify the main steps and activities. These flow charts can be developed further to provide a concise representation of the bridge management process and a basis for a systematic examination of the various sources of uncertainty. Based on experiences gained from bridge management, as well as from similar activities in the field of offshore structures, the paper then discusses the different sources and types of uncertainty such as physical uncertainty, modelling uncertainty, human errors etc. and how these can be managed more effectively to reduce the impact on decision making. The use of reliability techniques in assessing the relative significance of the various sources of uncertainty is then examined. Based on these preliminary investigations the paper outlines a possible methodology that can be developed to provide the bridge authorities, research workers and industry in general with an overview of the currently perceived problems and their relative importance so that resources could be employed in a cost effective manner.

INTRODUCTION

Co-ordinated bridge related research in the modern sense started in the United Kingdom from the early part of this century and has been mainly funded by successive government departments responsible for transport and war efforts. The research has been conducted also mainly by the government laboratories, namely the Building Research Station, the Transport Research Laboratory and the National Physical Laboratory. In recent decades, certain industry funded organisations such as the Cement and Concrete Association (now British Cement Association), the Welding Institute and the Steel Construction Institution have also carried out significant bridge related research. The leading universities have of course been involved in this research throughout this period.

Apart from any in-house R&D carried out by construction companies for testing and developing new methods and materials, publicly funded bridge research so far has gone through a number of distinct phases, which are shown in Fig.1.

1920-1960	Early research on behaviour of masonry bridges and other older bridge types
1960-1980	Analysis methods and tests
1970-1985	Design codes and limit state formulations
1980-1995	Assessment methods and standards
1990-	Bridge management, safety levels and reliability

Figure 1. Main themes of bridge research this century

In the closing years of the century, as the existing stocks of bridges are getting older and beginning to suffer from general deterioration and other problems, the attention of the bridge authorities and the profession is now focused on maintenance issues. As a result, in recent years, there has been a significant increase in research activities in the area of bridge management involving repair techniques, non-destructive testing methods, deterioration processes, assessment methods and bridge management systems (Das, 1997, Institution of Civil Engineers & Highways Agency, 1998).

Bridge management involves structural assessments and decisions which are aimed at prioritising maintenance work in a way that it will make the most effective use of resources. The information required for financial planning in this area has a considerable amount of uncertainty, the effect of which can be either wastage or inappropriate use of scarce resources. As bridge maintenance is becoming a major part of the highway authorities' expenditures, it is considered an appropriate time to take an overview of what the current uncertainties are in the process of bridge management and how future research work can be directed to address the resulting deficiencies.

Clearly, the aim should be that the level of resources employed in individual areas of research should reflect the accuracy that is warranted by the end use. For instance, there would be no point in producing more and more sophisticated methods of analysis if the basic parameters contain a high degree of uncertainty. This paper is intended to provide a broad outline of the decision process involved in bridge management and to explore the possibility of quantifying the relative needs for research in the various component activities involved.

The paper contains the preliminary observations from a joint project being undertaken by the Highways Agency and the University of Surrey aimed at prioritising research efforts in the area of bridge management. The project is taking on board experiences gained in research and development in the offshore industry where there is a large number of ageing platforms requiring significant maintenance resources (Onoufriou, 1997, Onoufriou 1999). In particular, application of the reliability analysis technique in selecting the potentially most fruitful areas of research is being considered.

The paper first outlines the bridge management activities and goes on to examine the various uncertainties affecting the decision process. The use of reliability techniques in assessing the relative significance of different problem areas is examined and recommendations are made for taking these important issues further.

THE BRIDGE MANAGEMENT PROCESS

The process of bridge management involves a variety of activities carried out at different levels (Figure 2). The maintaining agent inspects (and performs tests on) the bridges, generally at regular intervals,

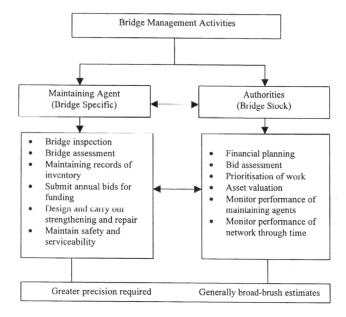

Figure 2: Bridge Management Activities and
Relative Accuracy Required

and assesses their structural adequacy as and when necessary. He also keeps records of inventory, inspection and assessment, and furthermore submits bids for funds on an annual basis. The owner or authority ensures that the maintenance of the bridges conform to the technical and other requirements of the network and also assesses the bids and allocates funds for any repair or strengthening works considered necessary (Das, 1998). The authority also, to varying degrees, monitors the performance of the maintaining agent, and of the network, through time. In practice, many of the agent's tasks are sub-contracted out while the authority's tasks are carried out at different levels, from the minister responsible down to the network area manager.

In general a higher level of accuracy is required on the bridge specific activities when compared to the planning activities relating to the whole stock of bridges. This is an important consideration, as the level of resources in individual areas of research should reflect the accuracy that is warranted by the end use. Both types of activities, however, can benefit from an assessment of the relative significance of key parts of the process in order to prioritise research effort and achieve more effective use the available resources.

All the activities involved in bridge management are primarily aimed at answering two questions (see Figure 3). The first question is what, if anything, needs to be done to the bridges now. The second is whether the maintenance is being carried out effectively. The first involves a complex array of activities, which will be discussed later. The second question is considered first.

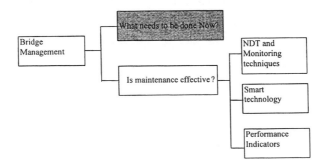

Figure 3. Overall aims of bridge management

Monitoring of Performance

The bridge authority needs to determine that the maintenance activities are doing what they are intended to do, so that he can conclude that either the current regime is satisfactory or the activities should be redirected in some way. For this purpose many investigative techniques involving non-destructive test (NDT) methods and 'smart' technology are being developed. Also being developed, performance indicators which can give an on-going picture of the state of the network of structures. These developments, although in their early stages, are extremely important for bridge management and research effort should be vigorously pursued to facilitate the wider use of these techniques in practice.

Funding Decision

The various activities shown in Figures 4 and 5 are carried out to answer the second question i.e. to determine if anything needs to be done to individual bridges in the bid year (next year) either from the consideration of safety or for whole life economy or for some other reason.

Figure 4 shows the activities involved in the assessment of the current structural adequacy of a bridge. These range from the investigation of existing information such as drawings, inspection and testing and structural analysis. The assessed capacity is then compared with the minimum acceptable criteria to determine if any work is immediately needed. The engineering criteria are themselves intended to

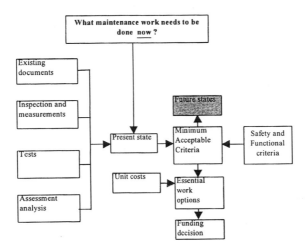

Figure 4. Funding decision for essential work

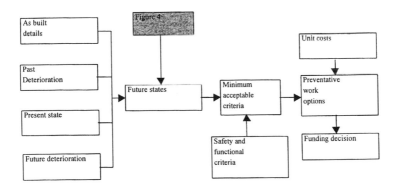

Figure 5. Funding decision for preventative work

represent client or societal expectations from the structure. If essential work is required, work options are considered using any available cost data.

If work is not essential at present, consideration of future adequacy is then necessary (Figure 5). This again involves the information gained from the present-state assessment, and additionally, information of past conditions, prediction of future deterioration and the consideration of possible preventative work options.

Research is at present carried out in varying degrees in the subject areas of many of the component activities, for example, inspection methods, analysis techniques, safety criteria etc.

UNCERTAINTIES

The activities shown in Fig.4 and 5 involve many uncertainties in the process. The final result of funding any maintenance work therefore can turn out to be very different from the intended objective. For effective management, it is imperative that these uncertainties are identified and reduced as far as possible. In examining the various sources of uncertainty there are a number of issues that need to be addressed such as the ones summarised below:

- Recognise and identify the various sources of uncertainty
- Recognise the different types of uncertainty e.g. physical or modelling uncertainty, human factors, gross errors etc.
- Address the question of how to manage uncertainty more effectively to reduce scope for error in decision making, i.e. reduce or better quantify uncertainty
- Examine how the various uncertainties affect the bridge management process, i.e. what can go wrong?
- Examine how reliability models can be used to study the various elements of uncertainty in bridge management and their relative significance with the aim of prioritising research efforts

Sources of Uncertainty

The flow charts presented above are an attempt to map out the main activities involved in the bridge management process. These can be expanded further to a more detailed level that will identify individual steps of each process and enable individual sources of uncertainty to be identified. For example in the area of inspections, which is one of the key inputs identified in Figure 4, the overall uncertainty is a compounded effect of the uncertainties resulting from each step in the process as identified in Figure 6.

The key components of uncertainty can be then be identified as follows:

- The uncertainties associated with the inspection planning process
- The probability of detection of a defect associated with different inspection methods used
- The uncertainty in sizing the defect
- The uncertainty in reporting and interpretation of the measurement obtained to a parameter that enters the strength assessment formulations
- Further uncertainty associated with the assessment analysis itself

Developing a more detailed mapping of the bridge management activities, as shown in the above example, can form a basis for a systematic examination of sources of uncertainty.

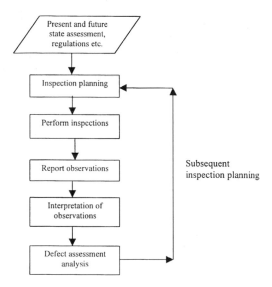

Figure 6: Inspection planning and execution

Types of Uncertainty

It is also important to recognise the different types of uncertainty that can be present as different types of action need to be taken in order to manage these more effectively. These include physical and modelling uncertainty which can be incorporated in a formal reliability analysis. The physical uncertainty represents the physical variability associated with a particular parameter such as for example the long term loading distribution on a bridge or the various material strength parameters. Modelling uncertainty on the other hand is associated with the idealisations used in assessing a particular effect which may have a bias and a distribution associated with it which will result in variability in the assessment prediction. In addition to physical and modelling uncertainty which can be modelled in a formal reliability analysis there are other sources of uncertainty associated with human factors, gross errors etc which also need to be given some consideration within the context of managing the uncertainties in the overall process.

Managing Uncertainty

Having identified the sources and types of uncertainty, the next question is what actions can be taken to reduce their impact on funding decisions. When it comes to physical uncertainty the objective should be to better quantify the distribution of the relevant parameters. This can be achieved by improving the data sets used to derive the statistics of these parameters. In practice this may require obtaining more and better quality data on loading histories and material parameters for example. To address modelling uncertainty it would be necessary to develop and validate improved assessment models, and provide guidelines which will reduce the scope for different assumptions associated with different users which may also have different levels of competency. Additional measures would have to be taken to address human factors and gross errors etc.

Effect of Uncertainties - What Can Go Wrong?

Having recognised the sources of uncertainty, it is important to understand how these affect the process of bridge management and what can go wrong as a result.

There are two main types of maintenance work for which funds are allocated i.e. 'essential work' and 'preventative work'. 'Essential work' is necessitated by structural inadequacy (for instance inadequate load carrying capacity), while 'preventative work' is intended to prevent essential work being required prematurely in the future, and is justified on the basis of whole life cost.

Maintenance funding can go wrong in a number of ways. These are as follows.

Unnecessary strengthening

Bridge assessment methods generally contain an adequate safety margin and hence are very conservative. Many bridges found to 'fail' assessment will in fact not fail in reality even if nothing is done. Such 'unnecessary' work is carried out because it is not possible to identify precisely which bridges are actually going to fail. Hence it can be said that it is an uncertainty (or error) in the decision process which results in the unnecessary work being scheduled.

Missed Essential Work

Essential work due at any point in time is identified through 'present state' assessments. If 10 % of the bridges are assessed annually, then in a period of 10 years all the essential work arising in that period should be identified. In reality, however, this will not happen as there are many uncertainties in the process. As shown in Figure 4, present state assessments make use of information from existing documents such as drawings, or suffer from the lack of them, use information from inspections and tests, and it finally employs calculation-based, or sometimes, judgmental assessment.

All these processes are imprecise and ultimately a number of the bridges with 'actual' essential needs will be missed through over-estimation of their structural adequacy, resulting from the over-estimation during the individual component activity.

Wasted Preventative Work

The purpose of preventative work is to postpone essential work on bridges which may become sub-standard in the near future, say within a time horizon of 40 years. Preventative work can be wasted for three reasons. First, it can be applied to bridges which will not become sub-standard in the next 40 years or so, and secondly, the maintenance method used may not be effective on a percentage of the bridges. Furthermore, in some cases preventative maintenance may not be applied at an optimum time.

The first error may occur as a result of underestimating the adequacy of the bridges that will 'actually' survive beyond 40 years without essential work. The adequacy of bridges for the future is assessed through 'future state' assessments, generally known as whole life assessment. As shown in Figure 5, the future state assessment uses information from the present state assessment, and additionally from the as built details, information on past deterioration and predictions of future deterioration. Within a whole life performance assessment of a structure (Department of Transport, 1999) these uncertainties will be reflected in the performance predictions for the structure at the time of construction (I_o), the future predictions at time t (I_t) and the minimum performance level (I_{crit}) as shown in Figure 7. The combined effect may result in the final over-estimation or underestimation of the adequacy of a group of bridges. In the latter cases preventative maintenance will be applied unnecessarily or not applied at an optimum time. There is also uncertainty associated with the effectiveness of maintenance methods (highlighted in Figure 8) which may also result in ineffective use of resources.

Reliability Based Assessments

Uncertainties such as those shown in Figures 7 and 8 can be investigated using probabilistic methods. The use of reliability based assessments provides a framework for taking into account the various uncertainties associated with particular assessment parameters and provides a more consistent and rational basis for comparison of results. Furthermore, it provides a basis for 'managing' uncertainty more effectively. The parameters calculated using reliability analysis (i.e. reliability index and probability of failure) can be interpreted as a measure of the confidence in the structural component assessed and the assessment process. As the uncertainty increases so does the predicted probability of failure while the reliability index reduces. Where the uncertainty associated with specific parameters or methods can be reduced or better quantified this can be readily reflected in the reliability predictions resulting in higher level of confidence in the structural component being assessed. This would then provide justification for removing some of the conservatism previously built into the decision making process and provide a more rational and consistent basis for decisions in prioritisation of work.

In recent years considerable research has been undertaken in the area of reliability based assessments for bridge management and the use of these methods has been incorporated in new guidelines produced by some authorities (Canadian Standards Association, 1990, Department of Transport, 1997).

In particular, reliability analysis has been used as a framework for developing optimised inspection and maintenance strategies. Such methods make use of model updating on the basis of new information, which will reduce the predicted probability of failure as more precise knowledge is for example obtained through inspection measurements. Such techniques have been developed for bridges (Frangopol, 1998) and research work is continuing to provide methodologies and guidelines in this area. The benefits of such techniques have also been clearly demonstrated in other areas of application such as the offshore industry (Onoufriou, 1995) where significant safety and cost benefits have been achieved from their application.

The bridge management process can benefit from these developments and experience.

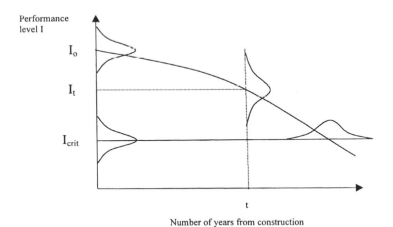

Figure 7. Uncertainty in whole life planning assessment

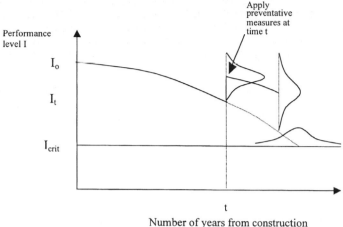

Figure 8. Uncertainty in the effectiveness of preventative measure

Reliability Analysis - Example

Reliability methods can be used to demonstrate and assess the sensitivity of the bridge management decision process to the various sources of uncertainty. An example of a simple reliability model is discussed here the purpose of which is twofold:

1. To demonstrate how different levels and characteristics of uncertainty can impact on the management decision process and ultimately on the way funds are used
2. The same model can be extended to examine the sensitivity of a component of the decision process to the various sources of uncertainty

To develop an appropriate model the following steps are performed:

- Model the governing limit state formula
- Introduce uncertainty variables to model parameter uncertainty
- Calculate the probabilities of failure and reliability indices corresponding to different levels and characteristics of uncertainty

In this example, the analytical model is that of a structural component, subjected to a loading 'L', which contains a defect 'd'. The effect of the defect is assumed to be a reduction in the intact strength 'R_i' given by a damage function 'S(d)'. The damaged strength 'R_d' is given by :

$$R_d = R_i - S(d)$$

Hence the limit state can be expressed as :
$$G = R_i - S(d) - L$$

We assume initially for the sake of this example that the main source of uncertainty is due to physical variability in the loading 'L' and that R_i and $S(d)$ can be assessed 'accurately'. Theoretically we can then predict the 'exact' probability of failure of this component which we can compare with an acceptable probability of failure of 1.0E-4 to decide whether this is sub-standard or above-standard. In this case the initial values are selected so that they give probability of failure of 1.1E-4 which makes this component just below standard.

Uncertainties associated with each of the terms in the limit state equation will affect the result and the maintenance decision on this component. In order to examine these effects we can model the uncertainty associated with each of the main parameters by introducing uncertainty variables 'e' as shown below:

$$G = R_i\, e_{Ri} - S(d)\, e_{sd} - L\, e_L$$

For example e_{Ri} is a normally distributed variable which is characterised by a mean value and a coefficient of variation (COV). This variable is multiplied by R_i to represent the uncertainty in the intact resistance.

Different combinations of mean and COV values are considered here which represent different levels and characteristics of uncertainty. For example a mean value of less than 1 would indicate a conservative bias while the coefficient of variation will indicate the degree of variability in the assessment (Table 1). Figure 8 shows how the reliability and probability of failure varies with different combinations of mean and COV in the uncertainty distributions for the intact resistance. These results highlight a number of possible outcomes of this assessment depending on the characteristics of e_{Ri}:

- If there is a non-conservative bias in e_{Ri} this will result in overestimating the intact strength R_i, and assessing this component as above standard (see points a and b on line A in Figure 9). As a result no essential work will be applied on this component which is in fact below standard.
- If there is a conservative bias on e_{Ri}, this will result in underestimating R_i and assessing the component as below standard (see points d and e on line A in Figure 9). Essential work will then be applied.
- The variability in the intact strength prediction represented by the COV in e_{Ri} will reduce the reliability prediction for this component (from line A to line B in Figure 9) and the component will be assessed as sub-standard and essential work applied.

Mean e_{Ri}	COV e_{Ri}	β	P_F	Characteristics of e_{Ri}
0.90	0	2.6	4.5E-3	conservative method
0.95	0	3.2	8.0E-4	conservative method
1.00	0	3.7	1.1E-4	Accurate
1.05	0	4.2	1.1E-5	non-conservative method
1.10	0	4.8	8.5E-7	non-conservative method
0.90	0.1	1.9	3.1E-2	combination of conservatism and variability
0.95	0.1	2.2	1.4E-2	combination of conservatism and variability
1.00	0.1	2.5	6.1E-3	variability in the assessment prediction
1.05	0.1	2.8	2.6E-3	combination of non-conservatism and variability
1.10	0.1	3.1	1.1E-3	combination of non-conservatism and variability

Table 1: Effect of Uncertainties in Intact Strength

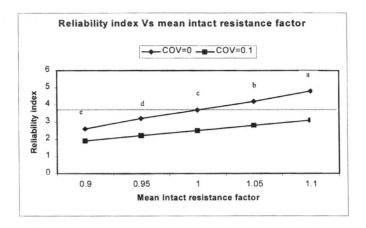

Figure 9: Effect of mean e_{Ri} on reliability index

The above example demonstrates how uncertainty and its characteristics can change the outcome of this assessment and in some cases result in incorrect funding decisions. Considering that such assessments are performed for a large number of components on a large population of bridges, the compounded effect of these may result in substantial errors in funding decisions and consequently wastage of resources. The maintenance expenditure for the Highways Agency's structures in the foreseeable future is likely to be of the order of £150m to £200m per annum. In view of the large sums involved, every effort must be made to reduce the sources of error in the decision process. To address this issue it is important to know the relative contribution of the various elements of uncertainty to the overall uncertainty in the assessments and ultimate funding decisions made.

The above model can be used to examine the effects of all contributions of uncertainty represented in this model (e.g. e_{Ri}, e_{sd}, e_L). The model can also be extended to represent the individual uncertainty contributions in the damage function S(d) for example. This may include components relating to the inspection, interpretation and assessment of the effect of the defect on the component. This model can then be used to calculate sensitivity measures. The sensitivity of the reliability index, β, to various parameters can be obtained from the probabilistic analysis in the form of $d\beta/d\theta$ (where θ is the parameter in question). Furthermore, importance factors can also be calculated which provide a measure of the relative contribution of each variable to the overall uncertainty in the assessment. Based on such an examination it would be possible to identify the areas where it is worth spending more effort to reduce or better quantify uncertainty which in turn will have the most beneficial effect on the bridge management decisions.

Further Work

Various issues relating to uncertainty and its effects on bridge management process were discussed in the previous sections of this paper and ideas put forward on how these issues may be addressed. Further work is clearly needed to progress this area. Based on the previous discussion a possible methodology is being proposed which can be summarised as follows:

- Develop a detailed flow chart of the bridge management process
- Identify individual sources of uncertainty
- Use of reliability models to examine sensitivity
- Investigate feasibility of developing system models to represent the process or sub-process
- Examine the relative contribution of uncertainties to component parts and the system as a whole
- A combination of quantitative and qualitative approaches need to be developed
- Identify and prioritise future research needs

RESEARCH NEEDS

The proposed methodology can be applied to obtain a better understanding of uncertainties and their relative effects on the bridge management process in order to prioritise research efforts accordingly. Some of the known key areas where further research is required and which need to be prioritised using the proposed methodology are discussed briefly below.

Existing Documents

The collection and maintenance of exiting drawings, details and reports relating to the structures is clearly an administrative matter. Where there is room for improvement, the necessary steps should be taken. Although it is good practice to store all information, invariably some choice has to be made. The most important items directly related to the decision process should be identified and carefully indexed and stored for future reference.

Inspection, Measurement and Testing

This is a fertile area of activity for errors, approximations and resulting uncertainty. For any particular investigation, usually there are a number of methods. The more precise they are, the higher are the costs of using them. Studies are now taking place to examine the probability of detection of some of these methods vis-à-vis costs. Such studies should continue. However, research efforts should be concentrated on the processes involved with the parameters important for the decision process. The inspection process should be optimised to provide the essential items accurately and with minimum use of resources. Research should be aimed at developing such methods.

Assessment Analysis

The methods of structural analysis used in assessments are now well established. However, there are certain areas where further work is still needed. These are

- Translation of inspection and test findings to structural strength parameters
- Uncertainty levels in the shear failure criteria
- Assessment methods for risk factors such as dynamic impact, steel and concrete fatigue, scour, pre-stressed concrete structures, thaumasite affected structures etc.

Other such areas of research need to be identified. It should however be borne in mind that the sophistication of the resulting methods and the magnitude of the efforts should be consummate with the likely impact on the final decision. In view of the other gross approximations only simplified methods of analysis may be justifiable.

Minimum Acceptable Criteria

These criteria are very important since they determine whether any maintenance work is necessary, and if so, its type and extent. However the criteria commonly used, such as a target probability of failure or a target reliability index, are only operational tools derived through comparison with existing codes and standards. They are deemed to satisfy the client's or society's requirements regarding safety and functional use. However, whether they actually do so has never been investigated and should be done. If these criteria are too onerous, they will unduly justify preventative work which will result in wastage of resources. On the other hand, if they are too lenient, 'actually' essential (urgent) work will be missed resulting in potentially dangerous backlogs for the future. Hence research should be aimed at examining whether the minimum acceptable criteria can be more focused on the client's and society's needs.

Safety and Functional Criteria

The main responsibility of the bridge manager is to maintain the structures in a safe condition with the minimum disruption to functional use. In practice, neither the safety nor the functional requirements are clearly stated anywhere. The ultimate measure of success of the funding decisions can only be in terms of the performance in these two aspects. Hence performance indicators need to be developed both for safety and functional use and consensus established.

Work Options

Once the assessment indicates that work is necessary, the type and extent of work have to be considered for the funding decision. Both are problematic until full investigations are carried out and the work is designed, which will not take place until funding is available. The uncertainty involved in predicting the work type and extent should be investigated.

Unit Costs

Predicting costs of yet to be implemented work is a problematic matter and involves a large amount of uncertainty. However, it directly affects the costing of the options for maintenance work, and hence, the final decision. Work should be carried out to identify the levels of uncertainty and to reduce them as far as practicable.

Deterioration

For taking funding decisions for preventative work, estimates of deterioration are necessary, both that which is present in the current state of the structure as well as that which will occur in the future. Much work is being carried out in this area and should continue. However, it should be borne in mind that the decision process involves cost discounting using a discount rate (6% in the UK). The accuracy of calculation of deterioration therefore becomes less important as the years become more distant.

CONCLUSIONS

This paper has been aimed at taking a closer look at the bridge management process with the objective of identifying and understanding better the various sources of uncertainties. The objective is to develop a rational method for focusing research in the areas where it will be most fruitful. It is hoped that with a rationally prioritised research programme, while recognising that uncertainties cannot be eliminated completely, it will be possible in future to considerably reduce their impact on the bridge management decision process and thus promote more effective use of limited resources.

The use of reliability models to examine the relative sensitivity of key steps in the bridge management process and thus to identify research needs has also been discussed in the paper. Models such as the one used in this paper can be refined and developed further to examine the relative significance of the various uncertainties introduced at the key stages of bridge assessments. Further development work is required in this area to investigate the feasibility of developing system models that represent the bridge management process. Such models should have the ability to examine the various component parts of the process with the objective of identifying the relative significance of the uncertainties that come into play.

Not all maintenance activities are amenable to precise quantification or mathematical representation. Hence, the examination of the bridge management process will require a combination of qualitative and quantitative approaches. It is hoped that through the process outlined it would be possible to identify the areas where better quality data, more reliable assessment models, or improved guidelines are needed to reduce or manage uncertainty more effectively. Given the potential impact on the effective use of limited resources it is important to be able to prioritise these topic areas for future research. It is hoped that the discussion in this paper sets the basis for taking these important issues further.

ACKNOWLEDGEMENTS

This paper is being presented with the kind permission of Mr Lawrie Haynes, Chief Executive of the Highways Agency.

REFERENCES

Canadian Standards Association, Supplement No. 1-1990 Existing Bridge Evaluation to CAN/CSA-S6-88, Ontario, Canada, 1990.

Das P.C., 'New Developments in Bridge Management Methodology', Structural Engineering International, Volume 8, Number 4, 1998.

Das P.C., 'Safety of Bridges', Thomas Telford, 1997.

Department of Transport, BD 21/97 : The assessment of highway bridges and structures. Design Manual for Roads and Bridges. The Stationery Office, London, 1997.

Department of Transport, Draft BD: Whole Life Performance Assessment of Highway Structures and Structural Components', 1999.

Frangopol D., Estes A., 'Optimum Design of Bridge/Inspection Programs Based on Reliability and Cost', Conference on Management of Highway Structures, London, 1998.

Institution of Civil Engineers, Highways Agency, 'The Management of Highway Structures', Conference Proceedings, London, 1998.

Onoufriou T., 'Experience with Reliability Based Inspection Planning of Offshore Structures'. Invited paper, Seminar on Industrial Applications of Probabilistic Methods, Electricite de France, Paris, 1995.

Onoufriou T., Forbes V., 'Developments in Structural System Reliability Assessments Of Fixed Offshore Platforms', Submitted for publication to the Journal of Reliability Engineering & System Safety, 1999.

Onoufriou T., "Overview Of Advanced Structural And Reliability Techniques For Optimum Design Of Fixed Offshore Platforms", Journal of Constructional Steel Research, Vol. 44 No.3, December 1997.

Lake Koocanusa Bridge rehabilitation

G DAVID BRIERLEY-GREEN C.Eng, P.Eng, P.E.
World Wide Infrastructure Ltd Olympia, WA USA

HISTORY

Koocanusa Bridge in Northern Montana was constructed in 1969 as part of the Libby Dam Project and is situated approximately fifty miles up steam of the dam. The bridge was built across the Kootenai River before completion of the dam and flooding of the river valley. The river elevation at the time of construction was approximately 2244 ft and after construction and flooding of the valley the high water level rose to 2459 ft. The supporting piers adjacent to the central spans are 235 feet in height. The bridge superstructure is of conventional form constructed using a steel truss system varying in depth from 35 ft at mid span to 55 ft at the piers. The deck was constructed using lightweight concrete having a 28-day compression strength of 4000 psi. The bridge elevation can be seen in Figure 1.

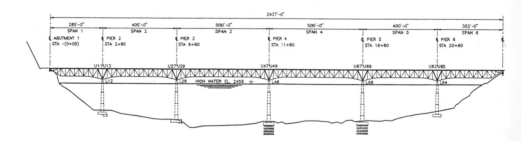

Figure 1. Bridge Elevation

The bridge performed well for the first twenty years except for minor problems with the expansion joints and the resulting deterioration of the paint system and steel in the vacinity of the intermediate deck joints. However in the mid to late eighties the deck concrete started showing signs of deterioration. Deck water contaminated with road salt was beginning take its toll. Over the next four years the deck surface delaminated and the embedded rebar started to corrode. In 1992 it had become apparent that the deck would have to be replaced and the bridge painted. It was also decided at this point in time to upgrade the bridge seismically so that it would comply with more recent stringent code requirements. The lateral design capacity of the structure had been based on wind loading, seismic loading had not been considered. A typical section of the original bridge can be seen in Figure 2.

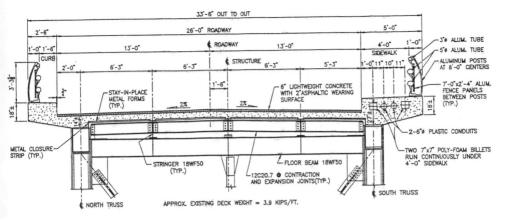

Figure 2 Typical Bridge Cross Section

FIELD SURVEY

In 1993 Arvid Grant Associates, Olympia, Washington was commissioned to provide an in-depth evaluation of the structure, determine vulnerable components under seismic loading, and make recommendations for rehabilitation. A thorough field survey of the bridge was deemed necessary before analytical work could commence. Experienced deep-sea divers performed a tactile survey of each pier and produced videotape of their findings. Paint samples were also taken for analysis and paint adhesion tests carried out at random locations. Various paint systems were also applied on prepared surfaces to determine the most suitable product for re-painting the structural steel. The structural steel was also closely inspected for damage, deterioration, and evidence of fatigue cracking. It was already known that the deck concrete had failed but never-the-less core samples were taken and the concrete evaluated. A general survey of the structure was also performed to determine if there had been any settlement or movement of the structure since completion of construction.

RESULTS OF SURVEY

Overall the structure was found to be in good shape. There was no fatigue cracking in the structural steel and the existing paint system had good adhesion to the steel over the vast majority of the structure. Some peeling of the paint was found at intermediate deck joint locations but the structural steel was not badly corroded. Pier concrete was also found to be in good condition with no sign of deterioration. The deck concrete was found to have a compression strength of 5000 psi and an adequate air entrainment level over the majority of the deck area but lower than the specified amount in the sidewalk and outside edges of the deck. One unusual finding was the alignment of the rocker bearings on piers 4 and 6. At mean temperature the top and bottom sole plates were not parallel. This may have been due to rotation of the pier since construction or an incorrect alignment when the bearings were installed. There is no way to determine which of these is correct.

EVALUATION OF THE STRUCTURE
During field survey work a 3D mathematical model of the bridge, using the finite element program SAP90, was prepared. In the mathematical model the chords of the truss have been treated as continuous beams. All other truss members are represented by pin ended struts. The deck has been considered as non-composite and is represented by applied loads along truss chords and longitudinal stringers. Piers 3, 4, and 5 carry all longitudinal forces applied to the structure and the transverse force is shared between the piers and the abutments. The reinforced concrete piers have been modeled down to foundation level using beam elements.

A site-specific response spectrum was developed for use in the seismic analysis. A peak ground acceleration of 0.13g has been used for the Design Earthquake (DE) and 0.18g for the Maximum Credible Earthquake (MCE). Fortunately, Montana is not very seismically active and these are relatively low values. The hydrodynamic effect of the water on the piers during a seismic event has also been included in the analysis.

The live and dead loading applied to the structure and the design check criteria is as prescribed by the AASHTO Specification for Highway Bridges. HS20-44 being the loading selected for the check and for rehabilitation.

On conclusion of the field survey and the analysis the following items were considered necessary for rehabilitation of the structure and to prevent damage under the forces generated by the Maximum Credible Earthquake.

- Replace all bearings
- Increase bearing seats at abutments and piers 2 and 6
- Replace deck concrete
- Strengthen top of piers 3, 4, and 5
- Strengthen floorbeam connections
- Make deck stringers continuous
- Clean and paint all structural steelwork
- Remove and replace drainage system

The truss was found to have no vulnerable components.

CONFIRMATION OF ANALYTICAL MODEL
Retrofitting large complex bridge structures is extremely expensive and in some instances the cost of retrofit can well exceed the replacement cost of the bridge. It is therefore very important to ensure that the analytical model on which recommendations are to be based should truly represent the real bridge. If this confirmation is not obtained, there can be no sound basis on which to justify recommendations for retrofit.

Dynamic field testing of Koocanusa Bridge was performed on completion of the field survey. A typical time domain reading and measured frequencies for the first longitudinal, lateral, and vertical motions of the bridge can be seen in Figures 3, 4, 5, and 6 respectively. A comparison between the analytically derived values and field-measured values can be seen in Table 1.

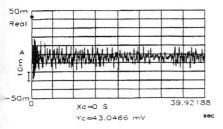

Figure 3. Typical Time History

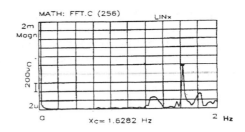

Figure 4. First Longitudinal Frequency

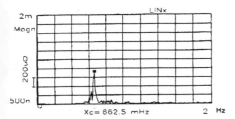

Figure 5. First Lateral Frequency

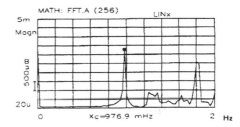

Figure 6. First Vertical Frequency

Table 1. Comparison of Field Measured and Analytically Derived Frequencies (Before Rehab)

Predominant mode And Direction	Measured Frequency (Hz)	Calculated Frequency (Hz)	% Difference in Values
1. Lateral	0.66	0.61	-8
2. Vertical	0.98	0.91	-7
3. Vertical	1.27	1.20	-6
4. Vertical/Longit	1.32	1.36	+3
5. Vertical/Longit	1.62	1.55	-4

It can be seen from the above table that there is good agreement between analytically derived and field measured values. The greatest difference is in the transverse direction, which is not surprising as the analytical model excludes the lateral stiffness of the deck when working compositely with the truss. Composite action will always occur regardless of design assumptions used. This effect would be limited by the action of the deck joints but never the less contribute some additional stiffness.

RETROFIT EVALUATION

Having determined which components needed to be replaced or strengthened and the analysis validated the mathematical model could now be revised. Changes to be made to the analytical model reflect the differences between the original structure and the proposed retrofit structure. The revised cross section can be seen in Figure 7

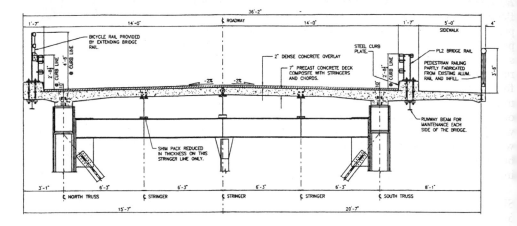

Figure 7. Typical Bridge Cross Section after Rehabilitation

Changes to the model and the structure included:

Stringers - The original deck was constructed with intermediate deck joints at approximately 120 ft centers these deck joints had proved a maintenance problem and hence they were removed. The new deck was made continuous, therefore, the longitudinal stringers supporting the deck had also to be made continuous.

Deck - Lightweight concrete had proved to be unsuitable at this location so regular weight concrete was selected for construction of the new deck. The structure being straight and of constant width for the majority of its length suggested the use of precast prestressed concrete panels, however, five different deck systems were designed and associated cost determined:

- Cast-in-place concrete with precast concrete forms $36 / Sqft
- Cast-in-place concrete with regular forming $36.
- Reinforced precast concrete panels $36
- Prestressed precast concrete panels post-tensioned longitudinally $31
- Concrete filled steel grid $41

Costs are in 1994 US dollars

Precast prestressed concrete proved to be the most economical and would also provide a high strength durable concrete deck that could be quickly and efficiently placed. The use of regular weight concrete added additional self weight loading to the bridge, the model loads applied to the stringers and the chords had to be revised to account for this increase.

Bearings - To reduce force carried by the central piers 3, 4, and 5 under seismic loading a flexible medium had to be placed between the steel superstructure and the pier top. There are proprietary products that can be used that are designed individually and specifically for this purpose. However, this can prove to be an expensive solution. A more economical route was sought, one that introduced more competition for selection of this item. A number of computer runs using the analytical model with varying longitudinal, lateral, and vertical spring stiffnesses for the bearings quickly provided a range of stiffness that would be acceptable and keep the max/min pier forces within member capacities. Having established this range it was clearly laid out in the specification and the field was then open to any manufacturer to provide a bearing that would fall within the stipulated parameters whether proprietary or generic. The retrofit analysis was based on the use of an elastomeric bearing having stiffness within the specified range, and being 44 inches wide, 50 inches long, and 13 inches in depth. These bearing allowed +/- 8 inches of movement in the longitudinal direction.

Sliding bearings were placed at all other pier and abutment locations that provided +/- 21 inches of longitudinal movement which was the estimated movement in that direction due to Maximum Credible Earthquake plus 50% of wind and temperature loading. Arriving at this design movement does require engineering judgement as codes give little guidance on this topic but it was felt that it was unreasonable to expect the earthquake only to occur at mean temperature and with no wind loading. Other engineers may see it differently, of course, but this was our best judgement at the time using our knowledge of site conditions.

Central Piers - The forces carried by the piers had been reduced considerably by the selection of the replacement bearings. However, some additional confinement of the concrete at the top of piers 3, 4, and 5 was necessary. This was due to increased bursting stresses resulting from the increase in dead load. Post-tensioned high strength steel bars in the transverse and longitudinal directions combined with the axial loading provided a state of triaxial compression in the top of the piers which proved sufficient to neutralize the bursting stresses.

Painting Structural Steelwork – This portion of the work is generally the least interesting to the structural engineer but it certainly proved to be a challenge. Lead base paint was commonly used for the protection of structural steel until quite recently. It is now regarded as a toxic material and as such it must not be allowed to contaminate the environment. This certainly poses a problem when a structure of this size over a reservoir needs to be re-painted. An enclosure had to be erected that completely encapsulated the structure over the length being cleaned and painted. There had to be a negative pressure inside the encapsulated area so that air from outside would be drawn in at leakage points. Strict limits on the amount of lead released at the site during this operation were in effect. Sediment samples along the length of the bridge had to be taken before and after painting to determine if there was any change in lead content that could be attributed to the painting process. Needles to say this proved to be a major cost item of the total rehabilitation cost of $8,500,000, painting the structure amounted to $3,500,000.

Abutments – Both abutments required minor modification. The backwalls had to be replaced to accommodate the slight shift in alignment and the new expansion joints, and the bearing seats had to be extended. The main complication regarding this work was keeping the bridge open to one lane of traffic during rehabilitation.

Guard Rails – Guardrail design has changed significantly over the last decade. Only systems that have been crash tested can be used on major highway routes or on Federally funded projects. This being the latter, even though it carries a small volume of traffic, had to be fitted with such guardrails. The existing aluminum guard rails, though in good condition, could not be reused unless crash tested. New steel guardrails were, therefore, used throughout and the existing aluminum rails and infill panels re-used for the pedestrian railing.

One complication surfaced from the Public Participation Meetings. The total rejection of closure of the bridge during the rehabilitation phase. Even though the bridge carried a light traffic volume the link was considered critical to the local community and the logging industry.

REHABILITATION
A contract for rehabilitation was let in the fall of 1995. Work in the field commenced in the spring of 1996. Careful consideration had to be given to the sequence of work to ensure safety and continued use of the structure. The following sequence was adopted:

- Remove one lane of existing deck in sections starting at one end of the bridge and working to the other, also work performed on abutments. Then complete adjacent lane while allowing traffic to use the lane just completed. With the exception of a few organizational problems at the start of panel placement this proved to be a fast and efficient method of deck replacement.

- On completion of the deck replace bearings on piers 3, 4, and 5. This required some preparatory work before the bridge could be lifted and the existing bearings removed. Mis-alignment of the top and bottom sole plates of the fixed bearings at mean temperature may have suggested that a longitudinal force existed at the pin location and that when the structure was jacked the top of the pier might move longitudinally relative to the truss. Or even topple the jacks and cause a catastrophe. The question to be answered was what restraint force should be provided for. After much discussion and a few sleepless nights a value of 150 kips along each chord line was determined. The restraint system the contractor selected utilized high strength bars connecting each bottom truss chord to the pier top on both sides of the pier. It was only necessary to raise the bridge two inches vertically to remove the existing bearings. This movement increased the tension in the bars and this combined with the initial pretension was sufficient to hold the two components firmly together. All bearings were replaced without incident. Finally the tops of the piers were post-tensioned.

- Cleaning and painting the steelwork could now commence. The enclosure developed for painting comprised an aluminum gondola, which was floated out then, raised by winch to the underside of the truss. The gondola was slightly wider than the truss allowing for tarpaulins to be raised each side. Raising and lowering the tarpaulins had to be done when the wind speed was less than 5 mph or it became very difficult to move those situated on the windward side. Once the tarpaulins were in place joints were sealed, a vacuum generated using filtered extract fans, and the painting could commence. Once painting had been completed within the enclosure the enclosure was moved forward to the next section. Insufficient time remained to complete painting before the end of the first construction season; hence. Painting was completed in the summer of 1997.

In accordance with the philosophy outlined above regarding confirmation of analysis the structure was again field-tested after completion of the rehabilitation. The first ten modes taken from the analysis and the corresponding field measured values are given in Table 2.

Table 2. Comparison of Field Measured and Analytically Derived Frequencies (After Rehab)

Predominant mode and Direction	Measured Frequency (Hz)	Calculated Frequency (Hz)	% Difference in Values
1. Longitudinal	0.2750	0.2731	+6.7
2. Lateral	0.2964	0.2971	-2.4
3. Lateral	0.3245	0.3337	-2.7
4. Lateral	0.4556	0.4436	+2.7
5. Lateral	0.5305	0.5083	+4.4
6. Lateral	0.6210	0.6363	-2.4
7. Lateral	0.6865	0.6865	0
8. Vertical/Longit	0.7614	0.7210	+5.6
9. Lateral	0.7239	0.7847	-7.7
10.Lateral	0.8019	0.7976	-5.4

It can be seen that good agreement exists between the two sets of data and we are confident that the structure will behave as predicted when subjected to the prescribed loading. Such in-depth testing is unusual but we believe necessary for the more complex structures. The cost of testing, before and after rehabilitation amounted to $40,000 or 0.5% of the cost of rehabilitation. A small price to pay for confirmation and reassurance.

ACKNOWLEDGEMENTS
I would like to acknowledge the support received from Dr. Arvid Grant, Dr. JJ Lee, Dr. Florian Neuner, Dr. Mark Rogge, Tom Whiteman, Steve Hall, and Raj Bharil during the field survey, analysis and design phase of the work and from Marco Vencroso, and Marion Barber of the Federal Highway Administration during the construction phase. The success of this project is a direct reflection of the dedication and effort of these individuals.

Sharavathi Bridge rehabilitation, India

M. A. SOUBRY, BSc(Eng), ACGI, MICE, MIStructE, CEng.
Hyder Consulting Limited, Guildford, UK
S. G. JOGLEKAR, BSc, BE, MSc, FIE(India)
STUP Consultants Limited, Mumbai, India

Abstract
Co-operation between British and Indian engineering led to successful and economical
prolongation of the life of the prestressed concrete bridge in the tidal zone of the Sharavathi
River. A deck overlay and external prestressing beneath the spans were added. A novel
engineering solution was devised using steel brackets to re-distribute part of the load away
from deteriorating pier cantilevers and half-joints.
Keywords: Bridge, concrete, cost effective, external cable, gap slabs, grouting, half joints,
India, innovative solution, marine environment, prestress, rehabilitation, steel, test load.

1 INTRODUCTION

National Highway 17 crosses the 1 km wide estuary of the River Sharavathi. The 34 span
bridge was originally constructed in 1970. 20 years later the bridge was showing significant
deterioration and traffic restrictions were in force. The spans (each just over 30 m) are of pre-.
cast, pre-stressed post-tensioned, concrete girders supported through half-joints.

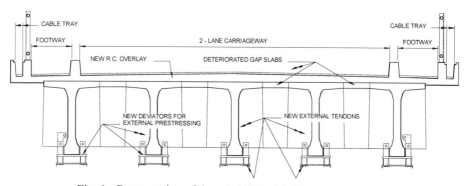

Fig. 1. Cross section of the rehabilitated bridge superstructure.

The consultants, STUP and Hyder, won the competitive tender for the investigation and
design for the rehabilitation of the Sharavathi Bridge. The group included, as sub-consultants
to Hyder, the Transport Research Laboratory (TRL), and STATS Limited. The contract was
with the Government of Karnataka, acting as agents for the Ministry of Surface Transport
(MOST).

The project was divided into three phases:

      I   Investigation

      II   Design of the rehabilitation (with alternative cost/life extension scenarios)

      III   Construction

2   PHASE I - INVESTIGATION AND SCHEME OPTIONS

2.1   INVESTIGATION

The project scheme was to initially investigate only three of the spans and the scheme selection was based upon the outcome of this. Great care was taken to ensure that the spans selected for investigation were representative of the whole bridge. The temptation to target particular areas where deterioration was particularly apparent was resisted. Following a general inspection of the bridge and a quantitative review of the PWD's previous inspection reports, a span was chosen to represent each of the worst, the least and the average amount of perceived levels of deterioration.

Expatriate staff and equipment from the UK specialist testing consultant, STATS, were mobilized to India for the detailed investigation and testing. The work was carried out together with the local consultant's personnel in a carefully planned programme. This ensured the transfer of technological knowledge, the last of the 3 spans being done by the local technicians with only advice from the specialists.

As the evidence was assembled, TRL provided advice and input on the state of the art in investigative measures, stress determination and rehabilitation methods.

To ensure the approach to the solutions was logical the investigation sought to answer the following questions for each element of the structure:

(a)   What was the nature of the defect?

(b)   What was the cause?

(c)   Could the cause be eliminated so that the progress of deterioration was halted or sufficiently reduced?

(d)   What was the extent of the defect, i.e.:

     (i)   had the strength reduced unacceptably, or what was the residual strength?

     (ii)   what type of remedial work was necessary, i.e. was it repairable or in need of structural alteration?

(e)   Could the remedial work to the particular element be better combined into an overall solution to the rehabilitation?

The wide ranging problems identified at the bridge were consolidated in tabular form (see Fig. 2). The table listed the defects found, their extent and severity, their causes and possible remedial measures. This allowed preparation of integrated and comprehensive schemes and ensured, as far as possible, that on-going deterioration would not be built-in to the rehabilitated structure. The choice of remedial work was developed from the tabulated results to suit the needs of the whole of the bridge. Apart from one particularly badly deteriorated span that was treated with emergency repairs under a separate contract, the investigation showed that spans were suffering to a generally similar extent. This was contrary to the

Sharavathi Bridge Rehabilitation - Phase I

	ELEMENTS	DEFECTS – Nature	Severity	Extent (> %)	CAUSES – Nature	Controllability	REMEDIAL MEASURES – Activities Required	Recommendation and Comment
SUPERSTRUCTURE	Main girders	Corroding tendons:	2	50	Water in ducts, chlorides:	Possible	External tendons, protect:	Implement
		Failed web:	4	*	Insufficient cover:	*Not possible	*Add shear plates:	*Implement
	Half joints in main girders	Cracks	2	50	Corrosion of reinforcement	Very difficult	Remove girders for access	Change structural system
	Girder half-diaphragms (cross girders)	Corroding reinforcement, spalling concrete	3	50	Chlorides	Very difficult	Remove girders for access	Change structural system
	Insitu deck slab concrete	Corroding tendons, slack tendons, spalled concrete	3	50	No grout, unstressed, chloride	Difficult	Add r.c. overlay slab	Implement
	Insitu diaphragm	Corroding reinforcement, spalled concrete	2	50	Chlorides	Possible	Cut back, prepare and repair, cosmetic only	Implement
	"Extra" counterweight diaphragms	Corroding reinforcement	2	50	Poor workmanship	Possible	Cut back, prepare and repair, cosmetic only	Implement
	Footway cantilevers and connections	Cracking and spalling concrete	3	50	Chlorides, poor detailing	Not possible	Replace all	Change to cantilever slab
	Footway planks and bedding	Corroding reinforcement cracked & spalling concrete	4	50	Chloride, poor construction	Not possible	Replace all	Change to cantilever slab
	Deck surfacing	Wear and damage	2	30	Insufficient maintenance	Possible	Strip, waterproof and re-surface	Add overlay slab also
	Expansion joints	Blocked and failing	3	30	Inadequate maintenance, poor detailing	Not possible in most cases	Replace with sealed system	Use ThormaJoint
SUBSTRUCTURE	Bearings	Corroded:	1	50	Insufficient maintenance:	Yes	Replace and grease box:	Implement
		Irregular inclinations:	2	30	Inclination cause unknown:	Possible	Realign where necessary:	Implement
	Cantilevers (hammerheads)	Corroding tendons	2	50	Chloride, poor design detail	Very difficult	Remove girders for access	Change structural system
	Half joints in cantilever	Corroding tendons, cracks	2	50	Chlorides, poor design detail	Very difficult	Remove girders for access	Change structural system
	Cantilever half-diaphragms (cross-girders)	Corroding reinforcement, spalling concrete	2	30	Chlorides	Very difficult	Remove girders for access	Change structural system
	Columns	Corroding reinforcement	1	30	Chlorides, carbonation of concrete	Possible	Monitor	Implement
	Counterbalance at land spans	Alkali silica reaction:	1	30	Reactive aggregates, chlorides	Difficult	Change structural system	Implement
		Corroding tendons:	2	-				

Key to severity codes:

0 None
1 Slight (need monitoring)
2 Developing (will affect strength, needs planned attention)
3 Severe (needs urgent repair)
4 Unacceptable (safety in doubt)
* Indicates one girder in structure

Fig. 2. Tabular summary of investigation results

perception when selecting spans, which was based upon the external, visually observable, deterioration.

A full report on the Material Investigation and Testing and the Repairs and Strengthening Schemes was prepared by the Consultants.

The tabulated remedial measures for each of the elements of the bridge indicated that extensive work in dismantling the bridge would have been necessary to return the elements to a satisfactory state. Therefore, alternative means of repair were sought, particularly by seeking suitable modifications to the structural system of the original bridge.

2.2   SCHEME OPTIONS - SUSPENDED SPANS

The potential weakness of the prestressing cables in the main girders was a primary consideration. While much of the deterioration of the prestressing tendons observed and measured during the inspections was superficial, there were significant lengths of cable which were inadequately grouted and severe corrosion damage was noted in some places. Coupled with observations that relaxation losses were potentially higher than originally designed, the considered judgement of the Engineers was that up to 20% of prestress could have been lost in some of the beams. Replacement of the lost prestress was therefore a favoured solution and the addition of external tendons to each of the girders was introduced into the scheme (see Fig. 3). Some of the prestressing was expected to have retained its effectiveness but there was no practical, non-destructive, way of establishing which. It was therefore necessary to carry out sensitivity analyses as the scheme was developed to ensure that overstresses did not result from the additional tendons in either extreme of deterioration, i.e. zero, or the 20% adopted.

Fig. 3. View of external prestressing

The weakness of the slabs between the girder flanges, the gap slabs, where transverse prestressing had become ineffective, was another primary consideration. The road surface was punching through into holes. Replacement of the gap slabs would have been difficult and with limited probability of success. An overlay slab, with the convenience of using the existing flanges and ineffective gap slabs as the soffit shuttering, was likely to be the most cost effective solution if its extra weight could be carried.

In this populated area pedestrians regularly use the bridge to cross the estuary. Failure of the concrete footway planks had led to alarming and dangerous conditions. The footway system was beyond repair and an alternative system needed to be found.

Additional prestress to the main girders was at risk of overloading the original concrete cross section. The addition of an overlay slab would serve to balance the additional stresses, both by its additional dead weight, and the increase in the second moment of area which would result from its composite action. Extension of the overlay at the edges would then be able to provide new cantilever footways and the fundamental form of the repair scheme for the suspended spans was confirmed.

2.3   SCHEME OPTIONS - PIERS AND LOAD BEARING JOINTS

The prestressing cables of the hammerhead cantilevers were in a similar condition to those in the main girders. Poor accessibility precluded repairs to the existing members and strengthening by additional external prestressing was impractical. One of the weakest parts of the load path was the half-joints. An alternative load path was sought.

Two primary options of providing support to the ends of the suspended spans, by-passing the half-joints, were investigated. One was to provide direct support upwards from the top of the caisson and the other would use brackets and tendons at the top of the pier. Several versions of each were considered. The direct support would have been more straightforward and provided long term durability but would have used more materials and have interfered slightly with navigation clearances. The brackets were a more complicated engineering solution, but their flexibility allowed variations of load transfer to be controlled. They were also the most economical solution.

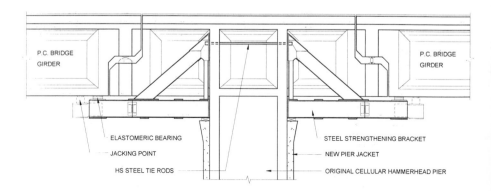

Fig. 4   Arrangement of pier strengthening bracket

3   PHASE II - DESIGN OF SELECTED SCHEME
3.1   THE SCHEME

The detailed design of the rehabilitation was prepared in India for the start of construction mid 1995. A choice of scheme had been made available to the Client taking account of the Client's intention to procure an additional new bridge nearby in the future. A short term, but nevertheless comprehensive, rehabilitation was instructed by the Client. The scheme fundamentally involved:

(a)   The addition of an overlay slab to eliminate problems with
    (i)   gap slabs,
    (ii)   deck waterproofing,
    (iii)   footway cantilevers.

(b)   The incorporation of external prestressing of the girders to
    (i)   support the additional overlay slab,
    (ii)   provide for potential loss of original prestress.

(c)   The addition of strengthening brackets to the pier tops to account for
    (i)   the deterioration and potential loss of prestress in the hammerhead,
    (ii)   potential problems at the half joints,
    (iii)   the additional weight of the superstructure overlay slab.

(d)   The enclosure of the cellular columns in a new reinforced concrete jacket to

    (i)   protect the columns against further attack by chlorides from marine spray,
    (ii)   provide positive bearing for the steel pier top brackets strengthening the hammerheads.

The brackets introduced in (c) above were designed to provide load relief and, for economy, were not designed to carry the full bridge superstructure loading.

Once the new strengthening brackets were incorporated at the pier tops, replacement of (original) bearings was to need a temporary supporting frame to ensure that the pier top brackets were not overstressed. Although provision was made, it was concluded during the rehabilitation that no bearing needed replacement.

The reinforced concrete pier jackets were designed to carry the full loading, even if the original pier concrete became ineffective.

Many general repairs and additional detailed components of the rehabilitation scheme were also necessary.

3.2   EXTERNAL PRESTRESSING AND PROTECTION OF EXISTING CABLES

The additional prestressing of the main girders was introduced by passing the cables up through holes cut in the top flanges of the T girders to new anchor plates cut and bedded into the top, the anchorages being finally covered by the new deck overlay concrete. The cables follow down each side of the girder web until they are turned parallel to the flanges over a fabricated steel deviator block that is saddled across the soffit of the girder at the quarter diaphragms.

Only 4 cables are provided per girder under this rehabilitation project, but ducts and deviators were put in place at the same time so that an additional 2 cables per girder can be threaded through and stressed if needed in the future. To access the anchorages a hole would need to be cut in the surfacing and the overlay to expose a new anchorage.

Protection against corrosion is a most important part of the viability and durability of the scheme. HDPE (high density polyethylene) ducting was provided continuously from one anchorage, bending round the deviator blocks, to the other anchorage. Grouting points were provided at the low points in all the ducts and vent points at high points at the anchorages so that grouting with cement grout of all the cables from low to high level could be seen to be complete.

Since investigations had indicated the existence of poor grouting of existing embedded cables the repair scheme included attention to this. Holes were drilled at strategic locations from outer surfaces to reach the cable ducts, and they were re-grouted to the extent that they would accept grout. Considerable intake of grout was observed.

3.3   PIER BRACKETS

As the scheme was developed the introduction of pier brackets was seen as a cheaper way of providing support to the ends of main girders. Additionally, the inherent flexibility of the brackets offered a different way of thinking of the support provided. Whereas the rigidity of the direct support option (see section 2.3) required struts to be designed to carry the full design load, the flexibility of the brackets offered the opportunity to adjust the reaction they carried. The brackets were initially introduced with the intention of limiting their capacity to the difference between the designed strength of the bridge and the potential degraded strength. This had the disadvantage that, should the pier hammerheads or half joints deteriorate in future to the extent that they can no longer support the bridge under live load, there was an inherent risk of collapse when that condition was reached. The strength was, therefore, tuned to a more cost effective solution as follows.

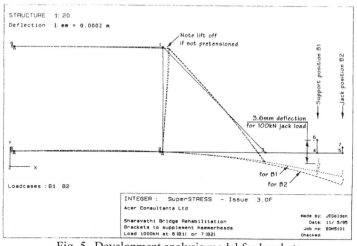

Fig. 5. Development analysis model for brackets

The ultimate capacity of the brackets was kept large enough to carry the characteristic dead load of the suspended deck. The philosophy of this is that the (unlikely) risk of complete collapse outlined in the previous paragraph is eliminated. Clearly, the bridge would be showing signs of distress at such a stage and live load could no longer be carried, but safety aspects would be satisfied and remedial action taken in a controlled way, rather than as an emergency.

The concrete bridge is relatively stiff compared with the brackets and, for the brackets to function at all, a pre-load needed to be introduced at the bearing. This was achieved by designing the bracket with a separate jacking position. Load was applied with the jacks and held while the permanent remedial works elastomeric bearings were set in position and shimmed to exactly the right thickness to fill the space between bracket and underside of girder.

The required jacking force to introduce the bracket pre-load needed to take account of the stiffness of the hammerhead (as it would deflect upwards as the load was relieved from it) and the compression of the elastomeric bearing as it picked up the load. It was calculated by plane frame analysis taking particular care with regard to the end constraints. The load was limited to ensure that the applied load did not lift the girders off the existing steel bearings

High strength steel tie rods passing completely through the cellular pier provided the main tensile member of the bracket. The ties were at first designed to be pre-loaded to ensure that the bearing plates did not lift off the concrete face as the ties extended under load. The analysis model indicating this lift-off is shown in Fig. 5. Without adequate pre-load the lengthened bar would allow the bracket system to "rock" However, it was found to be not possible to pre-load the rods but, in the process of transferring loads, jacking of brackets was carried out simultaneously on two sides of each pier. Thus the pre-load was balanced on the two sides of the common tie and shimming the gap prior to release of the jack load ensured a snug fit. This behaviour was taken into account in the final analysis.

4   PHASE III - CONSTRUCTION
4.1   WHOLE BRIDGE SURVEY
Completion of inspection and materials testing of the spans, other than the three representative spans originally selected by the Consultants, was carried out by the Contractor and by the

Consultants' representative posted at site during construction. Supervision of repair work was by the Client supported by technical advisors from the Consultants.

To confirm the behaviour of the bridge spans one full span was test loaded. As load was applied using sand bags the girder bottom bulbs were closely observed to establish the onset of cracking. A fundamental purpose of the test on these simply supported spans was to establish the effective net cable force. This was possible by back calculation once the tensile strength of the concrete had been overcome and made to zero at the crack locations. Release of the load and re-load ensured the repeatability of the test and confidence in the result. The test also allowed the

Fig. 6. View of pier and typical outer support bracket

stiffness of the concrete hammerhead portion of the pier to be confirmed and give greater confidence to its influence on the load sharing between the steel bracket and hammerhead.

4.2   CONSTRUCTION DIFFICULTIES

The bridge is an important link and, although traffic restrictions were in force, it was not considered acceptable to close the bridge entirely. Any diversion would have been unacceptably long. During the progress of the Works heavy vehicles were diverted to a ferry nearby while light vehicles were permitted to cross the bridge, one way at a time past the Works in progress, controlled by signals and traffic police. One difficulty was the continuing passage of prohibited vehicles across the bridge particularly at night, with consequent likely damage to partially complete work, or to freshly poured concrete.

The positions of the caissons supporting the bridge piers had been constructed with a very large tolerance on the nominal span lengths. The hammerhead lengths had been varied to match the standard length of pre-cast girders. Therefore, all pier brackets needed to be individually fabricated to match the as-measured dimensions and could not be manufactured in batches. In one case the caisson was out of position to the extent that a corbel needed to be cast to reach the correct alignment and pick up the load.

4.3   COST

The cost of the repair work amounted to about Rs. 10 Crores (equivalent to approximately £Sterling 150,000). The new bridge, expected to be constructed in due course, has an estimated cost of about Rs. 60 Crores at 1998 values.

4.4   PROGRAMME

The actual rehabilitation took place over a period of 3 years, during which the particularly heavy monsoon rains in this region take out 4 months of the effective working time each year.

5   CONCLUSIONS

At the start of the project, restrictions on traffic load and volume were in force. The bridge now allows all normal traffic to freely cross the river.

The work on this important transportation link has been completed successfully as a result of the good team spirit amongst all concerned. The Contractor satisfactorily overcame the logistical problems created by the 1 km length of the bridge over the deep water estuary.

The life of the bridge has been prolonged for a significant number of years for a reasonable outlay. The projected second bridge can be developed with proper planning and without unreasonable programme pressure.

6   ACKNOWLEDGEMENTS

The authors are grateful to the engineers in the Ministry of Surface Transport and Authorities in Karnataka PWD for the successful execution of the work, and for the contribution to the design from the Engineers of our respective firms.

Bibliography
Banerjee, A.K., Somasekharappa, Joglekar S.G. and Manjure, P.Y. (1998). *Rehabilitation of Sharavathi Bridge at Honnavar, Karnataka.* Indian Roads Congress.

A seismic retrofitting method of existing steel bridge piers

M. MATSUMURA, Research Associate, Department of Civil Engineering, Osaka City University, Osaka, Japan, T. KITADA, Professor, Department of Civil Engineering, Osaka City University, Osaka, Japan, and T. KAGAYAMA, Maintenance and Facility Design Department, Hanshin Expressway Public Corporation, Osaka, Japan

INTRODUCTION
After the Hyogo-ken Nanbu Earthquake, many energetic investigations [1] have been carried out for developing ductile steel bridge piers, which can support superstructures without increasing their elastic strength against such a strong earthquake as the Hyogo-ken Nanbu Earthquake. On the basis of the results of these researches, the seismic design method in the Japanese Specifications for Highway Bridges (JSHB) was revised in December 1996 [2]. In this seismic design method, the composite bridge piers of which steel columns are filled with concrete inside them are recommended as one of the most effective and economical ones. No recommendable structures seem to be specified except for the rectangular cross section with corner plates to make steel cross sections ductile (see. Fig. 1(a)).

Design guidelines [3] were drafted in the Hanshin Expressway Public Corporation (HEPC) for retrofitting the existing steel bridge piers by referring to Ref. [2] and the recent research results [1]. According to the guidelines, the HEPC takes the policy to retrofit the component stiffened plates without filling the column member with concrete in a steel bridge pier with box section in the case that the ultimate strength of the column member exceeds more than that of the basement structure consisting of anchor bolts and footing concrete by adopting the concrete filling method and consequently the basement structure collapses. And this failure mode is not allowed in JSHB. Among 72 steel bridge piers on the No.3 and 5 lines in Hanshin Expressway in Osaka, 56 ones are under the case and need to be retrofitted by the method treated in this paper and not using concrete-filled column members.

Described in this paper are the concepts of this seismic retrofitting method in the HEPC and the validity of the seismic retrofitting method not filling the column members with concrete through the gradually increased cyclic horizontal displacement tests using 5 specimens. Its validity is evaluated by focusing on the ultimate strength and ductility of the specimens retrofitted according to the method.

CONCEPTS OF RETROFITTNG
The basic policies of the seismic retrofitting methods in the guidelines are as follows:
1. The method filling column member with concrete (hereafter referred to as the concrete filling method) has superior priority for economical and operational reasons.
2. The weakest part controlling the ultimate strength of the bridge piers is not their basement structures.

On the basis of these policies, the seismic retrofitting methods for the existing steel bridge piers are mainly divided into two as follows:

• One is the concrete filling method.
• The other is the method to stiffen still more the stiffened plates not using the concrete filling method.

The former method has much more advantageous economically and operationally than the latter one. However, in the case that the ultimate strength of the concrete-filled column member exceeds more than that of the basement structure in a steel bridge pier, the excessive increment of the ultimate strength of the column member is apt to cause the collapse of the basement structure. The latter way is applicable to the case that the concrete filling method cannot be adopted.

To obtain expected ductility in existing column members not filling them with concrete, the restrictive conditions concerning the plate slenderness parameters of their cross-sections are regulated in Ref. [4] as follows:

• The slenderness parameter of plate panels between longitudinal stiffeners R_R $(= \sqrt{\sigma_Y / \sigma_{cr}}$, σ_{cr}: elastic buckling stress of plate panels) is less than 0.4.
• The slenderness parameter of overall stiffened plates, R_F $(= \sqrt{\sigma_Y / \sigma_{crg}}$, σ_{crg}: elastic buckling stress of overall stiffened plate panels) is less than 0.4.
• The slenderness parameter on the local buckling of longitudinal stiffeners, R_S is less than 0.5.

The following stiffening methods are concretely adopted for the stiffened plates and the existing longitudinal stiffeners in steel bridge piers to be retrofitted to satisfy the restrictive conditions mentioned above (see. Figs. 1 and 2, as well as Photos. 1(a) and (b)):

• The plate panels between the existing longitudinal stiffeners are stiffened still more with additional longitudinal stiffeners.
• The existing longitudinal stiffeners are still more stiffened with additional flange plates.

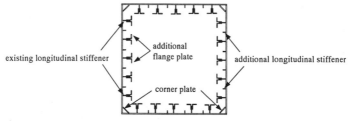

(a)Cross section

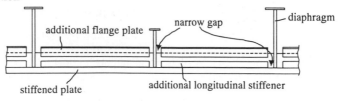

(b)Side elevation
Figure 1. Example of a retrofitted steel bridge pier

Moreover, narrow gaps, as shown in Fig. 1(b) and Photo 1(b), are introduced in the additional longitudinal stiffeners and the additional flange plates at the intersections between their additional members and the transverse stiffeners or diaphragms [5] to mitigate the ultimate strength increment of the column members themselves due to the retrofitting.

additonal flange plate

additional longitudinal stiffener

(a)Retrofitted stiffened plate

existing longitudinal stiffener   additional flange plate

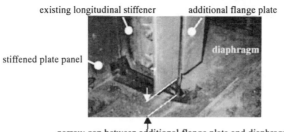

stiffened plate panel

narrow gap between additional flange plate and diaphragm

(b)Narrow gap at end of additional flange plate

Photo 1. Actual example of a retrofitted stiffened plate (HEPC)

EXPERIMENTS

The gradually increased cyclic horizontal displacement tests are carried out for 5 specimens, to investigate the validity of the seismic retrofitting method through the improvement of the plate slenderness parameters by adding the additional longitudinal stiffeners and flange plates to the stiffened plates of the original cross sections of 4 of the 5 specimens.

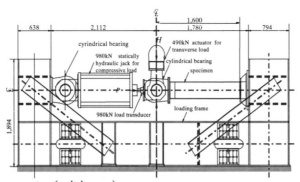

Figure 2. Loading apparatus (unit in mm)

Loading apparatus: The loading apparatus [6] adopted for the tests is designated so as to introduce a transverse load (or displacement) and a compressive load due to the dead load of the superstructure onto the specimen, simultaneously. The compressive load is applied to the specimens at their top by using a static hydraulic jack with a capacity of 980 kN, and the transverse load or displacement is generated by an actuator with a capacity of 490 kN. The loading apparatus is depicted in Fig. 2.

Characteristics of Specimens: 5 cantilever short column specimens with steel box cross sections are used in the tests. One of them aims at simulating the seismic behavior of a existing steel bridge pier and the others are corresponding to the retrofitted it.

Listed in Table 1 are the plate slenderness parameters on the cross sections of the tested specimens together with the corresponding restrictive conditions in the parenthesis and the cross sections before and after being retrofitted. The plate slenderness parameters of the specimen CEO-1-0.11 corresponding to the existing steel bridge pier are decided by making reference to actual steel bridge piers on the No.3 and 5 lines in Hanshin Expressway in Osaka.

Table 1. Plate slenderness parameters of cross sections and classification of specimens

Type of specimen / No. of specimen	R_R	R_F	R_S	Cross section(unit in mm)
Existing steel bridge pier model				
CEO-1-0.11	0.616	0.471	0.564	
Retrofitted steel bridge pier models				
MER-2-0.11 CER-3-0.11 CER-4-0.18 CER-5-0.30	0.396 (0.4)	0.232 (0.4)	0.446 (0.5)	

It is confirmed that the plate slenderness parameters on the cross sections after being retrofitted are within the restricted values by stiffening the original cross section with the additional longitudinal stiffeners and flange plates. Note that the last 4 letters of the specimen number mean the ratio of the applied axial compressive load, N divided by the squash load, N_{PS}. And "C", "E", "M", "O" and "R" mean cyclic loading, monotonous loading, existing bridge pier, original cross section and retrofitted cross section, respectively. For example, CER-4-0.18 is the 4th specimen retrofitted according to the seismic retrofitting method, subjected to gradually increased cyclic horizontal displacement keeping the equal compressive load, N the constant value of 0.11 N_{PS}.

Loading: According to Ref. [2], the gradually increased cyclic horizontal displacement is given on the basis of the yield horizontal displacement δ_{yo} at the tops of the specimens, as shown in Fig. 3. The applied axial compressive load caused by the weight of superstructures is kept at the constant value of 0.11, 0.18 or 0.30 N_{PS} (N_{PS}: the squash load of the cross section before being retrofitted, that is, the cross section of the specimen CEO-1-0.11) during the tests. Note that the specimen MER-2-0.11 is subjected to the monotonously increased horizontal displacement in only one direction. Moreover, the value 0.11 and 0.18 N_{PS} are decided by referring to the average and maximum values in actual cantilever-type steel bridge piers, respectively and 0.30 N_{PS} is decided by considering the maximum axial compressive load in column members of rigid framed bridge piers.

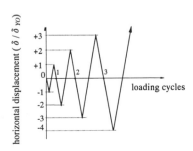

Figure 3. Relationship between loading cycles and horizontal displacement

Test Results: Figs. 4 and 5 illustrate the hysteresis curves at the tops of specimens, where the ordinate is the applied horizontal load, H divided by the yield horizontal load, H_{YO} of the specimen CEO-1-0.11 and the abscissa is the horizontal displacement, δ_o divided by the yield horizontal displacement, δ_{YO} of each specimen.

In the following considerations, the increment ratio of the ultimate horizontal load, β_u and the ductility factor, μ_u in Ref. [2] are used as indexes to evaluate the seismic performance of steel bridge piers and they are defined as shown in Fig. 4.

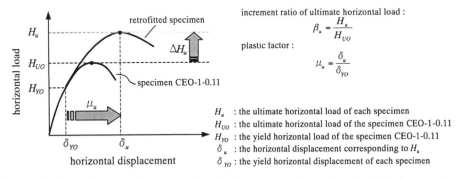

increment ratio of ultimate horizontal load :
$$\beta_u = \frac{H_u}{H_{UO}}$$

plastic factor :
$$\mu_u = \frac{\delta_u}{\delta_{YO}}$$

H_u : the ultimate horizontal load of each specimen
H_{UO} : the ultimate horizontal load of the specimen CEO-1-0.11
H_{YO} : the yield horizontal load of the specimen CEO-1-0.11
δ_u : the horizontal displacement corresponding to H_u
δ_{YO} : the yield horizontal displacement of each specimen

Fig. 4 Definition of terminology in relationship between horizontal load and displacement

Moreover, Fig. 7 shows the envelope curves connecting the average points of the upside-down peak points in the third quadrant and the peak points in the first quadrant of the hysteresis curves in Figs. 5 and 6. And Table 2 summarizes the increment ratio of ultimate horizontal load, β_u and the plastic factor, μ_u in the specimens.

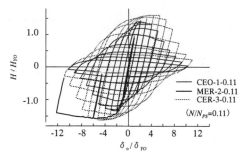

Figure 5. Hysteresis curves of specimens CEO-1-0.11, MER-2-0.11 and CER-3-0.11

The value of β_u and μ_u can be compared between the specimen CEO-1-0.11, the scaled-down model of the existing bridge pier and the specimen CER-3-0.11, the scaled-down model of the retrofitted bridge pier through Figs. 5 and 7 as well as Table 2. It can be seen that the specimen CEO-1-0.11 takes only 3 in μ_u, while the specimen CER-3-0.11 does 5.5 in μ_u and 1.2 in β_u, then the existing bridge piers retrofitted according to the seismic retrofitting method not using the concrete filling method has 1.8 times in μ_u and the increment of β_u by only 20 percent. The increased ultimate horizontal load is still a little smaller than the horizontal load decided by the fully plastic state of the retrofitted cross section.

And also the specimen MER-2-0.11, the scaled-down model of the retrofitted bridge pier and subjected to the monotonously increased horizontal displacement in only one direction, takes almost the same value in β_u and μ_u as those of the specimen CER-3-0.11 with the same axial compressive load of 0.11 N_{PS} but subjected to the gradually increased cyclic horizontal displacement.

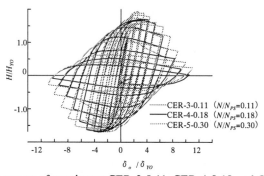

Figure 6. Hysteresis curves of specimens CER-3-0.11, CER-4-0.18 and CER-5-0.30

The difference of the plastic factor and the increment ratio of ultimate horizontal load due to the applied axial compressive load N can be estimated by using Fig. 7 and 8 and comparing the specimen CER-3-0.11 of $N=0.11\ N_{PS}$, CER-4-0.18 of $N=0.18\ N_{PS}$ and CER-5-0.30 of $N=0.30N_{PS}$ with the others. The comparisons show a tendency that the ultimate horizontal load increases and the plastic factor decreases as the applied axial compressive force increases.

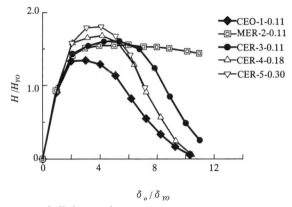

Figure 7. Envelope curves of all the specimens

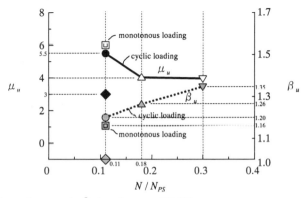

Figure 8. Variation of μ_u and β_u by change of N/N_{PS}

Table 2. Ultimate horizontal load and plastic factor

Specimens	Increment ratio of ultimate load β_u	Plastic factor μ_u
Existing steel bridge pier model		
CEO-1-0.11	1.00	3
Retrofitted steel bridge pier models		
MER-2-0.11	1.16	6
CER-3-0.11	1.20	5.5
CER-4-0.18	1.26	4
CER-5-0.30	1.35	4

CONCLUSIONS

The Hanshin Expressway Public Corporation has drafted the guidelines for two seismic retrofitting methods for the existing steel bridge piers. The seismic retrofitting method not filling the column members with concrete, one of the two methods proposed in the guidelines, is introduced in this paper. And the validity of the seismic retrofitting method is investigated through the gradually increased cyclic horizontal displacement tests.

Main conclusions obtained in the experimental investigation are summarized as follows:
 · The specimen CEO-1-0.11, the scaled-down model of an existing bridge pier, takes 3 in the plastic factor, which is the ultimate horizontal displacement, δ_u corresponding to the ultimate horizontal load divided by the yield horizontal displacement, δ_{yo}. While the specimen CER-3-0.11, the scaled-down model of the existing bridge pier retrofitted according to the method, takes 5.5 in the plastic factor and the increment of the ultimate horizontal load by only 20 percent.
 · The increment of the axial compressive load causes the increase of the ultimate horizontal load and the decrease of the plastic factor.
 · The sufficient ductility can be obtained in the existing steel bridge piers retrofitted by the seismic retrofitting method to stiffen still more the stiffened plates with a little increment of the ultimate horizontal load.

ACKNOWKEDGEMENTS
The authors would like to express their gratitude to the members of the Committee on Seismic Retrofitting Methods for Existing Steel Bridge Piers in the Hanshin Expressway Public Corporation.

REFERENCES
[1] Seismic Design WG, Committee on New Technology for Steel Structures : A Proposal for Seismic Design of Steel Bridges, Japan Society of Civil Engineers (JSCE), 1996 (in Japanese).
[2] Japan Road Association : Japanese Specifications for Highway Bridges (JSHB), Part V Seismic Design, December 1996 (in Japanese).
[3] Hanshin Expressway Public Corporation : Guidelines for Seismic Retrofitting Methods for Existing Steel Bridge Piers (Draft), March 1998 (in Japanese).
[4] Japan Road Association : Japanese Specifications for Highway Bridges (JSHB), Part I Generals and Part II Steel Bridges, December 1996 (in Japanese).
[5] Kitada, T., Nakai, H., Kagayama, T. and Matsumura, M. : A Seismic Design Method and Trial Design for Stiffened Plates in Existing Steel Bridge Piers, Memoirs of the Faculty of Engineering, Osaka City University, Vol.39, pp. 39-51, September 1998.
[6] Nakai, H., Kitada, T. and Nakanishi, K. : Hybrid Test for Simulating the Seismic Behavior of Steel and Composite Bridge Piers, Memoirs of the Faculty of Engineering, Osaka City University, Vol.37, pp. 49-60, September 1996.

Inspection, rating and rehabilitation of the Connecticut River Bridge

MICHAEL J. ABRAHAMS, P.E., Parsons Brinckerhoff Quade & Douglas, Inc., New York, USA
JOHN BRUN, P.E., National Railroad Passenger Corporation, Philadelphia, PA, USA
JOHN C. DEERKOSKI, P.E., Parsons Brinckerhoff Quade & Douglas, Inc., New York, USA
DEBRA L. MOOLIN, P.E., Parsons Brinckerhoff Quade & Douglas Inc., New York, USA.

I.   Introduction

The National Railroad Passenger Corporation's (Amtrak's) Springfield line connects the cities of New Haven and Hartford, Connecticut with Springfield, Massachusetts and provides both passenger and freight service. This rail line is one of three branch lines acquired during the formation of Amtrak and the Northeast Corridor some 28 years ago. The rail line

crosses the Connecticut River and the Enfield Canal in the Town of Windsor Locks, Connecticut. Railroad service to this area was established in the 1840's when the first river crossing was built at the site of the existing bridge.

The present bridge, Amtrak No. 49.73, was built in two stages in 1903 and 1904. The bridge is a two-track structure with 17 spans and a total length of 1,516 feet. Fifteen of the spans are constructed of built-up steel deck girders that vary in length from 24'-8" to 88'-0". The two remaining spans are constructed of skewed trusses; one the 135 foot long riveted steel through truss, and the other a 177 foot long eye-bar through truss. The bridge in service today replaced an existing single track bridge with a new two track steel structure with a higher load capacity. The new bridge was constructed in stages allowing the railroad to continue operating while it was built. The existing masonry piers were reused and widened eliminating the need to construct an expensive temporary trestle. The portion of the bridge that was constructed in 1903 consists of a 135 foot long riveted double intersection Warren through truss, over the Enfield Canal, and a deck girder span over the canal towpath. This portion of the bridge is believed to have been built by the canal owners as the truss design was entirely different from the remainder of the bridge including the 177 foot long eye-bar through truss, otherwise known as a sub-divided Pratt (Baltimore) truss.

The bridge carries two tracks; however, one track is currently abandoned and dead ends just past each abutment. Passenger trains are the primary users of the rail line although freight traffic also utilizes the route to service industries to the north. Approximately 15 Amtrak passenger trains and ten 50 to 60 car freight trains use the bridge each day.

Bridge Management 4, Thomas Telford, London, 2000.

This paper concentrates on the inspection, load rating, load testing and emergency repair of the 177 foot long eye-bar truss.

At the time of the start of inspection there was a 10 mph speed restriction for freight trains and a 35 mph speed restriction on passenger trains using the structure due to the condition of the eye-bar truss. Interim repairs had been made to some of the eye-bar truss pins and lateral bracing.

In April of 1998, Amtrak retained Parsons Brinckerhoff Quade & Douglas, Inc. to perform a detailed inspection and load rating of the bridge. The scope of inspection work included a hands on inspection and load rating of the steel superstructure, and inspection of the top of the piers and the abutments. In addition, the ultrasonic testing of all 42 pins of the eye-bar truss was performed.

II.   Inspection

The main focus of the work was to perform an in-depth, hands on, structural inspection, to provide Amtrak with a complete and thorough documentation of the overall condition of the bridge and describe the specific structural deficiencies identified during the inspection.

For the eye-bar truss access to the truss bottom chords, bottom chord connections, the lower portion of the vertical members, floor framing and pins for the ultrasonic testing was achieved by utilizing a hi-rail capable underbridge inspection unit reaching between the truss members.

Latticed web members were climbed for a full hands-on inspection. Truss web members with eye-bar configurations were accessed along the full length for hands-on inspection by utilizing a high rail mounted bucket truck.

The inspection of the substructure was limited to a visual inspection of the pier and abutment caps in the vicinity of the bridge bearings. These were accessed from the underbridge inspection unit and by ladder.

The inspection methods that were used on the bridge were basic inspection methods for steel structures. A hands-on, in-depth inspection was performed on all primary members, and a hands-on, in-depth inspection was performed as appropriate for secondary members.

During the inspection each eye-bar was shaken to observe the relative tensioning of the individual eye-bars and to record vibrational characteristics of each eye-bar. The vibrational characteristics were recorded in accordance with Section 15.8.8.2 "Method of Shortening of Eye-bars to Equalize the Stress" of the 1994 American Railway Engineering Association (AREA) Manual. This allowed the dead load force in each eye-bar to be calculated during the rating. The method calls for attaching a paper to the eye-bar and shaking the eye-bar to induce vibration. While the eye-bar is vibrating, a line is drawn along the paper for a given interval of time. The period of vibration can then be calculated and correlated to the force in the member by using a formula and chart in the AREA Manual.

Ultrasonic testing was conducted on each of the 42 truss pins. Pins were tested from both ends to verify the ultrasonic findings, and were tested in at least eight discrete positions on the pin faces. Extra care was taken to properly prepare the pin faces prior to testing. Notes identifying the test locations at each pin tested and any indications noted from the ultrasonic test along with all pertinent test data were recorded. Any location with a significant indication was recorded on a hard copy of the screen with that particular indication.

At several locations, markings were made on the side of the pin by inspectors to monitor for rotation of the pin in relation to the nut and the pin and nut in relation to the eye-bar. When observed under live load, no rotational movement of the pins was noticed. The results of the ultrasonic testing showed the pins to be in good condition with only very minor wear grooves. Based upon the minimal wear grooves in the pins it was concluded that any wear or play in the joints must be attributable to wear of the eye-bar holes or original fabrication tolerances.

In general, the condition of the truss was very poor. The top and bottom tie plates connecting the top horizontal lateral bracing to the top chords had cracks at three locations due to the considerable movement of the truss under live load. The pins at these three locations were moving in an elliptical motion and rotating under live load as the pin holes had become elongated. In addition, a number of eye-bars were loose and appeared to not be carrying any load. There were typical areas of pack rust and minor deterioration of the gusset plates and the connection details.

The webs of the top chords of the truss had been reinforced at several joints where the pins were loose. The pins at these joints were observed to rotate and translate in an oval fashion approximately 3/8" allowing the truss vertical and lateral bracing to move relative to the top chord under live load. At these locations the top and bottom lateral plates at midspan of each top chord were cracked. The cracks were previously welded by Amtrak personnel as a temporary repair. Shortly after the weld repairs were performed they cracked. Under live load, the top lateral plates moved up and down relative to one another at the crack. When observed under load, the truss deflected both vertically and laterally with the passage of each axle of the train.

The conditions found at the upper chord pins were of concern as they allowed significant movement at the joints resulting in distortion and cracking of the lateral plates and causing unknown redistribution of forces. In addition, the number of loose and possibly non-participating eye-bars was of concern as some eye-bars may have been carrying considerably more load than the load for which they were designed. The excessive deflections of the truss under live load add to the live load impact by increasing the component of the impact attributable to rocking effect.

III. Load Rating

The load rating analysis was performed in accordance with the requirements of the AREA Manual for Railway Engineering, 1996 Edition and Amtrak Rating Policy. Only the steel superstructure was rated. The superstructure members were rated on the basis of available plans as verified by field measurements, without losses due to deterioration (as-built) and with losses due to deterioration (as-inspected).

Rating spreadsheets were prepared for all primary members of each span of the structure using Microsoft Excel, and STAAD III structural analysis program. Electronic copies of the spreadsheets and input files containing three-dimensional models of each truss were furnished to Amtrak. STAAD III files containing the two dimensional models of the girder spans that were used to generate the maximum forces at the girder cover plate cutoff points were also furnished. These files were formulated to allow Amtrak engineering staff to check the structures for other possible loadings and/or changes to section properties.

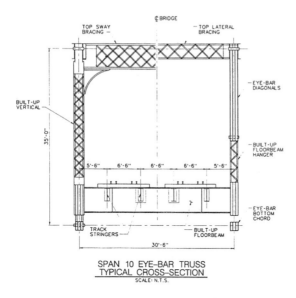

SPAN 10 EYE-BAR TRUSS
TYPICAL CROSS-SECTION
SCALE: N.T.S.

The available drawings indicated that the structural steel was soft open-hearth steel. As the minimum yield strength was not indicated on the drawings, it was taken as 30,000 psi for open-hearth steel per the AREA Manual.

Built up riveted members and eye-bar members containing four or more eye-bars were considered internally redundant and were not considered Fracture Critical for the Fatigue Ratings.

The tensile capacity of the eye-bars was evaluated at the center of the member and at the eye bar head, whichever area had the smaller net cross sectional area. As all of the joints appeared to be free to rotate the eye-bar members were evaluated as carrying axial forces only. The eye-bar members were assumed to transmit tension forces only and due to their slender nature were modeled to be ineffective in compression.

The three dimensional analysis of the structure revealed that the end supports for each floorbeam were such that under a given axle load one side of the floorbeam would deflect more than the other. As is the case with heavily skewed truss bridges the floorbeams frame into different panel points of each side of the truss. In the above cross section it is apparent that one side of the floorbeam frames into a truss vertical (a more rigid support). The other side frames into a hanger and two eye-bar truss diagonals (a more flexible support). The effect is that as a train passes over the structure, the bridge and train start to rock back and forth as the floorbeams alternately deflect more on opposite sides. This live load induced motion can be readily felt as one stands on the bridge. Unfortunately, the motion cannot be prevented as it is due to the basic bridge geometry. This motion has most likely accelerated the wear of the eye-bar holes and therefore increased the bridge deflections.

The following load ratings for one track loaded and a train speed of 10 mph were provided to Amtrak in the final inspection report along with the inspection findings.

Truss Member	Normal As-Built Rating	Normal As-Inspected Rating	Normal Fatigue Rating
Eye-bar Diagonals	E-65	E-65	E-26

Fatigue ratings of built up members were calculated using stress Category C requirements as the field inspection verified that the rivets were tight and appeared to have developed adequate clamping forces. Although the fatigue ratings for all members were below Cooper E-60 (a normal design load in the United States) the 95 year old members exhibited no signs of cracking or distress other than described herein.

The above ratings did not take into account the variation in tension of individual eye-bars and were therefore considered unconservative. Since the as-inspected load ratings of the eye-bar truss would be influenced by the elongated pin holes, the pin movements observed at the upper chord of the south truss, and the existence of numerous loose/non-participating eye-bars that could not be modeled or estimated. It was recommended that strain gauge load testing be performed to better assess the actual live loads carried by the members and individual eye-bars.

IV.   Load Testing & Analysis

A load test program including both strain gaging and deflection measurements was developed and conducted to more accurately determine the following:
1. The actual load distributions for members framing into the truss joints that were experiencing excessive pin movements and fatigue cracks.
2. The actual stress distribution in eye-bar members where there was evidence of unequal tension of eye-bars.
3. The stress in the members that are governing the rating to verify the member stresses and to verify the computer model used for the analysis.
4. Actual truss deflections for comparison to the computed deflections.

The load test was conducted under the direct supervision of Parsons Brinckerhoff Quade & Douglas, Inc.. Subconsultant, Specialty Measurements, Inc., installed the strain gauges and collected all of the strain data generated during the test. PB set up a deflection monitoring system and recorded all deflection measurements. Over a period of six days 60 strain gages were installed on the bridge. As it was not practical to gage each and every member and eye-bar of the truss, selected members were identified to be tested: members framing into loose joints, eye-bar members with several loose eye-bars, members with normal ratings below Cooper E-80. All eye-bars in selected members were fitted with strain gages to verify the load distribution to each bar.

One day was required for the computer equipment setup, wiring and testing of the data acquisition system was completed the day before the load test. The test itself took one day to perform. The strain measurements were taken while the bridge was loaded with two of Amtrak's General Motors F40PH locomotives weighing a total of 262,500 pounds. The locomotives were on site at the test location for approximately one hour.

The strains were recorded at a rate of 10 measurements per second as the dual locomotives crossed the entire structure in both directions at a speed equivalent to a slow walk. The continuous monitoring produced strain influence lines for the gaged members and provided the maximum and minimum strains for the load configuration. The strains were then measured with the locomotive stationary at a specific axle configuration. Each test was performed three times to ensure data repeatability. In addition, the testing team waited on site after the test locomotives were released to record the member strains during the passing of a typical Amtrak passenger train. At the conclusion of the load test the strain gauges and wiring were protected and left in place so that they could be reused to monitor the changes in truss strains during and after the completion of interim repairs.

Deflection measurements were recorded while the locomotives were stationary and centered on the bridge. Deflections were measured using 90 pound test monofilament fishing line stretched the length of each truss. The line was pulled taut and anchored to the end diagonals of the trusses just above the bearings with c-clamps to provide a stationary reference point. Steel machinist rules glued to mirrors were firmly attached to each vertical. Measurements were obtained by aligning the line with its reflection and recording the reading on the rule. Baseline measurements were taken without load then measurements were taken with the train in position. A minimum of three readings were taken for each measurement, readings were consistently within 1/32" in variation.

Subsequent to the field testing PB performed an analysis of the eye-bar truss using the loading of the two F40PH locomotives. The results of the analysis were compared to the calculated loads based upon the strain measurements.

An analysis of the test data yielded the following results:
- The live load deflections of the trusses as measured in the field were approximately twice that predicted by the computer analysis (0.70 inches vs 0.31 inches).
- The relative live load distribution in the individual eye-bars of a member as measured by the strain gaging was very close to that as determined by hand plucking the bars and calculating the relative tension based upon the field measured oscillations.
- In general there was fair to poor correlation between the measured and computed stresses. The poor correlation was judged to be the result of redistribution of loads due to the loose joints. In general the computer generated stresses were higher than those field measured.
- The strain influence lines generated showed that the eye-bars members were participating in compression more than assumed in the analysis model. This acted to help reduce load in other members.
- Correlation between the field measured loads and the calculated loads was improved slightly when the rotational restraint of the bearings was stiffened to reflect the effect of frozen bearings.
- The eye-bar diagonals that were previously governing the load ratings were shown to not be the critical members due to their low level of measured stress under live load.
- The load rating as adjusted based upon the strain gaging was calculated to be E-84 for one track loaded and a speed restriction of 10 mph. It was recommended that the load

restrictions remain in effect due to the unpredictable nature of the truss and the still excessive deflections.

V. Interim Truss Repairs

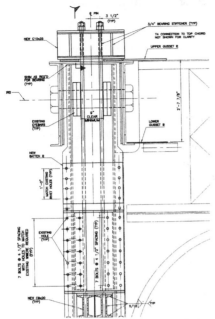

The overall recommendation of the initial inspection report was that the Span 10 truss be replaced. Until that time, the bridge needed to be closely monitored and repaired on a priority basis. Consideration was given to either repairing the loose pins locally or by adding redundant members. PB and Amtrak decided that it would be faster and easier to add a saddle and rod assembly on top of the top chord of the truss at each loose joint.

The saddle and rod assemblies were provided to minimize the relative movement between the vertical member and the top chord at the loose pins. The relative movement around the loose pins was a condition that appeared to be the cause of the cracking in the gusset plates.

The repair for the loose pins included the addition of four 1-1/2 inch diameter high strength rods connected to the vertical and the top chord member. The rod assembly strengthened the loosened pin connection and prevented excess deflection of the vertical member relative to the top chord by bypassing the pin and providing a direct connection between the vertical member and the top chord. The assembly was designed to carry the full dead and live plus impact load forces in the vertical member.

On February 9, 1999, the rod installations at the four top chord panel points were jacked to complete the top chord pin interim repairs. At that time, PB's strain gage testing was conducted to monitor the structure during jacking and after the completion of the load transfer. During jacking, the strain gauges were monitored for any increasing strain that would be the result of a member being loaded unexpectedly due to the deflections incurred by jacking. No build up of strain was noted at any member during any of the jacking operations.

The rods were jacked to the minimum load required to result in minimizing the movement of the pins and the relative movement between the top chord and the vertical member. This was accomplished at a level of approximately 24 kips per rod or 96 kips per member. The load was transferred to the rods incrementally, checking the movement of the pin under live load after each increment. At 24 kips per rod, movement of the pin under live load was no longer apparent. The minimum amount of load necessary was jacked onto the rods to minimize any unexpected effects on adjacent parts of the structure and to allow for further jacking of the rods in the event that the joints loosen in the future.

Deflection of the vertical members was also monitored during the jacking operation. At the first increment of approximately 12 kips per rod, 3/16 inch of upward movement was noted at the vertical, and at the second increment to 24 kips per rod, approximately 1/32 inch of upward movement was noted. This decrease in movement at an equal increase in load suggested that at this point, the vertical member had ceased movement and begun to compress.

At the completion of jacking all four rod assemblies, a final strain gauge measurement was taken under live load to insure that there was no unexpected change in the action of the bridge. After the completion of jacking at all four top chord locations, each location was observed hands-on under live load to watch for movement of the pin and relative movement between the top chord and vertical members. All four locations were found to show no movement of the pin and no relative movement. After this final inspection, the top chord lateral bracing gusset plates were replaced.

VI.   Conclusion

The Connecticut River Bridge No. 49.73 in Windsor Locks is a bridge that has been in service for nearly 100 years without undergoing a major rehabilitation. The structure shows the effects of being exposed to the elements and heavy railroad loads. The observed section losses to the structural steel can be attributed to deterioration due to the accumulation of debris on the horizontal surfaces of the bridge structure.

While the one eye-bar truss span is clearly nearing the end of its useful life, its companion riveted truss span is in generally good condition and should have considerable remaining service life. Given the similar span length, skew, and usage one may conclude that there is an inherent advantage to the more robust riveted construction.

The strain gage testing and subsequent analysis conducted as part of this study demonstrated that restraining effects of continuous welded rail and frozen or non-working bearings can be significant in evaluating the load distribution in existing structures. The testing also revealed that the eye-bars were in fact capable of resisting some compression forces and provide a slight reduction in the load in adjacent members.

It was also demonstrated that a very simple method of hand plucking of a series of eye-bars and timing their oscillations can produce relatively accurate load distribution evaluations, which are comparable to the much more accurate measurements made with strain gages.

Amtrak is currently planning a rehabilitation of the entire bridge structure that would include replacement of this 177 foot long eye-bar truss. The replacement scheme would consist of installing a new pier in the river channel at approximately midspan of the existing truss and installing two 88'-6' deck girder spans either new or borrowed from the adjacent girder spans. To facilitate railroad operations the construction will be performed while maintaining one track in service.

In an era of reduced funding to Amtrak every project is evaluated based upon passenger safety and the service requirements of the railroad. At this time, based upon Amtrak's Capital funding projections and considering the condition of the structure, the eye-bar truss span of this bridge is scheduled for replacement in the immediate to near future. Until such time, the interim repairs in conjunction with regularly scheduled (weekly) inspections and frequent maintenance by Amtrak forces will extend the life of this bridge until the capital funds are available to replace it.

Rehabilitation of Tsing Yi South Bridge Hong Kong

PC WONG and CY WONG, Highways Department (Hong Kong SAR)
F KUNG, NR HEWSON and K MORRIS, Hyder Consulting Ltd. (Hong Kong & UK)

INTRODUCTION
This paper describes the recent investigations and studies carried out by Hyder Consulting Ltd. on behalf of the Hong Kong SAR Government Highway's Department for the rehabilitation of the Tsing Yi South Bridge in Hong Kong. The work included a detailed inspection of the structure, extensive materials and load testing, assessment of the load capacity, design of hinge replacement and proposals for rehabilitation.

Fig 1 General location of bridge

Fig 2 View of Tsing Yi South Bridge

HISTORY
Tsing Yi South Bridge is an important structure being the first major bridge sea crossing in Hong Kong. It was constructed between 1971 and 1974 and carries 2 lanes of road traffic and two footways between Kwai Chung and Tsing Yi Island over a busy shipping channel. The bridge also carries key utility services to the residents of Tsing Yi Island and from the power station on the island including water and gas mains, telephone cables and high voltage electric cables.

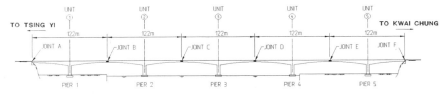

Fig 3 General arrangement showing notation

The deck is a six span prestressed concrete box girder constructed as five 122m long balanced cantilever units built symmetrically from each pier. The cantilevers are linked together with metal roller hinges at midspan. The ends of the bridge are supported on reinforced concrete bank seat abutments. The reinforced concrete piers are supported on a combination of spread footings and large diameter bored pile foundations. The water is up to 18m deep and the minimum headroom for the navigation channel is 17m.

The bridge has been the subject of various previous inspections, deflection monitoring, load assessment and materials testing work: In 1983 Mott, Hay and Anderson carried out an extensive study of the bridge, particularly for deflections of the cantilevers. They recommended additional external prestressing work that was implemented in 1990. In 1991 Ove Arup and partners were commissioned to carry out a survey on the concrete spalling and cracking of the bridge pier and deck. Rusting of reinforcement and spalling of concrete was located and made good in 1994. The previous studies of the bridge included materials testing, load testing and radiography investigation of the prestressing cables.

The structure has been subjected to a number of collisions from barges. Most of them are striking the bridge deck by derrick booms of the barges. The worst in a typhoon in 1992 caused damage to 52m length of footway and an 8m section of the main deck edge cantilever.

Serviceability Issues
Significant vibrations in the bridge deck are experienced under normal traffic loading particularly at midspan. This is mainly due to the flexibility design of the deck and is exacerbated by the movements across the damaged hinges. The ends of the cantilever units have deflected visibly due to long term creep of the concrete. Additional external prestressing tendons installed in 1990 raised the joints by about 100 mm on application which reduce to 80 mm and appears to be stable.

Fig 4 View of road alignment

Fig 5 View of internal tendons

INSPECTION AND ACCESS
The inspection was carried out in accordance with the UK Highways Agency Guidance for Inspections for Assessment[1]. A variety of access techniques and safety precautions were required to inspect the different parts of the bridge safely. These included: with high visibility jacket for all roadside inspections; with harnesses for the high level inspection; using the Highways Department's underbridge inspection vehicle and the abseiling method for the inspection of the deck soffit and the piers. Of particular note was the requirement of wearing full asbestos protection equipment to inspect inside the box, which is contaminated with asbestos dust. The China Light and Power cables inside the box are coated with asbestos materials which has degraded into dust on the floor of the chamber. The China Light and Power is an electric company supplying electricity to Kowloon and the New Territories. A specialist diving contractor was appointed to carry out the inspection of the piers and substructure below water level.

Fig 6 View inside box   Fig 7 Rope Access

Inspection Findings

The bridge was generally found to be in good condition with isolated areas of concrete spalling. The midspan hinges are all in a serious state of distress with extensive corrosion. The hinge mechanism is a roller that is designed to bear on the upper and lower bearing surfaces of the joint. This design requires a small tolerance in the joint to allow the rolling movement. This tolerance gap has increased with wear and now allows significant vertical movement across the joint that can be seen both at road level and at the hinge itself.

At the end of each T units, the concrete at the bottom slab of the box shows significant spalling which is aggravated by rain water leaking through the access manhole at the deck surface. The top of the bottom slab is in poor condition near the piers in areas where the cover is minimal and water has been standing. This does not cause immediate concern as the loss of concrete section is not a significant percentage of the deck cross section but it should be repaired. The abutment provides both reaction support and uplift restraint. The original bearings are located at both top and bottom of the web extensions of deck.

Fig 8 View of manhole access   Fig 9 View of exposed rebar in bottom slab

MATERIAL TESTINGS

Concrete Testing

Concrete tests were carried out to determine the strength and condition of the concrete and investigate the level of chloride contamination and corrosion of reinforcement. These included: (i) concrete cores for crushing, aggregate cement ratio, original water content and modulus of elasticity laboratory tests; (ii) drilling dust samples for chloride content and sulphate content; (iii) carbonation depth in holes used for dust sampling; (iv) hammer tapping and pull off tests of mortar to identify delamination; (v) covermeter survey to check the depth of cover to the steel; (vi) half cell potential and resistivity testing to check for the presence of corrosion; and (vii) concrete breakout to expose the reinforcement for examination.

The results showed the deck concrete to be generally in good condition with little risk of durability problems except in isolated areas. The pier concrete results indicate that the repair mortar placed in 1994 is resistant to chemical ingress though it has uniform high levels of sulphate which are likely to be from a mix additive. The chloride profiles indicated a possible back migration from the contaminated older concrete. The mortar pull off tests on the repair concrete around the piers showed the bond to be weak but no delamination was identified by the tapping survey.

Insitu Stress Measurement

The residual stress measurements can provide useful information on the structural behavior of a bridge. The insitu strains in the concrete were measured using the stress relief coring method and were used to calculate the actual stress in the concrete section at various positions along the deck. The measurement involved taking cores near the extreme fibres of the concrete section and measuring the change in strain around the hole.

The accuracy of the residual stresses determined using this method is highly sensitive to effects caused by the misalignment of the drilled hole, presence of reinforcement, local thermal effects and global temperature differences. Every precaution was taken during the tests to minimise these effects or to carry out additional measurements to correct these effects. However, due to congestion of reinforcement and variation of temperature, the residual stress results obtained from the site tests were difficult to compare with the analysis predictions.

Stress in External Tendons

A tendon pull-off test was carried out in order to measure the force in the external tendons. A jack was placed onto the tendons at the jacking end and a force applied until the end plate has seen to lift off the anchorage. This gives a measure of the actual force in the external tendons. An average of 3% loss of initial jacking force was measured indicating that the external tendons are performing well.

Special Investigation of Post-tensioning

Internal post tensioning systems are well known to be vulnerable to inadequate grouting which leaves voids in the tendon ducts. These voids can contain water which then leads to corrosion and eventual failure of the tendons. A series of radiographic photographs of the internal tendons in the top slab were taken to inspect the tendons at various cross sections. The radiographs showed where there is insufficient grouting to about 10% of the internal prestressing tendons but did not indicate any significant damage or corrosion. About 10% of the voids were investigated by drilling into the tendon duct and inspecting by boroscope. The tendons inspected were generally in good condition with minor surface corrosion. No water was encountered in the voids.

LOAD TESTINGS

Static Load Tests

A series of load tests were carried out on the bridge using the China Light and Power heavy vehicle loaded with concrete blocks. The objective of the load testing was to study the behavior of the structure for comparison with the assessment. The static load test involved strain and deflection measurements caused by a large test vehicle positioned at various locations along the bridge. Strain gauges were located in the longitudinal direction near the extreme fibres inside the bridge and the readings were electronically recorded. Horizontal and vertical deflection measurements were taken using traditional surveying techniques.

The site measured deflections were compared with the predicted deflections assuming a free or a partially restrained hinge using the assessment methods as described in the later sections. It was generally suggested that there were some jamming effects on all of the hinges, except for Hinge C. The jamming effects could be due to the condition of the hinges and the fact that the shim plates that have been placed to limit the vertical deflections are likely to induce some jamming.

The measured and the predicted strain readings were also compared. Based on the section properties including the concrete upstands and footways, the predicted strains are in close agreement with the measured strain.

Dynamic Response Measurements

Dynamic response measurements were taken under traffic loading to estimate the modes of vibration, the mode shapes and the corresponding damping ratios. The dynamic testing was carried out using a series of accelerometers placed along the length of the bridge. The vibration monitoring was carried out under normal traffic flow to measure the mode shapes for the vibration of one cantilever unit and compare the general movements of the other units. The principle modes of vibration and a measure of the amplitude and frequencies of those vibrations were determined and the results showed eight distinct mode shapes. The first eight natural frequencies were within the range of 0 to 2.5 Hz. The lowest natural frequency was 0.703 Hz which corresponded to the pier torsion dominated mode. The highest natural frequency was 2.461 Hz which was an out of plane pier bending and deck bending mode. No deck torsional mode was identified in the first eight natural frequencies.

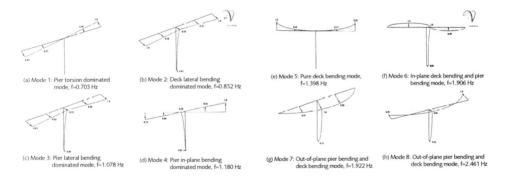

(a) Mode 1: Pier torsion dominated mode, f=0.703 Hz

(b) Mode 2: Deck lateral bending dominated mode, f=0.852 Hz

(e) Mode 5: Pure deck bending mode, f=1.398 Hz

(f) Mode 6: In-plane deck bending and pier bending mode, f=1.906 Hz

(c) Mode 3: Pier lateral bending dominated mode, f=1.078 Hz

(d) Mode 4: Pier in-plane bending dominated mode, f=1.180 Hz

(g) Mode 7: Out-of-plane pier bending and deck bending mode, f=1.922 Hz

(h) Mode 8: Out-of-plane pier bending and deck bending mode, f=2.461 Hz

Fig 10 Vibration Modes and Relative Amplitudes

The measured damping ratios were between 1% and 5%. The lowest damping ratio was 1.18% which corresponded to the 3rd out-of-plane mode (Mode 7), and the highest damping ratio was 4.79% which corresponded to the 1st in-plane mode (Mode 1). It was concluded that the east end of unit 2 and both ends of unit 3 move more freely than the others. The measurements also confirmed that the lateral vibrations at the base of unit 3 were greater than those of unit 2.

The analytical assessment as described in the later section suggested that Modes 1 to 4 of the site measurements were in close agreement with the predicted. The remaining 4 modes showed some variation, but as those modes had relatively lower amplitude it would not be considered any further.

ANALYTICAL ASSESSMENT

The assessment was generally based on the principles of BD 21/97 and BD 44/95 but to the current design loading given in the HK Structures Design Manual for Highways and Railways[2]. The scope of the assessment was: (i) to determine, in terms of vehicle loading, the maximum load that the structure will carry without suffering serious damage so as to endanger any persons or property; (ii) to predict the residual life of the structure based on the results of the materials testing; and (iii) to investigate the feasibility of rehabilitation options, i.e. to predict the behavior of the bridge for proposals which change the form of the structure.

The structure was modeled using three different analytical models in order to ensure that the structural behavior of the bridge is correctly assessed. The models were verified against historical records and recent testing. The analysis included the following work:

- a basic check on maximum service limit state stress levels at critical points and a calculation of the ultimate limit state load carrying capacity.
- a sensitivity analysis to check for variations in prestress level, dead load, superimposed dead loads (SDL), temperature effects.
- a calculation of the natural frequencies of the structure to compare with the site testing and normal design limitations.
- a prediction of the long term deflection of the structure.
- a check of the structural behavior and effect of proposed remedial measures.

SuperSTRESS [3] model

A simple 2-D plane frame model of the entire Tsing Yi South Bridge was established using the SuperSTRESS program. For the model all hinge joints were free for longitudinal and rotational movements. The abutment joints were free in the horizontal direction and fixed in the vertical direction. The spread foundations for Pier Nos. 1 and 4 were modelled as rigid supports. The piled foundations for Pier No.'s 2, 3 and 5 were modelled as spring supports.
The model was used to calculate the ultimate load capacity, to assess the global effect of the rehabilitation options and to check the output from the ADAPT model.

ADAPT [4] Model

A 2-D plane frame model of the entire bridge was generated using the ADAPT program. The model was built up in stages to match the construction sequence and strengthening works. The hinges were idealised as members with released longitudinal movement, axial force and in-plane bending moment. The program can predict the time dependent deflections of the structure during each stage of construction. The analysis can be projected forward to predict the future behavior. The model was used to determine the serviceability limit state stresses during construction and the current situation. The deflection results were compared with the historic survey data.

SAP2000 [5] Models

Two separate models were generated using the SAP2000 program. The first model of one cantilever section from the face of the pier to the hinge point was generated using solid elements. This was used to study the transverse effects on the box girders and the possible effect of shear lag. The second model of the entire bridge was generated as a line model. The hinges were idealised as members with released longitudinal movement, axial force and in-plane bending moment. The model was used to determine the natural frequency of the bridge and to compare with the site tests.

Assessment Findings

The bridge satisfies the serviceability limit state stress limitations even with a 30% additional loss of internal prestress. The structure can still sustain 150% of the current ultimate design loading with this assumed further loss. The effect of this additional loss is a further deflection of the cantilever tips of 120mm at Hinges C & D and 70mm at Hinges B & E.

The site deflection measurements are more than those predicted from the ADAPT model and, as indicated from a sensitivity analysis, they are very sensitive to temperature gradient. The discrepancy could be due to differences in the elastic modulus or greater than predicted relaxation of the tendons.

The analysis of various rehabilitation options has shown that the possibility of modifying the articulation of the structure is significantly restricted. Providing concrete stitches across all the joints would cause overstress on the piers and foundations due to longitudinal temperature effects. This option would require major strengthening work to the substructures. If just Hinge joints B & E were stitched then the strengthening work would only be necessary at each of these stitched joints.

The analysis and testing has highlighted the following features:
- The downward deflection at Hinges B & E are up to 20% greater than at Hinges C & D. This is caused by the tilting of the end units as the downward creep of the end cantilevers is restrained by the abutment supports.
- The joints or hinges at B, D & E are experiencing some jamming.
- The footways and upstands not included in the structural model may be contributing to the live load capacity.
- The dynamic analysis and the dynamic response measurements both indicate that in accordance with SDM, the structure does not cause significant discomfort to drivers though very noticeable to pedestrians

REHABILITATION AND HINGE REPLACEMENT DESIGN

The options for rehabilitation included the following:
- do minimum
- carry out hinge replacement and remedial works to specific areas in poor condition.
- remove the hinges by replacing them with stitched joints.

The first option was not appropriate as the serviceable life in the current condition is too short. Hinge replacement in the second option was considered to be essential in the short to medium term as the hinges are in a very poor condition. The third option to stitch the bridge deck, as discussed in the analysis above, is limited by the effects on other parts of the structure. The possibility of removing the maintenance problems of two of the hinge joint areas is attractive but proves to be a costly alternative. Solutions to raise the midspan to level out the vertical alignment proved to be uneconomic. There is limited space for additional prestress. Improving the road alignment at road level by resurfacing imposes additional load that will deflect the cantilevers further.

A major issue for either the short term or long term works to Tsing Yi South Bridge is that of traffic management and road diversions, including disruption to the navigation channel. Despite the fact that there are new bridges on either side, the area is highly sensitive to proposed developments and traffic growth being part of the link to the airport. The proposed rehabilitation works were designed to minimise any disruption to the existing traffic.

The existing hinge comprises a vertical roller mechanism in each web and a horizontal roller in the top slab. The replacement hinge has been designed to allow longitudinal and rotational movements across the joint thereby maintaining the original articulation. It was decided to replace the vertical web roller system with a purpose made pot bearing. The moving parts shall be easily inspected and will be replaceable without extensive concrete breakout. The top horizontal bearing does not require such a large rotational capacity so a separate simple guided bearing will be installed to prevent differential transverse movements across the joints. Combined with these works will be a complete rehabilitation of the hinge area and replacement movement joints.

Fig 11 View of replacement hinge

Fig 12 View of existing hinge from inside

CONCLUSIONS
The inspection and testing were successfully completed without any incident. The investigation revealed that Tsing Yi South Bridge is generally in good condition though there are localised areas in very poor condition particularly the midspan hinges. The bridge is considered to be structurally sound and does not have defects of structural inadequacy.

This project however highlights the importance of considering the durability and maintenance of a structure at the design stage. Items subject to wear such as bearings, joints and hinges need to be easily inspected and replaced. The life of these items is much less than the design life of a major structure and failure should be anticipated. In recent years the guidance to designers has been to avoid joints and bearings altogether where possible and design integral bridges thereby avoiding the problems that have occurred with the hinges on this structure.

References:
1 UK Highways Agency Guidance for Inspections for Assessment.
2 The Hong Kong Structural Design Manual for Highways and Railways.
3 SuperSTRESS: an elastic analysis computer program by Integer date 1989 – 1997.
4 ADAPT: a computer program to model time dependant effects such as creep by Adapt Structural Concrete Software System date August 1997.
5 SAP2000: a finite element computer program by Computers and Structures date July 1997.

Acknowledgments:
The authors would like to express thanks to the Director of Highways, Highways Department of Hong Kong SAR Government for permission to publish this paper and to all those people who have assisted in making this a successful project. Particular thanks are extended to Hong Kong University for advice on the dynamic testing, Testconsult for advice on materials testing and Taywood Engineering the testing contractor.

Structural behaviour and rehabilitation of fatigue cracked box girder bridge

KENTARO YAMADA, Professor, Ph.D., Nagoya University, TATSUYA OJIO, MS, Research Associate, Nagoya University, SHIGENOBU KAINUMA, Dr.Eng., Research Associate, Nagoya University, and KENICHI ITOH, Graduate Student, Nagoya University, Nagoya, Japan

INTRODUCTION

Rapid economic growth which Japan had experienced since 1960's caused also great increase in traffic volume and weight on major highway bridges. Such heavy traffics caused some damages to bridge members, first to cracking and deterioration of concrete decks. Then, fatigue cracks were observed in some steel members of various highway bridges. They were fatigue cracks at upper end of vertical stiffeners where cross frames were attached, at upper end of connection plates between upper flanges of cross beams and upper flange of main girders, and front ends of sole plate of plate girder bridges. Whenever such fatigue cracks were observed, investigation was carried out to find cause of the cracks and way to retrofit the members and the structures. Appropriate retrofitting measures, obtained from thorough investigation, can increase the life expectancy of the structures, and may be able to avoid future problems.

When fatigue cracks were observed in bridge members, owner of the bridge often invite experts to form a committee to investigate cause of the cracks and the most appropriate retrofitting procedures. When cracks were found in 1987 at diaphragm corners of three span continuous steel box girder bridge (Minato shin bashi) in Nagoya[1], the Ministry of Construction decided to form a committee. It was chaired by Prof. Nishino. Through measurements of stresses and vibration and structural analysis, possible cause of cracks and retrofitting measures were recommended. The repair works followed the recommendations. The cracked diaphragm corners were strengthened by high strength bolted splices, and truss members were added inside and outside of all diaphragms and every other cross ribs. Some cracks were left un-stiffened. They were monitored periodically to confirm the effectiveness of the additional stiffening trusses.

In 1994, relatively large cracks were found near the end supports of the same bridge. The cracks initiated from the front end of the sole plates, and propagated to lower flange and web of the box girder bridge. Temporary supports were installed under the diaphragms, which was then stiffened by poring concrete at around lower part to avoid any catastrofic damages associated with the extent of the cracks. Since Kobe Earthquake occurred in January 17, 1995, any investigation and repair works were postponed until more time and fund became available.

Bridge Management 4, Thomas Telford, London, 2000.

In the following fiscal year, the second committee was formed, which was chaired by Prof. Yamada. After the Kobe Earthquake the Ministry of Construction recommended to replace conventional metal supports to rubber one to increase resistance against earthquake. The Minatoshin-bashi Bridge followed the recommendation. In order to install much wider sole plates for the rubber shocs, the cracked lower flanges could be stiffened by high strength bolted splices.

One of the tasks, which the committee was asked to tackled, was to investigate the way to evaluate the previous retrofitting measures carried out about ten years ago and estimate remaining life of the bridge. This paper describes an example of durability evaluation of cracked and stiffened diaphragm corners of the Minatoshin-bashi Bridge.

DESCRIPTION OF THE BRIDGE

The Minato shin bashi Bridge[2], opened to traffic in 1964, spans over Horikawa Canal near Nagoya Port. It carries very heavy traffics, over 55,000 vehicles in one direction, since it situates in heavy industrial district. Recent statistics showed that more than 50 percent were trucks. Some of them are believed to be overloaded. Through the investigation supported by the committee, weight of wheels in service was also measured with newly developed WIM system using longitudinal ribs of orthotropic steel decks[3,4]. It also showed that very heavily overloaded trucks often passed on the bridge.

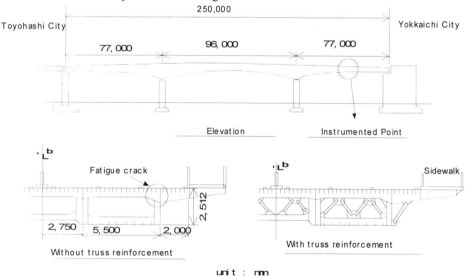

Fig. 1 Minato shin bashi bride and diaphragm details with and without truss member.

The bridge is three span continuous steel box girder bridge of 250 m long with orthotropic steel deck, as shown in Fig. 1. Two steel boxes are connected with cross beams, which are also stiffened by truss members in 1988 and 1989. The bridge carries six lanes of traffics, three in each direction. Each box is stiffened by the diaphragms at every 7.7 m and the cross beams at every 1.54 m. The orthotropic steel deck of 12 mm thick, stiffened by longitudinal ribs, supports the road way with about 80 mm thick asphalt pavement.

DURABILITY EVALUATION PROCEDURE

As mentioned earlier, one of the tasks to investigate the effectiveness of the previous retrofitting measures and evaluate the remaining life of the structures. In this context, we have proposed the following stress measurements and structural analysis near the diaphragm corners. They are;

1) Stress measurement with test truck of known weight,
2) Measurement of stress range histograms using histogram recorders, and
3) Stress analysis of section model by finite element method for truck loading.

Stress measurement was carried out in 1997. In order to simulate the condition before the stiffening trusses were installed to the diaphragms and the cross beams, high strength bolts at the ends of the trusses near the measured section were temporarily removed. In this way we could compared the measured stresses with and without stiffening truss due to the test truck with the same strain gages. The structural analyses were also carried out accordingly.

Comparison was made first between the measured maximum stresses near the diaphragm corner with and without stiffening trusses. The same comparison can be made through the finite element stress analysis. These results show how much stresses due to the test truck can be reduced by adding the stiffening trusses.

For the durability evaluation in term of fatigue, measured stress range histograms near the diaphragms can be used. Once the stress range histograms are obtained, we can use Miner's rule or equivalent stress range concept to evaluate the fatigue life of the diaphragm corners. However, two difficulties arises. The first difficulty is that fatigue strength of the diaphragm corner is not well defined. Since no fatigue test data is available for such details, we have to rely on S-N curves determined with some engineering judgement.

The second difficulties is how to estimate the traffic condition in the past and in the future. The stress range histograms are obtained through the measurement of particular date, for example 72 hours in weekdays for standard measurements carried out in Japan, it is not easy to apply the data to the past and the future estimations. In this study assumption were made in such a way that we could use the data of traffic volume recorded in the past. We assume that the increase in the traffic volume in the past is related to the increase in the number of cycles obtained by the stress range histogram measurement. The maximum weight of trucks in service is assumed unchanged, since no such data is available. For the future estimation, we assume that the traffic condition will not change so much because the number of traffics passes on this bridge is already to the maximum capacity, and present traffic volume seems saturated. If one can estimate future traffic condition accurately, it can be use as the future traffic conditions.

STRESS MEASUREMENT

Stress measurement was carried out on various structural details including the diaphragm corner at side span center. As mentioned earlier stiffening truss members of two sections were temporarily removed in order to simulate the stress condition before 1988-89 retrofitting.

Fourteen strain gages were attached near the diaphragm corner, as shown in Fig. 2. Stress gradient toward weld beads at the diaphragm corner was measured by a series of gages, such as gage 3 to 2 in lower flange of the cross beam, 7 to 5 in vertical flange, and 11 to 13 in web of the cross beam.

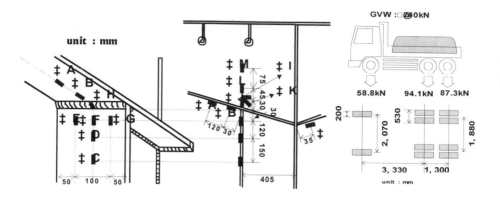

Fig. 2 Diaphragm corner and strain gages attached.   Fig. 3 Test truck and its dimension.

Stresses were measured dynamically with test trucks of 240 kN, as shown in Fig. 3 under normal traffic condition. The test truck ran in slow lane and middle lane, since these running positions caused largest stresses in the diaphragm corner. Stress range histograms were also measured for 24 hours with and without stiffening trusses.

Finite element stress analysis was carried out for the section model which include the diaphragm corner of interest. Shell elements were used for the model and the smallest mesh size at the diaphragm corner was 25x30 mm. Wheel loads corresponding to truck of 240 kN was applied to the model.

STRESSES WITH AND WITHOUT STIFFENING TRUSSES
Examples of measured stresses due to the test truck without the stiffening trusses are shown in Fig. 4. When the test truck runs in the middle lane, compressive stresses are observed at the diaphragm corner. However, when the truck runs in the slow lane, it will cause tensile stresses, which are somewhat smaller than the previous case. The maximum stress is observed when the rear-rear wheel passes at the diaphragm.

Stresses became much smaller after the stiffening trusses are installed, as shown in Fig. 5.
The effect of stiffening trusses is clearly shown by the gage 9 attached at weld bead, which showed the maximum stress of 65.4 MPa without stiffening truss was reduced to about 10 MPa with the stiffening trusses. The maximum stresses observed in the measurement and the analysis due to the test trucks are summarized in Fig. 6. Analytical results show that stress of -73.3 MPa occurs in the web of 30 mm upward, which is reduced to about -11 MPa by installing stiffening trusses. The other points also showed stresses less than 10 MPa with stiffening truss.

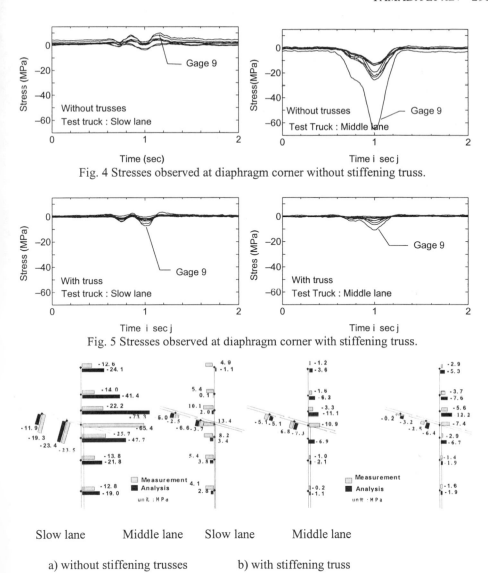

Fig. 4 Stresses observed at diaphragm corner without stiffening truss.

Fig. 5 Stresses observed at diaphragm corner with stiffening truss.

Slow lane   Middle lane   Slow lane   Middle lane

a) without stiffening trusses   b) with stiffening truss

Fig. 6 Comparison of maximum stresses measured and analyzed.

DURABILITY EVALUATION OF DIAPHRAGM CORNER

Stress range histograms were also measured for 24 hours with and without the stiffening trusses, for example, as shown in Fig. 7. Stress ranges were drastically reduced by installing the stiffening trusses. Note that for 24 hours in service large stress ranges, which was about three times larger than that due to the test trucks, were observed. It implies that simultaneous loading and/or overloaded truck caused much higher stress ranges than the single test truck of 240 kN. Actually another measurements, carried out by BWIM System using orthotropic steel deck, indicated that frequent overloaded trucks were observed on this bridge[3], [4].

Fatigue life of the diaphragm corner is evaluated based on the JSSC Fatigue Design Recommendations published in 1993[4]. The procedure is based on Miner's rule or equivalent stress range concepts. As mentioned earlier, stress category of structural detail at the diaphragm corner of interest is not well defined. Since any test data is available for such detail, we estimated the stress category as follows.

First, we decided to use strain gage measurements nearest to welds, such as gages 3,7,9 and 11. For gage 3, fillet weld perpendicular to the stress direction is categorized as stress category E. However, behind the fillet welds there exist the web of vertical column of the diaphragm, which causes global stress concentration to the fillet weld. Therefore, the stress category of this detail is categorized as stress category G. For gages 7 and 11, stress category F is adopted for toe cracks and H is adopted for root cracks. Both types of cracks were observed in this detail. For gage 9, which was attached over the weld bead, stress category C is used by assuming as if it is flame cut edge.

Number of vehicles passed on this bridge is estimated from the annual traffic data, as shown in Fig. 8. Note that rapid increase in traffic volume, as well as truck traffic, is observed since around 1970. The number of trucks, when the histogram data was obtained, was about 39,000. The stress range histogram was modified according to the traffic volume to find the stress range histogram in the past. Then, the fatigue damage is computed using Miner's rule, as follows,

$$D = \sum \frac{n_i}{N_i} \qquad (1)$$

where n_i is the number of cycles of corresponding stress ranges obtained in the stress range histograms, and N_i is the number of cycles obtained from given S-N curves of the structural detail of interest. One can assume that fatigue life elapses when Miner sum exceed unity.

As shown in Fig. 9, with different assumption of the structural detail, wide range of fatigue life is obtained. Fatigue life is reached as early as 1977, or after 14 years of service. Actually quite a few large fatigue cracks were found in 1987, and we can not decline the possibility that small fatigue cracks were already formed at much early stage.

After the stiffening trusses were installed in 1988-89, Fatigue damages do not increase. It implies that the stiffening all diaphragms and every other cross ribs reduced the stresses at the diaphragm corner, and no further damage will be expected in such detail, if the present traffic condition will continue in the future.

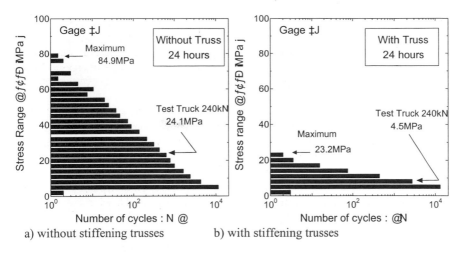

a) without stiffening trusses   b) with stiffening trusses

Fig. 7 Stress range histograms obtained by gage 11 with and without stiffening truss.

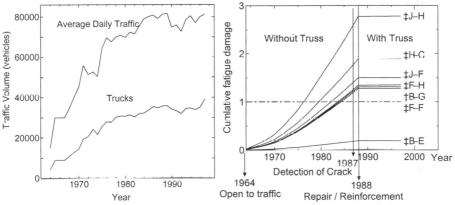

Fig. 8 Traffic volume data in the past. Fig. 9 Damage accumulation of diaphragm corner since the bridge was open to traffics.

CONCLUSIONS

Many Japanese highway bridges were constructed after the World War II as the economy recovered from the post war eras and grew rapidly in 1960s and 70s. These bridges were designed and fabricated with economy bases because of the limited fund available at that time. Economic growth was often associated with the rapid increase in traffic volume and weight, and that caused some damages to the highway bridges. The Minato shin Bashi Bridge, described in the paper, is one of those bridges. The bridge exhibited fatigue cracks at the diaphragm corner, which were retrofitted about ten years ago. The stiffening truss members were also installed to every diaphragms and every other cross ribs.

Stress measurement was carried out with and without the stiffening trusses to evaluate the effectiveness of the previous retrofitting measures. The measurement was carried out with the test truck. Then, the fatigue life of the diaphragm corner was also evaluated using the measured stress range histograms. With some assumptions the analytical results predicted occurrence of fatigue cracks within reasonable accuracy. Because the diaphragms and the cross ribs were stiffened by the stiffening trusses, the remaining fatigue life of the detail became satisfactorily long, and the previous retrofitting measures was proved satisfactory.

AKNOWLEDGEMENT

This study was carried out under the guidance of the committee works formed by Nagoya Highway Works of Ministry of Construction. The field tests were carried out with the help of Toyo Ghi-ken Consultant, and students of Nagoya University. The authors are grateful to those who supported this study.

REFFERNCES

1) Japan Road Association(1997) : " Fatigue of Steel Bridges" (in Japanese.)
2) Kamiya, H. and Kamijyo, S. (1965): " Design and Construction of Minatoshin-bashi ", Bridges and Foundation,(in Japanese)
3) Ojio, T., Yamada, K., Kainuma, S., Obata, T. and Furuichi, T. (1998) : "Estimation of Traffic Loads using Strain Recording in Orthotropic Steel Deck", Journal of Structural Engineering, JSCE, Vol.44A, pp.1141-1151. (In Japanese)
4) Ojio, T., Yamada, K., (1997) "Measurement of Wheel Loads Using Orthotropic Steel Deck", Proceedings of EASEC-6, Volume 1, pp.411-416.
5) JSSC (1993) : " Recommendations on Fatigue Design of Steel Structures", Gihodo-Shuppan. (in Japanese.)

Rehabilitation of the Monongahela Connecting Railroad Bridge

DAVID A. CHARTERS, JR., P.E.
Parsons Brinckerhoff Quade & Douglas, Inc., Pittsburgh, Pennsylvania, USA

BACKGROUND

The existing Monongahela Connecting Railroad Bridge (Mon Conn Bridge) in Pittsburgh, Pennsylvania, USA was constructed circa 1900 to carry steel mill trains across the Monongahela River (see Figure 1). Until the early 1980s, trains—typically consisting of two ladle cars—carried molten iron produced in blast furnaces on the north side of the river across the bridge to open-hearth converters and furnaces on the south side. The bridge was the lifeline of the steel plants on each side of the river, with over 25 utility pipelines and electrical power lines crossing the river on the bridge. With the demise of the US steel industry in the early 1980s, many of this steel mill's facilities have been inactive or demolished.

Figure 1.   Monongahela Connecting Railroad Bridge

Pittsburgh is undergoing a rejuvenation, replacing antiquated facilities with state-of-the-art high technology industries. In the late 1980s, the Urban Redevelopment Authority of Pittsburgh (URA) developed the north side of the river near the bridge into a high technology office park that includes representatives from local universities and businesses. In the early 1990s, the URA purchased inactive steel company property on the south side of the river near the bridge to be developed into a 53-hectare multi-use development including commercial, retail and residential properties. This required that the Mon Conn Bridge be rehabilitated to support highway vehicles instead of the defunct steel mill trains. The renovated bridge will serve as a vital link between the new south side development and the central business district of Pittsburgh.

The existing structure is a 340-meter-long seven-span steel bridge, consisting of five through-truss spans, one deck truss span, and one deck girder span (see Figure 2). The main span over the river measures 99 meters. The bridge site actually contains two companion bridges supported by the same substructure units—the Main Bridge that carried two railroad tracks for the supply trains (gondola cars), presently being reconstructed for vehicular traffic, and

Bridge Management 4, Thomas Telford, London, 2000.

CURRENT UK PRACTICE
In the UK prior to 1992, the majority of bridges were similarly match cast jointed with epoxy joints and internally prestressed. Outside the UK this is still a robust, economic and popular construction method, the author has been responsible for a number of designs [8]. For many locations particularly with a tropical or semi tropical climate or for railway structures where there is no use of de-icing salts durability is not such a problematic issue.

Since 1996 only two major structures have been constructed in the UK using segmental construction, both use external prestress so that ingress of chloride contaminated water at joints would not affect the prestress (both used epoxy joints). The first bridge the Second Severn Approach Viaducts [9] used external tendons protected by an external HDPE duct and a petroleum wax grout. The wax grout together with an extended end cap and uncropped strands means that the tendons could be re-stressed if required for replacement or demolition. The second bridge (figure 2) by the author [10] again uses external tendons but this time protected by an external HDPE duct with a cement grout. This system means tendons cannot be de-stressed but, they can be replaced by cutting as all tendons are double sleeved at deviators and anchorages. Although the externally (unbonded) prestressed segmental bridge has advantages for inspection and maintenance, it is in the authors view structurally less robust than an internally (bonded) prestressed structure.

Figure 2. The A13 Viaduct during construction. The structure uses epoxy jointed segments, HDPE ducting and external prestress.

FUTURE PRACTICE
Based on the Ynys-y-Gwas collapse and subsequent research it is clear that segmental bridges with mortar joints should not be used. Existing structures with mortar joints will require careful monitoring until they are demolished.

For the existing stock of segmental bridges monitoring by routine inspection could be supplemented by strain gauge monitoring. The monitoring of new structures like the Second Severn Viaducts and A13 Vviaduct should be considered by the maintenance authority as a means of monitoring structural safety. With the ever growing power of computers it may soon be possible to process the data in real time giving warning of unusual strains immediately to the maintenance team.

The US research could be interpreted as confirming that internally prestressed segmental bridges with epoxy joints are as durable as any other prestressed concrete structure. However, the results are ambiguous and the dual protection system recommended by TR47 is a sensible precaution, as is the use of plastic rather than steel ducting. Consequently for UK highway bridges designers have three options for their future segmental structures:

- The use of externally prestressed structures
- The use of internally prestressed structures with non-corrodable tendons i.e carbon or aramid fibres
- The use of internally prestressed structures with continuous internal ducts.

The first has been tried and tested and is likely to be used for the majority of structures of this type in the near future. The second option has possibilities, however, the cost of the non corrodable tendons are currently prohibitively expensive and in general they are more suited to external prestressing applications. The third option requires a continuous watertight duct within the concrete section by the use of a coupling devise or the use of a secondary duct. Again neither of these options are currently developed to a point where they have been used.

REFERENCES

1. Woodward RJ, Williams FW, Collapse of Ynys-y-Gwas bridge, Proc ICE, Part 1, August 1988.
2. Woodward RJ, Wilson DLS, Deformation of segmental post-tensioned precast bridges as a result of corrosion of the tendons, Proc ICE Part 1, April 1991.
3. Maguire RJ, Condition monitoring of prestressed concrete bridges, Construction Repair, November 1996.
4. The Concrete Society, Technical Report 47, Durable post-tensioned concrete bridges, 1996.
5. Miller Maurice D, Durability survey of segmental concrete bridges, PCI Journal, Vol 40, May-June 1995.
6. Poston RW, Waters JP, Durability of Precast Segmental Bridges – NCHRP Project 20-7/Task 992. Abstracts of ASBI 1998 Convention, November 1998.
7. West J, Vignos R, Breen T, Laboratory Durability Study of internal tendons in precast segmental bridges. Abstracts of ASBI 1998 Convention, November 1998.
8. Collings D, Bennett D, Railway system with a staring role in Malaysia, Concrete Quarterly, No 185, Winter 1995.
9. Kitchener JN, Mizon DH, Second Severn Crossing viaduct superstructure and piers, Proc ICE, Second Severn Crossing 1997.
10. Collings D, The A13 Viaduct: Construction of a large monolithic concrete bridge deck, (submitted to ICE proceedings for publication).

The performance of in-situ weathering steel in bridges

M McKENZIE
Transport Research Laboratory, Crowthorne, Berkshire, England

INTRODUCTION

Weathering steels are low-alloy steels containing up to 3% of alloying elements such as copper, chromium and phosphorus. Under suitable conditions, the atmospheric corrosion rate of such steels is lower than the rate for mild steel - in fact low enough for weathering steels to be used unprotected in structures and allowed to corrode. This is an attractive concept as it saves the costs of both initial and maintenance corrosion protection treatments such as painting. Weathering steels were used initially in the USA where exposure trials showed that corrosion rates fell to negligible levels after a few years exposure. However corrosion trials in the UK showed that corrosion rates were generally higher than in the USA and, although lower than for mild steel, were still significant as time passed. This led to the concept of using thicker steel to allow for any corrosion over the life of a structure. Corrosion trials carried out by the Transport Research Laboratory (TRL) and British Steel provided information on the corrosion rates of weathering steel in a variety of UK environments (Kilcullen and McKenzie 1979) and the results were incorporated in Department of Transport Standard BD 7/81 (Department of Transport 1981). This gave the additional thickness needed for different environments to allow for corrosion over the life of a bridge and indicated environments which were not suitable for the use of weathering steels - notably where they would be exposed to excessive chlorides. Since that standard was published further corrosion tests using weight loss coupons have been carried out by TRL (McKenzie 1990). In view of the fact that weathering steels had now been used in bridge construction in the UK for over 20 years it was decided carry out a more general review of the in-situ performance of weathering steel in highway bridges. A questionnaire was circulated to County Councils, Local Authorities and other weathering steel bridge owners to establish the number, age, location, condition and problems encountered on their weathering steel bridges. This was supplemented by site inspections on a number of bridges showing particular features or problems.

QUESTIONNAIRE ON UK PERFORMANCE

The questionnaire was intended to establish inventory type information - eg age, location, type of structure etc - along with details of any problems encountered and remedial measures taken. This provided data on 141 bridges. These ranged from small footbridges to large multispan viaducts. The main type of construction (over 80%) was a composite concrete deck with plate girders or rolled sections. There were 4 box girder bridges and 2 small bridges with wooden decks. Although weathering steel bridges are, by definition, designed to be used without further corrosion protection, some 14% had been part painted either for aesthetic reasons - eg the Royal Fine Arts Committee in Scotland had insisted that fascia girders were painted in one bridge - or as part of remedial treatments where corrosion was excessive. Five bridges had been enclosed either from the outset or as a remedial treatment.

Most bridges had been constructed since 1980 and the use of the material was increasing (see Figure 1). Bridges were located throughout the UK but were more popular in certain areas - notably Scotland and northern England.

The majority of bridges were built in what were described as rural areas (Figure 2). In terms of the classification used in BD 7/81 to decide additional steel thickness required to allow for corrosion, about 80% were in a mild environment with only 11% classed as severe. In over 8% of the bridges there had been no additional steel thickness to allow for corrosion, and in a further 28% it was not known what, if any, allowance had been made (Figure 3). However the higher additional thickness had been added in 26% of the bridges - rather higher than expected considering the BD 7/81 classification. The type of weathering steel used was mainly Corten B or grade WR50C.

Weathering steel bridges had been constructed mainly to cross watercourses or railways (Figure 4). Only 14% crossed roads, reducing the scope for assessing the direct effects of traffic passing under a bridge on corrosion performance.

PERFORMANCE

The overall assessment of performance was very good with over 80% of bridges being classed as having no defects (Figure 5). Only 5% were classed as having defects affecting steel corrosion generally. These included bridges built in inappropriate environments such as marine locations and what was perceived as excessive corrosion in some positions on a structure - eg pitting on a top flange and corrosion on a lower flange thought to have resulted from snow retention. The majority of defects identified could be attributed to specific causes rather than any general inadequacy in corrosion performance.

The main cause of defects was leaking expansion joints allowing salty water to flow over the steel at the end of beams leading to rapid, excessive corrosion. Some aspects of detailing also led to corrosion problems; open timber decks allowed salty water to flow directly over the steel, drainage pipes allowed water to blow onto adjacent steel. In one case a box girder footbridge was constructed with access hatches in the upper, salted surface. This allowed salty water to leak into the box leading to very heavy corrosion; the structure was eventually demolished and replaced with a concrete bridge.

Relatively more defects were identified on older bridges but there was no marked difference in the extent of defects on bridges crossing different types of obstacle. However the small number of structures in some categories - notably in bridges crossing roads where traffic effects are of interest - reduced the scope for assessment. There was also no marked difference in defects identified on structures in different environments. There were differences in the extent of defects identified depending on the degree of salting on the bridges with high salting leading to more defects (Figure 6). The assessment of salting covered both that applied to any road crossed by the bridge as well as salting to the road carried by the bridge. The obvious danger in the latter case is of leaking joints allowing water to flow directly over the steelwork.

Although the general corrosion performance is good, this must be based on assessment of the rust condition; the majority of bridges - about 60% - do not have a formal quantitative corrosion monitoring scheme in place. This is so even for recent bridges despite the requirements of BD 7/81. Where monitoring is being carried out, at least some section thicknesses are available for each bridge but repeat measurements, so that corrosion rates can

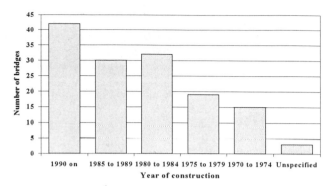

Figure 1 Age distribution of weathering steel bridges in the UK

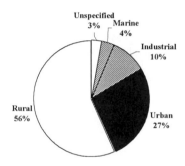

Figure 2 Descriptive environment

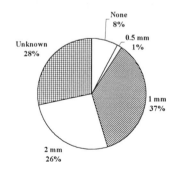

Figure 3 Additional steel thickness used

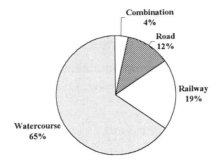

Figure 4 Type of obstacle crossed – 'Combination' implies multiple obstacles

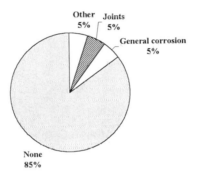

Figure 5 Defects identified on weathering steel bridges in the UK

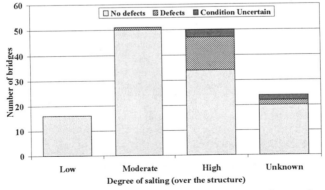

Figure 6 Defects identified on weathering steel bridges according to degree of salting

be estimated, are only available for about 20% of those monitored. The majority of these are in Scotland and have been reported elsewhere (Halden 1991). It is not possible to give general estimates of corrosion performance based on direct residual steel thickness measurements for bridges in the rest of the UK although some bridges do have corrosion rates from coupon measurements, usually as part of TRL research programmes.

REMEDIAL TREATMENTS
Where there is a specific cause for inadequate performance, eg a leaking joint or inadequate drainage, treatment of the cause is the preferred action. However if corrosion has by then progressed too far or the problem is expected to reoccur, the standard procedure is to apply a paint coating to the steelwork in the affected zone. A range of systems and degrees of surface preparation have been used including complete steel cleaning via grit blasting then application of a multicoat paint system, to hand cleaning and application of surface tolerant coatings - in some cases Hammerite was used. Because of the small sample sizes and variation in treatment it was not possible to assess the general effectiveness of such different approaches. In one case where there was an overall corrosion problem on a bridge in a marine environment, the entire bridge was blast cleaned and painted with a multicoat system. Another bridge in a marine environment was wet blast cleaned then enclosed as part of a TRL research project. As has been mentioned earlier, demolition and replacement was used in one case.

APPEARANCE OF RUST COATINGS
The rust coating that develops on a weathering steel bridge can show a wide range of colours and textures. Colour ranges from a lightish brown to very dark brown and is dependent on the lighting conditions and the descriptive powers of the observer.   Texture also varies considerably ranging from very fine grained adherent rust through larger grained but still adherent rust, flaking outer layers of rust over an underlying adherent coating, to very thick layers of rust which can be prised away from the surface. In general, attractiveness declines as the grain size, layer thickness and flakiness increase but this is definitely a case where beauty is in the eye of the beholder. There are also differences in the appearance of rust that forms on different parts of a bridge. The top surfaces of bottom flanges tend to have thicker rust layers than webs with more evidence of flaking rust. The undersides of upper flanges tend to have light brown rust spots, presumably from condensation, superimposed on a darker brown coating. Fascia girders, particularly those facing south, can have very adherent rust layers on the webs with thicker but still adherent layers on the flange. This is probably a result of more wet/dry cycles.   Sheltered girders and north facing fascias tend to have flakier rust.   The presence of traffic on the roads under a bridge can have a significant effect on the appearance of rust giving a generally flakier and less attractive appearance. This probably results from deicing salt and other traffic debris being thrown up onto the steel. The effect is most obvious on viaducts crossing both trafficked and untrafficked sections where differences in the appearance of the rust above the different sections are quite apparent.   By far the most unattractive rust is that formed when salty water, usually from a leaking joint, flows over the steel. This leads to very thick layers of variegated rust, around 5 mm in some cases, which can be seen to be peeling from the surface. This is most prevalent on the flanges but does spread up the base of the web. When such rust is peeled off, localised corrosion of the steel is often seen. There can also be man made effects on the appearance; the surface invites graffiti.

EFFECTIVENESS OF REMEDIAL TREATMENTS
The most common remedial treatment is to paint the steel either completely for generally unacceptable corrosion, or just in the affected parts, say in the case of a leaking joint. The main problem is surface preparation. Even if full grit blasting is possible this is unlikely to be

fully effective if corrosion has been caused by chlorides. Weathering steels tend to pit more than standard steel and it is difficult to fully clean to the base of pits even with wet blasting. This implies that subsequent paint life is likely to be poor. This effect was clearly seen on a bridge in a marine environment which resulted in unacceptable levels of corrosion. The bridge was grit blasted then painted but paint breakdown was evident after only 3 years. Similar problems will be encountered where only local areas are painted with the added problem of the join between the painted area and the unpainted steel. Moreover restricted access to the ends of beams can make surface preparation and repainting difficult; if only hand cleaning is possible and only surface tolerant coatings can be used paint life could be even shorter leading to an ongoing maintenance requirement.

PERFORMANCE ASSESMENT

Many bridges are not being monitored to provide quantitative estimates of the corrosion rates of the steel, and from those that are, there is only limited information on corrosion rates based on direct residual steel thickness measurements. Assessment of performance is presumably being made mainly on the appearance of the rust coatings which form on the steel. It is probable that rust appearance is related to the underlying corrosion rate. The thick, laminar layers of rust that form when salty water flows over the steel are a good indication of unacceptable performance, whilst the adherent, small grained rust coatings do indicate good performance. However the underlying corrosion rate could still vary considerably for what, visually, would be considered quite acceptable rust coatings. The larger grained, flakier rust coatings which fall between these extremes are even more difficult to relate to underlying corrosion rates. It would be useful for inspection guidelines to indicate what to expect when viewing rust coatings, and what is considered cause for concern. This is already done in some standards. This could be based on colour, thickness, grain size and flakiness perhaps related to representative photographs. However, although the appearance of rust coatings is important and informative, this is no substitute for quantitative measurements of residual steel thickness throughout the life of the bridge.

It was noticed that bridges crossing roads do develop flakier, larger grained rust, where the steel crossed trafficked sections compared with where it crossed untrafficked sections. From coupon studies this is known to lead to higher corrosion rates through not necessarily too high to allow the material to be used. Only a limited number of bridges have been built over trafficked roads in the UK and, although no general problems have been identified, any move to relax the current restrictions which limit this type of structure to high clearances, would have to be carefully considered. Such consideration is worthwhile as there would be considerable economic advantages in avoiding traffic delay costs by using a maintenance free material such as weathering steel for such bridges even if a greater corrosion allowance were needed above trafficked zones.

It has already been emphasised that regular measurements of residual steel thickness are desirable. Digital ultrasonic instruments, which measure residual steel thickness through the rust are now available which simplify such monitoring. These use the time between echoes in the steel rather than the usual total transit time for the ultrasonic signal thereby giving only the steel thickness ignoring the thickness of the rust

CONCLUSIONS

A questionnaire circulated to weathering steel bridge owners has provided a database giving inventory and performance data on 141 weathering steel bridges in the UK. Overall UK weathering steel bridges are considered to be performing well with over 80% being classed as having no defects. The main cause of problems seems to be associated with the effects of deicing salt - notably where leaking joints have allowed salty water to flow over the steel.

Many bridges are not being monitored to provide quantitative estimates of corrosion rate based on residual steel thickness measurements. Of those that are being monitored, general corrosion rates are available for only about 20%. These are mainly in Scotland and corrosion rates are considered acceptable.

ACKNOWLEDEMENTS
This work was carried out on behalf of the Highways Agency. Particular thanks are due to Awtar Jandu.

REFERENCES
Department of Transport (1981). *Weathering steels for highway bridges.* Departmental Standard BD 7/81, HMSO, 49 High Holborn, London WC1V 6HB

HALDEN, D (1991). *Design and performance of weathering steel bridges on Scottish trunk roads.* Proceedings of the Institution of Civil Engineers, vol 90, pp447-62, Part 1, April.

KILCULLEN, B AND M MCKENZIE (1979). *Weathering Steels.* Proceedings of the conference 'Corrosion in civil engineering' 21-22 February, 1979, Institution of Civil Engineers, London

MCKENZIE, M (1990). *The corrosion of weathering steel under real and simulated bridge decks.* Department of Transport TRRL Research Report 233, TRL, Crowthorne, Berks.

Stainless steel reinforcement for concrete – a solution chloride induced corrosion of reinforced concrete structures

GRAHAM GEDGE, Associate, Arup Research and Development, 13 Fitzroy Street London W1P 6BQ, United Kingdom.

INTRODUCTION

It is now well established that chloride induced corrosion of carbon manganese steel reinforcement is a factor limiting the durability of reinforced concrete structures. There are many methods available to combat this problem for both new and existing structures. However, few if any of these can be considered as inherent solutions to the problem. Some aim to delay the initiation of corrosion, others have a finite life and or require regular maintenance. Stainless steel reinforcement represents a potential maintenance free solution to the problem that makes use of the inherent properties of both the concrete and steel.

In addition to providing a potential solution to corrosion stainless steel also represents a material that requires minimal changes to established design procedures in terms of mechanical properties, bar size, availability and bending schedules. In the UK it is possible to specify stainless steel according to *BS6744*[1] to ensure that these properties are achieved.

This paper will review the current understanding of the performance of stainless steel in chloride-contaminated concrete and show that research over the last 30 years provides confidence that the materials can prevent chloride induced corrosion. The paper will also consider how design changes in relation to the specification of concrete could lead to the more economic use of stainless steel in reinforced concrete structures.

HISTORICAL USE OF STAINLESS STEEL IN REINFORCED CONCRETE

Stainless steels are not new materials and the use of stainless steel as reinforcement is not new. The first known use of stainless steel reinforcement was on the Progresso Pier on the Yucatan peninsula in the Gulf of Mexico[2]. This was constructed in the late 1930's and is still in service today some 60 years after construction and there is no evidence of corrosion to the steel. The reinforcement to the arches of the structure was made from stainless steel; the composition of the steel is comparable to a modern grade 1.4301.

Figure 1 Progresso Pier (background) 1998

Bridge Management 4, Thomas Telford, London, 2000.

In the 1970's it became apparent that carbon steel reinforcement did not always perform well as reinforcement where chloride contamination occurred to the level of the bar. This prompted research into methods to reinstate damaged structures and to prevent corrosion of new structures. One of the more overlooked areas of research initiated at this time was the work carried out in the UK by the Building Research Establishment (BRE) into the possible use of stainless steel. The BRE undertook a number of long term exposure trials[3,4,5] on stainless steel in concrete that was intentionally contaminated in chloride at the time the concrete was cast and also on samples without added chloride but which were exposed to environments where chloride ingress was likely to occur. The exposure sites were at Beckton, East London, Hurst Castle on the UK South coast and at Langstone Harbour near Portsmouth in the UK.

The materials investigated included grades 302, 304, 315 and 316 stainless steels placed in both prisms and stressed beam samples. The samples were exposed for up to 22.5 years in a range of marine environments. Periodically samples were recovered and corrosion assessed and compared to carbon steel. In all cases no corrosion of stainless steel was found despite chloride levels in some cases exceeding 3% and with cover to the reinforcement as low as 10mm.

The data obtained by these BRE studies represents a very significant body of track record data on the use of stainless steel reinforcement. As such it provides confidence that the use of stainless steel will result in durable structures.

More recent examples of the use of stainless steel reinforcement have been given in the Concrete Society[6] Report 51. Some of the examples from Report 51 are discussed in more detail below.

PERFORMANCE OF STAINLESS STEEL IN CHLORIDE CONTAMINATED CONCRETE

The work of the BRE represents an important and substantial body of exposure data to real environments that supports the limited track record, on actual structures, of the materials use as reinforcement. However, this work was not intended to understand why stainless steels perform as they do in concrete. The main contributors to this understanding have been the Italians working at the Politechno di Milano[7].

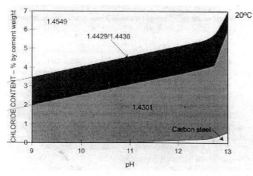

The work undertaken aimed to understand how stainless steel behaved at various pH levels with differing percentages of chloride present. The key part of the studies was to estimate the onset of corrosion by establishing the pitting potential of different grades of stainless steel in solutions of varying pH and chloride concentration.

Figure 2 Chloride concentration to pitting v pH.

The use of pitting potential to measure the onset of corrosion is well founded and an established technique[8]. Stainless steel protects itself by the formation of a stable passive film on the surface. The pitting potential of the steel can be taken as the value at which local breakdown of the passive film occurs in a given solution. It is also well known that stainless steels can, under some circumstances, suffer from pitting in chloride environments. Therefore by measuring the pitting potential of stainless steels in various pH solutions at a range of chloride concentrations it is possible to show a critical concentration for corrosion initiation.

The work of the Italians can be summarised in graphical form as shown in figure 2. An important feature of the graph is how the tolerance to chlorides, or the stability of the passive film, increases with increasing pH for all grades of stainless steel. More importantly as the typical pH of concrete is approached (pH 12 to 13.5) this tolerance shows a very marked increase.

If 1.4301 steel is considered the graph would predict that the onset of passivity breakdown would occur at a chloride concentration of the order of 4% at the level of the reinforcement. Comparison of this value with those reported on real structures, such as published data from the UK Bridge stock[9], shows that this concentration is unlikely to occur at the level of the bar within normal structural design lives.

For 1.4436 steel the chloride concentration is increased to 5% providing even greater confidence that loss of passivity will not occur.

This work demonstrates two important points regarding the use of stainless steel as reinforcement:

- The tolerance of stainless steel to chloride induced corrosion is high.
- This tolerance is due to the high pH of the concrete surrounding the steel.

The data also indicates that it is important to maintain concrete quality, it can be seen that as the pH decreases so does the tolerance to chloride. This indicates that although stainless steel is beneficial in combating the effects of carbonation, the combined effect of chlorides and carbonation could result in corrosion of the steel. It is therefore important to maintain concrete quality to deal with carbonation.

ECONOMICS OF USING STAINLESS STEEL
The development of the market for stainless steel has been hampered by:

- The perceived high initial cost of the material.
- The desire on the part of clients, specifiers and contractors for lowest initial cost structures.

In the former case at least part of the problem has been that suppliers regarded stainless reinforcement as a speciality, niche market that should attract a significant premium. Even allowing for this it has to be accepted that corrosion resistant alloys such as stainless steel will always be more expensive than carbon steels because of the alloy content. Against this must be balanced the improved performance and reduction/elimination of maintenance resulting from the use of stainless steel.

Current price comparisons related to carbon steel reinforcement range from 3.5 to 6.2. for 1.43xx and 4.5 to 8.5 for 1.44xx up to 40mm diameter compared to carbon steel. The differences in price arise from two sources:

- Different alloy compositions. Grade 1.43xx is approximately 25% less than Grade 1.44xx.
- Differing manufacturing routes for differing bar diameters.

The important point to note is that 15 years ago the comparison would have shown a minimum differential of a factor of 15. The reasons for the very significant reduction in first cost can be explained by a change in attitude by producers who now see the potential market as more significant and have altered the manufacturing route minimise production costs.

In considering the use of stainless steel it is also important to use the material appropriately and to compare costs in a sensible and meaningful manner. It may be appropriate on some structures to consider the global use of stainless steel to replace carbon steel. In many other instances it is more appropriate to make use of carbon steel generally and only use stainless steel in areas susceptible to corrosion. Fortunately it is often possible to predict with a high degree of confidence where corrosion problems will occur on a structure and therefore focus where stainless steel is used. This approach limits the overall tonnage of stainless steel used and reduces the overall cost impact. The benefits of this approach in terms of initial costs are discussed below.

In the case of minimising initial costs the folly of this approach is now well established. Increasingly it is recognised that durable maintenance free structures represent a far better investment than cheaper structures requiring regular maintenance. Life cycle costing has shown that over the life of a structure the use of stainless steel represents the most cost effective solution where there is a risk of chloride induced corrosion[6].

BRIDGE STRUCTURES
The different approaches to the use of stainless steel mentioned previously can be seen from actual structures; for example comparison between the Mullet Creek Bridge[10] in Canada and The Schaffhausen[11] Bridge in Switzerland. In the case of Mullet Creek the design was based on a straight substitution of Grade 316LN stainless steel for the entire bridge deck. This resulted in an increase in the initial cost, compared to epoxy coated bar, of approximately 20%. It has been estimated[12] that this cost could have been reduced to 5% by the use of Grade 304. It should be noted that the costs refer to the total as built structural costs, not simple reinforcement costs.

For the Schaffhausen Bridge[11] a different approach was adopted that utilised stainless steel where the risk of chloride contamination was highest. The remainder of the reinforcement was retained as carbon steel. Two grades of stainless steel were used, in the splash zone on the cable stay pylons a duplex stainless steel was this allowed the designers to utilise both the inherent corrosion resistance of duplex stainless steel and its improved strength. Grade1.43xx steel was used on the parapet edge beams that would be exposed to road de-icing salts. This targeted use of stainless steel increased initial construction costs by an estimated 0.5%.

In general the approach to date on using stainless steel has been cautious. There are very few examples of taking advantage of the inherent properties of both the concrete and stainless steel to reduce the requirements for durability to be found in most design codes and specifications. One example where a move towards this occurred was on the strengthening of the Highnam Bridge in Gloucestershire, UK[12].

For reasons of construction sequencing and difficulties in traffic management on the bridge during the work it was decided to omit the waterproofing membrane normally placed on UK highway bridges. Omission of this membrane would almost guarantee that the chloride would become contaminated with high levels of chlorides from road de-icing salts.

The bridge designers and the UK Highways Agency therefore specified that the beams of the bridge were constructed using stainless steel reinforcement. The grade selected was 316 and a total of 7.8 tonnes were used.

MATERIAL SELECTION

An important element in controlling costs when selecting any material to resist corrosion is the selection of the correct material or material grade. This applies equally to initial capital expenditure and future maintenance cost. The aim is to select a material that is resistant to corrosion in the anticipated environment without over-specification leading to unnecessary additional cost with little or no benefit in terms of corrosion resistance. Examples of the use of stainless steel in the literature and guidance commonly available to engineers indicate that as far as stainless steel reinforcement is concerned materials selection is, at best, confused.

In many cases the use of stainless steel has been as a straight substitution for carbon steel. It would seem that in this substitution process the properties of the concrete are all too often forgotten. This is particularly so with respect to the concrete pH.

In uncarbonated concrete the pH will be of the order of 12.5, Figure 1 shows that at this pH the resistance to chloride induced corrosion is high for all grades of austenitic and Duplex stainless steel. This data in conjunction with published data for actual chloride levels in concrete can be used as a basis for material selection taking in to account the pH of the concrete. This basis for selection can be applied to any type of structure where the probable chloride concentration is known.

For example, consider a UK Bridge structure exposed to road de-icing salts and assume that stainless steel is substituted for carbon steel without any other design for durability requirements being altered. Over the life of the structure chlorides can be expected to ingress to the level of the bar. It is not yet possible to accurately model the transport of chloride through concrete with time to predict a chloride level at the end of the design life. However, data is available in the Maunsel report from the existing bridge stock that can be used to select material grades.

This data indicates that it is extremely unlikely that the chloride concentration at the bar, assuming normal cover of the order of 50mm, will approach the 4% threshold required for the initiation of corrosion on a type 1.4301 steel. This steel is therefore appropriate for the application. This conclusion is also supported by the track record data from the Progresso Pier and the BRE studies.

It is possible to identify and tabulate the materials selection process for various different options when using stainless steel reinforcement.

DISCUSSION

This paper has outlined the tack record and use of stainless steel reinforcement and also the technical understanding of how it performs well. Despite this information being in the public domain the use of stainless steel to combat chloride induced corrosion remains the exception rather than the rule. The primary reason for this lack of use would appear to be the material cost of the reinforcement.

Although the use of stainless steel can be justified on whole life costing few clients seem prepared to accept an increase in initial cost of the order of 20% as reported for the Mullet Creek Bridge deck. If stainless steel is to be used more widely then a new approach is needed. This approach should aim:

- To use the most appropriate grade of steel for a particular application.
- To use stainless steel on those parts of the structure at risk of chloride induced corrosion. The remainder of the structure can remain as carbon steel reinforcement.
- To take benefit of the stability of the passive film on stainless steel in high chloride environments and reduce the durability design requirements associated with carbon steel. Thereby obtaining a cost saving on the concrete to offset any increase in reinforcement cost.

Changes in design for durability requirements could include one or more of the following:

- Omission of water proofing treatments such as Silane.
- Omission of water proofing membranes.
- Reduction in cover requirement for particular applications.
- Reduction in crack width control steel.

It must be emphasised that the use of stainless steel is not an excuse for the use of poor materials or poor workmanship, if stainless steel is to be used successfully it requires good quality concrete and high pH around it.

The data generated in the work carried out by the BRE indicates that the type of relaxation to design requirements can be undertaken with confidence. This research was undertaken without using any type of waterproofing on samples exposed to marine environments that are generally regarded as some of the worst that naturally occur. In addition the samples were deliberately cast with very low levels of cover, much lower than would now be used for structures operating in comparable locations. Despite not using any of the methods now regarded as normal practice to achieve acceptable durability none of the samples showed evidence of corrosion to the reinforcement.

A reduction in durability design requirements could be beneficial to the use of stainless steel, as it would reduce the overall cost of construction by:

- Elimination of processes such as Silane treatment resulting in a direct cost saving on materials and labour and a possible saving through a reduction in overall construction time.
- Reducing the structural weight of components such as bridge decks and edge beams through reducing the cover and therefore the volume of concrete. The production of lighter and thinner components on the superstructure will also have a beneficial effect in the possible reduction in foundation design. Again this would result in a possible cost saving in both money and programme terms that could easily offset the additional cost of stainless steel reinforcement.

These changes in conjunction with a more rigorous assessment of construction costs, one that compares total build costs rather than comparing the cost of carbon steel and stainless steel per tonne, may indicate that stainless represents a viable first cost option. In addition the owner of the structure will have the confidence that the maintenance burden on such a structure due to corrosion of the reinforcement will have been almost entirely eliminated.

A further benefit of the use of stainless in either new construction of for repair and strengthening works is that, unlike other corrosion resistant materials, the changes to design codes, design approach and design process are minimal. In effect stainless steels can be regarded, for structural design, as being the same as carbon steels. Stainless steels are available with the same mechanical properties as carbon steels and can be used in the same way. This is a significant advantage as it does not require the designer to learn new procedures to use a different material.

CONCLUSIONS

There is a considerable body of data and evidence indicating that stainless steel represents a solution to the problem of chloride induced corrosion of concrete. However, the material has not been widely used for this purpose and the main reason for this is the perception that it is expensive. This is unfortunate and if the material is to be adopted more widely this perception needs to be addressed. The use of stainless steel could be more economic if the following are more widely adopted:

- Materials selection procedure that ensures the correct grade of material is chosen for a given application. This selection procedure should take account of the inherent stability of the passive film in the high pH environments found in concrete.
- Serious consideration should be given to the targeted use of stainless steel on areas of a structure susceptible to corrosion. As opposed to the global substitution with stainless steel. This limits the impact of overall cost increase.
- Benefit should be taken from the inherent properties of both the concrete and stainless steel to reduce or eliminate some accepted requirements for durability when using carbon steel. For example the omission of waterproofing membranes/treatments and reduction in cover requirements.

By adopting this approach in combination with costing whole components or structures, rather than simply considering material costs of stainless bar, it is probable that stainless steel reinforcement will represent a zero cost increase over conventional reinforcement.

Some of these aspects require additional research particularly with respect to cost assessment of as built structures. It is hoped that such research will be undertaken in the future.

References
1) BS6744:1986 *"Austenitic stainless steel bars for the reinforcement of concrete"*, British Standards Institution, 1986.
2) *"Pier in Progresso, Mexico. Inspection report evaluation of stainless steel"*, ARMINOX, Denmark, March 1999.
3) Proc.Instn.Civ.Engrs, Part 1,1989, **86,** Apr., 305-331
4) *"The resistance of stainless steel partly embedded in concrete to corrosion by seawater"*, Flint and Cox, Magazine of Concrete Research, Vol.40 No.142, March 1988.
5) *"The long term performance of austenitic stainless steel in chloride contaminated concrete"*, Cox and Oldfield, Corrosion of reinforcement in concrete construction, Society of Chemical Industries, London 1996.
6) *"Guidance ion the use of stainless steel reinforcement"*, Technical Report No.51, Concrete Society 1998.
7) *"Stainless steel in reinforced concrete structures"*, Bertolini et al, Proc.International Conference, Tromso, Norway June 1998
8) *"Corrosion Volume 1"*, Editor Shrier et al, Butterworth-Heinneman, 3rd Edition,London 1994.
9) Maunsel Report
10) *"Application of stainless steel reinforcement for highway bridges in Ontario"*, Ip and Pianca, Proc. Of the Nickel-Cobalt 97 International Symposium – Vol 4, Ontario, Canada 1997.
11) *"Guidance ion the use of stainless steel reinforcement"*, Technical Report No.51, Concrete Society 1998.
12) *"Steel corrosion solution"*, C. Abbott, Concrete Engineering International, Vol 3 No.4, May 1999.

Hot dip galvanizing of bridge beams for long term corrosion protection

W. SMITH, Galvanizers Association, Sutton Coldfield, UK.

INTRODUCTION
Steel is the most widely used metal in construction today. However, in order for it to maintain its strength and integrity it must be protected from corrosion. Hot dip galvanizing is one of the oldest and the most durable forms of corrosion protection. It is used in a highly diverse range of applications including motorway crash barriers, street lamps, radio masts, bridges, manhole covers, rubbish bins….the list is lengthy.

The hot dip galvanizing industry has grown almost continuously since the process was first used to protect corrugated iron sheets almost 150 years ago. In fact, in 1998 more steel was galvanized than in any previous year. Continuous growth in the industry is the result of the simplicity of the process and the unique advantages of the coating.

This paper will illustrate the galvanizing process; give an insight into the performance of galvanizing as a corrosion protection system and outline the tremendous benefits to be accrued by maintenance engineers through the use of galvanizing in bridge applications. Galvanized steel is commonplace in temporary permanent bridges (Bailey bridges) so why not for permanent road or rail bridges? Two case histories involving the use of galvanizing for road and rail bridges will be set out.

THE GALVANIZING PROCESS
The process principally involves the formation of an impermeable layer of zinc which is firmly alloyed to the steel substrate (figure 1).

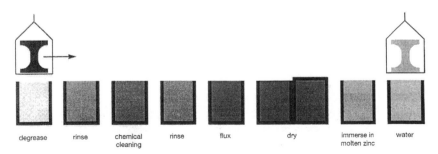

Figure 1 The galvanizing process

Bridge Management 4, Thomas Telford, London, 2000.

Since galvanizing will only "take" to a chemically clean surface, the steel must be cleaned in preparation for galvanizing to remove any surface contamination such as rust, old paint or grease. The clean steel is then immersed in the galvanizing bath - a large bath of molten zinc. Whilst in this bath the zinc wets the surface and a reaction takes place to form tough, hard wearing zinc alloy layers (figure 2).

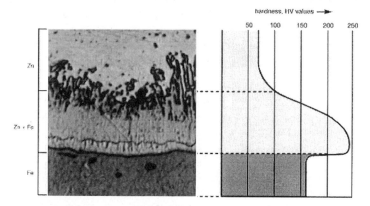

Figure 2 Toughness and abrasion resistance of a galvanized coating

As the work is removed from the bath some molten zinc remains on top of the alloy layers. The result of this process is a coating that is metallurgically bonded to the base metal. Normal galvanizing temperatures are 445°C - 455°C and the average time for immersion is about three or four minutes.

GALVANIZED COATINGS
Zinc coatings protect steel in three ways; (i) the zinc coating weathers at a slow rate giving a long and predictable life, (ii) the coating corrodes preferentially to provide cathodic (sacrificial) protection to any small areas of steel exposed, ie through drilling or cutting, and (iii) if the damaged area is larger, sacrificial protection prevents the sideways creep of rust which can undermine paint coatings.

CORROSION OF GALVANIZED COATINGS
It may be refreshing to find that there are products around which last longer now than they did before. Not only that, one of them lasts longer for entirely natural reasons, not because it has been redesigned and redeveloped. The product , hot dip galvanizing, is not a mystery but the answer to why this 150 year-old method of corrosion protection is lasting longer than ever is, quite literally, "blowing in the wind".

ZINC CORROSION RATE MAPS
In about 1965 the Central Electricity Generating Board (CEGB), as it then was, became interested in mapping England and Wales in order to determine a maintenance policy in relation to their galvanized steel transmission towers. CEGB conducted their first evaluation with zinc cans placed on existing towers 5m above the ground and exposed for 18 months. By measuring the amount of zinc loss a measure of the corrosivity of the atmosphere was gained. This led to a major trial in 1969 when they adopted a 2-year exposure test and followed it up with a repeat test in 1971. Ideally,

there was a site within every 10 square kilometre of England and Wales including some sites in Scotland. However, the trials lapsed in 1973 because CEGB decided not to fund it any more.

Several other bodies showed interest in these corrosivity figures. British Rail used existing data as the basis for predicting maintenance schedules for their overhead line gear - an application very close to the CEGB's. The Automobile Association (AA) set up their own series of tests in the mid-70's in order to try to confirm the results of an earlier survey on vehicle corrosion. This was an ironical application of the data derived from zinc since, in those days, very few cars used zinc for corrosion protection - the exact opposite being true today. By the early 80's, the Ministry of Agriculture had become interested in using the results in order to assess the rate at which farm buildings would corrode in different parts of the country so that they could prioritise grants for maintenance.

The Agricultural Development Advisory Service (ADAS) started to publish this information in the form of Corrosivity Maps which were updated every four years (see Page 14 of the Engineers & Architects Guide to Hot Dip Galvanizing). Because the data was based upon the corrosion rate of zinc, users and potential users of galvanized coatings were then able to extrapolate the corrosion rate of the coating from the maps. In effect, this gave a method of forecasting how long a galvanized coating would last at a particular point on the map. Obviously, this did not include site specific conditions (such as salt spray from motorways, or chemical plants) but it did give an authoritative basis on which to make general predictions of coating life.

ADAS produced maps in 1982, 1986 and 1990 which were reproduced in Galvanizers Association literature. Interestingly, when the information from the maps was consolidated onto a graph showing the distribution of corrosivity, a very clear trend emerged. By the time of the last map (1990), the average lifetime of a typical 85μm hot dip galvanized coating had risen to somewhere between 40 and 50 years - a very long time in terms of non-exotic surface coatings.

The general level of corrosivity across the country was declining and the lifetime of a given thickness of galvanized coating was therefore increasing. The cause of this improvement had been established in the late 60's when the relationship between sulphur dioxide (SO_2) levels in the atmosphere and the corrosion rate of zinc was confirmed.

The levels of SO_2 in the atmosphere have been declining since the early 60's following a series of Clean Air Act's but have received further impetus in the 90's with the Environmental Protection Act. Unfortunately, following the privatisation of ADAS, government funds are no longer available to update the survey. Figure 3 shows illustrative SO_2 trend data for Stockholm.

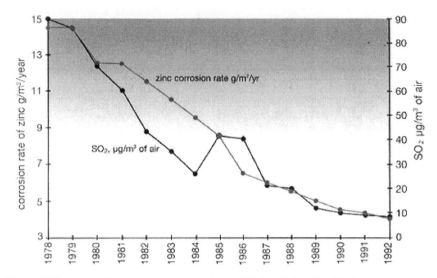

Figure 3 Zinc corrosion rates and atmospheric SO$_2$ levels for Stockholm

Clearly, with declining SO$_2$ levels, it is anticipated that the lifetime for a given thickness of galvanized coating in atmospheric exposure will increase.

THE MILLENNIUM MAP PROJECT

Following discussions with ADAS, Galvanizers Association are sponsoring a new survey and has launched a two year project to update the corrosivity data and to produce a new map in the year 2000. A large number of zinc can reference samples are being exposed to the atmosphere for a total period of two years at nominated sites around the country. The samples were cleaned and weighed before exposure. Upon retrieval, the samples will be cleaned and weighed again. The corresponding weight loss can then be converted into corrosion rate data for each sample. From this a picture of the average background corrosion rate for zinc can be built up, which is expected to confirm the overall downward trend in zinc corrosion rates (particularly in industrial environments) and also ensure that information is available to coating users regarding atmospheric corrosion rates in the future.

The performance of galvanizing in atmospheric exposure, even in aggressive local environments, can best be demonstrated through the use of case histories. Following are two such histories where galvanizing has been chosen for use in bridge applications.

THE STAINSBY HALL BRIDGE

This bridge, which is located on the A174/A19 Teesside Parkway interchange in Cleveland, is a four span concrete and steel composite access bridge to Stainsby Hall Farm. The steel beams of the bridge had been protected by hot dip galvanized zinc coatings. The bridge was erected in 1974 and was therefore approaching approximately 25 years exposure in an industrially polluted environment to heavy salt spray during the winter months at the time of it's last principal inspection (late 1998).

Figure 4 General view of the Stainsby Hall Road Over-bridge

The original hot dip galvanizing had been, on average 150 microns thick on the galvanized beams The galvanized beams had remained unpainted and no maintenance had been undertaken during the 25 year life of the bridge. As coating thickness generally increases with steel section thickness for hot dip galvanizing, such coating thicknesses are not uncommon for structural steel work, such as bridge sections.

Cursory spot inspections had been carried out in 1979, 1989 and 1997 which indicated that although zinc salts were forming on unwashed areas, that insignificant amounts of the zinc metal had in fact been lost. A coating thickness survey by the Zinc Development Association in 1987 had recorded coating thicknesses still in excess of those originally required by BS 729 and suggested that the bridge would not require maintenance for "at least 10 years". However a brief Department of Transport survey carried out in 1990 indicated that there was "significant loss of zinc and voluminous zinc corrosion products". It was therefore requested that a detailed "hands on" survey be carried out to establish categorically the condition of the protective zinc coating and the condition of the steel.

Steel Protection Consultancy (SPC) Carried out a completely independent survey of the four spans of the steel beams at the time of the 1998 principal inspection, on 17th and 18th June 1998.

The survey was undertaken jointly by both David H Deacon, Director and Alan Taylor, Senior Consultant of SPC. The site visit arrangements were made between David Deacon of SPC and Anthony Walker the Designer's Site Representative from Scott Wilson Kirkpatrick & Co Ltd acting for Amey - Robert McAlpine - Taylor Woodrow, A19 Joint Venture.

The site survey had been arranged for 18th June 1998 and William Smith of Galvanizers Association attended as an observer. Representatives of Scott Wilson Kirkpatrick were also present.

SITE SURVEY
The bridge has two main spans which cross the A174 Parkway interchange near the A19 Trunk Road. A general view of the bridge from the westbound carriageway can be seen in Figure 4. Access was obtained using a hydraulic mini lift on the westbound

span. This form of access had enabled a detailed examination of all exposed surfaces with the use of both hands for instruments and testing. Examination of the surfaces using, x10, x30 and x50 magnification had been carried out as well as removal of swabbings and scrapings of contamination for laboratory examination.

Further access had been obtained on the two embankments with the use of a ladder. This accessibility had enabled approximately two thirds of all the surfaces of the beams on all exposed faces to be carefully examined at arms length.

A range of destructive and non-destructive tests were carried out on different surface faces of the beams at a number of locations. These tests had included :

- Careful visual examination, with a convex mirror where appropriate.
- Visual examination using x10, x30 and x50 illuminated magnifiers.
- Film thickness measurements using a Positector 6000 SF3 Model 4.
- Destructive abrasion tests to remove contaminants to examine the zinc.
- Destructive tests using a Paint Inspection Gauge.
- Adhesion/cohesion tests using a hammer and chisel.
- Removal of scrapings for subsequent laboratory examination and analysis

SITE EXAMINATION
A general visual examination of the main beams showed that the underside of the lower flanges, as seen from ground level, exhibited varied amounts of what appeared to be zinc salts. A band of heavy deposits of zinc salts was evident on the underside of the internal unwashed surfaces of the beam on the east side of the westbound carriageway, i.e. the inner faces.

More general zinc salts on the westbound beam showing what appeared to be orange contamination within the matrix of the zinc salts together with general white zinc salt contamination on the underside of the flange above the north embankment area.

The external webs of the beams exhibited a dull zinc spangle over at least 80% - 90% of the area from the top of the webs down. On the internal web surfaces this dull spangle could be seen on approximately 50% of the surface area, with the lower half being more heavily covered with zinc salts.

The ends of the beams where they were in contact with the concrete in all areas showed no sign of rust or rust breakdown, or on any areas where the concrete bridge decks were in contact with the upper flange surfaces even where slight concrete laitance was present on the edge of the flange.

Figure 5 External web on galvanized beam still showing spangle after 25 years

After abrading the surfaces in ten separate areas to obtain clean zinc coating measurements, the zinc thickness on the underside of the upper flange and on the webs ranged from 94 - 159 microns. On the upper surface of the lower flange and on the internal face after removal of the heavy zinc salt deposits zinc thickness readings were measured at between 89 and 113 microns. On the outer face after the same removal of the zinc salts, figures ranging from 72 - 121 microns were measured.

On the underside of the lower flange after removing the zinc salts, figures ranging from 81 - 112 microns were obtained. The lower figures were in the areas with originally heavier deposits (with greater requirement for scraping and abrading to remove them) so it is likely that some zinc was removed and therefore the lower figures should be considered to be absolute minimum's.

Figure 6 Build up of salts on trailing edge of lower flange on galvanized beam

On the upper surface of the bottom flange readings ranging from 470 - 535 microns before scraping were recorded and on the same areas, after scraping, readings of 102 - 128 microns were measured. This test indicated that there was approximately 315 - 410 microns thickness of zinc salts on these heavily contaminated areas. Other spangled areas on the beam showed minimal salt development (10 - 25 microns) over sound galvanized coating.

Measurements were taken on a total of 10 different locations, picked at random covering all types of visual contaminants. Figure 7 shows test area 7, on the upper surface of the inner flange, where heavy surface contamination was present.

Figure 7 Inspection site on top surface of lower flange of galvanized beam. Note considerable zinc salt build-up with sound galvanized coating underneath

This area showed that 390 microns of contaminant was present over the surface with 102 microns of zinc still remaining. Other test areas showed salt build up of up to 530 microns over sound galvanized coatings (thickness not less than 72 microns).

Tests in each one of the areas examined showed that the adhesion of the zinc metal whether of original spangled condition, or whether under excessive amounts of surface contaminants and zinc salts, either light grey, dark grey or light brown in colour, showed that the adhesion was completely unaffected and it was not possible to remove any of the zinc from the steel surface using a sharp hardened cold steel chisel and hammer.

There were a number of areas on edges of the beams and web faces where grout spillage from the bridge deck had occurred on the surface of the zinc coating. No delamination or degradation of the underlying zinc metal had occurred in these areas. Splashes of surface grout from the casting were seen to run down the face of the beam. It was clear that attempts had been made to wipe or brush this away with wire brushes which had scored through to the zinc, but there was still over 100 microns of zinc on the surfaces in this area.

Only one minor area, approximately 8 - 10 cms long could be detected as showing visible substrate rust. On close examination and testing it was clear that this damage was caused during the construction and erection of the beam. It was located on the edge of the inner flange of the east side beam (trailing edge west bound carriageway) and therefore could not have been caused by a high loaded vehicle or flying debris. Examination of this small area of mechanical damage on the underside of the flange edge which was visibly rusting had deformed the metallic zinc adjacent to the area of damage - which was proof that the area had received a significant impact prior to or during erection. The adhesion of the zinc around the mechanically damaged areas after even 24 years showed that there was

- no corrosion occurring under the metal coating and
- no loss of adhesion.

This isolated area demonstrated quite clearly the sacrificial protection afforded by the zinc, which had prevented any rust pitting or loss of edge section.

LABORATORY TESTING
During the site survey samples of the zinc salts and surface contaminants were removed for laboratory examination and analytical examination. The samples removed from the various test areas when examined with high power binocular microscopy were virtually identical in appearance although with variations in colour traces.

Three of the sample scrapings were submitted to the laboratory and were tested qualitatively by x-ray diffraction to determine the principle compounds present. All the samples tested showed that over 90% of the removed salts were a complex zinc compound $(Zn_{12}(OH)_{15}Cl_3(SO_4)_3.5H_2O)$. In addition the remaining 10% of the sample comprised gypsum $(CaSO_4.2H_2O)$ and quartz (SiO_2) which were also present. The complex zinc compound was not readily soluble in water.

It was clear from the close and careful examination of over two thirds of the hot dip galvanized bridge beam surfaces that significant quantities of protective zinc metal coating still remained. Indeed over 75% of the surface area still exhibited the original zinc spangle without any significant zinc salts.

The condition of this zinc coating was adherent and continuous and there were no signs of corrosion or rusting on the edges of the beam surfaces in contact with the bridge deck or in the areas where the beams entered the concrete or any of the typical areas where a painted bridge of that age would exhibit those types of failure.

The significant areas still exhibiting the original zinc spangle were surprising and it was clear that very little loss of zinc has occurred. On areas of zinc unwashed by rain where chlorides and industrial sulphates had built up a corrosion poultice, the surface has initially reacted to form a large volume of zinc salts. Interestingly, the resulting complex zinc salts which had formed were protecting the remaining zinc coating layer, which was still considerable, from any rapid rate of deterioration. The zinc salts present were insoluble and could be removed to reveal a sound zinc base for maintenance in the future when this became necessary.

It is clear from this examination that protection to the steel beams would continue to be provided for a further 20 - 25 years and maybe longer, with the reduction in the industrial atmospheric pollutants in this area, bringing the life to first maintenance of the structure to over 50 years.

It was recommended that a further detailed examination be made in approximately 12 years and a general examination be carried out at the Principle Inspection stages. A painting feasibility trial on a typical area over the embankment were possible to establish how future painting programmes of weathered zinc beams could be effectively implemented. Additional useful technical information would then be available. It was considered that repainting of the structure would unlikely for at least another 20-25 years.

At that time, preparation of the weathered zinc surfaces would be the most important stage of ensuring the success of long term protection of the zinc metal. Surface preparation should then be carried out with low pressure wet/dry abrasive blast cleaning to remove all soluble and insoluble surface contaminants and exposing a clean roughened zinc surface with a fine profile when assessed with an ISO comparator. The paint system used at that time might be based on a high solids VOC compliant two-pack epoxy paint with a re-coatable polyurethane finish, total dry paint thickness of 225 - 250 microns in 2-3 coats.   Alternatively a water based coating system could be considered.

THE BURN REW BRIDGE

This bridge is a composite bridge, with a concrete deck over supporting hot dip galvanized beams. The bridge was erected in 1974. Steel Protection Consultancy carried out an independent inspection of the bridge as part of the Principal Inspection undertaken by engineers from the operators, Owen Williams Railways. The same techniques and testing regime used to examine the Stainsby Hall Overbridge were adopted for the inspection of this bridge

Figure 8 General view of the Burn Rew Rail Over-bridge

The bridge provides access to a food processing plant across the main Paddington to Penzance railway line. With the permitted rail closure, a scaffold tower was erected to aid close inspection of the surfaces of the beams and underside of the bridge decking. The night-time rail closure meant that the inspection was carried out with artificial light provided by the opertor's. General views of the structure were taken in the daylight which followed the inspection.

A view underneath the bridge (Figure 9 below) shows a build up of salts on the surface of the bottom flanges to the galvanized beams. These salts were readily removed to reveal sound galvanized coatings beneath.

Figure 9 Underside view of the Burn Rew Rail Over-bridge

The condition of the galvanized coating overall was extremely good, with over 95% of the surface of the beams showing the original zinc spangle, with a coating thickness of between 120 and 140 microns. Figure 10 below shows a representative area of the coating.

Figure 10 Internal web of galvanized beam still showing spangle after 25 years

Figure 11 below shows a single area of heavy salt penetration - the result of leakage through the fibreglass layer between the deck and the upper beam surface. The stalactites have formed over many years of salt percolation through this leaking joint. Some of the salts had run down the web face and ponded on the upper surface of the lower flange. While there was a significant build up of salts in this area, the presence of the zinc-iron alloy layers of the galvanized coating had prevented any serious pitting of the steelwork beneath, leaving still approximately 70 microns of alloy layers to protect the steelwork.

If salt had penetrated the bridge were it coated with metal spray and/or had it been painted without the benefit of hot dip galvanizing, the inspection suggests that serious attack of the protective coating resulting in under-film corrosion and loss of steel section would have been likely to occur.

Figure 11 Build up of salts resulting from deck leakage

Other areas showed lighter areas of salt formation, with no heavy build up as described above. Representative sections of these areas were mechanically abraded to remove the surface salts and coating thickness tests carried out to determine how much of the galvanizing remained. Readings of between 72 microns and 106 microns were recorded for these cleaned areas.

Those parts of the coating unaffected by surface contamination gave readings which ranged from 128 to 146 microns.

The cross bracing's which were secured with fasteners clearly had used hot dip galvanized bolts - which had performed very well in this exposure - together with either un-galvanized nuts or nuts with inferior corrosion protection. The nuts had rusted quite significantly (see Figure 12 below)

Figure 12 Fasteners used on the Burn Rew Bridge, showing sound galvanized coatings on the bolt heads (left), and corroded, poorly protected nuts - not galvanized (right).

It was recommended that the bridge deck surfaces are examined and a water proof membrane or mastic sealant used to prevent further penetration of the salts from the decking above the beams. The areas affected by the salt penetration should be cleaned and coated with three coats of Highways Agency or Railtrack RT 98 Surface Tolerant Epoxy primer and over-coated with a two-pack epoxy intermediate and finish coat containing MIO without aluminium flake pigment.

CONCLUSIONS

Case history information remains a useful basis for prediction of life for a galvanized structure, although the likely favourable changes in environment over time usually mean that they lead to conservative estimates of when information is extrapolated to new structures.

With galvanizing offering in many environments, for footbridges and smaller highways bridges, a realistic life to first maintenance of 50 years plus, bridge owners will undoubtedly be looking to galvanizing to significantly reduce their whole life costing's for their bridge applications.

Particularly in this smaller bridge market where perhaps presently pre-cast concrete has high market share, bridges using galvanizing as the sole means of corrosion protection will allow steel construction to challenge historical views in this area.

The design and use of temporary cladding materials to prevent the spread of harmful dusts etc, falls, and to improve working conditions

MALCOLM JAMES
MJ Consultancy.

1. Introduction.

This paper describes some recent research that has been undertaken on temporary cladding materials. It comments on the possibilities suggested by this work, for providing an inexpensive means of helping to control air born pollution, supporting waste control measures, and in providing higher standards of fall protection. The paper suggests that the research carried out so far encourages the idea that temporary cladding could also enhance the working environment and facilitate the development and use of advanced construction processes

Temporary cladding materials are widely used on scaffolding, although there is a great variation in the way they are erected and maintained. They have generally originated from two sources. One group being derived from materials used as horticultural screens, where they were intended to provide either shading from strong sunlight or reduce the air velocity within the shaded area. The other has been specifically produced for the construction industry, developing out of the use of tarpaulins used to protect the public from grit blasting, façade washing and similar operations. Both groups have evolved within by the construction industry to give a means of reducing the wind speeds across a work area, minimise both heat loss and rain or snow penetration. They have also been used to reduce the spread of debris from the site and provide a more aesthetically pleasing external appearance of the construction works.

Research into these materials, however, indicates that there could be other substantial benefits in the appropriate use of suitable temporary cladding materials. These would include helping to achieve full environmental control both on and off site, in particular in creating healthier working conditions through temperature and humidity control, control the distribution of fume, dusts and falling objects, and providing an enhanced means of fall prevention. In addition there is also the possibility that they could help reduce the wastage or the excessive use of expensive materials and create a better environment for the use of automated systems and robotics.

It is expected that temporary claddings could give these benefits by influencing the air flow across a work area, either encouraging or discouraging it as necessary, by having sufficient inertia to absorb the energy of flying debris, or by filtering dusts and debris from the air stream. There is also the possibility they could help provide some degree of protection against flame spread through the judicial use of materials with different melting temperatures, although this aspect is outside the scope of this paper.

While the intention behind the current use of temporary cladding is to give protection both to the site works and to the surrounding neighbourhood is worthwhile, it is undermined through a lack of care in selecting and maintaining appropriate materials. In addition as there has been no standard for temporary cladding, the materials have been selected only on price with the result they have become progressively lighter and less effective. A further problem is that there has been no design information available that would help users to select the most appropriate materials, design their installation for the best advantage and appreciate how to maintain them.

These difficulties are being tackled through research and the publication of BS 7955 'Containment nets and sheets on construction works – Specification for performance requirements and test methods'. In addition a draft Code of Practice for 'The use of Containment Nets and Sheets on Construction Works for Environmental and Safety Control', is being prepared and will be available for public comment shortly. While neither of these documents deals with the full potential of these materials, or provide comprehensive design information, they are a useful first step to better and more effective practice.

The research work that has been undertaken and described below, has been used to give valuable support the above two standards even though much of it has only been an initial study of the various aspects.

2. Impact resistance.

While the impact resistance of temporary cladding is important both from the point of view of preventing the fall of persons and for the control the fall of debris, it is only a subsidiary, although very useful, function of these materials.
As temporary claddings are relatively lightweight, typically $100 - 300$ g/m^2, and have individual components with relatively low strengths. Their ability to withstand impact loads depends on the ability of the various components to share and distribute these loads. Research has shown that this is possible and can give very significant levels of protection.
Again because of their lightweight nature these materials could be vulnerable to tearing or splitting and it is therefore essential for them to contain tear inhibitors. This is a requirement for materials manufactured to BS 7955.
As a result of these two attributes, temporary cladding can be used to enclose a wide variety of spaces in the comforting knowledge that, when correctly erected, these will give high levels of personal protection. They can provide this protection to awkward shaped areas where conventional protective components cannot be properly fitted.

The standard BS 7955, requires that temporary cladding should withstand an impact from a 50kg dropped from a height that gives an impacting energy of 150Nm (approx. 306mm), into a 2m x 1m specimen fixed on all four sides into a frame. The test sample should have two initial 50mm cuts at right angles to each other, at the point of impact, before the test, and these should not be found to have spread to more than 100mm afterwards. In addition separate 2m x 1m test samples are to be subjected to an impact from a 63mm diameter steel ball weighing 1kg and dropped 2m into the sample. In this case the sample should not be damaged.

These tests are intended to give confidence that the materials can successfully withstand low impact energy levels such as those that occur when a person trips or stumbles. Research carried out at BRE Cardington, however, has shown that when appropriately rigged, these materials can withstand much higher loads. This research involved swinging a 100kg steel cylinder 1m long x 380mm diameter, into three different types of cladding samples at 4.0m/second, i.e. at a fast running speed. In these tests the samples were rigged to completely cover a bay size of 9.0m x 3.9m high, and were supported on horizontal rails or ropes etc, at head and foot as well as at an intermediate level 1.075 above the floor. The test cylinder struck the samples 0.75m above the floor at the centre of the bay.

A variety of types of support were used in this test such as steel scaffold tubes, wire ropes, nylon ropes and webbing and a summary of the results are given below. Generally the materials failed by them tearing at their fixings.

Sample type and description		Number of impacts survived when fixed on:-			
		Scaffold tube	Steel wires.	Nylon ropes.	Webbing.
A	Polyethylene sheet with 1500 denier polyester reinforcing. Weight 300g/m^2.	3	3	3	3
B	Polypropylene woven sheets. Weight 140g/m^2.	Nil	nil	3	3
C	Fine polyethylene netting. Weight 110g/m^2.	N/a.	nil	1	Nil

Note: The samples type A, B and C referred to in the above table can also be described as being none permeable for type A sheets, semi permeable for type B and permeable for type C.

The research work on wind loading referred to below, has also shown that type B materials can be so tightly fixed to a scaffold frame, that the tubes will bend in high winds rather than the cladding rupture. This indicates that while these materials have the potential for supporting quite high loads, the more rigid the support the more likely it would be for the cladding to tear at such places. Also when compared to the minimal test requirements in BS 7955, it is probable that cladding arranged in large areas is more able to support higher impact loads. This could be a significant consideration when designing temporary cladding installations.

Where temporary claddings are required to support high loads, the nature and frequency of the fixings as well as the area of cladding supported by each one, is of critical importance The research work carried out has shown that it is more probable the fixing will tear the cladding where fixings are rigid.
This information reflects on the cladding installation in two different ways. In the first case the loading transmitted through the cladding onto the supporting frame, can be limited by using fixings of a known failing strength placed at an appropriate spacing. This arrangement is probable most appropriate in dealing with excessive wind loads.

In the second case fixings with possibly elastic properties, can be placed at close centres on flexible supports, and arranged so that any loading in the net is spread through as many of them as possible. This is the arrangement that will be most relevant where the cladding must support impact from persons or objects.

BS 7955 recognises these possibilities and requires that ties should have a stated strength which should be at least 500N, so that a person designing a temporary cladding layout will be able to specify a correct number to achieve the desired outcome.

3. Wind load and wind speed control.

It is often found that wind loads on temporary claddings can cause major problems by either imposing excessive loads on the supporting structure so it is damaged or collapses, or by tearing the cladding so badly, that contained dusts, debris and fume are allowed to escape. This may then lead to complaints from the public or harmful contamination of the neighbourhood.

The effect of wind damage can readily be seen on many temporary claddings erected in our cities, especially after a storm, when many are left in tatters. This is need not necessarily be the case as can be illustrated by one case where the cladding was so securely fastened they suffered little damage even though the wind force was strong enough to lift off a 30 tonne temporary roof.

While it is possible to undertake wind design calculations on temporary cladding on the assumption that it is a solid building, this may not always give the worse case and leaves a significant area of uncertainty which may trap the unwary. It could also be needlessly more expensive. Consequently wind studies were carried out on a full scale structure at the Health and Safety Executives, Health and Safety Laboratory (HSL), at Buxton between 1989 and 1997. The research was done in two phases and investigated the values of the form factors associated with different types of cladding when backed by walls having various void ratios. The information was intended to help designers calculate the wind loads on cladding for various wind speeds, areas and permeability of the backing walls.

In phase 1, which was completed in 1993, a 13.7m long plywood wall, 13.2m high, was constructed facing into the prevailing wind. A scaffold was built behind it on which was attached temporary claddings of types A, B and C. The plywood wall was built so that it could have various evenly distributed openings with void ratios of 100%, 50%, 25% and 12%. Load cells were attached to the ties between the scaffold and the wall, and could then monitor the magnitude of the wind loads on cladding to the lee side of the wall. In addition various configurations were tried where the cladding did not extend to the ground and where it was hung with 50% excess material across the wall.

Phase 2 studied the wind loads on a type B temporary cladding, arranged around an 18.4m x 17.8m x 17.5m maximum height scaffold and plywood box.. This box was covered with a monopitch corrugated steel roof and the wind loads were recorded for the wind blowing around 180^0 onto it. This phase therefore gave results for the loads on the cladding to the windward of the supporting structure. All these results are then summarised below.

Phase 1 results below, show the average form factors for the cladding when placed to the lee of a wall with varying void ratios.

Type of cladding	Void ratio in backing wall as the percentage of voids to the total wall area.					
	12%	25%	50%	100%		
				normal	1.49m gap at ground level	Set slack
A	0.5	0.46			0.27	0.19
B				0.51		0.56
C	0.04	0.07	0.12	0.34		

They show how the load characteristics are altered by varying the arrangement of the cladding as well as by the void ratio in the backing wall.

Phase 2 results below, show the average form factors for cladding type B, when arranged to the windward of a structure with varying void ratios.

	25% void ratio, 1m gap at roof level and rear wall open.	50% void ratio, rear wall open.		50% void ratio, rear wall closed	
		No gap at roof level.	1m gap at roof level	No gap at roof level.	1m gap at roof level
Form factors.	0.34	0.22	0.32	0.30	0.21

Note: it is thought that where wind penetration through a semi permeable cladding to windward of a structure, is very restricted, then the wind will flow around the cladding showing an apparent drop in the form factor values.

Other work has been done on wind loading on temporary cladding, Gylltoft in Sweden measured the forces on the cladding around a building and Schnabel in Germany studied the loads generated on a model in a wind tunnel. This latter project used fine wire mesh to model the cladding and it has been thought that as this mesh would not flex very much, the results obtained might not be a close approximation of what happens on site.

All the research work carried out does show how the void ratios in the backing structure or wall will have an important effect on the wind loads on the cladding. The results also suggest that the airflow between the cladding and the backing wall can be encouraged giving significant useful effects from the point of view of environmental control.

4. Dust and fume control.

Much of the dust and fumes created on construction and maintenance site is harmful to some degree or other and may also be unacceptable to the amenity of the area if scattered widely. It could well be that the control of dusts etc. will become more of a problem in the future. Temporary cladding can clearly be used as a means of controlling or limiting this problem to within acceptable limits, and it is therefore very desirable that a clear understanding should be developed on how this can be done.

For instance the production of an acceptable design will require a knowledge of the following three issues.

- The extent that a particular cladding arranged in any particular way, will filter or block the escape of dusts etc.
- How this is affected by wind speed and particle size.
- The acceptable limits of pollution for that area and the type of dusts being generated.

Any design however, would also have to consider the issues of impact strength and wind load. For instance some forms of temporary cladding can be provided that are fume tight, these are used to cover sewage works etc. However such materials may be more easily damaged by storm conditions, especially if exposed at a great height. It may well be more practical to use a semi permeable material and accept that some dusts will be able to escape to a certain degree. This may not be a problem if the subsequent pollution is still below acceptable limits.

The following results were obtained from an experiment at the HSL site at Buxton. In this concrete blocks were cut with a grinding machine on a scaffold platform in an enclosed scaffold. The cladding was type B, and the work done was on a day when the winds were relatively light.

The results summarised below, show how the dust drifted down the 1.2m wide by 1.8m high corridor formed by the scaffold platforms above and below, with the solid plywood wall on one side and the cladding on the other. They also show the extent that the cladding prevented the escape of the dusts even immediately adjacent to the work area. However, it is possible that a different set of results would have been obtained if a study had been made when different wind speeds occurred and the dusts had different particle sizes.

At source of cutting.	3m downwind of source.	12m downwind of source.	Above source on a higher platform.	Adjacent to source but outside the cladding	4m upwind of source.
916 mg/m^3	212 mg/m^3	2.15 mg/m^3	29.8 mg/m^3	31.8 mg/m^3	Nil.

The above results may be compared with observations also made at the Buxton site following a severe blizzard blowing directly onto one face of the cladded scaffold. After this storm the platforms on that face were covered with a light coating of fine snow which had been driven through the cladding.

The results obtained for different arrangements of the cladding during the wind load research and the observations made during the grinding of the concrete blocks etc, suggest that there is a strong likelihood that dusts can be controlled to some extent by varying the air flow within a contained area. Controlling the air flow direction and its velocity could enable dust levels both inside and outside the cladding to be maintained at an acceptable level.

5. Environmental control (1.0)

Exposure to the vagaries of the weather results in the specification for materials and finishes to be increased to maintain satisfactory levels of strength, finish, and life effectiveness. Similarly exposure to the weather also can have create health problems if it is not suitably controlled. Being exposed to changing weather patterns is hardly conducive in obtaining high productivity levels or for the efficient working of sensitive electronic equipment unless specially (and expensively) protected. This 'control' is generally currently provided through the use of waterproof clothing etc.

However due consideration does not lead to the conclusion that this is satisfactory, or represents an approach that matches our present technological development. It is interesting to note the requirements of regulation 24 of the Construction (Health, Safety and Welfare) Regulations, that after having regard for the purposes of the work, and the protective clothing or equipment provided, a worker should be protected from inclement weather.

It is therefore clear that the costs associated with construction and maintenance work from being exposed to the weather are very high although generally well hidden. This suggests that enclosing the work will be a useful step in dealing with these problems.

However, the inappropriate use of a particular type of cladding can give adverse effects in terms of humidity and temperature gain. The results of these can often be seen in hot weather where the claddings have been deliberately cut to allow fresh air to reach those working inside the cladding. During the work at Buxton it was found that substantial levels of condensation accumulated inside the scaffold tubes used to support the plywood box, and during periods of heavy frost, the condensate froze and the tubes split.

6. Proper use of temporary cladding materials.

Because of the variable quality of cladding materials coming into the country from abroad, as well as some made here to compete on price, it is essential that only materials conforming to BS 7955 should be used. This comment applies both to the fixings as well as the cladding.

The cladding arrangement should be designed to encourage those properties desired at any particular site can be done by taking into account such matters as the following -

- the properties of the actual materials, such as weight, permeability etc.
- the arrangement of the cladding, its height, slackness and gaps etc.
- the fixing of the cladding, the type and frequency.
- whether only fire retardant materials should be used.

All these matters interact with each other making it possible to provide an extensive range of performance arrangements to deal with a wide range of site problems.

Having identified the desired properties of the enclosure, and undertaken a design to produce these, it is clearly important that the cladding should be installed to the design and maintained in that condition until the end of the contract or any particular phase in the contract. It is also clear that the cladding should be properly maintained and repaired as soon as possible after any damage has occurred.

At the end of each contract the cladding materials should be disposed of safely. It is not advisable to burn them and it is unlikely it will be economic to re-cycle them. This means that the best option would be to arrange their disposal at a licensed tip. However if the cladding is contaminated with harmful dusts etc, as it certainly could be, then depending on the nature of the contaminents, other precautions might be required for disposal.

7. Conclusions.

Where construction and maintenance works etc, are carried out in the open there are serious problems though the vagaries of the weather, in maintaining productivity and quality, minimising waste, keeping to programme, preventing pollution and ensuring that the health of workers does not suffer too much. In addition open sites are hardly the place for sophisticated electronic control systems or for the minimal specification of compounds and similar materials without exposure to the elements having serious consequences.

Therefore it is very desirable, particularly in a time of rapid climate change, that sites should as far as is possible, be enclosed and maintained at a constant temperature and humidity. Developing and designing temporary cladding systems to give such control will therefore be of a major cost, quality and safety benefit. From the research done so far, it seems clear that this is not only possible but that the costs of installing suitable systems to help achieve this, will repay their installation costs many times over on the contract itself. There will then be even bigger benefits through extending the lifetime of the structure and through reducing maintenance costs. Effective cladding could also provide substantial safety benefits.

While there is no doubt that a large amount of research is still required to provide information on how an appropriate design can be specified, the costs of such research should be recoverable within a few years.

8. References

1. K.Gylltoft 'Wind loads on sheeted scaffolds. Afield study'
   Swedish National Testing Institute. Technical Report SR-RAPP 1986:27 1986.

2. P.Schnabel 'Zwischenbericht zum Forschungshaben Stromungstechnische Modellversuche zur Ermittlung der Windlasten auf mit Netzen oder Planen bekleadete Fassadengeruste.
   LGA Munchen Report A 18/91. 1992.

3. E.J.Hollis
   Debris netting project
   HSL Field Engineering Reports IR/L/FE/93/1 – 15. 1993.

4. Construction (Health, Safety and Welfare) Regulations.1996.
   Statutory Instruments 1996 No 1592

5. E.J.Hollis and D.Allinson
   Wind loading on a temporary roof
   HSL Field Engineering Reports FE/97/010 – 016. 1997.

6. E.J.Hollis, B.GUIVER, E.J.Brueck and T.Ward
   Dust measurements in a clad scaffold
   HSL Field Engineering Report FE/OT/NV/97/001. 1997.

7.   N.I.Fattuhi and L.A.Clark.
        Research project on the use of different types of safety nets undertaken at BRE
        Cardington by Birmingham University.
        HSE contract 3706/R33.050. 1997.

8.   Loss prevention standard LPS 1215: Issue 2 July 1998.
        Flammability requirements and tests for LPCB approval of scaffold cladding materials.

9.   BS 7955. 1999 'Containment nets and sheets on construction works – Specification for
        performance requirements and test methods'.
        British Standards Institution.

10.   Draft Code of Practice 'The Use of Containment Nets and Sheets on Construction Works
        for Environmental and Safety Control' available for public comment in 2000.
        British Standards Institution.

Bridge isolation – a further technical innovation

Eric COSTES, Jarret company, 198, Av. des Grésillons 92602 Asnières, France
Technical Director

INTRODUCTION
JARRET has studied and manufactured pressure spring dampers devices for more than 35 years. These devices use the rheological properties, these are the properties of a fluid when it flows, of a silicon fluid. The compressibility of the fluid is used for the spring function (absorption and restitution of the energy) and the viscosity of the fluid is used for the damping function (dissipation of the energy). The applicabilities of these devices are varied, with applications in many fields, such as : civil engineering, the iron and steel industry, mechanical handling , military, nuclear and railway sectors.

The first applications intended for the protection of bridges and structures, using the JARRET technology, were carried out in Italy in the mid-Eighties. Since then, the number of applications have multiplied all around the world.

The objective of this article is to demonstrate how « Bridge Isolation » is a further technical innovation that has practical (and cost-effective) solutions and benefits, when compared to the traditional way of protecting bridges and structures. This paper will describe the behavioural laws of the main current devices used for seismic protection, while relating then to one of the principle recent achievements, in term of size, namely, the viaduct of Saint André.

MAIN CHARACTERISTICS OF THE BRIDGE
The viaduct of Saint-André, located on the Maurienne motorway between Praz and Modane, is a prestressed concrete structure, some 902 meters in length. Its deck, with a winding layout, consists of 11 spans out of unicellular caissons. It is built between two abutments and has ten low height piles.

The viaduct, located in a designated S2 seismic zone (nominal acceleration = 1.5 m/s^2), is subjected to bridge regulation by the Association Française pour le génie Parasismique (AFPS). However, its heavy deck (55,000 tons) has no joints and the very stiff piles pose problems with respect to its seismic behaviour. It was thus necessary to devise seismic devices to decrease the loads on the piles and abutments which appear during a seism.

Transversely, two preloaded spring dampers, placed between the deck and each pile, fulfill a double role at the time :

- under the normal service (thermal expansion, wind and traffic effects), the spring dampers
  maintain the deck on the axis of the piles due to their preload,

- under seism, the spring dampers allow, due to their low stiffness, a displacement of the deck which decreases the loads at the head of pile and consequently at the level of their foundations

Transverse displacements are limited by the function damping device which reduces a great part of the seismic energy transmitted to structure. At the end of a seism, the transversal preloaded spring dampers return and pre-centre the deck on its axis. This is « Bridge Isolation » in practice.

Longitudinally, elastomeric bearings located at the head of the two central piles ensure the role of the longitudinal fixed point. The other piles, as well as the abutments are, equipped with multidirectional elastomeric sliding bearings. Six longitudinal dampers are placed in each abutment to reduce the displacements and the loads generated by the longitudinal component of a seism. The dampers develop only a very low load under the displacements at a low velocity related to the normal conditions of thermal expansion, shrink and creep.

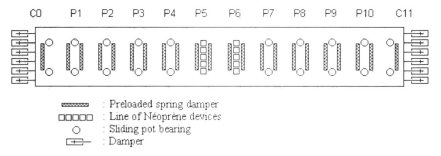

Figure n°1: Bearing scheme chosen

CHARACTERISTICS OF THE SEISMIC DEVICES

Longitudinal seismic devices
The selected longitudinal devices are pure dampers. They develops a nominal force of 2450 kN under a velocity of 0.68m/s. The stroke is ±300 mm and the mass is approximately 1800 kg. Figure n°2 illustrates the principle of this device. The main parts of this device are the reservoir filled with a viscoelastic silicon fluid and the crossing piston. One of the ends of the piston, on the left-hand side on the figure, constitutes one of fastenings, for example, on the abutment. The reservoir constitutes the other fastening, on the right-hand side on the figure, for example, on the deck. The fastenings are equipped with ball joints which avoid the lateral load into the device. The ball joint has the added benefit of having a multi-dimensional rotation. A length adjustment is possible with a screw and nut system, which facilitates the installation of the device. The design of the JARRET damper does not comprise any external devices, such as a pipe or an expansion tank, which have to be protected. The fixing of the devices is carried out, classically, by ties and bolts in the case of a metallic structure or by ties and embedded steel plates which distribute the stresses on a large surface in the case of a concrete structure.

The piston comprises a part of a larger diameter - i.e. the head of the piston - adjusted to the internal diameter of the reservoir in order that only one small annular gap remains. This gap creates a lamination of the silicon viscoelastic fluid during the displacement of the piston. It is noticed that the crossing piston allows the creation of a pure damper device, with no restitution of energy, because the displacement of the piston does not change the volume of the reservoir. Consequently, the pressure of the fluid involves only a flow of fluid from one side to another, across the head of the piston. At rest, the initial pressure exerted on the seals ranges from 200 to 600 bar, and, in operation, it can be varied up to 3000 / 4000 bar. This high pressure allows the design of compact devices.

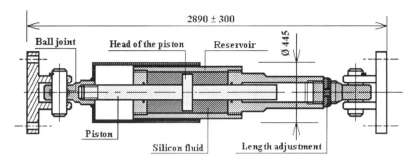

Figure n°2 : Cross-section of a JARRET damper

The force developed by a damper depends directly on the viscosity of the fluid and the velocity of the piston. To put this viscosity into perspective, liquids have a viscosity ranging from 50 - 500 cSt, whereas the Jarret silicon fluid can be varied between 10 million and 20 million cSt. During the working of a damper, the pressure loss induced by the annular gap around the head of piston generates a difference of pressure between its two faces. The force developed by the damper follows from this variation of pressure. The behaviour law of such a damper is:

$$F = C . X^{\circ \alpha} \tag{1}$$

With: F = force developed (N), C = viscosity coefficient $(N/(m/s)^{\alpha}$, X°= velocity of the piston (m/s) and α= exponent velocity close to 0.1 for a JARRET device, but with oil, the exponent velocity would be 1.0.

The behaviour law can be expressed in a more explicit way by introducing the concept of nominal force (Fn) and nominal velocity (Vn). These values depend : for Fn, on the mechanical capacities of loading the structure, where the damper is fixed; for Vn, on the results of the expected seismic response. Fn and Vn define the setting point of the damper. They correspond to the maximum force and velocity of the device during the seism.

$$F = \frac{F_n}{V_n^{\alpha}} . X^{\circ \alpha} \tag{2}$$

It is noticeable that this behaviour is not linear since velocity is affected by the exponent α. This peculiarity requires a direct, step-by-step, « time history » calculation to perform a seismic study of the dynamic response of the structure. The classical methods of modal analysis are not so applicable (cf. DATRY & A.A. 1998).

Figure n°3 underlines two of the principle advantages of the JARRET dampers. The first relates to the rapidity of the response of the damper due to the strong pressure of the silicon fluid, as mentioned earlier. The force quickly reaches a higher level, which guarantees an optimal effectiveness from the begining of the seism. However, the force developed at very low velocity, under thermal expansion, shrink and creep for example, remains weak.

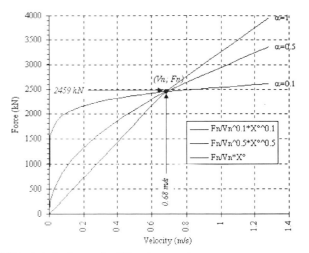

Fn and Vn define the setting point of the damper. The figure n°3 presents the force versus the velocity of dampers with Fn = 2450 kN, Vn= 0.68 m/s and the exponents α of 1 (linear), 0.5, and 0.1 (JARRET).

Figure n°3: Behaviour law Forces = f(Velocity) of a damper for different exponents α.

Consequently, the damper does not oppose the dimensional changes of the structure. The second advantage is related to the quasi-asymptotic character of the developed force. Even if the working velocity due to the seism is higher than that envisaged, for example if the seism is more important than that retained at the time of the seismic study, the level of the force developed by the device remains on a level close to Fn. This constitutes an important advantage since the damper will not subject the structures (deck, abutment, pile) to overloads. Thus, the dimensioning of the structures can be optimised.

Figure n °4 shows that the non-linear character of the behaviour law of the JARRET damper leads to a dissipation of energy more important than that obtained with an exponent α higher than 0.1. This dissipation is characterised by the area of a loop force Force = f(Stroke). The more the shape of the loop force tends towards a rectangle, the more important is the dissipation. A rectangular loop force - as that produced by a friction damper - is however not desirable because of the effect of the threshold. It is impossible to move a friction damper with a small load even if the velocity is small. If the load which is applied on a friction damper is lower that the nominal force, the stroke is nil. As soon as the load reaches the nominal force of the damper, a movement is made.

The figure opposite presents the loops developed force versus the stroke under the effect of a sine movement of dampers which have an exponent α of 1 (linear), 0.5, and 0.1 (JARRET). The excitation has an amplitude ±300 mm at the frequency of 0.36 Hz in order to reach Vn= 0.68 m/s.

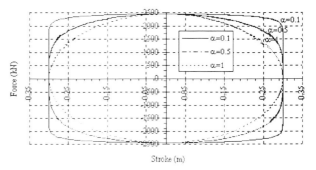

Figure n°4: Comparison of the response under sine excitation of dampers with different exponents α.

This behaviour leads to discontinuities which are undesirable. A friction device opposes its nominal force at any displacement. When a low reaction force is required for the low velocities, such as thermal expansion, shrink and creep, the use of a friction device is not satisfactory.

Transversal seismic devices
The selected transerval devices are preloaded spring dampers. They develop a nominal effort of 1950 kN under a stroke of 80 mm and a velocity of 0.47 m/s ; the unit mass is approximately 850 kg.

Figure n°5 illustrates the principle of the preloaded spring damper. The device consists of a guide fixed to the under-side of the deck. In this guide, there is a device made up of a reservoir, a plunger piston (non-crossing piston) and a wedge. The reservoir and the wedge are pushed by the intermediary of a pad (PTFE) on slide plates fixed to the head of the pile. At the time of the longitudinal movements, such as those resulting from thermal expansion, shrink and creep or even from a longitudinal seism, the whole preloaded spring damper moves along the slide plates without any movement of the piston in the reservoir. At the time of the transverse movements, the piston enters into the reservoir and causes a decrease in the volume of the silicon fluid. This decrease of the volume brings about an increase in the pressure of the silicon fluid. The pressure acting directly on the area of the piston is the origin of the spring

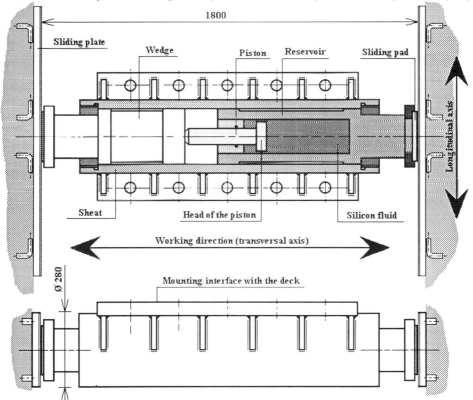

Figure n°5: Cross-section of a preloaded spring damper

characteristic. The more the pressure pushes on the piston, the more the force, developed by the device, increases. The fluid is initially under high pressure to have a sufficient preload to maintain the deck in a centered position. Like the damper, the piston is provided with a head tolaminate the silicon fluid when the device operates.

The force developed by a preloaded spring damper can be sub-divided into two parts. These are:

- the static component, which depends directly on the compressibility of the silicon fluid and thus of its pressure. This static component results from the initial pressure, when the device is at rest, and from the stroke of the piston, which creates the change of the pressure. The initial pressure defines the preload of the device.

- the dynamic component, which depends on the viscosity of the silicon fluid and the velocity of travel of the piston. At the time of the excitation of the device, the pressure loss created by the annular play around the head of piston generates a gap of pressure between its two faces. The force developed by the device ensues directly from this gap of pressure. This dynamic component is, at any point, comparable with the force developed by a damper (cf. §2).

The behaviour law of such preloaded spring dampers is:

$$F_{static} = F_p + K.X \tag{3}$$

$$F_{dynamic} = C.X^{o\alpha} = \frac{F_n}{V_n^\alpha}.X^{o\alpha} \tag{4}$$

$$F = F_{static} + F_{dynamic} = F_p + K.X + \frac{F_n}{V_n^\alpha}.X^{o\alpha} \tag{5}$$

With: F = force developed by the device (N), F_p = preload (N), K = stiffness (N/m), X = stroke (m), C = viscosity coefficient (N/(m/s)$^\alpha$), X^o= velocity (m/s) and α= velocity exponent close to 0.1 for an JARRET device.

Figure n°6 presents the behaviour law of a preloaded spring damper answering to the above formulae (5) with the parameters of the devices used for the viaduct of Saint Andre: F_p=800 kN, K=8200 kN/m, F_n=620kN, V_n=0.47 m/s and α= 0.1, maximum stroke=80 mm.

The graph opposite shows each component and the total behaviour of a preloaded spring damper.

The dynamic component represents the response under a sine excitation which has an amplitude of ±80 mm at the frequency of 0.935 Hz in order to reach Vn= 0.47 m/s. The velocity is nil at the maximum strokes (+ and - 80 mm).

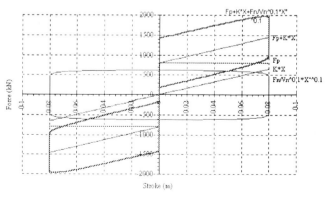

Figure n°6: Static and dynamic components of a preloaded spring damper.

The graph opposite presents the loops of the force developed versus the stroke under the effect of a sine excitation of a preloaded spring damper with the velocity exponent α of 1 (linear), 0.5, and 0.1 (JARRET).

The sine excitation has an amplitude of ±80 mm at the frequency of 0.935 Hz in order to reach Vn= 0.47 m/s.

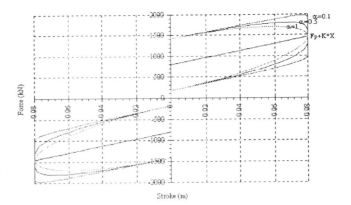

Figure n°7: Comparison of the response under sinusoidal excitation of a preloaded spring damper with different α.

The remarks made for the damper device are also applicable to the self-resetting preloaded spring damper.

CONCLUSIONS

Design For Durability and Protection

These preloaded spring dampers ensure the design provides the optimum solution. The device functions as a protective unit and the technical, innovative simplicity ensures the enhanced durability of performance within a specially hardened steel tube that contains the visco-elastomer under pressure.

The self-resetting spring damper devices, as used for the seismic protection of the viaduct of Saint André, have been studied and manufactured by a company (JARRET) that has specialized for several decades in the development of cushioning devices that use a visco-elastic fluid to protect bridges and structures of all types. The very simple design, without surge tank and/or corrector pressure units, makes the JARRET devices particularly robust and without any specific maintenance needs.

Bridge isolation

The viaduct of Saint Andre illustrates the successful use of these devices that operate equally well in normal situations and when there is a seism. The implemented solution allows two goals to be achieved :
- the perenniality of the bridge is improved ;
- the dimensioning of the structures remains economic, in terms of support, thermal expansion joints and the reinforcements of the piles and the abutments.

JARRET has developed a dedicated software pprogram, which permits the determination of the seismic response of structures equipped with seismic devices with non-linear behaviour laws. This software makes it possible to try many solutions, quicly and easily, to resolve the means of achieving optimum energy dissipation. Thus, this ancillary technical innovation directs the choice towards the better solution of « Bridge Isolation », which is adapted to resolve the situations.

REFERENCES

Datry, Marracci, Krief, Neant. (1998). parasismic Protection by viscoelastic cushioning devices, the viaduct of Saint Andre. AFPC-AFREM XIII congress of the FIP 98 (ISSN N °1286-4781).

Numerical and experimental analysis of jointless precast prestressed hollow-core beam slabs with cast-in-place continuity

Prof. F. GASTAL, PhD.
Co-ordinator – PPGEC
gastal@ufrgs.br

R. A. BARBIERI, MSc.
PhD. Candidate, Assoc. Researcher
barbieri@cpgec.ufrgs.br

R. S. REZENDE
Graduate Student
rafarez@ufrgs.br

Federal University of Brasil at Rio Grande do Sul - UFRGS
Graduated Program in Civil Engineering - PPGEC
Laboratory of Structural Testing and Modelling - LEME

INTRODUCTION

A hollow-core slab is a precast, prestressed concrete member with continuous voids provided to reduce weight. Primarily used as floor or roof deck systems, hollow-core slabs have also been applied as wall panels, spandrel members, bridge deck units and parking-deck elements (PCI, 1998).

The members are usually made in a long line bed, cut to the desired length and primarily used as simply supported elements, as shown in figure 1. The transmission of forces between adjacent elements provides the bi-dimensional behaviour expected from a slab area. The lateral connections are produced by the casting-in-place of longitudinal keys among the elements (Stanton, 1992).

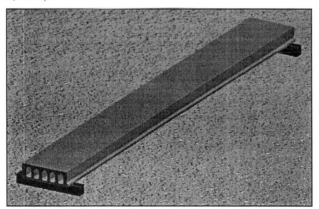

Figure 1 – Prestressed precast hollow-core beam slab

Bridge deck joints are a persistent and costly maintenance problem. Water leaking through the joints is a major cause for the deterioration of bridge girder bearings and supporting structures. Debris accumulation in the joints restrains deck expansion and causes damage to the bridge. Furthermore, joints and bearings are expensive to install and maintain. Therefore, the cost of construction and maintenance for a bridge can be greatly reduced if the number of deck joints in multi-span bridges are minimised (Carner & Zia, 1998; Gastal & Zia, 1989).

The objective of this work is to study the behaviour of hollow-core slabs when connected to full continuity, consisting in an alternative for jointless deck bridge construction. Two different procedures to achieve continuity are investigated: one considers the use of a concrete topping in which reinforcing bars are inserted over the connection zone; the other consists of simply casting in place the inner support slab zone and filling its voids with reinforcing bars.

To compare the effectiveness of these different continuity solutions, a numerical and an experimental study were carried out using full-size elements loaded up to failure. These are described herein.

DESIGN FOR CONTINUITY

When simply supported beam slab elements are used, there is no continuity and service load limits (q_{single}) are defined by the moment capacity (M_{cap}) of the single element cross-section, as seen in figure 2. When designed for full continuity there is a significant increase on the structure's load capacity (q_{cont}), once an adequate reinforcement is provided over the zone connecting two adjacent single elements.

Considering the elastic relation (M_{cont}/M_{span}), between the negative moment over the connection zone and the positive span moment, and imposing a limit due to the M_{cap} of the single element, the amount of reinforcement necessary to resist the connection moment is easily determined.

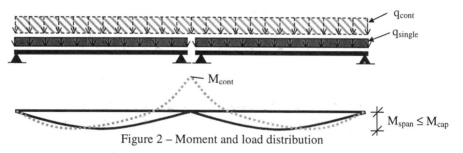

Figure 2 – Moment and load distribution

The relation between the amount of reinforcement provided to a continuous element and the total area of reinforcement necessary to resist the elastic negative moment over the connection zone, is hereby defined as the "rate of reinforcement" and may vary from 0 %, in simply supported elements, to 100 %, in fully continuous ones.

Reinforcement bars are placed on top of the continuity zone on either of the connection solutions, as sketched in figure 3. To compare the behaviour of both types of continuity and to relate them to the use of single elements, the prototypes in study maintain the same external dimensions and position of the continuity reinforcement. The rate of such reinforcement, however, is varied to investigate its influence on the overall behaviour of the structure.

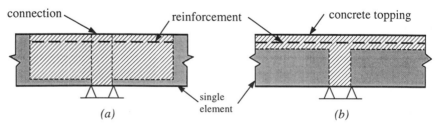

Figure 3 – (a) Connection by insertion of reinforcement; (b) connection by reinforced concrete topping

NUMERICAL MODEL

The numerical model used in this work was implemented by Gastal (1986) for the analysis of jointless bridges. Reinforced or prestressed concrete beams or steel girders, topped by a concrete deck slab, may compose such structures. Construction sequences and techniques may vary, as can cross-sectional shapes and material properties. Similarly to the structure in study, they present a composite structural behaviour and share some similarities.

For the numerical analysis, the structure is modelled by distinct elements: a two nodded isoparametric beam element is used to represent the cross-sectional and material properties of the girders and deck, whereas the connection zone is modelled either by a two-nodded uniaxial spring-like element or by a combination of both, depending on the type of continuity to be represented.

Both elements have their stiffness matrices modified to account for a variable nodal position, imperative condition in representing the actual supporting conditions of a general beam. The presence of pre- or post-tensioned tendons is modelled by a matrix superposition and three levels of mild reinforcement are also considered.

Time-dependent analysis considers the effects of concrete aging, differential creep, differential shrinkage and pretressing steel relaxation. A time incremental procedure is assumed and the suggested models of ACI Committee 209 (1971) and PCI Committee on Prestress Losses (1975) are adopted for both concrete and steel properties, respectively. Different loading sequences and construction stages may be predefined and solved in one single analysis, for a general beam type of any number of spans.

The beams in study have been modelled as shown in figure 4 and the material properties are described below.

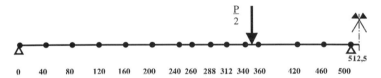

$\frac{P}{2}$

512,5

0   40   80   120   160   200   240 260 288 312 340 360   420   460 500

Figure 4 – Finite element modelling of the continuous beams and nodal co-ordinates, in cm

EXPERIMENTAL PROGRAM

The experimental program comprised six full-scale tests where two different hollow-core beam slabs with two different types of continuity were analysed.

Three tests were carried out with a 16 cm thick and 60 cm wide cross-sectional, 5 m long hollow-core beam slabs. In the first test, two separated elements have been used working as simply-supported members. In the second and third tests two continuous 10.25 m long members, with 40 % and 100 % rate of continuity, respectively, have been analysed. The reinforcement was inserted into the voids, as shown in figure (3.a). The cross sectional properties of the 16 x 60 cm hollow-core slabs were: $y_x = 8.11$ cm, $A_c = 574.5$ cm^2 and $I_x = 16624.6$ cm^4, representing the centroid, the concrete cross sectional area and the concrete cross sectional moment of inertia, respectively.

The same experimental program, for the other three tests, was carried out with 12 cm thick and 60 cm wide cross-sectional, 5 m long slabs, topped with a 4 cm thick layer. The continuous members have been produced with the union of two hollow-core slabs, with the reinforcement being positioned in the concrete topping layer, as shown in figure 3.b. The section properties of the 12 x 60 cm, 4 cm topped hollow-core slabs were: $y_x = 8.81$ cm, $A_c = 697.42$ cm^2 and $I_x = 17562.9$ cm^4. Both sections are shown in figure 5.

Continuity and the concrete topping were cast *in situ*, after the elements had been placed onto the testing supports. Both the hollow-core slabs and the cast-in-place concrete had a compressive strength of 30 MPa and a modulus of elasticity of 30000 MPa. In the numerical analysis, the ultimate compressive and tensile strains were considered as –0.35 % and 0.014 % respectively.

The prestressing steel used in these hollow-core slabs had an ultimate strength of 2000 MPa, at a strain of 7.1 %. The elastic range was limited by a strain of 0.734 % and a stress of 1430 MPa. The strands were submitted to an initial stress of 1520 MPa. Both sections contained 6 prestressing strands, two on the top and four on the bottom of the cross-section. The cover depth of 1.7 cm was maintained, but varied the diameter of strands. In the 16 x 60 cm members, the upper strands had a 3 x 3 mm cross-section and the lower ones a 3 x 4.5 mm cross-section. In the 12 x 60 cm topped cross-section slabs, all strands had a 3 x 3 mm cross-section.

The mild steel used in the continuity reinforcement consisted on 20 mm bars with a tensile strength of 500 MPa at a strain of 1 %. The elastic range was limited by a strain of 0.238 %. In the 40 % rate of continuity members two bars were used while in the 100 % rate of continuity members, five bars were used, all having a length of 4 m.

Figure 5 – Hollow-core slab cross-sections

The loading was applied through two concentrate forces and the prototypes have been tested to failure. Figure 6 shows a member under loading.

The specimens were instrumented with electrical strain gages bonded to the top and bottom surfaces of the element at sections near the maximum positive moment region, over the spans, at the connection sections and in two of the reinforcement bars, in each test. Strain gage signals as well as the load cell response were received by a computerised data acquisition system.

Displacements were controlled by mechanical dial gages positioned over the supports, along the spans and at the extreme sections. The formers to indicate any undesired support settlement or prestressing strand slippage, respectively. None of these actually occurred in any of the tests. Figure 7 shows the position of strain and dial gages.

Figure 6 – A simply supported member under loading

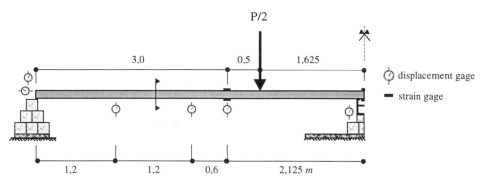

Figure 7 – Loading scheme and instrumentation of the elements

NUMERICAL AND EXPERIMENTAL RESULTS

The numerical analysis was conducted, for each of the specimens, considering the actual physical properties of the materials, as determined at the laboratory. Geometrical characteristics of the cross sections and the position of the prestressing strands of each element were measured and considered in the numerical model. Prestressing forces were assumed to act at the centroid of the prestressing strands.

Time effects were included in the analysis considering the age of the precast elements, from time of production up to final testing. Test loading was considered as an instantaneous effect.

Numerical and experimental results, for displacements and strains, are shown and compared in figures 8 to 11, both for the simply supported and the continuous elements, organised according to the type of continuity and rate of continuity reinforcement.

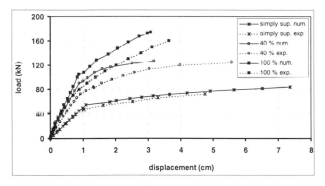

Figure 8 – Results for the 16 cm thick slabs with continuity reinforcement placed in the voids: displacement in the spans for different situations of continuity

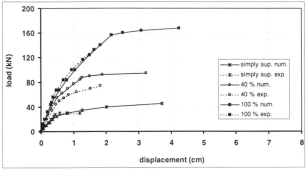

Figure 9 – Results for the 12 + 4 cm thick slabs with continuity reinforcement placed in the concrete topping: displacement in the spans for different situations of continuity

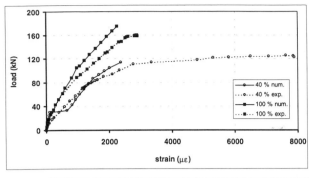

Figure 10 – Results for the 16 cm thick slabs with continuity reinforcement placed in the voids: mild steel strain.

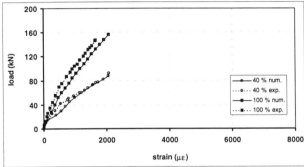

Figure 11 – Results for the 12 + 4 cm thick slabs with continuity reinforcement placed in the concrete topping: mild steel strain.

LOAD CAPACITY x RATE OF CONTINUITY REINFORCEMENT

To determine the potential increase in load capacity of the continuous elements, based upon the rate of continuity reinforcement provided, a numerical analysis was performed considering a 10.25 m long, double span, uniformly loaded (q_{cont}) continuous element as shown in figure 2.

For both types of continuity, theoretical design cross-sections were considered and material properties were the same for both cases, as previously defined. The results obtained are shown in figures 12 to 16.

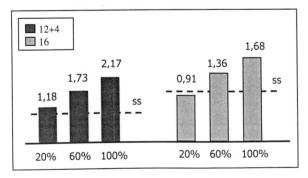

Figure 12 shows the improvement in load capacity provided by the two types of continuity. It can be seen that members with reinforcement in concrete topping have a higher load improvement when compared to the members with reinforcement in the voids.

Figure 12 – Relationship between the ultimate load of continuous and simply-supported members

Figure 13 shows the variation of displacement with load, for 16x60 cross sectional members and different rates of continuity reinforcement. It can be seen that the continuous elements present a good increase in their stiffness. In the elastic range, before the cracking of the spans, all continuous elements show a similar behaviour, regardless of the reinforcement rate. After cracking occurred in the spans, the amount of reinforcement determines the behaviour of the members. In the slab with 20% rate of reinforcement, failure happens in the steel, without reaching the hollow-core beam slab span capacity (0.91 factor in fig. 12). The same general behaviour is presented by the 12 + 4 x 60 cm topped cross section, with the ultimate load directly related to the reinforcement rate, as shown in figure 14. Differently from the previous section, the ultimate load of the 20% continuous member is higher than the ultimate load of the simply supported element.

Figure 15 shows the strain in reinforcement steel for 16x60 cross sectional members with 20%, 60% and 100% reinforcement rate. It can be seen in all cases that steel has yielded, at the point where the numerical curves ended. The initial stiffness, before the connection cracking, is identical for all reinforcement rates.

Figure 16 shows the relationship between moments in the continuity zone and at the middle span region against the load. It can be clearly noted the contribution of the continuity to the resistance of the member, at different load stages. The initial rate between the moments is similar for all cases and close to the elastic value. After cracking of the continuity element, this ratio decreases drastically. If the amount of reinforcement is sufficient, the relation increases with cracking of the spans and decreases again when the steel yields.

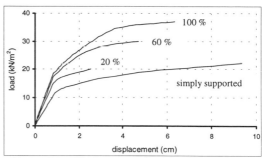

Figure 13 – Displacement for 16 x 60 members

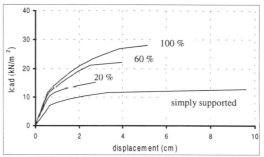

Figure 14 – Displacement for 12 + 4 x 60 members

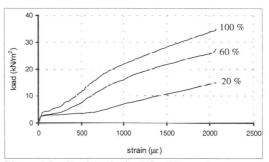

Figure 15 – Steel strain for 16 x 60 members

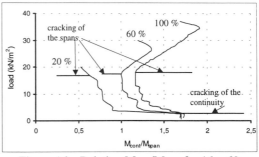

Figure 16 – Relation M_{cont}/M_{span} for 16 x 60 members

CONCLUSIONS
Considering the numerical and experimental results presented above, the following conclusions can be drawn.

Continuity in hollow-core precast elements may be adequately obtained either by providing a reinforced concrete topping or by inserting additional reinforcement bars within the original cross-section, along the continuity zone. In both cases, a short diaphragm connection element ought to be cast between the two adjacent simply supported elements.

Both procedures to achieve continuity are simple and can be executed at low cost, yet there is an enormous increment in load capacity as compared to the single simply supported elements. Once certain continuity reinforcement is provided, the elastic stiffness of the cross-section and the cracking load limits are greatly increased, regardless however on the amount of such reinforcement.

Within the plastic range of behaviour of the continuous element, the increase on the amount of continuity reinforcement does not only reduce the reinforcement strain levels, but also elevates the structural strength. Those advantages, nevertheless, seem to be limited by the use of an amount of reinforcement close to the obtained by the suggested design.

The adopted numerical model has shown to be adequate to represent the overall behaviour of such structures and consists of an important tool for future studies.

Low continuity reinforcement rates may be used to improve the member's stiffness before cracking of the spans, without producing any increase in load capacity.

ACKNOWLEDGEMENTS
The authors gratefully acknowledge the support of PRECONCRETOS ENGENHARIA, who provided the facilities and the specimens for the undertaking of this work.

REFERENCES
1 ACI Committee 209, **"Designing for the effects of Creep, Shrinkage and Temperature in Concrete Structures"**,SP.27, ACI, Detroit 1971.
2 BARBIERI, R. A. **Análise Numérico-Experimental de Elementos Pré-Moldados Protendidos com Continuidade Posterior por Adição de Armadura**. MSc. thesis, UFRGS, Brasil, Jul. 1999.
3 CARNER, A.; ZIA, P. Behaviour and Design of Link Slabs for Jointless Bridge Decks. **PCI Journal**, Chicago, PCI. v.43,n.3, p. 68-81, May-Jun. 1998.
4 GASTAL, F. **Instantaneous and Time-Dependent Response and Strength of Jointless Bridge Beams**. PhD. dissertation, NCSU, USA, Dec, 1986. p. 289.
5 GASTAL, F.; ZIA, P. Analysis of Bridge Beams with Jointless Decks. In: International Symposium on Durability of Structures, 1989, LISBON. **Conference Report...** Zurich: IABSE: ETH Hönggeberg, 1989.
6 NILSON, Arthur H. **Design of Prestressed Concrete**. New York: John Wiley and Sons, 1978.
7 PCI-Precast/Prestressed Concrete Institute. **Manual for the Design of Hollow Core Slabs**. Chicago, 1998.
8 PCI Committee on Prestress Losses, "Recomendations for Estimating Prestress Losses", **PCI Journal**, V.20, n.4, Jul-Aug 1975.
9 STANTON, John F. Response of Hollow-Core Slab Floors to Concentrated Loads. **PCI Journal**, Chicago, PCI. v.37,n.4, p. 98-113, Jul.-Aug. 1992.

Bridge maintenance strategy and sustainability

K. N. P. STEELE
Surrey County Council Environment, Highway House, 21 Chessington Road, West Ewell,
Epsom, Surrey, KT17 1TT.
G. COLE
Surrey County Council Environment, Highway House, 21 Chessington Road, West Ewell,
Epsom, Surrey, KT17 1TT.
Dr. G. A. R. PARKE
Department of Civil Engineering, University of Surrey, Guildford, Surrey, GU2 5XH.
B. CLARKE
Department of Civil Engineering, University of Surrey, Guildford, Surrey, GU2 5XH.
Prof. J. E. HARDING
Department of Civil Engineering, University of Surrey, Guildford, Surrey, GU2 5XH.

Abstract

Bridge infrastructure within the County of Surrey is coming under increasing pressure as traffic volumes continue to grow. Policy changes and diminishing funds have resulted in a reduction in highway construction over the past decade. The result is that the County bridge network continues to age and structures are expected to have ever greater load capacities. It follows that the economic implications of asset maintenance are of great importance. Within the industry much research is focused on these issues and environmental topics are often neglected.

Surrey County Council is the principal owner of highway infrastructure within its boundaries and is fully aware of these constraints. In the interests of sustainable development and environmental improvement, it is looking at ways to monitor the environmental consequences of maintaining the bridge network.

It is believed that important environmental implications are associated with the different maintenance strategies demanded by various structural bridge forms. To test this hypothesis, the potential of life cycle assessment technique is being evaluated.

Findings will be integrated into a data system which will allow for the objective comparison of different maintenance strategies. In the long term, this will lead to the development of a framework of best practice maintenance procedures. The existence of such procedures will support future decision-making, and encourage sustainable development through reduced environmental impact.

1. Introduction

Bridges are single purpose structures that provide an essential function in society today. It is imperative that they are maintained and managed such that they continue to provide the necessary links within local and national road networks. To achieve this the structural integrity and general maintenance state of a bridge network is an essential, yet an often over looked issue which is of great importance to society. Bridge networks must be managed in a coordinated and methodical fashion. Structures should be regularly inspected and assessed, and appropriate maintenance should be carried out in an economic fashion. Structures should be kept in a safe acceptable condition throughout their design life. A traditional management view will consider these economic, technical and social issues, but what about associated environmental issues?

2. Sustainabilty - A Bridge Manager's Perspective

The Local Transport Plan (LTP) process was unveiled in the Government's White Paper, *A New Deal For Transport: Better For Everyone*. Its goal; the development of a more integrated and environmentally sustainable UK transport system. Amongst other things, the LTP process attempts to develop the transport system by taking a more holistic approach. Engineers have a responsibility and commitment to society to provide development in a sustainable manner. The big question is "How do we achieve this?" We can start with the question of how we define "Sustainability," and what it mean to the bridge engineer?

Sustainability means many things to many different people. Subjectively it can mean anything from achieving greater social equality, to conserving the stock of natural assets. Classic terminology can be adopted:

"Sustainable development meets our present needs without compromising the ability of future generations to meet theirs." Brundtland Commission (1987).

Perhaps this definition is too vague for the purposes of the bridge engineer, however, the scope of the quote should not be forgotten. In today's society, unless a project is financially competitive, then it will struggle to find a place in cost conscious economies despite its environmental potential. Sustainable decision making and sustainable development will only be achieved if one considers the economic, social and physical implications of what is being examined. Within the overall goal for sustainable development, these broad areas are considered relevant to decision making as practical constraints on human activities[16]. These three areas can be envisaged as overlapping lobes. The area in which all three lobes overlap, where all three broad sets of constraints are satisfied, represents the area in which decisions promote sustainable development, Figure 1. These three areas can be defined under the headings:

- Scientific - Physical laws and relationships that shape ecosystems.
- Economic - Business systems and economic relationships.
- Social - The social structure and issues that shape society; a reflection of peoples values.

Environmental issues are not a core function of the scientific lobe, rather they encompass all three lobes. The attainment of environmental engineering solutions, lies at the centre of all three lobes.

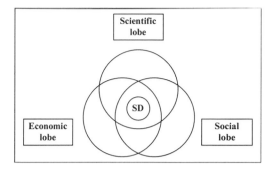

Figure 1. Sustainable decision making = SD

The DETR White Paper is helping to encouraging this form of thinking. In reality, as engineers we are already bound by social and economic constraints. Roads must be kept open, safety is of over all importance, value for money must be achieved, and we must work within budgets. It is the area of scientific consideration which now requires a focusing of energies. To deliver environmental improvement, one must examine bridge management strategy, and the resulting interactions it produces for bridges and their surrounding environments.

It is standard practice for bridge managers to employ the use of a bridge management system (BMS). A BMS provides the base from which bridge managers can make planned and holistic judgment on maintenance and management strategy. By definition, a BMS considers the traditional socio-economic needs for structural management, it is the environmental issues associated with bridge network management that are neglected, Figure 2.

2. Bridge Management Systems

Bridge management systems are valuable tools enabling bridge managers to make informed decisions on bridge networks. As a tool, a BMS must provide recommendations for structural maintenance and strategy which are consistent with the policies, long term goals, and budgetary constraints of the bridge network owners. The structure of a BMS will therefore vary considerably depending on the specified requirements[8]. The basic components of all management systems are essentially the same. A model system which details the traditional core components is illustrated in Figure 2.

Environmental conservation and sustainability issues are essentially not a part of current bridge inspection and management processes. This is accentuated through design codes and bridge inspection and maintenance practice. The situation is no different at Surrey, and the Surrey County Council (SCC) BMS makes no provision for environmentally based decision making.

With these issues in mind, SCC is funding research to develop a sustainable bridge management system.

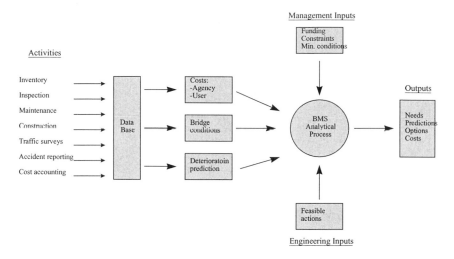

Figure 2. Structure of model bridge management system.

4. Environmental Policy

SCC has a comprehensive environmental policy which serves as a holistic objective for all the services that the authority provides to the county[4]. Sustainability and the issues of best value, quality and efficiency, form a core part of this policy. In the Infrastructure Group, which oversees the management of all highway bridge procedures, this policy is translated into the following form:

- To provide, and efficiently maintain a stock of bridges and other highway structures in a safe condition compatible with their likely use.
- To comply with Highway Agency guidelines, advice and technical standards whenever these can be achieved.
- To pursue the most environmentally friendly solutions practicable.
- To provide the safest achievable environment on the highway and its associated structures for all road users, and make particular provisions to encourage cyclists and pedestrians, including disabled people whenever possible[5].

Environmental issues are recognised as important. The hindrance to sustainable decision making and the adoption of the "most environmentally friendly solutions practicable", is that the burden placed on the Surrey environment by the options available to bridge managers are not known or presently quantifiable. This makes it difficult to compare and contrast in environmental terms the different management strategies available for different bridge types. The objective of this research is to develop a methodology which will provide a means to compare various bridge forms on an environmental basis. This will be undertaken as a life cycle study, such that all inputs and outputs, from construction through use and maintenance, to final demolition might be considered. It is the use phase which is important, not only because of the long time spans involved, but because it is the phase which presently offers the greatest opportunity for implementing change.

Government policy changes and diminishing funds have meant that the programme of new highway construction has been drastically reduced over the last five years. This has brought about a shift in function for the Infrastructure Group at SCC. The focus of the Group is now on bridge maintenance and management and no longer bridge construction.

5. Bridge Management in Surrey

The system presently used by SCC in its bridge management activities is COSMOS. The "COmputerised System for the Management Of Structures" programme is the interface between the user and an Oracle database system which stores all structure information. COSMOS is a fully working bridge inventory and management system which encompasses many of the activities and inputs as defined in Figure 2. Surrey County Council has further developed the COSMOS system to include modules which generate site instructions and inspection notices, monitor expenditure, archive anti graffiti work, prompt pump maintenance patrol works, and allow the publication and circulation of information on a full featured Intranet system.

The problems associated with structures on the network are unique to each and as such, there is not a definitive rule for all structures. There are numerous complex issues behind the construction and management of bridges. Until now function (a social issue) and cost (an economic issue) have dominated. The issue of environment has been of less importance. However, as society becomes increasingly aware as people have sufficient funding, time and interest to be concerned with their living space, the environmental implications of bridge management strategy are having ever greater bearing. The COSMOS system provides the facilities for which it was designed, yet it is always in a constant state of further development. To meet the demands of the Surrey community, SCC would like to expand the system such that it incorporates an environmental theme.

6. The Surrey Bridge Network

In Surrey, as in the rest of the United Kingdom, bridge infrastructure is being increasingly stressed. Surrey primary and principal route networks have some of the highest traffic flows in the country. Across all classes of road, Surrey's highways carry over twice the national average of traffic flows[6]. Maintenance and incidents on the busy motorway and trunk road network can have significant effect on the county network. In consequence, traffic restrictions of any sort, even on unclassified roads can have direct and substantial adverse effect on local communities. Significant socio-economic pressures can result. SCC, as Highway Authority, is directly responsible for the safe operation and structural maintenance of some 2,133 structures on county roads, including road bridges, footbridges, viaducts, aquaducts, tunnels, subways, culverts and retaining walls. Within the 2,133 registered structures, approximately 750 were built between 1880 and 1920.

7. Bridge Form and Structural Classification

Bridges can be classically divided into distinct structural forms. The three basic bridge forms are arch, beam and cable, and these in turn can be broken down into various subgroups[9]. Each structural form can be constructed from a variety of materials, however within the Surrey bridge inventory, patterns emerge where individual bridge forms are typically constructed of a specific material; masonry arch bridges are a good example. The following proportions provide a good indication of the content of the Surrey bridge inventory[10]:

Brick masonry arch	37%
Reinforced concrete	30%
Lattice plate and steel girders	12%
Other	21%

The Surrey bridge network can thus be broken down into three distinct structural categories which comprise the majority of structures; masonry arch, reinforced concrete, and steel bridges. Each will display different deterioration characteristics and thus have different maintenance requirements. A specific maintenance strategy is therefore needed for each structural form, each of which can be expected to have a different burden on the Surrey environment.

8. Environmental Considerations

Bridges have long life spans. It is reasonable to suspect due to the lengths of time involved, that significant environmental burden will occur during the "use" phase as well as the "construction" and "demolition" phases of a structure's life. Policy and funding also dictate that structural maintenance is where the greatest need for knowledge lies. The crucial point for the implementation of an environmentally beneficial management strategy, is that it is not known by bridge managers which options or solutions are of environmental best practice or offer the most sustainable engineering solution.

What are the effects of concrete coatings or steel paint-systems on local water ways? Is it better to use superplasticizers during the construction of a reinforced concrete structure or rely on concrete repair technology during maintenance? Is it environmentally better to use steel or carbon fibre plates in the strengthening of bridge beams? What are the environmental consequences of transport congestion or traffic diversions resulting from the re-saddling of masonry arches? Would it be environmentally better to use radial pinning or anchoring techniques for strengthening of masonry arches? Are the excess pollutants produced by traffic re routing over lengthy detours of such significant environmental burden, as to merit the consistent adoption of the fastest maintenance option?

Presently the positive merits of one strategy are not known or quantifiable against an other. Apart from stated environmental policy, there is very little guidance within design codes or maintenance manuals for the bridge manager to make informed environmental decisions. In short, the effect that these strategies might have on the Surrey environment is not known.

SCC has no database of preferred materials or catalogue of most efficient maintenance options for common structural defects. To provide data, information is presently been amalgamated into a data system. This is of matrix form, and will illustrate the key structural faults of a bridge form and/or material bridge construction (under the three structure types listed above), and the typical maintenance options adopted to fix the problem. Information is being sourced from:

- Bridge file records.
- COSMOS.
- Site instructions actioning maintenance works.
- A summary archive of bridge strengthening works.
- The knowledge of the bridge inspectors and maintenance managers.

To supplement this codified system for classifying maintenance options, a material database of all the materials used through bridge maintenance activities is being compiled.

9. Life Cycle Assessment and Bridges

Life cycle assessment (LCA) or cradle to grave analysis is a methodology which will enable objective comparison of various bridges forms to be undertaken. Although still not widely used, LCA methodology is gaining momentum in the construction industry. Evidence of this is illustrated by the fact that no fewer than 37 presentations of LCA and construction issues were staged at the Second International Conference on Buildings and the Environment[11]. However, in application to bridges, it appears that very little work has been conducted, and any research will be highly original.

Studies completed to date have been limited in scope[12,13]. This is due to the complex nature of LCA, and a lack of accurate data with which to perform assessments. Very little information has been correlated and validated which defines the environmental consequences of construction sector materials. This brings many uncertainties into play. An enormous number of materials are used within the construction industry. A bridge may comprise of many hundreds of components, and be exposed (through maintenance practice and interaction with the surrounding environment), to potentially hundreds of different chemicals over its life time. Each of these could have detrimental environmental consequences. The problem is accessing information that measures or ranks the magnitude of the polluting effects for this wide cross section of materials. Through the compilation of a materials database, the structured mapping of the maintenance strategies used by SCC, and the use of reliable and accurate LCA software, it is hoped that these problems can be over come. Life cycle assessment methodology can then be applied to different bridge forms and their relative maintenance strategies. Developing a LCA methodology which is suited to "Surrey factors" will be a challenge.

10. The future

LCA is a useful tool in monitoring environmental performance or making decisions towards environmental improvement. It offers a way to examine and validate environmental impacts. It allows comparisons between competing systems and approaches. It can provide validation that improvements in environmental performance at one stage of a life cycle, are not offset by greater environmental impacts elsewhere. Very little research has been conducted into the application of life cycle assessment to bridges. This is due to the complex nature of bridge network management, and the fact that bridge network managers are only beginning to develop an environmental awareness.

For the purposes of this research, once a coherent understanding of Surrey County Council bridge maintenance strategy has been developed, findings will be integrated into a data system comprising of life cycle assessment methodology. This will allow for the objective comparison of different maintenance strategies and structural forms. In the long term, it is hoped that this will lead to the development of a framework of best practice maintenance procedures that can be linked into the COSMOS bridge management system. Potential may exist for the implementation of ISO 14001 environmental management systems[14,15]. The existence of such procedures will support future decision-making, and encourage sustainable development through diminished environmental impact in the County of Surrey

Acknowledgments

The first author would like to acknowledge the support of Callum Findlay, Head of Engineering at Surrey County Council, for the research programme and for permission to publish this paper.

References

(1) Young, M., 1998, Bridge management the local authority perspective. The Management of Highway Structures Conference, Thomas Telford, London,
(2) The National Audit Office, 1996, Highways Agency, The bridge programme, HMSO London.
(3) Das, P., 1998, Development of a comprehensive structures management methodology for the Highways Agency. The Management of Highway Structures Conference, Thomas Telford London,
(4) Sustainable Surrey Forum, Summary Report, March 1999.
(5) Surrey County Council Transport Policies and Programme 1999/00, Bridge Strengthening and Assessment, pp. 82-121.
(6) Wootton, N., 1999, Management of bridge maintenance: a local authority perspective, 8ᵗ International Conference: Structures, Faults and Repair.
(7) Darby, J. J., Brown, P., Vassie, P. R., 1996, Bridge management systems the need to retain flexibility and engineering judgment, Bridge Management 3, inspection maintenance assessment and repair, Edit, J. E. Harding, G. A. R. Parke, M. J. Ryall, E & FN Spon London, pp. 212-218.
(8) Road transport research, 1992, Bridge management, OECD, Paris.
(9) Cracknell, D., W, 1999, Kempes Engineering Year Book, Miller Freeman, London.

(10) Palmer, J., Cogswell, G., 1990, Management of the bridge stock of a UK County for the 1990s, Bridge Management , inspection maintenance, assessment and repair, Edit, J. E. Harding, G. A. R. Parke, M. J. Ryall, Elsevier Applied Science, London, pp. 39-50.

(11) CIB Task Group 8, 1997, Environmental assessment of buildings, Second International Conference on Buildings and the Environment, Paris.

(12) Horvath, A., Hendrickson, C., Sept 1998, Steel verse steel-reinforced concrete bridges: Environmental assessment, Journal of Infrastructure Systems, ASCE, pp. 111-117.

(13) Widman, J., June 1998, Experiences from life cycle assessments on steel and concrete composite bridge, Construction and the environment, CIB World Building Congress, pp. 871-878.

(14) Harris, J., Griffin, S., 1999, Environmental management system - Implementation within the County Council, Internal report for discussion.

(15) BS EN ISO 14001 : 1996, Environmental management systems - Specification with guidance for use.

(16) Cowell, S. J., Hogan, S., Clift, R., Positioning and Applications of LCA, Centre for Environmental Strategy, University of Surrey, Guildford, UK.

Maintenance and management of bridges in Osaka City

T.KITADA, Professor, Osaka City University, Osaka, Japan, M. NAKANISHI, Dupty General Manager, Project Management Department, Kansai International Air Port Land Development Co., Ltd., Osaka, Japan and Tadamasa ITO, Assistant Manager, Bridge Department, Public Works Bureau, Osaka Municipal Office, Osaka, Japan

INTRODUCTION

Described in this paper are the present conditions, main maintenance and repair works, bridge management system, adjustment works due to environmental reasons, main working expenses and issues on management of the bridges constructed and managed by the Public Works Bureau, Osaka Municipal Office.

The present situation of the bridges in Osaka City was almost completed by the replacement of old bridges with new ones and the construction of new bridges for the first auburn planning enterprise stated from 1921. These bridges were designed by using new technologies at that time and by considering fire protection too because of the knowledge of Kanto Earthquake (1923). Moreover, some bridges in Osaka after World War were newly constructed and retrofitted for the quick development of motorization and the holding of the Senri International Exposition (1970) in cooperation with the economic revival and high development in Japan.

PRESENT CONDITIONS OF BRIDGES

The number of the bridges located in Osaka City is 873 and the number of the bridges which Osaka City government is responsible to is 761. The bridges mean the later bridges hereafter. The number of bridges constructed every year is predominant soon before the Exposition (1970), as shown in Fig.1. There are two toll bridges. As illustrated in Fig.2, the number of the bridges was 196 in 1889, increased to 395 for the first project of expanding city area in 1897, and substantially increased to 1,629 for the second project of expanding city area in 1925. Then, the number was decreased to about 1,200 in about 1935 because of the reclamation of canals, but increased again to 1,470 for the third project of expanding city area in 1955. The number decreased again to about 800 from 1970 for the reclamation due to the construction of sewerage systems. Then, the construction of elevated bridges and large or long span bridges started and consequently the average bridge length and the area of the decks of the bridges increased although the number of the bridges decreased. This tendency in Osaka City is very predominant compared with the other big cities in Japan, as shown in Fig.3. Through these progress, the bridges attained to the current situation. The total length of the bridges is now 46.8 km and the total area of the decks of the bridges is 697,000m^2. The average age of the bridges is about 31 years and the oldest bridge is the Honmachi Bridge of 86 years old, constructed in 1913.

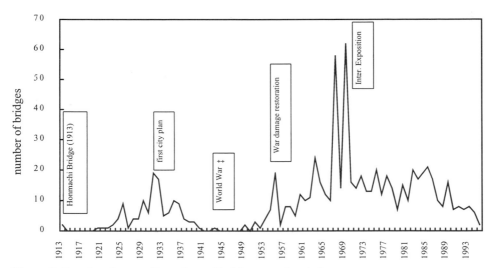

Figure 1. Secular change of number of bridges constructed every year

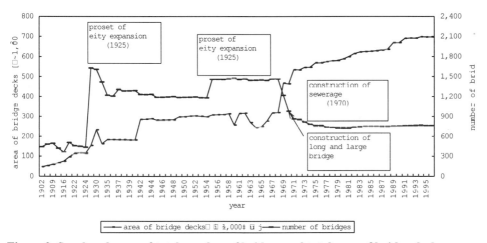

Figure 2. Secular change of total number of bridges and total area of bridge decks

MAINTENANCE AND REPAIR

The main maintenance and repair works are the periodical repair works of expansion joints, railings, main structures, drains and painting, the retrofitting of concrete slabs due to the increased design live load specified in the revision of the Japanese Specifications for Highway Bridges (JSHB, see Ref.1), seismic retrofitting due to the revision of the seismic design method in JSHB (see Ref.2), fireproof improvement of soundproof walls, and strengthening due to the increase of the design live load in JSHB, and the daily, emergency and detail inspections as well as the inspection using false works for painting.

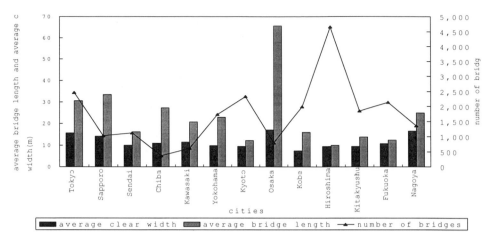

Figure 3. Comparison of number of bridges, average bridge length and average clear width among big cities in Japan

Repair of Expansion Joints : The total length of the expansion joints is 27 km, consisting of 6 km rubber joints, 10 km simplified dummy joints and 11 km steel dummy joints. The rubber joints are repaired every 15 years, and the simplified dummy joints every 10 years. The steel dummy joints are scheduled to repair soon before the extent of the damage under control reaches the limit state not permitted.

Repair of Concrete Wall Railings : The constant repair by using glass fiber cloths or paint for concrete is executed to prevent the fall of covering concrete of wall railings and curbs due to the exfoliation by the corrosion of reinforcements and the secular deterioration of concrete because the thickness of the covering concrete is thin.

Retrofitting and Repair of Railing : The height raising works are constantly carried out in railings because the minimum required height of the railings was changed from 90 cm to 110 cm in 1992, and repair is also carried out in the railings in which the possibility of fall of children from the bridge can not be neglected because the space of lateral members is wider than 15 cm, or dangerous deterioration or damage can be observed. The bridges with these railings waiting for retrofitting or repair are 390 in number and 21 km in total length.

Repair of Drainage Systems : The drainage systems of the elevated bridges are 21 km in total length, and are inspected and repaired if necessary ever year.

Painting : The painted area in the steel bridges (608 bridges) is 1,920,000 m^2. Polyurethane resin paint has been used from 1990 to increase the period of repainting and steel bridges are generally repainted every 15 years.

Retrofitting : Almost all the concrete slabs are retrofitted because the maximum weight of heaviest vehicles specified in JSHB was changed from 200 kN to 250 kN in 1993. The concrete slabs of about 80 bridges of which area is 18,000 m^2 are scheduled to be repaired for the reason of the increased live load. For the same reason, the load carrying capacity of 200 bridges located in the trunk roads were checked and it is planed that the necessity of strengthening and the strengthening methods are examined if necessary after the detailed investigation and stress measurement of 160 bridges of which load carrying capacity may be insufficient.

The seismic retrofitting of 145 very important bridges among 352 bridges necessary to be retrofitted started from 1996 and will be finished until 2000 because the Seismic Design Part of JSHB (see Ref.2) was revised in 1996 on the basis of the results of many investigations after the Hyogo-ken Nanbu Earthquake. The strengthening of RC bridge piers has been almost finished and the retrofitting of steel bridge piers, the setting of new types of bridge restrainers to prevent bridges from falling down, and the connection works of neighboring simply supported bridges to change them to continuous bridges are now carried out.

The soundproof walls in 18 elevated highway bridges have been replaced with new ones from 1997 as a five-year plan by using about 920 million \ (Japanese Yen, 100 \ is about one US $) because a soundproof wall made of flame-resisting polycarbonate in a existing bridge flamed by a car collision.

BRIDGE MANAGEMENT SYSTEM

A bridge management system called OBAS (Osaka City Bridges Administration System) as shown in Fig.4 is developed by the Bridge Department, Public Works Bureau, Osaka Municipal Office. Data on bridges can be separated into three parts, one is the property data on the dimensions of the bridges etc., next is the maintenance data on inspection, damage and repair, and third one is the drawing data. The property data are controlled by the main system and the maintenance data and the drawing data by the Karte (card) system. OBAS has a net work system as shown in Fig.5. The main data base is in the office of the Bridge Department and connected to the sub-data systems in the many satellite offices for maintenance or construction.

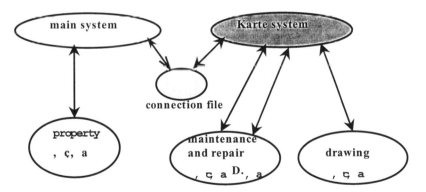

Figure 4. Compassion of OBAS

BRIDGE MODIFICATION DUE TO ENVIRONMENTAL ADJUSTMENT

The beatification of bridges, for example, the beatification of elevated bridges with decoration plates underneath them, decoration on historic environment, illumination, and 100 monumental bridges can be mentioned specially (see Refs.3 and 4).

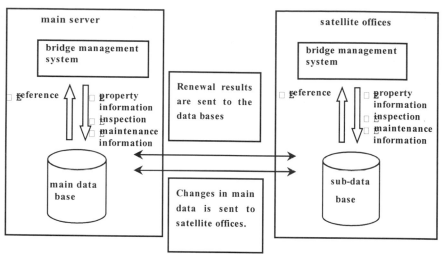

Figure 5. Network of OBAS

MAIN WORKING EXPENSES AND ISSUES OF BRIDGE MANAGIMENT

The work on bridges is categorized into the following four groups. These are new construction/reconstruction, improvement of function, seismic retrofitting and repair. The secular change of working expenses in the recent 7 years is illustrated in Fig.6. The total working expenses on bridges are 8,600 million \, and its items are 2,400 million \ for the repair, 1,900 million \ for the construction/reconstruction, and 4,300 million \ for the improvement and the seismic retrofitting. The working expenses for the construction/reconstruction have, however, a tendency to decrease. The expenses for the seismic retrofitting are temporary for the Hyogo-ken Nanbu Earthquake. The expenses for the repair are almost constant every year although repair work is increasing year by year.

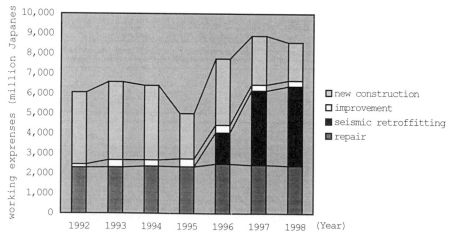

Figure 6. Secular change of working expenses for bridges

Improvement of Bridges : The improvement of existing bridges is for the following four reasons. That is to say, insufficient function on the number of lanes, load carrying capacity, etc., adjustment due to riparian works, adjustment due to city planning, and deterioration or insufficient load carrying capacity.

Fig.7 shows the secular change of working expenses per year for reconstruction and its accumulated expenses. The total working expenses is almost 1,100 billion (US) \. About 11 billion \ (110 million US $) will be required every year for the reconstruction, if it can be assumed that the average life of the bridges is about 100 years old. The bridges older than 50 years and constructed before World War are 10,800 m^2 in the area of the bridge decks and almost 15% all the bridges. The number of the bridges constructed every year before World War has a peek in about 1932 for the first auburn planning enterprise. These bridges will be about 100 years old after about 30 years.

The secular change of the average life and the oldest life during the recent 50 years are illustrated in Fig.8. The average life is prolonged from 12.5 years in 1941 to 30.8 years in 1994. This means the increase of about 2.5 times in the average life for about 50 years. Roughly speaking, it may be predicted from the data that the average life after 50 years becomes about 60 years old, which is two times the 30 years old. The strategies to prolong deliberately the life of the bridges constructed before World War should be investigated by separating them into three categories, the bridges to be replaced, the bridges to be repaired on a large scale, and the bridges to be repaired partly. The seismic retrofitting is also important for these bridges. The rational and economical inspection and repair methods are required for the large bridges or long span bridges constructed in 1960's.

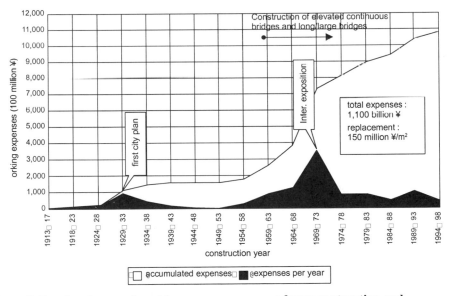

Figure 7. Secular change of working expenses per year for reconstruction and accumulated one

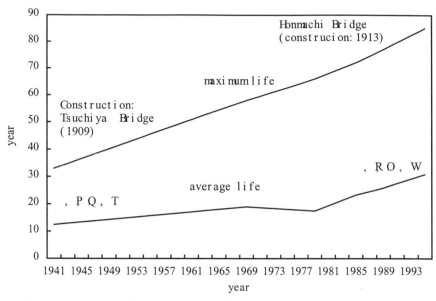

Figure 8. Secular change of average and maximum lives

Table 1. Representative lives of replaced bridges in 1927 and 1995 (unit in years)

Lives	1927	1995
Maximum life	----	82
Average life	13	66
Minimum life	17	32
Average life of all the bridges	8	26

CONCLUSIONS
The present situation, maintenance and management on the existing bridges in the Public Works Bureau, Osaka Municipal Office are briefly described in this paper. The bridges to be replaced or repaired are increasing year by year, while working expenses for bridge management are limited always. Economical and rational bridge management methods based on the total or life-cycle cost should be investigated to keep the bridges safe at all the times although it may be very difficult to predict the bridge life because it is greatly influenced by the economical situation, transportation system, living style etc. in the society for long years. Conversely, the situation of the bridges may decide the economical situation, transportation system, living style etc. in the society if I may be allowed a little exaggeration. I would like to conclude this paper by emphasizing the importance of the bridge management.

This paper was written out on the basis of data prepared by the second and third authors while the second author, Mr. Nakanishi was working for the Public Works Bureau, Osaka Municipal

Office as the head of Bridge Department.

REFERENCES
1. Japan Road Association : Japanese Specifications for Highway Bridges, Part   General, Pert   Steel Bridges, November 1993
2. Japan Road Association : Japanese Specifications for Highway Bridges, Pert   Seismic Design, December 1998
3. Bridge Department, Public Works Bureau, Osaka City Government : Bridges in Osaka, September 1998.
4. Bridge Department, Public Works Bureau, Osaka City Government : Bridges in Osaka Port, September 1998.

BWIM systems using truss bridges

TATSUYA OJIO, MS, Research Associate, Nagoya University, KENTARO YAMADA, Professor, Ph.D., Nagoya University, and HIDEMASA SHINKAI, Graduate Student, Nagoya University, Nagoya, Japan

INTRODUCTION

Unlike railroad bridges, traffic loads, passing on the highway bridges in service condition, is not well known in the design stage. The design truck is changed time to time to adjust the actual traffic loads in service. The last modification of design trucks for Japanese highways was carried out in 1994, where truck weight is increased from 196 kN to 245 kN. Weigh measurements of trucks in service, for example using scales buried under the traffic lanes or using weigh stations, were often carried out to support such modification of design trucks.

In order to evaluate durability of existing bridges and pavement, for example in terms of fatigue, measurements of traffic condition in service are extremely important. For this purpose several methods have been proposed to measure traffic loads in service without using weigh stations. Since they often use bridge members as a scale, they are called Bridge Weigh-In-Motion (BWIM) system. The method was first proposed by Moses[1] in 1979, where main girders of plate girder bridge is used as the scale. Miki et al. used practically the same technique in their BWIM systems[3]. Matui et al. used opening displacement of cracks existing in RC slabs[2]. The former two procedures can estimate the whole truck weight, and the latter one can estimate axle weight as well.

The authors developed the method to estimate wheel loads of vehicles using orthotropic steel deck[4,5]. In this method ten strain gages were attached to five longitudinal ribs made of bulb plates at two sections, about 1.54m apart. Comparing analytical stress waves obtained by FEM stress analysis of orthotropic steel deck, we could estimate not only weight of wheels, but also speed of trucks, wheel spacing, wheel positions in the lane, and type of trucks.

These BWIM systems, which were developed before[1-5], are generally based on reverse problem, where wheel load or weight of trucks is estimated from response of bridge members. Therefore, they need influence lines or influence surfaces of members which the sensors are attached. It also need complex calculation procedures for estimation.

Method described in this study has a simple concept based on forward problem like a calculation of response from load. Gross vehicle weights of trucks can be estimated from strain recordings by a simple calculation procedure. The concept was applied to BWIM system using hangers and stringers of Warren truss bridge.

ESTIMATION PROCEDURE

Responses induced by trucks running through a member of a bridge are a linear composition of influence line. When an influence line of response of a bridge member is indicated as $y = f(x)$, where x is the position of the moving load W, the response wave R can be expressed as

$$R = W \cdot f(x) \tag{1}$$

When N axles of a truck run through, the response wave is described as a liner combination of influence lines for each axle,

$$R = \sum_{n=1}^{N} Wn \cdot f(x - Xn) \tag{2}$$

where x is the position of the first axle, Xn is the distance from the first axle to each axle, and Wn are the axle loads. When a truck runs over the bridge, influence area A can be expressed by integrating R, as follows.

$$A = \int_{-\infty}^{+\infty} R \, dx = \int_{-\infty}^{+\infty} \left[\sum_{n=1}^{N} Wn \cdot f(x - Xn) \right] dx \tag{3}$$

The equation can be further transformed as follows.

$$A = \sum_{n=1}^{N} \left[Wn \cdot \int_{-\infty}^{+\infty} f(x - Xn) \, dx \right] = \sum_{n=1}^{N} Wn \cdot \int_{-\infty}^{+\infty} f(x) \, dx = GVW \cdot \int_{-\infty}^{+\infty} f(x) \, dx \tag{4}$$

In Eq. 4, the integral term depends only on the measuring points and the term A is not depended on wheel spacing, Xn. Therefore, term A is an indicator of gross vehicle weight GVW of the truck.

If a truck with known gross vehicle weight $GVWc$ runs over the bridge, an influence area Ac can be obtained. When a truck with unknown weight runs on the same bridge, we can measure influence area A. Then, gross vehicle weight GVW of the unknown truck can be estimated using Eq.5.

$$GVW = \frac{A}{Ac} GVWc \tag{5}$$

Responses induced by moving loads are usually measured as time-history. When a truck run with a constant speed V, a position x can be expressed as $x = V \cdot t$, the influence area can be described as,

$$A = \int_{-\infty}^{+\infty} R \, dx = \int_{-\infty}^{+\infty} R \left[V \cdot dt \right] \tag{6}$$

When a truck enters the bridge at $t = T_1$, and exits at $t = T_2$, with constant speed of V, the influence area of the truck is described as,

$$A = V \cdot \int_{T_1}^{T_2} R(t)\, dt \tag{7}$$

APPLICATION OF THE THEORY

In order to fully utilize the above mentioned concepts, a few caution is necessary. The system needs structural response (time-history) of members where the sensors are attached, and velocity of truck. In order to keep certain accuracy sensor points should be sensitive to traffic loads. The influence line should not be too long, because trucks have to run with constant speed over the influence line. It is also necessary that only a truck runs over the influence line at one time. Considering such requirement, suitable sensor points in bridge members are hangers, struts, stringers, cross beams, short span girders, and so on.

At least, two strain gages (or displacement meter) attached at the two sections are needed, which have same influence line. From comparison of measured time-histories of two sensors, velocity of the truck can be calculated. If only one sensor is available, velocity have to be measured by other instrument, such as optical sensors, as described later.

It is also necessary to use known truck for calibration. The gross vehicle weight of a test truck ($GVWc$) should be measured in a weigh station. The integral area of the test truck (Ac) can be measured from several test runs on the test sections.

In order to use this system with accuracy the following factors should also be considered.
(1) Effect of simultaneous loading : When the length of influence line is about 10m or less, little effect of simultaneous loading may be observed.
(2) Effect of vibration of structure or any impact induced by roughness of bridge surface : Because the indicator of weight in this method is calculated by integrating dynamic response, any small vibration is cancelled out. The estimation is considered to be not sensitive to dynamic effects.
(3) Effect of velocity : Any error associated with the estimation of velocity affects the weight estimation (see Eq.7). Make sure that trucks run over the bridge with constant speed. The accuracy associated with velocity should be verified by test runs using the test trucks.
(4) Discrepancy of running position in the instrumented lane : Trucks tend to run right or left side of a lane. It is advised to select sensor points which are insensitive to the running position in transverse direction. The sensitivity of the sensor points to the measurement should also be verified by test runs.

In truss bridges, stringers of floor system and vertical hangers of Warren truss are normally sensitive to traffic loads. The length of influence line of bending moment of stringers is considered as the span of the stringer. For the vertical members of Warren truss the length of the influence line is considered as their panel length. In this study, a Warren truss bridge was selected as a test bridge. Two vertical hanger and four stringers were used as sensors of weight in the experiment.

BWIM SYSTEM USING WARREN TRUSS BRIDGE

Measurement was carried out on Yoshida-Bashi Bridge, which was constructed in 1961 with span of 42m, as shown in Photo 1. It is a through-Warren truss bridge located in Kiso mountain area on a national route 19. The Chuo Express Way was constructed about 20 years ago and this route is now used as by-pass. Yet, it still carries average daily traffic of about 15,000 in two lanes. The elevation and plane of Yoshida-Bashi bridge is shown in Fig. 1. The truss flame has vertical hangers to suspend the floor system. The floor system is consisted of cross beams, stringers, RC slabs and pavement.

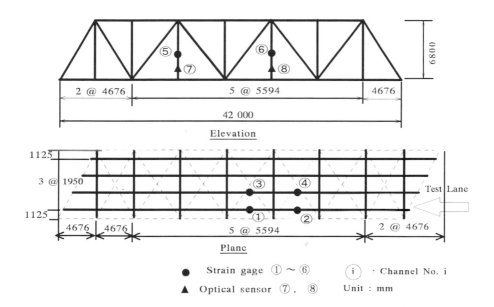

Fig. 1   Yoshida-Bashi Bridge and measuring points

Photo 1   Yoshida-Bashi Bridge and stress measurement

Stress measurement was carried out in October, 1998. Two strain gages, 5 and 6, were attached at two vertical hangers (influence line = 11.2m). Four strain gages, 1,2,3 and 4, were attached at the bottom flanges of four stringers (influence line = 5.6m). Two optical sensors were clumped at two vertical hangers of 11.2m apart, which are numbered as channel 7 and 8. The optical sensors react when any vehicle runs in front of them, and thus velocity of vehicle can be calculated. Strain waves and responses of optical sensors were measured for 15 seconds by a dynamic digital recorder for each loading case. The sampling interval was 0.005 second.

Theoretically only one truck is necessary for calibration. However, two test trucks, which has different wheel base, as shown in Fig.2, were used in this test. One was a dump truck loaded with gravel. The other was a truck loaded with a bulldozer. The test trucks ran over the bridge with speed between 30 and 50 km/h during the test. The running positions in transverse direction of test trucks were recorded by video camera. Examples of response waves in vertical members and optical sensors are shown in Fig. 3.

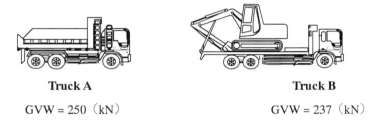

Truck A   **Truck B**

GVW = 250（kN）   GVW = 237（kN）

Fig. 2   Test Trucks used for calibration

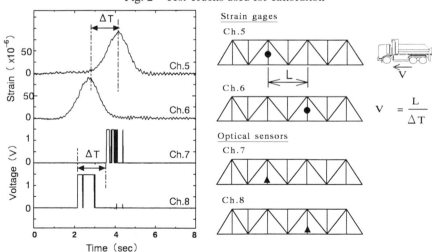

Fig. 3   An example of response wave

Each of the test trucks ran six times without other vehicles in the experiment. Velocities of trucks were calculated from the difference of arrival times between two sensor points. In this method, three possible ways to calculate the truck speed. First, the speed was calculated using optical sensors. Then, the speed was calculated by comparing strain waves from stringers or vertical hangers. Fig. 4 shows the comparison between velocities calculated by strain recordings and those of optical sensors. They were in good agreement with each other.

As shown in Fig. 6, the sensitivity of accuracy to the running speed of trucks are not too large. The test trucks actually ran left and right side of the test lane over the range of about 50cm. As shown in Fig.7, the accuracy were not too sensitive to the running position.

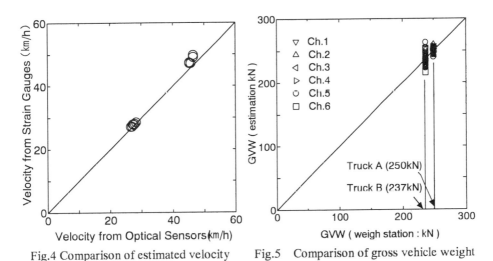

Fig.4 Comparison of estimated velocity   Fig.5   Comparison of gross vehicle weight

Table 1 Influence areas calculated from test trucks

Sensor Point		Truck A GVW= 250 kN			Truck B GVW= 237 kN		
		Maximum Response (x10⁻⁶ strain)	A* (x10⁻⁶ strain • m)	A*/GVW (x10⁻⁶ strain • m / kN)	Maximum Response (x10⁻⁶ strain)	A* (x10⁻⁶ strain • m)	A*/GVW (x10⁻⁶ strain • m / kN)
1	Stringer	82	454	1.816	67	440	1.826
2	Stringer	89	496	1.984	75	481	1.996
3	Stringer	104	594	2.376	88	570	2.365
4	Stringer	82	468	1.872	71	449	1.863
5	Hanger	79	693	2.772	70	678	2.813
6	Hanger	81	748	2.992	72	716	2.971

A* : Average of six running.

The influence areas were calculated by integrating the strain recordings and the velocities.

The influence areas for unit load (A/GVW) are summarized in Table 1. The influence area per unit load, calculated from the truck A, are in good agreement with that from the truck B at every measuring point. This indicates that the influence area does not depend on wheel spacing .

The calibration value of influence area for unit load was obtained by averaging the results of these tests. Estimated gross vehicle weights in every test running were calculated using the calibration value. The relationship between estimated gross vehicle weights and real weights measured in weigh station are shown in Fig.5. Here, the accuracy of estimation for test truck is defined as follows.

$$Accuracy = GVW(estimated)/GVW(weigh\ station)$$

The estimated truck weight is generally in good agreement with the real weight measured in weigh station. We can say that the error of estimation was within 10%.

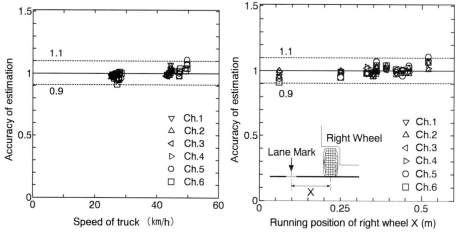

Fig.6 Accuracy for running speed of truck Fig.7 Accuracy for running position

CONCLUSIONS

In this study, a new concept of Bridge Weight-Motion (BWIM) system is introduced and its applicability is demonstrated by stress measurement using stringers and vertical hangers of a Warren truss bridge. The method is based on equality between a moving load and a distributed load on bridge surface. Integration of the response of moving loads, such as a stress history of a truck of unknown weight, is linearly correspond to the gross vehicle weight of the truck. Therefore, if one know the stress history and its integration of a test trucks, one can estimate weights of trucks in service. This method needs only time history response and velocity of each truck.

In the test of a Warren truss bridge, the vertical hangers and stringers, which had relatively short influence lines, were successfully used for calibration. The estimated weight of the test trucks were in good agreement with their actual weights measured at the weigh station within 10 percent accuracy at every sensor point. Factors affecting the weight estimation are also discussed.

AKNOWLEDGEMENT

This study was carried out as a cooperative research between Nagoya University and Chubu Technical Research Office of Ministry of Construction in Japan. The field experiment was carried out with help of Kiso Work Office of Iida Highway Work Office, Ministry of Construction, Japan Engineering Consultants Co., Ltd. and Tokyo Sokki Kenkyujo, Ltd. The authors are grateful to those who supported this study.

REFFERNCES

1) Moses, F. (1979) "Weigh-In-Motion System Using Instrumented Bridges", Transportation Engineering, Proceedings of ASCE, Vol.105, No. TE3.

2) Matui, S. and El-Hakim, A. (1989) "Estimation of Axle Loads of Vehicle by Crack Opening of RC Slab", Journal of Structural Engineering, JSCE, pp.407-418.(In Japanese)

3) Miki, C., Murakoshi, J., Yoneda, T. and Yoshimura, H. (1987) "Measurement of Weight of Traffics in Service ", Bridge and Foundation, pp.41-44. (In Japanese)

4) Ojio, T., Yamada, K., Kainuma, S., Obata, T. and Furuichi, T. (1998) : "Estimation of Traffic Loads using Strain Recording in Orthotropic Steel Deck", Journal of Structural Engineering, JSCE, Vol.44A, pp.1141-1151. (In Japanese)

5) Ojio, T., Yamada, K., (1997) "Measurement of Wheel Loads Using Orthotropic Steel Deck", Proceedings of EASEC-6, Volume 1, pp.411-416.

Whole life costing in bridge management – a probabilistic approach

S.R. RUBAKANTHA, WS Atkins Consultants Ltd, Epsom, UK, and G.R. PARKE University of Surrey, Guildford, UK

INTRODUCTION

Over the last few years, issues related to the validity and applicability of whole life costing (WLC) to the bridge management have been discussed extensively. At present, even though some scepticism remains, there seems to be a wider realisation of the importance of WLC based decisions because of the fact that it accounts for the long term cost implications as opposed to the initial cost alone. Particularly, the merits of WLC in bridge management are evident in the applications related to cost evaluation of durable designs for new bridges, selection of cost effective protective maintenance methods and financial evaluation of Design Build Finance Operate (DBFO) schemes. In addition to these potential applications, another avenue will emerge with the introduction of whole life assessment standards, which requires the maintenance agents to submit alternative maintenance strategies together with their whole life costs to the Highways Agency so that the cost effective strategy can be selected for funding.

In any whole life cost exercise, the primary difficulty is to formulate the necessary database of costs and timing for the future maintenance, and invariably one has to use the data that can not be obtained with certainty. In the basic form, cost and life data can be best described by probabilistic distributions of their actual variation and uncertainties. In the conventional approach, the best estimates are determined for the life and timing data and a deterministic discounting is adopted to evaluate the WLC. The probabilistic approach offers the assessor with the benefit of being approximately right in the WLC estimates rather than being exactly wrong with comparison using deterministic values that are uncertain. WLC appraisals to compare alternative options will also indicate the chances of one option being preferable to the other and the possibility of cost over runs above a threshold value, providing a better basis to select the option that suits the long term maintenance strategy. This paper proposes a probabilistic whole life cost procedure that can be used in the place of present deterministic discounting exercises. Discussing the sources of costs and timing, it identifies the already existing stochastic data and other areas where such data formulations is necessary. The benefits of a probabilistic approach are also specified for various applications of WLC.

CURRENT APPLICATION AND USE

Guidance on whole life costing procedure of bridges is specified in the Departmental Standards BA28/92 and BD36/92. In their current form, they cover the data and required assumptions for costing alternative maintenance options. Revised documents are being produced to include updated data for costing alternative designs of new bridges, maintenance management options and to provide essential data for the cost implications of whole life assessment options which will promote the use of WLC in related applications.

Bridge management strategies can be planned at two levels: essential work that is carried out when the structure reaches the critical state in the absence of regular maintenance and the other is to carryout preventative maintenance at a regular intervals to avoid structural deterioration. Preventive maintenance of new bridges begins at the design stage where improved durability provisions can be considered such as epoxy coated reinforcement, integral abutments, continuous deck, etc. Such measures could increase the up front cost of bridges but can minimise the costs and frequency of future preventive maintenance. New designs need to account for the optimum trade off between money spent on the initial durability provisions and subsequent maintenance needs which can be best achieved by the use of whole life costing. Use of WLC for the maintenance planning of existing bridges needs some additional information such as the remaining life of bridges with different age and type as well as the effectiveness of various preventive maintenance methods.

There is a potential for using WLC for prioritisation of maintenance and rehabilitation works. One of the primary functions of any authority responsible for highway maintenance is the selection of critical projects among a large number of works competing for funding. Decisions on prioritisation and which scheme to fund can be hardly justified on the basis of whole life costs alone, especially for the bridge stock in urban areas. In such situations, the other aspects such as risk of structural failure, social impact and limitation on weight and width restriction of the specific routes may influence the decisions. Whole life costs must be treated as an element of wider considerations of other deciding factors, and its relative importance have to be established in the effective integration of all the relevant factors and criteria by using a prioritisation system.

COST DATA

Building up the necessary cost database is a prerequisite for any WLC procedure. Actual future costs of maintenance are dependent on the new developments in repair techniques, improved access to bridge components, inflation, etc. The influence of inflation can be incorporated either by using the construction price index to multiply the future costs or using effective discount rates that allow for the depreciation. Discounted costs have a diminishing effect with time and the activities beyond certain duration need not be included, hence it is important to select the costing period based on the magnitude of the costs associated with all the components of future expenditure. If traffic delay costs are included into the WLC, their increase with the growth of traffic flow have to be taken into account which may give a significant discounted value say even after 50 years. There are other factors that may decide the period of analysis depending on the requirements of the application. For instance, a WLC of a DBFO project lasting 30 years needs to include the costs over this period and any costs incurred after this period is irrelevant to the contract.

Cost data of bridge maintenance can be formulated using a number of ways. The most common method is to systematically analyse the costs of typical previous maintenance and repair contracts.

Past cost information can be obtained from a number of sources, NATS database, maintenance agents and local authorities are some examples. Statistical analysis of such data will give probabilistic cost estimates or costs expressed in distributions rather than single values.

These variations can be represented by simple normal distributions. In the absence of past data for some new form of protective maintenance, expert opinion can be used to formulate the costs. It involves carefully prepared group sessions organised among the experienced engineers to seek objective data on costs. It is more convenient and meaningful to request costs data from the engineers on the basis of minimum most likely and maximum, than as point estimates. These estimates may not be sufficient to formulate a descriptive distribution and a simple triangular distribution can be assumed of their variation.

Models to predict costs are the "easy-to-use" methods to determine the costs of maintenance, however creation of such models demands extensive collection, analysis and correlation of past cost data. An added difficulty is to relate the costs of maintenance to damage condition and extent as the actual cost of repair will vary according to the access and ease of applying maintenance. When describing the costs in terms of probability distributions, variations arising from such differences can be included.

MAINTENANCE TIMING
Intervention time to treat the bridges under a protective maintenance regime determines the frequency of applying the costs. The first maintenance time is decided from the deterioration of the structure whereas the subsequent treatments are the function of the effective life of maintenance. Advanced modelling of reinforcement corrosion in concrete structures provides the time of depassivation of steel, onset of corrosion and the appearance and cracks. For the deterioration prediction to be inclusive of changing conditions, this modelling needs to be probabilistic to accommodate the variation in exposure conditions and rate of corrosion, etc. The distributions describing these variations have to be established for the bridge population. Use of probabilistic prediction of element life will give the probable range of life with the attached probabilities demonstrating that the elements of similar types can have different lives due to changes in the exposure conditions. The components that have reached the critical level may not receive the preventative maintenance when required, due to lack of funds and the delays caused can be included in the modelling uncertainty in maintenance timing.

To account for the probability of maintenance occurring, the current BD36/92[1] adopts a simplified method which recommends the maintenance and traffic delay costs to be multiplied by a factor of 0.02 to reflect the 20% chance of the assumed maintenance actually taking place. A detailed treatment of probability of the actual occurrence of maintenance can only be given in the probabilistic WLC whereby probability distributions are used for the lives.

Reinforcement corrosion is the main type of deterioration affecting bridge decks and the critical time for maintenance has to be decided from the profiles showing the depassivation of steel against time. The lives of other components such as bearing and bridge joints are estimates from past records and engineering judgement. The component life profiles are included in the whole life performance assessment document that is under preparation by the Highways Agency. This will assist finding the maintenance frequencies of other components. Instead of the performance indicators based on single values, the actual uncertainties attached to the profiles will offer timing for maintenance in probabilistic terms (Figure 1).

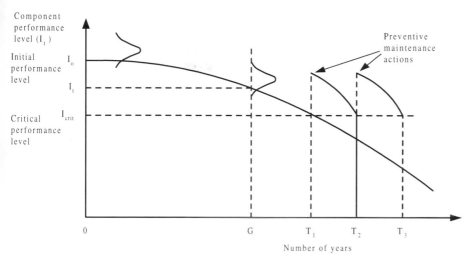

Figure 1: Maintenance timing of bridge components

In the absence of lives and timing of maintenance data, expert opinion can be sought organising group sessions similar to that used for collection of cost data. Engineers experienced in repair contracts have to extrapolate their knowledge to similar work and the costs in the future. Views expressed by the participants will be more realistic if the data are collected for the minimum, most likely and maximum figures for the durable lives for the bridge elements, which can be modelled into distributions of maintenance timing.

PROBABILISTIC APPROACH AND BENEFITS
The probabilistic approach involves the generation of WLC estimates for alternative maintenance options utilising the stochastic data for the cost, maintenance timing and other uncertain data (Figure 2). It has to begin with the process of collecting costs and maintenance free data in terms of probabilistic estimates, be the source a theoretical model, a deterioration prediction or an expert opinion. In the conventional approach, the most likely values of cost and maintenance free lives are estimated for the range of values or probability distributions, this cast doubt about the results when the inputs deviates from the most likely values.

Process of probability based WLC procedure involves formulation of uncertain data into probability distributions, creating maintenance strategy (timing and extent of treatment), defining the correlation between the inputs and execution of simulation using Monte Carlo methods. Selection of distribution types for uncertain data is decided by the extent of data available of the variation: limited data of minimum, most likely and maximum can be assigned with triangular distributions. If sufficient data are available, a curve fitting exercise can be carried out to determine the distributions of normal and lognormal types. Detailed methods of curve fitting are unnecessary for most of the inputs, unless the input has a decisive effect on the WLC, as the shape of the distribution for most inputs to WLC procedure was found to have less impact on the resulting WLC distribution.

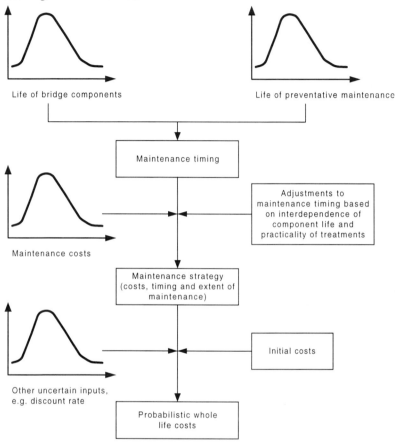

Figure 2: Whole Life Costing Based on Probabilistic Inputs

Different components of the bridge end their durable life at different points in time. The durable life of some elements are interdependent, for example a leaking deck joint will accelerate the deterioration in the column head, bearings, abutment and deck ends. Randomness in varying their lives need to take account of the dependency of element lives by specifying the appropriate correlation coefficients.

Probabilistic WLC procedures are easy to implement by the assessor. Creation of probabilistic models fundamentally needs the underlying deterministic model for which the inputs can be replaced with distributions and software to carry out the simulation. WLC models can be developed using spreadsheet applications and there is a number of ready made software available capable of carrying out a simulations on such models. However, sufficient expertise is required in formulating a realistic maintenance strategy in deterministic models for the bridge in question.

CONCLUSION

The probabilistic approach is essential in representing the uncertainties in the input data for the whole life costing procedure. It recognises the fact that basic data used for the exercise can vary and presents the effect of variation on the whole life costs with the associated probabilities. Besides, it overcomes the major impediment to the validity of whole life cost exercise, that is "the whole life costing is built on uncertain data". Rather than searching for more accurate data, the method provides a means to compare the options in the middle of the inherent uncertainties present in the basic data.

In addition to these general benefits, the method enables more realistic comparisons to be made based on the probabilistic results acknowledging that components to the same specifications can have varying lives due to the difference in exposure conditions and extent of damage that will also have different maintenance costs.

For the cost evaluation of DBFO projects, it will show the probability of costs deviating from the given value and a better perspective for cost overruns which helps apportioning the risk between the parties involved.

The overall approach must begin with modelling the basic data collected as a probabilistic distribution. Means of costs and timing data, i.e. theoretical models, analysis of past records, expert opinion sessions, need to be designed to provide probabilistic data. In using deterioration models, some background information is necessary such as the variation of exposure conditions, rate of deterioration, degradation of protective maintenance, etc for the bridge population of interest. A risk-based WLC can be implemented with the available tools – a WLC model can be created by incorporating stochastic inputs and Monte Carlo simulation method in the conventional deterministic WLC models.

REFERENCES
1. BD 36/92 Evaluation of Maintenance Costs in Comparing Alternative Design for Highways Structures, DOT, UK.
2. BA 28/92 Evaluation of Maintenance Costs in Comparing Alternative Design for Highways Structures, DOT, UK.

Bridge management based on lifetime reliability and whole life costing: the next generation

DAN M. FRANGOPOL[1], JUNG S. KONG[2], and EMHAIDY S. GHARAIBEH[2]
[1] Professor, and [2] Graduate Research Assistant, Department of Civil, Environmental, and Architectural Engineering, University of Colorado, Boulder, CO 80309-0428, USA

ABSTRACT
The next generation of bridge management systems has to use powerful reliability, optimization, and life-cycle engineering tools to predict the performance of bridges and the costs associated with bridge interventions including both agency and user costs, to find the optimal lifetime strategy based on benefit/cost methodologies, to implement this strategy at both network and project levels, and to update the optimal strategy as more information becomes available. This paper describes those portions of the current effort which have direct bearing on the bridge performance prediction from the viewpoint of a reliability approach and on the integration of whole life costing with lifetime reliability. A particular emphasis is placed upon reliability-based bridge management. Examples of solutions for reliability-based bridge performance predictions and benefit/cost analyses are presented.

INTRODUCTION
In bridge engineering, the financial resources do not keep pace with the growing demand for bridge maintenance, rehabilitation, and replacement. For this reason, it is imperative that the best possible use of existing resources should be achieved. The current bridge management systems try to meet this goal. However, as indicated in Frangopol and Das (1999) and Frangopol (2000), currently available bridge management systems have important limitations. One of the most severe limitations is that bridge reliability is not directly incorporated in bridge management. Consequently, so far, the best possible use of financial resources has been only a subject of continuing concern to bridge engineers and a dream for bridge managers. This dream can only come true by using reliability-based bridge management (Frangopol and Das 1999, Thoft-Christensen 1999, Das 2000, Frangopol *et al.* 2000, Frangopol 2000).

The next generation of bridge management systems has to use powerful reliability, optimization, and life-cycle engineering tools to predict the performance of bridges and the costs associated with bridge interventions including both agency and user costs, to find the optimal lifetime strategy based on benefit/cost methodologies, to implement this strategy at both network and project levels, and to update the optimal strategy as more information becomes available. This paper describes those portions of the current effort which have direct bearing on the bridge performance prediction from the viewpoint of a reliability approach and on the integration of whole life costing with lifetime reliability. A particular emphasis is placed upon reliability-based bridge management. Examples of solutions for reliability-based bridge performance predictions and benefit/cost analyses are presented.

Bridge Management 4, Thomas Telford, London, 2000.

BRIDGE RELIABILITY STATES

Currently available bridge management systems, including Pontis (Thompson *et al.* 1998) and BRIDGIT (Hawk and Small 1998), use condition states to define condition of bridge elements at any given point in time. These condition states, largely based on visual inspections, indicate relative health of bridge elements but do not identify their specific reliability. Recently, Frangopol and Das (1999) and Thoft-Christensen (1999) proposed five bridge reliability states. These states are indicated in Fig. 1 along with the associated reliability indices. The service life of bridges is a progression of reliability states from excellent ($\beta \geq 9.0$) to unacceptable ($\beta < 4.6$). As indicated in Wallbank *et al.* (1998), Das (1999), and Frangopol *et al.* (1999), the justification for carrying out essential maintenance (such as major repairs) is that without it the element will be unsafe, and the justification for preventive maintenance (such as painting, silane treatment) is that if it is not done at the time it will cost more at a later stage to keep the element from becoming critical. The attributes for reliability and associated maintenance actions are also indicated in Fig. 1. It should be mentioned that preventive maintenance work should be considered as a package of actions (such as silane treatment, deck waterproofing, expansion joint replacement, extraction of contaminants). It is expected that the cost of these packages (called options in Fig. 1) should increase with the decrease in the reliability state of the bridge.

RELIABILITY STATE

5	4	3	2	1

RELIABILITY INDEX

$\beta \geq 9.0$	$9.0 > \beta \geq 8.0$	$8.0 > \beta \geq 6.0$	$6.0 > \beta \geq 4.6$	$4.6 > \beta$

ATTRIBUTE FOR RELIABILITY

EXCELLENT	VERY GOOD	GOOD	FAIR	UNACCEPTABLE

MAINTENANCE ACTION

PREVENTIVE 5	PREVENTIVE 4	PREVENTIVE 3	PREVENTIVE 2	ESSENTIAL 1
OPTION 5a OPTION 5b OPTION 5c • • •	OPTION 4a OPTION 4b OPTION 4c • • •	OPTION 3a OPTION 3b OPTION 3c • • •	OPTION 2a OPTION 2b OPTION 2c • • •	OPTION 1a OPTION 1b OPTION 1c • • •

Figure 1. Definition of Reliability States, Attributes, and Maintenance Actions

Using this approach, maintenance actions are selected in response to distinct changes in the reliability states. In this manner, bridge reliability is directly incorporated in bridge management and all limitations associated with current Markovian-based bridge management systems can be relaxed.

FIRST REHABILITATION TIME

In order to estimate the number of bridges of a particular type requiring first rehabilitation (i.e., first essential maintenance) each year in the future it is necessary to capture the propagation of uncertainties from the time of construction to the time of failure (i.e., time at which the reliability index β downcrosses the target level 4.6). This time is called first rehabilitation time or rehabilitation rate. As indicated in Fig. 2, there are two cases to be investigated: (a) first rehabilitation time assuming no preventive maintenance has been done, and (b) first rehabilitation time assuming preventive maintenance has been done. The random variables associated with no maintenance action scenario are: initial target reliability index B_o, time of damage initiation T_I, and reliability index deterioration rate A. Five additional random variables have to be introduced in order to characterize the preventive maintenance scenario: time of first application of preventive maintenance T_{PI}, time of reapplication of preventive maintenance T_p, duration of preventive maintenance effect on reliability T_{PD}, deterioration rate of reliability index during preventive maintenance effect Θ, and improvement in reliability index (if any) immediately after the application of preventive maintenance Γ. The two distinct events (i.e., reliability profile with and without preventive maintenance actions) shown in Fig. 2 are identified through the particular values $\beta_o, t_I, \alpha, t_{PI}, t_p, t_{PD}, \theta$, and γ of the random variables B_o, T_I, A, T_{PI}, T_p, T_{PD}, Θ, and Γ, respectively.

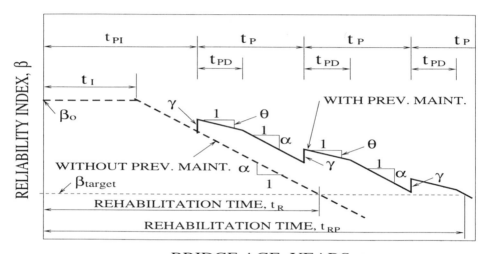

Figure 2. Time Variation of Reliability Index with and without Preventive Maintenance

The probability density functions (PDFs) of the initial reliability index (i.e., at time $t = 0$) for moment and shear of steel/concrete composite bridges are shown in Fig. 3. Both PDFs are assumed log-normal with mean values 7.5 and 8.5 and standard deviations 1.2 and 1.5 for moment and shear, respectively. The density distributions of all the other random variables for steel/concrete composite bridges are indicated in Kong and Frangopol (1999).

FIRST REHABILITATION TIME AFTER NO MAINTENANCE

Using Monte-Carlo simulation it is possible to generate the PDF of the reliability index of a group of steel/concrete composite bridges at any point in time. For example, Fig. 4 shows the time variation of the PDF of reliability index for moment assuming a high deterioration rate of reliability index $\alpha_{max} = 0.10$ years^{-1}. Fig. 5 compares the PDFs of the first rehabilitation time (i.e., rehabilitation rate) for steel/concrete composite bridges provided by experts in 1997 [i.e., triangular distribution (20, 35, 50) where 20, 35, and 50 represent the lowest, mode, and highest age (in years), respectively], in 1998 (i.e., the logistic distribution characterized by the parameters 35.9 years and 6.2 years), and the one obtained through complex reliability analysis computations using Monte-Carlo simulation and quadratic fitting. As indicated in Frangopol and Das (1999) it is interesting to note that: (a) the target reliability index was not specified in 1997 and 1998; (b) in the reliability analysis carried out in 1999 the target reliability index was specified as 4.6; (c) the modes of the three distributions in Fig. 5 are approximately the same.

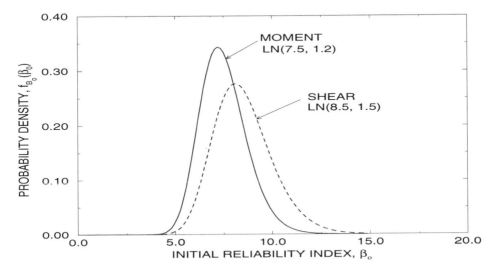

Figure 3. Probability Density Functions of the Initial Reliability Index for Moment and Shear in Steel/Concrete Composite Bridges

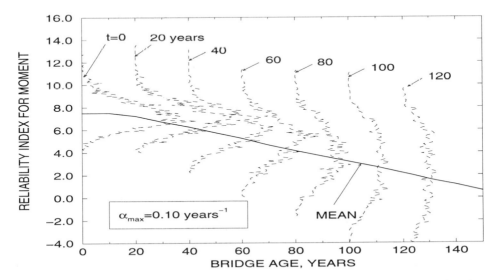

Figure 4. Time Variation of the Probability Density Function of the Reliability Index for Moment in Steel/Concrete Composite Bridges Assuming No Maintenance and High Reliability Deterioration Rate

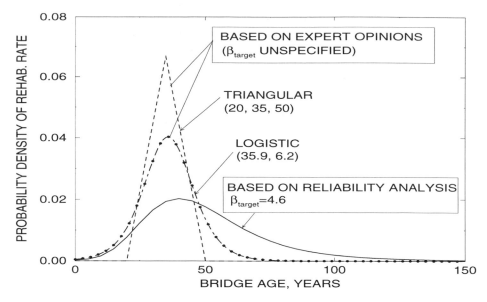

Figure 5. Probability Density of First Rehabilitation Time for Steel/Concrete Composite
Bridges Assuming No Maintenance

FIRST REHABILITATION TIME AFTER PREVENTIVE MAINTENANCE

The reliability-based procedure developed for finding the PDF of bridge rehabilitation rate
for steel/concrete composite bridges assuming no preventive maintenance was extended to
the case of preventive maintenance by adding the random variables associated with the
preventive maintenance actions. The values of the cumulative distribution function (CDF)
and PDF for the two cases (i.e., without and with preventive maintenance) of first
rehabilitation time of steel/concrete composite bridges are indicated in Table 1. The results
are based on simulation followed by quadratic fitting.

EFFECT OF PREVENTIVE MAINTENANCE

The beneficial effect of preventive maintenance is clearly demonstrated on a stock of 100
steel/concrete composite bridges by examining the results presented in Table 2. As indicated,
the numbers of bridges in reliability states 1 ($\beta < 4.6$), and 1 and 2 ($\beta < 6.0$) decrease
considerably when preventive maintenance has been done.

In order to describe in economic terms the benefit of using preventive maintenance, the
benefit-cost ratio $R = B/C_{PM}$ has to be computed. The cumulative unit benefit B (i.e., the
difference between the present value of expected unit maintenance cost assuming no
preventive maintenance and the present value of expected unit maintenance cost assuming
preventive maintenance) does not include the present value of the expected cumulative unit
cost of preventive maintenance C_{PM}. The benefit-cost ratio is shown in Fig. 6 for post-
tensioned and prestressed (i.e., pre-tensioned) concrete bridges using two values of the
discount rate r. The benefit of using preventive maintenance is decreasing with increasing the
discount rate, and is substantially higher for post-tensioned as compared to prestressed
concrete bridges. The ratio R is also influenced by the user cost (Frangopol 2000). The results
in Fig. 6 are based on data presented in Maunsell Ltd. and Transport Research Laboratory
(1998).

Table 1. Cumulative Distribution Function (CDF) and Probability Density Function (PDF) of First Rehabilitation Time of Steel/Concrete Bridges: (a) without Maintenance, and (b) with Preventive Maintenance. Results are Based on Simulation followed by Quadratic Fitting.

Bridge Age (1)	Without Maintenance		With Preventive Maintenance	
	CDF (2)	PDF (3)	CDF (4)	PDF (5)
5	0.0045	0.0006	0.0073	0.0006
10	0.0125	0.0016	0.0118	0.0009
20	0.0708	0.0077	0.0302	0.0022
30	0.2156	0.0166	0.0680	0.0044
40	0.4139	0.0203	0.1313	0.0070
50	0.5999	0.0177	0.2194	0.0094
60	0.7395	0.0127	0.3242	0.0108
70	0.8322	0.0082	0.4341	0.0110
80	0.8903	0.0051	0.5387	0.0102
90	0.9260	0.0031	0.6312	0.0089
100	0.9481	0.0020	0.7086	0.0074
110	0.9621	0.0012	0.7710	0.0059
120	0.9712	0.0008	0.8200	0.0046
130	0.9772	0.0005	0.8578	0.0035
140	0.9813	0.0004	0.8866	0.0027

Table 2. Expected Number of Bridges in Reliability States 1 ($\beta < 4.6$), and 1 and 2 ($\beta < 6.0$) from a Stock of 100 Steel/Concrete Composite Bridges with the Same Age: (a) without Maintenance, and (b) with Preventive Maintenance. Results are Based on Simulation.

Bridge Age (1)	Without Maintenance		With Preventive Maintenance	
	State 1 (2)	States 1 and 2 (3)	State 1 (4)	States 1 and 2 (5)
5	0	14	0	14
10	1	15	1	15
20	2	25	2	23
30	10	49	4	33
40	30	72	9	45
50	54	86	16	59
60	72	93	26	70
70	82	96	38	79
80	88	97	49	85
90	92	98	60	90
100	95	99	69	93
110	96	99	76	95
120	97	99	81	96
130	97	99	85	97
140	98	99	89	98

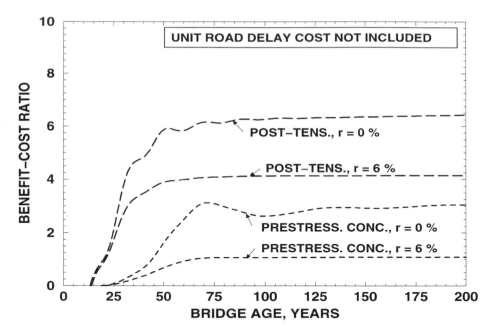

Figure 6. Benefit-Cost Ratio for Post-Tensioned and Prestressed Concrete Bridges

CONCLUSIONS

Bridge management based on lifetime reliability and whole life costing is considered to be the next generation of bridge management systems. These systems are based on reliability states instead of condition states. Maintenance actions are selected in response to distinct changes in reliability states. In this manner, bridge reliability is directly incorporated in bridge management. Application of benefit/cost analysis to reliability-based bridge management decision making guides the selection of the optimum program strategy in the face of uncertainties and fiscal constraints.

ACKNOWLEDGEMENTS

The partial financial support of the U.S. National Science Foundation through grants CMS-9506435 and CMS-9522166, of NATO through Grant CRG. 960085, and of the U.K. Highways Agency is gratefully acknowledged. Fruitful discussions with Dr. Parag Das of the U.K. Highways Agency are also gratefully acknowledged. The opinions and conclusions presented in this paper are those of the writers and do not necessarily reflect the views of the sponsoring organizations.

REFERENCES

Das, P.C. (1999). "Prioritization of bridge maintenance needs," *Case Studies in Optimal Design and Maintenance Planning of Civil Infrastructure Systems*, D.M. Frangopol, ed., ASCE, Reston, Virginia, 26-44.

Das, P.C. (2000). "Reliability based bridge management procedures," *Proceedings of the Fourth International Conference on Bridge Management*, Guilford, U.K., Thomas Telford.

Frangopol, D.M. (2000). "Life-cycle management of highway bridges: Past, present, and future," *Proceedings of the 16th Congress of IABSE*, Lucerne, Switzerland (under review).

Frangopol, D.M. and Das, P.C. (1999). "Management of bridge stocks based on future reliability and maintenance costs," *Proceedings of the International Conference on Current and Future Trends in Bridge Design, Construction, and Maintenance*, Singapore, October 4-5, Thomas Telford Publ., London.

Frangopol, D.M., Gharaibeh, E.S., Kong, J.S., and Miyake, M. (2000). "Optimal network-level bridge planning based on minimum expected cost*,"* *Fifth International Bridge Engineering Conference*, Transportation Research Board, Tampa, Florida.

Frangopol, D.M., Thoft-Christensen, P., Das, P.C., Wallbank, E.J., and Roberts, M. (1999). "Optimum maintenance strategies for highway bridges," *Proceedings of the International Conference on Current and Future Trends in Bridge Design, Construction, and Maintenance*, Singapore, October 4-5, Thomas Telford Publ., London.

Hawk, H. and Small, E.P. (1998). "The BRIDGIT bridge management system," *Structural Engineering International* , IABSE, **8**(4), 309-314.

Kong, J.S. and Frangopol, D.M. (1999). "Bridge life-cycle safety management," University of Colorado, Boulder (in progress).

Maunsell Ltd. and Transport Research Laboratory (1998). "Strategic Review of Bridge Maintenance Costs: Report on 1997/98 Review," *Draft Report*, The Highways Agency, London.

Thoft-Christensen, P. (1999). "Estimation of bridge reliability distributions," *Proceedings of the International Conference on Current and Future Trends in Bridge Design, Construction, and Maintenance,* Singapore, October 4-5, Thomas Telford Publ., London.

Thompson, P.D., Small, E.P., Johnson, M., and Marshall, A.R. (1998). "The Pontis bridge management system," *Structural Engineering International*, IABSE, **8**(4), 303-308.

Wallbank, E.J., Tailor, P., and Vassie, P.R. (1998). *Strategic Planning of Future Maintenance Needs*, Seminar, Management of Highway Structures, ICE, London, June.

Annualized life-cycle costs of maintenance options for New York City bridges

DR. R. B. TESTA, Professor and Chair, Department of Civil Engineering & Engineering Mechanics, Columbia University, New York, NY 10027, USA

DR. B. S. YANEV, Director, Bridge Inspection & Bridge Management, Division of Bridges, Department of Transportation,
Adjunct Professor, Columbia University, New York City

INTRODUCTION

New York City Department of Transportation was in charge of 770 bridges with 4706 spans in 1998. Their average age was 72 years and the total bridge deck area is approximately 1.4 ml m^2 (15 ml ft^2).. (Roughly 100 culverts were eliminated from the total count reported in previous years and earlier publications.) Typical recent annual expenditures range from \$450 to \$600 ml. for bridge rehabilitation, \$20 to 30 ml. for component rehabilitation, \$20 ml. for maintenance and \$25 ml. for repairs and hazard mitigation. The average bridge condition over the last 10 years has remained at a rating of approximately 4.5 on the New York State Department of Transportation scale (such that 7 signifies new and 1 – failure). The City bridge managers must determine whether these annual expenditures of over \$0.5 billion (\$366/m^2 or \$34/ft^2 of bridge deck area) could be more effectively distributed among the above (or other) bridge related operations in order to achieve better bridge performance and reduce costs. To this end the life cycle performance of bridge components has been continually studied. The model developed in Yanev (ref. 1) obtains the steady state equilibrium equation for the New Yok City bridges and demonstrates that reduction of the bridge deterioration rate "r" would be the most effective way of improving overall bridge condition.

The problem is thus reduced to influencing r in a cost-effective manner. It is commonly assumed that bridge deterioration rates (in an otherwise relatively stable environment) are directly related to bridge maintenance. There is no known relationship, however, between specific maintenance operations and rates of deterioration for bridges or their components. This is the main difficulty in estimating the cost and the type of the maintenance, capable of reducing the rate of bridge deterioration r and, consequently raising the overall bridge condition rating "R" of a large group of bridges, as is the one in New York City. Consequently, funding requests for bridge maintenance are lacking in their cost-benefit justification. As part of a project for updating the New York City Bridge Maintenance Manual (ref. 2), a procedure is proposed for obtaining the required relationships between bridge condition ratings, the respective maintenance and its cost.

DEFINITION OF PARAMETERS

Bridge inspection condition ratings according to New York State Department of Transportation Inspection reports are consistently available since 1982 for the bridges and all their components Conditions are graded as follows: 7 - new, 1 – failed and 3 – not functioning as designed

Bridge Management 4, Thomas Telford, London, 2000.

Limitations of such condition ratings have been discussed at length, for instance in Yanev (ref. 3), concluding that, as long as design and construction remain generally adequate, they can be useful for policy decisions pertaining to the entire stock of bridges.

In complience with Federal and State laws, all bridge and their components are rated by the above system at least once every two years. A weighted average fomula is applied to the lowest condition ratings Ri of the thirteen bridge components listed in Table 1 to obtain the overall bridge condition rating R as follows:

$$R = \sum_{i=1}^{13} k_{ei} R_i \qquad (1)$$

Table 1 contains the weights w_i used for the "Bridge condition" formula and their normalized values k_{ei}. Also listed are the minimum life L_{io} of the 13 components obtained from New York City inspection records and the maximum life L_{il} recommended by (ref. 2). Depending on the importance of the component, its failure rating R_{ic} is assumed equal to or greater than 1 as shown in Table 1. It is further assumed that L_{io} has been obtained at negligible maintenance level and that the relationship between the level of maintenance and the bridge component life is directly proportional. Condition ratings are assumed to deteriorate linearly with time. Actual deterioration histories, such as the ones reported by Yanev (ref. 3) show linear or bi-linear patterns at 0 or minimal maintenance. Deterioration at the highest recommended maintenance level at this time can only be assumed.

Influence of Maintenance Tasks on Component Ratings. (Ref. 2) identifies 15 routine maintenance tasks for the New York City bridges. They are listed in Table 2. The extent to which each maintenance task affects the rate of deterioration of a component is not easily quantified. Yet such a relationship is needed for the estimate of the maintenance benefits in terms of bridge life-cycle costs. For this study, the relative influence of each maintenance task j on the rating of each component i is estimated from experience and assigned a subjective value from 0 to 1 as shown in the "infuence matrix" of Table 2. The rows j refer to the fifteen maintenance tasks and the columns i correspond to the thirteen bridge components. The values are normalized so that the sum of each column is 1. A matrix of influence coefficients k_{ji} results. These influence coefficients k_{ji} of each maintenance task on the component ratings, and the earlier defined k_{ei} defining the influence of components i on the overall bridge rating R, one may interpret the sum of the products of the k_{ei} with the k_{ji} as the influence of each maintenance task on the bridge rating, defined as the influence coefficients k_{mj} and expressed as:

$$k_{mi} = \sum_{i=1}^{13} k_{ei} k_{ji} \; / \sum_{j=1}^{15} \sum_{i=1}^{13} k_{ei} k_{ji} \qquad (2)$$

maintenance level. For the fifteen tasks of Table 2, a "full maintenance" frequency is specified in ref. 2). The total cost of the full maintenance is shown in Table 3 and it can be entered as data the model. Table 3 also introduces the concept of the level of maintenance ($0 < M_j < 1$) and the calculation of the cost for that level. Thus the overall bridge maintenance level M can be defined as:

$$M = \sum_{j=1}^{15} k_{mj} M_j \qquad (3)$$

Table 1 : Bridge Components and Properties

	COMPONENT	Lio	Li1	Rio	Ric	weight	Kei	Kei * Rio
1	Bearings	20	120	7	1	6	0.083	0.5833
2	Backwalls	35	120	7	1	5	0.069	0.4861
3	Abutments	35	120	7	2	8	0.111	0.7778
4	Wingwalls	50	120	7	1	5	0.069	0.4861
5	Bridge seats	20	120	7	1	6	0.083	0.5833
6	Primary members	30/35	120	7	2	10	0.139	0.9722
7	Second. Members	35	120	7	1	5	0.069	0.4861
8	Curbs	15	60	7	1	1	0.014	0.0972
9	Sidewalks	15	60	7	1	2	0.028	0.1944
10	Deck	20/35	60	7	2	8	0.111	0.7778
11	Wearing surface	15/20/30/35	15/20/30/35	7	1	4	0.056	0.3889
12	Piers	30	120	7	2	8	0.111	0.7778
13	Joints	10	30	7	1	4	0.056	0.3889

Lio = expected life with no maintenance
Li1 = expected life with full maintenance
Rio = rating at start
Ric = rating for component failure
Weight = influence of component i on bridge rating
Kei = normalized influence of component i on bridge rating

Ro =	7.00

Table 2. Normalized Influence of Maintenance Tasks on Ratings

K_{ij}	Brgs	BkW	Abut	WgW	Seats	Prim	Sec	Curbs	SW	Deck	Wear	Piers	Joints
Debris rem	0.068	0.0649	0.0303	0.0278	0.08	0.0485	0.0495	0.1013	0.1404	0.0825	0.1071	0.0116	0.053
Sweeping	0.0194	0.013	0.0152	0	0.05	0.0485	0.0495	0.1266	0.1404	0.0928	0.119	0.0116	0.0662
Clean Drain	0.0874	0.1169	0.1364	0.2222	0.1	0.0971	0.099	0.1266	0.1754	0.1031	0.119	0.0581	0.0662
Clean abut/piers	0.0971	0.1299	0.1515	0.25	0.1	0.0777	0.0792	0	0	0.0515	0.0595	0.1163	0.0331
Clean grating	0.0971	0.0649	0.1061	0.0278	0.1	0.0971	0.099	0.0127	0.0175	0.0825	0.119	0.1163	0.0596
Clean exp jts	0.0971	0.1039	0.1515	0.1389	0.1	0.0971	0.0792	0.0633	0.0877	0.0928	0.1071	0.1047	0.0662
Wash deck etc	0.0485	0.039	0.0303	0	0.06	0.0388	0.0396	0.1266	0	0.1031	0.119	0.0465	0.0662
Paint	0.0971	0.0649	0	0	0.05	0.0971	0.099	0	0	0.0412	0	0.1163	0.0331
Spot paint	0.0971	0.0649	0	0	0.05	0.0971	0.099	0	0	0	0	0.1163	0
Sidewalk & curb	0	0	0	0	0	0	0	0.1266	0.1754	0.0103	0.0119	0	0.0331
Pavmt & curb seal	0.0971	0.1299	0.1515	0.1389	0.1	0.0971	0.099	0.1266	0.1754	0.1031	0.119	0.0581	0.3311
Elect device maint	0	0	0	0	0	0	0	0	0	0	0	0	0
Mech Comp	0.0971	0.0649	0.0758	0.0556	0.1	0.0971	0.099	0.1266	0	0.0515	0	0.1163	0.0662
Repl wear surf	0	0.013	0	0	0.01	0.0097	0.0099	0.0633	0.0877	0.1031	0.119	0.0116	0.0662
Wash underside	0.0971	0.1299	0.1515	0.1389	0.1	0.0971	0.099	0	0	0.0825	0	0.1163	0.0596

The total cost for the NYC system of each task performed at the full level of maintenance is also entered in Table 3.

LIFE-CYCLE MODEL
The model for projecting a bridge life can now be constructed as a function of the bridge component deterioration, subject to the prescribed maintenance level. The deterioration rate r_i of a component i can be expressed as:

$$r_i = - dR_i/dt = (r_{il} - r_{io}) (\sum_{j=1}^{15} k_{ji} M_j) + r_{io} \qquad (4)$$

where: $r_{io} = (7 - R_{ic})/L_{io}$ is the component deterioration rate at maintenance $M_j = 0$
$r_{il} = (7 - R_{ic})/L_{il}$ is the component deterioration rate at maintenance $M_j = 1$ and
R_{ic} is the prescribed rating at which the component is assumed to have failed.
The overall bridge deterioration rate r can be derived from the component ones as:

$$r = - dR/dt = \sum_{i=1}^{13} k_{ei} r_i \qquad (5)$$

Component repair. Another facet of bridge maintenance involves repair and/or replacement of individual components when they have failed or when they reach some critical level of deterioration before their failure value (R_{ic}). This is distinguished from overall reconstruction or rehabilitation of a bridge when all components are restored. The main parameters associated with this portion of the input to the model for optimal maintenance are:
dR_{ri} = increase in component rating as a result of repair.
R_{ri} = rating of component i used as a guide to need for repair.

Among the computations performed in the model, the time of each component repair is determined using the life-cycle calculation described above. The current version of the model specifies component repairs at the rating of $R_{ri} = 2$, but this can be selected at will. In addition, the components are grouped for repair at intervals of 5 years in the life of the bridge, with non-critical elements repaired at the next 5 year repair stage from the time they reach a rating R_{ri}, while a critical element is repaired at the 5 year repair stage preceding the time when it would reach a rating of R_{ri}. All of these options and values can be selected in the model to explore various strategies.

A specified maintenance level together with a regimen for component repair thus permits construction of the entire history of component and bridge rating. From that history, the expected life of the bridge, that is, the time at which a failure rating is reached, is determined. The result is the expected life L output for the specified overall maintenance level M, where:

$$L = (L_1 - L_o) M + L_o \qquad (6)$$

is this expected life that will be used in determining the various annualized costs over the useful life of bridges.

LIFE-CYCLE COSTS
Unit Costs. The costs of maintenance tasks are determined as part of the scope of (ref. 2) (Table . For the model presented herein all costs are reduced to a cost per square metre of bridge deck area.and they are expressed in present day dollars (U.S.D.) for the present year. It is

understood that in any succeeding year the annual cost will be higher in terms of that current year's dollars, rising each year with the inflation rate, but it will have the same value in today's dollars for a constant level of maintenance. These maintenance costs are designated by the following, per square foot of bridge in today's U.S.D.:

C_{m1j} = the annual cost of maintenance task i when carried out at level $M_j = 1$.
C_{mj} = $M_i \, C_{mlj}$ is the annual cost of maintenance task i at level M_j.

$C_m = \sum_{j=1}^{15} C_{mj}$ is the annual cost of all maintenance ta level M.

Other costs associated with deterioration of a bridge are expenditures at specific times in the life of the bridge rather than annual costs, but they are still estimated today, in today's dollars, so that at the time of the actual expenditure the cost in the then current dollars will be higher in accordance with the rate of inflation. The following costs, given per square foot of bridge, are identified and included in the model:

C_R = cost of total rehabilitation or replacement of the bridge at the end of its life,
   currently assumed = $4840/m^2$ ($450/ ft^2$).
C_C = total cost of corrective repair of components, currently assumed = $1614/m^2$
   ($150/ft^2$) although various portions of this cost occur at several times during the
   life of the bridge.
C_U = cost to users due to closures caused by bridge deterioration, closure for repairs,
   and other disruptions. These occur at various times during the life of a bridge but
   are expressed in today's costs. Similar user costs associated with bridge
   reconstruction are already included in C_R.
C_{NY} = non-explicit cost to New York City, identical in all respects to the user cost
   described above (C_U) in that it is associated with disruptions of service because of
   deterioration and repairs. It is listed separately so that its effect on the optimal
   strategy may be investigated independently even though both costs are handled in
   the same manner in the model.

Annualized unit costs.
The choice to use costs annualized over the structural intended life is made based on considerations discussed in a number of texts, such as (ref. 3, ref. 4). As argued by M. B Leeming in (ref. 5), discounting may still have some use in comparing maintenance or repai alternatives over a limited timespan.

RESULTS AND FUTURE DEVELOPMENT
An example of results is shown in Fig. 1. As expected, the model obtains a minimum life-cyc! cost at full maintenance. It is noteworthy that the most significant cost benefit is derived from th user costs which are not commonly included in life-cycle cost considerations during a projec selection. Also available from the model output is a comparison of annualized life-cycle cos! due to different maintenance task funding allocation, but adding up to the same reduced fundin level. Once refined, this feature is likely to have frequent application under the routine budg constraints. All benefits due to the recommended full maintenace are obtained by the develope model as a direct result of the introduced assumptions. While the latter are based on substanti engineering experience, field data would have to confirm them. The argument in favor intensified maintenance is generally recognized as valid. The difficulty in its implementatic

TABLE 3. Maintenance Activities, Annual Frequencies, Costs and Influences

j	Activity	Frequency (100%) fixed / moveable		k_{mj}	M_j	Annual Cost [S]	C_{mlj} [S/m^2]
1	Debris Removal	12(52*)	26	0.061	1.0	2,319,653	1.61
2	Sweeping	26	26	0.052	1.0	613,071	0.43
3	Clean Drainage	2	2	0.106	1.0	863,804	0.65
4	Clean Abut., Piers	1	12	0.081	1.0	2,776,013	1.94
5	Clean Grating	1	2	0.070	1.0	55,490	0.40
6	Clean Joints	3(26*)	3	0.091	1.0	3,262,730	2.26
7	Wash Deck, Etc.	1	1	0.051	1.0	1,455,198	1.01
8	Paint	0.083	0.083	0.042	1.0	36,041,997	17.68
9	Spot Paint	0.25	0.25	0.037	1.0	23,743,128	11.67
10	Sidewalk/curb Rep.	0.25	0.25	0.025	1.0	1,328,182	0.93
11	Pavmnt/curb Seal	0.5	0.5	0.122	1.0	2,334,466	1.61
12	Electric Maint.	12	12	0.000	1.0	1,107,143	0.75
13	Mech. Maint.	12	12	0.067	1.0	1,010,502	0.75
14	Wearing Surface	0.2	0.125	0.035	1.0	1,390,305	0.97
15	Wash Underside	1	1	0.159	1.0	13,189,518	9.24
				1.000		91,491,200	

* East River Bridges

k_{mj} - normalized influence of maintenance task j on overall bridge rating

M_j - level of performance for maintenance task j

C_{mj} - annual cost/ sq. meter of bridge for maintenance task j at level M_j

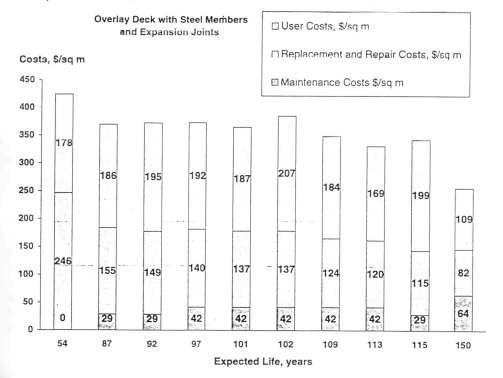

Fig. 1. Comparison of annualized costs for different maintenance strategies and levels of funding.

stems from the ever present funding limitations, compounded with the reconstruction needs which would be reduced only after an enhanced maintenance policy has been in effect over a significant period. Higher maintenance levels imply immediate budget increases and delayed benefits, rendered even less attractive by discounting.

Work is currently in progress at New York City Department of Transportation, Bureau of Bridges, and at Columbia University to broaden the scope of the study in improve its precision. Most significant are the following areas under consideration:

The "levels of maintenance M_{ij}". It is realized that the "level of maintenance" is typically discontinuous and frequently consists of either performing a certain operation or not, although cycles of varying length are also possible, for instance in painting and washing of bridges. The continuous "level of maintenance" will eventually be replaced by discrete maintenance levels known to be of practical significance. A sensitivity study can easily be performed on the existing model in order to obtain an optimal level of maintenance when "full" maintenance is not affordable.

Structural diversity. Not only concrete versus steel but also different structures within those larger categories will be distinguished. The unique needs of the 25 City moveable bridges and the four East River bridges, Brooklyn, Williamsburg, Manhattan and Queensborough will be taken into account.

Types of maintenance. The effectiveness of the 15 maintenance operations is subject to review and optimization. Other maintenance operations may be considered as well. Extending the results reported by Yanev in (ref. 3), bridge components with a significant contribution to the overall bridge condition and a consistent history of early failure, such as joints and scuppers are identified as targets for more intensive maintenance or, depending on the cost analysis, early replacement. It is noted that some of these components, paint, scuppers and wearing surface in particular, do not appear explicitly in the bridge condition formula (Eq. 1 herein). Implicitly they are represented by the influence coefficients k_{ij}, which are established in a highly subjective manner. The next step is to better quantify these factors. Particularly significant are the current developments in bridge painting strategies.

The proposed method is a first step in establishing a more rigorous relationship between bridge maintenance expenditures and bridge conditions. It also allows for continuing refinement and optimization. The optimal maintenance strategy for a system of bridges will not necessarily mean the same level of maintenance for all bridges in the system. In order to develop a strategy that attempts to optimize the cost of routine maintenance, as well as the periodic costs of component repair and eventual replacement or rehabilitation, each bridge or class of bridges must be assessed in light of its current condition. Moreover, the assessment should include some allowance for the much less tangible costs of disruptions during repairs or reconstruction. The result for each group of bridges can then be combined into an optimal strategy for the whole system.

ACKNOWLEDGEMENT
The work presented herein is part of the continuing research at the Bridge Management & Research/ Development Unit, New York City Department of Transportation. The work on the use of Table 1 was conducted by Mr. Herbert Pechek during his internship with the Uni

Elizabeth Watson obtained the results from the developed software. The views stated in this article are those of the authors and do not represent the position of any organization or agency.

REFERENCES:

1. YANEV, B. S. The Management of Bridges in New York City. *Engineering Structures*, Vol. 20, No. 11, Nov. 1998, pp.1020 – 1026.

2. VAICAITIS, R. et al. Preventive Maintenance Management System for New York City Bridges.Technical Report No. 98-1, May 28, 1999. The Center of Infrastructure Studies, Department of Civil Engineering and Engineering Mechanics, Columbia University, New York.

3. YANEV, B. S., Bridge Management for New York City, *Structural Engineering International*, SEI Vol. 8, #3,August 1998, pp. 211 - 215.

4. YANEV, B. S. 1994. User costs in a bridge management system. Characteristics of *Bridge Management Systems*:pp. 130 - 138, Transportation Research Circular #423, April 1994, ISSN 0097-8515.

5. LEEMING, M. B. , The Application of Life-Cycle Costing to Bridges. *Bridge Management 2* Proceedings, pp. 574-583, Thomas Telford, London, 1993.

Structural evaluation methods for concrete bridges

C. ABDUNUR, Ph.D
Laboratoire Central des Ponts et Chaussées, Paris, France.

INTRODUCTION

Whether damaged or apparently sound, bridges are or should be periodically inspected to allow early assessment and, eventually, optimum strengthening.

For a given type of construction, structural investigation methods are usually chosen for their aptitude to evaluate the required parameters and for their suitability with regard to the mechanical system, constituent materials and the observed or assumed type of damage or deficiency.

The development of these methods, towards better performance and new concepts, did not always follow the chronological order of the structural designer's usual checking procedure. Stress, for example, the first parameter that conventional calculations usually yield, was one of the last made accessible to direct measurement on real structures.

A REMINDER OF CONVENTIONAL METHODS

For monitoring the general or local deformation of a bridge and the movement of its bearing points, the adopted methods are now well known. They are based on topographic and geometric variation measurements.

Under dead load, topographic monitoring of the structure's geometry helps to evaluate its general condition. It may disclose support settlement or movement, a modified cyclic response to temperature reflecting gradual damage, irreversible deformation of a concrete deck owing to under-estimated creep and other random effects.

As to dead load forces and stresses, they remained inaccessible by this procedure.

Under test loading, conventional measurements are mainly geometric variations. Deflections, rotations, and strains are measured along the bridge spans and compared with their theoretical values. Traditionally, deflections under static test-loads are among the first data obtained for the acceptance of a bridge. They set a reference for future assessment. Rotation measurements along the span improve these deck deformation data under bending and extend them to the supports and other adjacent members. Strain profiles, measured in certain sections, estimate the corresponding load-induced *variation* of stress and of local curvature.

However, for most bridges with transverse cracks and other singularities, strains are too restricted and deflections too global to permit early detection of such defects and appropriate analysis of the actual response to loading.

MORE RECENT METHODS

It took some time to go beyond the above-mentioned limits of geometric monitoring and set or adopt additional concepts for the assessment of bridges. Specific methods thus emerged:
- to directly measure and monitor the actually acting forces and existing stresses, using hydrostatic pressure as an additional metrological element,
- to better investigate structural discontinuities along a bridge, using a more balanced metrological approach as provided by inclinometry.

To attain these objectives, several developments were necessary:
- specific force-displacement devices for measuring support reactions and tendon forces,
- surface inserted or completely embedded thin flat jacks for direct concrete stress monitoring,
- coupled inclinometers with differential readings for monitoring individual cracks,
- the moment-curvature method, facilitated by recent mobile cuvaturemeters, for investigating
   multiple discontinuities and determining the actual structural system.

MEASUREMENT OF FORCES ACTING ON THE BRIDGE

Support reactions provide the external equilibrium of the bridge. Embedded post-tensioned tendons maintain the concrete stress profile within the allowable limits. The force, acting through each of these structural elements, has its own specific metrology.

Support reactions

In statically indeterminate bridges, support reactions are periodically measured to evaluate the developing forces during construction and, mainly, their *redistribution* throughout service life.

The measurement procedure consists of inserting a set of jacks around the existing bearing and using them to lift the deck. The force-displacement curve is plotted for the whole lifting-lowering cycle. The average curve thus eliminates frictional effects, f, (figure 1-a). The first part of the graph marks the release of the original bearing. The second part, usually a straight line, represents the deck bending deflection; the slope gives the flexural stiffness. The reaction is deduced by extrapolation to zero displacement. However, this value closely follows the cyclic variations of thermal gradients. In fact, the present metrology was the first to reveal the intensity of these effects on continuous bridges, sometimes attaining the equivalent of full live load configurations. Consecutive parallel measurements of the reaction and thermal gradient are hence obligatory for at least 24 hours to establish a relationship, as shown in figure 1-b, and finally obtain the corrected reaction value at zero thermal gradient.

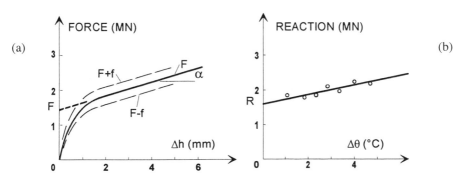

Figure 1 - (a) force-displacement curve giving a preliminary value of the support reaction.
(b) Final value after correction for temperature gradient.

As the *total* reaction can only be measured, high precision instruments are imperative to reasonably detect differential values, of greater monitoring interest. A 1% error already limits reaction measurements to the abutments, where loading is much reduced and variations more detectable. Figure 2 shows two versions of the equipment in use at present: the flat piston jack and the double flat jack. To put these devices in position, a 15-cm minimum clearance should be available between the deck and the bearing shelf. The bearing strength of the involved concrete surfaces must also be checked. Now it is also possible to install bearing systems directly equipped with permanent force-measuring devices.

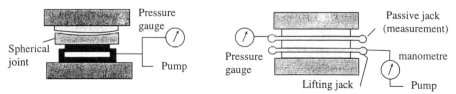

Figure 2 - Flat piston jack (left) and double flat jack (right) for measuring support reactions

In continuous spans, temperature-corrected reactions respond to dead load developments, differential settlement, stress profile modification in the deck (e.g. post-tensioning or differential creep) and specific or random consequences of various defects. Support reaction metrology can hence be used to detect abnormal dead load distribution, explain certain observed defects and, more basically, verify or modify structural design assumptions.

Forces in embedded tendons
The actual prestressing force, determined through the measured concrete stresses in a section, can also be evaluated by direct action on the tendons, using the Crossbow method. This is based on the simple fact that the effort necessary to deflect a tight rope is proportional to its axial tensile stress. Figure 3 outlines the procedure, after carefully clearing the adjacent concrete, duct and grout. The resulting disturbed tendon length, 2(l+x), is 60 cm approximately. A controlled perpendicular force P, coupled with a displacement sensor, then successively deflects prestressing wires through a distance f, limited to 4 mm. Theoretically, the tensile prestressing force F in each wire may be deduced by the formula:

$$P = 2 (F + k) (f/l) + K (f/l)^3 , \text{ where k and K are given constants.}$$

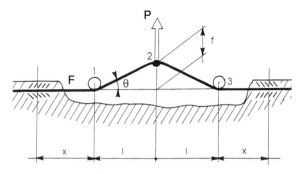

Figure 3 - Principle of measuring the force in a prestressing wire.

In practice, the parasitic effects of friction, flexural stiffness, over stretching and random bond failure necessitate prior calibration tests on simulation models in the laboratory, using the same type of wires and the exact disturbed length 2(l+x). A family of P = g (f) reference curves are traced for different F-values. The curves established in-situ could then be interpreted by direct comparison, leading to the actual prestressing force F in each wire. These data should undergo statistical treatment before they become reliable experimental results.

STRESS MONITORING

The evolving state of stress is considered as an essential parameter of structural safety. While the existing stress profile can now be directly and accurately measured at a given time by the release method, a further and equally important step would be to monitor its cyclic and irreversible variations.

At a given point in a concrete structure, stress monitoring may be defined as the time-dependent hydrostatic pressure of an incorporated thin flat jack, constantly ensuring the same local normal strain variation that would occur if the medium were free of any inclusion.

However, both the monitoring equipment and the initial stress data may differ considerably, depending on whether the concrete structure is already existing or not yet cast.

Existing structures

If concrete is already cast, the initial instantaneous stress is first directly measured by the release method shown in the upper shaded area of figure 4. This consists of cutting a tiny slot normal to the stress direction, then inserting and pressurising a special semicircular thin flat jack to cancel the surface displacements across the slot. The cancelling pressure p_c is assumed to be the initial average stress over the depth of the slot. The initial stress profile is traced after repeating the operation at closely successive depths and treating the data numerically.

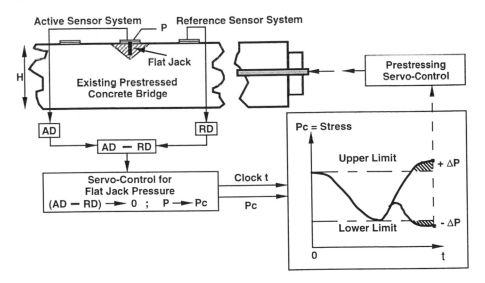

Figure 4 - Processes of stress monitoring and eventual re-adjustment
for an already existing concrete bridge

Once the initial stress is determined, the sensors over the flat jack now form the « active system », measuring the time-dependent response of the surface displacement field in the artificially reconstituted continuum. Along the longitudinal axis of the active system, a second group of identical sensors, forming the « reference system », is placed slightly beyond the range of the jack to monitor all displacement field variations in a close but undisturbed section. A specific regulator then automatically adjusts the jack pressure so that the displacement *variations* measured in the active system always equal those detected in the reference system. The time-dependent pressure p_c, thus obtained, gives the normal stress evolution in the equipped section. Monitoring can be remote-controlled and last for the lifetime of the bridge.

Structures under construction

Before concrete casting, a different monitoring equipment is posed to be completely embedded. The flat jack is circular and both active and reference systems are adapted to internal measurements (figure 5). As soon as the concrete hardens, all sensors are set to zero. Stress monitoring then starts and proceeds in the same way as in existing structures. Though restricted to future structures, this configuration has the additional advantage of tracing the stress history of early-age concrete, where the actual mechanical effects of the physico-chemical formative mechanisms are needed to evaluate their impact on durability.

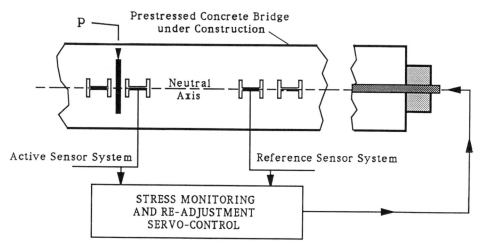

Figure 5 - Embedded stress monitoring system posed before casting a concrete bridge

Stress re-adjustment

In certain cases, such as post-tensioned concrete bridges, direct stress monitoring may lead to corrective measures. If the flat jack pressure p_c clearly indicates a lasting stress outside the allowable limits, as shown in the two lower shaded areas of figure 4, then a selective action on external tendons may restore it to its optimum level.

INVESTIGATION OF TRANSVERSE DISCONTINUITIES

Inclinometry, through rotation or curvature measurements, proved particularly suitable for detecting and monitoring transverse discontinuities in a bridge, such as flexure cracks or opening segment joints with inaccessible residual sections. Specific instruments are now accurate, robust and mobile, supplying principal angular deformations and sieving out local disturbances.

Preliminary approach

Figure 6 illustrates the behaviour of a detected rotation discontinuity $\Delta\theta$, monitored by differential inclinometer readings on either side of an existing flexure crack, then plotted versus increasing bending moment M. The linear part of the curve $M = f(\Delta\theta)$ represents the opening of the crack without any further growth i.e. within its initial tip height limit z_0 before loading. The non-linear part marks the crack growth $z_{0+}\Delta z$, under a greater moment.

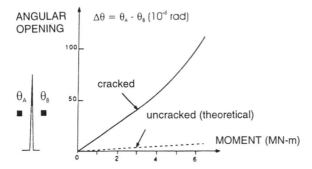

Figure 6 - Influence of bending moment on the angular opening of a bridge girder crack, measured by differential readings of two inclinometers A and B.

The moment-curvature method

The development of mobile curvaturemeters led to a structural evaluation method, based on the measured load-induced curvature distribution. The method detects transverse discontinuities and predicts the actual flexural response to applied loading.

The bridge is equipped, throughout its spans, with inclinometry instruments supplying curvature variation data. Under convenient test load configurations, the measured curvature variation diagrams θ' are plotted for the whole length of the bridge.

If these diagrams are regular and reasonably follow the theoretical ones, then the flexural adequacy of the structure is probably maintained.

If, on the contrary, a sharp curvature redistribution appears at certain points, then transverse cracks or other discontinuities should be suspected. In this case, the structural system may be represented as illustrated in figure 7: The cracked sections and their disturbed vicinities are assimilated to a series of elastic or plastic hinges H, alternating with sound beam segments B and jointly setting up a new system in equilibrium.

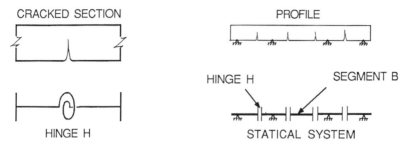

Figure 7 - Modelling assumptions of cracked sections of a bridge.

The main difficulty is the realistic determination of the relative residual flexural stiffness $[EI]_H$ of the hinge where several variables and assumptions are often involved. The proposed evaluation is hence experimental and based on the simple relationship between the applied moment M (x), the resulting measured curvature θ' (x) and the flexural rigidity EI (x). It proceeds as follows and as shown in figure 8.

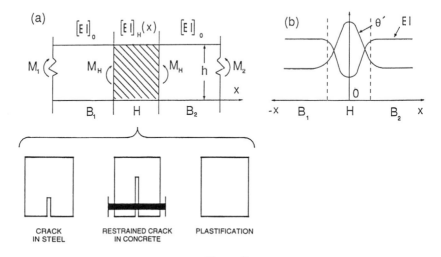

Figure 8
(a) Moments M and stiffnesses EI in cracked sections H and sound segments B.
(b) Redistribution of curvature θ' and stiffness EI.

Under a test moment, as already stated, the resulting curvature diagram θ' (x) is known throughout the spans and in particular:
θ'_H (x), over a cracked section or hinge H and its short influence zones,
θ'_B (x), in the parts close to the crack of both adjacent sound beam segments B_1 and B_2.

Moreover, all sound segments B are assumed to conserve their initial given stiffnesses $(EI)_0$.

For each module $B_1/ H / B_2$, the beam equation $M = EI . \theta'$ is applied at successive sections, more closely spaced in the hinge H zone where the moment M_H remains almost constant but θ' varies considerably and the local stiffness EI inversely to it (figure 8-b). The moment M_H is applied by the sound beam segments B_1 and B_2. Hence, at any section x within the H length,

$$[EI]_H (x) . \theta'_H (x) = M_H = [EI]_0 . \theta'_B$$
or
$$[EI]_H (x) = [EI]_0 . \theta'_B / \theta'_H (x)$$

$[EI]_H (x)$, thus determined, is plotted versus x, as shown in figure 8-b, giving the required equivalent residual rigidity of the damaged section or hinge. This quantity varies with x for an elastic hinge and with both position x and moment M_H for a plastic hinge. The hinge area H, in figure 8-a, can have many possible configurations, among which the three illustrated, without modifying the above reasoning.

Most often, the *relative* residual rigidity $(EI)_H / (EI)_0$ is preferred for estimating the fractional remaining capacity of the bridge. Its *inverse*, $(EI)_0 / (EI)_H$, is usually chosen for graphical representation, as in the example shown in figure 9.

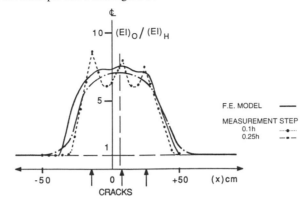

Figure 9 - Theoretical and experimental redistribution curves of the inverse relative flexural stiffness over a beam segment with three cracks.

The residual rigidities of the damaged sections H and those of the sound segments B now define the new actual structural system of the bridge. This enables:
-the prediction of the real flexural response of the damaged structure to any given loading,
-the evaluation of the residual load-bearing capacity and, eventually, the optimum needs for strengthening.

The whole procedure can be repeated at a later stage to verify the effectiveness of eventual repairs or simply to monitor a time-dependant mechanical change.

Furthermore, it may be imagined that, in the not too distant future, inclinometry could fully accompany deflectometry in the acceptance procedures for newly constructed or strengthened bridges.

CONCLUSION
Investigation methods, presented or not in this paper, must be selected with experience-based judgement. Many of them have to be used on distressed bridges because defects have been detected too late. Their cost is very high compared to that of a "normal" assessment policy. Hence, it is never too early for a first detailed inspection, for other periodic ones and for appropriate action as soon as defects appear.

REFERENCES
1. Chabert A., Ambrosino R., "Pesées des réactions d'appui", *Association Française des Ponts et Chapentes*, National Conferece, theme n° 3, pp 31-46, Paris, 1983.
2. Abdunur C., "Stress redistribution and structural reserves in prestressed concrete bridges", 3rd International Bridge Management Conference , Guidford, UK, 1996.
3. Abdunur C., "Monitoring the influence of transverse cracks in a bridge", *Intelligent Civil Engineering Materials and Structures*, American Society of Civil Engineers, 1997.

Bridge management in Europe (BRIME): structural assessment

Rolf Kaschner and **Dr Peter Haardt**, Bundessanstalt für Strassenwessen, Germany

Dr Christian Cremona, Laboratoire Central des Ponts et Chaussées, France

Dr David Cullington and **Dr Albert F Daly**, Transport Research Laboratory, UK

INTRODUCTION

The objectives of the BRIME project are to develop a management system framework applicable to bridges on the European road network and to examine the inputs required for such a system. An overview of the project is described by Woodward *et al* (2000). It consists of seven workpackages divided between partners in six participating countries. The work described in this paper belongs to Workpackage 2: Structural assessment. It is led by BASt (Bundessanstalt für Strassenwessen, Germany) and LCPC (Laboratoire Central des Ponts et Chaussées, France) with support principally from TRL (Transport Research Laboratory, UK). This paper, and the workpackage, deals with the structural assessment input to the management system, ie, the calculation of the safe load carrying capacity of bridges.

The European road network is called on to carry steadily increasing heavy goods traffic, and, from time to time, increases in legal vehicle or axle loads. It contains many bridges built before modern design standards were established. Furthermore, as bridges grow older, deterioration caused by heavy traffic and an aggressive environment becomes increasingly significant resulting in a higher frequency of repairs and a reduction in load carrying capacity.

It is important that assessment standards are safe but not over-conservative, because this will cause network operating costs to rise unnecessarily as bridges are strengthened or replaced and traffic restrictions are imposed. The BRIME project provides an opportunity for the partners to review standards and procedures in the participating countries and pursue the development of efficient assessment methods suitable for adoption in Europe.

The first stage of the work has been to collate information on bridge assessment from the participating countries. In the later stages, the aim is to provide guidelines that reflect the current best practices, a methodology that will allow for future development and sufficient flexibility to accommodate variations in national priorities as the process of harmonisation is pursued.

SCOPE OF PROJECT

The following tasks fall within the scope of the project:

- a review of current procedures and standards used for bridge assessment in Europe;

- the development of models for taking into account the bridge specific traffic conditions and material properties;

- the use of reliability methods based on a probabilistic approach for bridge assessment including the use of measurements for updating the reliability of structural elements;

- the provision of recommendations for methods and procedures that can be adopted for the assessment module of the management framework highlighting where further development will be beneficial.

All partners contributed to the provision of information on current national procedures, the review being carried out by BASt, LCPC and TRL. LCPC is principally responsible for the work on reliability methods including provisions for material strengths and BASt for the development of traffic models. TRL is responsible for introducing and interpreting current UK procedures for assessment, which are the most comprehensive in the BRIME partner countries.

ASSESSMENT PROCEDURES AND STANDARDS

The review confirmed that there are significant differences in procedures and methods used for bridge assessment by the partner countries (Kaschner *et al*, 1999b). This includes the reasons for initiating a bridge assessment, which can be summarised as:

1) when there is a need to carry an exceptional heavy load

2) where the bridge has been subjected to change such as deterioration, mechanical damage, repair or change of use

3) where a bridge is of an older type built to outmoded design standards or loading and has not been assessed to current standards.

In most of the partner countries, assessment is only carried out on specific structures that have to carry an exceptional load or have been affected directly and significantly by change (reasons 1 or 2). In the UK, reason 3 has predominated since 1987 when a comprehensive programme of bridge assessment started.

Bridge assessment in the partner countries generally relies on orthodox structural calculations in which the load effects are determined by structural analysis and the corresponding resistances are determined by code-type calculations. Reliability calculations are beginning to be introduced in which a target reliability index is the governing factor.

Currently, the rules used in bridge assessment are provided mainly by design standards with additional standards relating to testing methods including load testing. The design standards used are normally the current standards, although in some cases, previous versions of the codes are referred to in order to deal with non-conforming details. For assessment, only current design loading specifications can be used, although these can be modified specifically for assessment and can include a reduced load level based on restricted traffic conditions. Additional requirements can be given regarding exceptional traffic loading.

Design standards are mainly based on two alternative approaches:

- allowable stress design as prescribed in the German codes

- partial safety factor design as prescribed in the French, UK and Eurocode documents.

In the UK, assessment standards have been developed by modifying the design standards. The modifications provide more realistic formulae for member resistance, allowances for non-conforming details and imperfections, and methods for incorporating insitu material strengths in calculations. Advice on the management of sub-standard bridges is also given (Highways Agency 1998). This provides for five levels of assessment of increasing sophistication that may be applied when a simple assessment (Level 1) indicates that the bridge is sub-standard (*provisionally sub-standard* in the parlance of the document). Advice is given on monitoring sub-standard bridges so that they can remain in service until they can be strengthened.

In the other partner countries, there is minimal official documentation for assessment although Germany has an assessment standard for bridges in the former states of East Germany. Norway has provision for assessment loading and Slovenia has been developing a reliability method for assessment. Spain uses the design documents for assessment, as does France, but in the latter case there is flexibility to reduce partial factors or improve structural or resistance models with the help of laboratory investigations or site measurements.

STRUCTURAL PRINCIPLES OF ASSESSMENT

Allowable stress approach

In the allowable stress approach, the principle is to verify that the maximum load effect, S, calculated in any section of any part of a structure under the worst case loading, remains lower than a so-called allowable resistance value, $R_{allowable}$. This value is derived from the failure load effect, R_f, of the material divided by a safety factor, K, set conventionally. The aim of the structural assessment is to verify that:

$$S \leq R_{allowable} = \frac{R_f}{K} \tag{1}$$

The allowable stress design approach formed the basis of all engineering throughout the first part of this century. In its application, two questions were raised (Calgaro 1991):

♦ Can the allowable stress criteria be replaced by other criteria (limit states)?

♦ Can the introduction of structural safety concepts be carried out in a rational way?

Probabilistic approach

The concept of probabilistic structural safety started at the 3rd IABSE congress, Liege in 1948, although it was in the sixties that many developments were made. Cornell (1967) introduced the reliability index from which Lind (1973) developed the idea of safety coefficients. This initiated the emergence of the semi-probabilistic approach to structural safety.

In a probabilistic approach, the load effect S in a structural element, and the variable characteristic of the strength R of this element, are randomly described because their values are not perfectly known. If the verification of the criterion related to the limit state results in

$$R < S \tag{2}$$

then the limit state is exceeded. The probability P_f that the event $R < S$ characterises the reliability level of the component with regard to the limit state being considered:

$$P_f = \text{Prob}(R < S) \tag{3}$$

Details regarding the methods and techniques of structural safety are given by Melchers (1999).

Semi-probabilistic approach

The semi-probabilistic approach used in many design codes replaces this probability calculation by the verification of a criterion involving characteristic design values of R and S, denoted R_d and S_d, and partial safety factors γ_R and γ_S. This may be represented in the following form:

$$\gamma_S S_d \leq \frac{R_d}{\gamma_R} \tag{4}$$

The partial safety approach is described as semi-probabilistic because statistics and probability are applied to determine the design values, the formulation of assessment criteria, and the values of load and resistance factors.

The semi-probabilistic approach allows simple calculations to be made. The partial safety factors must cover uncertainties induced by parameters otherwise omitted from the process. They are also calibrated to provide designs that are not too different from designs made by former design methods. Their essential purpose is to cover the variabilities in structural behaviour and loading.

Limits in using design methods for structural assessment

It is important to note that the rules set down in design codes constitute a set of prescribed rules that are only valid in a certain context. For assessment, situations often exist which render design codes inapplicable either because of existing structural condition or because of the presence of non-conforming details. This is particularly important for older bridges and current design codes have to be interpreted carefully before being used. It is clear that the establishment of principles and procedures to be used for the assessment of existing bridges is needed because some aspects of assessment are based on an approach that is substantially different from new design, and requires knowledge beyond the scope of design codes. In addition, bridge assessment should be carried out in stages of increasing sophistication, aiming at greater precision at each higher level. In order to save structures from unnecessary rehabilitation or replacement (and therefore to reduce owners' expenditure), the engineer must use all the techniques, all the information available in an effective way. Simple analysis can be cost-effective if it demonstrates that the bridge is satisfactory. If it does not, more advanced methods should be used. Reliability theory based on a probabilistic description of resistance and load variables is one such approach.

The modelling of uncertainty is required for a rational approach to the evaluation of structural safety. This has many advantages over more traditional deterministic techniques for the following reasons:

♦ the evolution of loads with time is not considered in deterministic methods;
♦ the properties of materials can change with time, eg, through corrosion, fatigue, etc;
♦ the combination of multi-component load effects is not properly considered (such as the combination of normal and moment effects);
♦ real elements are often different from specimens on which design performance is based,
♦ studies on the sensitivity to errors in modelling the behaviour of structures are generally omitted;
♦ poor workmanship which is statistically inevitable;

♦   construction requirements discovered when the works are being carried out may lead to alternative solutions which can alter the behaviour of the structure.

Probabilistic methods constitute an alternative to allowable stress design or partial safety factor approaches. They are based on:

♦   the identification of all variables influencing the expression of the limit state criterion;

♦   a study of the statistical variability of each of these variables, often considered to be stochastically independent;

♦   a derivation of probability functions for each variable;

♦   the calculation of the probability that the limit state criterion is not satisfied;

♦   a comparison of the probability obtained to a previously accepted limit probability.

Although extremely useful, probabilistic reliability theory is sometimes limited.   Some parameters are difficult to measure and probability calculations quickly become impractical. These considerations are critical in determining what can be expected from probabilistic theory. Probabilities suffer from the fact that they are only estimates of frequencies (sometimes not observable) based upon an evolving set of partial data. Conventional distributions are often used for convenience and, of course, the outcome of a probabilistic approach depends strongly on the assumptions that are made about the uncertainties associated with the variables. If these assumptions are not founded on adequate and appropriate data, estimates of safety will be misleading. Indeed, probabilistic methods are often misused when the variables are not carefully modelled. It is therefore essential that the quality of data and validity of assumptions are borne in mind when using a probabilistic approach to make decisions about the apparent safety of a structure.

ASSESSMENT STEPS WHEN USING A PROBABILISTIC APPROACH

The essential parameters which characterise structural resistance or applied loads cannot be defined solely in terms of characteristic values reduced by partial safety factors: they must be considered as random variables. Choice of stochastic models for resistance variables such as yield strength and modulus of elasticity can be based on information from a number of sources:

♦   *Experimental measurements:* Based on such data statistical methods can be used to fit probability density functions (see below). The main problem with fitting probability density functions on the basis of experimental results is that usually most of the data are obtained in the central part of the density function, whereas the most interesting parts from a reliability point of view are the tails. For a resistance variable the lower tail is of interest while for a load variable it is the upper tail.

♦   *Physical reasoning:* In some cases it is possible on the basis of the physical origin of a quantity modelled as a stochastic variable to identify which stochastic model in theory should be used. Three examples of this are the normal, the lognormal and the Weibull distributions. When a stochastic model can be based on physical reasoning the above mentioned problem of tail sensitivity is avoided.

♦   *Subjective reasoning:* In many cases there are not sufficient data to determine a reasonable distribution function for a stochastic variable and it is not possible on the basis of physical reasoning to identify the underlying distribution function. In such situations subjective reasoning may be the only way to select a distribution function. In the future it is anticipated that code based rules will be established for selecting appropriate distribution types for the most frequently occurring stochastic variables.

Strength modelling

The uncertainties associated with strength properties (and some stiffness properties) will be considered in this section. To adequately describe the resistance properties of structural elements, information on the following is required (Melchers 1999):

♦ statistical properties for material strength and stiffness;
♦ statistical properties for dimensions;
♦ rules for the combination of various properties (as in reinforced concrete members);
♦ influence of time (eg, size changes, strength changes, deterioration mechanisms such as fatigue, corrosion, erosion, weathering, marine growth effects);
♦ effect of "proof loading", ie, the increase in confidence resulting from prior successful loading;
♦ influence of fabrication methods on element and structural strength and stiffness (and perhaps other properties);
♦ influence of quality control measures such as construction inspection and in-service inspection;
♦ correlation effects between different properties and between different locations of members and structure.

Except for the first three of these, relatively little information is available in statistical terms. A useful summary of statistical properties for reinforced and prestressed concrete members, metal members and components, masonry and heavy timber structures is given by Ellingwood *et al* (1980).

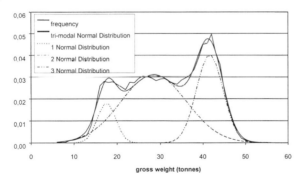

Figure 1 Tri-modal normal distribution of vehicle gross weight.

Traffic modelling

At the simplest level of assessment, load models from design codes or assessment codes, eg, BD 21/97 (Highways Agency 1997) can be used. Unfortunately, simple load models do not satisfy the requirements of higher level methods of assessment. Artificial traffic, which represents actual conditions with sufficient accuracy as a basis for simulating load effects, can be described using the stochastic simulation or Monte Carlo method, provided sufficient traffic and structural data are available. In principle, structural and traffic data, as well as a description of the composition of traffic including dynamic parameters, have to be provided.

For example, the frequencies of gross weight can be approached by bi- or tri-modal normal distributions as shown in figure 1.

The calculation method and the results of extensive investigations for different traffic and structural data are described in detail by Kaschner *et al* (1999a). As an example results are given for relevant axle load measurements in the German highway network. In this case simulation of load effects was carried out with the help of the computer program REB (Geißler 1996) on a single-span and a two-span beam and slab bridge. The distribution of daily extremes result from 900 random tests by an adaptation of an asymptotic distribution of extremes type I (Gumbel), given by:

$$F_x(x)_d \quad = \quad \exp\left[-\exp\left\{-a\left(x-u\right)\right\}\right] \tag{5}$$

The application of this equation results in a forecast of annual extremes. The 98% fractile of load effects corresponds to an average return period of 50 years. This period is used in many national codes for the characteristic load value. Figure 2 shows a comparison of simulated load effects for different spans with load effects calculated by taking the EC1 load model into account. In this figure, Mf1 is the bending moment at midspan (single-span beam), Mf2 is the maximum field moment (two-span beam) and Ms is the bending moment at the inner support. In these plots, Sim_flow/EC1 represents load effects resulting from a traffic flow of 10,000 HGVs per day.   The "traffic jam" situation (Sim_jam/EC1) was calculated using 1% of the daily HGV traffic (ie, 100 vehicles) travelling in a closely spaced convoy. The simulated mid-span moments generally show distinct reserves. However, the EC1 values of support bending moment (Ms) are exceeded, particularly for the "traffic jam" situation for spans more than 40m.

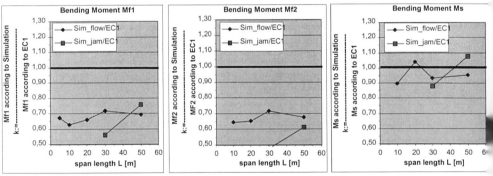

Figure 2 Comparison between simulation and EC1.

As an alternative to the method of stochastic simulation, the fractile values of relevant load effects can be approximated for single-span and multi-span beams with spans up to about 30m. The method is described in detail by Kaschner *et al* (1999a). The basic idea is that for those static systems where only one HGV is required per lane due to the shape of the influence line and the geometric extent, the extreme loading effects can be determined by using the extreme vehicle gross weights. This presupposes that the HGVs in lane 1 and 2 are of the same type and are positioned as a "vehicle packet". Then, the gross weights for vehicles on two lanes can be approximated by taking into account the distribution of loads in the transverse direction. For this a fractile value can be calculated from the sum of vehicle gross weights by taking into account the following equation:

$$P(Z < z) \quad = \quad F_Z(z) \quad = \quad \int_0^z F_X(z-y)\, f_Y(y)\, dy \qquad (6)$$

where $F_X(x)$, $F_Y(y)$ and $f_X(x)$, $f_Y(y)$ are accompanying distribution and density functions for extremes of gross weight for both lanes. Figure 3 gives the results of a numeric solution to equation (6) for vehicles with the highest possible gross weight. For the support bending moments (Ms), it was assumed that an additional HGV was positioned in lane 1 of the neighbouring span. For comparison, the results of the comprehensive simulation previously discussed are also given.

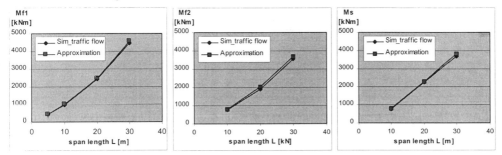

Figure 3   Comparison between simulation and approximation.

For shorter spans (up to 30m), the approximation makes it possible to calculate load effects resulting from traffic, which are necessary for structural assessment, without complex individual simulation, provided that fractile values of vehicle gross weight are known.

Evolution of the probability of failure

One advantage of the probabilistic approach resides in its effectiveness in including any additional information coming from inspections or tests. Indeed, the first objective is to improve the knowledge of the structure. This improvement can be carried out directly in a probabilistic approach by means of conditional probabilities. In comparison, it is impossible to incorporate inspection data into an assessment approach based on partial safety factors in a rational way. Another advantage of a probabilistic approach is that it is possible to estimate the level of safety through the probability of failure. In a deterministic approach, the classification of the reliability of a bridge is made in a binary way: the bridge is either safe or it is not.

The bridge reliability can also be assessed with respect to time. This is exactly what is necessary to provide more effective bridge management since that makes it possible to select the more economic interventions, like the next inspection, repairs, or simply no action.

Updating probabilities of failure

Qualitative or quantitative information can be obtained from inspections. Each inspection is an event, associated with an event margin H and an occurrence probability. Qualitative inspection results provide information on the detection (or non-detection) of an event related to a particular phenomenon. The information is expressed by:

$$H \leq 0 \qquad (7)$$

Quantitative inspection results correspond to measurements of an event related to a particular phenomenon. The information is expressed by:

$$H = 0 \tag{8}$$

Let us assume that we are studying the reliability of a component described by its safety margin M. Let us also assume that different qualitative and quantitative inspection results are available and described by a set of event margins $\left(H_{quant,i}\right)_{1 \le i \le n}$ and $\left(H_{qual,j}\right)_{1 \le j \le m}$. Then the probability of failure of the component when these qualitative and quantitative information are known is given by the conditional probability:

$$P_f^{updated} = \mathbf{Prob}\left(M < 0 \middle/ \left[\bigcap_i H_{quant,i} = 0\right] \bigcap \left[\bigcap_j H_{qual,j} < 0\right] \right)$$

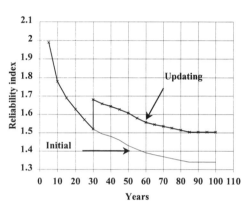

Figure 4 Probability updating of a prestressed section (SLS - Stress release method).

Cremona *et al* (1999) provides more details. Figure 4 provides an example for the assessment of a prestressed concrete structure. In this example, the reliability index has been updated on the basis of additional information regarding prestress losses (Cremona 1995).

Minimal safety levels

The evolution of the probability of failure over time has to be compared with a maximum probability prestressed concrete beam at the Serviceability Limit State, taking into account prestress of failure. The majority of the current standards provide such probabilities. In the Eurocodes, the probability of failure corresponds to $7.237 \; 10^{-5}$ (ie, $\beta = 3.8$) and is related to a 50-year reference period. This index aims to achieve the same level of safety of all the elements of a bridge whatever the consequences of the failure of these elements. Also, this level of reliability does not imply excessive costs when designing a bridge. Within the framework of a structural assessment, the modification of the reliability index can provide considerable cost savings. For example, the Canadian code for bridge assessment (CSA 1988) makes it possible to modulate the annual reliability index by taking account of factors such as the mode of failure of the overall structure, mode of failure of individual elements, the level of inspection, and the traffic conditions.

CONCLUDING REMARKS

This paper gives a brief description of the work carried out under Workpackage 2 of the BRIME project, the objective of which is to develop a framework for the structural assessment of bridges. The project is identifying methods of calculation that can be applied to bridge assessment as an input to a bridge management system. Semi-probabilistic methods are suitable for many assessments, but where these prove to be inadequate, or where deterioration of the structure is significant, reliability methods may be preferable. The continuing development of these methods is important to the cost-effective management of the European road network.

REFERENCES

CSA (1988). CAN/CSA-S6, 1988/Supplément, Canadian Standards Association.

Calgaro, J A (1991). *Introduction à la réglementation technique* Annales des Ponts et Chaussées, N.60.

Cornell, C A (1967). *Some thoughts on maximum probable loads*, and *Structural safety insurance*. Memorandum to ASCE Structural Safety Committee, MIT, Cambridge, USA.

Cremona, C (1997). *Probability-based optimisation of inspection intervals for steel bridges*. IABSE Workshop on evaluation of existing steel and composite bridges, Vol.76, Lausanne.

Cremona, C, R Kaschner, P Haardt, D W Cullington and S Fjeldheim (1999). *Experimental assessment methods and use of reliability techniques*. Deliverable D6, BRIME, PL97-2220 (to be published).

Cremona, C (1995). *Prestressed bridge reliability updating by prestress measurements*. ICASP 7, Paris.

Ellingwood, B, T V Galambos, J C MacGregor and C A Cornell (1980). *Development of a probability based load criteria for American National Standards A58*. NBS Special Publication, N.577, NBS, Washington D.C.

Geißler, K (1996). *REB - Rechenprogramm zur Restnutzungsdaueranalyse von Bahn- und Straßenbrücken, modifizierte Version mit Berücksichtigung der Extremwertauswertung*

Highways Agency (1998). *BD 79/98: The management of sub-standard highway structures*. Highways Agency, London.

Highways Agency (1997). *BD 21/97: The Assessment of Highway Bridges and Structures*. Highways Agency, London.

Kaschner, R, C Cremona and D W Cullington D (1999a). *Development of models*. Deliverable D5, BRIME PL97-2220 (to be published).

Kaschner, R, P Haardt, C Cremona and D W Cullington (1999b), *Review of current procedures for assessing load carrying capacity*. Deliverable D1, BRIME, PL97-2220 (to be published).

Lind, N C (1973). *The design of structural design norms*. Journal of Structural Mechanics, Vol 1.

Melchers, R.E (1999). *Structural reliability analysis and prediction*. Wiley.

Woodward, R J, P R Vassie and M B Godart (2000). *Bridge Management in Europe (BRIME): Overview of project and review of bridge management systems*. Paper presented at the 4th International Conference on Bridge Management, Surrey, UK.

Assessment at the serviceability limit state

R.J. LARK, Cardiff School of Engineering, Cardiff University, Cardiff, UK, and
B.R. MAWSON, Gwent Consultancy, Newport, UK.

INTRODUCTION

Bridge structures are subject to increasing deterioration due to both adverse environmental effects, resulting in degradation, and ever increasing volumes and weight of traffic. In the UK, although 91.5% of the current bridge stock has been constructed since 1955 with an intended service life of 120 years,[1] many structures have now been found to require some form of rehabilitation within less than a quarter of this design life.[2] The latest increase in allowable vehicle weights (from 38 to 40 tonnes) has already resulted in a 15 year assessment and strengthening programme, and in some places there is still a lot of work outstanding.

One of the main difficulties experienced in this assessment programme is that existing techniques result solely in pass / fail criteria with respect to the ultimate limit state (i.e. collapse of one element of the structure). Engineering judgement can, of course, be applied in reviewing the likelihood of this collapse but this makes it very difficult to ensure that the level of safety is consistent across a group of different bridges.

To address this issue the Highways Agency have co-ordinated a comprehensive programme of research into assessment techniques,[3] but to date, the majority of the published output of this work has still concentrated on the capacity of the structure at the ultimate limit state. Despite this, many difficulties still exist in realistically modelling ultimate limit state failure modes[4] and disquiet has been voiced over the selection of appropriate load factors.[5]

When approached from the standpoint of a 'one-off' assessment of safety these difficulties can be tolerated, but in the context of bridge management, what is of more interest is the response of the structure to 'real' loads and the variation of this response with time. Remedial action must be taken long before a structure approaches its ultimate limit state. This paper therefore proposes a serviceability limit state approach to bridge assessment which it is suggested is more consistent with current bridge management practice.

The paper will begin by reviewing current trends in bridge assessment and the relationship between the favoured techniques and the notion of whole life assessments. The reality of bridge management will then be identified and the concept of condition monitoring as a technique for both characterising the response of the structure to working loads and highlighting the onset of deterioration will be assessed. The paper will conclude with a proposal for a serviceability approach to assessment

BRIDGE ASSESSMENT

The assessment of structural adequacy and any consequential actions are of crucial importance to the bridge management process as they are the means by which it is ensured that a given bridge stock is always in a safe and serviceable condition.   As noted in BA 79/98,[6] if assessments are unduly conservative, structures will be unnecessarily restricted or strengthened, whereas if the rules are lax, some structures may be left in service without having an appropriate margin of safety against failure.

Current UK codes for the assessment of bridge structures adopt a deterministic, partial safety factor approach.[7, 8]   In theory this ensures a reasonably consistent margin of safety against failure when any bridge is subject to the maximum load expected within a 120 year return period.   In practice, however, this is not the case and both Hogg and Middleton[9] and Flaig[10] have shown that the level of safety implicit in a design carried out to current standards is a function of details as diverse as the bridge type, its span, the assumed mode of failure and the variability of its material properties.

In addition to this the national bridge assessment programme has revealed a large number of bridges which have failed the assessment but show little evidence of distress.   It is therefore widely accepted that a failed assessment does not necessarily mean that a bridge poses a safety risk or, as noted by the Concrete Bridge Development Group[11], that it is inadequate.   Indeed, in future codes it is probable that there will be a formal requirement for a plausibility check, such that, if a bridge fails an assessment but there is no evidence of distress or failure, the difference between the real condition of the structure and that predicted by analysis will have to be explained.

To address these problems it is possible to :

i)   use more advanced analytical techniques to address more realistic failure modes.

ii)   use bridge specific load data

iii)   adopt a more rational approach by computing the actual probability of failure or reliability of a structure by taking into account the uncertainties in the design variables and comparing this reliability with an acceptable target reliability.

These increasing levels of sophistication are the basis of the Highways Agency's staged approach to assessment (Levels 1 to 5), the rationale of which is described by Das[12] and the details of which are given in BA 79/98.[6]

There are however still problems with this approach not the least of which is the definition of a prescribed target reliability as required by levels 4 and 5.   The difficulty is that for any given structure this target is likely to depend on :

i)   the loading, which in turn will be a function of the chosen return period and the anticipated traffic flow;

ii)   the definition of the failure mode under consideration and the relationship of this failure mode with the chosen method of analysis;

iii)   any reserve strength or redundancy in the system;

iv)   the likelihood of there being a warning of failure and the frequency and thoroughness of the inspection and maintenance regime;

v)   deterioration;

vi)   whether it is a component, bridge, bridge type or network reliability which is being considered.

Allen[13] endeavours to address these issues by defining an overall target reliability of the form :

$$\beta_t = \beta_o + \sum \beta_{mi}$$

where β_t is the overall target reliability,

β_o is the base target reliability, and

β_{mi} are a range of modification factors to allow for the above effects.

Such an approach may be satisfactory for one-off assessments but in the context of bridge management, what is of more interest is the response of the structure to 'real' loads and the variation of this response with time and, in particular, deterioration.

One approach to deterioration is to develop reliability based whole life profiles,[14,15,16] but again such an approach is not without problems. For example, difficulties can be anticipated in :

i)   realistically modelling the relationship between deterioration and ULS failure modes, especially when this is non-linear and in situations where the failure mode changes with deterioration;

ii)   identifying a cut-off reliability (as above for target reliability);

iii)   allowing for the effects of deterioration associated with secondary and possibly non load-bearing items;

iv)   recognising the significance of a structure's load history;

v)   allowing for the effects of redundancy;

vi)   identifying an appropriate relationship with the condition indices which will typically be the output of the inspection process.

What is required to overcome these difficulties is a more positive link between assessment procedures and bridge management practice. A better framework for recording and incorporating the information that has allowed engineers in the past to assess a structure on the basis of engineering judgement is needed. It is felt by many practising engineers therefore, that a better idea of the health of a bridge structure would be gained by observing its behaviour under the application of loads which are representative of those that they are likely to carry in practice,[17] and it is precisely this which should be achieved by investigating its behaviour at the serviceability limit state.

BRIDGE MANAGEMENT

Shetty[18] defines bridge management as "the rational planning and implementation of all actions necessary to ensure the safety, serviceability and durability of a bridge throughout its service life". In practice, what this means for most bridge managers is that it is a process of balancing on one hand an ageing bridge stock with, on the other, funding shortages for maintenance and repair.

Faced with this challenge a number of management systems, typically referred to as Bridge Management Systems, have been developed.[10] In the early years these were mainly databases which stored information on the location of the bridge, road category, construction date and data collected during inspections. Then, with time, they have become more sophisticated enabling the user to undertake activities such as inspection planning, deterioration prediction and the economic evaluation of repairs.

To date, however, the practice of bridge management, whether computerised or not, is typically initiated by an inspection of the structure under consideration.   Engineering judgement is then applied to issues such as the consequences of failure, the importance of the road within the network, the results of a structural assessment, public perception and the co-ordination of MR&R works with work on adjacent roads and bridges.[19] This would suggest that, as might be expected, the condition of the structure at the time of the inspection is fundamental to the bridge management process. In current assessment procedures however, the only way of making use of bridge specific data is by means of the 'worst credible strength' concept.[7] Furthermore, although bridge specific data can be used to update a structure's reliability index, great care is needed to ensure that this process is not unduly optimistic because of notional improvements in both the mean and standard deviation of such data, and that sufficient samples are taken to yield reliable results.[20] As a result, procedures for the use and interpretation of such data in reliability based assessments are, as yet, ill defined.

The assertion of this paper is that, in reality, bridge management is about maintaining the serviceability of a structure. If a bridge can carry normal loads, including typical overloads, on a regular basis and without showing signs of distress, and as long as the ratio of the actual load at which collapse would occur to that at which this distress would be apparent provides an appropriate factor of safety, it is deemed to be satisfactory. This is acceptable because, even when components fail in shear, there is evidence to suggest that significant warning of total collapse can be anticipated.[21] Indeed, it is considered by many bridge managers that, subject to improvements in data capture, the use of regular inspection and monitoring techniques would provide a measure of a structure's serviceability and would enable the development of defects to be traced.[19] What is required, therefore, is that this should in some way be incorporated in a procedure for the assessment of the structure's integrity.

CONDITION MONITORING
The concept of condition monitoring is to make best use of past, present and future performance based data to verify the current integrity, and predict the future performance, of a structure. It is a technique which has been widely used in other engineering disciplines including geotechnical engineering for which much instrumentation has been developed [22]

In the context of bridge structures condition monitoring may comprise :

i)   environmental monitoring from which response models may be developed, e.g. diffusion of chlorides, rates of corrosion, seasonal and diurnal response etc.

ii)   material monitoring, e.g. creep, shrinkage, variation of strength and stiffness with time etc.

iii)   structural monitoring, e.g. stress, strain, deflections, vibrational response, load testing etc.

iv)   load monitoring, e.g. "weigh-in-motion" systems, vehicle counts etc.

iv)   health monitoring, e.g. visual inspections, defect monitoring, acoustic monitoring etc.

Clearly some of these can only be applied to new structures but all have their place in a rational inspection and maintenance regime and all represent different forms of monitoring under serviceability conditions.

If such data is to be used for assessments the output of condition monitoring should enable :

    i)   analysis of structures on the basis of measured strains and deflections;

    ii)   comparison of observed and predicted performance;

    iii)   assessment of the state of a structure on the basis of its loading history and a known, measured response,

and further research is now required to identify the most appropriate techniques for achieving this. Moss and Matthews[22] and McGowan et al.[23] have reviewed the art of in-service structural monitoring but, as yet, nobody has assessed them specifically in relation to the process of bridge management. To achieve this and to identify what data is needed to satisfy the above a framework for a serviceability approach to assessment is required. One aim of this paper is to identify a basis for such a framework.

SERVICEABILITY ASSESSMENT

The justification for proposing an alternative approach to assessment based on Serviceability Limit State criteria is that certification of a structure today is no guarantee of its performance tomorrow. What is required, therefore, is a procedure in which the everyday response of the structure to both the structural and environmental effects to which it is subject can be monitored and projected forward. In other words the proposal is one of "change management" whereby it is the *variation* of the reliability of the "in-service" structural and environmental response of the structure *with time* which is of interest.

To elaborate, it has already been shown that the reliability of a structure with respect to an environmental serviceability limit state such as the onset of corrosion can be expressed as a reliability index, the time-dependent variation of which can then be derived by assuming a deterioration model which, typically, is a function of the chloride levels in the concrete.[24] The latter is a measurable parameter, therefore by monitoring chloride levels the likelihood of corrosion being a problem can be anticipated and, by comparing the development of this likelihood with time with that predicted on the basis of the design parameters, a rational decision on the need for preventative action can be made.

Likewise, but not yet explored to the same extent, it is also possible to calculate the reliability of a bridge with respect to structural serviceability limit states. The variation of this reliability with either the applied loads or the material resistance can therefore also be determined and if these in turn are related to time, for example due to changes in loading regimes and the perceived likelihood of extreme loading events or to changes in material resistance due the effects of creep, shrinkage and / or deterioration etc., then the variation of this reliability with either these changes themselves or time can also be examined. As with the environmental example given above, the significant benefit of adopting an SLS approach is that the analytical procedure and limit state models adopted in this procedure can be verified by monitoring the response of the structure. Then, if found to be inappropriate, it can be modified to reflect the actual behaviour of the structure and, when satisfactory, can be updated in response to the past performance of the structure. In other words, the actual service of the structure can be used as a "proof" load and by monitoring the structure's behaviour the uncertainties associated with its assessment can be continually refined and reduced.

What is being proposed therefore, is a procedure which seeks to make best use of the recent advances in our understanding of the reliability of bridge structures applied in such a way that they can support a bridge manager's engineering judgement of the response of their stock and can be enhanced by both traditional and future bridge monitoring techniques. For example, strain, deflection and / or the nature of cracking patterns may currently be used to characterise the response of a structure to both loads and movements from which its structural reliability can be identified. Likewise chloride levels, moisture content, porosity and degree of corrosion etc. are measures of environmental reliability. In the future therefore, it can easily be envisaged that defect, vibrational and other NDT monitoring techniques may also provide suitable techniques for gauging the "in-service" reliability of a structure.

Whichever "measure" is adopted it is the variation of the resulting reliability from that anticipated at the time of design which is significant, with improvements representing an enhanced understanding of the structure's behaviour and reductions typically being due to changes in loading and deterioration, although they can also be due to changes in our understanding of the performance of the structure. What matters is that these variations in reliability are managed. Thus if the "in-service" reliability of a structure is reduced because of deterioration or an increase in loading, can it be restored through a knowledge of the actual behaviour of the bridge or by tighter control of the traffic using it, or are remedial works required? Unlike assessment at the ULS, the adequacy of the response to this question can be monitored and the bridge manager can be assured that a margin of safety against total collapse still exists. The opportunity therefore exists to manage the risk posed by the structure and in each period to address the changes which have occurred in both the response and loading of the structure, whether that be due to traffic, environmental or deterioration effects, to assess the consequences of these changes and, on the basis of this assessment, to implement the necessary monitoring, controls or remedial action.

CONCLUSIONS
This paper has sought to present a basis for a serviceability limit state approach to bridge assessment. Current trends in bridge assessment have been reviewed and the reality of bridge management has been identified as being about the maintenance of the serviceability of a structure. Techniques for calculating, gauging and monitoring the "in-service" reliability of a structure have therefore been suggested and a process of "change management" is proposed as the basis of a serviceability approach to bridge management and continual assessment.

Detailed work is still needed to precisely define health monitoring models capable of describing the structural serviceability and safety of our bridges. More research and field trials are also required to improve our knowledge of the ratio of the collapse load of a structure to that at which distress is first apparent. Nevertheless, it has been shown that such an approach to bridge assessment, and thereafter bridge management, is clearly feasible. By formulating serviceability assessment criteria on the basis of the actual behaviour of structures, it is suggested that the process of what is being proposed is no different to that by which codes of practice have been developed in the past. Unlike models of ultimate limit state behaviour, however, the adequacy of serviceability limit state models can be continually reviewed and, where necessary, modified on a bridge specific basis. This makes this proposal a readily applicable and yet truly flexible approach.

REFERENCES
1.   Rochester, T.A. (1997)   *Trunk road bridges - current needs for design and maintenance*. Safety of Bridges, Ed. Das, P.C., Thomas Telford, London.

2.   Rubakantha, R.S. and Vassie, P.R. (1996) *Risk-based approaches to economic appraisal, load assessment and management of bridges : A review.* Bridge Management 3, Ed. Harding, J.E. and Parke G.A.R., E&FN Spon, London.

3.   National Audit Office (1996) *Highways Agency: The Bridge Programme.* HMSO, London.

4.   Middleton, C.R. (1997) *Concrete bridge assessment: an alternative approach.* The Structural Engineer, Vol. 75, No. 23 & 24.

5.   Panel Session (1999) *Safety Criteria for Buildings and Bridges.* Thomas Telford Conference, London.

6.   BA 79/98 (1998) *The Management of Sub-Standard Highway Structure.* Design Manual for Roads and Bridges, Vol. 3, Section 4, Part 18. HMSO, London.

7.   BD 44/95 (1995) *The Assessment of Concrete Highway Bridges and Structures.* Design Manual for Roads and Bridges, Vol. 3, Section 4, Part 14. HMSO, London.

8.   BD 21/97 (1997) *The Assessment of Highway Bridges and Structures.* Design Manual for Roads and Bridges, Vol. 3, Section 4, Part 3. HMSO, London.

9.   Hogg, V. and Middleton, C.R. (1998) *Whole life performance profiles for highway structures.* The Management of Highway Structures. Thomas Telford Conference, London.

10. Flaig, K.D. (1999) *Development of a reliability based bridge management system.* Ph.D. Thesis, Cardiff University.

11. Frostick, I. (1997) *Report of Working Party on The Assessment of Concrete Bridges.* Concrete Bridge Development Group.

12. Das, P.C. (1997) *Development of bridge specific assessment and strengthening criteria.* Safety of Bridges, Ed. Das, P.C., Thomas Telford, London.

13. Allen, D.E. (1992) *Canadian highway bridge evaluation: reliability index.* Canadian Journal of Civil Engineering, Vol. 19.

14. Flint, A.R. and Das, P.C. (1997) *Whole life performance based assessment rules, background and principles.* Safety of Bridges, Ed. Das, P.C., Thomas Telford, London.

15. Nowak, A.S. (1997) *Revised rules for steel bridges.* Safety of Bridges, Ed. Das, P.C., Thomas Telford, London.

16. Thoft-Christensen, P., Jensen, F.M., Middleton, C.R. & Blackmore, A. (1997) *Revised rules for concrete bridges.* Safety of Bridges, Ed. Das, P.C., Thomas Telford, London.

17. National Steering Committee for the Load Testing of Bridges (1998) *Guidelines for the supplementary load testing of bridges.* Thomas Telford Publications, London

18. Shetty, N.K. (1997) *Conceptual framework of a comprehensive risk-based BMS.* Unpublished communication.

19. Flaig, K.D. and Lark, R.J. (1999) *The development of UK bridge management systems.* Proceedings of the Institution of Civil Engineers, Transport. Submitted for publication.

20. Kersken-Bradley, M. et al. (1990) *Estimation of structural properties by testing for use in limit state design.* Joint Committee on Structural Safety, IABSE-AIPC-IVBH, Zurich.

21. Ricketts, N.J. and Low, A.McC. (1993) *Load tests on a reinforced beam and slab bridge at Dornie.* Transport Research Laboratory Report 377, Crowthorne, Berkshire.

22. Moss, R.M. and Matthews, S.L. (1995) *In-service structural monitoring, a state of the art review.* The Structural Engineer, Vol. 73, No. 2.

23. McGown, A. et al. (1997) *Integrated, multilevel condition monitoring of structures.* The Structural Engineer, Vol. 75, No. 6.

24. Gehlen, Ch. and Schiessl, P. (1999) *Probability-based durability design for the Western Scheldt Tunnel.* Structural Concrete, Vol. P1, No. 2.

Updating traffic load models for the assessment of existing road bridges

SIMON F. BAILEY Swiss Federal Institute of Technology, Lausanne, Switzerland
EUGENE J. O'BRIEN University College Dublin, Earlsfort Terrace, Dublin 2, Ireland
ALEŠ ŽNIDARIČ Slovenian National Building and Civil Eng. Institute, Ljubljana, Slovenia

INTRODUCTION

The key to the assessment of existing structures is the use of updated models of loads and resistance derived from site specific data. In comparison to the use of notional models defined in codes, the consideration of site data reduces uncertainty and facilitates a more accurate assessment [1]. In the case of road bridges, the updating of traffic load models presents perhaps the greatest scope for reducing uncertainty and thus verifying adequate performance that might otherwise be put in doubt by the use of a general traffic load model. The most appropriate approach to gathering site specific traffic data is certainly the use of Weigh-In-Motion (WIM) techniques, based either on the instrumentation of a bridge or the installation of pavement sensors. Such techniques were recently reviewed and developed within the European 4th Framework research project, WAVE and the concerted action, COST 323, of which some of the authors were members.

This paper begins with a summary of the COST 323 research into WIM technology and its application to bridge assessment. Case studies illustrating three approaches to updating traffic load models are then presented: an assessment load model; the use of site specific WIM data and the direct measurement of traffic action effects. The first case involves the development of a load model for a bridge in Slovenia that was derived from WIM data. This model has been used to justify the opening of a third traffic lane on a deteriorated prestressed concrete bridge that was heavily congested. A case study from Switzerland compares the application of WIM data from strain measurement and a traffic survey using pavement sensors. The third case treats a long-span bridge in Northern Ireland where deflections due to traffic load were measured directly and statistical techniques were subsequently used to extrapolate the measured values and estimate the characteristic deflection. All examples clearly illustrate that considering updated traffic action effects is an important element of bridge assessment and of significant benefit to the efficient management of road bridges.

OVERVIEW OF COST 323: WIM TECHNOLOGY

The development of Weigh-In-Motion (WIM) has been ongoing for the last 20 years in Europe, initially in France and the UK. However, in the early 1990's WIM systems still had limited accuracy and/or excessive cost, and in many cases their durability was questionable or unknown. Therefore, in 1992 the Forum of European Highway Research Laboratories (FEHRL) proposed a concerted action on WIM that was supported by the European Commission General Directorate for Transport. The result was COST 323, which extended over 6 years from November 1992 and involved 18 countries. The goal of the COST 323 action was to promote the development, implementation and application of WIM techniques and systems throughout Europe.

The principal results of COST 323 were the following:
- A glossary of WIM terms was developed to facilitate communication between WIM researchers, users and manufacturers. It contains more than 830 entries and 160 definitions, in 10 languages: English, French, German, Italian, Spanish, Portuguese, Dutch, Hungarian, Slovenian and Slovakian [2]. It is available on the Internet (http://www.zag.si/wim/).
- A pan-European database of WIM sites with more than 350 entries from 15 countries was developed. Site descriptions cover the road, pavement, WIM sensors and system, the purpose of the measurements, etc.. This database has become the first part of a comprehensive database of WIM [3] that will include detailed vehicle by vehicle records.
- A specification on WIM was developed containing a detailed procedure for WIM system accuracy classification. An appendix gives simplified requirements, and proposes some standard test plans that are easy to apply. The specification is already in use in many countries as an unofficial standard or for vendor-customer contracts.
- Four large-scale field trials of WIM systems were carried out in the context of COST 323.
  - The first test took place on an urban road in Zürich from 1993 to 1995 [4]. It involved 4 complete WIM systems, 2 combined systems and 4 sensors, representing most of the existing technologies. The gross weights of more than 2000 (small) lorries were measured, both in motion and at rest.
  - In June 1996 a three day trial was carried out in France involving four portable WIM systems supplied by their owners and one multiple sensor WIM system installed for the OECD/DIVINE project [5].
  - A large scale European Test Program started in 1997 involving two further tests. The Cold Environment Test, started in June 1997 in Northern Sweden while the Continental Motorway Test, started in March 1997 on a heavily trafficked motorway in Eastern France. Results of these tests are presented in [6] and [7].
- Four reports were drafted about the application of WIM to traffic management, bridges, pavements and enforcement. They present both a state-of-the-art summary of the use of WIM data and the future perspectives, tools and benefits.

The research carried out within the framework of COST 323 has provided the basis for the practical application of WIM techniques, which are illustrated by case studies in the following three sections.

RAVBARKOMANDA VIADUCT, SLOVENIA

Live load modelling based on WIM data
There are several approaches to using WIM data for modelling traffic loads [8]. Moses for example proposed the following load model [9] to predict the maximum load effect Q:

$$Q = a \times W_{.95} \times H \times m \times I \times g \qquad (1)$$

where a is a deterministic influence coefficient relating load effect to vehicle weight, $W_{.95}$ is the characteristic vehicle weight (95th percentile), H is the headway factor, representing the probability of multiple presence of vehicles on the bridge, m is a factor reflecting the variation of load effects from random heavy vehicles, I is the impact factor and g is the girder (lateral) distribution factor.

A basic statistical evaluation of WIM data is used to derive deterministic values of $W_{.95}$, m, I and g. Calculation of the headway factor H requires prediction of the expected maximum load effects, either by full simulation or by a simpler convolution method [9, 10]. Since the latter assumes that the maximum load effect is caused by two vehicles crossing side by side over the critical section of the span, it is only applicable when the probability of having more than two trucks on the bridge at the same time is acceptably low. It has been observed that for most bridges shorter than 30 m with less than 2000 heavy vehicles/day the probability of occurrence of such events can be neglected.

Example

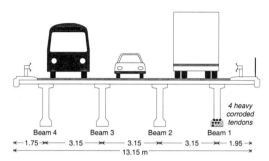

Figure 1. Proposed traffic lanes on the
Ravbarkomanda viaduct

The derivation of an updated traffic load model is illustrated for a heavily deteriorated motorway viaduct between Ljubljana and Italy. The viaduct comprises two parallel bridges, each consisting of seventeen 36.3 m simply supported spans. During reconstruction of one bridge, the owner wanted to decrease traffic jams on the other (open) bridge by introducing a third traffic lane (Figure 1). However, a detailed inspection showed that 4 of 11 tendons in beam 1 were heavily corroded in one of the spans, thus reducing the resistance by about 35%.

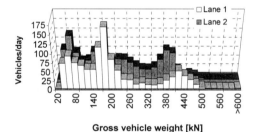

Figure 2: Daily gross weight histograms

Seven days of WIM measurements were performed in order to evaluate heavy traffic weight in both lanes (Figure 2). The convolution method was then applied to determine the maximum expected combined weight of two vehicles on the bridge ($W_{.95} \times H$) for different time periods (Figure 3).

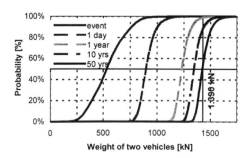

Figure 3. Combined weight of two vehicles

A 5-axle semi-trailer of 420 kN was used in load tests to determine the factors a and m. Due to the well-defined loading and measured impact and load distribution factors, the live load factor was reduced from 1.8 (design code) to 1.4. The expected maximum bending moment for the next 50 years was therefore predicted to be 44% lower than the design moment (Table 1). Subsequent analysis revealed sufficient safety levels, allowing the bridge owner to introduce the third traffic lane for cars, thus considerably reducing congestion.

Table 1. Bending moments due to load models from two design codes and from WIM data

Loading scheme	Bending moment		
	Unfactored	Partial load factor	Factored
Bridge-specific from WIM data	2 102 kNm	1.40	2 943 kNm
DIN 1072	2 901 kNm	1.80	5 222 kNm
Eurocode 1, Part 3	3 881 kNm	1.35	5 239 kNm

Table 1 illustrates the significant reduction in traffic action effect that was justified by site investigation. This reduction was achieved by adopting both a site specific load model and a lower live load factor justified by the fact that uncertainty about traffic action effects was greatly reduced by measurements. Both of these steps were needed in order to justify the introduction of the third traffic lane for cars.

BRIDGE ON THE RIVER AAR, AARWANGEN, SWITZERLAND
The results of testing carried out on a two span steel-truss composite bridge in Aarwangen, Switzerland has been used in order to compare WIM data collected from pavement sensors and bridge strain measurement. Simultaneous measurements with temporary WIM sensors placed adjacent to the bridge and strain gauges on bridge elements were used to study vehicle characteristics and traffic load effects. The results of this study provide valuable information about the effects of actual traffic on bridges and the use of WIM data for bridge evaluation.

Strain measurements
Static and dynamic load tests were carried out during a bridge closure using two 25 tonne 3-axle test vehicles. Continuous dynamic measurements under normal traffic conditions were subsequently made over two periods totalling 18 days [11]. Strain measurements during the static load test enabled the behaviour of the bridge to be assessed and showed that traffic load effects were lower than assumed during design, thereby illustrating the advantage of carrying out measurements on a structure as opposed to using a default structural model. Strain data from continuous measurement under normal traffic was processed in order to identify vehicle loading events and to determine the peak static effect of traffic actions during each event. Figure 4 shows a histogram of peak stresses obtained over a period of three days.

WIM Traffic survey
A vehicle survey was carried out using temporary pavement WIM sensors installed 200 m from the bridge [12]. The portable WIM system consists of inductive loops which measure vehicle speed and length combined with WIM strip capacitive sensors that weigh half of each axle. More than 30 000 heavy vehicles were measured using the WIM system over a period of 18 days. However, the system is sensitive to vehicle speed and experiences problems measuring axle groups. Erroneous measurements were therefore later filtered out and approximately 10,600 heavy vehicles were retained for further analysis. A histogram of heavy vehicle total weight is shown in Figure 5.

Discussion
The two approaches to updating traffic load models involve different amounts of effort and yield results of different quality. The continuous measurement of strains in a bridge requires expensive equipment and the use of expert technicians as well as complex data post-processing. On the other hand, the use of temporary pavement WIM sensors is relatively straightforward and real-time processing produces vehicle data that is easy to use.

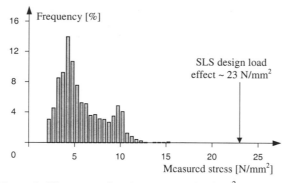

Figure 4. Histogram of peak stresses (> 2 N/mm^2) measured under normal traffic conditions

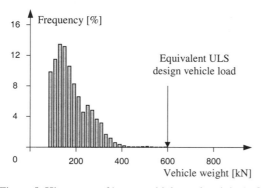

Figure 5. Histogram of heavy-vehicle total weight (> 80 kN)

In terms of financial investment in equipment and labour, strain measurement is about twice as expensive as the use of temporary pavement WIM sensors. A comparison of costs should in turn be balanced against the value of the data collected. In this respect, this study has shown that strain measurement yields precise data about traffic action effects and bridge behaviour, whereas vehicle data collected using temporary pavement WIM sensors is subject to significant errors. A decision about which approach to adopt must be made by considering the purpose of measurement. If vehicle counting and a general idea of average vehicle loads is required then temporary pavement WIM sensors suffice; otherwise a more expensive permanent WIM installation is necessary for weight data of greater accuracy. If dynamic traffic action effects and bridge behaviour are of interest then strain measurement is preferable. The measurements carried out clearly demonstrate the relative merits of the two approaches to updating design traffic load models for bridge evaluation, but that in both cases a site-specific traffic load model is significantly lower than the design load model.

THE FOYLE BRIDGE, NORTHERN IRELAND
The Foyle Bridge carries the A2 trunk road across the River Foyle about 3 km downstream from the centre of the city of Londonderry in Northern Ireland. The structure is 866 m long and comprises three main spans in steel, totalling 522 m, together with a 344 m approach viaduct in prestressed concrete. The bridge consists of two independent parallel structures, each carrying a single carriageway. The main steel structure consists of twin steel box-girders

that vary in depth from 3 m at midspan to 9 m over the intermediate supports. It is on the primary trunk route between North Donegal and Northern Ireland. It was, therefore, designed to carry the full highway loading specified for trunk roads at the time of its design in 1980. However, as it serves a primarily rural area, the frequency of heavy loads is very much less than would be the case in a heavily industrialised region where a large proportion of lorries may be expected to be near or to exceed the maximum legal weight.

For several years the steel portion of the bridge has been the subject of ongoing monitoring [13]. A computer controlled remote access monitoring system has been installed which allows a range of stresses, deflections, temperatures and wind parameters to be observed [14, 15]. The equipment includes a deflection measuring system which is shown diagrammatically in Figure 6. In each box, two lasers were fixed at midspan and each projects a spot of light on to a target fixed over an intermediate support. Any movement of the lasers causes the position of the light spots to vary, and this movement is tracked by computer controlled cameras. The resulting apparent movements are then combined to give the translational and rotational movements at the midspan section. Measurements were recorded at a rate of 8.3 Hz with typical scans consisting of 4096 readings taken over a period of about 8 minutes. Full details can be found in [14, 15]. The maximum traffic-induced deflection from a 48 minute daily sample was treated as a random variable and was found to conform well to the Type I extreme value distribution as illustrated in Figure 7.

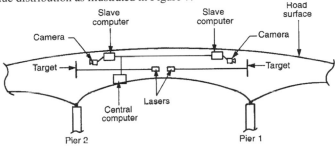

Figure 6. Deflection measurement system

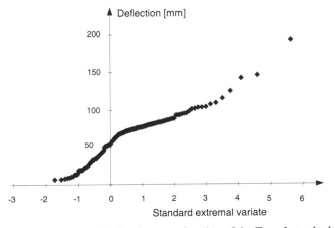

Figure 7. Measured deflections as a function of the Type I standard extremal variate

For a 1000-year return period (as recommended for design purposes in EC1) there are 10.96 x 10^6 48 minute samples, which corresponds to a probability of $1/(10.96 \times 10^6) = 91.26 \times 10^{-9}$. Equating the cumulative distribution function for the Type I extreme value distribution to a probability of $(1-91.26 \times 10^{-9})$ gives the following expression for the 1000-year deflection:

$$\delta_{1000} = c + m\{-Ln [-Ln(1 - 91.26 \times 10^{-9})]\} = 420 \, mm \qquad (2)$$

where m and c are the slope and intercept respectively of a line fitted to the points of Figure 7. Samples from 155 daily maximum deflections were used to calculate characteristic deflections for the bridge. The 1000-year characteristic value was found to be 420 mm. Allowing for uncertainty due to the sample size [16, 17], the representative value was determined to be 451mm.

A design value for deflection was calculated for comparison purposes. Taking two notional lanes of HA loading in accordance with BD37/88, an unfactored design deflection of 1042 mm was found. It should be noted however, that this value would correspond to jammed conditions whereas the characteristic deflection of 451 mm is based on free flowing traffic. Both jammed and free flowing conditions must be considered for the design of bridges.

CONCLUSIONS
The aim of this paper was to demonstrate that design load models are often conservative. Considering actual traffic during bridge assessment can therefore often lead to the avoidance, or at least reduction, of unnecessary strengthening or traffic restriction that might otherwise be suggested by the use of a design traffic load model.

The use of a temporary pavement WIM system has shown that measurements can be subject to significant error. However, these errors are known to be much lower if a permanent pavement WIM system is used. The prediction of extreme traffic loads, in particular, is therefore subject to high uncertainty. Furthermore, traffic action effects calculated using traffic data are dependent on an assumed bridge model, thus introducing further uncertainty.

Deflection or strain measurement provides accurate information about the actual traffic action effects on a structure. Furthermore, direct measurement provides information about load distribution within a structure as well as the combined effect of vehicles and dynamic effects.

It is worth emphasizing that the cost of deriving a bridge-specific traffic load model using weigh-in-motion measurements is in most cases marginal compared to the cost of strengthening or rehabilitation of an existing road bridge.

REFERENCES
[1] Bailey, S.F. and Hirt, M.A., 'Site Specific Models of Traffic Action Effects for Bridge Evaluation', in *Recent advances in Bridge Engineering, Evaluation, Management and Repair* CIMNE, Barcelona, 1996, pp. 404-425.

[2] COST323, *Glossary of terms for WIM*, final draft, ed. LCPC Paris/ZAG Ljubljana, 1998.

[3] Siffert, M., Dolcemascolo, V. and Henny, R., 'European database of WIM', in *Pre-Proceedings of 2nd European Conference on Weigh-in-motion of road vehicles*, Lisbon, eds. E. J. OBrien & B. Jacob, Sept 14 - 16, European Commission, Luxembourg, 1998, pp. 97-105.

[4] COST323, *Test of WIM Sensors and Systems on an Urban Road, Zürich 1993-95*, Report EUCO-COST/323/3/97, ed. M. Caprez, ETH, Zürich, 1997.

[5] COST323, *Test of Four Portable and one MS-WIM Systems (Trappes, June 1996)*, Report EUCO-COST/323/98, ed. R. Blab, ISTU, Vienna, March 1998.

[6] Stanczyk, D. and Jacob, B., 'Continental Motorway Test of Weigh-in-Motion Systems: Final Results', in *Post-Proceedings of 2nd European Conference on Weigh-in-motion of road vehicles*, Lisbon, eds. E.J. OBrien & B. Jacob, Sept 14 - 16, European Commission, Luxembourg, 1998, pp. 51-61.

[7] Jehaes, S. and Hallstrom, B., 'Accuracy Analysis of WIM Systems for the Cold Environmental Test', in *Post-Proceedings of 2nd European Conference on Weigh-in-motion of road vehicles*, Lisbon, eds. E.J. OBrien & B. Jacob, Sept 14 - 16, European Commission, Luxembourg, 1998, pp. 63-72.

[8] COST 323, 'Bridge applications of WIM', Draft No. 2, UCD Dublin,1998

[9] Moses, F. and Verma, P., 'Load Capacity Evaluation of Existing Bridges', *National Cooperative Highway Research Program (NCHRP) - Report N° 301*, 1987

[10] Žnidarič, A. and Moses, F., 'Structural safety of existing road bridges', *7th International Conference on Structural Safety and Reliability*, Kyoto, 1997

[11] Schumacher, A. and Blanc, A., *Stress Measurements and Fatigue Analysis of the New Bridge at Aarwangen* ICOM Report No 386, November 1998, 34 pp.

[12] Schumacher, A., *Aarebrücke Aarwangen - Mesures du trafic sur la RC 244* ICOM Report No 685-2, December 1998, 22 pp.

[13] Sloan,T.D., Kirkpatrick, J., Boyd, J.W. and Thompson, A.,: 'Monitoring the inservice behaviour of the Foyle Bridge', *The Structural Engineer*, April 1992, 70, No. 7, pp130-134

[14] Leitch, J.G., Tompson, A. and Sloan, T.D., 'A novel deflection measurement system for a major box-girder bridge', Proc. 4th Int. Conf. on Civil and Struct. Eng. Computing, London, Civil-Comp Press, 1989, Vol. 2, pp 301-306.

[15] Sloan, T.D. and Thompson, A., 'Development of an automatic data collection system for a major box-girder bridge', Proc. 4th Int. Conf. on Civil and Struct. Eng. Computing, London, Civil-Comp Press, 1989, Vol. 2, pp 313-318.

[16] Goda, Y., 'Uncertainty of design parameters from viewpoint of extreme statistics', Journal of Offshore Mechanics and Artic Engineering, Transactions of the ASME, 114, May 1992, pp 76-82.

[17] OBrien, E.J., Sloan, T.D., Butler, K.M. and Kirkpatrick, J., 'Traffic load 'fingerprinting' of bridges for assessment purposes', *The Structural Engineer*, 73, No. 19, 3rd October 1995.

The determination of site-specific imposed traffic loadings on existing bridges

S. GRAVE, Post-graduate researcher, Trinity College Dublin, Ireland
E.J. OBRIEN, Professor of Civil Engineering, University College Dublin, Ireland
A.J. O'CONNOR, Lecturer, Trinity College Dublin, Ireland

INTRODUCTION

A great deal of attention has been centred in recent years on the assessment of the load carrying capacity of existing bridges. In contrast, the traffic loadings which bridges are required to carry are often notional and, consequently, in less heavily trafficked regions, they can be excessively conservative. BA79/98 [1] on the management of sub-standard structures, recommends detailed load modelling for the accurate determination of the level of reliability of existing structures. Such modelling requires knowledge of advanced statistical concepts and thus is not practical for a majority of engineers. This paper describes the development of a simple statistical approach to determine the critical loadings to which a structure will be subjected without the requirement for complex statistical analysis. The background to the development of the simple statistical approach for the assessment of single lane, bi-directional, highway bridges is presented. The simple approach is validated by comparison with a more complex statistical approach. Finally, the robustness of the approach is demonstrated for a number of different sites with different traffic flow characteristics.

TRAFFIC SIMULATIONS

Traffic data obtained from Weigh-In-Motion (WIM) systems is required to generate traffic simulations. Using WIM data, the distributions of gross vehicle weight (GVW), speed and spacing between vehicles may be determined and modelled by statistical distributions - see for example, Figure 1. The modelled distributions are validated using statistical goodness-of-fit tests such as the Kolmogorov-Smirnov (K-S) test [2]. For each traffic file the site-specific proportion of trucks (i.e., vehicles with GVW > 3 500kg) per class of axles may also be obtained where a class of axles contains all trucks having the same number of axles.

In this study the sample files of real data used have been obtained from WIM stations on the A6 motorway at Auxerre (France) and the RN10 near Angers (France) [3]. Both sites have four lanes (2 in each direction) but traffic was recorded in the slow lanes only. A full week of data was recorded at both sites (from the 26/05/1986 to the 02/06/1986 for Auxerre and from the 07/04/1987 to the 14/07/1987 for Angers).

Full traffic simulations are performed using the modelled distributions of GVW per class of axle, speed and of time interval between consecutive trucks. Significant differences may exist in the GVW distribution moments for the opposing directions. Consequently, directional distributions were derived. In addition, for each class a typical truck was derived with fixed axle spacings and proportions of weight carried.

The program developed to generate traffic files theoretically allows for any time period, but due to time and memory limitations the simulations usually represent from 1 week to 2 months. A period of 1 month was selected for this study.

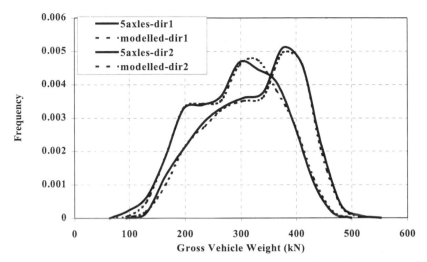

Figure 1 - Probability Density Functions of GVW for 5-axle trucks - Angers 1987

DETERMINATION OF LOAD EFFECTS

Two-lane simply supported and two span continuous bridges ranging from 20 m to 40 m (i.e. 'short' bridges) are considered. The extreme loading cases are governed by the meeting of two heavy trucks within a critical influence zone on the structure. For bridges exceeding 40 meters, meeting events with more than two trucks begin to influence the extremes. In this study, four load effects are considered:

- 1: bending moment at mid-span of a simply supported bridge,
- 2: moment at central support of a two-span continuous bridge,
- 3a, 3b: shear on left-end and right-end support of a simply supported bridge,
- 4a, 4b: shear on left-end and right-end support of a two-span continuous bridge.

An influence line can be defined as a diagram which shows the variation of a load effect (i.e., bending moment, shear force, etc.) at a given position in a structure as a unit load travels across it [4]. The influence lines corresponding to the load effects considered are illustrated in table 1. In this study the notion of influence line was extended from that of a unit load, to represent the effect induced by a 'unit truck'. Indeed it is extremely useful to show the variation of the load effect at a given position in the structure as a 'unit truck' (i.e., a truck with GVW of 1) travels across it. This function has been termed the *'characteristic response'* of the structure.

The characteristic response can be determined by adding the effects of all the axles when the spacings between the axles and the proportion of GVW carried by each axle are known. Figure 3 shows the characteristic response of a *'unit 5-axle truck'* with the properties shown in table 2 and illustrated in figure 2. Irregularities in the curve correspond to the arrival/departure of one of the axles on/off the bridge. This concept of characteristic response is very useful in developing the simple approach described in this paper.

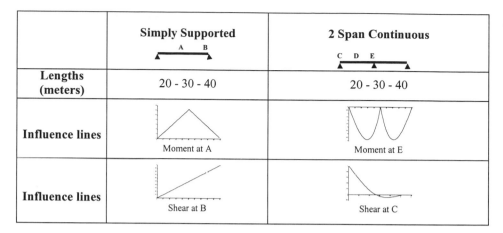

	Simply Supported A B	2 Span Continuous C D E
Lengths (meters)	20 - 30 - 40	20 - 30 - 40
Influence lines	Moment at A	Moment at E
Influence lines	Shear at B	Shear at C

Table 1 - Influence lines

	Axle 1	Axle 2	Axle 3	Axle 4	Axle 5
Proportion of GVW carried per axle	13 %	30%	19 %	19 %	19 %
	Spacing 1-2	Spacing 2-3	Spacing 3-4	Spacing 4-5	
Axle-spacing between consecutive axles	3.1 m	5.0 m	1.1 m	1.1 m	

Table 2 - Characteristics of a 5-axle truck modelled using real data from Auxerre

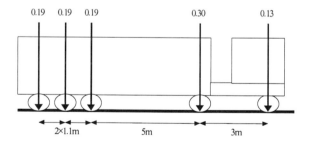

Figure 2 - Typical 5-axle truck used in the simulation for Auxerre

In figure 3 the X-axis goes from 0 to 30.2 meters since the 1st axle of the 10.2 m long truck is located at $X = 30.2$ when the 5th axle of the truck is at the end of the 20 m long bridge. For the moment at mid-support of a two-span continuous bridge, the maximum value induced by a 5-axles truck (with characteristics given in table 2 and GVW of 1) is 0.89 and occurs when the 1st axle of the truck is located at $X = 15.5$ meters.

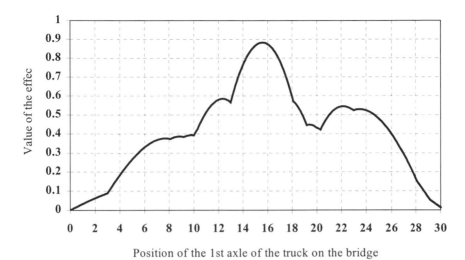

Position of the 1st axle of the truck on the bridge

Figure 3 - Characteristic Continuous Support Moment Response, Total Length 20m

MONTE CARLO SIMULATION AND LOAD EFFECT CALCULATION

Monte Carlo simulation is used to generate traffic records, with corresponding truck characteristics (GVW, speed and arrival time) generated using the probability distributions obtained form WIM records [5]. The simulations performed represent one month of traffic. The critical loading events are determined through identification of a meeting event in the traffic file, i.e., two trucks meeting on the bridge. Events where only one truck is present on the structure are not considered as critical. For every critical loading event identified, the load effects are calculated repeatedly as the trucks move at 0.1 m intervals along the bridge. Only the maximum value for each load effect is recorded.

The maximum values recorded for each of the critical loading cases form a statistical distribution of extreme values. This distribution can be fitted to a known distribution of extreme values: Gumbel (type I), Fréchet (type II) or Weibull (type III) [6]. A program is used to calculate a first estimate of the parameters characterising the distribution and then to optimise these parameters using a weighted least-squares method.

Generally the distributions of extreme values of the load effects have been found to be well described by a Weibull (type III) distribution. Following optimisation of the parameters of the distribution, the characteristic value (i.e., value with a specified probability of exceedance, α, during the design lifetime of the structure) can be determined by extrapolation to an appropriate return period [7], R :

$$R = R_{y\alpha} \cong \frac{-T}{Ln(1-\alpha)} \cong \frac{T}{\alpha} \quad if \quad 0 < \alpha << 1 \tag{1}$$

The characteristic values for extrapolation of the traffic effects, were determined such that the probability of exceeding the load effect during the specified design life of $T = 50$ years was calculated with a 0.95 fractile probability that y_α is exceeded during T. From equation (1), a return period of $R = 1000$ years is required. The characteristic value changes slightly from one simulation to another as the traffic is never exactly the same. Therefore a number of full simulations were run in order to get a statistical distribution for the characteristic value.

DEVELOPMENT OF A SIMPLE APPROACH

The aim of the simple approach is to provide an accurate estimation of the characteristic value of load effect due to imposed traffic loading on existing bridges without performing large scale simulation which requires a lot of computation time.

The critical loading situations for each of the load effects for the bridges considered is given by the meeting of two 'heavy' trucks within a critical influence zone. It is easier to analyse the situation in terms of characteristic response than in terms of influence line since the 'heavy' trucks are long (9 to 10 meters) and the bridges considered are short (20 to 40 meters). The load effect is highly sensitive to the location of the truck on the bridge and thus cannot be easily anticipated. An examination of the characteristic response allows the identification of the critical location(s) of the truck (i.e., the location(s) of the truck that induces the maximum effect).

The simple approach developed requires a knowledge of:
- the number of axles and axle spacings for each 'heavy truck',
- the critical location(s) for the 'heavy trucks'.

In order to answer the first question, the distribution of gross vehicle weights of all the 5-axle trucks going in the same direction is used. By obtaining this WIM data and modelling it using a Weibull type distribution, typical site specific weights for 'heavy trucks' can be generated. To obtain the critical locations corresponding to each load effect, the characteristic response for that particular effect and span length needs to be examined.

Figure 4 shows the Weibull distribution of GVW for all 5-axle trucks in 'Angers 1987'. The best-fit estimation of the parameters of the distribution was done for the right tail of the curve as only the extreme values are of interest.

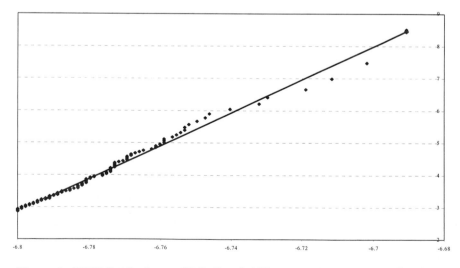

Figure 4 - GVW distribution on Weibull probability paper, Angers 1987 – direction 2

Table 3 shows the characteristic gross vehicle weights for different probabilities, sites and directions. The difference from one site to another and from one lane to another can be seen.

There are two possible ways to place the trucks on the bridge in order to induce the maximum effect:

1. The trucks can be placed at one particular location (such as the critical point on the influence line) and will induce a given load effect. This is the traditional approach.
2. The trucks can be placed in a zone and will induce a different load effect depending on their location inside that zone.

In the first case the load effect is known once the location is fixed (as the axle-spacings and the axle-weights are fixed). The value of the load effect is equal to the product of the characteristic response and the GVW of the truck. In the second case the load effect will lie somewhere in a fixed range defined by the zone in which the truck is located.

The first approach was initially used to determine an estimate of what the maximum load effect is when the two trucks are located at the critical location of the bridge. The second more accurate approach was then used to obtain a distribution of values and determine the mean and variance of that distribution.

	Direction	Parameters of the Weibull distribution			0.98	0.985	0.99	0.995
		Lambda λ	Delta δ	Beta β	X (y=0.98)	X (y=0.985)	X (y=0.99)	X (y=0.995)
Angers	Dir1	528	189	6.12	428	433	439	448
	Dir2	1748	1351	83.4	459	463	470	480
Auxerre	Dir1	1941	1583	36.91	517	528	544	570
	Dir2	873	422	10.23	585	593	604	622

Table 3 - Extrapolated values (kN) for gross vehicle weight for different probabilities and sites/directions

For a given load effect and span length, there is a location on the bridge for which the load effect is maximum (i.e., the maximum ordinate on the characteristic response plot) - see Figure 5. Different zones have been identified using the characteristic responses: zones in which the load effect induced would be at least X% of the maximum value. By decreasing the value of X, the mean of the load effect distribution is decreased and the spread of the distribution increased. Typical values of 80%, 85% and 90% were used in the simple approach.

RESULTS OF SIMPLE APPROACH

100 full simulations were run for each site and for each bridge length (20m, 30m and 40m). The distribution of the 100 characteristic values obtained was modelled as a Normal distribution, with optimised moments and the mean was found. The simple approach used 1000 critical meeting events (i.e., two trucks of fixed GVW placed randomly in the critical zone) for each site and bridge length. The distributions obtained from the 1000 values were found to be well fitted by a Normal distribution; the mean critical value was used.

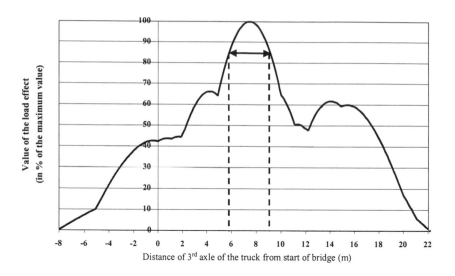

Figure 5 - *X*=85% zone for the characteristic response of Central Support Moment in Two-span continuous bridge, 20 m long

Table 4 shows the comparison of mean values of the characteristic load effects listed in table 1 for the 2 sites, 'Angers 1987' and 'Auxerre 1986'. The results presented in table 4 have been obtained using an 85% zone for the characteristic response and a value of *Y*=0.99 for the fixed gross vehicle weight. The 'SM' columns show the relative difference (in %) between the mean characteristic value from the full simulations and the corresponding value calculated by the simple approach. For the 3 span lengths considered, the results corresponding to effects 1 and 3a/3b are very similar for the full simulation calculation and the simple approach calculation. Results for effects 2 & 4a/4b demonstrate the sensitivity of the simple model to the shape of the characteristic response. However, the simple approach is still within 14.3% of the full simulation for all cases. The root mean square difference for all spans, effects and data is 7.34%.

Length (m) and Data Source	Effect 1			Effect 2			Effects 3a & 3b			Effects 4a & 4b		
	FS kNm	SM (%)	TR (%)	FS kNm	SM (%)	TR (%)	FS kN	SM (%)	TR (%)	FS kN	SM (%)	TR (%)
20 Angers	2920	1.7	11.8	740	8.1	17.6	605	7.4	20.2	385	14.3	32.5
Auxerre	3550	5.6	24.0	930	8.1	26.1	765	2.0	28.4	495	11.1	39.2
30 Angers	5200	0.4	10.6	1040	-7.7	8.1	685	6.6	18.1	490	10.2	24.5
Auxerre	6500	1.4	19.2	1340	-9.7	13.1	875	5.1	24.8	620	9.7	32.9
40 Angers	7450	0.1	9.9	1645	-10.6	-1.9	730	2.7	16.0	560	7.1	21.1
Auxerre	9350	0.8	18.0	2075	-10.8	4.8	965	-0.5	18.4	710	5.6	29.0

Table 4 - Results of Full Simulation (FS), Simple Method (SM) & Turkstra's Rule (TR)

The effects were also calculated using Turkstra's rule, i.e., by putting a 1000 year (return period) truck alongside a 20 week (return period) truck at the critical location for each effect considered. The results obtained from Turkstra's rule range from 1.9% below to almost 40% in excess of those obtained from the full simulation. The root mean square difference for all spans, effects and data is 21.6%.

CONCLUSIONS

This paper describes the development of a simple statistical approach to determine characteristic load effects in short- and medium-span bridges without the requirement for complex statistical analysis. The simple approach has been validated by comparison with a more complex statistical approach based on direct simulation. The accuracy of the approach has been tested for two different WIM sites with different traffic flow characteristics and for a range of spans and load effects. The results obtained demonstrate the sensitivity of the approach to the shape of the characteristic response. Overall, the simple approach is shown to be quite accurate, always being within 14.3% of the result of a full simulation. The conventional Turkstra's Rule, on the other hand, is significantly less accurate with errors of up to 40%.

REFERENCES

[1] BA79/98 (1998), *Design Manual for Roads and Bridges*, Department of Transport, London.

[2] ANG, A.H.S., Tang, W.H., *Probability concepts in engineering planning and design*, Volume 1, Wiley and Sons, 1975.

[3] JACOB, B.A. et al., *Traffic data of the European countries*, Report of WG2, Eurocode 1 part 3, March 1989.

[4] GHALI, A., *Structural Analysis*, E&FN Spon, 1997.

[5] O'CONNOR, A.J., Jacob, B., O'Brien, E.J. and Prat, M., 'Approche probabiliste des valeurs extremes de charges sur les ponts routiers', Seconde Conference Nationale, 'Fiabilite des Materiaux et des Structures', Champs sur Marne, France, Hermes, Nov. 1998.

[6] CASTILLO, E., *Extreme Value Theory in Engineering*, Academic Press, 1991.

[7] JACOB, B.A., 'Methods for the prediction of extreme vehicular loads and load effects on bridges', Report of Subgroup 8, Eurocode 1.3, LCPC Paris, 1991.

Boyne Railway Viaduct – structural assessment

Authors:
Rónán Gallagher, Post-graduate, Trinity College Dublin, Dublin 2, Ireland, Dermot O'Dwyer, Lecturer, Trinity College Dublin, Dublin 2, Ireland and Michael Hartnett, National University of Ireland, Galway, Ireland

Figure 1. Railway Viaduct crossing the river Boyne at Drogheda

ABSTRACT

This paper describes an analysis of the Boyne Railway Viaduct, as shown in Figure 1. The Boyne Viaduct is a three span simply supported structure, constructed in 1932 to replace the original wrought iron bridge. The Viaduct is constructed of riveted mild steel lattice girders and comprises one 81m and two 43m spans. The bridge carries both passenger and freight rail traffic. The assessment of the bridge involved both static and dynamic analyses.

The project's first stage involved the development of computer code and a three dimensional model to analyse various static load cases and axle positions. Validation of the computer code was carried out by comparing the results from the static analysis with those from previous work obtained from Strudl software. The second stage required the modification of the computer code to enable a dynamic analysis to be carried out. These dynamic analyses yielded values for the natural frequencies and mode shapes and gave the response of the structure to moving train loads. The developed dynamic analysis software was checked by comparison with the ANSYS Finite Element package.

Bridge Management 4, Thomas Telford, London, 2000.

Validation of the model was achieved by instrumenting the bridge and monitoring the bridges response to both static and dynamic loads. Comparison and comment on the correlation between the two sets of theoretical predictions and the measured values for deflections and natural frequencies is made in the paper.

Finally, an assessment of the factor of safety of the bridge from both a maintenance and an operational perspective is described.

THE BRIDGE AND ITS HISTORY

Iranród Éireann's Dublin Belfast railway line crosses the river Boyne at Drogheda, 32 miles north of Dublin by means of the Boyne Viaduct. The crossing originally carried double track and was opened on 4th April 1855. The original structure was approximately 536m in length, consisting of 15 semi-circular masonry arch spans and 3 girder spans. There are 12 masonry arches on the south approach and 3 on the north approach, each of about 19m clear span. The 3 girder spans comprise two 43m spans between bearings and one central span of 81m. These girders span over the waterway, quays and public road. The whole structure is supported on masonry piers with the underside of the centre girder being 30m above high water[1,2]. The masonry used throughout is of good quality limestone, solidly constructed and still in excellent condition.

The original structure was conceived by Sir John MacNeill, F.R.S. and was designed by James Barton, chief engineer of the railway operator, the Great Northern Railway[3]. The original girder spans were constructed wholly of wrought iron, the main girders being of the riveted double web multiple intersection latticed type, having flanges approximately 1m wide, 7m deep, and continuous over the piers, with and overall length of 167m. The original Boyne Viaduct was the first application of the principle of multiple-lattice construction to a bridge of this span. Due to the efficient design of the structure, it was capable of carrying with safety the heavier loads and higher speeds in use in 1931 when it was replaced. For about 45 years prior to its renewal, traffic over the Viaduct was restricted to one track and a speed limit of 25 miles per hour with a single locomotive having a maximum axle load of 17 tons. These restrictions had the effect of reducing the live load carried by each main girder to about 75% of the amount which would have been placed upon them had both tracks been occupied simultaneously.

The 3 arch spans on the North side and the North arch on the South side of the river are devoid of filling over the arch rings. The tracks were carried on 4 longitudinal single web double intersection latticed type wrought iron girders, having brick jack arching between them, on which the ballast was laid. On the remaining 11 arches the original filling between the arch ring and the underside of the ballast is still in position, but on the four arches referred to this was removed and the girders substituted in 1858. At that date, it was found that the abutment piers carrying the ends of the wrought iron spans were being trust towards the river. As the pier on the North side was not founded on solid rock, the engineers considered it advisable to remove the filling over the arches, but the work was never actually completed. After the filling over the four arches mentioned, had been removed a position of equilibrium had been apparently arrived at, and, while the North and South piers remain about 7 and 2 inches respectively out of plumb, no movement has occurred since.

In 1929 the Great Northern Railway Company intended to introduce locomotives having axle loads of 21 tons, as against the permissible axle of 17 tons over the Viaduct. As a result of these new planned loadings, and having regard for the age and condition of the wrought iron structure, tenders were invited both for, repairing and strengthening the old structure, and replacing it. Due to the excessive estimations for the repair work, a contract for the replacement of the bridge was awarded in 1930.

The new super-structure was designed to carry a single track and 18 units of British Standard Loading. The new bridge was to follow, in so far possible, the overall design of the existing structure. The design of the new structure was influenced by the necessity to construct it within the limits of the existing bridge while the existing bridge carried traffic. The upper chord of the centre span was given the appearance of being well curved, while sloping end posts were avoided for aesthetic reasons. The sloping end posts were also avoided in order to facilitate erection by cantilevering one span from that next to it. It was decided to incorporate similar expansion arrangements in the new structure as the old one, this resulted in the new three spans being bolted together on the neutral axes of the bottom chords of the main girders.

CONDITION OF BRIDGE

The bridge is in excellent condition with very little corrosion or deterioration and its present state can be accredited to routine maintenance. Figure 2 shows a detail of one of the bridge's bearings. It can be seen from the photograph that the bearing accommodates both longitudinal and rotational movement.

Figure 2. Bridge bearing allowing longitudinal movement and rotation

From an inspection of the bridge there is evidence of a system for lubrication, that is a vertical pipe located at each bearing. However, it was not clear from inspection whether the rotational slider is still free to move. This is because the magnitude of the rotation at this support is so small as to be insignificant. In the 1930s when the bridge was built, any calculations that needed to be done were carried out by hand. Consequently, the rotational calculations at the supports may not have been done and a rotational bearing was probably installed as a precaution.

REASONS FOR ASSESSMENT
Iranród Éireann, the current owner of the bridge commissioned and funded this assessment of the Viaduct to establish;

(i)   The effect on the structure of increasing the axle loads
(ii)   The effects on the structure of increasing the running speeds
(iii)   The peak stresses caused by rolling stock with different dynamic characteristics
(iv)   The effects of vertical track alignment irregularities, such as joints, on the peak stresses in the structure

Answering these questions has required the development of a mathematical model of the structure and a full dynamic analysis.

ANALYSES
The information required from the analyses was the peak stresses, the deflections and the interaction between the bridge and a passing train. To enable this to be undertaken both a static and dynamic analyses were carried out. In attempting to model the bridge two systems were identified, the first was two parallel trusses and the second was a system of cross beams. The first step taken was a simple 2D model which yielded the deflections of the main nodes of the bridge. However, in order to get a full picture a full 3D model needed to be developed.

Most of the members on the main girders arc of lattice type. As a result each lattice member had to be modelled individually. By modelling these compound members, using ASNSYS, and applying unit axial deflections and end rotations their equivalent stiffness could be calculated. A simplified structural model of the overall structure was developed with the complex lattice elements being represented by simple beam elements with equivalent properties. This procedure simplified the overall structural model. This simplified model was used to calculate nodal displacements. The stresses in the lattice members were then calculated by applying the calculated nodal displacements to the compound lattice member models.

i)   Model each lattice member
ii)   Develop equivalent cross sectional area and stiffness
iii)   Use these in overall model of structure
iv)   To obtain member stresses, apply model displacement from overall model to individual lattice model

The above steps were coded in a separate routine to enable the member stresses to be calculated independently from the main programme. This makes it easier to track the peak stresses in each member and also facilitates calculating the effects of localised corrosion.

In the static analysis two types of model were use, the first had all the joints fully fixed and the second was pin jointed. The real structure's joints probably behave in a semi-rigid manner. By carrying out analyses on the two extreme situations it was possible to establish whether the bridge's behaviour was sensitive to the rigidity of its joints. It was found that the overall deflection was virtually the same for both cases. This similarity also carried through to the dynamic analysis, where both the natural frequencies and eigenvectors were comparable.

DYNAMIC MODEL

The equation of motion of dynamic system may be defined as:

$$[m]\{\ddot{x}\}+[c]\{\dot{x}\}+[k]\{x\} = f(t) \tag{1}$$

Where m is the global structural mass matrix, c is the global structural damping matrix and k is the global structural stiffness matrix[4,5]. All of the matrices are of the order (n*n), where n is equal to the number of degrees of freedom of the structure. The forcing vector is of the order of (n*1). The damping matrix c is extremely difficult to evaluate accurately, because it simplifies the analyses and allows modal de-coupling c is taken to be linearly related to the stiffness and mass matrices.

$$[c] = a*[k]+b*[m] \tag{2}$$

Where a and b are real constants, and, k and m are the stiffness and mass matrices respectively.

The first computation carried out in the dynamic analysis was the evaluation of the natural harmonics of the system, that is, the first, lowest, natural frequencies and their corresponding eigenvectors. These values were arrived at by solving the equation of motion for free vibration. The expression for free vibration for a system is obtained by ignoring the damping in the general equation of motion and by setting it's right hand side, that is, the forcing function to zero.

$$[m]\{\ddot{x}\}+[k]\{x\} = 0 \tag{3}$$

Two different techniques were used in solving for the deflections of the bridge as a train passed over it. The first of these was a direct time-stepping method, which used the general equation of motion with all of the matrices of dimension (n*n), where n is the number of degrees of freedom of the system, and was then solved for each degree of freedom. The second method employed was the mode superposition method. In this technique, the first say, p, eigenvectors are written in matrix form [ϕ] of order (n*p), i.e. each column representing an eigenvector. The transpose of this matrix was also calculated and was of the order of (p*n). To enable the equation of motion to be reduced, a new variable z had to be introduced.

$$\{x\} = \{z\} * [\phi]$$
$$\Rightarrow \{\dot{x}\} = \{\dot{z}\} * [\phi]$$
$$\Rightarrow \{\ddot{x}\} = \{\ddot{z}\} * [\phi]$$

(4)

Each term in the equation of motion is then pre-multiplied by the matrix of eigenvectors and post-multiplied by the transpose of the matrix of eigenvectors.

$$\left[\phi^T\right] * [m] * \{\ddot{z}\}[\phi] + \left[\phi^T\right] * [c] * \{\dot{z}\}[\phi] + \left[\phi^T\right] * [k] * \{z\}[\phi] = \left[\phi^T\right] * f(t)$$

(5)

$$[M] * \{\ddot{z}\} + [C] * \{\dot{z}\} + [k] * \{z\} = \left[\phi^T\right] * f(t)$$

(6)

where

$$[M] = \left[\phi^T\right] * [m] * [\phi]$$

(7)

$$[C] = \left[\phi^T\right] * [c] * [\phi]$$

(8)

$$[K] = \left[\phi^T\right] * [k] * [\phi]$$

(9)

Where [M] is the modal mass matrix, [C] is the modal damping matrix and [K] is the modal stiffness matrix, each being of the order (p*p). This new equation is hence solved for {z}, and the actual displacements {x} are calculated by post-multiplying by [φ]. The advantage of using the modal system matrices is that the number of degrees of freedom necessary to describe the system's dynamic behaviour can be reduced significantly. The basic model has hundreds of degrees of freedom but five mode shapes, degrees of freedom, describe the overall behaviour of the structure adequately. This reduction in problem size significantly reduces the computation time required when a direct time stepping technique such as the Runge Kutta method is employed [4].

The output from the computer code output the eigenvectors and deflections in numerical form. This made it difficult to interpret the results and identify errors in the model. To overcome this a series of routines were written which allow the data to be passed to the drafting package AutoCAD. The data on the structure was transferred to AutoCAD as a data (.Dxf) file. This allowed the data to be viewed graphically and the graphical images manipulated. The first two eigenvectors of the bridge centrespan are shown in figure 3.

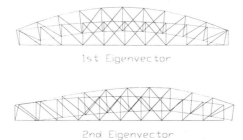

Figure 3. The first two eigenvectors of the centre span of the Boyne Viaduct.

MODEL CALIBRATION

Calibration of the model and the computer code used was done by comparing results given by the program with existing calculations. These existing calculations included results produced from hand calculations and from the M.I.T. Strudl software. The Finite Element package ANSYS was also used to validate the model. All the results found good agreement with each other. For example, shown in table 1 are the first five natural frequencies for the centre span of the Viaduct computed by both ANSYS and the program.

	Program (Hz)	Ansys (Hz)
1st natural frequency	2.8874	2.8923
2nd natural frequency	4.8014	5.1369
3rd natural frequency	4.9581	5.5729
4th natural frequency	5.3531	5.9477
5th natural frequency	5.6507	6.2914

Table 1. The first five natural frequencies of the centre span of the Boyne Viaduct.

DYNAMIC INTERACTION

The dynamic analysis of the interaction between the Viaduct and the rolling stock was tackled using direct time stepping [4,5,7,8]. This direct method has the disadvantage that it is computationally expensive. However, it has the very important advantage that it allows non-linear effects to be modelled. In order to model the dynamic interaction it was necessary to generate a simple dynamic model for the rolling stock [6,8].

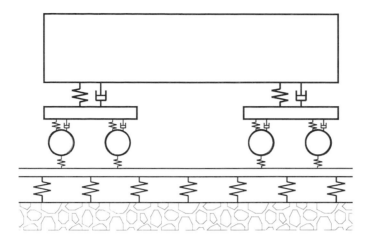

Figure 4. Simple two dimensional rail vehicle model.

The dynamic model of the bridge, which included the track support system, was treated in isolation from the rolling stock model. The interaction between these models was achieved as follows.

The time stepping model uses very short time steps so that the forces and accelerations can be assumed to remain constant during the time step. The Runge-Kutta integration technique uses the nodal displacements and velocities at the start of the time step along with the constant accelerations during the time step to calculate the nodal displacements and velocities at the end of the time step. At the start of the next time step these new displacements are use to calculate new values for the forces and accelerations and so the cycle continues.

In the Viaduct analysis the interaction forces between the bridge and the rolling stock were calculated at the start of each time step. The dynamic response to the two independent models where then calculated independently of one another. At the start of the next time step the relative positions of the bridge and rolling stock where again compared and the interaction forces recalculated. The two models are linked via the Hertzian contact springs [6]. This approach enabled non-linear effects to be modelled with relative ease. If a wheel was found to have lifted from the rail then no interactive forces would be applied to the wheel and at that location on the rail. Similarly by treating the two models separately it was relatively easy to add additional forces due to wheel flats and vertical rail mis-alignment.

EXPERIMENTAL MEASUREMENTS

The next step to take is to validate the model with measurements taken on site. This needs to be done for both static and dynamic load cases. The technique that will be employed involves a laser pointer situated at a location remote from the bridge and sighted on a target on the bridge. Sited on the bridge will be a digital camera that will record the relative movement of the laser pointer on the target. Strain gauge and accelerometer readings will also be taken. From these recordings the vertical acceleration, velocity and displacement of the bridge can be calculated.

CONCLUSIONS

The analyses that were carried out showed that there was little difference in modelling the structure as being pin jointed or having rigid joints.

Modelling the dynamic interaction between the bridge and the rolling stock can be carried simply using direct methods and small time steps. This allows the modelling of non-linear effects but requires very short time steps and hence is computationally expensive. Modal superposition is computationally efficient and can be used to give the overall dynamic behaviour with considerably fewer computations but non-linear effects remain problematic.

REFERENCES

1. Howden, G. B., *Reconstruction of the Boyne Viaduct, Drogheda*, Proceedings of the Institute of Engineers of Ireland
2. Houston, Terry, *Motherwell Bridge, The First Hundred Years*, M & M Press, 1994
3. Cox Ronald, C., Gould, M., H., *Civil Engineering Heritage : Ireland*, Thomas Telford, London, 1998
4. Thomson, William, T., *Theory of Vibrations with Applications*, 4th Edition, Chapman & Hall, London, 1993
5. Craig, Roy R., *Sturctural Dynamics, An Introduction to Computer Methods*, John Wiley & Sons, Chichester, 1981
6. Esveld, Coenraad, *Modern Railway Track*, MRT Productions, Duisburg, West Germany, 1989
7. Frýba, Ladislav, *Vibration of Solids and Structures under Moving Loads*, 3rd Edition, Thomas Telford, London, 1999
8. Frýba, Ladislav, *Dynamics of Railway Bridges*, Thomas Telford, London, 1996

Assessment of reinforced concrete slab bridge decks with circular voids

D W CULLINGTON and M E HILL
Transport Research Laboratory, Crowthorne, Berkshire, RG45 6AU, UK

ABSTRACT

When the assessment standard for concrete bridges was first issued, the method it contained for calculating the shear resistance of reinforced concrete voided slab decks was based on the design rules. A number of bridges were found to be sub-standard for shear and a programme of tests was started to review the method of assessment. Nineteen tests were carried out on models with a scale of 0.42:1. Several were used to investigate the effect of shear span. Others had variations in the voids, concrete strength, and reinforcement. After the first phase of tests, the Highways Agency promulgated a revised calculation that increased the resistance by 50% in typical configurations. The calculation derives the shear resistance of a voided slab from that of a solid slab, reducing it by a factor depending on the ratio of void diameter to distance between void centres. Further tests have been done to refine the method.

INTRODUCTION

When the rules for assessing the load carrying capacity of concrete bridges were prepared in the late 1980s (BD44/90, Department of Transport, 1990) they were based on the design standard with modifications based on available knowledge. There was little or no new research. It was intended to study individual topics as they arose in assessment and introduce modifications to the standard where possible.

One mode of assessment failure encountered in practice was longitudinal shear in reinforced concrete bridge decks containing circular voids. The assessment standard made no special provision for this type of deck, the beam rules being used as in design. Clause 5.3.3.2 of the standard states that the in the absence of shear reinforcement, the ultimate shear resistance V_u of a section is given by

$$V_u = \xi_s v_c b_w d$$

where,

ξ_s is the depth factor defined in Clause 5.3.3.2;
b_w is the breadth of the section which, for a flanged beam is taken as the rib width;
d is the effective depth to tension reinforcement;
v_c is the ultimate shear stress found using Clause 5.3.3.2.

This was thought to be conservative for shear in slabs with circular voids for two reasons:

- the shear capacity was based upon the minimum web thickness between adjacent voids ignoring a significant proportion of the section

- the area of reinforcement was generally limited to the bars contained within the minimum web width.

In addition, close to the interface between solid and voided sections there was a possibility that a form of enhancement could take effect. In an assessment, a section such as A-A in Figure 1 would be treated as voided. In practice two possibilities could be envisaged for the failure plane that could increase the resistance assigned to the section under the load.

(i)   situated wholly within the voided section - in this case the failure plane is forced to rise at a steeper angle than would otherwise be expected, as if there was a support at the interface

(ii) passing through the voided and solid sections – in this case having a resistance depending on the proportion of the plane in each section.

Figure 1. Possible shear failure planes

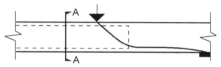

(i) Failure plane wholly within voided section

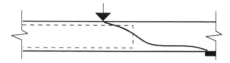

(ii) Failure plane crosses solid-voided interface

In view of the potential to increase the assessed resistance, the Transport Research Laboratory was commissioned by the Department of Transport to obtain experimental data on the subject and propose a revised method of assessment.

SPECIMENS AND TESTING METHOD
Nineteen load tests have been carried out on model voided slabs. The test slabs were based on an existing sub-standard voided slab bridge built to a scale of 0.42:1. They were cast with the tension steel in the top flange (top-cast bars) to reproduce the conditions that would be found in the hogging region of a bridge[1]. The slabs were turned over prior to testing so that the bearings supporting the slab represent vehicle loads and the loading jack represents the pier. The typical slab geometry and test configuration is shown in Figure 2.

Brief descriptions of the test specimens are given in Table 1 together with their geometric and material properties. Initially six slabs (VS1 to VS6) were tested to determine the basic shear resistance and investigate enhancement near the solid-voided interface. The slabs were of the same cross-section with the exception of VS4 which was solid.

In Table 1, VS5 is described as the reference specimen, the main difference from the others in the initial series being a detail change at the end of the transverse reinforcement. Descriptions such as "reduced web" and "increased tension steel" are relative to the reference slab.

---

[1] The effectiveness of tension steel located at the top of a reinforced concrete section at the time of casting is known to be inferior to members in which the tension steel is located near the bottom of the mould. The loss of effectiveness is caused by low quality concrete surrounding the steel (Clark and Thorogood, 1983).

Figure 2. Test arrangement

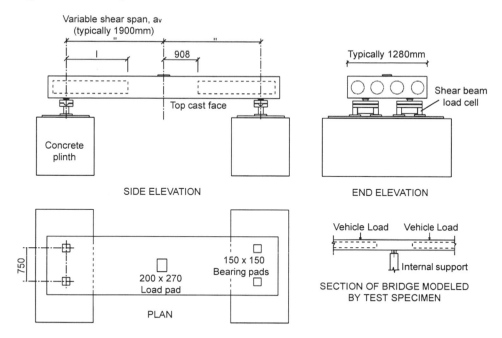

Voided slabs VS7 to VS12 were designed to examine the effects of changing the cross-section properties, such as void size and proportion of tension reinforcement. As far as possible only one variable was changed in each test. The final set of tests (VS13-VS18) was used to check the resistance model at extremes, and included a full-size specimen.

Apart from VS16 and VS17, which are beams, the models were slabs simply supported on four bearings. A hydraulic actuator operated through a servo-control panel and reacted off a cross-head provided loading.   The instrumentation consisted of eleven displacement transducers set along the longitudinal and transverse centrelines of the slab.

Load cells were located between the actuator and specimen to measure the applied load, and under the bearings to check that the load was evenly distributed. Load was initially applied in load control before being transferred to displacement control toward the end of each test. Records were made of the loads at which flexural and shear cracking occurred and crack propagation was regularly mapped throughout each test. All the specimens failed in shear. Where possible, load was reapplied after failure to determine the residual strength.

TEST RESULTS
The ratio of the test results to BD44/90 values are given in Table 1 where resistance is calculated using partial safety factors of unity and an allowance has been made to account for the self-weight of the specimens. The test results show that BD44/90 is conservative, the average ratio being 1.97 ignoring the solid slab results and high value from VS15 that arises from an extreme lateral distribution of the tension reinforcement.

Table 1. Details of voided slab test specimens

	No. of voids	Overall depth, h (mm)	Effective depth, d (mm)	Overall width (mm)	Distance between void centres, b (mm)	Void diameter, ø (mm)	% tension steel in full section	% tension steel in net section	Links	Depth of comp. flange (mm)	Concrete strength, f_{cu} (N/mm²)	Shear span, a_v	l	Ratio of test result to BD44/90 (2)	Description
VS1	4	353	309	1280	320	213	1.37	2.15	(1)	64	40	5.2d	2.2d	1.83	Standard slab tested at x=2.2d
VS2	4	341	309	1280	320	213	1.37	2.15	(1)	64	39	6.2d	3.2d	2.05	Standard slab tested at x=3.2d
VS3	4	341	309	1280	320	213	1.37	2.15	(1)	64	44	4.1d	1.1d	2.94	Standard slab tested at x=1.1d
VS4	0	341	309	1280	n/a	n/a	1.37	n/a	(1)	n/a	40	3.7d	n/a	0.88	Solid slab, first end
VS4b	0	341	309	1280	n/a	n/a	1.37	n/a	(1)	n/a	45	3.7d	n/a	1.04	Solid slab, second end
VS5	4	341	309	1280	320	213	1.37	2.15	None	64	41	6.2d	3.2d	2.01	Reference slab
VS6	4	341	309	1280	320	213	1.37	2.15	None	64	44	7.2d	4.3d	1.92	Standard slab tested at x=4.3d
VS7	4	341	311	1280	320	213	1.37	2.12	None	44	40	6.2d	3.2d	1.88	Reduced compression flange
VS8	4	341	311	1280	320	253	1.37	2.76	None	44	42	6.2d	3.2d	2.01	Increased void diameter, ø
VS9	4	341	311	1080	270	213	1.37	2.34	None	64	39	6.2d	3.2d	2.66	Reduced web width, b_w
VS10	4	341	311	1280	320	213	2.27	3.54	None	64	39	6.2d	3.2d	2.01	Increased tension steel, A_s
VS11	4	341	311	1280	320	213	1.37	2.12	0.07%	64	39	6.2d	3.2d	1.62	Shear reinforcement included
VS12	6	341	309	1280	213	142	1.37	2.72	None	100	42	6.2d	3.2d	2.10	Reduced void diameter
VS13	4	341	309	1280	320	253	0.92	1.86	None	44	20	6.2d	3.2d	2.23	Weak slab
VS14	2	341	309	1066	533	213	2.83	3.53	0.07%	64	88	6.2d	3.2d	1.01	Strong slab
VS15	5	341	302	1280	256	142	1.00	1.29	None	150	30	6.2d	3.2d	9.37	Random changes
VS16	1	341	309	320	320	213	1.37	2.15	None	64	40	6.2d	3.2d	1.76	Beam as reference slab
VS17	1	812	738	762	762	507	1.38	2.16	None	153	31	4.8d	3.2d	2.28	Full-scale, otherwise as V16
VS18	2	341	305	1066	533	213	1.49	1.90	None	64	65	6.2d	3.2d	1.28	VS14 reduced A_s, f_{cu}, no links

(1) The transverse reinforcement in VS1-VS4 was anchored by bends that contributed to the shear capacity. The effect of this detail was quantified by calculation and verified by the test on VS5. Results from the four test results have been adjusted to allow for this.

(2) In the calculation of resistance the depth factor has been applied using the actual depth of the model. Partial factors being unity.

With the exception of the two solid slab tests (VS4 and VS4b) and VS14 which had only two small voids and therefore was close to solid, the lowest ratio of applied shear force to assessed shear capacity is 1.62. The ratio for the solid slab VS4 is 0.88 and that for VS14, which has a small void area, is 1.01. These low values are believed to be caused by the top-cast bar effect.

It was found that the measured resistance was independent of the distance between the support (load) and the solid-voided interface from 4.3d to 2d (ratios 1.92, 2.01, 2.05, and 1.83 in descending order of distance). In general, the failure plane was principally in the voided section but with the support at a distance of d it was partially in the solid section. There was an increase in resistance (ratio 2.94) but it was still substantially less than the solid slab strength.

DEVELOPMENT OF RESISTANCE MODEL
The resistance model was initially devised from the results of tests VS1 to VS6, ignoring the increase in resistance occurring at d from the voided-solid interface. It is shown with all test results in Figure 3.

The model takes a different approach to BD44/90. The shear resistance of a voided slab is now found from the resistance of a solid slab having the same overall dimensions by applying a multiplying factor of less than unity to account for the voids. It was assumed that the reduction in resistance was attributable to:

- a reduction in aggregate interlock caused by the loss of concrete area in the voided region
- a reduction in dowel action caused by loss of effectiveness of the reinforcement in the flange region
- a reduction in compression zone shear caused by encroachment of the voids into the compression zone.

A fit was sought with the experimental data using an expression of the form:

$$V_{cv} = [1-(M. \ (ø/b) +N. \ (ø/b)^{x})].V_{c'} \tag{1}$$

where

$V_{c'}$ is the resistance of the solid slab calculated to BD44, ignoring the presence of voids
V_{cv} is the resistance of the circular voided slab
$ø$ is the diameter of the void
b is the distance between centres (or the width of the "voided beam" element)
$M.(ø/b)$ accounts primarily for the loss of aggregate interlock
$N.(ø/b)^{x}$ accounts primarily for the loss of dowel action and compression shear
x is a constant used to fit the curve to the data.

Although the expression was simply intended to fit the experimental data, some consideration was given to the relative magnitude of the components of shear resistance in its formulation. This was desirable because in the initial tests ø/b was constant. A fit was therefore sought with the constants M and N in an acceptable ratio. Using this principle, the experimental data from VS1 to VS6 led to values of M=0.4, N=0.6 and x=2.5, thus giving the expression:

$$V_{cv} = [1-(0.4 \, (ø/b) +0.6 \, (ø/b)^{2.5})]V_c.$$ (2)

Other constraints were that V_{cv} should (a) tend to V_c as $ø/b$ tends to zero – i.e. when the voids diminish to nothing the slab becomes solid and (b) tend to zero when $ø/b$ tends to unity. The resistance model represented by expression (2) was first issued to Regional Bridge Engineers in the form of notice BE21/14/011 (Chakrabarti, 1995) and later adopted in the Advice Note BA44/96 (Highways Agency et al, 1996).

Figure 3. Test data plotted against new and modified assessment methods.

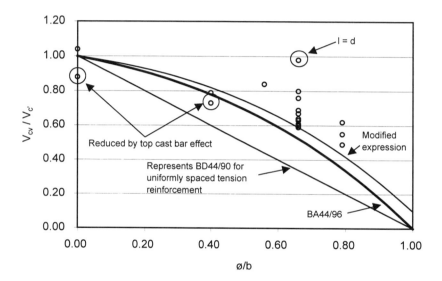

For application in assessment, the following restrictions were proposed, based on the model dimensions used in the tests:

a) $ø/b$ not to be greater than 0.7
b) $ø/h$ not to be greater than 0.65, where h is the overall section depth
c) thickness of the compression flange not to be less than 0.4(h-ø): i.e. the voids should be near the centre of the slab.

Outside these limits the original resistance model in BD44/90 was retained. In the initial test series the slabs had no shear reinforcement, but this was not part of the restrictions. It was checked in later specimens. The increase in resistance close to the interface was not accounted for because it was limited to a small region of the slab and not generally at the position of maximum shear.

Tests VS7 to VS18 were carried out to corroborate the resistance model over a range of geometries and if possible to reduce the restrictions and improve the assessed capacity. Based on these tests it was proposed to modify the restrictions as follows:

a)   ø/b not to be greater than 0.80 – based on VS8 and VS9
b)   ø/h not to be greater than 0.75 – based on VS8
c)   the thickness of the compression flange not to be less than 0.35(h-ø) – based on VS7 and VS8
d)   when ø/h is 0.4 or less an increase of 15% is permitted in the net resistance of the slab based on VS12 and VS15
e)   the component of resistance provided by shear reinforcement and calculated in the normal way may be added to the net resistance of the voided slab but is limited to 50% of it – based on VS11.
f)   the longitudinal tension reinforcement used in the calculation of the solid slab resistance must be limited to 3% of the net area of the voided slab

The result from VS12 demonstrated the presence of a significant reserve of strength over the assessed capacity obtained from expression (2). However, a single test result is insufficient to devise an alternative model with varying ø/h and so the increase outlined in (d) was suggested as a simple of modification for bridges with small voids. Expression (2) is aimed at sections with typical-size voids in a central position. It is adequate for other configurations but as it depends only on the ratio ø/b and not ø/h, it is a simplification that could probably be improved with further tests and a more complicated resistance model.

In addition to these relaxations it is possible to modify the constant terms in the expression for V_{cv} but the improvement is small (see Figure 3) and does not apply to near-solid slabs where top-cast bars reduce the experimental resistance. In the modified expression, M becomes 0.3. For this to be conservative, the longitudinal tension reinforcement used in the calculation of the solid slab resistance has to be limited to 3% of the net area of the voided slab as given in (f). This is instead of applying the limit to the area of a solid member as in BD44/90. It is based on the result of VS10, which was heavily reinforced, and gave a non-conservative result when compared with the assessed capacity obtained from the modified expression. It is not strictly necessary with expression (2). It has nevertheless been used to derive the assessed resistance for VS10 as plotted in Figure 3.

The average ratio of measured to calculated resistance using BA44/96 is 1.33 compared with 1.97 using BD44/90, and the improvement ratio is therefore 1.48.

CONCLUSIONS
- The results of nineteen tests on voided slabs show that the assessment standard BD44/90 is conservative for this type of deck.
- A new method of calculation promulgated in BA44/96 finds the resistance of the voided section by applying a reduction factor to the resistance of a solid section.
- A number of restrictions on the use of the new method were imposed to exclude its use outside of the geometrical range used in the test programme.
- Since the issue of BA44/96 further testing has been completed that suggests the rules restricting the application of the method could be relaxed.

ACKNOWLEDGEMENTS
The work described in this paper was carried out under a contract from the Highways Agency and has been prepared with their permission and that of the Transport Research Laboratory. The views expressed are not necessarily those of the Highways Agency or TRL. Thanks are

due to Professor L A Clark for his valued advice and recommendations, and to the members of staff at TRL who carried out the testing work.

REFERENCES

CHAKRABARTI S (1995). *Longitudinal shear resistance of voided slab bridge decks.* BE21/14/011, Bridges Engineering Division, Highways Agency, London.

CLARK L A (1992). *Longitudinal shear strength of voided slab bridges.* Report prepared for the Transport Research Laboratory, Crowthorne.

CLARK L A and THOROGOOD P (1983). *Flexural and punching shear strengths of concrete beams and slabs.* TRRL Contractor Report 60, Transport and Road Research Laboratory, Crowthorne.

DEPARTMENT OF TRANSPORT (1990). Departmental Standard *Standard BD44/90 The assessment of concrete highway bridges and structures.* HMSO.

HIGHWAYS AGENCY, THE, THE SCOTTISH OFFICE DEPARTMENT, THE WELSH OFFICE and THE DEPARTMENT OF THE ENVIRONMENT FOR NORTHERN IRELAND (1996). *Advice Note BA44/96 The assessment of concrete highway bridges and structures.* HMSO.

Application of extreme models for the evaluation of long term stress redistribution in repaired bridge girders

Prof. FRANCO MOLA,
Department of Civil Engineering, Politecnico di Milano, Italy
Eng. GABRIELE MEDA,
Structural Engineer, INPRO Cons. Eng., Milano, Italy
Eng. ALESSANDRO PALERMO,
Specialisation School for R.C. Structures "F.lli Pesenti", Politecnico di Milano, Italy

Abstract
The long term analysis of repaired bridge girders collaborating with rebuilt slabs is presented. The problem is approached in a general way or recurring to extreme creep models able to reduce the computational efforts and to maximise the redistribution of stresses affecting the girder and the slab. The proposed algorithms, applied to a case study, allow to discuss the basic aspects of the long term behaviour of repaired bridge girders.
Keywords: girder,slab, creep, shrinkage, stress redistribution.

1. Introduction

A widely adopted procedure for repairing bridge girders consists in the replacement of the slab with a new one. In this way it is possible both to eliminate the deck damaging and to modify the slab geometry allowing to increase the load bearing capacity and to enlarge the deck width. When this procedure is prescribed it has to be taken into account that the repaired structure will exhibit a high degree of rheological nonhomogeneity so a significant stress redistribution between the slab and the girder will take place under the effects of sustained loads or imposed deformations. Observing that the slab has a markedly lower age with respect the girder, different stress redistributions can be present when considering the effects due to sustained actions representing the self-weight of the slab, the self weight of the girder, the loads connected to the finishing works and differential shrinkage. In fact if we proceed in rebuilding the slab without propping the girder, the slab behaves as a delayed restraint for the girder and collaborates in equilibrating the sustained loads due to the structural self-weight. On the contrary the slab behaves as a preexisting restraint for the girder when the effects of the loads due to the finishing works and differential shrinkage are investigated. The stress redistribution connected to these various actions are different as for delayed restraint they are essentially dependent on the residual creep of the girder while for preexisting restraint they are connected to the time-dependent behaviour of the slab. In order to refinedly evaluate the stress redistribution which takes place in repaired bridge girders it is necessary to carefully describe the rheological behaviour of the slab and of the girder. Referring to linear viscoelasticity it is possible to proceed using Mc Henry Principle of Superposition. In this way the compatibility relationships governing the problem assume the form of a system of Volterra integral equations which can be solved by means of standard numerical techniques. This way of proceeding, even though theoretically exact, is affected by various uncertainties, in particular the ones connected to the evaluation of the parameters of the creep models. In order to obtain reliable results it is convenient to adopt extreme models to describe the time dependent behaviour of concrete, able to maximise the stress redistribution between the slab and the girder. This aim can be achieved assuming for the aged girder a viscoelastic behaviour according to the Kelvin-Voigt non ageing model and for the slab the rate of creep or Dischinger model. As a particular case the sectional arrangement considering a purely elastic girder can be considered. In the present work the basic aspects of the sectional analysis of repaired bridge girders adopting a general approach according to Mc Henry Principle of Superposition or assuming extreme creep models are discussed. The derived procedure is then applied for the analysis of a significant case study.

2. Problem formulation

Referring to Fig.1 and indicating by $X_1(t)$, $X_2(t)$ the statically indeterminate unknowns acting between the slab and the girder, the application of Mc Henry Principle of Superposition [1] allows to write the subsequent matrix compatibility equation when the slab and the girder are collaborating before the application of loads or imposed deformations

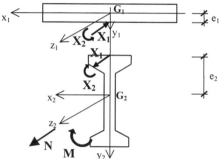

Fig.1 Repaired bridge girder

$$\int_0^t \left[\underline{\underline{C}}^{(1)}E_0J^{(1)}(t,t')+\underline{\underline{C}}^{(2)}E_0J^{(2)}(t,t')\right]d\underline{X}(t')+\underline{\underline{D}}^{(2)}E_0\underline{Q}J^{(2)}(t,t_0)+\underline{\lambda}^{(1)}(t)+\underline{\lambda}^{(2)}(t)=0 \qquad (1)$$

In Eq. (1) $\underline{\underline{C}}^{(1)},\underline{\underline{C}}^{(2)}$ are the flexibility matrices of the slab and of the girder referred to the conventional elastic modulus E_0, $\underline{\underline{D}}^{(2)}$ is the flexibility matrix of the girder connected to the external load vector $\underline{Q}^T=|N \quad M|$, $J^{(1)},J^{(2)}$ are the creep functions of the slab and of the girder and $\underline{\lambda}^{(1)}(t),\underline{\lambda}^{(2)}(t)$ are the vectors of the imposed deformations.

For $\underline{\underline{C}}^{(1)},\underline{\underline{C}}^{(2)},\underline{\underline{D}}^{(2)}$ we respectively have

$$\underline{\underline{C}}^{(1)}=\frac{1}{E_0A_1}\begin{vmatrix}1 & \dfrac{e_1}{r_1^2}\\[2mm] \dfrac{e_1}{r_1^2} & \dfrac{1}{r_1^2}\end{vmatrix}, \quad \underline{\underline{C}}^{(2)}=\frac{1}{E_0A_2}\begin{vmatrix}1 & -\dfrac{e_2}{r_2^2}\\[2mm] -\dfrac{e_2}{r_2^2} & \dfrac{1}{r_2^2}\end{vmatrix}, \quad \underline{\underline{D}}^{(2)}=\frac{1}{E_0A_2}\begin{vmatrix}1 & 0\\[2mm] -\dfrac{e_2}{r_2^2} & \dfrac{1}{r_2^2}\end{vmatrix} \qquad (2)$$

At initial time Eq (1) gives

$$\left[\underline{\underline{C}}^{(1)}\frac{E_0}{E_1(t_0)}+\underline{\underline{C}}^{(2)}\frac{E_0}{E_2(t_0)}\right]\cdot\underline{X}(t_0)+\underline{\underline{D}}^{(2)}\frac{E_0}{E_2(t_0)}\cdot\underline{Q}+\underline{\lambda}^{(1)}(t_0)+\underline{\lambda}^{(2)}(t_0)=0 \qquad (3)$$

so the initial value $\underline{X}(t_0)$ of the unknown vector can be easily calculated.

When the slab is made collaborating with the girder at time $t_0^* > t_0$ with t_0 time of application of loads and imposed deformations, Eq. (1) assumes the following form

$$\int_0^t \left[\underline{C}^{(1)}E_0 J^{(1)}(t,t') + \underline{C}^{(2)}E_0 J^{(2)}(t,t')\right] d\underline{X}(t') + \underline{D}^{(2)}E_0 \underline{Q}\left[J^{(2)}(t,t_0) - J^{(2)}(t_0^*,t_0)\right] +$$
$$+ \underline{\lambda}^{(1)}(t) - \underline{\lambda}^{(1)}(t_0^*) + \underline{\lambda}^{(2)}(t) - \underline{\lambda}^{(2)}(t_0^*) = 0 \tag{4}$$

expressing that no relative deformations between the slab and the girder are allowed after t_0^*.

From Eq. (4), at time t_0^* we immediately deduce the following relationship

$$\underline{X}(t_0^*) = 0 \tag{5}$$

giving the initial condition for Eq. (4).

Eqs. (1) (4) represent two matrix Volterra integral equations which can be solved by means of standard numerical methods. In particular, adopting the mid – point technique for evaluating the integrals Eq. (1) can be written in the following numerical form

$$\left[\underline{C}^{(1)}b_{kk}^{(1)} + \underline{C}^{(2)}b_{kk}^{(2)}\right]E_0\Delta\underline{X}_k = -\underline{D}^{(2)}\underline{Q}E_0 J^{(2)}(t_k,t_0) - \underline{\lambda}^{(1)}(t_k) - \underline{\lambda}^{(2)}(t_k) -$$
$$- \sum_{i=1}^{k-1}\left[\underline{C}^{(1)}b_{ik}^{(1)} + \underline{C}^{(2)}b_{ik}^{(2)}\right]E_0\Delta\underline{X}_i \tag{6}$$

where according to fig.2 $\Delta\underline{X}_i = \underline{X}(t_i) - \underline{X}(t_{i-1})$ is the variation of the unknown vector between the two generical times t_i, t_{i-1} assumed for subdividing the interval $(t = t_k) - (t_1 = t_0)$ with $\Delta\underline{X}_1 = \underline{X}(t_0)$ and

$$b_{ik} = \frac{1}{2}\left[J(t_k,t_i) + J(t_k,t_{i-1})\right] \tag{7}$$

The same algorithm applied to Eq. (4) drives to the subsequent expression for $t \geq t_0^*$

$$\left[\underline{C}^{(1)}b_{kk}^{(1)} + \underline{C}^{(2)}b_{kk}^{(2)}\right]E_0\Delta\underline{X}_k = -\underline{D}^{(2)}\underline{Q}E_0\left[J^{(2)}(t_k,t_0) - J^{(2)}(t_0^*,t_0)\right] - \left[\underline{\lambda}^{(1)}(t_k) - \underline{\lambda}^{(1)}(t_0^*)\right] -$$
$$- \left[\underline{\lambda}^{(2)}(t_k) - \underline{\lambda}^{(2)}(t_0^*)\right] - \sum_{i=1}^{k-1}\left[\underline{C}^{(1)}b_{ik}^{(1)} + \underline{C}^{(2)}b_{ik}^{(2)}\right]E_0\Delta\underline{X}_i \tag{8}$$

with $\Delta\underline{X}_1 = 0$.

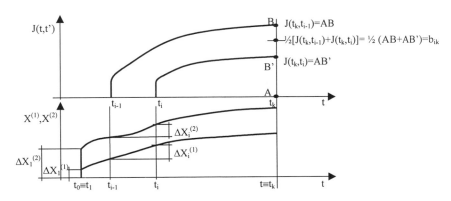

Fig.2 Numerical evaluation of superposition integrals

The two algebraic systems (6) (8), solved in sequential way, i.e. putting k=1,2.....n with n arbitrarily large, allow to calculate the time evolution of the state of stress and deformation in repaired bridge beam sections. The accuracy of the solution depends on the reliability of the assumed creep functions $J^{(1)}, J^{(2)}$. At this subject many refined models are nowadays available, in particular the one introduced by CEB MC90 [2] allows to derive quite good solutions. Even though the models are intrinsically well refined the solutions of Eqs. (6) (8) can be significantly made unreliable by an incorrect choosing of the main parameters of the models. Taking into account that many uncertainties affect the model parameters, especially the age of the girder, the times of loads application and the time of the connection between the girder and the slab, it is convenient to introduce some considerations in order to investigate about the limiting values of the unknowns $X_1(t), X_2(t)$. We start observing that the girder is considerably more aged with respect the slab so its residual creep deformations are quite small.

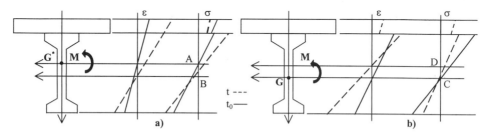

Fig. 3 Time evolution of strains and stresses in the repaired bridge girders

Referring to Fig.3a when the loads are applied after the connection between the girder and the slab, the neutral axis, initially located at the centroid G^* of the transformed section, shifts downwards in time consequently in the girder, exhibiting smaller creep deformations, the stresses increase while in the slab they decrease. When the loads are applied before the connection between the girder and the slab we have the situation of Fig.3b, where at initial time the neutral axis lies at the centroid G of the beam and shifts upwards in time. Consequently the stresses in the girder decrease while in the slab increase with zero value at initial time. Regarding Fig.3a, when the girder and the slab exhibit the same creep function $(J_1 = J_2)$ the section is rheologically homogeneous, the neutral axis does not move in time and the state of stress in the whole section remains constant so that no redistribution of stress takes place between the slab and the girder.

The maximum value of the segment AB representing the shifting of the neutral axis associated to the maximum redistribution of stresses takes place when the girder and the slab present the maximum rheological inhomogeneity i.e. for an elastic beam and a slab exhibiting a rate of creep behaviour. This assumption gives the upper bound for the stress redistribution values. Taking into account that a residual creep is always present in the girder a better approximation can be achieved adopting for the girder a non ageing creep behaviour assuming as final creep coefficient the one obtained from the CEB MC90 model connected to very high times of loading ($t_0 \cong 5 \cdot 10^3 \div 10^4$ days). When dealing with the problem of Fig.3b the maximum shifting CD of the neutral axis is obtained for homogeneous sections while considering the girder elastic no dislocations of the neutral axis are possible. In this second case also, considering the marked difference in age between the slab and the girder the adoption of a rate of creep model for the slab and of a non ageing model for the girder drives to suitable results.

For the evaluation of the solution of the problem, referring to the well refined creep model of CEB MC90 and to the limiting creep models now introduced we obtain the subsequent relationships for the creep functions J(t,t')

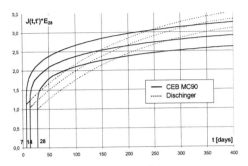

Fig.4 Slab rate of creep model   Fig.5 Girder non ageing model

CEB MC90 Model, [2]

$$J(t,t') = \frac{1}{E(t')} + \frac{\varphi_0(t')}{E_{28}} \beta_c(t-t') \tag{9}$$

RATE OF CREEP MODEL (Dischinger Model), [3]

$$J(t,t') = \frac{1}{E_D(t')} + \frac{\varphi_\infty^{(D)}}{E_{28}} \left(e^{-\beta(t'-\bar{t}_0)} - e^{-\beta(t-\bar{t}_0)} \right) \tag{10}$$

$\overline{t}_0$ = hardening time

NON AGEING MODEL (Kelvin-Voigt Model), [4]

$$J(t,t') = \frac{1}{E_k} \left[1 + \varphi_\infty^{(K)} \left(1 - e^{\frac{t-t'}{\tau^*}} \right) \right] \tag{11}$$

According to the hypothesis previously discussed we can assume for the slab the creep function given by Eq. (10) putting , $\varphi_\infty^{(D)} = \varphi_0(t_0)\dfrac{E(t_0)}{E_{28}}$, $E_D(t') = E(t')$. For the girder we on the contrary assume $\varphi_\infty^{(K)} = \varphi_0(\overline{t}_0)\dfrac{E(\overline{t}_0)}{E_{28}}$, $E_K = E(\overline{t}_0)$ with $\overline{t}_0$ age of the girder at time of loading. Finally for both models we can put $\beta = \dfrac{1}{\tau^*} = 0,005$ days^{-1} obtaining the creep functions reported in Figs. 4,5. Remembering Eqs. (10) (11) for the coefficients $\beta_{ik}^{(1)}, \beta_{ik}^{(2)}$ we respectively write

$$b_{ik}^{(1)} = \frac{1}{2}\left[\frac{1}{E(t_i)} + \frac{1}{E(t_{i-1})} \right] + \frac{1}{2}\frac{\varphi_\infty^{(D)}}{E_{28}}\left[\left(1 + e^{-\beta\Delta t_i} \right) \cdot e^{-\beta(t_{i-1}-\bar{t}_0)} - 2e^{-\beta(t_k-\bar{t}_0)} \right] \tag{12}$$

$$b_{ik}^{(2)} = \frac{1}{E_K}\left[1 + \varphi_\infty\left(1 - \frac{1}{2}e^{-\left(\frac{t_k-t_{i-1}}{\tau^*}\right)}\left(1 + e^{\frac{\Delta t_i}{\tau^*}} \right) \right) \right] \qquad \Delta t_i = t_i - t_{i-1} \tag{13}$$

$$J^{(2)}(t_k,t_0) = \frac{1}{E_k}\left[1 + \varphi_\infty^{(k)}\left(1 - e^{-\left(\frac{t_k-t_0}{\tau^*}\right)}\right)\right]$$

(14)

$$J^{(2)}(t_k,t_0) - J^{(2)}(t_0^*,t_0) = \frac{1}{E_K}\left[\varphi_\infty^{(k)}\left(e^{-\left(\frac{t_0^*-t_0}{\tau^*}\right)} - e^{-\left(\frac{t_k-t_0}{\tau^*}\right)}\right)\right]$$

(15)

When for the girder the limiting case of elastic behaviour is assumed Eqs. (13) (14) (15) have to be substituted by the following expressions

$$b_{ik}^{(2)} = \frac{1}{E_K} \qquad J^{(2)}(t_k,t_0) = \frac{1}{E_K} \qquad J^{(2)}(t_k,t_0) - J^{(2)}(t_0^*,t_0) = 0$$

(16)

and the last Eqs. (16) shows that no redistribution of stresses can be expected for loads applied before the connection between the slab and the girder. It is noteworthy to observe that the hypothesis of assuming limiting creep models to describe the long-term behaviour of repaired bridge girders allows to simplify the mathematical aspects of the problem. In fact the constitutive laws of Dischinger and Kelvin-Voigt models are espressed as linear first order differential forms so that Eq.(1) can be reduced to a system of two linear differential equations with variable coefficents. For the solution we can proceed by means of the finite difference method which, not requiring to evaluate the load history represented by the sum at second member of Eqs.(6)(8), allows to strongly reduce the computational efforts.

3. Case study

Let us consider the bridge girder $\boxed{2}$ of fig.6 to which the new slab $\boxed{1}$ substituting the previous highly degraded one is connected. At time of the slab replacing the age of the girder is $7 \cdot 10^3$ days (~ 20 years) and a residual shrinkage $\varepsilon_{sh2} = -10 \cdot 10^{-5}$ is expected for it. Regarding the slab an age of 7 days is assumed together with a final shrinkage $\varepsilon_{sh1} = -60 \cdot 10^{-5}$. The creep functions for the two parts assuming the CEB MC90 Model or the simplified Dischinger-Kelvin-Voigt approach are sketced in Fig.7.

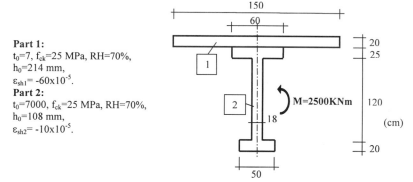

Part 1:
$t_0=7$, $f_{ck}=25$ MPa, RH=70%,
$h_0=214$ mm,
$\varepsilon_{sh1}= -60 \times 10^{-5}$.
Part 2:
$t_0=7000$, $f_{ck}=25$ MPa, RH=70%,
$h_0=108$ mm,
$\varepsilon_{sh2}= -10 \times 10^{-5}$.

Fig.6 Repaired bridge girder

The repaired girder has been studied for the two effects respectively connected to the application of a constant bending moment M=2500KNm and to differential shrinkage assuming the CEB MC 90 creep Model or the Dischinger Kelvin-Voigt approach. In fig.8 the evolution in time of the stresses in the slab and in the girder are reported. We observe a reduction of the compressive stresses in the slab accompanied by a corresponding increase of the stresses in the girder.

The different approaches give very close results while for the curvatures, illustrated in fig.9, the results are close only for long times with maximum variations of about ± 30% for intermediate times.

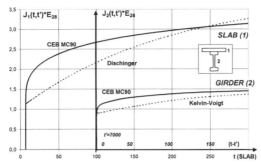

Fig.7 Creep functions for slab (1) and girder (2).

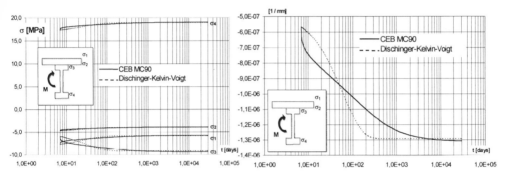

Fig.8 Stresses due to bending moment   Fig.9 Curvatures induced by constant moment

The stresses due to differential shrinkage are reported in fig.10. The maximum stress σ_2 in the slab and the maximum one σ_3 in the girder present at final time the maximum divergence and analogous considerations can be made regarding the curvature, sketched in fig.11 where a difference of about 15 % can be observed for long times. In any case the approach based on the limiting models give results on the safe side so it can be used for design purposes, enphasizing the possibility of proceeding by means of simpler algorithms for long term analysis of aged girders made collaborating with rebuilt slabs.

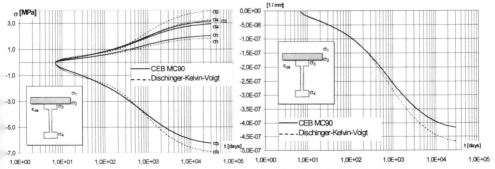

Fig.10 Stresses due to shrinkage   Fig.11 Shrinkage induced curvature

4. Conclusion

The long term general analysis of repaired bridge girders collaborating with re-built slabs, represents a quite difficult problem as the rheological inhomogeneity of the two parts requires to approach the problem recurring to the Mc Henry Superposition Principle and to solve a system of Volterra integral equations. The simplified approach based on the assumption of the Kelvin-Voigt non ageing creep model for the aged girder and of the Dischinger rate of creep model for the slab significantly reduces the computational efforts and drives to reliable results lying on the safe side. For this reason this simplified approach can be recommended for practical proposes in order to reach, without making cumbersome calculations, reliable results allowing to define the basic prerequisites of the repaired girders. Nevertheless, taking into account the importance of assuring for the repaired bridge girder a reliable behaviour together with a good durability, a second step analysis based on the application of the general procedure related to Mc Henry Principle of Superposition should be in any case mandatory.

5. References

1. Mc Henry, D. (1943), *A New Aspects of Creep in Concrete and its Application to Design*, Proc. ASTM 43.
2. CEB – FIP, (1993), *Model Code 1990, Design Code*, Thomas Telford, London.
3. Dischinger, F. (1939), *Untersuchung über die Knicksichereit die Elastische Verformung und das Kriechen des Betons bei Bogenbrücken*, Der Bauingenieur, H.33/34.
4. Nowacki, W. (1965), *Theorie du Fluage*, Eyrolles, Paris.

Active control of bridge loads

T. ATKINSON, PAT-GB, Wolverhampton, UK, P. BROWN, Oxfordshire County Council, UK, J. DARBY, Consultant, Abingdon, UK, T. EALEY, Buro Happold, Bath, UK, J. LANE, Transport Research Laboratory, Crowthorne, UK, J. W. SMITH, and Y. ZHENG, University of Bristol, UK

INTRODUCTION

Under European Union legislation, heavy goods vehicles of up to 40 tons gross weight were allowed on UK highways from January 1st 1999. The Highways Agency was already implementing government policy to upgrade the UK bridge stock by a programme of assessment and strengthening. There are approximately 155,000 bridges in the UK of which about 13,000 on major trunk roads and motorways are the responsibility of the Highways Agency. The remainder are in the care of local authorities, Railtrack and private organisations (Leadbeater, 1997). It is estimated that more than 20% have strength deficiencies in some form (Williams, 1997).

Bridges that fail their assessments have to be repaired, strengthened, demolished or subjected to load limits. Any form of structural rehabilitation or complete replacement is likely to be very expensive and a financial burden on the bridge owner. Useful data on the costs of maintenance and major rehabilitation work has been published by Wallbank, Tailor and Vassie (1998). Significant repair and strengthening of a typical dual two lane concrete overbridge may cost between £150,000 and £400,000, excluding the cost of traffic delays.

Often it is necessary to post load limits to ensure public safety for a bridge that is significantly below the required strength. This may be satisfactory as a temporary measure, especially for any bridge which is not heavily trafficked. For a bridge which carries a high volume of heavy vehicles, the cost penalty of having to make lengthy and time consuming detours may be substantial. Another cheap and immediate measure is to reduce the number of lanes on the bridge. Most bridges carry two lanes of traffic and therefore reduction to single lane operation will reduce the bridge loading by half. Unfortunately, the capacity will also be cut by half and the effect on traffic delays and queueing may be unacceptable at busy times.

A feasible alternative is to install a system of active load control. The technology for weighing vehicles in motion is currently well developed (COST 323, 1995; McCall and Vodrazka, 1996). Weigh in motion (WIM) has already been used as a means of monitoring traffic loads on a bridge (Opitz, 1993). The bridge loading was predicted from WIM sensors in real time and displayed on a computer screen. The data could be used by the operators to control load by changing the permitted lanes for heavy vehicle access. In the current project, which is funded by the DETR through the LINK Inland Surface Transport Research Programme, the objective is to go one step further and control bridge loading actively by a combination of WIM and traffic signals. This concept is analogous to single lane operation.

The major difference is that, in a two lane highway, both lanes will be open most of the time while one lane will be closed briefly on the rare occasions when a combination of heavy lorries is predicted to exceed the bridge load limit. There are some uncertainties about the performance of such a system in practice, including traffic delays, driver reaction and public perception. It is the purpose of this project to address these by means of a field trial and by computer simulation.

CHOICE OF SITE FOR TRIAL SYSTEM

A trial active load control system has been installed on an under-strength bridge in Oxfordshire. The site was chosen from a number of bridges where load control might be a feasible alternative to strengthening. The following bridges were considered:

A) Steel truss carrying two lanes over a river: this bridge already has a posted load limit and is not heavily trafficked.
B) Three span bridge over a railway: the central span is a brick arch with a capacity insufficient for 40 ton vehicles according to simple masonry arch theory. The outer two spans are short reinforced concrete slabs found to have insufficient reinforcement. Although the traffic flow here is not great, a nearby waste disposal site, a dairy and a bus route provide a regular stream of heavy vehicles. This bridge was rejected for two reasons. Firstly, there were access roads so close to the bridge that it would have been difficult to install WIM units with sufficient braking distance for safe load control. Secondly, the structural form was not ideal for strain monitoring which was a secondary part of the project.
C) Two span steel girder bridge over a railway: there are five parallel riveted plate girders. An intermediate support divides the girders into two assymmetric spans. The bridge failed the assessment due to a problem with localised lack of strength. It was not considered suitable since load control would not be beneficial.
D) Single span steel girder bridge over a railway: this bridge was built in 1928 and was similar to (C), comprising five parallel steel riveted plate girders with curved top chords attaining maximum depth at mid-span. The five girders are connected with brick jack-arches carrying rubble infill and a concrete deck. The girders were assessed to be overstressed in bending under full HA load, although transverse distribution was uncertain. The bridge has sufficiently long approaches and, due to the potential for transverse distribution, was considered suitable for the project. Strengthening of the bridge would cost of the order of £250,000 and replacement may cost as much as £800,000. On the other hand, active load control is estimated to cost about £40,000 for this site.

A location plan of Bridge (D) is shown in Fig 1. The bridge is on a two lane rural highway which carries a significant amount of heavy trucks due to the presence of a number of industries in the area. Prior to installing the system, the traffic flow at the site was as follows:

Total weekday 16 hour count (both lanes)	= 5,201 vehicles
Heavy goods vehicles (HGV)	= 9.46 %

Bending plate weigh-beams were installed on the approaches as shown, and are currently monitoring vehicle loads on the highway. The weigh-pad array on the northern approach is 115m from the bridge. A road leading to the yard of a large builders merchant enters from the left just north of the array. There is a farm entrance just south of the array but this leads only to the farmhouse. The entrance for heavy farm traffic is north of the array. The southern weigh-pad array is also 115m from the bridge and there are no entrances to the highway prior to the bridge, apart from an old entrance to a disused station, now fenced off.

The bridge is shown in elevation in Figure 2. The road profile rises significantly as it goes over the railway. In order to assess the dynamic effects of speed and vehicle bounce, the bending stresses in the girders are being measured with strain gauges for structural verification and to provide data on transverse distribution. This need not be a standard part of a normal load control system.

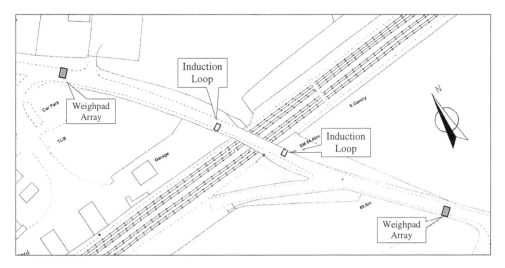

Figure 1. Plan view of active load control bridge site

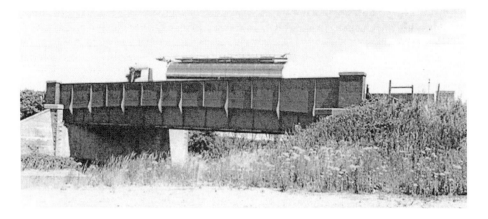

Figure 2. Elevation view of bridge

LOAD CONTROL SYSTEM

The load control system proposed here consists of sets of bending plate weigh-beams and vehicle detector loop arrays at the approaches to the bridge in question. These are shown schematically in Fig 3. An electrical induction loop embedded in the road surface detects the presence of a vehicle. The bending plate weigh-pad measures each axle load in turn and the second induction loop detects when the vehicle has completely passed over the array. The vehicle axle load distribution, gross weight and class may be interpreted from the data. An algorithm calculates the expected bending moment or other stress resultant that will be applied to the bridge. This is added to the effects of load detected by the weighpad array in the other lane at the far side of the bridge. If the expected combined bending moment exceeds a pre-calculated threshold, then traffic signals on the bridge are set to stop one of the approaching heavy vehicles. Note that there is an induction loop on the far side of the bridge in each lane. This is used to count vehicles off the bridge so that the signals may be set to green again when it is calculated to be safe to do so.

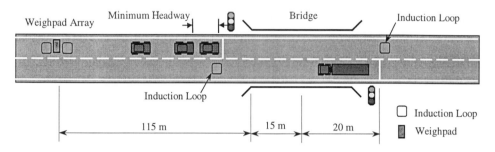

Figure 3. Schematic arrangement of weigh-pad and induction loops

COMPUTER SIMULATION OF LOAD CONTROL SYSTEM

A computer simulation of the load control system has been developed. Its purpose was to test a variety of conditions under which the load management system may operate, including: normal weekday traffic; short bursts of traffic containing a high proportion of heavy vehicles; very dense traffic with the normal proportion of heavy vehicles; and various transient states leading to queuing. The simulation comprises a length of highway which includes the bridge and weigh-pad arrays on both approaches, as shown in Figure 3.

It is assumed that vehicle arrivals at the weigh-pads in each lane conform to the Poisson distribution and therefore that the headway distribution can be described by the negative exponential. Hence, the probability of the next headway being of duration t is given by

$$P = e^{-qt} \tag{1}$$

where q is the mean traffic flow in vehicles per second. A headway conforming to this distribution may be generated randomly be re-arranging Equation 1 as follows:

$$t = \frac{1}{q} \ln\left(\frac{1}{P}\right) \tag{2}$$

where t is the headway in seconds, and P is the probability of the headway distribution that can be generated by selecting a random number between 0 and 1. Since the minimum headway between successive vehicles is approximately one second, the above formula for headways should have one second added in order to allow for a practical minimum headway. By accumulating headways generated as above, traffic flow is obtained.

So far this headway flow is only an empty time sequence with no specification for each vehicle. The next step is to assign each vehicle a specific description, including vehicle weight, axle spacing, and vehicle travel speed. The distribution of heavy goods vehicles (HGV) in the traffic flow is assumed to be random. Each vehicle in the traffic stream is specified either as an HGV or a light vehicle by selecting a random number and comparing it with the proportion of HGVs in the traffic flow. The specific classes of HGV are assigned by a similar process with respect to the frequency of these classes.

The commercial vehicle load spectrum used for this purpose is similar to that published in BS5400 (1980). However, in recent years the number of five axle and six axle vehicles has grown significantly and these have been included, based on recent classification counts (Rahgozar, 1998). Vehicles with seven or more axles need special permits and have been excluded from the load spectrum. Hence, the data used for generating vehicle classes are shown in Table 1. As soon as sufficient loading data have been accumulated from the site trial, a more accurate load spectrum will be used. This will reflect actual traffic conditions at the site.

Table 1. Commercial vehicle load spectrum

Vehicle Designation	Total Weight (kN)	No. in each group per million vehicles	
		Data from BS 5400: Part 10	Data obtained from recent count
6A-H (new)	440	Not included	52500
5A-H (new)	380	Not included	85900
5A-H	630	280	4100
5A-M	360	14500	21400
5A-L	250	15000	7900
4A-H	335	90000	29000
4A-M	260	90000	29000
4A-L	145	90000	10450
4R-H	280	15000	36700
·4R-M	240	15000	36000
4R-L	120	15000	21100
3A-H	215	30000	10200
3A-M	140	30000	9400
3A-L	90	30000	6900
3R-H	240	15000	31600
3R-M	195	15000	30800
3R-L	120	15000	18600
2R-H	135	170000	183500
2R-M	65	170000	182500
2R-L	30	180000	191500

It is assumed that there is no overtaking and that traffic adopts a follow-the-leader flow model.
The model adopted in this project is based on the concept that the acceleration of the following vehicle will be influenced by the speed of the lead vehicle, modified by the vehicle spacing as follows:

$$a = \frac{v_{i-1}^2 - v_i^2}{2(x_{i-1} - x_i)} \tag{3}$$

where v_{i-1} and v_i are the speeds of leading vehicle and following vehicle respectively, and $x_{i-1} - x_i$ is the distance between successive vehicles.

The load control operates as follows: (i) an eastbound vehicle is predicted to create a bending moment, m_e ; (ii) a westbound vehicle is predicted to create a moment, m_w , such that $m_e + m_w$ exceeds the allowable moment; (iii) if these events occur within a time slot during which both vehicles could be on the bridge together, then the signals are activated to stop the second vehicle; (iv) when the first vehicle has been counted off the bridge, the second vehicle is released.

The traffic simulation software runs under the MATLAB environment and requires data for traffic flow, percentage of HGVs, bridge span, load spectrum and duration of simulated run. The animated simulation of traffic flow is shown in Figure 4.

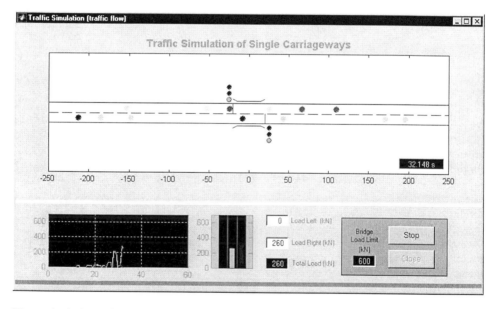

Figure 4. Animated traffic simulation window

An important consideration is the form of the signal. One possibility is to adopt the style of level crossings or fire stations with flashing lights. However, these are appropriate when there is a long warning period followed by a visibly significant event. Bridge load control resembles single lane operation and the normal 3 colour sequence, possibly with a simple explanatory notice, is thought to be more suitable.

RESULTS OF COMPUTER SIMULATION
The results of the simulation can be presented in terms of a time history of total load on the bridge, as shown in Figure 5. This is a 60 second snapshot of loading on the bridge and

demonstrates how vehicle loads on opposite lanes combine to give the total load on the bridge. It may be seen how the loads in each lane are added to provide a total load on the bridge. The percentage of heavy vehicles has been increased for the purposes of this illustration only.

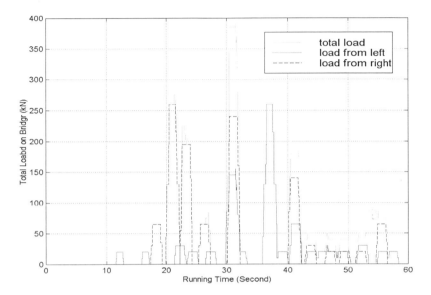

Figure 5. Total load on the bridge

It is of interest to enquire how often the control signals will be activated by overloading events. In Figure 6 the frequency of simultaneous meetings of commercial vehicles on the bridge is shown. It is evident that the frequency increases with total traffic flow. Although it may be seen that two commercial vehicles may meet on the actual bridge about 25 times per day, a minority of these meetings are likely to be combinations that will exceed the allowable bridge load. In the event of queuing traffic, vehicles will be closely spaced as is assumed for HA loading. Queuing may arise for a number of reasons, such as a major public event, an accident, or a broken down or slow vehicle. The simulation was used to assess the frequency of critical levels of total load on the bridge in these circumstances. It was found that critical queuing load events will be very rare.

CONCLUSIONS
1. Active control of traffic loads on bridges is a viable alternative to the posting of load limits or reducing the number of lanes. Cost estimates for strengthening or replacing bridges have shown that active load control would be a highly cost effective alternative.
2. A simulation study has shown that overloading events are likely to be infrequent and, with active load control, the overall delay to traffic will be minimal.
3. Measurements of vehicle loads at the bridge site will enable a more accurate load spectrum to be obtained.

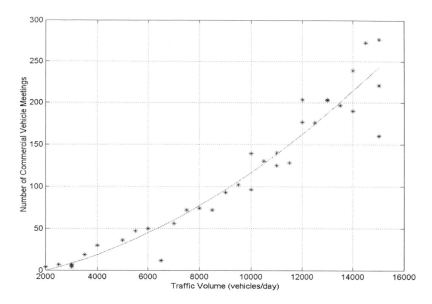

Figure 6. Number of HGV meetings versus traffic flow

REFERENCES

BS 5400 (1980). *Steel, Concrete and Composite Bridges.* Part 10: *Code of Practice for Fatigue*, British Standards Institution, London.

COST 323 (1995). *Post-proceedings of the First European Conference on Weigh-In-Motion of Road Vehicles.* ISBN 3-9521034-0-3.

Leadbeater, A (1997). Bridge assessment and strengthening - local authority perspective. *Safety of Bridges*, Parag C Das (ed), (12 - 19), Thomas Telford, London.

McCall, W and Vodrazka, W (1996). *States' Best Practices WIM Handbook.* Center for Transportation Research and Education, Iowa State University.

Opitz, R (1993). BCS 200 Active Bridge Control System. *Bridge Management 2*, Thomas Telford, 989-997.

Rahgozar, R (1998). *Fatigue Endurance of Steel Structures Subjected to Corrosion.* PhD Thesis, Department of Civil Engineering, University of Bristol.

Wallbank J, Tailor P and Vassie P (1998). Strategic Planning of Future Structures Maintenance Needs. *Conference on the Management of Highway Structures*, Institution of Civil Engineers, London.

Williams, T K (1997). Bridge management problems and options. *Safety of Bridges*, Parag C Das (ed), (131 - 137), Thomas Telford, London.

Characteristics of composite concrete beams

J. HULATT[*], L. HOLLAWAY[*ψ] and A. THORNE[*]
University of Surrey, Guildford, UK

ABSTRACT
The use of advanced polymer composite materials (APCs) in the civil engineering industry, although increasing, has not yet reached its full potential. APCs have proven to be a useful material in a number of applications (Hollaway, 1990), including plate bonding (Hollaway & Leeming, 1999), due to their lightweight, durability and ease of handling. With the economic advantage becoming more apparent, the use of APCs in further applications is set to grow. This paper discusses the use of APCs in combination with concrete to form a 'duplex beam' as an example of a typical structural element. The beam is of a rectangular cross-section with GFRP composite webs extending into the compressive zone to form permanent shuttering for the concrete. The flange is constructed from interleaved layers of GFRP and CFRP. With careful design, the high strength, lightweight properties of APCs can be fully utilised when constructed in this way. Consequently, the concrete and APC elements are used in the compressive and tensile regions of the beam respectively, where they are most effective. The method of manufacture of the duplex beam will be discussed and a description of two geometrically different beams tested will be given; viz. the concrete in one of the beams will be fully encapsulated by the GFRP composite whilst in the other beam it will be enclosed on three sides only, the top being exposed to the atmosphere. Characteristics of the beam will also be given. Initial results show that the encapsulated beam has a much improved failure strength.

INTRODUCTION
Advanced Polymer Composite Materials (APCs) have become increasingly more widespread within the civil engineering industry. This has lead to an increase in research within this area to identify potentially suitable applications and uses for the material (Bodamer, 1998). Already carbon fibre reinforced polymer (CFRP) has shown to be an ideal replacement for steel when used for strengthening of structures (Meier & Kaiser, 1991, Norris et al, 1997, Garden and Hollaway, 1997). The lightness of the material, along with its inherent strength and stiffness, allows for faster and easier erection. This, as well as its enhanced durability properties, makes it an ideal material for this type of application.

Many of the western world's reinforced concrete transport infrastructure is affected by corrosion of reinforcement as a result of the use of de-icing salts. Large amounts of research is being undertaken in US, Japan and Europe to overcome this; these are discussed in "Recent advances in bridge engineering, 1997, NSF, 1993 and Non-metallic (FRP) reinforcement for concrete structures, 1997". Research carried out by Triantafillou and Meier (1992) has led to the concept of a "duplex beam". Thus to use two durable materials in the tensile and

---

[*] Composite Structures Research Unit, Department of Civil Engineering
[ψ] To whom correspondence should be addressed in the first instance

Bridge Management 4, Thomas Telford, London, 2000.

compressive regions is obviously attractive. For APCs to be used effectively, the materials must be utilised in the correct manner. This duplicity incorporates the best features of both APCs and a standard material such as concrete. With the concrete located in the compressive region of a beam, where it is most effective, and the APCs in the tensile region, an efficient beam can be manufactured. As part of two ongoing programmes of work[#][φ] two such "duplex beams" have been manufactured and tested for comparison and are reported in this paper. One beam has concrete that is fully encapsulated and the other has the top surface of the concrete exposed to the atmosphere.

The objectives of this paper are to determine the effects of fully encapsulating the concrete section of the beam when comparing it to a conventionally constructed beam. The experimental results for the two beams are compared with predictions made by numerically modelling them in a Finite Element Package. The two beams were tested experimentally using 4-point bending and FE analysis was carried out using ABAQUS v5.7 (1998), with pre- and post processing by PATRAN v8.5 (1999).

METHOD OF MANUFACTURE

The two 1.5m span beams were manufactured using the vacuum bag technique. Figure 1 shows details of the various layers of the conventional composite beam as it is layed up. Initially, a male tool was prepared from Medium Density Fibreboard (MDF) to the correct profile. A glass reinforced self-adhesive PTFE film was applied to the surface of the tool for ease of demoulding the composite. The material, supplied by Advanced Composites Group (ACG), was in the form of a prepreg material and this was used for both beams. Two layers of +/- 45° GFRP prepreg[1] were placed by hand onto the mould. This was followed by a layer of unidirectional (UD) CFRP[2] in the flange, Figure 1. Each of the three layers was rolled flat before placement of the next layer.

To ensure good compaction of the three layers, and to reduce voidage, the material was debulked. In this operation a pinprick sheet was placed over the third layer of prepreg and the whole system was wrapped in a breather blanket. The wrapped system was placed in a polythene bag, sealed and a vacuum pump used to evacuate air to a vacuum of 1 atmosphere. The debulking was continued for 5 minutes after which time the pump was switched off, the mould removed from the bag and the breather blanket and pinprick removed. An adhesive film was then used to bond the pre-cut foam sections to the vertical sides of the composite (Figure 1). The foam was added to form a sandwich construction to improve the buckling resistance of the webs. A further two layers of GFRP were laid over the previous prepreg layers and the foam sections with one CFRP layer being interleaved with the GFRP prepregs in the soffit of the beam; the whole system was again debulked using the procedure described above. The final layers were added, debulked and the beam was placed in the oven for 16 hours at 60°C under a vacuum of 1 atmosphere to affect the correct curing regime. The final beam had seven layers of +/- 45° GFRP and four layers of UD CFRP (Figure 1).

A GFRP plate to act as the base of the permanent shuttering and the two diaphragms to prevent local buckling at the support positions were manufactured (Figure 2). The plate was made from 6 plies of 0/90° GFRP and was constructed in the same way by debulking and

[#] COMPCON Project – European Community under the Ind. & Material Technologies prog. (Brite-Euram III)
[φ] EPSRC Built Environment Prog. – Optimisation of Polymer Composite/Concrete Structural Units 1996-2000

[1] GFO100 (390 g/m2 E glass 2 x 2 twill)/LTM 26-50% fibre volume fraction
[2] HTA 12k 150 g/m2 LTM 26-68 g/m2 60% fibre volume fraction

curing at 60°C. The diaphragms were made from 10 plies of 0/90° GFRP each, again in a similar fashion. They were bonded in place with 3M 9323/2 two part epoxy adhesive. Once the adhesive had cured, the concrete was poured into the permanent shuttering. The concrete was designed to have a 28 day strength of 40 N/mm². A number of ways in which the concrete is bonded into the beam had been investigated previously, including resin injection and mechanical means (Canning et al, 1999). In the present investigaton, the concrete was removed from the beam after it had cured, grit blasted and bonded back in place using the 3M 9323/2 two part epoxy adhesive. The beam was tested at 28 days; in addition three concrete cubes, to confirm the concrete 28 day compressive strength, were also tested.

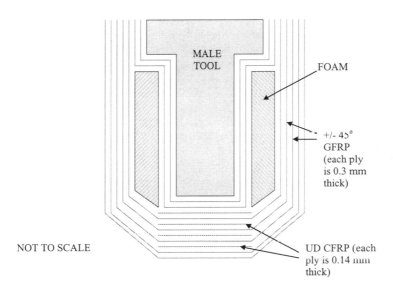

Figure 1: Order of layers for the composite beam from the male tool outward (exploded view)

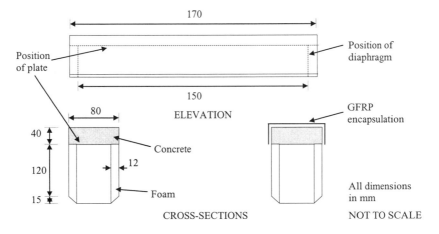

Figure 2: Sections and elevations of the standard and encapsulated beams

Figure 2 shows the completed beam with some basic dimensions. The encapsulated beam was made in exactly the same way. However, prior to the 28 day testing, a separate cap was made by laying six layers of GFRP over a male tool. The cap extended to the bottom of the permanent shuttering on either side of the beam (Figure 2). This was then bonded in place using the same 3M 9323/2 adhesive.

EXPERIMENTAL WORK

The two beams were tested in 4-point bending. The load points were located at 600mm from each end of the beam which gave a length of 300mm in pure bending. The load was applied via a hydraulic jack in a number of steps and cycles. The load was incremented in a 2kN step, with strain and load recorded, up to 10kN. At this point the load was released. The load cycling was repeated to 16, 25, 32kN before loading to failure. Strain gauges were located on top of the beams, the flange and down one web of the beams. The strain gauges, along with the loading rig, can be seen in figure 3. The strain and load were recorded using an ORION datalogger.

Figure 3: Photograph of the general beam loading arrangement and strain gauge positioning

Figure 4: Failure of the flange in the encapsulated beam (beam inverted)

The experiments resulted in a different failure mechanisms for each beam. The standard beam, with the top of the concrete exposed to the atmosphere, failed at a load of 51.1 kN with the concrete failing at the midsection with buckling outwards of the permanent shuttering. However, it was unclear which came first, the buckling of the permanent shuttering which caused the concrete to fail, or the concrete failing which caused the permanent shuttering to buckle. The encapsulated beam however, failed at a load of 56.8 kN due to tensile failure of the carbon in the flange which resulted in local plane buckling in the GFRP at the midspan. Sectioning the beam after failure showed that the concrete was completely intact. Figure 4 shows the failure of the flange.

Figure 5 shows a plot of load against deflection for both experiments. It can be seen that the two lines are nearly coincident up to a load of 30kN. With further increase in load the encapsulated beam shows a stiffer load deflection response.

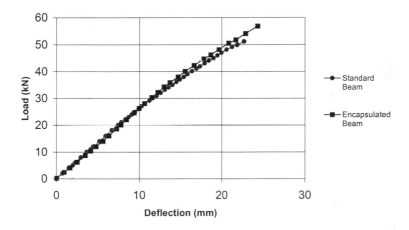

Figure 5: Plot of load against deflection for the standard and encapsulated beams

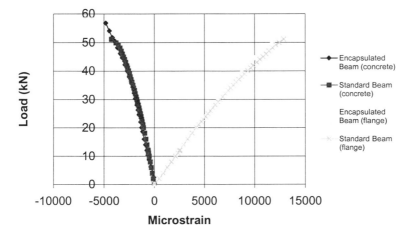

Figure 6: Plot of the load against strain response for the two beams

Figure 6 shows a plot of microstrain against load for the two beams at two different locations; on top of the concrete and on the underside of the flange. A similar response for both beams is evident with the encapsulated beam again showing a marginally stiffer response.

Figure 7 shows a plot of flexural rigidity (EI) against applied load. The two lines again converge at approximately 30kN which shows the stiffness of the two beams to be very similar. The EI value was calculated using a spreadsheet based on an equation for flexural rigidity (Triantafillou & Meier, 1992). With a known deflection for a given load, the flexural rigidity could be found.

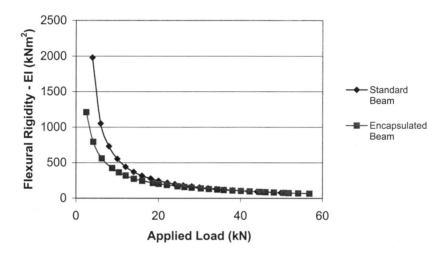

Figure 7: Plot of the flexural rigidity against load for the two beams

FINITE ELEMENT ANALYSIS

Finite Element (FE) analysis of the two beams was carried out using ABAQUS v5.7 and PATRAN v8.5 as a pre and post-processor. The material properties used were derived from experimental coupon tests and are presented in Table 1. The analysis was run as a linear static analysis with a serviceability load of 25kN applied in 4-point loading as mentioned previously. The serviceability load was chosen as it gave a factor of safety of 2 on the ultimate failure load. The model was meshed using 20-node hexahedron brick elements that allowed a greater accuracy when compared with the 8-node elements which are more commonly used.

Material	$E_{(longitudinal)}$ (GPa)	$E_{(other\ directions)}$ (GPa)
UD CFRP	140	10
+/- 45 GFRP	12	10
Foam	0.005	0.005
Concrete	28	28

Table 1: Table showing material properties used in FE models

The results from the experimental work, the numerical FE analysis and a spreadsheet based on Triantifiilou's (1992) paper were compared (Table 2). Values were compared analysing the strain on the top of the concrete and on the underside of the flange.

			Method	
Beam Type	**Location**	**Experimental**	**FE Analysis**	**Spreadsheet**
Standard	Microstrain in concrete	1390.8	1330	1391
Standard	Microstrain in flange	5359.1	5325	5660
Encapsulated	Microstrain in concrete	1324.8	1264	~
Encapsulated	Microstrain in flange	5001.0	5059	~

Table 2: Table comparing values of strain at two locations using three different methods

DISCUSSION

The results show that the concept of the "duplex beam" is a satisfactory structural unit. The two beams compared well to each other with similar load deflection responses and strain plots. All plots show the last cycle of loading which was from zero kN to complete failure of the beam.

Of particular interest was the way in which the beams failed. Encapsulation of the beam resulted in a change in the failure mechanism. By encapsulating the concrete the failure load was increased to 57 kN and the failure mode of the beam was by tensile failure of the carbon fibre composite in the flange, at a strain of 0.0137, resulting in local plane buckling in the GFRP at midspan; no concrete crushing occurred. Results from Canning et al (1999) describe six number beams similar to the standard beam. These have all failed by compression failure of the concrete resulting in buckling of the permanent shuttering at a load of approximately 50 kN. The strain in the flange of the beams was approximately 0.0130. In all cases the strain in the flange and concrete are close to failure and hence an economic and balanced section has been produced.

The duplex beams were originally designed to have a comparable ultimate load to that of a RC beam of section with similar depth and breadth. This has resulted in a composite/concrete beam which deflects unacceptably for many construction situations, such as primary load carrying units in buildings where the deflections are too great for the secondary finishings. Current work is being undertaken on a stiffness basis where deflection is the primary concern. Increasing the carbon content in the flange will increase the flexural rigidity. However, the design as a whole needs to be considered to maintain a balanced section. Increased carbon content not only provides the increase in stiffness but additionally results in a unit that has a much higher ultimate failure load, thus increasing the factor of safety.

The manufacture of the beams was carried out with a minimal amount of effort and the technique could readily be applied to large span beams to the civil engineering industry. However, the bonding in of the concrete after it is cured is really not practicable when scaled up to engineering practices. Other methods of creating the necessary shear transfer between the composite and concrete have been investigated by Canning et al (1999) and developments in using an adhesive that cures when in contact with fresh concrete are currently being investigated.

The linear FE analysis of the two beams also compared well with experimental and theoretical investigations. However, a more accurate model could be created using non-linear analysis. A more time consuming non-linear analysis was not undertaken in this case as the objective of the FE work was to confirm numerically the difference between the encapsulated and standard beams. Further work could be carried out in this area.

CONCLUSIONS

The encapsulated beam was shown to have an increased failure strength when compared to a beam without a cap. The increase in failure load was approximately 11% at which time the carbon fibre composite in the flange failed. The failure mechanism for the standard beam was shown to be different, with the concrete crushing and permanent shuttering buckling outwards. It should be noted that the failure was of a brittle nature. However, as a result of meeting the stiffness requirement there is a very much larger factor of safety than other, conventionally used, brittle materials such as timber, cast iron and reinforced glass.

REFERENCES

ABAQUS Version 5.7, Hibbitt, Karlsson & Sorensen, Inc, Pawtucket, RI, 1997

Bodamer, D. (1998)
A Composite Sketch, *Civil Engineering – ASCE, Vol. 68, No. 1, Jan. '98, pp 57-59*

Canning, L., Hollaway, L. and Thorne, A. (1999)
Investigation of shear transfer for composite action in an innovative polymer composite/concrete structural unit, *submitted to Journal of Construction and Building Materials*

Garden, H. N. and Hollaway, L. C. (1997)
An experimental study of the strengthening of reinforced concrete beams using prestressed carbon composite plates, *Proc. 7th International Conference on Structural Faults and Repair, University of Edinburgh, July 8-10th '97, Vol. 2, pp 191-199*

Hollaway, L. (ed.) (1990)
Polymers and Polymer Composites in construction, *Thomas Telford Ltd, London*

Hollaway, L. and Leeming, M.B. (1999)
Strengthening of Reinforced Concrete Structures using externally bonded FRP components in structural and civil engineering, *Woodhead Publishing Ltd, Cambridge*

Meier, U. and Kaiser, H. (1991)
Strengthening of Structures with CFRP Laminates, *Proc. Conf. on Advanced Composite Materials in Civil Eng. Structures, ASCE, pp 224-232*

MSC/PATRAN, version 8.5, MacNeil-Schwendler Co., 1999

"Non-metallic (FRP) reinforcement for concrete structures", *Proc. of the Third International Symposium on Non-metallic (FRP) Reinforcement for Concrete Structures, Vols. 1 & 2, October 1997, Pub. Japan Concrete Institute*

Norris, T., Saadatmanesh, H. and Ehsani, R. (1997)
Shear and flexural strengthening of R/C beams with carbon fiber sheets, *Journal of Structural Engineering, ASCE, Vol. 123, No. 7, July '97, pp 903-911*

NSF (1993)
NSF 93-94, Engineering Brochure on Infrastructure, US National Science Foundation, Arlington, VA

"Recent advances in bridge engineering – Advanced rehabilitation, durable materials, non-destructive evaluation and management", *Ed. by Urs Meier and Raimondo Betti Proc. Of the US-Canada-Europe Workshop on Bridge Engineering, Dubendorf, Switzerland, July 1997*

Triantafillou, T. C. and Meier, U. (1992)
Innovative design of FRP combined with concrete, *Proc. Advanced Composites in Bridges and Structures, pp 491-498.*

Analysis of R/C beams externally strengthened with carbon fibre sheets

A.J. Beber, MSc.   A. Campos Filho, Dr.   J.L. Campagnolo, MSc.   L.C.P. Silva Filho, PhD
PhD. Candidate   Associate Professor   Associate Professor   Lecturer

Laboratory of Structural Tests and Modelling – LEME
Civil Engineering Graduate Program – PPGEC
Federal University of Brazil at Rio Grande do Sul – UFRGS

INTRODUCTION

The service life extension of our infrastructure towards the third millennium, confronts the construction industry with a distinctively challenge of infrastructure renewal along with increasing economical constraints. Several research centres world wide have been studying numerous materials and methods for the repair and strengthening of reinforced concrete structures. The use of steel plates externally bonded onto concrete elements is one of the best techniques for strengthening or repairing deteriorated concrete elements. This is a simple and quick to apply technique and it also allows significant increases in strength and stiffness without major changes in the element's geometry. Since the 1960's, there have been several field applications of epoxy-bonded steel plates in the strengthening of reinforced concrete structures, especially in Europe, USA and Japan. Generally, this technique is used to repair or strengthen concrete elements with insufficient load carrying capacity due to structural damage, functional changes and frequently reinforcement corrosion. The principle of this strengthening technique is fairly simple: steel plates are epoxy-bonded to the tension flange of, for instance, a concrete beam, increasing its strength and stiffness [Saadatmanesh & Ehsani, 1990].

Despite all these advantages, under certain environments, steel plating has been demonstrated to be vulnerable to corrosion, especially at the steel epoxy interface [Pilakoutas, 1997]. According to Pilakoutas (1997), the Transport and Road Research Laboratory has carried out a number of durability experiments in order to investigate the long-term behaviour of strengthened structures. The test results clearly demonstrated that after a long-term exposure, corrosion of the steel plate occurs at the interface between the adhesive and the steel plate. This type of corrosion adversely affects the bond at the steel plate/concrete interface and is quite difficult to monitor during routine inspections. There are also operational problems involving this technique. The plates are difficult to shape in order to fit complex profiles. In addition, the weight of the plates makes them difficult to transport and handle on site, particularly in areas of limited access, and can cause the dead weight of the structure to be increased significantly after installation. Welding is not permitted, hence there are limitations regarding to the length of the plates.

Recently, researchers all over the world have been seeking for alternative materials. Their attention has been drawn to the use of non-metallic composite materials as a substitute to the steel plates in order to overcome these shortcomings.

Fibre reinforced polymers (FRP) offer an excellent alternative to steel plate bonding because of their high tensile capacity, non-corrosive nature and lighter weight. Unlike steel, FRP's are unaffected by electrochemical deterioration and can resist the corrosive effects of acids, alkalis, salts and similar aggressive materials under a wide range of temperatures [Hollaway & Leeming, 1999]. They are perfectly bonded to the concrete surface avoiding the use of lifting devices (low self-weight). Many researchers have studied the mechanical properties of the FRP's. However, very few studies have been carried out on the combination of fibre composites and traditional construction materials [Norris et al, 1997].

FIBRE REINFORCED POLYMERS (FRP)

In the past, the fibre reinforced polymers have been used almost exclusively in the aerospace and defence industries; however, they are now frequently used in the manufacturing industries such as the automotive and sports industries. These materials present high strength, low self weight, high durability and the ability to form complex shapes. They are generally constructed of high performance fibres such as carbon, aramid, or glass which are placed in a resin matrix [Norris et al, 1997].

According to Taylor (1994), the main reason for reinforcing polymers is to increase their stiffness, although useful increases in strength, impact resistance, fatigue resistance and creep resistance should also be obtainable. By selecting among the many available fibres, geometries and polymers, the mechanical and durability properties can be tailored for a particular application.

The FRP offer unique advantages in several applications where conventional materials fail in providing satisfactory service life. The high strength-to-weight ratio and excellent corrosion resistance made the FRP very attractive to structural applications. The mechanical properties of advanced composites vary with the amount and orientation of the fibres in different directions.

The carbon fibres are a new and highly promising materials based on the strength of the carbon-carbon bond in graphite and the lightness of the carbon atom [Taylor, 1994]. The carbon fibre reinforced polymers (CFRP) are the most appropriate for strengthening reinforced concrete beams because of its extraordinary mechanical properties (figure 1) allowing significant increases in strength and stiffness.

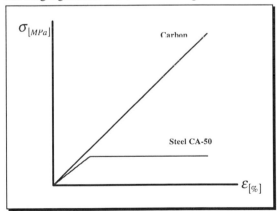

There are currently three types of CFRP strengthening systems [Robery & Innes, 1997]: (i) pultruded plates; (ii) tension wound carbon fibre strands and (iii) pre-impregnated (prepreg) sheets (figure 2).

Prepreg epoxy sheets are pre-impregnated with epoxy resin, which holds the fibres together. The prepreg sheets are very flexible and can be easily cut by means of scissors or any other cutting tool. At application, the prepreg sheets are impregnated again with epoxy resin and the chemical reaction is started.

Figure 1. Stress-strain relationship

EXPERIMENTAL PROGRAM

Characteristics of beams

Ten rectangular beams were tested aiming the evaluation of the effect of externally bonded carbon fibre sheet on the flexural capacity of reinforced concrete beams.

Figure 2. Prepreg sheet

The experimental study consisted of casting ten reinforced concrete beams as shown in figure 3. The beams were divided in five groups: control beams and beams strengthened with one, four, seven and ten CFRP sheet layers, respectively.

TABLE 1 – Beams designations

Group	Beam code	Designation
I	VT1 – VT2	control
II	VR3 – VR4	one CFRP sheet layer
III	VR5 – VR6	four CFRP sheet layers
IV	VR7 – VR8	seven CFRP sheet layers
V	VR9 – VR10	ten CFRP sheet layers

All ten beams had a span length of 235 cm (beams were 250 cm long) and cross-sectional dimensions of 12 x 25 cm. The flexural reinforcement ($\rho = 0,0052$) consisted of two 10 mm CA-50A tension and two 6 mm CA-60B compression steel bars. The shear reinforcement consisted of twenty-two 6 mm stirrups equally spaced (11 cm) along the beam length.

Materials

In situ concrete was used for all beams. For each group of beams, nine 10 x 20 cm concrete cylinders were cast and tested to determine the concrete compressive strength. The average compressive strength was 33.58 MPa.

Samples of each type of steel rebar were tested under uniaxial tension. The average measured yield stress for the 10 mm bar was 565 MPa and 738 MPa for the 6 mm bar.

According to the supplier, the CFRP sheet characteristics are listed in table 2.

TABLE 2 – CFRP characteristics

Tensile Strength	3400 MPa
Modulus of elasticity	230000 MPa
Cross section per unit width	1,11 cm^2/m
Fibre weight per unit area	200 g/m^2
Ultimate strain	1,48%

CFRP sheet bonding procedure

The tension face of the beams to be externally strengthened were blasted down to aggregate by a disk grinder. All the dust was removed by air blowing. The primer was applied onto the previously prepared concrete surface uniformly using a brush. While the primer was curing, the carbon fibre sheets were cut to size. After the primer had cured, the resin undercoating was applied in order to lay down the first carbon fibre sheet layer. The resin functions as an adhesive to bond the carbon fibre sheet to the concrete surface and also to form a composite by impregnating into the sheet. Special attention should be given while laying down the carbon fibre sheet in order to avoid air encapsulation. Once finished, the over coating resin completes the composite material moulding. After the carbon fibre sheets were bonded, seven day resin curing was carried out at ambient conditions.

Instrumentation

The strains in the concrete, steel rebars, and the CFRP sheet in the section at midspan of each beam were measured by electric resistance strain gages. For measuring the steel strain, one strain gage was mounted in each tension rebar. The concrete strain was measured by means of two strain gages placed on both sides at the top of each beam. For measuring the CFRP sheet strain, two strain gages were placed on the sheet surface at midspan. Deflection was measured by means of displacement gages, placed vertically under each load application point and midspan and, horizontally at the support. Load was measured by a load cell.

Test Set-Up and Test Procedure

All beams were simply supported on a clear span of 235 cm, and they were subjected to two concentrated loads symmetrically placed about the midspan. The loading points were 78.3 cm apart. The beams were incrementally loaded to failure in steps of 4 kN (control and strengthened with one layer beams) and 5 kN to all the rest. After each increment of load, the strains were measured by an automatic data acquisition system.

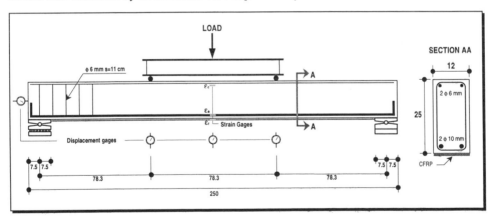

FIGURE 3 – Cross section and Test Set-Up of beams

TEST RESULTS

Ultimate load

There is a significant increase in the ultimate load for the strengthened beams. These results are shown in Table 3.

The theoretical results for the ultimate moments, in accordance with the Brazilian code for reinforced concrete (NBR6118), were predicted with reasonable accuracy. In order to predict the ultimate moment, the cross section was verified for the following failure modes: (i) classic R/C failure due to steel yielding or concrete crushing; (ii) peeling off, i.e., sheet debonding due to high tensile and bending stresses and (iii) shear failure due to insufficient shear reinforcement.

Based on the experimental and theoretical results of this study, it was observed that the increases in the ultimate load are only possible when other failure modes do not interfere such as the capacity to carry higher shear and sustain higher strains in carbon or steel without bond failure.

It is interesting to note that the weakest point for the strengthened beams was the concrete/internal steel interface. This type of failure was verified in almost all strengthened beams. It is due to high shear stress concentration at the end of the CFRP sheets (figure 5).

TABLE 3 – Comparison of ultimate loads and failure modes

Beam code	VT1	VT2	VR3	VR4	VR5	VR6	VR7	VR8	VR9	VR10
Experimental failure load [kN]	47.4	47.0	65.2	62.0	102.2	100.6	124.2	124.0	129.6	137,0
Theoretical failure load [kN]	46.4		66.5		120.8		122.2		121.9	
Difference [%]	+2.2	+1.3	-1.9	-6.8	-15.4	-16.7	+1.6	+1.5	+6.3	+12.4
Failure mode	Type 1	Type 1	Type 1	Type 2	Type 3	Type 3	Type 3	Type 3	Type 3	Type 3
Increase in failure load [%]	-		+35		+115		+163		+182	

Type 1 – steel yielding
Type 2 – CFRP sheet rupture
Type 3 – CFRP sheet debonding

Deflection

Deflection of a beam primarily depends upon the loading, span, restraints, sections geometry and materials properties. Strengthening a beam with CFRP sheet results in a significant increase in cross sectional stiffness. This increase is evident in figure 4, which shows the comparison among the *load vs. displacement* curves for all tested beams.

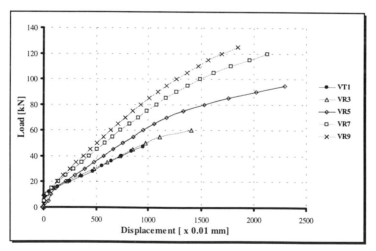

FIGURE 4 – Comparative deflections

Rebar strains

The CFRP sheet attached to the reinforced concrete beam share the applied load with the steel rebars. Therefore, a decrease in rebar stress is expected for the strengthened beams at a given load. The strengthening also reduces large plastic strains in rebars, since most of the increase in load-carrying capacity was obtained after the rebar yielding.

The rebar yielding for the control beams occurred at a load of 44 kN. As expected, the rebars yielding, for the strengthened beams, occurred at higher loads. Still, when the strain results for the control beams at a load of 44 kN are compared with the strengthened beams it is observed a decrease in the strain values. It was observed the strains in the carbon sheets at failure were lower (around 1.0%) than the ultimate tensile strain (1.48%).

TABLE 4 – Comparison of strain rebars results

Beam code	VT1	VT2	VR3	VR4	VR5	VR6	VR7	VR8	VR9	VR10
Load at the beginning of steel yielding [kN]	44.0	44.0	47.9	48.0	60.0	60.1	80.1	85.1	90.0	95.0
Difference [%]	-	-	+9	+9	+36	+36	+82	+93	+105	+116
Strain of steel bar at 44 kN load [10^{-6} mm/mm]	2634	2688	2447	2295	1907	1840	1350	1275	1096	1052
Difference [%]	-	-	-8	-14	-28	-31	-49	-52	-59	-60

FIGURE 5 – Concrete delamination

Cracking

The crack patterns developed similarly for all the strengthened beams. First cracking usually occurred at a slightly higher loads than the control beams. This is attributed primarily to the increase in the moment of inertia of the strengthen section also to the slight increase in the concrete tensile strength due to confinement. Initially, the cracks were vertical, as would be expected for flexural cracks, but later they would bend over in the shear regions. Generally there were more cracks, more closely spaced, more uniformly distributed, and not opened up as wide on the strengthened beams.

NUMERICAL RESULTS

With the aim of showing the possibilities for the numerical analysis of structures strengthened with CFRP sheets it was used a computational program based on the finite element method. This program was developed by Campos Filho (1987) for the analysis of reinforced concrete structures subjected to plane stress states. The concrete is represented using a quadratic isoparametric quadrangular finite element. Ottosen (1977, 1979) proposed the failure criterion and the constitutive relations for concrete. Steel bars and the CFRP sheets are represented by an embedded model. Comparisons between the experimental and numerical results validated the use of a FEM in the analysis of strengthened beams. Despite the fact that after the concrete cracking, the numerical and experimental strain results may differ, the computer program provides very accurate results. These comparisons are presented in figures 6 to 9.

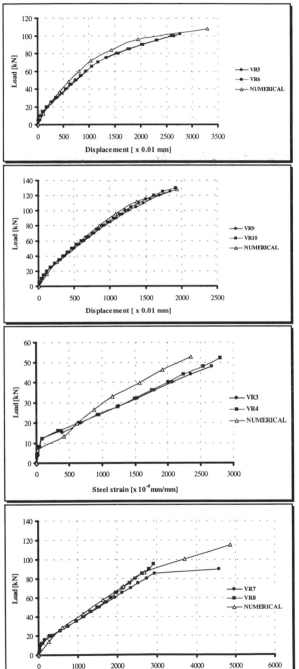

FIGURE 6 – Beam strengthened with four layers

FIGURE 7 – Beam strengthened with ten layers

FIGURE 8 – Rebar strain/numerical vs. experimental (beam strengthened with four layers)

FIGURE 9 – Rebar strain/numerical vs. experimental (beam strengthened with seven layers)

CONCLUSIONS
The use of CFRP sheets for the strengthening of reinforced concrete beams has been demonstrated a feasible way to increase the load carrying capacity and stiffness of existing structures. It is very simple, easy and quick to install, requiring essentially unskilled labour although it is necessary an appropriate site supervision to ensure its quality. The thin CFRP sheets allow very little cross-sectional increases while providing the concrete element significant increases in strength and stiffness.

The ultimate load increases were significant although these values tend to a limit determined by the CFRP sheet debonding. Peeling off is found to be a typical failure mode. It is associated with high shear stress concentration at the ends of the CFRP sheets

Stiffness increases are evident. At a given load, rebar stresses and deflections were found to be smaller when compared to rebar stresses and deflections of control beams.

Ultimate load capacities for the strengthened beams can be predicted with reasonable accuracy by conventional beam-bending theory (in accordance with the Brazilian code for concrete – NBR6118).

The purely linear elastic behaviour of the CFRP sheets prevents the use of the composite's full tensile strength. According to the experimental results, the maximum strain at the CFRP sheet was 10.6‰ i.e. 28% less than its full strain potential.

REFERENCES
ASSOCIAÇÃO BRASILEIRA DE NORMAS TÉCNICAS. **Projeto e execução de obras de concreto armado: NBR 6118**. Rio de Janeiro, 1980.

BEBER, Andriei José. **Avaliação do desempenho de vigas de concreto armado reforçadas com lâminas de fibra de carbono**. Porto Alegre: CPGEC/UFRGS, 1999. 108 p. Dissertação de Mestrado em Engenharia.

CAMPOS FILHO, Américo. **Análise teórico-experimental de elementos de concreto armado para obtenção de modelo matemático**. São Paulo: Escola Politécnica da USP, 1987. 293 p. Tese de Doutorado em Engenharia.

COMITÉ EURO-INTERNATIONAL DU BETON. **CEB-FIP model code 1990**. Lausanne, 1993 (Bulletin d'Information, 213).

NORRIS, Tom; SAADATMANESH, Hamid; EHSANI, Mohammad. Shear and flexural strengthening of R/C beams with carbon fiber sheets. **Journal of Structural Engineering**, New York, ASCE. v.123, n.7, p.903-911, July 1997.

OTTOSEN, N. S. A failure criterion for concrete. **Journal of the Engineering Mechanics Division**, New York, ASCE, v. 103, n. 4, p. 527-535, Aug. 1977.

OTTOSEN, N. S. Constitutive model for short-time loading of concrete. **Journal of the Engineering Mechanics Division**, New York, ASCE, v. 105, n. 11, p. 127-141, Feb. 1979.

PILAKOUTAS, Kypros; HE, Jin Hong; WALDRON, Peter. CFRP plate strengthening of RC beams. In: INTERNATIONAL CONFERENCE ON STRUCTURAL FAULTS AND REPAIR, 7., 1997, Edinburgh. **Proceedings...** Edinburgh: Engineering Technics Press, 1997. 3v. v.1, p.119-127.

ROBERY, Peter; INNES, Craig. Carbon fibre strengthening of concrete structures. In: INTERNATIONAL CONFERENCE ON STRUCTURAL FAULTS AND REPAIR, 7., 1997, Edinburgh. **Proceedings...** Edinburgh: Engineering Technics Press, 1997. 3v. v.1, p.197-208.

SAADATMANESH, Hamid; EHSANI, Mohammad. Fiber composite plates can strengthen beams. **Concrete International**, Detroit, ACI, v.3, p. 65-71, Mar. 1990.

SCHWARTZ, Mel M. **Composite materials handbook**. New York: McGraw-Hill, 1984.

TAYLOR, Geoffrey. **Materials in construction**. 2.ed. London: Longman Scientific & Technical, 1994. 284p.

Stressed and unstressed advanced composite plates for the repair and strengthening of structures

JOHN DARBY, SAM LUKE and SIMON COLLINS
Mouchel Consulting, West Byfleet, Surrey, UK

INTRODUCTION

The benefits of advanced composite materials have now been recognised by many sectors of the civil engineering industry, largely as a result of the publication of research results demonstrating the scope for economic and durable strengthening. Much of this research has been aimed, in particular, at application of unstressed plates applied to concrete bridge structures.

This paper considers some of the results of this research and some practical applications, together with more recent developments that have extended the technique to encompass stressing of the composite reinforcement and application to metal structures.

Mouchel have been involved in much of this early work, both as lead partner in the ROBUST project described below, and in the development of stressing techniques subsequently used on Hythe Bridge.

THE ROBUST PROJECT

Understanding of plate bonding in the UK was advanced by the work of the consortium that undertook the ROBUST project. This consortium included specialists from all parts of the industry. Led by Mouchel Consulting, other Partners included Balvac Whitley Moran, University of Surrey, Oxford Brookes University, Techbuild composites (now Fiberforce composites), James Quinn Associates, Sika , Vetrotex, and Oxfordshire County Council.

A major impetus behind the research was the need to strengthen many UK highway bridges. This resulted from the assessment programme undertaken in preparation for the introduction of more 40 tonne vehicles on UK roads as a result of the EC directive. However, bridge strengthening represents only a very small proportion of the opportunities for application of advanced composite materials.

The ROBUST project, funded under the DTI-LINK structural composites programme, provided the opportunity to study the effects of different reinforcement upon beams of length from 1.0m to 18.0 m. The project included 130 flexural tests on beams within the laboratory, and 110 tests on FRP materials to characterise the materials and examine long term durability, stress relaxation and fatigue. Tests included the load testing of 10 beams of 18m length, which were strengthened after they had been removed from a bridge following concern over corroded post-tensioned tendons. These beams enabled demonstration of the techniques developed, both at full scale and under representative application conditions.

Three-dimensional non-linear finite element static analysis was carried out, and demonstrated very good correspondence with the experimental beams. Fatigue tests on beams demonstrated that the life of the system was limited by the fatigue life of embedded steel reinforcement, and was not limited by the composite plate bonding.

A significant development within the project was the use of peel ply technology to improve the reliability of bond and application under site conditions. The ROBUST project therefore provided a sound basis for the design of strengthening systems, significantly increasing the understanding of individual materials and their composite behaviour.

SUBSEQUENT APPLICATIONS

Greater understanding of the behaviour of composite bonding has led to an increase in the number of applications of composite plate bonding. Mouchel have undertaken the design of a number of applications of composite plate bonding, both on bridges and industrial structures.

A typical example of these applications is Haversham Bridge, shown in Figure 1. In this case the three span bridge was strengthened on the top surface over supports to increase the flexural capacity. 768m of CFRP (Carbon Fibre Reinforced Polymer) plate were bonded, during which time traffic flow was maintained in adjacent lanes. The durability of the CFRP was of particular benefit on the top surface, due to potential exposure to de-icing salts, but the solution was also of lower cost than steel plate bonding.

Figure 1. Strengthening of Haversham Bridge, Milton Keynes, UK.

GENERAL BENEFITS OF PRESTRESSING COMPOSITE PLATES.

The benefits that arise from prestressing structures are very well known. When located such that the prestress opposes stresses due to dead and live loads, there are benefits to durability as well as the load capacity of the parent structure. It is therefore aspects such as the cost of prestressing, and durability of the prestressing itself, that most strongly influence appropriate applications. Pretensioning is both durable and relatively low cost, and is hence competitive for small items such as concrete lintels. Post-tensioned structures using currently available techniques are economic only for larger spans. There is therefore potential for retrofitting of prestress, as a strengthening technique, providing the cost and durability issues are fully addressed.

Following the retrofitting of prestress, the degree of strengthening of most existing structures is likely to be limited by the tensile stress within constituent materials. The additional

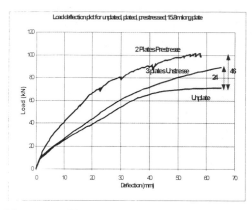

capacity that is potentially available will therefore be particularly high for those structures with significant dead load, the effects of which may be effectively counteracted. An example of the benefits of adding prestress is demonstrated by Figure 2, which shows the economic benefit of prestressed CFRP plates compared with unstressed CFRP plates demonstrated during ROBUST. Two prestressed plates will increase load capacity of an 18m beam by 46%, compared with 24% for three similar unstressed plates.

Figure 2. The general benefits of prestressing demonstrated during the ROBUST project.

Composite plates are ideal prestressing tendons that meet these needs. The excellent durability of carbon fibre composite plates in exposed conditions has been demonstrated, and the flat profile meets the needs of limited headroom and visual intrusion. Limitation was the absence of techniques to both stress and anchor such plates under site conditions in an economically viable way. This problem has been overcome in a method developed by Mouchel Consulting, and first applied to Hythe Bridge in Oxford.

DESCRIPTION OF HYTHE BRIDGE IN OXFORD

Hythe bridge, shown in Figure 3, was constructed in 1874. It carries a busy city centre road over a backwater stream of the River Thames in two square clear spans of 7.80m. The decks comprise 8 inverted Tee section Cast Iron beams, with Cast Iron channel section edge beams supporting a decorative parapet. Brick jack arches between the beam flanges, and infill material, support a 6.8m carriageway. The structure also carried a substantial number of congested services that would be very expensive to divert or disrupt. The original granular infill material had been replaced by 30 N/mm2 concrete during an earlier roadwork contract. This took advantage of the opportunity for limited local access, during which time the bridge remained open to traffic.

The structure was assessed to support only Group 2 Fire Engines (7.5 tonnes), compared with a required capacity of 40 tonne. In addition, some of the cast iron beams had also suffered cracking. Due to the location, it was important that any strengthening methods adopted did not cause traffic disruption

THE PRE-STRESSING TECHNIQUE DEVELOPED

The proprietary technique developed by Mouchel Consulting for Hythe Bridge requires anchorages to be fixed to the structure at a distance apart dictated by the pre-stress required. These anchorages may be fixed by bonding, friction or mechanical means, or a combination of each. 'End tab' plates are bonded to each end of the carbon fibre reinforced polymer (CFRP) tendons, and provide a means of attaching jacking equipment and anchoring the tendon when extended.

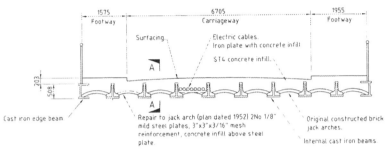

TYPICAL SECTION THROUGH EXISTING DECK

Figure 3. Hythe Bridge, Oxford.

The tendons are stressed by a hydraulic jack, which reacts against a jacking frame temporarily fixed to the anchorage. The stressed tendon is secured after extension by a shear pin that transfers load from a keyway in the end-tab to the anchorage. The tendons are bonded to the beam by epoxy resin in addition to the end anchorages. The anchorage itself is surrounded by a protective casing and fully grouted

STAGES IN DEVELOPMENT OF THE TECHNIQUE

Oxfordshire County Council and Mouchel Consulting adopted a partnership approach to enable the development of this new technique. The work was divided into stages, with progression to each stage depending upon a continuing expectation of ultimate success.

FEASIBILITY STUDY

The first stage was a feasibility study undertaken by Mouchel Consulting on behalf of Oxfordshire County Council in December 1997. The objective was to assess alternative methods of increasing mid span bending capacity to support the full 40 tonne loading.

Solutions involving beam replacement or additional beams were feasible, but had already been rejected by Oxfordshire County Council on grounds of the disruption required. Studies therefore concentrated upon forms of plate bonding that could be undertaken from the soffit. The feasibility study evaluated the material required to add strength with either steel plate bonding, unstressed composite plates, or stressed composite plates. The results are shown in Table 1.

The advantages and disadvantages of each method were then evaluated, and in particular the risks associated with adopting methods unsupported by test results. The conclusions are summarised below.

Steel plate bonding:
- Thickness of 135mm would restrict headroom and impose high additional load.
- Such thickness would require several layers, and is beyond previous experience. Extensive testing would be required before the effectiveness could be relied upon.
- Difficult handling and fixing, with drilling of Cast Iron not advisable.
- Expensive to undertake, and final system would require continuing maintenance.
- Risk of fatigue failure of cast iron remains, although reduced.

Items	Steel Plates	Unstressed Composite Plates	Stressed Composite Plates
Basis of method	Reduces extra stresses due to live load	Reduces extra stresses due to live load	Relieves existing dead load stresses
Plate dimensions	135mm x 450mm	70mm x 350mm	3 no 90mm x 4mm or 2 layers of 3no 90mm x 2mm
No of laminations	Say 6+	30	1 or 2
Need for fixing: • **Weight of plate** • **End Peel** • **Live Load anchorage** • **Pre-stress anchorage**	High Yes Yes No	Low Yes Yes No	V Low No Yes Yes
Headroom Loss	135mm + straps	70mm + straps	5mm, except at anchorages
Risks due to innovation	Multi layer effects	Multi layer effects	Stressing device CI local effects
Future Maintenance	Painting of complete system	None if clamps and straps stainless steel	Protection of end anchorage required.

Table 1. Comparison of Strengthening Methods

Unstressed Composite Plate Bonding:

- Multiple layers required, increasing labour and material costs beyond that normally associated with plate bonding.
- Behaviour of such high build multi-layer systems providing such a high degree of strengthening is unproven. Extensive testing would be required before the effectiveness could be relied upon.
- Strapping may be required for such a thick system to prevent peel forces at maximum live load, unless proven not to be required by representative tests.
- Risk of fatigue failure of cast iron remains, although reduced

Stressed Composite Plate bonding:

- No system for stressing composites in this situation was available.
- Stressing of Cast Iron could reveal weaknesses in the underlying material.
- Compressive stress very beneficial for resisting future fatigue failure.

It was concluded from detailed consideration of these factors that the stressed composite solution was most likely to offer a satisfactory solution to the problem. However, a feasibility trial was required to demonstrate the effectiveness of any proposed prestressing system before any main contract could be let.

The Feasibility Trial

The feasibility trial took place in May 1998, utilising cast iron beams saved from a smaller span structure that had been demolished earlier. These beams were of similar cross section to Hythe Bridge, but of length 4.70m and flange width 305mm. Three beams were erected in a works yard belonging to Oxfordshire, as illustrated in Figure 4. Note that the beams were erected with the flat soffit uppermost to facilitate the observations, testing and instrumentation.

Figure 4.   Cast Iron Test beam during the Feasibility Trial

The trial set out to investigate the following criteria: -

- The friction anchorage contributed by the high strength friction grip bolts.
- The bond strength between anchorage and Beam.
- The bond strength between end tabs and CFRP plate.
- The performance of the jacking system.
- The procedures for application of the anchorage, tendons, and temporary clamping.
- The transfer of load to the beam as demonstrated by instrumentation.

The trials proved to be a complete success. The tendons were pre-stressed to a load of 25 tonnes, and cycled without sign of failure. The information required for design of Hythe Bridge was obtained, and valuable information gained to enable procedures to be effective on a larger scale.

DESIGN OF THE STRENGTHENING

The desired capacity was achieved by stressing each beam with four CFRP (Carbon fibre reinforced polymer) tendons per beam, each stressed to a total of 18 tonnes. Allowance was made for beam restraint, and local effects examined by finite element analysis. Three of the cast iron beams were found badly cracked, and figure 5 shows a typical profile. It is believed that the most serious of these cracks may well have originated from damage incurred many years ago when services were installed. These cracked beams were treated slightly differently from the undamaged beams, with strengthening provided by a combination of steel plate bonding and pre-stressing. The cracks were first stitched together by the 'Metalock' system. Additional plating was then applied, also shown in Figure 5. The stressing force was reduced to 12 tonnes per plate on these beams. The primary strengthening of these beams was still provided by the pre-stress. Although the steel plating contributed to bending capacity, its main purpose was to provide edge restraint and containment of the cracked sections.

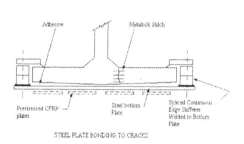

STEEL PLATE BONDING TO CRACKS

Figure 5.   Diagram of beam soffit showing crack profiles, and cross section of additional plating applied.

MAIN CONTRACT
The main contract was won by Balvac Whitley Moran, following competitive tender by three specialist firms judged to meet the quality standard. The site work to Hythe Bridge took place between December 1998 and April 1999. The main material components, CFRP plates and anchorages, were also obtained competitively in advance of the main contract, and supplied to the successful contractor. The jacking equipment was developed and owned by Mouchel Consulting, and made available to the successful contractor. General supervision was by Oxfordshire County Council, with supervision of specialist composites aspects by Mouchel Consulting.

The procedures were further developed during the main contract, with close co-operation between contractor, client and consultant. Work took place despite sub-zero temperatures, and there was no disruption to traffic that continued to use the structure. Experience was gained in surface preparation techniques, system tolerances, and temperature control and sensitivity.

Figure 6.   Photograph of an end anchorage box and beam with four tendons in place

CONCLUSIONS

- Composite plate bonding has passed through the early experimental stage. Applications may now be made with confidence, but there is a need to demonstrate that applications remain within the limits demonstrated by representative tests.
- The Hythe Bridge contract demonstrated that stressing of CFRP could be successfully undertaken under extreme site conditions to achieve a considerable increase in load capacity.
- Pre-stressing of composite materials offers a means of strengthening structures that can be more economic than unstressed composite materials or steel plate bonding.
- The system developed may be applied to steel, concrete and masonry, as well as cast iron as demonstrated by Hythe Bridge.
- The system is appropriate in most structural situations where tensile stress is the factor that limits load capacity.
- The system is durable, with negligible maintenance requirements.
- The development benefited from close co-operation between client, consultant and contractor, enabling the process to be developed and difficulties to be overcome as they arose.
- Reliable bonding at high stress is a specialist area requiring trained operatives and expert design and supervision. If these pre-requisites are satisfied, a reliable system results with considerable economic advantage to the strengthening and repair industry.

Reinforced concrete beams upgraded with externally bonded steel or FRP plates

PROF. M. RAOOF, and M.A.H. HASSANEN MSc, Civil and Building Engineering Department, Loughborough University, Loughborough, Leicestershire, U.K.

INTRODUCTION

With the development of strong epoxy adhesives back in the 1960's, bonding external steel or fibre reinforced plastic (FRP) plates to the tension side of reinforced concrete beams, Fig.1a, has proved very attractive for increasing the flexural strength of beams and/or slabs in both buildings and bridges, in a number of countries. Using this technique, the work can be carried out relatively simply and quickly, even while the structure is still in use, and its application causes minimal changes in the member dimensions (including overhead clearance), and negligible increases in the self-weight. It also does not alter the configuration of the structure. With the external plates supplementing the area of internal tension reinforcement, their application leads to reduced cracking and deflections under service loads, and also increases the flexural ultimate strength.

Reinforced concrete beams strengthened by externally bonded steel or FRP plates are commonly designed for flexure on the basis of conventional ultimate limit state procedures as recommended by various codes of practice, assuming the presence of full bond between the external plate and concrete up to the ultimate load, and using the plane-section bending assumption and a concrete stress block at failure. However, as repeatedly reported in the literature, the designer should also check that premature anchorage failure (Fig.1b) caused by peeling and debonding of the plate (which initiates at its end) does not occur prior to the beam achieving its full flexural strength. The premature failure mode shown in Fig.1b involves the plate and concrete cover becoming separated as a unit, from underside of the main reinforcing bars, which is the most commonly reported type of plate peeling phenomena, and forms the subject of the present paper. The occurrence of plate peeling at the plate/glue or concrete/glue interface has experimentally been found to be a rare case, and occurring as a result of bad workmanship: a treatment of this case is outside the scope of the present paper.

In recent publications by the first author and his associates [1-3], a semi-empirical model (backed by extensive test data as reported by others) was reported for predicting the potentially dangerous (largely brittle) premature plate peeling failures of simply supported reinforced concrete beams strengthened in flexure by gluing external steel plates to their tension sides. The purpose of the present paper is to report an extended version of this semi-empirical model which is applicable to cases when fibre reinforced plastic (as opposed to steel) plates are used for upgrading R.C. beams in flexure. Similar to steel plated beams, it will be argued that, due to large variations in the spacings of stabilised cracks within the concrete cover zone (by a factor of, say, 2), a unique solution for FRP plate peeling load does not exist, and one needs to resort to theoretical upper/lower bound solutions, with the lower bound being the appropriate (i.e. safe) one for design purposes.

Finally, it will be shown (both theoretically and by using large scale test data) that, ironically, adding externally bonded steel plates to R.C. beams in order to strengthen them in flexure may (if one does not guard against premature plate peeling failures) significantly reduce their flexural failure load below that of the corresponding unplated (i.e. original) beam.

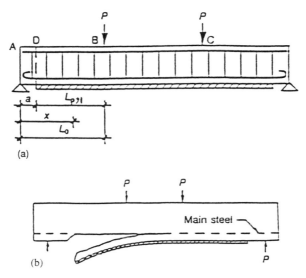

Fig. 1. Reinforced concrete beams strengthened with externally bonded plate: (a) general configuration; (b) plate peeling failure.

THEORY
The first author and his associates [1-3] have recently suggested a mode of failure (Fig. 2a) which is controlled by the characteristics of the individual teeth in-between adjacent stabilised cracks in the concrete cover, and forms the basis of the proposed model which very much depends on the size of stabilised crack spacings.

By considering the behaviour of an individual plain concrete tooth formed between two adjacent stabilised cracks (Fig.2b), a semi-empirical model is developed for steel plated beams [1-3] which enables one to estimate the magnitude of axial tensile stresses, σ_s, in the externally bonded plate at the critical instance of plate peeling failure. The model is based on an extended version of some well-established formulations for determining the minimum and maximum stabilised crack spacings, with the maximum crack spacing being assumed to be twice that of the minimum crack spacing. With the stabilised crack spacings calculated, it is, then, possible to estimate reasonable values for the critical tensile stresses at point A, in Fig. 2b, σ_A , assuming that the concrete tooth may be treated as a cantilever bending under the action of tangential (shear) stresses, τ, at the interface between the externally bonded steel plate and concrete, with the depth of the cantilever equal to either minimum or maximum stabilised crack spacings. It is further assumed that the brittle plate peeling failure initiates at the instance when σ_A reaches the concrete tensile strength, f'_t, i.e. at the critical state: $\sigma_A = f_t'$. Assuming an effective length of the plate L_p over which an equivalent uniform state of shear stresses, τ, exists at the plate/concrete interface within the critical shear span, where the plate terminates (Fig.1a), it is simple to calculate the lower and upper bounds to the magnitudes of plate tensile stresses $\sigma_{s(min)}$ and $\sigma_{s(max)}$ (directly under the point load nearest to the support in Fig.1a) corresponding to the minimum and maximum stabilised crack spacings, respectively, where $\sigma_{s(max)} = 2\,\sigma_{s(min)}$ and [2]

$$\sigma_{s(min)} = 0.154 \frac{L_p h_1 b^2 \sqrt{f_{cu}}}{h' b_1 t \left(\sum 0_{bars} + b_1\right)} \tag{1}$$

In Equation (1), t = plate thickness, b_1 = width of the steel plate, b = width of the beam, h' = net height of the concrete cover(= length of the cantilever), f_{cu} = concrete cube strength, $\Sigma 0_{bars}$ = sum of main reinforcing bar circumferences, and $2h_1$ is the height of the assumed region of concrete in tension as shown in Fig.3. For steel plated beams, the value of the effective length, L_p, in Equation (1) is given by the lower value of L_p as calculated from Equations (2), $L_{p,2}$, and the actual length of the plate as shown in Fig. 1a, $L_{p,1}$ (whichever is smaller), where [2]

$$L_{p,2} = \ell_{min}^p \left(21 - 0.25\,\ell_{min}^p\right)\,,\quad \ell_{min}^p \leq 72\ \text{mm} \tag{2a}$$

or

$$L_{p,2} = 3.0\,\ell_{min}^p \qquad,\quad \ell_{min}^p > 72\ \text{mm} \tag{2b}$$

In the above, ℓ_{min}^p = estimated minimum stabilised crack spacing, with its value given by

$$\ell_{min}^p = \frac{A_e\,f_t'}{u\,\left(\Sigma 0_{bars} + b_1\right)} \tag{3}$$

where, u = steel/concrete average bond strength = $0.28\,\sqrt{f_{cu}}$, f_t' = cylinder splitting tensile strength of concrete = $0.36\sqrt{f_{cu}}$, and A_e = the effective area of concrete in tension, Fig.3, given by $A_e = 2h_1 b$, with the other terms in Equation (3) as defined previously. Note that the units of u, f_{cu}, and, f_t' are in MPa, with the units of L_p and ℓ_{min}^p in Equations (2a,b) both in mm.

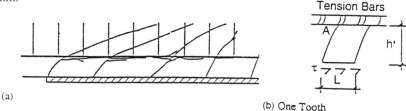

(a)

(b) One Tooth

Fig. 2. Assumed mode of failure due to premature plate peeling: (a) pattern of stabilised concrete cracking; (b) behaviour of an individual tooth in the concrete cover.

In situations where FRP (as opposed to steel) plates are externally bonded to the soffit of a R.C. beam, it has been shown [4] that one may still use Equation (1) for estimating reasonable values of the axial tensile stresses in the FRP plate, with the proviso that the magnitude of L_p in Equation (1) should now be estimated using the following semi-empirical relations (instead of Equations (2a,b))

$$L_{p,2} = \ell_{min}^{P} \, (24.0 - 0.5 \, \ell_{min}^{P}) \qquad , \qquad \ell_{min}^{P} \leq 40 \text{ mm} \qquad (4a)$$

or

$$L_{p,2} = 4.0 \, \ell_{min}^{P} \qquad , \qquad \ell_{min}^{P} > 40 \text{ mm} \qquad (4b)$$

with the values of ℓ_{min}^{P} in Equations (4a,b), again, given by Equation (3). The appropriate L_p for FRP plates would, then, be the lower value of L_p as calculated from Equations (4a,b), $L_{p,2}$, and the actual length of the plate within the critical shear span as shown in Fig. 1a, $L_{p,1}$ (whichever is smaller). If the plate is only positioned within the constant moment zone (for beams under symmetrical four-point loading), the value of L_p in Equation (1) may only be estimated by either Equations (2a,b) (for steel plates), or Equations (4a,b) (for FRP plates).

Finally, with the magnitudes of $\sigma_{s(min)}$ and $\sigma_{s(max)} = 2 \, \sigma_{s(min)}$ directly under the point load nearest to the support estimated, it is, then, simple to predict the corresponding lower and upper bounds to plate peeling moment, $M_{peel,\ell}$ and $M_{peel,u}$, respectively, at this location, by resorting to traditional methods based on the plane-section bending assumption, taking the effect of concrete tensile stresses below the neutral axis into account [2,4].

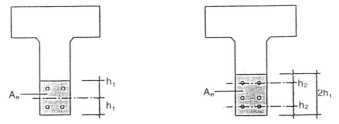

Fig. 3. Assumed concrete region in tension: (a) even distribution of reinforcing bars; (b) uneven distribution of reinforcing bars.

RESULTS AND DISCUSSION

Figs. 4a and 4b present correlations between the experimental data of Oehlers and Moran [5] and Oehlers [6], respectively, and predictions of the bounding values of plate peeling moment at the point load for 83 R.C. beams with externally bonded steel plates. Unlike all the theoretical results reported in Ref. [3] which assumed a trilinear axial stress-strain relationship for the reinforcing bars but a bi-linear behaviour for the externally bonded steel plates, the plates and reinforcing bars relating to all the theoretical plots in Figs. 4a and 4b are assumed to have a bilinear (elasto-plastic) axial stress-strain characteristic after BS8110 [7]. The semi-empirical model is found to successfully predict the lower and upper bound solutions, $M_{peel,\ell}$ and $M_{peel,u}$, respectively, to all the extensive set of large scale test data, M_{exp}, with the specimens having widely varying design parameters and external plate bonding techniques. In the tests on 57 beams as reported by Oehlers and Moran [5] the specimens had steel plates bonded to their soffits within only the constant moment zone with the beams subjected to symmetrical four-point loading, while in the tests on 26 beams as reported by Oehlers [6], the beams had steel plates terminated within the shear span with the specimens again subjected to symmetrical four-point loading. As discussed by Oehlers [6], for beams with plates terminated within the shear span, the ultimate loads (at which

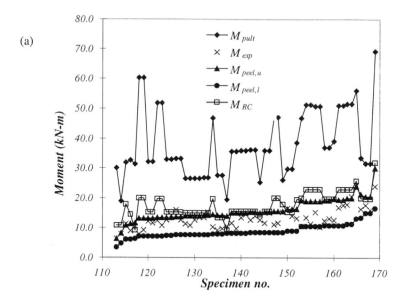

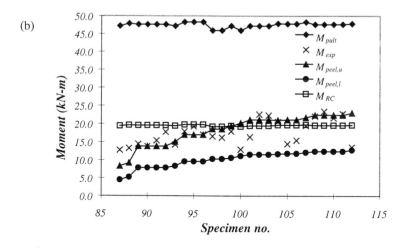

Fig. 4 Correlations between the theoretical upper and lower bound solutions and experimental data after: (a) Oehlers and Moran [5], and (b) Oehlers [6].

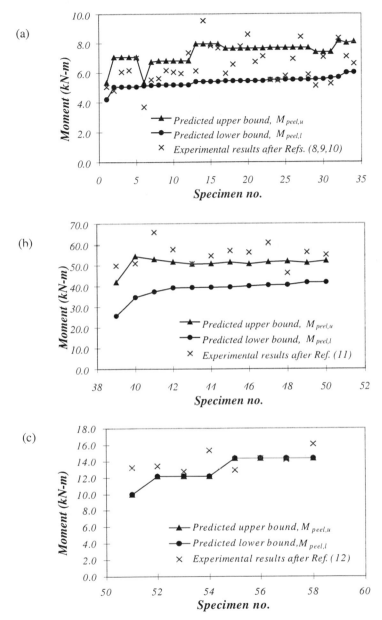

Fig. 5 Correlations between upper and lower bound theoretical predictions of FRP plate peeling bending moment and test data: (a) test data after Refs. [8,9,10]; (b) test data after Ref. [11]; (c) test data after Ref. [12].

total collapse occurred) were experimentally found to be somewhat greater than the associated initial plate peeling loads: the test data in Fig. 4b all relate to initial plate peeling moments. The results in Figs. 4a,b also include plots of M_{pult} and M_{RC} for all these beams, where M_{pult} = absolute maximum bending moment for the plated section based on BS8110 [7] with material partial safety factors set equal to unity, and M_{RC} = maximum ultimate bending moment for the corresponding unplated (i.e. original) R.C. beams, with material partial safety factor of 1.15 for the steel plates and reinforcing bars, and material partial safety factor of 1.5 for the concrete according to BS8110 [7]. It is most interesting to note that, in the vast majority of cases in Figs. 4a and 4b, values of M_{RC} are greater than the associated experimental data relating to the plate peeling moments, M_{exp}, and also their corresponding theoretical upper and lower bound predictions, $M_{peel,u}$ and $M_{peel, \ell}$, respectively: in other words, it is ironical that adding externally bonded steel plates to R.C. beams in order to strengthen them in flexure may (if one does not guard against plate peeling failures) significantly <u>reduce</u> their failure load in flexure. Such largely brittle failures can obviously have serious implications in practice, where this method has been used extensively for upgrading both bridges and buildings, in a number of countries. The very large differences between calculated values of $M_{peel, \ell}$ and M_{pult} are also believed to be of significant practical importance.

As regards externally bonded FRP plates, Fig. 5a presents comparisons between the upper and lower bound theoretical predictions, $M_{peel,u}$ and $M_{peel, \ell}$, respectively, and the extensive set of small-scale test data of Hollaway and his associates as reported in Refs. [8-10], where the correlations between the present theoretical bounding solutions and test data is very encouraging. Similar correlations between the theoretical bounding solutions and (this time) large scale test data are presented in Fig.5b, where the test data is after Ref. [11], with the beam specimens having a clear span of 2.4m. In the vast majority of the tests carried out in Ref. [11], the test specimens had various forms of plate end anchorage arrangements which could well have been instrumental in producing experimental results which lie above (although quite close to) the theoretical upper bound. Finally, in Fig. 5c, the theoretical lower and upper bounds for premature FRP plate peeling bending moments are found to be equal in magnitude with $\sigma_{s(min)} = \sigma_{s(max)} = \sigma_u$ (where, σ_u = yield strength for the FRP material) for all the initially pre-cracked R.C. beams (prior to plating) tested in Ref. [12] (as opposed to the initially uncracked (i.e. as cast) beams of Refs. [8-11]): the correlations between theory and test data, is again found to be very encouraging. In the theoretical results of Figs. 5a-c, a bi-linear (elasto-plastic) axial stress-strain relationship as recommended by BS8110 [7] has been assumed for the embedded steel bars, while the FRP plate is assumed to be linearly elastic with a brittle fracture at ultimate axial load. The assumed stress-strain law for concrete in Figs. 4a,b and 5a-c is as that recommended by BS8110 [7].

CONCLUSIONS
The present paper reports the salient features of a semi-empirical model which throws some light into the appropriate mode of steel or FRP plate peeling failure, and the outcome of which is supported by an extensive set of mainly large scale test data from other sources. Based on this model, the premature plate peeling failure is controlled by the spacings of stabilised cracks in the concrete cover zone, and due to large variations (by a factor of, say, 2) in such crack spacings in practice, it is concluded that a unique solution for the plate peeling failure load does not exist, and one has to resort to upper/lower bound approaches, with the lower bound predictions being the appropriate (i.e. safe) ones for design purposes. Furthermore, in view of the lack of a unique solution, it may be argued that the previous general practice among various researchers who have (over a number of years) carried out

purely experimental parametric studies has been fraught with difficulties and uncertainties with the wide scatter problem making any conclusive deductions based on purely experimental comparisons very difficult if at all possible. Finally, it has been demonstrated both theoretically and by using extensive large scale test data, as reported by others, that if the practising engineers do not guard against the potentially dangerous brittle steel plate peeling phenomenon, the plate peeling failure moment can be significantly lower than the ultimate (failure) moment of the corresponding unplated (i.e. original) R.C. beam which has been designed according to ultimate limit state code recommendations even when (unlike the plated beam) material partial safety factors are included in the design calculations for the unplated beam. This observation may have significant practical implications, particularly when one considers that the method of strengthening R.C. beams with externally bonded plates has already been used extensively, in a number of countries, for upgrading both bridges and buildings.

ACKNOWLEDGEMENTS
The authors are grateful to Professor Hollaway and Dr. Garden of the Civil Engineering Department, University of Surrey, for providing them with some crucial experimental data on FRP plated beams, which led to the derivation of Equations (4a,b) in the present paper.

REFERENCES
1. S. Zhang, M. Raoof, and L.A. Wood, "Prediction of Peeling Failure of Reinforced Concrete Beams with Externally Bonded Steel Plates", Proc. Instn Civ. Engrs, Structures and Buildings, Vol.110, Aug., 1995, 257-268.

2. M. Raoof, and S. Zhang, "An Insight into the Structural Behaviour of R.C. Beams with Externally Bonded Plates", Proc. Instn Civ. Engrs, Structures and Buildings, Vol.122, Nov., 1997, 477-492.

3. M. Raoof, J.A. El-Rimawi, and M.A.H. Hassanen, "Theoretical and Experimental Study on Externally Plated R.C. Beams", Engineering Structures, Vol. 22, no.1, Jan., 2000, 85-101.

4. M. Raoof, and M.A.H. Hassanen, "Peeling Failure of R.C. Beams with FRP or Steel Plates Glued to Their Soffits," Submitted for Publication.

5. D.J. Oehlers, and J.P. Moran, "Premature Failure of Externally Plated Reinforced Concrete Beams", J. Struct. Engng, ASCE, Vol. 116, no. 4, 1990, 978-995.

6. D.J. Oehlers, "Reinforced Concrete Beams with Plates Glued to their Soffits", J. Struct. Engng, ASCE, Vol.118, no. 8, 1992, 2033-2038.

7. British Standards Institution. Structural Use of Concrete. London: BSI, 1985, BS8110.

8. R.J. Quantrill, et al., "Experimental and Analytical Investigation of FRP Strengthened Beam Response: Part 1", Magazine of Concrete Research, Vol. 48, no.177, Dec., 1996, 331-342.

9. R.J. Quantrill, et al., "Predictions of the Maximum Plate End Stresses of FRP Strengthened Beams: Part 2", Magazine of Concrete Research, Vol. 48, no.177, Dec., 1996, 343-351.

10. H.N. Garden, et al., "A Preliminary Evaluation of Carbon Fibre Reinforced Polymer Plates for Strengthening Reinforced Concrete Members", Proc. Instn Civ. Engrs, Structures and Buildings, Vol. 123, May, 1997, 127-142.

11. P.A. Ritchie, et al., "External Reinforcement of Concrete Beams Using Fibre Reinforced Plastics", ACI Structural J., July-Aug., 1991, 490-500.

12. A. Sharif, et al., " Strengthening of Initially Loaded Reinforced Concrete Beams Using FRP Plates", ACI Structural J., Mar.-Apr., 1994, 160-168.

The development of design charts for column strengthening using FRP

DENTON, S.R., PB Kennedy and Donkin Ltd, Bristol, UK, CHRISTIE, T.J.C., PB Kennedy and Donkin Ltd, Bristol, UK, POWELL, S.M., PB Kennedy and Donkin Ltd, Bristol, UK.

INTRODUCTION

Many highway bridge supports in UK have been found to have insufficient capacity to satisfy the requirements for vehicle impact loading set out in BD48/93[1]. One possible method for strengthening columns to resist vehicle impact is through the use of fibre reinforced plastics (or polymers) (FRP).

FRP can be applied to existing columns with fibres aligned with the axis of the column, in the "longitudinal" direction, principally to increase their flexural strength, or with the fibres aligned around the column, in the "hoop" direction, to increase their shear strength and to confine the concrete. Confining the concrete provides a dual benefit: firstly the strength of the concrete is increased and secondly its strain capacity is increased.

The UK Highways Agency has commissioned research to investigate the structural behaviour of concrete columns strengthened using FRP. This research has lead to the development of a design method which has been validated by tests[2]. This design method has been incorporated in an Interim Advice Note (IAN)[3], to be promulgated by the Highways Agency.

Design charts for circular reinforced concrete columns have been developed on the basis of this design method to enable the required thickness of "longitudinal" or axial FRP, to strengthen these columns in flexure, to be established rapidly by hand. These charts will form part of the IAN, in which their application is explained and examples are presented. The purpose of this paper is to describe the theoretical basis of these design charts, particularly those issues not dealt with in the IAN.

The development of these design charts presented numerous challenges, particularly because the total number of charts required had to be limited to a manageable figure whilst catering for all common ranges of column size, reinforcement arrangement, material properties, fibre type, fibre properties etc. Furthermore, the effect of adding FRP cannot be considered in isolation since the contribution to the capacity of the column made by the steel reinforcement and concrete is strongly dependent on the neutral axis depth which is itself dependent upon the amount of FRP provided.

In this paper, emphasis is placed on the fundamental theory underpinning the development of the charts rather than the supplementary steps taken to limit the number of charts required. Such steps included taking advantage of key non-dimensional parameters and making some safe approximations, and are discussed in detail by Denton and Christie[4].

Bridge Management 4, Thomas Telford, London, 2000.

SIGN CONVENTION AND COLUMN GEOMETRY

Throughout this paper compressive strains and stresses are positive. Tensile strains and stresses are negative. The assumed column geometry is shown in Figure 1.

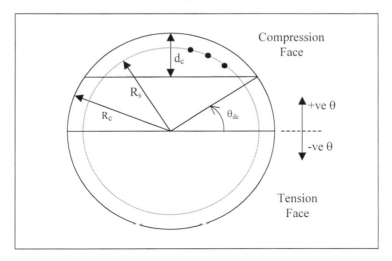

Figure 1 : Column Geometry

THEORY

The design approach set out in the Interim Advice Note (IAN) is based on the assumptions that plane sections remain plane and that the ultimate axial force and bending moment capacities of an FRP strengthened column about its centroid axis may be expressed as follows:

$$N_u = N_{uc} + N_{us} + N_{uw} \qquad (1)$$
$$M_u = M_{uc} + M_{us} + M_{uw} \qquad (2)$$

where (N_{uc}, M_{uc}) are the contributions to the coexistent axial load and moment capacities made by the concrete, (N_{us}, M_{us}) are the contributions made by the steel and (N_{uw}, M_{uw}) are the contributions made by the FRP.

Concrete and Steel Reinforcement Contributions

For any ratio of neutral axis depth to column radius, d_c/R_c, and compressive strain in the extreme fibre of the concrete, ε_{cp} (which may be greater than the BS5400 Part 4 limit of 0.0035 due to confinement), the contributions of the concrete and the steel reinforcement to the axial load and moment capacity may be calculated using standard methods.

In the IAN a rectangular concrete stress block of magnitude $0.6f_{cw} / \gamma_{mc}$ is used, where f_{cw} is the effective concrete "cube" strength and may be greater than f_{cu} due to confinement. The stress-strain curve specified in BS5400 Part 4 is used for the steel reinforcement.

FRP Contribution

Since the purpose of the design is to determine the FRP thickness, the contribution of the FRP to the axial load and moment capacities cannot be established at the outset as can effectively be done for the concrete and reinforcement. As the provision of charts for various standard thicknesses of FRP would lead to an unmanageable number of different charts an innovative technique has been developed to allow the required FRP thickness to be determined.

The key step in understanding how the FRP contribution is determined using the design charts stems from considering the ratio of the contributions to the axial load and bending moment capacities made by the FRP. If the radius to the FRP is conservatively assumed to be equal to the column radius and the compressive strength of the FRP is neglected then, using expressions given in the IAN or by integration, it can be shown that

$$\frac{R_c N_{uw}}{M_{uw}} = \frac{\left[(\frac{d_c}{R_c}-1)(\frac{-\pi}{2}-\theta_{dc})+\cos\theta_{dc}\right]}{\left[\frac{1}{2}(\frac{-\pi}{2}-\theta_{dc})+\frac{1}{4}(\sin 2\theta_{dc})+(\frac{d_c}{R_c}-1)\cos\theta_{dc}\right]} \tag{3}$$

where, as shown in Figure 1, $\qquad \theta_{d_c} = \sin^{-1}(1-\frac{d_c}{R_c})$ \hfill (4)

From these expressions it can be seen that the ratio N_{uw}/M_{uw} is constant with respect to differing FRP thicknesses for any given values of d_c and R_c.

One further relationship that is used in the development of the design charts enables the contribution of the FRP to the axial load to be related directly to FRP thickness, t_1. Again, either from the expression in the IAN or by integration, it may be shown that

$$t_l = \frac{-k_1 N_{uw}}{R_c \varepsilon_{cp}(E_{wt}/\gamma_{mwE})} \tag{5}$$

where E_{wt} is the stiffness of the FRP, γ_{mwE} is the partial safety factor applied to this stiffness and the FRP "thickness-factor", k_1, is given by

$$k_1 = 0.5\left[(1-\frac{R_c}{d_c})(\frac{-\pi}{2}-\theta_{dc})+\frac{R_c}{d_c}\cos\theta_{dc}\right]^{-1} \tag{6}$$

DEVELOPMENT OF THE DESIGN CHARTS
A range of design charts have been developed for high yield and mild steel reinforcement, confined and unconfined concrete and various mechanical reinforcement ratios, ω, defined as:

$$\omega = \frac{A_s(f_y/\gamma_{ms})}{\pi R_c^2(f_{cw}/\gamma_{mc})} \tag{7}$$

where A_s is the area of axial steel reinforcement, f_y is the steel yield strength and γ_{ms} the corresponding partial safety factor.

Each design chart is made up of three "sub-charts" as follows (see Figure 3 at end of paper):

(i) Upper Section : N-M Interaction Sub-chart

(ii)   Central Section   : FRP Strain Sub-chart

(iii)   Lower Section   : FRP Thickness-factor Sub-chart

These sub-charts have the same horizontal axis, but the vertical axes differ. Their development and purpose are described in the following sections.

N-M Interaction Sub-chart
This chart enables the contributions to the axial load and moment capacities provided by the FRP and the combined contributions provided by the concrete and reinforcement to be established. Furthermore, it enables the neutral axis depth of the strengthened column to be established, although for simplicity this is not done explicitly.

The horizontal axis of the N-M Interaction Sub-chart represents axial load and the vertical axis represents moment. In the actual design charts these are non-dimensionalised in the forms $N/R_c^2(f_{cw}/\gamma_{mc})$ and $M/R_c^3(f_{cw}/\gamma_{mc})$ respectively. A complete design chart is shown in Figure 3, and on the N-M Interaction Sub-chart two families of curves are plotted: the first family are curved solid lines; the second are straight dotted lines with varying gradient.

The curved solid lines represent the combined contributions of the concrete and steel to the axial load and moment capacities of a column for four different ratios of effective steel radius to column radius, R_s/R_c. Each point on these curves corresponds to a different value of d_c/R_c. Combinations of ultimate applied axial load and moment (N,M) that lie below the relevant curve can safely be carried by the column without strengthening using axial FRP. However, if the column is required to sustain combinations of loading outside this region axial FRP is required. These curves are similar in concept to the charts for designing reinforced concrete columns included in, for example, CP110[5].

The development and purpose of the straight dotted lines are best explained by considering a single column and the effect of adding varying amounts of FRP to this column. For this column, an N-M curve is first plotted showing the combined contribution of the concrete and reinforcement to the axial load and moment capacities for different values of d_c/R_c, as shown diagrammatically in Figure 2.

If a specific value of d_c/R_c is selected then this corresponds to a unique point, denoted Point B, on the N-M curve, as also shown in Figure 2. For this point the coexisting ultimate axial load and bending moment are denoted (N_B, M_B).

If a layer of axial FRP (of thickness t_{L1} say) is added while keeping this value of d_c/R_c constant and N_u and M_u are recalculated for the strengthened column, N_u will have increased from N_B to $(N_B + N_{uw1})$ and M_u will have increased from M_B to $(M_B + M_{uw1})$. If this new point (A_1 say) is plotted on the N-M diagram, then the gradient of the line BA_1 is M_{uw1} / N_{uw1}.

If a further increase in the FRP thickness to t_{L2} (say) is made while still keeping d_c/R_c constant, so that N_u and M_u increased to $(N_B + N_{uw2})$ and $(M_B + M_{uw2})$, and this point (A_2 say) is plotted on the diagram, then the gradient of BA_2 is M_{uw2} / N_{uw2}. But it is shown in Equation 3 that, if d_c/R_c is constant M_{uw}/N_{uw} is also constant. Thus $M_{uw1} / N_{uw1} = M_{uw2} / N_{uw2}$ so the gradient of BA_1 equals the gradient of BA_2.

It follows therefore that if any thickness of FRP were to be applied to the column whilst keeping d_c/R_c constant then the resulting coexistent values of N_u and M_u of the strengthened column, denoted Point A, (N_A, M_A), must lie on a straight line passing through points B, A_1 and A_2.

In general therefore, if the FRP is thickened keeping d_c/R_c constant, then the point (N_A, M_A) on the interaction diagram denoting the coexisting ultimate axial load and moment capacities of the strengthened column migrates upwards along a *straight* line of constant gradient equal to M_{uw}/N_{uw}, as shown in Figure 2. If we repeat the above procedure for different values of d_c/R_c we obtain a series of lines which can be considered as "d_c/R_c contours". These are *not* parallel and are plotted on the N-M Interaction Sub-chart as a family of dotted lines.

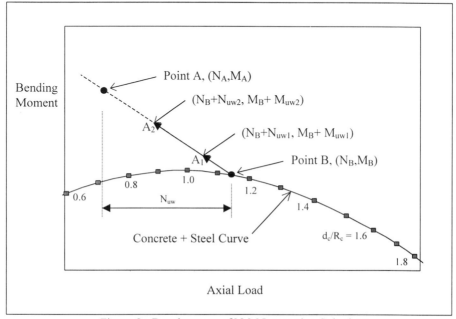

Figure 2 : Development of N-M Interaction Sub-chart

To use the N-M interaction chart in design the process described above is effectively reversed. At the outset of a design the required coexistent applied ultimate axial load and moment, (N,M) will be known, as will the properties of the existing column which are used to select the appropriate chart.

Point A, given by $(N_A, M_A) = (N,M)$, is first plotted on the selected chart. In principle, the value of d_c/R_c for the *strengthened* column can now be read directly from the "d_c/R_c contours". Furthermore, by constructing a line through Point A parallel to the nearest contour until it intersects the solid line corresponding to the concrete and steel contributions, at Point B, then the contribution to the axial load and moment capacity made by the concrete and steel, (N_B, M_B), for this value of d_c/R_c can be established. As illustrated in Figure 2, the difference between the points (N_A, M_A) and (N_B, M_B) corresponds to the contribution to the axial load and moment capacity made by the FRP, i.e. (N_{uw}, M_{uw}).

Thus, simply by plotting Point A and constructing a single straight line to find Point B, both the neutral axis depth of the *strengthened* column and the contribution of the FRP to the coexistent axial load and moment capacities are established.

To limit the number of design charts required, as explained previously, curves corresponding to the concrete and steel contributions are plotted on the N-M interaction sub-chart for four different ratios R_s/R_c. As a result the dotted lines no longer relate directly to specific neutral axis depths, resulting in some limited conservatism. However, once the point (N_B, M_B) has been identified by following a dotted "contour" from (N_A, M_A), the neutral axis depth is still implicitly established.

Although the identification of the neutral axis depth of the strengthened column is central to the design process, it is not necessary for its actual value to be established explicitly. Instead, its value is effectively "transferred" to the other sub-charts by constructing a vertical line through all three sub-charts from the point (N_B, M_B).

FRP Strain Sub-chart.
In developing the N-M Interaction Sub-chart, the strain capacity of the FRP was not considered. The FRP Strain Sub-chart enables the strain capacity of the FRP to be checked to ensure that it is not exceeded before the concrete reaches its required ultimate compressive strain.

The value of the strain in the extreme fibre of the concrete, ε_{cp}, is initially assumed to have a particular value (viz. 0.01 for a column with hoop wrapping and 0.0035 for a column without hoop wrapping). Thus, if d_c/R_c is known the maximum tensile strain in the FRP, ε_{wt}, can be calculated from the equation,

$$\varepsilon_{wt} = -\varepsilon_{cp}\left[\frac{2R_c}{d_c} - 1\right]$$

(8)

and the design is acceptable provided, $\left|\varepsilon_{wt}\right| < \left|\dfrac{\varepsilon_{wtm}}{\gamma_{mw\varepsilon}}\right|$

(9)

where ε_{wtm} is the maximum permissible tensile strain in the FRP and $\gamma_{mw\varepsilon}$ is the corresponding partial safety factor.

From Equations (8) and (9) the limiting FRP strain requirement can be more conveniently rewritten as

$$\frac{-\varepsilon_{wtm}}{\gamma_{mw\varepsilon}\varepsilon_{cp}} > \left[\frac{2R_c}{d_c} - 1\right]$$

(10)

The Strain Sub-Chart has been developed to enable this limit to be checked graphically. The vertical axis of this Sub-Chart, which is directly related to the left-hand side of Equation 10, is the dimension-less parameter, $-\varepsilon_{wt}/\varepsilon_{cp}$.

In concept, the curves on this Sub-chart are constructed through the intersection points of vertical lines projected downwards from Points "B" on the N-M Interaction Sub-chart corresponding to a range of values of d_c/R_c and horizontal lines at values on the vertical axis for these same values of d_c/R_c given by:

$$\frac{-\varepsilon_{wt}}{\varepsilon_{cp}} = \left[\frac{2R_c}{d_c} - 1\right]$$

(11)

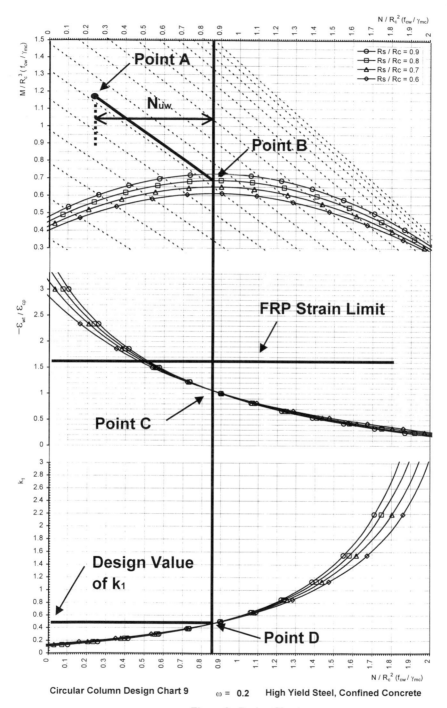

Circular Column Design Chart 9   ω = 0.2   **High Yield Steel, Confined Concrete**

Figure 3 : Design Chart

To use this Sub-chart, a horizontal line of value $-\varepsilon_{wtm}/(\gamma_{mw\varepsilon}\varepsilon_{cp})$ is first plotted, which corresponds to the maximum strain capacity of the FRP. Then, if the vertical line projected from the actual Point B, (N_B, M_B), identified on the N-M Interaction Sub-chart for a strengthening design, intersects the curve on the FRP Strain Sub-chart corresponding to the relevant value of R_s/R_c, denoted Point C, below this maximum strain capacity line then the strain capacity of the FRP is not exceeded. However, if this point lies above this line, the strain capacity of the FRP is exceeded and it will be necessary to add extra FRP or use an alternative fibre type. The appropriate steps to take in such an instance are explained in the IAN and by Denton and Christie[4].

FRP Thickness-factor Sub-chart
The FRP Thickness-factor Sub-chart is required to convert the value of N_{uw} found from the N-M diagram, as described above, into a required thickness of FRP, t_l. The vertical axis of the chart is the thickness-factor k_1 defined in Equation 6. Since this factor is a function of d_c/R_c only, a similar procedure to that used for developing the FRP Strain Sub-chart may be used to construct the curves on this chart.

The sub-chart is used by reading off the value of k_1 at the intersection point of the vertical line projected downwards from Point B on the N-M Interaction Sub-chart and the curve corresponding to the appropriate ratio of R_s/R_c. The value is then substituted into Equation 5, along with N_{uw} and other material property and geometric parameters which to give the required thickness of FRP.

APPLICATION
The following steps are undertaken to perform a design, as illustrated on Figure 3:

(1) The appropriate design chart is selected.
(2) Point A corresponding to the ultimate applied coexistent axial load and moment is plotted on the N-M Interaction Sub-chart.
(3) A straight line from Point A is drawn parallel to the nearest dotted contour line, to intersect the curve corresponding to R_s/R_c at Point B.
(4) From the horizontal distance between Points A and B, N_{uw} is calculated.
(5) A vertical line through point B is constructed to intersect the relevant FRP strain curve (central sub-chart) at Point C and the thickness-factor curve (lower sub-chart) at Point D.
(6) The horizontal line corresponding to the maximum FRP strain capacity is plotted on the FRP Strain Sub-chart. A check is made that Point C lies below this line. If this is not the case, see IAN[3] for required action, or Denton and Christie[4] for a full explanation.
(7) The FRP thickness factor k_1 is read from Point D.
(8) The required FRP thickness is calculated from Equation 5.

REFERENCES
1. The Highways Agency, *The assessment and strengthening of highway bridge supports*, BD48/93, 1993.
2. Cuninghame, J.R. and Sadka, B, 'Fibre Reinforced Plastic strengthening of bridge supports to resist vehicle impact', SAMPE Europe Int. Conf. *Advanced materials and processes, affordabilities for the new age*, Paris, April 1999.
3. The Highways Agency, *Strengthening Concrete Bridge Supports Using Fibre Reinforced Plastics*, Interim Advice Note, 1999
4. Denton, S.R. and Christie, T.J.C., "Design Charts for Strengthening Circular Concrete Columns using FRP", in preparation.
5. British Standards Institution, *The Structural use of Concrete*, CP110 Part 3, 1972

Selecting appropriate remedial measures for corroding concrete structures

P.R.Vassie. Transport Research Laboratory, Crowthorne, Berkshire, RG45 6AU, UK

C.Arya. University College London, Department of Civil and Environmental Engineering
Gower Street, London WC1E 6BT

INTRODUCTION

Reinforcement corrosion is the most common cause of deterioration of concrete structures. It is caused by carbonation and by sodium chloride in de-icing salt and sea water. In most cases it is important to stop the corrosion process and repair any damage to the concrete that has resulted. If corrosion is not stopped it could affect the safety of the structure by reducing bond strength and the cross-section of the reinforcement. The safety of users could also be affected by lumps of concrete falling of the structure. There are a number of methods for preventing/stopping corrosion in concrete structures and it is often difficult for the engineer to select the best method in particular circumstances.

This paper reviews the following remedial measures:

- silane
- paint
- inhibitors
- cathodic protection
- desalination
- realkalisation
- concrete repairs
- enclosures for catching debris and
- strengthening using fibre composite materials.

Each remedial measure is discussed in terms of:

- the condition of the structure
- the amount of pre and post treatment monitoring that is needed
- relative costs and benefits
- its practical applicability and the degree to which it may affect the users of the structure.

The discussion leads to some guidelines on when the use of different remedial measures is appropriate and the life and relative costs of different remedial measures.

REMEDIAL TECHNIQUES FOR STRUCTURES VULNERABLE TO REINFORCEMENT CORROSION

There are a number of types of remedial measures:

a)   preventative methods
b)   methods for stopping the corrosion process

Bridge Management 4, Thomas Telford, London, 2000.

c)   methods for repairing damaged concrete
d)   methods for preventing injury from spalling concrete
e)   methods for strengthening/replacement

The condition of the structure determines which types are appropriate. This is important because the wrong choice of remedial measure is likely to be completely ineffective. Types (a) and (b) are only likely to be effective when applied before corrosion has initiated. Type (e) would only be relevant if the load carrying capacity of the structure has been shown to be inadequate and would be combined with methods (a) to (d) to stop further corrosion and prevent it from reoccurring. Types (b) and (c) are normally used together since it is unusual to find corrosion without any concrete damage. The use of type (d) on its own would normally only be considered if the strength of the deteriorated part of the structure was insensitive to the steel-concrete bond and to the steel cross section.

The amount of testing work on the structure that is needed in order to be able to specify a durable repair depends on the type of remedial measure. Testing prior to repair work is often important. It's purpose is to ensure the correct repair/maintenance method and extent of repairs is used in order to achieve an effective and durable repair. For example preventative methods are often applied to the entire surface of a structural element so testing is only needed in order to decide the potential cause of deterioration. Sometimes, however, preventative measures are only applied to areas of the structure which have the most extreme exposure to corrosive substances in the environment or which are vulnerable to corrosion due to low concrete cover or because they are a sink where corrosive substances can accumulate. In this case testing and observations are also needed to locate the vulnerable areas.

Areas of damaged concrete can be determined by visual observation or by tapping the concrete surface with a hammer and listening for a low frequency audio response. Most techniques for stopping corrosion are often applied to the entire structural element hence pretesting is only required to determine the cause of corrosion. Concrete repairs are an exception in this respect because if they are to stop corrosion for a substantial period of time tests are needed firstly to decide whether or not there is macrocell corrosion. If there is no macrocell corrosion concrete repairs should be effective if the damaged concrete is replaced and preventative measures are used to avoid further deterioration. If there is evidence of macrocell corrosion then extensive testing prior to repairs is needed in order to determine the total extent of all the anode-cathode cells since this is the area of the structure where concrete repairs are needed. The repair of this area should remove most of the chloride ions generally responsible for macrocell corrosion and prevent the formation of incipient anodes, although the extent of repairs may be too high to be viable.

In order to determine the degree of strengthening required, testing is needed to find the strength of the concrete and steel and the cross section of steel and steel-concrete bond still available after corrosion.

PREVENTATIVE MAINTENANCE
Preventative methods increase the age of a structure when corrosion first initiates and may also subsequently reduce the rate of corrosion. Some preventative methods must be applied during construction, but all must be applied before corrosion initiates to be at all effective. The earlier they are applied the more effective they are. Some can be reapplied a number of times to prevent corrosion throughout the design life of a structure. Preventative maintenance is probably the most effective and reliable way of controlling corrosion in concrete structures; its main limitation is that it must be applied early in the life of the structure and although it may need replacing periodically, the frequency is unlikely to be much greater than for techniques designed to stop on-going corrosion.

The preventative techniques to be discussed in this section are barriers such as silane, paint and waterproofing membranes applied to the concrete surface and inhibitors added to the concrete mix. Techniques such as cathodic protection, desalination, realkalisation and migrating inhibitors primarily used for stopping on-going corrosion can also be used for preventative maintenance. They are generally more effective when used as preventative maintenance, but they have proved to be too expensive to have been used much for prevention of corrosion.

Silane

Silane, typically trimethoxy isobutyl silane, is a clear volatile liquid which is absorbed by the concrete surface. It acts as a pore liner making the pore surface hydrophobic, increasing the contact angle between the concrete surface and water, thereby preventing its absorption together with dissolved substances such as sodium chloride. Silane is most effective in preventing chloride ingress to concrete surfaces that are periodically wetted rather than to submerged concrete. The effective life of a silane treatment is not known but it is at least 10 years [1] and could be much longer. Silane is quick and easy to apply and does not require much surface preparation. It does not affect the appearance of the structure. It is not necessary to remove old silane when retreatment becomes necessary. Silane is one of the least expensive types of corrosion control maintenance. The main disadvantage of silane is that it is harmful to the environment because of its vapour and absorption of excess material by the soil in which the structure stands. Solutions of silane in water and immobile aqueous emulsions are now available which largely eliminate these environmental problems. Repeat applications that may interrupt the use of the structure could also be inconvenient and costly in some cases. Silane forms a barrier to the passage of water and aqueous solutions in the liquid phase while allowing free transport of water vapour. The humidity in concrete pores is usually higher than in the air outside the structure hence the silane should facilitate the concrete to dry until the relative humidity inside and outside the concrete eventually attains equilibrium.

Paint

Paints have been applied to some structures to improve the appearance and to prevent ingress of corrosive substances such as chlorides and carbon dioxide [2] and in these respects some paint systems have been effective. It is important that paint systems applied to concrete should allow the transmission of water vapour. A paint system that is in good condition is probably better than silane at preventing chloride ingress to concrete from sinks where salt solution can pond. The main limitations of paint systems are:

- considerable surface preparation of the concrete is needed before application
- their life is only 10 to 15 years
- old paint must be thoroughly removed before repainting
- the effectiveness and appearance of paint films is reduced later in their life
- paint systems take longer to apply than silane and generally cost a little more.

Waterproofing Membranes

These materials are sometimes applied to parts of structures where water or salt solution is known to accumulate [3] and is probably the most effective way of preventing chloride ingress from ponds of salt solution. Waterproofing membranes have a life of about 20 years but they are significantly more expensive than silane or paint. They require a considerable amount of surface preparation to the concrete before application which is often quite slow. The removal of old membranes prior to replacement has frequently proved to be difficult.

Inhibitors

There are two types of inhibitor that have been added to concrete mixes to provide protection against reinforcement corrosion: Calcium nitrite [4] and amino alcohol [5]. They provide protection against corrosion caused by carbonation and chlorides. The degree of protection provided depends on the amount of inhibitor added to the mix. It is thought that they can provide protection up to a total chloride concentration of about 2 per cent by weight of cement. These inhibitors function by interfering with electrochemical reactions occurring at the anodic and/or cathodic sites. They do not influence the ingress of corrosive substances and hence do not begin to function until these substances have reached the reinforced steel which may take several decades. It is possible that a certain amount of inhibitor could be leached from the concrete during this period reducing somewhat the protection provided. This is most likely to be a problem where water ponds on the concrete surface. Inhibitors have the advantage that they only need to be applied during construction so that no interference to users occurs during service; they are however, relatively expensive. They do not affect the appearance of structures.

METHODS OF STOPPING CORROSION AND REPAIRING DAMAGE TO CONCRETE

After the initiation of reinforcement corrosion there is an interval of time before damage to the concrete can be observed. This interval depends on the rate and type of corrosion. If corrosion is the macrocell type leading to the formation of isolated though intense pits the interval before concrete damage occurs can be large enough for substantial loss of reinforcing bar cross section to result without visual indications. This type of corrosion is fortunately rare. Normally macrocell corrosion occurs in conjunction with microcell corrosion so that there are signs of concrete damage before the cross section of the bars is too severely reduced. Microcell corrosion by itself normally results in concrete damage before the bar section has reduced significantly.

Structures are not monitored in service to find out when corrosion initiates hence by the time steps are taken to combat corrosion, damage to the concrete and significant reductions in reinforcement section have often already occurred. Maintenance both to stop the corrosion process and to repair the damaged concrete is therefore usually needed for structures where corrosion is well established. The following techniques for stopping corrosion will be discussed: cathodic protection, desalination, realkalisation and migrating inhibitors. Concrete repairs will also be discussed as a method of repairing concrete and as a method of stopping corrosion. Methods for stopping corrosion deal both with the situation already pertaining within the concrete and with the prevention of additional corrosive substances entering the concrete.

Patch repairs

Patch repair is the traditional and most widely used method of repairing structures suffering from corrosion. It is particularly useful where the entire structure is not affected or the repaired area is of minor importance. It usually involves cutting out the concrete to a depth below the main reinforcement from those areas where corrosion had caused disruption of the concrete cover [6]. Any rust on the exposed reinforcement is also removed. There is the option of applying an epoxy resin paint or acrylic rubber to the steel to provide it with a small measure of additional protection. Finally, the exposed area is filled with fresh concrete or a proprietary material such as a polymer concrete overlay and sealed at the surface with materials such as acrylate or latex.

Unfortunately, in practice, this method has not always provided us with a lasting repair solution. Experience has shown that although the actual repaired area usually performs satisfactorily, incipient anodes may occur causing corrosion to be initiated in the surrounding areas of concrete [7].

Incipient anodes arise where concrete repairs are used for damaged concrete resulting from macrocell corrosion. The damage occurs at a position roughly corresponding to the anodic sites. These anodes provide a degree of cathodic protection to the surrounding steel permitting higher than normal chloride levels to be tolerated without corrosion and producing the localised distribution of pits typical of macrocell corrosion. When these anodes undergo concrete repair this cathodic protection is lost and the incipient anodes within the originally protected area, but outside the repaired area, start to develop.

Concrete repairs, made to a structure suffering from macrocell corrosion, and designed to stop corrosion and repair the damaged concrete are sometimes likely to be so extensive as to be not economically viable. It is becoming more common to use concrete repair to deal with the damaged concrete and another method to stop the corrosion e.g. CP or desalination. Instead of applying CP to the entire structural element using an impressed current system it is possible to use a newly developed sacraficial zinc anode [8] situated just within the concrete repair, to provide protection against the development of incipient anodes. These anodes are easier to install and require no electrical equipment or post application maintenance, unlike the impressed current systems. Experience of these sacraficial anodes is currently very limited.

This type of repair is usually costly due to the labour intensive work of mechanically breaking out the concrete and the temporary support often required due to the loss of composite strength [9]. Another disadvantage is that it is often uncertain how much break out is necessary. Too much breakout leads to greatly increased cost and disruption whereas too little breakout leads to an indurable repair. A possible consequence of concrete repair is the induction of micro-cracks within the parent concrete as a result of the hammering action on the surface. It also has an impact on the environment due to excessive noise and which makes it unsuitable for repairing structures that lie within urban areas. Concrete can now be more effectively removed by high pressure water jetting although the water generated can be a problem in some circumstances. Preventative measures are often required to stop further ingress of corrosive substances after concrete repair. The life of concrete repairs is very sensitive to how well the work is done. A poor repair is unlikely to last for more than 2-5 years whereas a good repair will last more than 15 years.

Cathodic protection (CP)

This method involves passing a small direct current, typically less than $20mA/m^2$, between a permanent metal anode distributed across the surface of the concrete to be protected and the embedded reinforcing steel [10]. The impressed current lowers the potential of the steel bars thereby stopping corrosion, or reducing the rate to negligible values.

Cathodic protection has been used successfully for over twenty five years on concrete structures suffering chloride-induced corrosion. Like desalination, this process requires relatively little concrete preparation prior to installation. This reduces both the cost of the treatment compared with patch repairs and the resulting environmental and user disruption. It also means that temporary structural support is not necessary. Robery [11] has used whole life cost analysis to show that cathodic protection is more cost effective than concrete repair.

CP is generally considered to be unsuitable for prestressed concrete structures and structures susceptible to alkali-silica reaction although currents used in CP are much less than for desalinsation or realkalisation so this restriction may be less absolute for CP. More significant is the fact that CP systems must be continuously operated and maintained during their life time. Most anode systems have a certain life expectancy, typically 10 years for paint systems and 25 years for mesh systems at the end of which they will require repair or replacement. CP systems do not require additional preventative treatment to stop further ingress of corrosive substances unlike desalination and realkalisation. There is also a need to monitor the system in order to ensure that corrosion has not reinitiated. Although several criteria exist by which the effectiveness of CP can be judged their validity is still the subject of some debate. The track record of CP is better established than for desalination or realkalisation.

Realkalisation

Involves restoring the alkalinity of concrete and re-establishing the passive oxide film around the embedded steel reinforcement. This is achieved by passing a low voltage direct current between an external temporary anode surrounded by an alkaline anolyte, commonly sodium carbonate, and the steel reinforcement, which acts as a cathode. During treatment sodium ions migrate into the concrete, raising its pH, and hydroxyl ions are produced at the steel cathode, thereby restoring the passive oxide film [12].

There are many advantages to realkalisation. For example it is a silent treatment which is ideal for city locations. Also it can be carried out over a relatively short period of time, typically two or three weeks. Robery [11] has found using whole life costing, that realkalisation is more economical in the long term than other carbonation repair strategies.

Despite the advantages, realkalisation is not widely specified. In the early days this was mainly due to uncertainties regarding possible adverse effects of the treatment. These include an increased risk of alkali silica reaction [13], hydrogen embrittlement [14], loss of bond [15] and concrete microcracking. Work by Miller [16], however, has demonstrated that provided the current density remains between 1-2 A/m^2 during the treatment, none of the above will usually cause any structural damage. Sergi [17] also suggests that the risk of ASR can be minimised by using lithium carbonate instead of sodium carbonate. Nevertheless, the long term effectiveness of the treatment remains uncertain which perhaps prevents it being specified more widely. Realkalisation requires post treatment monitoring to confirm its effectiveness and a surface coating to prevent further carbonation. Its life is not known but the surface treatment will probably need replacing within 15 years.

Desalination

Desalination involves passing a direct current, typically around 1 to 5 A/m^2, between the embedded steel reinforcement, which acts as a cathode, and an anode which is surrounded by a liquid electrolyte (anolyte), usually tap water, and attached to the surface of the concrete [18]. The current results in chloride ions being drawn towards the anode where they eventually pass into the anolyte and are thereby removed from the concrete. Application of this technique also results in the generation of hydroxyl ions at the steel reinforcement which helps to restore the passive oxide film and further aids the cessation of corrosion.

Desalination has many advantages. For example, it is relatively quick - typical treatment time six to eight weeks - and easy to apply with no ongoing maintenance requirement. Monitoring is however needed to confirm the success of the treatment since its track record is short. The equipment is portable and, once running, it is a silent process, making it ideally suited to urban areas. The desalination process normally requires much less concrete breakout than concrete repair which means less noise, dust, vibration and environmental disruption. It also means that temporary structural support is not necessary. All these factors act together in making the treatment cost-effective.

The major drawbacks of this technique are similar to those of realkalisation and include its restriction to use on non prestressed structures and structures not susceptible to alkali silica reaction. Because it is relatively new, optimum operating conditions are not well established. Also, its effectiveness remains uncertain. On bridge decks, where possession time is critical, soffit desalination has been used in an effort to reduce traffic disruption but doubts exist regarding its effectiveness. Desalination treatment is usually followed by a surface coating to prevent further ingress of chloride and to improve the appearance of the concrete surface which can be affected by desalination. The life of a desalination is not known at present although the surface coating is likely to need replacement within 15 years.

Migrating Inhibitors
During the last few years a number of migrating corrosion inhibitors [19,20] have been introduced which are designed to migrate quickly through the concrete cover and stop ongoing corrosion when they reach the reinforcing steel. These materials are based on inhibitors that are known to be effective in stopping corrosion on bare steel hence the most critical feature of their performance on reinforced concrete is how quickly they migrate through a slightly porous material like concrete to generate a concentration at the reinforcement sufficient to stop corrosion. The evidence to date suggests that these inhibitors (a) can only migrate at a sufficient speed through poor quality concrete with a water:cement ratio greater than 0.6 and (b) cannot accumulate in sufficient quantities near the reinforcement to provide protection when the total chloride concentration exceeds about 1% by weight of cement.

It has not yet been established whether migrating inhibitors can readily stop or substantially reduce the rate of ongoing corrosion; they may only provide additional protection against corrosion initiation. It appears, therefore, that migrating inhibitors are only effective in a limited number of situations. They are however of relatively low cost, although their effective life is not known.

MAINTAINING THE SAFETY OF STRUCTURE USERS
Concrete structures are normally very safe. Occasionally partial or total failures occur where corrosion has dissolved some reinforcement, jeopardising the safety of users. Spalled lumps of concrete from corroding structures also present a hazard.

Where the assessed load carrying capacity of a structure is inadequate strengthening is required. Strengthening is generally carried out in conjunction with maintenance to stop the corrosion process and to prevent is reoccurrence. A frequently employed method of strengthening concrete bridge decks is plate bonding this method is discussed below. If the deterioration is very severe there may be an economic case for replacing elements. Decisions about whether to strengthen and rehabilitate or to replace are usually based on economics and the consequences relating to suspension in the use of the facility.

Spalled lumps of concrete particularly from higher levels of a structure are a hazard for people or traffic passing underneath. Dealing with this hazard forms a part of many repair schemes for corroding concrete structures. Loose concrete can be easily detected by hammer tapping; the loose concrete can then be safely removed. This procedure is clearly only a temporary measure and would need to be repeated at least annually until the corrosion process is stopped. Another approach is to fix netting or cladding to the area of spalling/delamination in order to catch the lumps of concrete which can then be safely removed. In situations where the loss of bond due to spalling and delamination has been shown to have insignificant structural consequences maintenance has been restricted to just catching the spalled lumps of concrete, thereby reducing costs.

Strengthening with FRP

Significant loss of steel cross-section due to corrosion will lead to a reduction in the strength and stiffness of concrete structures. Repair methods based on concrete repair and stopping corrosion will not restore structural capacity. Total replacement of the structure is clearly an expensive option and potentially very disruptive. The use of externally bonded steel plate can be far cheaper and, particularly in relation to structures where possession time is critical, can be made even more cost-effective by using fibre reinforced plastic (FRP) laminate, instead of steel [21].

FRP laminates consist of uni-directional carbon, aramid or glass fibres set in an epoxy-resin matrix. They offer an outstanding combination of properties such as low weight, high strength and good durability. The composite can be used to increase both flexural strength and shear strength. Installation involves carefully examining the surface of the concrete to make sure that it is free from any defects likely to have a detrimental effect on the transfer of stresses between the concrete and the FRP. Any spalled, delaminated or weak concrete must be repaired before plate bonding is started. The bonding surface of the plate is cleaned with a degreasing agent and adhesive applied to both the concrete surface and the FRP. The FRP is then simply pressed and rolled into position.

The use of FRP laminates for strengthening concrete structures has increased significantly over recent years. This is because the reduced weight makes handling and installation quick and easy. The work can be carried out from access platforms rather than scaffolding. In addition, FRP laminates are available in very long lengths which avoids the need for laps and joints. The main disadvantages of using FRP as external reinforcement are the risk of fire and accidental damage. Moreover, the technique is relatively new and there are currently no standards or specification for repair and strengthening with FRP. Work is currently being undertaken, however, by a number of organisations, including the UK's Concrete Society, to develop appropriate design guidance.

The Strength of Corroded and Repaired Elements

The strength of an element before and after repair warrants some consideration. Locally corroded reinforcing bars are probably more subject to fatigue than an uncorroded bar and account should be taken of this where there are significant live load stresses. Local corrosion is also likely to reduce the ductility of bars and there is a case for eliminating locally corroded bars from plastic analyses. General corrosion leads to a reduction in bond when the concrete cracks parallel to the reinforcement and a further reduction when spalling or delamination occur. The exact consequences of general corrosion and loss of bond have not been fully established. Mangat [22] suggests that the reduction in bond strength is sensitive to the degree of corrosion and that marked reductions of flexural strength can occur when there is a significant loss of bond. These factors should be considered when assessing the strength of structures and deciding on the need for temporary support during repair work.

There is considerable doubt as to whether the strength of an element is restored after the damaged concrete and reinforcement is repaired. Mangat [23] suggests that the structural performance of repaired beams depends on the characteristics of the repair concrete/mortar. A ductile repair material with a high coarse aggregate content gives the best structural performance. Work by Cairns [24] on the other hand indicates that the ultimate bending capacity of a beam will not be reduced by failing to relieve deadload stresses during repair, although the total deflection will increase.

SITE SPECIFIC FACTORS AFFECTING THE CHOICE OF MAINTENANCE METHOD

The design constraints affecting a structure must be considered and assessed because they may and location may influence the method of repair, access and temporary works. The transport undertaking carried or crossed by the structure must be considered; in particular the gradient and curvature of roads and the proximity of junctions may affect the management of traffic and hence the choice of repair method. The proximity of railways, rivers or canals may also be significant. The effect that remedial works can have on the users of the structure is a major consideration and influences the planning, design and construction of the work. For example on bridges traffic volumes, speeds and composition and their variation with time through a day, week and year will influence the management of traffic and possibly the choice of maintenance strategy. Maintenance work on bridges can produce congestion some distance away hence the planning of maintenance work at different places and times on the road network needs careful consideration if disruption is to be minimised. Restricting maintenance work to overnight and weekends and avoiding Summer and Christmas holidays can help to reduce disruption. Some forms of maintenance are sensitive to the prevailing environmental conditions such as temperature, humidity, rainfall, windspeed and direction. The effect of these variables on the condition of the concrete substrate often affects the timing and maintenance method selected. The proximity of saline water, river water, industrial pollutants and unusual soil conditions such as hazardous waste or mine workings often influence the design and choice of maintenance method. Health and safety must always be given careful thought in the design of remedial works. For work on bridges the use of traffic signs and barriers, safety zones, access to the works and the disposal of waste must be given priority. These site specific factors are often influential and can result in the selection of a maintenance method that does not correspond to the most technically feasible or cost-effective options.

The main factors, apart from the site specifics discussed above, affecting the choice of methods for preventing and repairing corrosion of reinforcement are summarised in Tables 1 and 2 which can be used as guidance when choosing a suitable maintenance strategy.

CONCLUSIONS

Preventative maintenance is probably the most effective strategy for maintaining most structures that are vulnerable to reinforcement corrosion. The costs are low compared with other methods and it has proven effectiveness provided it is applied early in the structures life. The main disadvantages of many preventative techniques is that repeat treatments are needed at intervals throughout the life of the structure and this could result in some disruption to users. Preventative maintenance is a blanket strategy whereby all structures in the stock are treated. There will always be some more durable structures that will not really need preventative maintenance, but until tests can be developed to identify these it will be necessary to treat all structures. Hence preventative maintenance is probably only justified when there is evidence that most of the structures in the stock will suffer from corrosion at some point during their life.

Concrete repairs have been used successfully to repair concrete damaged by corrosion although considerable care must be taken. It is much more difficult and costly to use the concrete repair method to stop corrosion. Most of the other techniques for stopping corrosion have a limited track record and their long term effectiveness is not known. Cathodic protection appears to be the most reliable of these methods although its application is less convenient.

REFERENCES

1. Vassie, P.R. and Calder, A. Reducing Chloride Ingress to Concrete Bridges by Impregnants in Controlling Concrete Degradation. Ed. Dhir, R.K. pp 133-148. Thomas Telford, London (1999).

2. Robinson, H.L. Evaluation of Coatings as Carbonation Barriers. Proceedings of the

P Vassie 13

	Protects against	Effectiveness for intermittent wetting	Effectiveness for Ponding	Aesthetics	Ease of initial application	Ease of replacement	Frequency of replacement	Comparative cost of prevention
Silane	Chloride	Good	Poor	Neutral	Easy	Easy	15*	Low
Paint	Carbonation & chloride	Very Good	Good	Improved	Moderate	Moderate	10-15 years	Low
WPM	Carbonation & chloride	Very Good	Very Good	Reduced	Moderate	Difficult	20-25 years	Moderate
Inhibitors	Carbonation & chloride	Very Good	Good	Neutral	Easy	Not needed	-	High

*Silane has a relatively short track record; there is evidence that it is fully effective for 10 years and its life is expected to be much longer

Table 1. Characterisation of Methods for Preventing Reinforcement Corrosion

	Pre repair testing	Propping during repair	Effective for corrosion caused by CO_2/Cl/both	Effects of Repairs on users	Post repair monitoring /maintenance	Preventative maintenance needed	Comparative Cost of repair
Concrete Repair	Extensive	Usually	Both	High	Monitoring	Yes	Very High
Impressed Current CP	Moderate	Unlikely	Both	Low	Monitoring and maintenance	No	High
Desalination	Moderate	Unlikely	Chloride	Low	Monitoring	Yes	Moderate
Migrating inhibitor	Limited	Unlikely	Both	Low	Monitoring	Yes	Low
Sacrificial CP	Moderate	Unlikely	Both	Low	Monitoring	No	Low
Realkalisation	Moderate	Unlikely	Carbonation	Low	Monitoring	Yes	Moderate

Table 2. Characteristics of Methods for Stopping Reinforcement Corrosion

15. Nustad, G.E. and Miller, J.B., Effect of electrochemical treatment on steel to concrete bond strength. NACE Conference: Engineering Solutions to Industrial Corrosion Problems, Sandefjord, Norway, 1993, Paper 49.

16. Miller, J.B., Structural aspects of high powered electro-chemical treatment of reinforced concrete. International Conference on Corrosion and Corrosion Protection of Steel in Concrete, Ed., R.N. Swamy, Sheffield, 1994, 1489-1498.

17. Sergi, G., Walker, R.J. and Page, C.L., Mechanisms and criteria for the realkalisation of concrete. 4th International Symposium on Corrosion of Reinforcement in Concrete Construction, Eds., C.L. Page, P.B. Bamforth and J.W. Figg, Cambridge, U.K. 1996, 491-500.

18. Arya, C., Sa'id-Shawqi, Q. and Vassie P.R.W., Factors influencing electrochemical removal of chloride from concrete. Cement and Concrete Research, Vol. 26, No. 6, 1996, 851-860.

19. Wiss, Janny and Elstner associates, Corrosion testing of organic corrosion inhibiting admixtures. Company report, 19p.

20. Zoltanetzky, P. Gordan, C. and Parnes, J., New developments in corrosion inhibiting admixture systems for reinforced concrete. International Conference on Corrosion and Corrosion Protection of Steel in Concrete, Ed., R.N. Swamy, Sheffield, 1994, Vol. 2, 825-837.

21. Meier, U., Deuring, M., Meier, H. and Schwegler, G., Strengthening of structures with advanced composites. Ed., J.L. Clarke, E & F.N. Spon, 1993, 153-171.

22. Mangat, P.S. and Elgarf, M.S. Bond characteristics of corroding reinforcement in concrete Beams. Materials and Structures, **32** pp89-97 (1999).

23. Mangat P.S. and Elgarf, M.S. Strength and Serviceability of repaired reinforced concrete beams undergoing reinforcement corrosion. Magazine of Concrete Research **51**, pp97-112 (1999).

24. Cairns J. An analysis of the behaviour of reinforced concrete Beams following deterioration and repair. First International Conference on Bridge Management, pp643-653, Guildford, UK (1990).

Second International Colloquium on Materials Science and Restoration.   Technische Adakemie, Essinger, Germany (1986).

3. Darby, J.J.et al. The Effectiveness of silane for extending the life of chloride contaminated reinforced concrete. Bridge Management 3. Ed J.E. Harding. E & F N Spon, London (1996).

4. Berke, N.S. and Rosenberg, A. Technical Review of Calcium Nitrite Inhibitor in Concrete. Transportation Research Record No 1211 pp18-27 (1989).

5. Phanasgaonkar, A. et al. Corrosion Inhibition Properties of organic amines in a simulated concrete environment: mechanism and time dependency of inhibition. Proc. Int. Conf. Understanding Corrosion Mechanisms in Concrete. MIT, Cambridge, MA, Section 6, pp6 (1997).

6. Read, G.E., Degradation of reinforced concrete structures due to corrosion attack. International Conference on Structural Improvement Through Corrosion Protection of Reinforced Concrete.   Institute of Corrosion, London, 1992, Conference Document E7190.

7. Vassie, P.R., Reinforcement corrosion and the durability of concrete bridges. Proceedings of the Institution of Civil Engineers, Part 1, 76, 1984, 713-723.

8. Sergi, G and Page, C.L.. Sacraficial anodes for cathodic protection of reinforcing steel around patch repairs applied to chloride contaminated concrete. Proceedings of Eurocorr 99, Aachan, (1999).

9. Wallbank, E.J., The performance of concrete in bridges: a survey of 200 highway bridges. Her Majesties Stationery Office, London, 1989

10. Ward, P.M., Cathodic protection: a user's perspective. Chloride corrosion of steel in concrete, ASTM STP 629, Eds., D.E. Tonini and S.W. Dean Jr., American Society for Testing and Materials, 1977, 150-163.

11. Robery, P.C., Gower, M. and El-Belbol, S., Comparison between cathodic protection and other electrochemical repair techniques. Proceedings 7th International Conference on Structural Faults and Repairs, Ed., M.C. Forde, Edinburgh, U.K. 1997, 231-244.

12. Banfill, P.F.G., Features of the mechanism of realkalisation and desalination treatments for reinforced concrete. International Conference on Corrosion and Corrosion Protection of Steel in Concrete, Ed., R.N. Swamy, Sheffield, 1994, 1489-1498.

13. Sergi, G and Page, C.L. The effects of Cathodic Protection on Alkali Silica Reaction in Reinforced Concrete. Contractor Research Report 310, Transport Research Laboratory, Crowthorne, Berkshire, UK (1992).

14. Law, D.W. and Green, J., NCT chloride extraction, chloride remigration, bond strength and hydrogen embrittlement tests. Report No. 1303/91/5390, Taywood engineering Ltd., R&D Division, Middlesex, England, 1991.

A143 Haddiscoe Cut Bridge – maintenance and strengthening

R. N. Welsford, Thorburn Colquhoun, Bedford, UK

HISTORY

Haddiscoe Cut Bridge carries the A143 Primary Road over the Haddiscoe New Cut Waterway, and the Norwich to Lowestoft Railway Line. With an overall length of 372 m it is the longest bridge in Norfolk.

It was constructed to replace a two leaf bascule lifting bridge, built in 1827 by the Norwich and Lowestoft Navigation Company, and a level crossing on the railway line. As well as causing traffic delays as a result of Railway and Waterway operations, the bascule bridge had been subject to a 5 ton weight restriction for many years (Norfolk County Council, 1961).

In 1957 the Bridges Section of the Highways Department of Norfolk County Council was commissioned to prepare a design for a scheme to bridge the New Cut Waterway and the Railway. A design was developed for a three span bridge with approach embankments, and was submitted for the approval of the Fine Arts Commission. Approval was received and a detailed design developed. The three spans were all to be simply supported and of steel concrete composite construction. Substructure components were to be constructed from insitu reinforced concrete. Two spill through type abutments would be formed, each with two rectangular section tapering columns and integral crossheads, and two piers, each with two circular section columns with integral crossheads. All substructure elements were to be founded on piles. Parapets comprising reinforced concrete walls topped with steel railings were to be constructed integral with the reinforced concrete deck edges. All the steel beams were to be supported on the crossheads by the use of laminated rubber bearings. The ensuing structure would accommodate a 7.3 m wide carriageway flanked by 1.85 m wide footways to either side.

Tenders were invited for the works, and of twenty two received that submitted by A Monk and Company Ltd. was accepted. A contract was let for the bridge to be constructed in an eighteen month period (Norfolk County Council, 1961).

Work commenced on site in March 1960. However, the Contractor found it was impossible to build the required full height of the approach embankments due to progressive settlement of the underlying fenland during construction. It was decided to limit the height of the approach embankments. In order to achieve this the overall length of the bridge had to increase. The design was modified to incorporate a further 10 No. simply supported steel concrete composite spans, and the approval of the Fine Arts Commission to the revised design obtained.

The elements intended to be spill through abutments became piers, and further piers comprising circular section columns with integral crossheads were constructed. The articulation arrangement for the central three span portion of the bridge was such that the two outer spans were free to expand over the spill through abutment style piers. Both outer spans had fixed bearing supports on the inner piers, and the central span had fixed bearing supports over one pier, and free bearing supports over the other. For both five span sections of the bridge to either side of the central three span section, fishplates were provided between the girder ends over the piers. These linked the spans for thermal and horizontal load effects, and were fixed at the neutral axis level of the girders to prevent inducement of moment effects. The abutment ends of the five span sections linked in this manner were fixed, with all other supports being free. There were thus three transverse joints on the bridge required to accommodate expansion/contraction effects, two over the spill through abutment type piers, and one over one of the central piers. A further eleven joints were required to accommodate rotation effects, albeit that these were of a buried joint style.

The ensuing thirteen span bridge was opened to traffic on 15 May 1961, within the original eighteen month period of the Contract. Following legal formalities associated with hand-over of responsibilities of the old bascule bridge from the British Transport Commission to Norfolk County Council, the old bridge was demolished.

However, settlement of the underlying ground to the approach embankments of the new bridge continued to develop, causing the vertical alignment of the road to become unacceptable Further approach structures were constructed to replace the embankments They comprised reinforced concrete deck slabs supported on reinforced concrete ring beams that were founded on piles. The structures were designed in a manner that allowed their construction in longitudinal halves, thereby enabling the road over the bridge to remain open during construction.

INSPECTION AND ASSESSMENT
In the early 1980s the bridge was re-waterproofed and re-surfaced. Norfolk County Council utilised the opportunity presented by this work to undertake a detailed inspection of the reinforced concrete deck surface. At that time the bridge was also used by the Transport and Road Research Laboratory as a test site for the pioneering work undertaken by Vassie and Cavalier into the use of Half Cell Potential measurements to non-destructively detect evidence of reinforcement corrosion (Cavalier & Vassie, 1981). Concrete and reinforcement repairs were also undertaken at that time.

In 1992 a Principal Inspection and Load Assessment of the main bridge was undertaken for Norfolk County Council. The Principal Inspection noted a number of defects that required attention, the primary ones being (W.S. Atkins, 1992):
- Leakage of water laden with road salts through the transverse joints, both expansion/contraction and rotation types, over the piers and abutments was causing chloride attack to the underlying reinforced concrete substructure elements, the ends of the steel beams and the reinforced concrete deck slab of the superstructure.
- The rubber surfaces to a significant proportion of the laminated bearings was perishing allowing the inner plates to corrode and expand, causing distress to the bearings.

The Load Assessment found the superstructure to be adequate to accommodate the effects generated by 40 tonne Assessment Live Loading, but noted the following deficiencies (W.S. Atkins, 1992):
- The parapets did not comply with contemporary standards.
- The central piers could not accommodate horizontal live load effects as promulgated by BD 37/88 (Department of Transport 1988) transferred via fixed bearings.

FEASIBILITY STUDY

Following consideration of the findings of the above exercises, in 1993 Norfolk County Council appointed Thorburn Colquhoun to undertake a study to investigate feasible measures to address the noted deficiencies. Part of this commission was to include a Principal Inspection and Load Assessment of the reinforced concrete approach structures to the bridge. These exercises noted the following salient points:

- The Principal Inspection noted evidence of percolation of water laden with road salt through the central longitudinal construction joint in the deck slab. This was causing chloride attack to the central elements of the ring beams that support the deck slab.
- The Load Assessment found the reinforcement that passed through the central longitudinal construction joint of the approach span deck slab to be inadequate to accommodate the hogging bending moments that would be generated in this section by 40 tonne Assessment Live Loading. It was considered that the apparent over-stress that would be induced due to this deficiency may well be promulgating cracking to the construction joint, thereby creating the water pathway noted in the inspection.

Identification of measures to address these deficiencies was then included in the brief for the feasibility study.

In investigating the reinforcement pattern in the vicinity of the parapets, Thorburn Colquhoun found that reinforcement in the portion of the deck that cantilevered out to support the parapet was subject to over-stress under the effects of the Accidental Wheel Load representation of 40 tonne Assessment Live Loading. The need to address this matter was also included in the remit for the feasibility study.

Four means of addressing the deficiencies of the parapets and the cantilever slabs that support them were identified:

1. Demolish the existing parapets and supporting cantilever slabs, form new insitu reinforced concrete cantilever slabs and install new metal parapets compliant with contemporary standards.
2. As item 1., but use precast reinforced concrete slab units to speed up the on-site construction programme.
3. Reduce the carriageway width and provide new parapets inboard of the existing.
4. Adjust the carriageway and footway widths and install safety fencing across the bridge to protect the parapets, and reduce the likelihood of vehicle over-run on to the edge cantilever portions of the deck slab.

Of the above options number four was far and away the most cost effective. However, the need to incorporate appropriate set backs between carriageway edges and new safety fences would reduce the effective width available to accommodate the carriageway and footways. It would not be possible to maintain a 7.3m wide carriageway and two footways of reasonable width. Following consultations within Norfolk County Council, with District and Parish Councils and with Norfolk Constabulary, it was agreed that a 50 mph speed restriction could be imposed on the road over the bridge. The carriageway width could be reduced to 6.55m, and only one 1.9m wide footway need be provided, thereby making full accommodation for use by the disabled. The imposition of the speed restriction allowed Norfolk County Council to accept the provision of set backs of 600mm between the carriageway edges and the safety fence faces. Figure 1 shows the ensuing pattern on the bridge deck cross section.

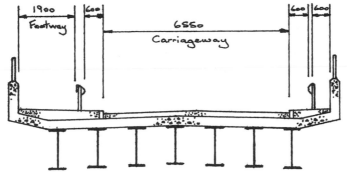

Figure 1. Proposed carriageway and footway arrangement

Altering the highway cross section over the bridge to that described above moved the crown of the carriageway. This in turn locally increased surfacing thicknesses over the deck. Norfolk County Council considered that they would be able to ensure that due to the nature of the works that would be implemented, overlay of the surfacing at a future date would not occur. They thus accepted that a reduced load factor for surfacing could be applied, as allowed in BD 37/88 (Department of Transport, 1988). Thorburn Colquhoun were then able to confirm that the ensuing revised loading pattern on the structure generated acceptable effects.

The laminated rubber bearings were in such poor condition that replacement was the only option. It was considered sensible that like for like replacements be generally deployed, thereby not altering support stiffnesses.

To reduce the noted deficiencies of the central piers to accommodate horizontal live loading effects transferred to them via the fixed supports with the central three spans, it was decided to alter the articulation and expansion/contraction arrangements. The spill through abutment type piers had significant capacity to accommodate horizontal loading effects. By making the bearings over the two central piers free, linking the three central spans together and providing fixed bearings to one end of this group of three linked spans, all horizontal forces on the three spans would be transferred to one of the spill through abutment type piers. This change in articulation arrangement is shown in Figure 2.

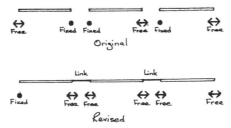

Figure 2. Original and Revised Articulation Arrangements for Central Three Spans.

With regard to the noted deficiencies in the approach structure deck slabs, it was accepted that whilst over-stress of the central longitudinal construction joint area was noted in hogging under live load effects, at the Ultimate Limit State the moments could redistribute to the sagging areas where there was sufficient over-capacity in the reinforcement. Cracking would be anticipated over the central longitudinal construction joint, which could be addressed by provision of a jointing medium and careful attention to the waterproofing detail.

CONNECTIVITY SLABS

It was very clear that the primary cause of deterioration to components of both the superstructure and substructure, was leakage of salt laden water through the joints over the piers. Both the expansion and rotation type joints were noted to allow significant percolation of water through the carriageway surfacing and then down on to the ends of the girders and decks slabs of the superstructure and the pier crossheads. The ensuing chloride attack was generating localised corrosion to the girder ends, and corrosion of reinforcement causing spalling of concrete cover. Whilst section loss to the girders and reinforcement was insignificant at that time, it was considered essential to stem this cause of deterioration and prevent future loss of strength of the affected components. Renewal of joints would address the percolation, particularly if asphaltic plug joints were used for the rotation only joints, rather than buried joints as had previously been the case. However, such provision would only be efficacious for a limited period, and would require maintenance renewals on a cyclic basis. Norfolk County Council asked Thorburn Colquhoun to explore the feasibility of structurally connecting the reinforced concrete decks over the pier tops where rotational effects only occurred.

In exploring this matter great attention was paid to ensure that the provision of any connection between the deck slabs did not generate significant composite action between the deck slab and the girders. This would cause hogging bending moments to be induced in the girder ends under live load effects, in turn generating compressive stresses in the lower flanges, which they were not sufficiently restrained to accommodate. Reference was made to the paper "Deck slab continuity in composite bridges" by A. Kumar (1994). It was considered necessary to ensure the "connectivity slabs" between composite deck slabs were debonded from the steel girders over a particular length at their ends. Additionally, it was appreciated that the provision of laminated rubber bearings at the supports would assist in reducing the tie forces induced in the debonded connectivity slabs under girder rotation effects. As the girders rotate under sagging of the spans, the bearings will tend to move towards the pier centre lines thereby reducing the strains and hence stresses induced in the connectivity slabs. This characteristic is illustrated in Figure 3. The connectivity slabs would also have to be designed to accommodate local wheel load effects and differential vertical movement between the bearings to adjacent spans as illustrated in Figures 4 and 5. The longer the length of debonding the greater the bending effects would be under local wheel loadings, but due to decreased stiffness of the slab the lower bending effects due to differential bearing compression. Analysis was undertaken to determine the optimum length of the debonding to minimise the combined effects of these two actions. This was determined to be 500mm over each girder end.

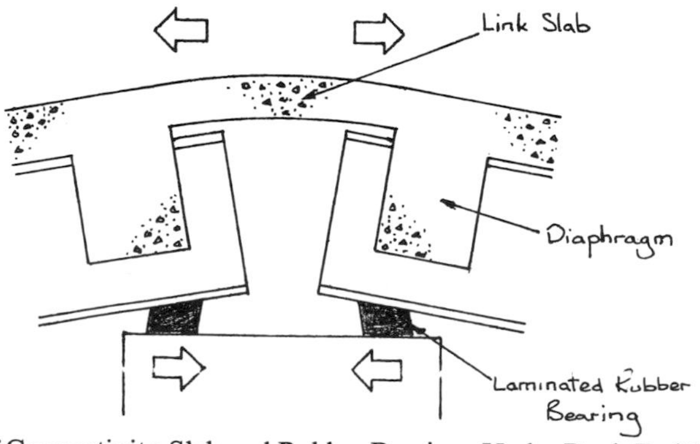

Figure 3. Action of Connectivity Slab and Rubber Bearings Under Deck End Rotations.

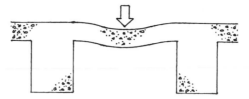

Figure 4. Local Wheel at Mid Span of Connectivity Slab

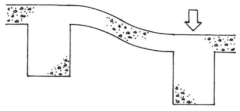

Figure 5. Local Wheel Load at Support

In order to anchor the tie forces induced in the connectivity slabs it was considered necessary to provide new diaphragm beams, which are also shown in Figure 3. These beams could be designed to have an added function of acting as jacking spreader beams, thereby assisting in the required bearing replacement operations.

Having identified that it was feasible to provide connectivity slabs over the ten piers where the spans would be linked, thereby removing 10 potential rotation joints, possible means of installing them were explored. In order to prevent loss of bond between reinforcement and concrete, vibrations had to be prevented from being induced in curing concrete. Thus the bridge would have to be closed to traffic whilst the concrete of the connectivity slabs and diaphragms cured. Norfolk County Council embarked on a series of consultation exercises with District and Parish Councils, Norfolk Constabulary and other Emergency Services to determine the least disruptive means of undertaking the work. It was concluded that it would be acceptable to close the bridge to traffic on two weekends between midnight on Friday and very early morning on the following Monday. Whilst, if an appropriate concrete mix was used, this would provide enough time to place, compact and cure the concrete to a number of slabs, it would not be sufficient time to break out the old slabs and fix reinforcement and formwork for the new components. It was found that the composite spans could accommodate the effects that would ensue if the ends of the existing deck slabs were demolished, and then over-spanned by short span temporary bridging units. Thus if traffic over the bridge were reduced to single line working under Traffic Light control, one half of the connectivity slab areas could be broken out and reinforcement and shuttering prepared, then over-spanned by temporary bridging units whilst the other half was similarly prepared. Placing of concrete could then take place in the pre-arranged weekend closures.

IMPLEMENTATION
A scheme design was prepared to accommodate the findings derived from the foregoing studies and tenders were invited to undertake the works. In April 1998 May Gurney were awarded the contract.

The system for construction of the connectivity slabs worked well. The form of temporary bridging unit used is illustrated in Photograph No.1.

Photograph No.1. Temporary Bridging Units

Originally it had been anticipated that the bearings could be replaced in short over night road closures thereby only requiring a jacking system capable of accommodating dead load effects. However difficulties were encountered in achieving curing of the bearing fixing adhesive, and in accommodating the new bearings in the gaps ensuing from removal of the old bearings. It transpired that the old bearings had been subject to non-uniform deterioration, thus their eventual thicknesses varied very slightly from one to another, and that the reinforced concrete deck slab had crept to follow the profile arising from the deteriorated bearings. It became necessary to install bedding mortar between the bearings and the crosshead substrates rather than adhesive. The jacking system was enhanced to accommodate live loadings from a single line of vehicles thereby enabling the longer curing time for such product to be accommodated with minimal disruption to day-time traffic over the bridge.

Works were completed in December 1998, within the period prescribed by the Contract.

CONCLUSION

The arrangement of carriageway, footway and safety fence over the bridge is shown in Photograph No.2. Provision of the safety fence between the carriageway and the footway has the added benefit of providing protection to pedestrians.

Implementation of the scheme has thus addressed the noted deficiencies of the parapets, cantilever slabs, central piers and approach structure slabs, and has improved pedestrian protection measures. However, perhaps of greater significance it has reduced the number of joints across the main bridge from fourteen to two, thereby significantly reducing the future maintenance liabilities associated with this structure.

Photograph No.2. View Over Bridge on Completion of Works

REFERENCES

Cavalier P.G. & Vassie P.R., TRRL (1981). ICE Proceedings, Part 1, Volume 70, August 1981, pp 461-480. London. Institution of Civil Engineers.

Department of Transport (1988). Departmental Standard BD 37/88 Loads for Highway Bridges. London. HMSO.

Kumar A. (1994). Locally Separated Deck Slab Continuity in Composite Bridges. Continuous and Integral Bridges, edited by B.P. Pritchard. London. E & FN Spon.

Norfolk County Council (1961). Haddiscoe Cut Bridge. Norfolk County Council.

W.S. Atkins (1992). Principal Inspection of A143 Haddiscoe Cut Bridge. W.S. Atkins East Anglia.

W.S. Atkins (1992). Load Assessment of A143 Haddiscoe Cut Bridge. W.S. Atkins East Anglia.

Refurbishment of town bridge, Northwich, Cheshire

T JENKINS,
Parkman Ltd, Ellesmere Port, South Wirral.

SYNOPSIS.
This paper describes the strengthening and refurbishment project, for the Grade II listed Town Bridge, owned and operated by British Waterways, who were the Clients for the project. This is a wrought iron swing bridge, carrying the A533 over the Weaver Navigation, in Northwich, Cheshire. Built in 1899 it is of particular historical significance, as the first fully electro-mechanically operated swing bridge, and partially floating structure of its kind in the country.

Inspection of the structure revealed significant breakdown of the paintwork and severe corrosion of certain elements. During the subsequent refurbishment contract, the corroded elements were replaced and elements which were inadequate to carry the 40 tonne assessment loading were strengthened. Having this significant historical interest, design of the strengthening works had to be particularly sympathetic with the original 1897 design.

LOCATION AND HISTORY OF TOWN BRIDGE
There has been a crossing of the River Weaver at Northwich since Roman times, although history does not record exactly when the first bridge was constructed to replace the earlier stepping stones. The "Old Bridge" was built in 1663 from masonry, and replaced an earlier wooden structure. This was followed in 1858 by a plate girder steel bridge. In turn, this was replaced 100 years ago by the present swing bridge.

Records of the previous bridge on the site show that, as a result of brine extraction in the area, settlement of the abutments occurred to the extent of nearly 2 metres over a 16 year period. That structure had to be raised by 500mm just to maintain adequate clearance to the navigation. As the chemical industry grew and tall fixed-masted vessels were used for trade along the river, it was clear that a swinging bridge was needed to allow these craft up to the local chemical works, notably ICI. Colonel Saner, Chief Engineer for the Weaver Navigation Trustees at that time, designed the replacement bridge.

BRIDGE DETAILS.
The structure is a complex assembly of elements, identified in figure 1 of the paper. The two span (nose and tail) bridge rotates on roller bearings within the pintle section of the bridge. Beneath the pintle are located semi-submerged pontoons which counteract some 70% of the dead weight of the bridge and reduce bearing loads accordingly. The superstructure comprises two main trusses interconnected by transverse girders. On this framework rests decking of various types, each generally associated with a previous strengthening of the structure.

The bridge comprises wrought iron construction throughout, except for the transverse and longitudinal deck beams, replaced in the 1920,s and 1978 with mild steel and the pontoons which are also mild steel. The rollers and track are cast iron.

Bridge Management 4, Thomas Telford, London, 2000.

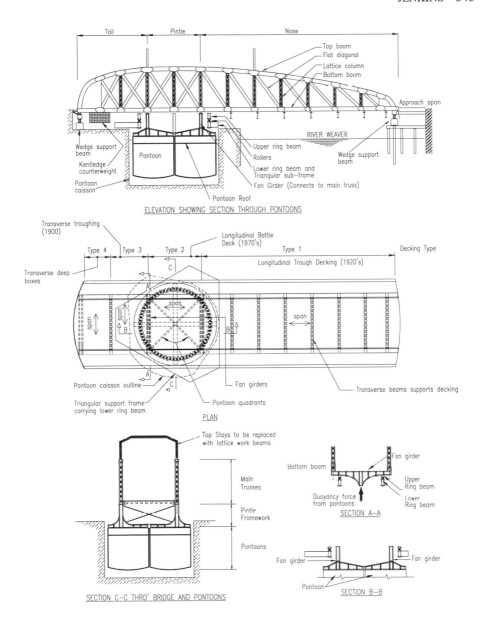

Figure 1 – General arrangement of bridge indicating elements referred to in the paper

Superstructure dead and live loads are carried down from columns in the truss, via the four projecting ends of the two fan girders, into the upper ring beam. These loads are then transmitted through roller bearings onto the lower ring beam, built into the triangular support frame which is founded on the caisson walls. Buoyancy forces from the pontoons are transmitted upwards to the centre of each fan girder, counteracting the effects of most of the dead load and reducing the loads on the bearings. This allows the bridge to be slewed more easily using electric motors.   With the bridge in the road open position, wedges are slid between bearing plates on the deck and the abutments, and convert the structure to a continuous two span bridge for live loading.

INSPECTION AND ASSESSMENT

Since its construction, the bridge has undergone a number of changes, which were required to deal with the steady increase in live loading over the period. The main alteration to the structure was in the 1920's when a strengthening programme included the addition of main diagonal compression members to the tail end of the structure and the replacement of sections of the decking plates. Also, around this time, the original wooden block and bitumen surfacing was replaced with asphalt. The structure then remained unchanged until 1975 when further sections of the decking were replaced and the bridge was generally refurbished.

In March 1993, Parkman undertook a Principal Inspection and an assessment of the bridge. The bridge was re-inspected in 1997, this time including the substructure and its assessment. The main areas identified in the inspections as requiring attention were:
- A break down of the paint system throughout the structure, with large areas exhibiting severe corrosion of the wrought iron, with the trough decking type 1 badly affected.
- Severe corrosion to the ends of the transverse girders supporting the decking type 1, at the connection into the bottom boom element.
- Severe corrosion of the projecting walls of the pontoon above roof level in the vicinity of the splash zone, with elements exhibiting thick laminations of rust, a common feature in all four of the pontoon quadrants.
- Severe corrosion inside the pontoon.
- Distortion of the wall plates in many corners of the quadrants, with inward deformations of the order of 75mm, and loss of plate thickness of the order 2 to 3mm.

The main areas identified in the assessment as requiring attention to achieve the 40 tonne assessment load were:
- The decking type 1 replaced in 1926 was inadequate in flexure.
- Several of the riveted joints of the main trusses required strengthening.
- The webs of the bottom boom member of the truss did not satisfy buckling criteria.
- The ends of the fan girders projecting beyond the upper ring beam required strengthening.

The pontoon structures are unaffected by vehicular live loading, as this is transmitted directly through the ring beams to the triangular sub-frame foundations. However, under flood conditions, river levels have been as much as 3 metres above normal operating level. At these times, average pressures on the pontoon will have doubled. Repairs were evident in the inspection, most notably to the bottom of one of the pontoons which had deflected inward on its bottom surface. The original stiffening of the plate work within the pontoons was very light in comparison with the loading to which the pontoons are normally subjected. This means that membrane action, with associated large deflections of the surfaces, is a principle means by which forces were transmitted around the pontoon shells. The fact that these structures were still intact after loading of twice normal working load has confirmed their basic adequacy. The corners, and large stiffening plates on the flat sides of the pontoons were identified as the critical elements of the tank structures ensuring their overall stability.

STRENGTHENING AND REFURBISHMENT

Prior to the detailed design and contract preparation, British Waterways commissioned Parkman to carry out a feasibility study to determine the viability and cost to strengthen and refurbish the bridge. British Waterways had available Mechanical and Electrical expertise and undertook this aspect of the work themselves. A number of issues were investigated, and the major ones are outlined below.

As the structure is of riveted wrought iron construction, possible problems in strengthening methods were identified. The two main restrictions were:

- Welding to the face of wrought iron plate. This practice was to be avoided if possible due to the problem of laminates tearing apart.
- Replacement of rivets with bolts. The replaced bolts must be capable of working with the rivet group from the onset of loading.

With these points in mind it was decided that all strengthening works would use bolted connections, and close tolerance bolts would be used when the fixings were acting in combination with the existing rivets. To be in keeping with the heritage of the structure, any bolt visible on the outside of the bridge was to be cup headed to give an appearance of the existing rivets. With the structure Grade II listed, the improvements were to be sympathetic with the original 1897 design.

An important aspect of the refurbishment work was how the contractor would undertake the work, bearing in mind the need for the pontoon removal. Three options were investigated:

(i) To crane out the bridge superstructure and other elements using a vacant area of land adjacent to the site for use as a compound and working area.

(ii) To move the bridge by jacking up and sideways onto a pontoon barge, and the barge moved to a location down stream for refurbishment. This would then free the substructure including pontoons for removal.

(iii) To raise the bridge vertically by jacking and remove the pontoons through a breech in the caisson wall. This was an option which it was felt provided the greatest risk due to the unknown nature of the ground and the construction of the caisson.

At the design stage, it was decided to replace the under-strength decking type 1, including the transverse supporting girders. This was an alternative to strengthening and repair for which unforeseen difficulties were likely to arise. Cheshire County Council was anxious to minimise the closure period of the bridge and by fabricating and painting the replacement members off site, this would assist in keeping the closure to a minimum. It also gave open access for the insitu refurbishment of the bottom boom and nose of the structure.

The strengthening and painting can be summarised as follows:

- the original nose decking and supporting transverse girders were to be replaced, and joints in the main trusses strengthened with close tolerance bolts;
- all steelwork was to be blast cleaned and a full four coat paint system applied, retaining the existing black and white colour scheme;
- the deck plates were to have spray waterproofing applied prior to the laying of a mastic asphalt surfacing, with the surfacing reduced to 40mm thick to keep to a minimum the dead weight of the bridge;
- defects in the pontoons were to be repaired, and the corners and main stiffeners strengthened, before waterproofing and painting externally using a glass flake resin;
- the bridge was to be re-trimmed on completion by adjustments of the ballast.

Included in the refurbishment was the overhaul of the mechanical and electrical equipment, much of which was original; the electric motors were to be replaced and the switch gear brought up to modern standards. Video cameras were to be installed to oversee the safe operation of the bridge.

The refurbishment and strengthening of the bridge was jointly funded by British Waterways and Cheshire County Council. Cheshire County Council were to fund the cost of strengthening to increase the capacity of the bridge beyond British Waterways legal obligation to provide for Construction and Use (C & U) Vehicles (defined in BE 3/73), up to the 40 tonne assessment loading, defined in BD 21/93 (now BD 21/97). They were also to fund one third of the refurbishment costs. A detailed exercise was undertaken to provide an accurate costing of the works with a clear break down of the costs of the works apportioned between British Waterways and Cheshire County Council.

CONTRACT PROCUREMENT

Six contractors were invited to tender for the project and each was asked to return with their tenders an acceptable method of carrying out the work. Three satisfactory methods of achieving the work were proposed.

The first method was to jack the superstructure vertically upwards and support it on a steel frame whilst the substructure was raised and slid out from underneath. The second method was to construct a working platform on piles over the river and slide the superstructure sideways before lifting the substructure to a working position. The remaining tenderers opted to crane lift the superstructure and substructure components onto adjacent waste ground and carry out the refurbishment at ground level on dry land.

One of the lowest tenderer's proposals to carry out the work contained insufficient detail, and showed a possible lack of understanding of the work content. To evaluate the tenders in terms of quality rather than cost alone, the two lowest acceptable tenderers were invited to interview to present their proposals and demonstrate previous experience in this field of work.

After detailed examination of the proposals, Kvaerner Construction Ltd were awarded the contract in December 1997, with site commencement in early 1998. Most of the temporary and permanent traffic management works were undertaken by Cheshire County Council in parallel to the main contract works.

CARRYING OUT THE REFURBISHMENT.

The project was split into five main sections:-   1- Preparatory Work, 2- Lift-Out, 3- Strengthening Work and Refurbishment, 4- Lift-In and 5- Completion.

Preparatory Work

Following the setting up of the site and the preparation of the compound to accept the crane and receive the components the electrical and mechanical drive system was disconnected and moved offsite for refurbishment. To reduce the weight of the superstructure for the lift the surfacing was removed from the deck and the steel deck over the pintle was removed to provide access below. Lifting frames were fitted to the superstructure and substructure and the two main components separated by removal of bolts and rivets and jacking apart. The superstructure was supported by steel stools erected on the caisson top until the deck was ready to be lifted.

Plate 1 Town Bridge being lifted from its support, where it has been since construction in 1899

Lift-Out

A 1200 tonne crane AK 912 (the largest mobile crane in Europe) from Baldwin was used to carry out the lift.   The crane arrived in sections and together with jibs hooks and counterweights took in total 34 trucks to bring it to site. It was erected on piled foundations during a three day period with the main components arriving at night due to restrictions on road access (a road closure was required).

Following assessment of weight and confirmation of the lifting centre, the superstructure was lifted into the compound and placed onto a prepared area over a six hour period.   The remaining pintle and pontoon components secured together were lifted out as a complete unit in a second lift. This assembly was placed onto a specially shaped bed of soft sand in the compound on which the dished bases of the pontoons were seated.

Over the following four days the pintle and pontoons were further separated and lifted onto their own stillages. All components were fully encapsulated in scaffolding and sheeting to allow the refurbishment to proceed.

Strengthening, Refurbishment and Painting

The electrical design and fabrication and the mechanical refurbishment was carried out offsite concurrently with the main works on site. The most substantial element of mechanical work was the testing and replacement of several of the cast iron rollers which allowed the bridge to swing. One roller had failed in recent years and it was feared that substantial numbers of the rollers could be cracked. Ultrasonic testing of the cast iron rollers proved inconclusive and subsequent testing by X-rays was adversely affected by scatter. Testing using photographic plates and a gamma ray source yielded satisfactory detail and enabled 18 weakened cracked rollers to be identified out of the 76 total. The cast iron roller track was drilled to measure the depth against the original records and it was found that there had been virtually no wear on the track over the 100 years.

The main components of the bridge and also the separate East Approach Span were wet abrasive blasted to remove existing coatings and the majority of the corrosion products. The 1920's trough decking was removed and all remaining components were examined for loss of section and other faults such as cracks and condition of rivets.

Deck

To enable complete access to all surfaces in the tail of the bridge, 90 tonnes of underslung kentledge were removed. A further 30 tonnes of pig iron ingots were removed from inside the truss tail boxes.

The hanger bolts for the kentledge were discovered to be seriously corroded by up to 50% of their cross sectional area. This corrosion was always very localised to the point where the bolt passed through a clearance hole in the metalwork and water was trapped. The most critical point was within the hole, and not visible without disassembling the hanger bolt. Corrosion like that found in the kentledge hanger bolts was also found in the hanger connections between bottom boom and transverse girders.

The revised centre of gravity of the bridge was determined, with the new deck fitted, and the kentledge adjusted accordingly, after refitting with new hanger bolts. Provision was made for the inspection and future replacement of these hanger bolts. Once the bridge was reassembled and resurfaced, a physical check using jacks, established the centroid of the bridge to be approximately 140mm off centre, well within acceptable limits.

Corrosion in lattice members of the main truss columns, not apparent at the time of inspection, was exposed by the initial blast clean. Again, this was in a water trap at the lower end of each lattice member at its connection to the corner angles. The extent of the corrosion was increased because of an excessive projection of the diagonal lattice member beyond the fixing rivet, allowing ingress of water and corrosion between the joined members. Numerous diagonals and sections of angles were replaced as additional works to the contract.

Pontoons

Full inspection of the four pontoons following the wet blast, showed the corrosion in certain areas to be far more extensive than had been initially believed. A band of corrosion 300mm deep was present at the splash zone and the plate in this area (originally $3/8$" thick) was cratered with holes. It was decided that the top of the pontoons above pontoon roof level be replaced in its entirety, a patch repair deemed not practicable. A fabrication subcontract was implemented for the components. (Only one metre further down the pontoon walls, the steel condition was in such good condition that the makers mark was still visible, probably safeguarded by oxygen depletion in the surrounding water, which was isolated within the caisson).

Following all the steel work operations the pontoons were dry blast cleaned inside and out, and painted. Externally a glass flake paint was used to maximise the longevity of the refurbishment – 25 years to first maintenance.

<u>Pintle and East Approach Span</u>

Additional works to these two components was limited to minor works except for the strengthening of the fan girders. After wet blasting, the top flange of these rivetted box members were lifted off to determine the constructional details. Strengthening beams were designed to fit internally within the ends of the fan girder, fabricated, fitted, and the girders close up again.

Plate 2 Completed Bridge (South Elevation)

Lift-In and Completion

Lift-in was carried out in the reverse order to the lift-out with the pontoons and pintle reassembled and secured as one unit, prior to the first lift. This was followed by the lift-in of the truss. The superstructure was reconnected to the top of the pintle with bolted connections. Angular steel girder top stays (previous replacements for the original stays following damage in a traffic accident in the 1970s) were replaced with new curved fabricated lattice girders to match the original stays.

The deck was waterproofed and resurfaced with mastic asphalt of 40mm thickness to minimise superimposed weight. The refurbished drive system was reconnected to the bridge. After fine adjustment and thorough testing of the electrical and mechanical works the road was reopened to traffic. The site compound was cleared and returned to the town as a car park for shoppers and commuters.

ACKNOWLEDGEMENTS
The works described in this paper were undertaken for the North West Region of British Waterways. The author wishes to thank British Waterways for giving permission for this paper to be published. Any views expressed are not necessarily those of British Waterways.

Bridge management in Europe (BRIME): modelling of deteriorated structures

DR ALBERT F DALY, BE, MESc, PhD, CEng, MICE
Project Manager Civil Engineering
Transport Research Laboratory
Crowthorne BERKS UK

ABSTRACT

This paper describes a study of the modelling of deteriorated structures which is being carried out at TRL as part of a project entitled Bridge Management in Europe (BRIME). A general overview of BRIME and proposed procedures for bridge assessment are given in separate papers presented elsewhere at this conference. The main objective of BRIME is to produce a framework within which a comprehensive bridge management system can be built encompassing all aspects of bridge engineering. This part of the project focuses on the common forms of deterioration found in bridges and the effects they have on assessed capacity. The ultimate aim is to identify models for the deterioration process which can be used as part of the general assessment procedure to enable a reliable estimate of load carrying capacity to be produced and used in an overall bridge management system. This paper identifies the need for assessment procedures for deteriorated structures, summarises the progress of the project to date and presents the main conclusions found so far.

INTRODUCTION

The increasing age of the bridge stock throughout Europe has highlighted the problems associated with deterioration in existing structures. Surveys have indicated that the main reasons for deterioration, besides normal wear and tear, are the increasing weights and volumes of traffic using the road network, and adverse environmental conditions such as exposure to chlorides and freeze-thaw attack. The magnitude of the capital investment in the European bridge stock requires that effective maintenance is practised to ensure that bridges are kept in service safely at minimum cost. The assessment methods currently being used do not normally include reliable techniques for the evaluation of the structural consequences of deterioration.

The first stage in the project was to identify the common forms of deterioration present in the European bridge stock and to determine the main causes. It is clear from previously published work that the corrosion of steel due to chloride contamination and the carbonation of concrete is a serious problem in bridges. Other forms of deterioration such as alkali-silica reaction, freeze-thaw action and sulphate attack are also common. In this project, models for taking these forms of deterioration into account are being developed for use within a general assessment methodology. Methods for identifying and quantifying the deterioration are also being investigated. These are an important aspect if models of behaviour are to be effective and practical.

CONDITION OF BRIDGES AND APPROACH TO ASSESSMENT

As part of this project, a questionnaire was circulated to the participating BRIME countries (France, Germany, Norway, Slovenia, Spain, UK) to obtain general information on the number and type of bridges in the national bridges stock, the condition of the bridges, and the main forms of deterioration present. Specific information was also obtained on the procedures used to assess bridges and the methods used to take account of deterioration.

The responses to the questionnaire are summarised in Table 1. The information obtained was very variable since each country is at a different stage in terms of bridge assessment and has different priorities. In most countries, existing data pertained only to the bridges on the national routes (ie, roads managed directly by government departments) and thus the questionnaire tended to focus on these. It is recognised that, in some countries, this represents only a small proportion of the national bridge stock. Other bridges are the responsibility of local authorities and private owners such as toll road concessionaires, railway operators, regional transport systems, national river authorities, etc. Thus the statistics may be biased towards the more recent, longer span structures with better maintenance regimes.

It was clear from the analysis of the responses that it is the same problems which are found in all countries in spite of the different traffic conditions and climate. The sources of deterioration can be sub-divided into three different groups:

- deterioration or defects arising from faults in design and construction: low cover, reinforcement congestion, badly located joints, poor drainage system, ASR susceptible aggregates, insufficient foundation capacity;

- defects due to construction method: poor quality concrete, bad compaction, inadequate curing, poorly fixed reinforcement, faulty ducting for post-tensioning systems, inadequate grouting, inadequate painting or coating;

- deterioration from external influences: overloading, vehicle impact, chloride attack, carbonation, poor maintenance, freeze-thaw action, dynamic loading.

CURRENT ASSESSMENT PROCEDURES

In the questionnaire, countries were asked to supply information on general assessment procedures and, in particular, how deterioration in bridges is taken into account. Again there are clear differences between countries. The UK is the only country that has adopted standards specific for bridge assessment and these are used in a formalised assessment programme. In other countries, the appropriate design codes are used, modified occasionally according to the specific requirements and according to the expertise and experience of the assessment engineer.

The current practice in all countries is to take account of deterioration in some way. In general, this means using actual sound section dimensions as measured on site or assumed, and modifying the material properties based on material tests or NDT methods. Except for the UK, taking account of deterioration is, in general, carried out on an *ad hoc* basis and depends on the knowledge and experience of the assessment engineer. In the case of the UK, there are assessment documents relating to deterioration arising from chloride induced corrosion and alkali-silica reaction. These documents tend to be general in nature and contain little quantitative guidelines.

Table 1 Estimated number of bridges and % sub-standard.

Country	Est. number of bridges	Number on national roads	% sub-standard	Main causes of deterioration
France	232,000	21,500	39% [1]	Corrosion of reinforcement Inadequate compaction Corrosion of prestressing tendons Defective grouting Inadequate water-proofing Inadequate design for thermal effects Alkali-silica reaction
Germany	34,800	19,550	42% [1]	Corrosion of reinforcement Design/construction faults Faulty bearings, joints, drainage, etc Overloading Vehicle impact Fire, flooding.
Norway	21,500	9,000	42% [2]	Corrosion of reinforcement Freeze-thaw damage Alkali-silica reaction Deterioration of paint, etc Corrosion of steel Construction faults, Shrinkage Use of sea water in mix Settlement of foundations, Scour
Slovenia	N/A	1,761	15% [1]	Corrosion of reinforcement Corrosion of prestressing tendons Failure of waterproofing Corrosion at abutments Freeze-thaw damage Corrosion of steel Defective expansion joints
Spain	N/A	15,000	N/A	Corrosion of reinforcement Corrosion of steel Inadequate waterproofing Defective expansion joints Impact from high-sided vehicles
UK	155,000	9,515	30% [3]	Corrosion of reinforcement Corrosion of prestressing tendons Impact damage Shrinkage cracking Freeze-thaw Alkali-silica reaction Carbonation

[1] Based on survey of national roads only.
[2] Based on inspection of 149 bridges
[3] Estimated from surveys on all bridges

The main conclusion is that, in general, while many countries have adopted general rules for investigating deterioration as part of condition assessment, there are few procedures available for taking deterioration into account in a structural assessment. Generally, the emphasis is on determining the presence of deterioration, and determining the condition of the bridge in terms of the extent, and possibly the rate, of progress of damage. This information is then used in formulating maintenance strategies and prioritising repair or rehabilitation works.

ASSESSMENT PROCEDURES FOR DETERIORATED BRIDGES
Bridge owners and managers are required to ensure that the structures for which they are responsible serve the purpose for which they were built in a safe and maintainable manner. They are also required to ensure that appropriate maintenance strategies are implemented in a cost effective way. For an ageing bridge stock, and this is the current state in most countries, maintenance of the existing structures is becoming more and more important in terms of the commitment of resources. As a result, the requirement to be able to identify the presence of deterioration and to quantify it in terms of its effect on serviceability and carrying capacity is increasing.

It is important to differentiate between the condition of a deteriorated bridge and the effect the deterioration has on carrying capacity. In this project, only the latter is being investigated. This difference is important as the condition of a bridge can be considered to be poor, but the effect on structural performance may be slight, or insufficient to bring the structure below the minimum acceptable performance level. On the other hand, the converse may be true and bridges which have been classified as being in good condition, based on visual examination, may still be grossly under-strength. An example is the local corrosion in the tendons of segmental post-tensioned members due to the ingress of chlorides through defective joints. In such a situation a small amount of localised corrosion can cause serious loss of strength even though a visual examination might suggest only minor corrosion. In the case of Ynys-y-Gwas bridge in Wales, the result was a dramatic and completely unexpected collapse in 1986 (Woodward and Williams 1988). This is not to say that condition assessment is not important: bridge users must have confidence in the structure and a structure will be considered a failure if its appearance prevents normal use. Similarly, the present and future condition of a bridge should be such that normal maintenance is sufficient to keep the bridge serviceable. This paper focuses only on the effect of deterioration on assessed capacity.

Minimum acceptable level of performance
The phases of the service life of a structure are dictated primarily by loss of strength, although loss of serviceability can be just as important. At some point, a minimum acceptable level of performance is reached and this defines the end of the service life. This is illustrated schematically in Figure 1 which gives as an example a structure which is deteriorating due to corrosion of reinforcement resulting from chloride ingress. In this example, the chloride concentration at the reinforcement is used as a measure of the condition of the structure. When the chloride level reaches some critical value, then corrosion will be initiated and loss of section will result. The objective of an effective maintenance strategy is to increase the service life at minimum cost. For example, routine maintenance such as painting, cleaning and minor cosmetic repairs can be used to slow down the rate of corrosion. Repair or rehabilitation work can be used to restore lost capacity as shown in the figure. This ensures that the structure does not go below the minimum acceptable level of performance - the level below which the structure must not be allowed to reach – as defined by the appropriate technical authority.

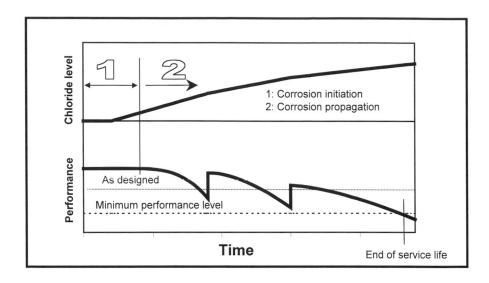

Figure 1 Corrosion loss and performance as a function of time.

For new construction, this level is defined by national codes and standards and it is likely that these will continue to be used for assessment, albeit with appropriate modifications. For a bridge, several levels of performance must be considered including:

♦ Serviceability: appearance and maintainability; deflection and deformation; crack widths, safety of bridge users;

♦ Ultimate limit state: using full highway loading, or restricted traffic if appropriate;

♦ Future state: future progress of deterioration, at possible increased rate due to increasing damage;

♦ Consequences of failure: in terms of both costs and potential loss of life.

In general, it can be argued that serviceability need not be considered in bridge assessment. This is because, if a bridge has already seen many years of service without serviceability problems, it can be assumed that the serviceability requirements have been satisfied. This approach has been generally accepted in the UK. Bridge owners may be willing to accept lower levels of serviceability based on the previous acceptable performance of the structure. Greater deflections or crack widths may be acceptable where aesthetics are of less importance because of the location and use of the bridge. There are other implications in accepting lower levels of serviceability and these should be fully considered and taken into account within an overall management strategy before being accepted. It may mean that the need for rehabilitation work is postponed which might increase the cost. Risks to user safety might also occur: for example, spalling concrete could fall onto a carriageway or member of the public.

In structural assessment, attention is focused on the ultimate limit state for the current condition of the bridge, and for the condition at some time in the future. It is envisaged that, in assessing bridge capacity, the engineer should make some allowance for future deterioration, ideally presenting an indication of the residual service life. Attention should be given to possible modes of failure particularly those unique to the deterioration process.

METHODS OF DEALING WITH DETERIORATION IN ASSESSMENT
There are a number of general methods which can be used to take account of the modified behaviour of a structure as a result of deterioration. These can be applied to most forms of deterioration and are described in the following paragraphs.

Reduced cross-sectional area
In determining the strength of a deteriorated or damaged structure the most common approach is to take account of the material loss by direct measurement of the remaining sound material. Measurements can easily be made of external concrete dimensions or the steel thickness of the steel beam but not so easily when deterioration of reinforcing steel or delamination within the concrete is suspected. In some cases, breaking out portions of the structure can be justified in order to determine the extent of material loss. A wide range of non-destructive testing (NDT) techniques are available which can be used to determine residual steel thickness, presence of corrosion in concrete structures, delamination of concrete and presence of faults as well as concrete strength. The information provided by these techniques is largely qualitative, which is useful but is unlikely to completely satisfy the requirements of practical strength assessment. However NDT techniques are continuously being developed and improved.

Condition factor
Where measurements are not possible or where there are other uncertainties in the determination of resistance, some codes suggest the use of a *condition factor* to take account of any deficiencies that are noted in an inspection but not allowed for in the determination of member resistance. The factor should be based on engineering judgement and should represent an estimate of any deficiency in the integrity of the structure or member. It may be applied to a member, a part of the structure or the structure as a whole. While this is an imprecise and subjective method of allowing for defects in the structure, it is often the only approach available due to the absence of data on the strength of deteriorated structures. In general, there is a lack of specific guidance and it is left to the experience and judgement of the assessment engineer to make an appropriate choice of condition factor.

Modified concrete properties
A knowledge of the in-situ material properties is required for a reasonable estimate of the capacity of a structure to be made. The most direct method of determining in-situ concrete strength is to take core samples from the structure and carry out compression tests. NDT techniques can also be used but careful calibration is required for a particular structure.

The assessment of the strength capacity of a bridge entails calculations pertaining to a number of different load effects such as flexure, shear, bond, bearing and deflection. A number of concrete properties are required, such as compressive and tensile strength, bearing strength, and elastic modulus. Most design codes use the compressive strength (cube or cylinder) as a reference for these concrete properties. This is a convenience which may bear little resemblance to the actual behaviour of concrete in a structure (Clark 1989). The equations were developed to correlate reasonably well with test data for non-deteriorated concrete. For deteriorated concrete these relationships may not be appropriate, since the various concrete properties may be affected differently by the deterioration. The solution is to avoid using the cube strength as a reference where possible and measure the properties with appropriate tests. These properties can then be used directly in the assessment.

Modified steel properties
The strength of corroded steel members can be taken into account by reducing the cross-sectional area to allow for the material loss. There is no evidence in the literature to suggest that the stress-strain relationship of structural steel is greatly modified by normal forms of deterioration. It is known that corrosion in concrete can have an effect on the properties of the reinforcement. Ductility can be reduced and fatigue properties can be affected.

Modified bond properties
For reinforced concrete to behave as a composite material, adequate bond between the concrete and reinforcement must be maintained. Bond is developed by chemical adhesion and mechanical interlock. For deformed bars, the main component of bond is the interlock between the deformations and the surrounding concrete. When cover is low and where no transverse steel is present, failure occurs by splitting of the concrete and bond strength depends on the tensile strength of the concrete. Where splitting of the concrete is prevented, either by adequate cover or the provision of transverse steel, the concrete between the bar deformations shears from the surrounding concrete and the bond strength is a function of the strength of the concrete in direct shear. Most codes simplify bond strength by presenting it as a function only of concrete strength and bar type.

Small amounts of surface corrosion can enhance bond strength by increasing the confinement of the reinforcing bars. In general, however, corrosion disrupts the interface between the steel and concrete thus reducing the bond strength and this can be the most significant effect of corrosion. Various equations have been proposed to model the relationship between bond strength and extent of corrosion. While these are based almost exclusively on pull-out tests on laboratory specimens, they provide a useful means of determining the effectiveness of reinforcement in naturally corroded members.

Modified structural behaviour
The approaches so far assume that neither the structural properties of the material nor the structural behaviour of the components are altered by the deterioration. Where substantial amounts of deterioration have occurred the mechanisms by which a structure resists load may be modified. For example, a reinforcing bar extensively corroded on one side only will develop bending stresses when subject to an apparently axial load. Complete breakdown of the bond between steel and concrete would result in a beam being incapable of behaving as a conventional beam. In fact, such a member could carry the load by acting as a tied arch, provided there was adequate mechanical anchorage at the bar ends. Shear strength can also be affected since the effect of dowel action could be lost. In most instances, impairment of performance is likely to be sensitive to detailing but no specific guidance can be given.

Where serious corrosion exists, there is always the possibility that the integrity of the structure is impaired. In such cases, it would be necessary for the assessment engineer to demonstrate that the assumed structural behaviour can actually be achieved. For deteriorated structures, advantage can be gained by using alternative methods which take account of restraints which are generally ignored in design. An example is compressive membrane action which can considerably enhance the strength of slabs. The implication is that, if traditional methods result in slabs which are over-designed, then the loss of steel due to corrosion may not be such a serious problem.

Additional stress
Some forms of deterioration result in additional stresses being imposed in the structure. For example, when deterioration due to ASR and the production of expansive gel occurs over a significant proportion of a structural component, overall expansion will occur. When this expansion is restrained either by internal reinforcement or external restraints, additional stresses will develop in the concrete and the beam is effectively prestressed (Clark, 1989). For a reinforced concrete section, these additional stresses can be taken into account by treating it as a prestressed beam. For prestressed beams, the additional ASR prestress can be included in the assessment. In principle, these stresses can be estimated from a knowledge of the free expansion due to ASR and the reinforcement details. In practice, however, complications arise since the stresses due to restrained ASR expansion depends on the stress history and on the expansion rate.

CONCLUSIONS
This paper describes the main forms of deterioration found in the European bridge stock and outlines straight-forward methods of taking account of deterioration in the determination of structural strength. How they can be applied to different forms of deterioration depends on the type of deterioration and how it affects structural behaviour. It is important to note that the assessment of the effects of deterioration and the determination of residual life depends on the correct diagnosis of the deterioration and the conditions causing it. Models for taking account of deterioration within a formal bridge assessment procedure are currently being developed and evaluated.

ACKNOWLEDGEMENT
The work described in this report has been part funded by the European Commission Directorate General for Transport, with balancing funds provided by the authorities responsible for the national road networks in the UK, France, Germany, Norway, Slovenia and Spain.

REFERENCES
Clark, L A (1989). *Critical review of the structural implications of the alkali silica reaction in concrete.* Contractor Report 169, Transport Research Laboratory, Crowthorne.

Kaschner R, P Haardt, C Cremona, D W Cullington and A F Daly (2000). *Bridge Management in Europe (BRIME): Structural assessment.* Fourth International Conference on Bridge Management, Surrey, 2000.

Woodward, R J and F W Williams (1988). *Collapse of Ynys-y-Gwas bridge, West Glamorgan.* Proceedings ICE, Part 1, Volume 84, August, pp 635-669.

The strength of deteriorated reinforced concrete bridges

I. L. KENNEDY REID.
WS Atkins Consultants Ltd., Epsom UK.

INTRODUCTION
This paper summarises a project carried out by WS Atkins Consultant Ltd for the Highways Agency (Mr Sibdas Chakrabarti) to carry out model testing leading to the production of an Advice Note and Technical Memorandum giving guidance on the assessment of reinforced concrete bridge structures.

SCOPE
The scope of the project principally covered concrete structures subject to carbonation and chloride attack, resulting in reinforcement corrosion and delamination of cover. It excluded prestressed and lightweight concrete and members subject to alkali-silica reaction and thaumasite attack.

Following a desk study of reasons for deterioration and a literature search covering previous work carried out into the strength of deteriorated reinforced concrete structures, the laboratory testing was carried out by Professor Paul Regan at Westminster University. The objectives of the testing were:

- To study the bond strength with corroded bars and with simulated delamination of cover, both in pull-out tests and in beams;

- To study the failure mode with delaminated cover of slabs, of rectangular and Tee-beams, and of beams with continuity over supports;

- To study punching shear in a slab and axial strength of a column, both with delaminated cover;

- To study shear strength in beams with the end parts of their stirrups corroded;

- To study the shear strength of beams which already had perpendicular cracks.

BOND PULL OUT TESTS
The test arrangement is shown in Figure 1 which shows clamping added.

Bond was found to vary with:

- Depth of cover, concrete strength, proximity of bar to corner of the member, and whether top or bottom cast;

- Main bar diameter and rib type, direction and orientation;

- Stirrup diameter, bar type, spacing, proximity of corner to main bar, and bend diameter;

- Whether there was a vertical crack in the cover longitudinally along the main bar;

- Transverse pressure.

Bond was found to be independent of:

- Anchorage length;

- Position of delaminated cover in an anchorage comprising part delaminated and part full cover.

Corrosion:

- The effect of corrosion on ultimate and residual bond stresses was found to be equivalent to losing cover to between half barrel level and level of top of main bar. See Figure 4.

SLAB TEST
Delaminated slabs, see Figure 2, were tested on differing bearing configurations of steel and rubber with continous strips, plates under the outer bars only and under all bars. It was found that:

- Bond stresses were very high over the supports;

- With mixed anchorages the strengths of the individual anchorages could not be fully aggregated.

BEAM TESTS
Four beams were tested with different proportions of the span delaminated; see Figure 3, typical.

- With main bars half exposed the beam behaved almost normally;

- With main bars fully exposed the beam acted as an arch;

- With the central half of the span exposed this part of the beam acted as an arch. With insufficient stirrups at the quarter points cracking would occur and the whole beam would act as an arch;

- With the outer quarter spans exposed the thrust lines are inclined in the outer quarters and then run horizontal in the central part, resulting in a need for adequate stirrups to resist the change in direction of thrust at quarter span.

BEAM SHEAR TESTS
Four beams were tested with the shear span having bars exposed to half barrel and no shear links. It was found that:

- There was enough bond to develop flexural cracks;

- These developed into shear cracks;

- Some of the beams failed in shear, see Figure 5.

T-BEAM TESTS
Two T-beams were tested with lower main steel bars exposed to half barrel in the shear span, and links provided throughout. It was found that:

- All beams failed in shear with a slight reduction in strength;

- Bond was enhanced due to tension in stirrups trapping bars.

FLEXURAL TESTS
Eight beams were tested with the main bars able to slide in tubes cast into the beam except over the end anchorage lengths. It was found that:

- The beams failed in flexural compression at a significantly lower load than beams with full bond;

- A formula has been developed for flexural failure of unbonded beams from the test results.

CANTILEVERS UNDER UDL
Four beams were tested with discrete loads on the cantilevers representing a uniformly distributed load (UDL). Bars were mixes of bars in and not in the corners of stirrups and clamped and not clamped by the load. (The differences between these pairs was found from the pull-out tests significantly to affect the bond strength). It was found that:

- The bond strengths of the individual bar configurations were not fully additive in developing the cantilever strength.

RESTRAINED BEAMS WITH PARTLY EXPOSED MAIN STEEL
Four beams were tested with a cantilever extending from an otherwise simply supported span. The moment/shear ratio at the end of the interior span was changed by altering the beam depth, the rib orientations of the bars were altered and the stirrups had u-bends and 90° bends over the main bars. Among the conclusions reached it was found that:

- Bond strengths were approximately 50% higher in a beam than in pull-out tests.

SHEAR STRENGTH OF RC BEAMS WITH DEFECTIVE STIRRUPS
Stirrups having less cover than main steel are particularly vulnerable to corrosion and can lose a large proportion of their cross section at their ends before the main steel loses much section.

Fourteen beams with and without cover were tested to assess the effect of defective stirrup anchorages. See Figure 6. It was found that:

- The crack patterns of those beams failing in shear were little changed by lack of stirrup end anchorages;

- Bond tests of short anchorages cast vertically demonstrated bond values more than twice that in BD44;

- The shear strength can be calculated using these values and stirrup anchorages either side of a diagonal shear crack;

- The loss of shear strength is equivalent to roughly half the loss of cross-sectional area of the ends of the links, provided the loss of area does not exceed two thirds, and provided half the beam depth is adequate to develop bond;

- It would be over-conservative to ignore stirrups with corroded end anchorages.

COLUMN TESTS
Three axial tests were carried out on columns two of which had delaminated cover. It was concluded that:

- With intact links the stress reduces by the compressive strength of the reduced cover;

- With corroded links the main bars could buckle and might have to be discounted.

PUNCHING SHEAR TESTS
Three slabs were tested, two of which had delaminated cover. It was concluded that:

- Punching strength is not affected by loss of cover to tensile reinforcement in an area 8 effective slab depths square around a concentrated support, provided the bars are fully anchored beyond.

SHEAR STRENGTH OF BEAMS WITH PERPENDICULAR CRACKS
Five beams were cast and perpendicular cracks induced by axial force. See Figure 7. The beams were then tested in shear. It was found that:

- A 3.5 mm wide crack was 20% weaker than an uncracked beam;

- A 9 mm wide crack was a further 20% weaker. Shear strength based on dowel action.

Beams were also cracked in flexure and then tested in shear. It was found that:

- A 2 mm wide crack was 24% weaker than an uncracked beam.

RECOMMENDATIONS
Draft recommendations include the following:

- The strength of sections (and the slenderness of columns) should be based on ignoring the cover when delaminated;

- Reinforcement strength should be based on the corroded cross-section;

- Sections a limited distance along a main bar from corrosion can be based on the corroded bar cross-section plus the strength provided by delaminated bond over the intervening length, the bond strength depending on whether or not delaminated;

- Compression bar effectiveness in beams should be reduced in proportion to corroded link section versus minimum link section. Outer compression bars in beams should be ignored if cover delaminated and there is significant link corrosion. Longitudinal reinforcement in columns should be discounted where corrosion and delamination reduces effectiveness of links to below minimum;

- Where main bar capacity is less than the force developed through moment and shear, the effectiveness of the links should be reduced;

- Where the ends of the links are corroded, their loss of effectiveness in shear may not be as much as the proportion of cross-section lost.

- Without shear links, the shear strength of a delaminated member should be reduced and is dependent on the extent of flexural cracking and whether delaminated bond strength is adequate;

- Shear enhancement of delaminated members can be based on strut and tie analysis to determine force in main reinforcement over support and thereby adequacy of anchorage using delaminated bond strength;

- Full allowance may not need to be made for delamination for punching shear in the punching shear zone other than the loss of cover, provided the reinforcement is fully anchored beyond the area considered;

- Anchorage strength. Force in main steel derived from moment and shear. Bond capacity based on a formula depending on cover from mid-barrel, stirrup spacing, diameter of main bar, area of stirrup leg adjacent to main bar, square root of concrete strength and degree of clamping pressure. Delaminated bond strength is very low.

- Rules given so that bond strength of bars in different situations i.e. in/not in stirrup corners/clamped/unclamped are not fully additive;

- Bond strength of corner bars exposed to half barrel on both faces is ignored;

- Perpendicular cracks with aggregate interlock will reduce shear resistance.

REFERENCE
1. Highways agency. Design Manual for Roads and Bridges. Volume 3. Highway Structures Inspection and Maintenance. Part 14. BD44/95. The Assessment of Concrete Highway Bridges and Structures.

2. Series of Reports on Model Testing for Deteriorated Structures project by P.E. Regan, Westminster University, for WS Atkins and the Highways Agency.

ACKNOWLEDGEMENTS
The author acknowledges the kind permission of the Client Mr Chakrabarti of the Highways Agency to publish the paper, the sterling work by Professor Regan in managing the testing, and the guidance and help of Mr Mike Chubb and other colleagues of WS Atkins.

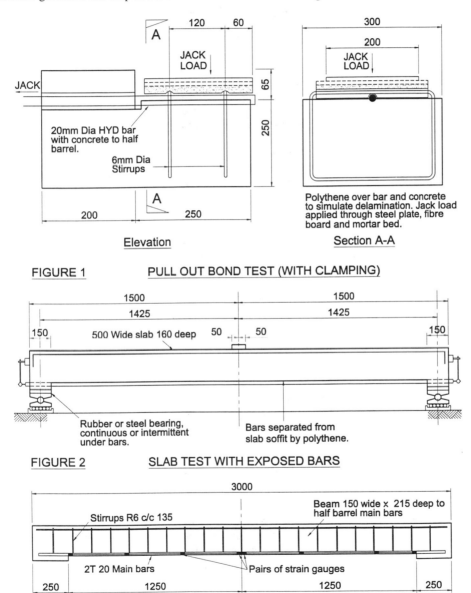

FIGURE 1   PULL OUT BOND TEST (WITH CLAMPING)

FIGURE 2   SLAB TEST WITH EXPOSED BARS

FIGURE 3   BEAM TEST WITH PARTIALLY EXPOSED BARS

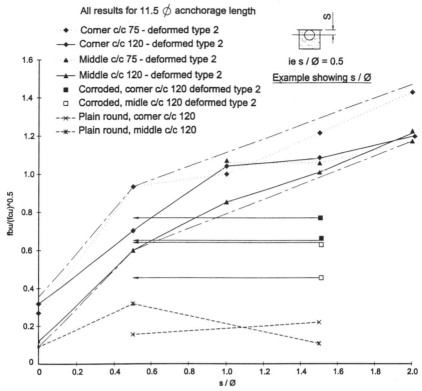

FIGURE 4 ULTIMATE BOND STRESS fbu AS A FUNCTION OF COVER,

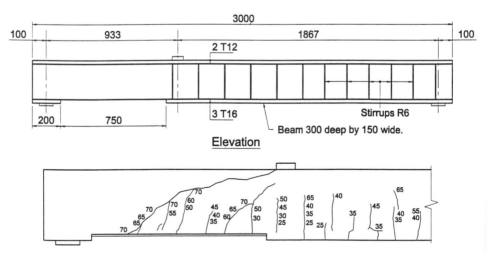

Crack pattern of one of the beams which failed in shear

FIGURE 5 BEAM SHEAR TESTS WITH PARTIALLY EXPOSED
REINFORCEMENT IN SHEAR SPAN

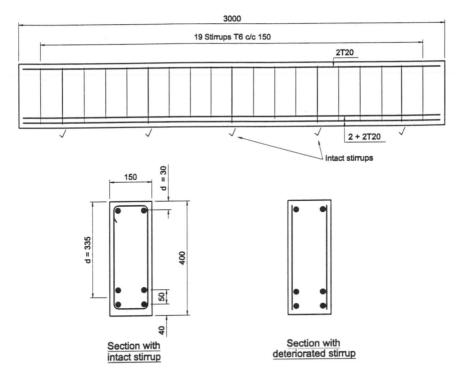

FIGURE 6   BEAMS WITH DEFECTIVE STIRRUPS FOR SHEAR TESTS

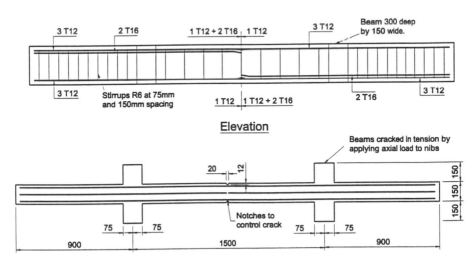

Plan showing only upper layer of top steel

FIGURE 7   SHEAR STRENGTH OF BEAMS WITH PERPENDICULAR CRACKS

Specifications for competitive tendering of NDT inspection of bridges

Dr P C Das, OBE
Highways Agency
St Christopher House
Southwark Street
London SE1 0TE UK

Dr M S A Hardy, Prof D M McCann & Prof M C Forde
University of Edinburgh
School of Civil & Environmental Engineering
Kings Buidings
Edinburgh EH9 3JN UK

KEYWORDS: Bridges; Scour; Radar; Conductivity; Tendering.

ABSTRACT
Frequently NDT of bridges has been undertaken on a negotiated basis, relying upon goodwill from both the employer and the NDT test house in order to achieve value for money and quality. When competitive tendering has been introduced into this arena, problems have arisen due to lack of a quantified basis for competitive tendering.   The objective of this paper is to address the issues of drawing up competitive tenders in order to obtain the best value for money price from competing test houses.

This paper focuses on NDT techniques to identify voiding in grouted post-tensioned bridge tendon ducts; and scour measurement at bridge sites - in order to illustrate the procedures for competitive tendering of an NDT contract.

The procedure for preparing tender documents has been given. In addition, guidance has been given to the client as to how to monitor the role of an independent consultant undertaking the tendering process.

INTRODUCTION
The role of NDT in the testing of structures in general is increasing at a significant rate within the international community. Against the background of increased use of NDT techniques, many clients are seeking to gain best value for money by introducing the concept of competitive tendering. However, there is increasing frustration from both the client base and the test house base upon (1) the quality of tender documents received by test houses and (2) the quality of the investigation undertaken by the test houses, respectively. There is thus a need for documentation to give guidance upon the procedure for drawing up competitive tenders, evaluating competitive tenders and finally evaluating the reports as a result of work undertaken from a competitive tendering environment.

In order to give specific examples, this paper will focus upon NDT techniques to identify voiding in grouted post-tensioned bridge tendon ducts; and scour measurement at bridge sites - in order to illustrate the procedures for competitive tendering of an NDT contract. The key issues with regard to bridge scour are that conventional techniques such as rodding and diving will only identify the existing river bottom profile as can be observed. It will not identify any previous scour history.

ROLE OF NDT IN BRIDGE EVALUATION
In recent years, a number of non-destructive testing methods have been used, mainly on trial basis, for a variety of structures related applications including some involving bridges. Although such applications have been sometimes intended for the precise location of relatively small features such as reinforcement bars and structural cracks, in general, the results from these methods are expected to be qualitative. This is because most of these methods rely on reflected electro-magnetic or acoustic signals or dynamic responses, which give a very complex but imprecise picture of the reality.

When assessing the load carrying capacity of an existing bridge, the calculations used are of a precise nature. The results also have to be fairly decisive, in that the bridge authority has to take specific actions based on the results of the assessments. In this context, at first sight, there would seem to be no place for "imprecise" qualitative testing methods in bridge assessment. Indeed, there have been attempts in the past to test the comparative "accuracy" of such methods for locating small structural features for assessment purposes, the results of which have not been encouraging.

Nevertheless, the current national bridge assessment and strengthening programme has highlighted a number of situations where the precision of the conventional assessment methods cannot be relied upon. In such cases, a few examples of which are given later, instead of the "pass of fail" type of results from the assessments, it may be more logical to allow a "pass, fail or monitor" type of result. The monitor option could be used when a bridge fails assessment by a small margin but may have features which are likely to lead to a reserve for strength, which cannot at present be rationally taken into account in the calculations.

When deciding whether to adopt a "monitor" option for an assessment-failed bridge, the bridge authority's confidence would be reinforced if some form of qualitative examination could confirm the possibility of there being a hidden reserve of strength. Non-destructive testing methods such as the sonar, radar or the dynamic methods could be of great value in arriving at such decisions. Furthermore, once the monitor option is decided upon, some of these methods could also be used for examining the bridge periodically to see if any significant structural deterioration is taking place.

The Highways Agency has for some time been actively examining the various NDT methods for use in different aspects of the management of the network. There is already advice available on the use of ground penetrating radar (GPR) for road pavement assessment (HA 72/94, Design Manual For Roads and Bridges, the Highways Agency). The present intention is to eventually provide similar guidance for bridge-related use. The purpose of this paper is to highlight the areas of bridge assessment and monitoring where non-destructive testing techniques could be fruitfully used. It also describes what is needed at present in terms of the specifications for equipment and test procedures, so that bridge authorities can make use of these methods with confidence.

NDT METHODS

A variety of methods have been used in recent times in a number of different ways. Those intended to be covered in this paper fall into the general categories of radar, sonar and dynamic testing methods.

The applications so far have been mainly exploratory or in the context of research and therefore competent specifications for commissioning such testing widely are not at present available. The common characteristics of all these methods are:-

(1) Most of the methods were originally developed for purposes other than structural testing, for example for quick testing of aircraft bodies or machine parts.
(2) Highly sophisticated equipment are available commercially.
(3) The instruments come with wide ranges of capability, a large part of which could be inappropriate for any particular application.
(4) Most equipment are portable, and the tests can be carried out quickly and conveniently without much disruption to traffic.

The general principles of such methods are that:

• Radar will propagate through most materials, including air, but not metals; however it is rapidly attenuated in saline conditions and in clays.
• Sonic will propagate through most materials including metals, but not air.

- The conductivity technique is essentially a high powered metal detector and will identify changes in ground conductivity. Shallow metal would shield deeper penetration due to its high conductivity.
- Dynamic testing methods generally involve the vibrational characteristics of the whole bridge.

BRIDGE ASSESSMENT AND TESTING
There are three aspects to any assessment and testing programme:
(a) Protecting the value of the bridge asset commensurate with its residual life (Das, 1996).
(b) Ensuring public safety (Menzies, 1999).
(c) Choosing the best value for money assessment and investigation procedure commensurate with a whole life costing programme (Frangopol & Hearn, 1996).

STRATEGY FOR BRIDGE INSPECTION
The general strategy for bridge inspection is normally as follows:

Phase 1: Visual inspection, but beware of the special case of post-tensioned bridges where this strategy may prove ineffective.
Phase 2: Analysis of load carrying capacity
Phase 3: Review need for further investigation - if none, then revert to routine visual inspection schedule, but beware of the special case of post-tensioned bridges. If further investigation is required, then proceed to Phase 4.
Phase 4: "Desk study" - before undertaking any more detailed field study, research needs to be undertaken of the origins of the bridge: who designed and built it and the possible construction methodology.
Phase 5: Cost effectively choose the most suitable strategy for further investigation. An NDT method would only be chosen when a direct physical measurement strategy was statistically inadequate, too risky or too expensive.
Phase 6: Implement the investigation technique.

GUIDANCE FOR NON-SPECIALIST CLIENT OR AGENT ENGINEER :
When it comes to choosing an NDT technique some commonly asked questions include:

When to consider each type of test?
Type of testing organisation to shortlist?
What to expect from the tests?
How to ensure consistent bids for work?
How to judge results or performance?
How to evaluate tenders?
How to use the results?

CONSIDERATION OF THE INVESTIGATION OPTIONS
Before one can address the above questions, an informed and detailed knowledge of investigation options is needed. This same knowledge is also needed before the "Desk Study" is undertaken.

An example of such a set of options with relative cost rankings is given in Table 1 for the identification of voiding in metallic post-tensioned bridge beam tendon ducts; and for bridge scour identification in Table 2.

Investigation Method	Cost of Method	Effectiveness of Technique
Visual Inspection	Low	Technique is ineffective as bridges rarely show distress before catastrophic failure.
Load Test	Relatively high	Ineffective procedure and dangerous as the structure could fail before any meaningful deflection response is obtained.
Stress/strain measurement	Relatively high	Generally ineffective as Cavell (1997) has shown that post tensioned bridge strain variations due to loss of pre-stressing can be similar to variations resulting from temperature gradients throughout the year. Thus this technique is not sensitive to the defects in post tensioned bridges.
Impulse radar	Intermediate	Effective with non metallic liners such as in the joints of segmental bridges and in the newer post tensioned bridges. Radar will not penetrate post tensioned metal ducts.
Impact echo	Intermediate	Potentially useful in identifying voiding in non metallic and metallic post tensioned ducts. Essential to ensure that impact frequency is sufficiently high to identify the defect.
Manual drilling of tendon duct with visual inspection using endoscope.	Intermediate	Statistically limited and potentially dangerous if the tendons themselves are drilled. Advantage is that a direct physical observation can be made.
Radiography	High	High powered radiographic techniques give good image of voiding but requires closure of the bridge and may not be used in urban areas due to risk of radiation.
Ultrasonic tomography	Intermediate	Promising technique that could identify voids by producing a 2-D or 3-D image of the beam cross-section.

Table 1: Techniques for the identification of voiding in metallic post-tensioned bridge beam tendon ducts

Field Technique	Maximum River depth limitation	Advantages	Disadvantages	Relative cost
Rodding from a boat	3 m	quick & cheap	Difficult to get an accurate profile of the river bed when a stony profile exists; no information on profile below river bed	low
Diving	no practical limit	quick	Diver may have little relevant experience	medium
Bridge mounted sonar	5 m	continuous record in time	Data at one point only; no information on profile below river bed	high
Ground Penetrating Radar (GPR)	30 m	continuous and accurate record of river bed; may give information on profile below river bed	site work slow and requires expensive instrumentation	high
Continuous seismic profiling (CSRP)	Maximum depth = 30m Minimum depth = 3m	continuous and accurate record of river bed; may give information on profile below river bed	site work slower than GPR and requires expensive instrumentation	high

Table 2 Summary of Current Methods for Bridge Scour

Thus GPR would not be a first choice for every routine bridge scour survey, but it may well be a serious choice where there is a possibility of infilled scour holes and thus a need to identify the depth of deep scour during storm conditions on critical structures. Depending upon the size of the bridge, a typical GPR survey would probably involve two days field work and around a week to interpret using a signal processing package.

THE DESK STUDY
Against the above background, each "Desk Study" will vary and will involve visits to the site and other sources of information. In order to illustrate the kind of information expected, the example below relates to a Bridge Scour investigation where radar is planned (hence the need to know the conductivity of the river):

1. The location of the structure.
2. Conductivity of the river water.
3. Engineering drawings, especially of foundations.
4. Sketches giving the pier and abutment numbers.
5. Direction of river flow and typical flow rates.
6. Parts of the structure which are below the high water mark.

7. Reference datum points.
8. Results of previous measurements of the bed profile with the survey date.
9. Local geology.
10. Access to the site
11. Permission constraints on access from land and river owners.

MODEL SPECIFICATION
In order to satisfy value for money criteria and technical consistency in the tendering process, a model specification for an NDT survey must be drawn up. The example below relates to a scour survey.

To ensure that the information produced by a radar survey is sufficiently accurate and reliable for the bridge engineer, a technical brief and quality control plan must be prepared and used in order to manage the survey.

In addition some radar contractors who do not have bi-static radar system facilities available, may have to undertake some coring of the river bed.

The technical brief should specify the following:

*   the precise location of the bridge, using ordinance survey co-ordinates
*   the name of the river to be surveyed
*   the date and time of the survey
*   the weather at the time of the survey
*   the conductivity of the river at daily intervals
*   the approximate level of the river
*   the trade name and model of the radar system (e.g. GSSI SIR System 10A or Pulse Echo 1000)
*   the name and model number of the antenna together with their frequency (e.g. GSSI 100MHz antenna used in bi-static arrangement or Pulse Echo 220MHz)
*   statement of whether the antenna is used in bi-static arrangement or monostatic arrangement
*   method of providing lateral location across the river (e.g. steel wire; dinghy with outboard motor and position fixes to be placed on radar trace; other means of position fixing may be considered appropriate by the survey team depending on the operational conditions at each individual site).
*   lateral location of radar traces across the river bed
*   number and names of personnel involved in radar site survey, including their qualifications.

The specifier should should make clear that the radar survey interpretation should include data to be given in three formats:

*   raw radar traces
*   filtered radar traces, if applicable
*   diagrammatic type interpretation (e.g. CAD type drawing)

It is essential that the radar contractor provide not only an interpretation of the river bed but the actual raw radar data in order that the specifying engineer has a clear view of the quality of the data print-out.

The radar contractor should give a clear indication of the accuracy of the survey based upon the choice of antenna and the radar hardware. This data should be provided in a format such that the specifying engineer can compare the claimed accuracy with the radar plots.

The specifier has to decide how prescriptive he/she is going to be with respect to a generic system. The specifiers must make themselves aware of available data on the accuracy of NDT systems (e.g. Martin, Hardy, Usmani, & Forde, 1998) A summary of the more commonly encountered generic radar systems and analysis procedures is given in Table 3 below:

GPR Method	Contractor's Equipment Spec.	Advantage	Relative advantage	Disadvantage	Cost implication
1	Analogue radar system	Low cost		Visual interpretation of raw data only	Low cost survey
2	Digital: single channel; mono-static bow tie antenna	Facility for both visual interpretation and post-processing	More flexible interpretation than analogue, method 1	Resolution of penetration of river bed limited	Medium cost
3	Digital: single channel bi-static	Facility for both visual interpretation and post-processing	Better resolution than monostatic, method 2		Medium cost
4	Digital: multi-channel	Facility for both visual interpretation and post-processing	Potential for higher resolution than single channel bi-static, method 3		Higher cost
5	Post-processing software	Facility for "cleaning-up" data; more effective on bi-static; and even more effective on multi-channel	May give better interpretation than methods 2-4 without signal processing	Requires extra processing, usually on a desk-top computer; requires more technical expertise	Extra over cost

Table 3.

MODEL OUTPUT (REPORT AND RESULTS)
The output from an investigation will vary depending upon the type of problem and investigation specified. Two typical examples are given below. The first relates to an impact echo survey of post-tensioned bridge ducts, whilst the second relates to the results from a radar survey of bridge scour.

Impact Echo Test Results and Presentation
Results from impact-echo tests should include at least a representative selection of raw impact echo frequency domain results - so that the Engineer can gain a feel for the clarity of the results.

Overall the data should be analysed in tabular format indicating where defects might exists.

A sensitivity analysis should be included indicating the minimum depth which can be resolved.

The test report should:

1. Include a reference to the specification or an Advisory Note (if more appropriate)
2. identify the location of the structure
3. give the dimensions of the structure
4. specify the size and location of the test area
5. identify and plot the measuring grid
6. state the make and model of instrumentation used
7. if a non-standard instrument is used, state the specifications
8. state the likely input frequency achieved and ball-bearing size
9. give the date and weather conditions (possibly temperature) during test

10. plot the results in tabular format or graphical format
11. Indicate the size or volume of the void if possible.
12. give the results of any complementary tests

The results can be used for a number of purposes primarily to identify voided ducts. Then the Engineer can make a considered judgement as to whether further inspection such as physical drilling and endescopic inspection is required.

Scour Report
The results from a radar survey of bridge must be presented in a format which can be clearly understood and also clearly and unambiguously related to the bridge crossing the river. Thus, the traverse lines across the river should be clearly marked and in addition, markers should be shown on the radar traces indicating lateral locations as the antennae are moved across the river. It may be appropriate to give a diagrammatic interpretation of the radar results - although this is not an essential requirement provided that the raw radar data traces are clear. The diagrammatic interpretation of the radar results should not replace the raw data.

The radar survey for the scour report should include:

- a written report giving conclusions, assumptions used to interpret the radar data, accuracy of measurement achieved, problems encountered and any other technical data.
- the raw radar data
- a graphical display of the survey results
- tables of data

The diagrammatic and radar output should contain a summary of the information outlined as a part of the survey. A typical radar plot from a scour survey might be as below – Figure 1.

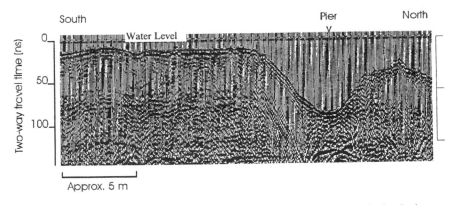

Figure 1 GPR data showing an "infilled" scour hole in a sandy riverbed

CERTIFICATION PROCEDURE
In many Civil Engineering applications, it is normal procedure to adopt a third party certification procedure. At the present time this is not normally undertaken in NDT surveys. However the adoption of such a strategy would add to the cost of the survey, but would dramatically improve the consistency of

work undertaken. It has to be appreciated that lowest absolute cost does not always equate to best value for money.

TENDER PROCEDURES

It is suggested that a two envelope tender system be used:

- "envelope 1" would contain the proposed equipment to be used, work and survey methodology, together with the qualifications of the staff to be employed.
- "envelope 2" would contain the competitive price for the survey.

Clearly the survey should be evaluated based upon envelope no. 1 containing an adequate minimum survey level, otherwise the targets for the bridge management programme would not be met. Envelope no. 2 would then be scrutinised in relation to bids which met the minimum technical target standard.

CONCLUSIONS

- There is a clear need to develop a consistent framework for bridge investigation.
- There is a need to develop a proper framework for specifying NDT surveys of bridges in order to improve the consistency of information obtained and also to obtain best value for money.
- Examples of specifications have been included with respect to identification of voids in post-tensioned bridge ducts and also GPR surveys of bridge scour.
- A two envelope tender system should be used: "envelope 1" would contain the proposed equipment to be used, work and survey methodology, together with the qualifications of the staff to be employed; "envelope 2" would contain the competitive price for the survey.
- Certification procedures should be considered as lowest absolute cost does not always equate to best value for money.

ACKNOWLEDGEMENTS

The authors acknowledge the provision of facilities by the University of Edinburgh. They also gratefully acknowledge the funding provided by EPSRC and The Highways Agency, London. Valuable discussions were held with Kevin Broughton.

REFERENCES

- Cavell, D.G. (1997) *Assessment of deteriorating post-tensioned concrete bridges*, PhD thesis, University of Sheffield.
- Das, P.C. (1996) "Bridge Management Methodologies", *Recent Advances in Bridge Engineering*, Eds. J.R. Casas, F.W. Klaiber & A.R. Mari, Barcelona, 1996, 56-65.
- Frangopol, D.M. & Hearn, G. (1996) "Managing the Life-Cycle Safety of Deteriorating Bridges", *Recent Advances in Bridge Engineering*, Eds. J.R. Casas, F.W. Klaiber & A.R. Mari, Barcelona, 1996, 38-55.
- Highways Agency (1994) *Design Manual for Roads and Bridges*, HA72/94, HMSO, London, 1994.
- Martin, J, Hardy, MSA, Usmani, AS & Forde, MC, (1998) Accuracy of NDE in bridge assessment, *Engineering Structures,* Vol 20, No. 11, 979-984.
- Menzies, J, (1999) Structural Risk During Extended Life, *Proc. 8th Int. Conf.: Structural Faults & Repair-99.* Engineering Technics Press, Edinburgh, CD-Rom: ISBN No: 0-947-64441-5

This paper was first published in *Current and future trends in bridge design, construction and maintenance,* by Das P.C., Frangopol D.M. and Nowak A.S. (eds), Thomas Telford, London, 1999.

Effect of corrosion on the ductility of reinforcement

L.A. CLARK, Y. DU and A.H.C. CHAN
School of Civil Engineering, The University of Birmingham, Birmingham, UK

INTRODUCTION

Reinforcement corrosion is a major cause of the deterioration of reinforced concrete structures. In recent years a considerable amount of research has been conducted on the strength of reinforced concrete members containing corroded reinforcement in order to provide guidance on the structural assessment of such members. Bending and shear capacity have been studied together with the effect of corrosion on bond strength. As a result assessing engineers are gaining confidence in the capacity of corroded structures. However, an area where there is still a lack of data is the effect of corrosion on the ductility of reinforcement per se and of reinforced concrete members. Ductility can be particularly important for corroded structures in seismic regions and/or where redistribution of elastic moments is necessary to justify structural capacity.

This paper gives an overview of the results of an investigation into the effects of corrosion on the ductility of bare reinforcing bars and of bars embedded in concrete cylinders. The effects of corrosion on the ductility of reinforced concrete beams were also studied, but are not reported here.

In all cases accelerated corrosion was induced in the laboratory by impressing a direct current on the reinforcement to be corroded. 3.5% sodium chloride solution was used as the electrolyte and the external cathodes were of stainless steel.

BARE BARS

Tension tests were performed on 75 bare reinforcing bars after they had been corroded to different extents. The variables were: bar type (plain and deformed); bar diameter (8, 16 and 32 mm) and amount of corrosion (0, 5, 10, 15 and 20% expressed as mass of bar). These amounts of corrosion were achieved in about 7, 14, 21 and 28 days, respectively, by impressing currents of 0.5, 1.0 and 2.0 mA/cm^2 on the 8, 16 and 32 mm bars, respectively.

The average amount of corrosion was determined by weighing the bars before and after corrosion. The variation of corrosion along the length of each bar was determined by progressively immersing the bar in water and measuring the volume of water progressively displaced. As a result the corrosion over 10 mm lengths of each bar could be determined. A typical corrosion distribution for an 8 mm plain bar with 20% average corrosion is shown in Figure 1. It can be seen that the corrosion attack penetration varied considerably along the length of a bar.

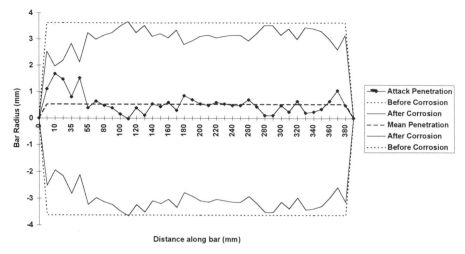

Figure 1   Typical corrosion distribution for R8 with 20% corrosion

The effect of corrosion on the force-extension curve of a reinforcing bar is shown in Figure 2 for a 16 mm plain bar. The effects illustrated were typical of all bars. It can be seen that the corrosion reduced the yield force, the ultimate force, the strain at maximum force, and the elongation.

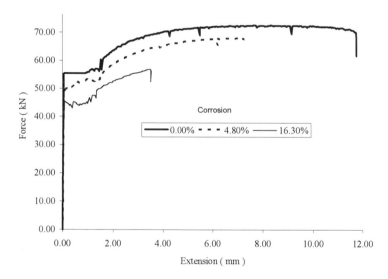

Figure 2 Effect of Corrosion on Force - Extension Curve of R16

An analysis of the results showed that the reductions in yield force and ultimate force were greater than those arising from the reduction of the average cross-sectional area of the corroded bar. The additional decrease was due to the concentration of stress caused by the local uneven corrosion attack penetration. It was found that the yield or ultimate force (F) of a corroded bar could be related to that (F_o) of an uncorroded bar as follows:

$$F = [1 - (0.01 + \alpha) Q_{corr}] F_o \qquad (1)$$

where Q_{corr} is the average percentage corrosion, and α represents the intensity of the local attack penetration.

Significance tests showed that, at the 5% significance level, α was the same for both yield and ultimate forces, and for either plain or deformed bars with diameters in the range 8 to 32 mm, and could be taken as 0.0043. Hence:

$$F = (1 - 0.0143 \, Q_{corr}) F_o \qquad (2)$$

A similar analysis of the yield and ultimate stresses of corroded bars (f) as a function of their uncorroded values (f_o) resulted in the relationship:

$$f = (1 - 0.0054 \, Q_{corr}) f_o \qquad (3)$$

If there were no effect of the local attack penetration then f would equal f_o. It has been suggested that the reason for f being found to be less than f_o is that the bar could contain an outer "shell" of stronger steel which is removed during the corrosion process leaving only the weaker interior material. This hypothesis was examined by reducing the cross-sectional areas of three 8 mm plain bars by 24% by machining. The ratios of the average yield and ultimate strengths of the machined bars to those of un-machined bars were 1.01 and 1.00, respectively. Hence, the observed reductions of yield and ultimate strengths of corroded bars were not due to the corrosion of an outer strong "shell". Furthermore, the results of the Authors' tests are in accord with those of Andrade et al [1] on 12 mm ribbed bars, as shown in Figure 3 for stresses.

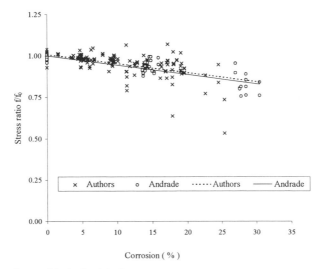

Figure 3 Comparison with Andrade's data

The effect of corrosion on bar ductility was determined in terms of (see Figure 4): (i) the ratio of ultimate to yield stress (f_u/f_y); (ii) the strain (ε_{sh}) at the onset of strain hardening; (iii) the ultimate strain (ε_u); (iv) the ductile area (A_d); and (v) elongation. It was found that corrosion had little effect on the ratio of ultimate strength to yield strength and on the strain at the onset of strain hardening. However, as a result of the uneven local attack penetration, corrosion had a significant effect on ultimate strain, ductile area, and elongation.

It was found from regression analyses that each of the latter ductility measures (ultimate strain, elongation and ductile area) were influenced by corrosion as follows:

$$D = (1 - 0.029 \ Q_{corr}) \ D_o \qquad (4)$$

where D is the corroded value of the ductility measure and D_o that of an uncorroded bar.

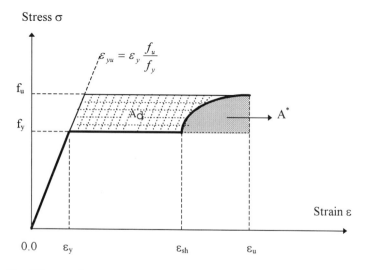

Figure 4   Ductility notation

It should be noted that 20% corrosion induced a 58% reduction of the ultimate strain of the uncorroded ribbed bars from about 0.12 to 0.05. The latter value is less than the value of 0.06 specified in Model Code 90 [2] for Class S reinforced which should be used where high ductility is required (e.g. in seismic regions). Hence, an assessing engineer should be wary of relying on moment redistribution in structures with reinforcement corroded to more than 20%.

Corrosion had a greater effect on the ductility measures than on either yield or ultimate stresses on forces.

The effects of corrosion on bar ductility observed in the Authors' tests can be compared with those of Andrade et al [1] performed on 12 mm ribbed bars corroded in either fresh water or 3% sodium chloride solution. However, Andrade's data permit only the stress ratio (f_u/f_y) and the elongation to be compared. Andrade's f_u/f_y data confirm that the stress ratio is not affected significantly by corrosion. With regard to elongation, as shown in Figure 5, Andrade et al observed a greater effect of corrosion on bars corroded in salt solution than those corroded in fresh water. Furthermore, their data from bars corroded in salt solution are reasonably close to those of the Authors corroded in salt solution.

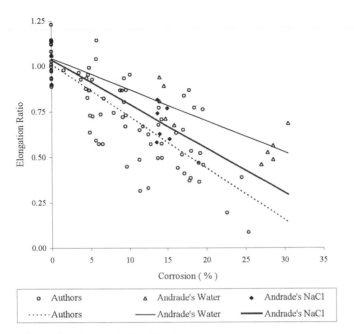

Figure 5 Comparison with Andrade's ductility data

Andrade's data suggest that more local attack penetration occurred with bars corroded in salt solution than those corroded in fresh water. However, they report that the bars corroded in salt solution showed a more general corrosion attack. This anomaly needs to be resolved.

BARS IN CONCRETE

Tension tests were performed on thirty 16 mm diameter plain (RC16) and ribbed (TC16) bars cast in 50 mm diameter concrete cylinders, with a 28 days compressive cube strength of about 35 N/mm². The concrete cylinders were placed within 60 mm diameter stainless steel tubes (cathodes) with an electrolyte of 3.5% salt solution. It was intended to test the specimens in tension at the same ages, and the same amounts of corrosion, as the bare bars described earlier. However, after corroding for more than 7 days, at which time about 5% corrosion had occurred as planned, it was found that the corrosion rate slowed significantly and corrosion eventually ceased. The reason for this was that the corrosion products initially dispersed into the concrete pores, then induced cracking of the concrete and then dissipated out of the concrete cylinder through the cracks and pores. They then started to fill the 5 mm annulus between the concrete cylinder and the stainless steel cylindrical cathode. The outcome was that a maximum corrosion of only about 8% was achieved for the bars cast in concrete rather than the 20% intended.

After corrosion, the cover concrete and corrosion products were removed from the bars, and it was found that deeper local corrosion attack penetration had occurred than was the case for the bare bars. It is suggested that this is due to the material components and the non-uniform pore distribution of the cover concrete resulting in less uniform corrosion inducing properties than the electrolyte around the bare bars.

In spite of the greater local attack penetration of the bars embedded in concrete, it was found that the effect of corrosion on the yield and ultimate forces and stresses, the stress ratio and the strain at the onset of strain hardening were, at the 5% significance level, the same as for bare bars. Hence, Equations 2 and 3 can be used for bars embedded in concrete. However, corrosion had a greater effect on ultimate strain, elongation (see Figure 6) and ductile area in the case of bars embedded in concrete than for bare bars, because of the greater local corrosion attack penetration of the former bars. It is necessary to replace Equation 4 with:

$$D = (1 - 0.044\ Q_{corr})\ D_o \tag{5}$$

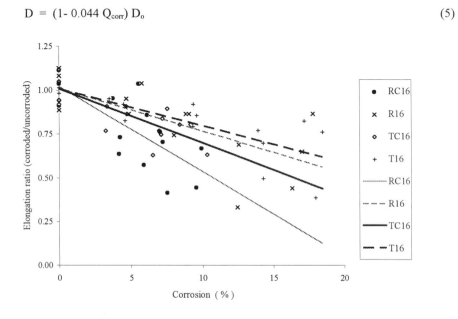

Figure 6 Effect of Corrosion on Bar Elongation

Hence, corrosion reduced the ductility measures of bars embedded in concrete about 50% more than bare bars. This suggests that at about 11% corrosion the ultimate strain could be reduced to less than the Model Code 90 Class S required value of 0.06.

The Authors' test results have been compared with those of Morinaga [3] for bars, extracted from actual structures, which had corroded due to chloride ingress. Morinaga found the constant 0.0143 in Equation 2 to be 0.0161 for yield strength and 0.0263 for ultimate strength, and the constant 0.044 in Equation 5 to be 0.0603 for bar elongation. In contrast Zhang [4] found values of 0.0104, 0.0105 and 0.0137, respectively, for bars extracted from carbonated structures. The latter bars would be expected to have less local attack penetration than either Moringa's or the Author's bars. Hence, although the laboratory data are in general accord with the field data, it is apparent that the nature of the corrosion attack has a significant influence on the extent to which corrosion affects the mechanical properties of bars embedded in concrete. Hence, caution is required when applying either laboratory or field data to a different corrosive environment.

CONCLUSIONS

1. Corrosion had an effect on the mechanical properties of reinforcement in excess of that due to the average reduction in cross-sectional area.

2. The yield and ultimate forces and stresses, the ultimate strain and elongation and the measure of ductility referred to as the "ductile area" were reduced by corrosion, but the ratio of ultimate to yield forces and stresses and the strain at the onset of strain hardening were not reduced.

3. Greater reductions of ultimate strain, elongation and "ductile area" occurred for bars embedded in concrete than for bare bars.

4. Regression analyses of test data resulted in equations to predict the mechanical properties of corroded bars. The predictions are in accord with the data of other researchers.

5. From comparisons of laboratory and field data it is apparent that the effects of corrosion on the mechanical properties of bars embedded in concrete are dependent on the corrosion conditions. Hence, care must be taken when applying either laboratory or field data to different situations.

6. Corrosion in excess of about 10% can reduce the ultimate strain of a reinforcing bar embedded in concrete to below the value required by Model Code 90 for high ductility situations. Hence, care should be taken when relying on moment redistribution in the assessment of structures with such corroded reinforcement.

REFERENCES

1. ANDRADE, C., ALONSO, C., GARCIA, D and RODRIGUEZ, J. Remaining lifetime of reinforced concrete structures: Effect of corrosion on the mechanical properties of the steel. Life prediction of corrodible structures, NACE, Cambridge September 1991, pp.12/1 - 12/11.

2. COMITE EURO-INTERNATIONAL DU BETON. CEB-FIP Model Code 90, 1991.

3. MORINAGA, S. Prediction of service lives of reinforced concrete buildings based on rate of corrosion of reinforcing steel. Special Report of Institute of Technology, Shimizu Corporation, Japan, No.23, June 1988, pp.82.

4. ZHANG. Private communication, 1995.

ACKNOWLEDGEMENT

The financial support for Mr Y Du from the British Cement Association is gratefully acknowledged.

Fatigue analysis of a cracked steel deck using measured stress spectra and full-scale laboratory tests

M. H. KOLSTEIN, Senior Research Engineer, and J. WARDENIER, Professor, Steel and Timber Structures, Faculty of Civil Engineering and Geosciences, Delft University of Technology, The Netherlands

SUMMARY
During visual examination of the condition of the surfacing of a seven years old bascule bridge cracks were found in the heaviest loaded traffic lane (1997). The cracks grew very fast and could cause a problem for the safety of the traffic. Preliminary repairs by grinding and filling the groove by a butt weld had to be carried out directly. At that time it was clear that further investigations were urgently needed to obtain information about this type of crack and to avoid unacceptable frequent repairs of this bridge. Further it had to be checked if this type of crack could occur or could be present in other bridges. Apart from a study of relevant literature, site measurements on the bridge as well as laboratory tests have been carried out to obtain information about stress spectra, stress distributions and fatigue strength of this specific detail. Finite element calculations as well as fracture mechanics calculations were done to verify the experimental observations. This paper highlights the fatigue analysis of the cracked steel deck using measured stress spectra and full-scale laboratory tests.

INTRODUCTION

Description of the bridge
The main span of the bascule bridge is 60 m and the total width with six traffic lanes is to 27 m. The bridge is shown in Figure 1. The shape of the longitudinal stiffener of the orthotropic deck structure is a so called Krupp profile FHK 2/325/6 with a structural height of 325 mm, a base distance between the outer side of the trough legs of 300 mm, bottom width of 105 mm and a plate thickness of 6 mm. The plate thickness of the cross beam web support of the continuous longitudinal stiffener is 10 mm. The surfacing on the 12 mm deck plate of the bascule bridge consisits of a 8 mm rather thin epoxy layer compared to thick wearing surfacings of 40 to 80 mm on fixed bridges. The number of trucks in the most heaviest loaded lane on this bridge amounts to be about 7000 trucks a day in 1997.

Description of the observed cracks
During visual examination of the condition of the wearing surface one longitudinal crack of 800 mm was found. The crack in the deck plate was located at the intersection of the continuous longitudinal trough stiffener and the crossbeam. As shown in Figure 2 and 5 the cracks initiated at the root of the stiffener-deck plate weld. They propagated through the total deck plate and wearing surface and grow in longitudinal direction parallel to the stiffener to deck plate weld. After detailed inspection by removing the wearing surface from the steel plate a second crack was found at the other side of the same longitudinal stiffener. These parallel cracks grow to each other. Further detailed inspection showed that more cracks

appeared to be present in the steel deck. Since the initiation of the crack was inside the trough no inspection from underneath was possible and they could also not be observed during regular visual inspection of the steel deck in the past. As long adjacent parallel cracks could result in a deep sag of the deck plate above the longitudinal trough, preliminary repairs by grinding and filling the groove by a butt weld had to be carried out directly. In order to obtain an insight in the traffic-induced stresses in the steel deck after repairing strain gauges were applied at several locations.

Fig. 1.The Van Brienenoord Bridges

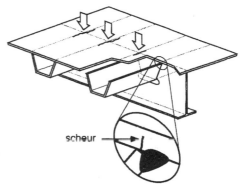

Fig. 2. Crack location and crack initiation [1]

INSTRUMENTATION

Most of the strain gauges on the existing bridge deck were positioned close to the welds between the deck plate and the longitudinal trough stiffeners in the offside wheel track of the heavy loaded slow lane (see Figure 3). The strain gauges at the topside of the deck plate (24 – 28 and 19 – 23) were applied after removing the wearing surface. All strain gauges measured the strain perpendicular to the welds.

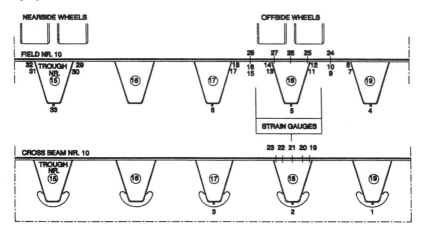

Fig. 3. Review location of strain gauges on the existing bridge

In this paper the strains measured at locations 19 and 23 on top of the deck plate and 25 mm from the root of the weld are of particular interest as they are positioned at positions close to the location of the observed cracks (see Figure 4 and 5).

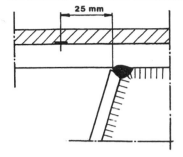

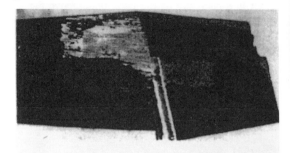

Fig. 4. Gauge position 19 and 23   Fig. 5.   Crack in the deck plate

MEASUREMENTS

Stress variations

One week measurements of stress variations under normal trafficing have been carried out. Typical examples of measured peak values at gauge position 19 and 23 are given in Figure 6. As shown compressive stresses as well as tensile stresses occur. Tensile stresses occur if the wheel load passes the cross beam in the centre line of the longitudinal trough stiffener. However the transverse position of the wheel loads deforms the deck plate in such a way that before and after passing the cross beam also compressive stresses occur at this particular point.

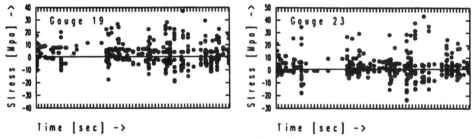

Fig. 6. Measured peak stresses at gauge position 19 and 23.

Stress spectra

The range pair cycle counting method has been used to convert the stress variations of a complex waveform into a sequence of identifiable cycles to enable Miner's rule to be used to calculate the fatigue damage. In Figure 7 the range pair counts are accumulated from the largest class to the lowest class. The results are divided by the total number of counts and plotted in modified relative frequency curves of range-pair counts. A one-week cumulative stress spectrum for gauge 23 is given in Figure 7. The total number of cycles larger than 5 MPa in this graph amounts to be $2,4.10^5$.

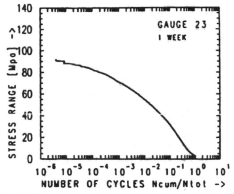

Fig. 7. Cumulative stress spectra gauge 23

Calibrated truck measurements

Influence lines are a well-known method of relating the load on the bridge deck to the stresses in the structural details. However to construct the influence lines from experiments, it is necessary to apply very concentrated loads of an appreciable magnitude at many points on the bridge deck. This presents a lot of practical problems. Instead a set of four points was used, i.e. a dual axle lorry with calibrated wheel loads. The test vehicle was placed in a number of fixed locations and at each position the stress at all the measuring points were recorded. To obtain an indication of the dynamic effect the lorry has been driven several times over the bridge. The lateral position of the lorry was measured by 10 mm wide measuring switches on the bridge deck. As an example the recorded analogue strain signals measured by strain gauge 21 and 23 for a vehicle speed of about 70 km/h are shown in Figure 8. The static and dynamic tests showed that depending on the location of the wheel load, compressive stresses or tensile stresses occur in the deck plate at the intersection of the longitudinal stiffener web and the continuous cross beam. The maximum measured static tensile stress under the front wheel load of 22 kN with a wheel print width of 200 mm and a length of 210 mm at location 23 was 37,7 MPa. The dynamic measurements resulted in an impact factor of about 1,20.

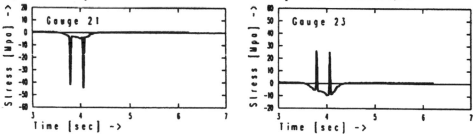

Fig. 8. Dynamic results of the calibrated truck measurements.

FATIGUE CALCULATIONS

Fatigue damage

Using the measured stress spectra fatigue calculations have been carried out as a function of the fatigue categories according to the Eurocode design rules. Two sets of fatigue calculations have been carried out. First calculations use fatigue strength curves with a cut-off limit at 10 million cycles. Secondly the fatigue strength curves have been applied assuming that there is no cut-off limit which means that also stress ranges beyond the cut-off limit contribute to the total fatigue damage (see Figure 9). In both cases the fatigue calculations have been made assuming that the measured one-week spectrum can be used directly to determine a year spectrum by multiplying the number of cycles by a factor 52. Up to class 90 the resulting calculated fatigue damage for strain gauge 23 appears to be almost the same for both definitions of the fatigue strength curve (see Figure 10). For higher fatigue strength classes the cut-off limit results in a lower fatigue damage.

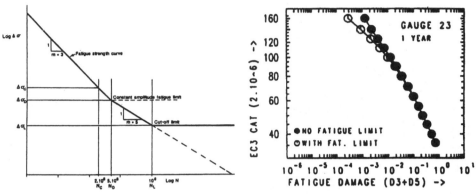

Fig. 9. Fatigue strength curve   Fig. 10. Fatigue damage at gauge pos. 23

Fatigue classification of the detail
Full scale laboratory tests have been carried on the welded detail as shown in Figure 2 and 4. The stress distribution of this structural detail as well as the fatigue strength at the top side of the deck plate have been determined [2]. To determine the fatigue design class and to evaluate the fatigue life of the joint the laboratory tests had to be combined with the site measurements on the bridge. Therefore a Stress Range Factor (SRF) has been calculated using the measured stress distribution on the laboratory specimens. In the analysis the SRF has been defined as the rato between the maximum stress and the stress 25 mm from the weld root (see Figure 12). Different wheel prints resulted in a maximum SRF of 4,07 and a minimum value of 2,90. The mean value of the SRF of 13 test results was 3,26.

The geometric stress of the detail was defined as the extrapolated stress at the weld root using the measured stress at a distance of 0.4 times and 1.4 times the thickness of the deck plate. Based on this stress the fatigue design classification according to the Eurocode appeared to be Class 124.

Fatigue design life
Using the measured spectra from Figure 6, the stress distribution from Figure 12 and a fatigue class of the welded detail equal to Class 124 several fatigue design life calculations have been carried out as shown in Table 1. In these calculations the measured stress spectrum has not been corrected for future changes in the traffic loading.

Table 1. Fatigue design life for different deck plate thicknesses

Deck plate thickness (mm)	12	18	20	22	24
Fatigue design life (yrs)	2.4	47	118	287	713

These results indicate very clear that the cracks as found on the bascule bridge could have been expected and a considerable thicker deck plate is required to obtain an acceptable fatigue design life.

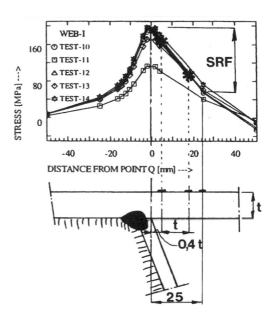

Fig. 11 Test set-up and test specimen

Fig. 12 Stress distribution at deck plate top-side

REPAIR PROCEDURE

To minimize traffic jams it was decided to replace the orthotropic bridge deck completely. Simulations including increase in traffic loads and frequency in the future [4] resulted for a design life of 50 years in a deck plate thickness of 28 mm in the heaviest loaded lane .

REMAINING QUESTIONS

Shear effect of the cross beam
A test specimen which has previously been used for a test in the past [10] to determine the stress distribution and the fatigue strength at the bottom side of the connection between the longitudinal trough stiffener and the cross beam was consired in more detail. In that research program a total of 3.5 million load cycles of 300 kN had been applied on this specimen. Non-destructive testing indicated cracks in the deck plate weld at the intersection of the stiffener web and the cross beam at locations which have not been directly loaded in the previous tests. From finite element calculations carried out [9] it can be seen that the shear deformations in the cross beam also effect the shape of the deck plate (see Figure 13). Additional finite element calculations showed that the shear deformations resulted in relative high peak stresses in the deck plate resulting in the cracks observed. This mechanism must be studied in more detail for different combinations of plate thicknesses.

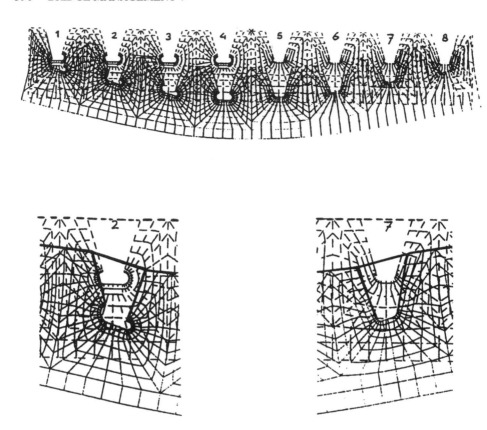

Fig. 13 Shear deformation in the cross beam affecting the stresses in the deck plate

Composite effect of the wearing course and the steel deck plate
The composite action between the steel deck plate of the orthotropic deck and the wearing course on the steel plate reduces the stresses considerable. This mechanism depends on the temperature and the condition of the wearing course. In the current European Standard only the dispersal plate of the concentrated loads (wheel contact area) through the wearing course and the steel deck is taken into account. The effect of the bending stiffness of the wearing course is still neglected. For the fatigue evaluation of existing structures this mechanism should be quantified in more detail and taken into account.

Repair solutions for deck plate cracks
This paper concentrated on the deck plate cracks of one particular movable bridge. Repair methods which decrease the stresses in the steel plate are limited since the thickness of the wearing course is only 8 mm. Except repairs by grinding and rewelding there is no experience for other repair methods. In case deck plate cracks occur in steel decks with a relative thick wearing course of 40 - 80 mm other methods of strengthening could be used and these should be studied and evaluated urgently.

CONCLUSIONS

Site measurements carried out on the orthotropic bridge deck of a heavyly loaded bascule bridge with a 12 mm steel deck and 8 mm epoxy on top confirmed the relative short fatigue life of the deck plate weld at the intersection of the continuous longitudinal trough stiffener and the crossbeam. The full scale laboratory tests showed relative high peak stresses at this point. The geometric stress being used in the fatigue calculation is the extrapolated stress at the weld root using the measured or calculated stresses at a distance of 0.4 times and 1.4 times the thickness of the deck plate. Based on this geometric stress the fatigue design classification at 2 million cycles according to the Eurocode gives 124 which nearly agrees with the class for the geometric stress approach of tubular joints (for 12 mm) [11].

The experience as described in this paper shows that there is a need to evaluate other existing structures and to reconsider the design of orthotropic bridge decks.

ACKNOWLEDGEMENT

The authors wish to thank the Civil Engineering Division of the Ministry of Transport and Public Works and Watermanagment for their permission to publish this paper.

REFERENCES

[1]   Van der Weijde, H.: *Repair Van Brienenoord Bridge Rotterdam*, Bouwen met Staal 146, 1999, Rotterdam, The Netherlands (In Dutch).

[2]   Kolstein, M.H., Wardenier, J., Van der Weijde, H.: *A New Type of Fatigue Failures in Steel Orthotropic Bridge Decks*, In Proc of Fifth Pacific Structural Steel Conference, 1998, Seoul, Korea.

[3]   Kolstein, M.H., Wardenier, J.: *Laboratory tests of the deck plate weld at the intersection of the trough and the cross beam of steel orthotropic bridge decks*, In Proc of the 2nd Eurosteel Conference, 1999, Prague, Czech Republic.

[4]   Vrouwenvelder, A.C.W.M.: *New Bridge Loadings*, In Syllabus Fatigue Steel Bridge Decks, Staalbouwkundig Genootschap, 1998, Rotterdam, The Netherlands (In Dutch).

[5]   Kolstein, M.H.: *Research Activities in Consequence of Observed Cracks in the Deck Plate of the Bascule Bridge – Van Brienenoord*, In Syllabus Fatigue Steel Bridge Decks, Staalbouwkundig Genootschap, 1998, Rotterdam, The Netherlands.

[6]   ENV 1993-2, *Eurocode 3: Design of Steel structures – Part 2: Steel bridges*, European Committee for Standardization, 1997, Brussels, Belgium.

[7]   Kolstein, M.H., Wardenier, J.: *Stress Reduction due to Surfacing on Orthotropic Steel Decks*, In Proc IABSE Workshop - Evaluation of Existing Steel and Composite Bridges, 1997, Lausanne, Switzerland.

[8]   Kolstein, M.H., Wardenier, J.: *Evaluation of Recently Observed Fatigue Cracks in the Stiffener to Deck Plate Joint of Orthotropic Bridge Decks*, 1999, Singapore.

[9]   Leendertz, J.S., Kolstein, M.H., Wardenier, J.: *Numerical Analyses of the Trough to Crossbeam Connections in Orthotropic Steel Bridge Decks*, In Proc Nordic Steel Conference, 1995, Malmö, Sweden.

[10]   Kolstein, M.H., Leendertz, J.S., Wardenier, J.: *Fatigue Performance of the Trough to Crossbeam Connections in Orthotropic Steel Bridge Decks*, In Proc Nordic Steel Conference, 1995, Malmö, Sweden.

[11]   Van Wingerde, A.M., Van Delft, D.R.V., Wardenier, J., Packer, J.A.: *Scale Effects on the Fatigue Behaviour of Tubular Structures*, In Proc IIW Int Conf Performance of Dynamically Loaded Welded Structures, 1997, San Francisco, USA.

Orthotropic steel bridges: management tools for live load and fatigue assessment

DEMPSEY, A.T. Department of Civil Engineering, University College Dublin, Ireland.
KEOGH, D.L., Department of Civil Engineering, University College Dublin, Ireland.
JACOB, B. Laboratoire Central des Ponts et Chaussees, Paris, France.

1. Introduction

Orthotropic steel decks were first conceived in the early 1930's for use in moveable bridges, but since then, their use has become more widespread typically, in long span bridges (Heins & Firmage, 1979). The steel plate is supported by longitudinal stiffeners, which span between transverse crossbeams. The advantage of this system over conventional bridge design at the time was that a light steel deck plate replaced the original concrete deck slab, thus reducing the dead weight of the bridge in comparison to a typical concrete deck. Another advantage of this type of construction is that it has the ability to act compositely with the main girders, transverse beams and stiffeners. In long span bridges, the bending moment due to the dead load comprises a large portion of the design moment. Therefore, a reduction in the dead load can significantly affect the economy of the bridge. As the bridge span reduces, the influence of the dead load moment in the design criteria also reduces and thus the advantage of orthotropic deck design over conventional bridge design reduces. In general, an orthotropic deck tends to be more economical than a concrete deck for bridge spans in excess of 100m.

As the ratio of live load to dead is greater for orthotropic bridges than for other highway bridges and the fact that there are numerous weldings (especially between the longitudinal stiffeners and the steel plate) on the structure, means that these bridge types are highly sensitive to traffic loads (Bignonnet et al., 1990). It has generally been found that bridges of this type fail under fatigue loading rather the ultimate load. Therefore, in the design and assessment of such bridges, the knowledge of traffic loads is important in predicting accurately the fatigue lifetime of such bridges. The collection of this traffic load data is described in section 2 where the development of a novel Bridge Weigh-In-Motion system (B-WIM) for orthotropic bridges is described. Normally, in fatigue assessment or design, the dynamic response of the bridge is accounted for by including a dynamic amplification factor in the load effect (response of the bridge). This paper also includes recommendations for a dynamic model of an orthotropic deck, so that the dynamic load effect of the bridge can also be calculated.

2. Fatigue Design and Assessment – Bridge Loading Codes

In the design and assessment of these orthotropic bridges, fatigue load models in either national standards such as BS 5400 (BS 5400 Part 10) or Eurocode 1 part 3 "Traffic Loads on Bridges" (ENV 1991-3, 1995) are used. Such conventional load models are designed and calibrated in order to provide stress variations on a large set of bridges and bridge details sensitive to fatigue that are similar, in terms of fatigue damage to those produced by real

Bridge Management 4, Thomas Telford, London, 2000.

traffic flows (Kretz & Jacob, 1991). When comparing the national codes, four different fatigue check procedures were found (Jacob, 1998):

1. Comparison of the maximum stress produced by a conventional load model with the fatigue strength of the detail considered,
2. Fatigue life or cumulative damage calculation using a simplified load model (fictitious truck or small set of trucks) and Miner's rule,
3. Fatigue life or cumulative damage calculation using a precise number of different truck types, which represent a realistic mix of real traffic and Miner's law,
4. Fatigue life or cumulative damage calculation using detailed traffic records and influence lines to generate stress variations and Miner's law.

Eurocode 1 part 3 provides a total of five fatigue load models, including all of the above fatigue check procedures. In designing such bridges fatigue load models 1 to 3 are normally used, e.g., the construction of the orthotropic steel box girder bridge over the river Foyle (Wex et al., 1984).

For the assessment of a particular orthotropic deck bridge, it should be borne in mind that for Eurocode 1.3, the first four load models were designed by using traffic details from a heavily trafficked French motorway (Jacob & Kretz, 1996). For the assessment of a particular bridge, which is less trafficked than these French motorway conditions, the fatigue lifetime calculation will indicate a shorter lifetime than is actually the case. Conversely, it is possible that the fatigue lifetime of a bridge could be overestimated for a bridge which was subject to particularly heavy truck loads. In order to allow for this eventual possibility, as stated above, Eurocode 1 part 3, fatigue load model 5 allows for the fatigue lifetime calculation for a particular bridge for traffic conditions on that bridge. Section 4 of this paper describes this process of using detailed traffic records for a particular bridge to assess the fatigue lifetime. These traffic loads would normally be calculated from pavement WIM sensors, but for the case of an orthotropic deck, the pavement is thin (due to limits on the dead load) and therefore, the installation of pavement sensors is practically impossible. Pavement sensors would have to be installed prior to the bridge on a flat portion of the road, and as the approach spans tend to be quite long, the pavement WIM system might have to be installed several kilometres away from the bridge. The WIM statistics regarding composition of traffic flow and headway, etc. could be very different at the pavement WIM station than on the bridge. It was with this point in mind, that the development of a B-WIM system for orthotropic decks was developed. If the static axle and gross vehicle weights as well as the other relevant truck parameters are calculated at a point where the bridge is being assessed for fatigue, this removes any errors or uncertainties relating to the composition of the traffic. This is the obvious advantage of a B-WIM system installed on the orthotropic deck, as opposed to using a pavement system installed prior to the bridge or indeed a set of traffic measurements for an entirely different road.

3. Orthotropic Bridge Weigh-in-Motion for Calculation of Live Load

B-WIM systems were developed in the late 1970's by Moses and others (Moses,1979) in the USA and by Peters in Australia (Peters, 1984). The basic principle of these systems is that the bridge is used as a large scales to weigh trucks. Strain gauges or transducers are attached to the soffit of the bridge beams, girders or slabs and axle detectors are installed on the pavement to provide information on the truck classification, axle spacings and velocity, which are needed by the B-WIM algorithm to calculate the axle and gross vehicle weights. The B-WIM algorithm itself is an inverse calculation in which the response of the structure (bending moment) is measured and the axle loads, which have produced it, are calculated. In matrix form, the inverse problem can be stated as the following equation:

$$F.A = M$$

where F is the matrix of influence line ordinates, A is the vector of axle loads and M is the vector of the measured moments. This, in essence, has been the concept behind all B-WIM algorithms developed prior to the commencement of the WAVE project (Weighing-in-motion of Axles and Vehicles for Europe) (Jacob & O'Brien, 1996). The WAVE project (1996-1999) consisted of four work packages of which work package 1.2 was concerned with B-WIM systems. The main objectives of this work package was to improve the accuracy of the calculated axle and gross vehicle weights by using new and improved algorithms, extending B-WIM to other bridge types and to improve the durability of the systems. The extension of B-WIM to other bridge types included orthotropic steel decks. One of the requirements in the development of such a system was that axle detectors were not allowed on the road surface. This has advantages in the fact that the system is completely undetectable from the road surface, installation of the system does not require any lane closure and the durability of the system is greatly increased. The weakest link, in terms of durability of the classical B-WIM systems was always the axle detectors (O'Brien et al., 1999). The fact that there were no axle detectors for this system meant that all of the truck parameters had to be determined from the measured strain underneath the bridge. Therefore, in order to calculate accurately the truck parameters, i.e, vehicle occurrence, number of axles, axle spacings and vehicle velocity, an Orthotropic Bridge Weigh-In-Motion (OB-WIM) system was developed. A brief description of this algorithm is now described. A complete description can be found in Dempsey et al. (1998 and 1999). The algorithm requires bridge instrumentation at three different longitudinal sections. The strain gauges (which only measure longitudinal strain) are placed on the soffits of the longitudinal stiffeners. The OB-WIM algorithm consists of two parts: firstly a vehicle event, the number of axles, the velocity and the axle spacings are calculated by the Free of Axle Detector (FAD) algorithm (Znidaric et al., 1999). This algorithm was found to be satisfactory in determining truck events. The number of axles on each truck and the calculated velocities and axle spacings were found to be sufficiently accurate for use in the optimisation algorithm which calculates the other truck parameters, i.e., axle weights. This optimisation algorithm (which can be based on a 1- or 2-dimensional bridge model) minimises an objective function to calculate the following truck parameters, velocity, axle spacings, axle weights and transverse location of trucks in lanes), using the input from the FAD algorithm (number of axles, velocity, axle spacings and transverse location) as initial guesses for the optimisation algorithm. This objective function can be described as the sum of squares of differences of the measured and modelled bridge responses. It is a well-known fact that the speed and convergence of optimisation algorithms is greatly increased, if a good initial guess of some (or all) of the optimisation parameters is known (Rao, 1984). The optimisation routine used is Powell's method or the conjugate directions method. This optimisation algorithm for the two-dimensional bridge model can be defined as:

With initial parameters estimates from the FAD algorithm:

$$y_0 = \left\{ n^*, v^*, L_1^*, L_2^*, \ldots L_{n-1}^*, t^* \right\}$$

and assuming: $A_1 = A_2 = A_n = 30kN$

Minimise :

$$O_{2D}(y) = \sum_{j=1}^{S} \sum_{x=1}^{K} [M_j(x) - M_j^{'}(x)]^2 + k|y - y_o|^2$$

where:
$$M_j(x) = A_1 I_j(x) + A_2 I_j(x) + \ldots\ldots + A_n I_j(x)$$

to find:
$$y = \{v, L_1, L_2, \ldots L_{n-1}, A_1, A_2, \ldots A_n, t\}$$

where

n is the number of axles,

y_o is the initial estimate of the parameters from the FAD algorithm,

n is the number of axles on the truck,

v is the velocity,

$L_1, L_2, \ldots L_{n-1}$ are spacings for a truck with n axles,

t is the transverse location of the truck,

$A_1, A_2 .. A_n$ are the axle weights for an n axle truck,

$O_{2D}(y)$ is the objective function for a 2-dimensional bridge model,

S are the number of instrumented longitudinal stiffeners,

K are the number of strain readings for a truck crossing,

$M_j(x)$ is the modelled bridge bending moment,

$M'_j(x)$ is the measured bridge bending moment,

$I_j(x)$ is the influence line ordinate for the jth longitudinal stiffener when the first axle is at a position x from some specified position on the bridge,

y are the optimisation parameters,

* indicates that the parameter has been calculated from the FAD algorithm,

This algorithm was tested extensively with experimental field tests, to determine the suitability of such an algorithm in determining truck events and trucks parameters. The accuracy of the algorithm in calculating axle and gross vehicle weights was determined by comparing the calculated WIM weights to the static axle weights, which were determined by randomly stopping trucks from the traffic flow and weighing their axles using a static portable scales. This test was conducted in conjunction with the Continental Motorway Test (CMT), which was a part of the COST 323 Action "Weigh-in-motion of Road Vehicles", (Stanczyk & Jacob, 1998). Figure 1(a) and (b) illustrate the accuracy of the system in calculating the axle and gross vehicle weights.

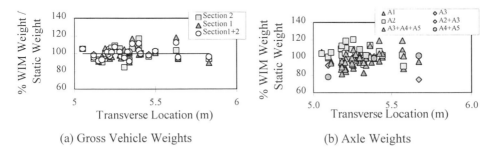

(a) Gross Vehicle Weights   (b) Axle Weights

Figure 1 - Accuracy of the OB-WIM algorithm in calculating weights (A1=Axle 1, etc.)

The accuracy class for this system, according to the COST 323 specification for WIM systems was found to be C(15). This means that for gross vehicle weights approximately 95% of the calculated weights (the exact percentage depends on the environmental and test conditions) are within ±15% of the static weights (Jacob & O'Brien, 1998).

4. Design and Assessment of Orthotropic Decks for Fatigue

As stated above, the codes give an option whereby the assessor can use site specific traffic loads to provide a more accurate prediction of the true fatigue lifetime of the bridge or the bridge detail. This section describes how detailed traffic records are used to assess the fatigue damage and remaining lifetime of a bridge or bridge detail.

One of the peculiarities of fatigue damage is that it is not related to the extreme traffic loads, i.e., the tail of the gross vehicle weight distribution curve but to the cumulative effects of the whole operating traffic loads, considering weights, geometry and frequency of occurrence. The ultimate load or extreme load case is concerned only with the small percentage of trucks in the tail of the gross weight distribution curve. Therefore, the first step in this fatigue assessment using detailed traffic records is the collection of traffic data. These traffic details are normally determined from weigh-in-motion (WIM) systems installed in the road surface or on bridges. WIM systems provide continuous traffic information, in particular, truck and vehicle classification, axle and gross vehicle weights, axle spacings, velocity, inter–vehicle distances and other traffic parameters. Ideally, this WIM data should be measured over a long time period, e.g., one year, however due to certain limitations with memory sizes of WIM systems and the resources necessary to deal with such large data files, this is not always possible. Instead, WIM data from a shorter traffic measurement period are used, but these must nevertheless be representative of longer ones. Therefore, measurement periods of shorter than a week should not be used, because of the weekly periodicity and weekend effect. Also, the choice of week in the year should be chosen carefully so as to be representative of a typical weekly traffic (Jacob, 1998). When the weekly traffic has been chosen, it is also necessary to conduct checks to determine; (1) if the WIM system operated continuously for the entire period and (2) if the ratio of unidentified trucks or measurement errors to the total number of trucks is minimal. The authors suggest that the OB-WIM system should be used to obtain these detailed traffic data files. Once, the weekly traffic file has been prepared and verified, the stress history for the detail required on the bridge is calculated using the influence line and the traffic data. Then, this stress history is converted into a statistical description of stress variations such as a stress amplitude histogram, using a cycle counting procedure, i.e., commonly the 'rainflow' counting method (Chabrolin, 1989). Various different methods of calculating the fatigue damage from this stress amplitude histogram may then be used, such as Miner's rule deterministic approach and probabilistic approach or fracture mechanics, deterministic or probabilistic approaches (Jaocb, 1998). Only Miner's rule deterministic approach is discussed here, as it is recommended for use in the national codes and the Eurocode.

Fatigue Damage Using Miner's Rule
Miner's rule states that the conventional damage d_i induced by a stress cycle of a given range $\Delta\sigma_i$, is proportional to the mth power of $\Delta\sigma_i$:

$$d_i = \Delta\sigma_i^m / C_m$$

where C_m is a constant the value of which depends on the value of m. Generally, for steel used in structural engineering and depending on the intensity of $\Delta\sigma_i$, m is between 3 and 5. The main hypothesis of this rule is that the individual damages of d_i of a series of stress cycles $\Delta\sigma_i$ during a reference period T, are independent of the order and time of occurrence of these stress cycles and the accumulated damage d_T is equal to the sum of the individual damages d_i (linear damage accumulation), (Jacob, 1998), i.e.,

$$d_T = \sum_{i=1}^{K} \frac{n_i}{N_i} = \left(\frac{n_1}{N_1} + \frac{n_2}{N_2} + \dots + \frac{n_K}{N_K} \right)$$

where

n_i are the specified number of repetitions of the various stress ranges which occur in the design life of the structure,

N_i are the corresponding numbers of repetitions to failure for the same stress ranges, obtained from the S-N curves.

The fatigue resistance of a detail depends on the geometry and quality of the weld and it is modelled by an 'S-N' curve which relates the number of cycles leading to failure, N_i, and the constant amplitude stress cycle $\Delta\sigma_i$. Conventional S-N curves can be found in the literature and in Eurocode 3 (ENV 1993-2, 1997) and the British Standard (BS 5400: Part 10, 1980).

The failure of a bridge detail is conventionally determined when the total accumulated damage equals unity. Therefore, if the time history of the stress variations during the reference period T, is representative of the whole stress variation time history (assumed to be stationary) for the duration of the bridge lifetime, this fatigue lifetime, D, is defined as:

$$D = \frac{T}{d_T}$$

Miner's rule is easy to implement either using recorded stresses or calculated stresses from detailed WIM data using a simulation programme, such as CASTOR-LCPC (Eymard & Jacob, 1989). In this simulation, the influence lines are calculated considering only the static response. The effect that these moving vehicle loads have on the stress variations due to the dynamic response of the bridge subject to moving loads is not considered. If the dynamic response of the orthotropic deck is to be considered, a dynamic orthotropic bridge model would have to be used. The development of a simplified procedure for calculating the dynamic response of orthotropic decks is given in the next section.

5. Recommendations for Dynamic Modelling of Orthotropic Bridge Decks.

The selection of the model type has a bearing on the amount of effort required for both modelling and interpretation of results. Many orthotropic bridges lend themselves to three dimensional models as the orthotropic plate is at a different (vertical) level to the longitudinal and transverse beams. This can be allowed for by modelling them seperately, at different levels, and connecting the two levels with rigid vertical members. Traditionally, two dimensional models were preferred as they were more economical on computer resources, but this is no longer an issue. The choice of model type will depend largely on the amount of effort required to determine the member properties and to interpret their output. The properties of the members in the three dimensional model are often easier to determine, as each part of the bridge, beams, plate etc., are individually discretised. For the two dimensional model, a member may represent, for example, a longitudinal beam and a portion of plate. Not only has the quantity of plate to be decided upon, but the resulting property is more difficult to determine. This should take account of shear lag effects (Hambly, 1991). Interpretation of results may be more labourious for a three dimensional model as there are more members and more degrees of freedom at each node. For the two dimensional model, it may be difficult to determine exactly what portion of the structure should be designed to resist the predicted moments.

With dynamic modelling it is the natural frequencies and mode shapes which are of primary concern. If the user is not concerned with design moments at this stage, a three-dimensional model may be the most appropriate. Such a model may lend itself to an easier description of both the stiffness and weight distribution of the bridge. In general the three dimensional model should resemble the real bridge as much as possible. This will mean locating longitudinal and transverse members and the plate at the same level as the centroids of the parts they represent. These should be connected at their ends by 'rigid' vertical members. Care should be taken to provide sufficient rigid vertical beams such that members on different levels behave compositely and do not bend individually between vertical beams

The plate is referred to as orthotropic because it possesses different stiffness in the two orthogonal directions. This is achieved by the provision of some form of plate stiffener in the longitudinal direction only. The plate is 'geometrically orthotropic' as it is the geometry that gives it different stiffnesses in the two directions. Most commercial finite element programs do not incorporate elements, which allow for modelling of this type of plate directly. Alternatively, they use 'materially orthotropic' elements. These assume the same stiffness in both orthogonal directions by adopting a single thickness for the plate. The orthotropy is accounted for by allowing the specification of different modulii of elasticity in the two directions. For dynamic modelling, it is important that the weight of the model be accurately allowed for as well as the stiffness. For this reason, it is convenient to choose the depth of the elements so as to give the correct cross sectional area of the plate (and hence correct weight). The modulus of elasticity in both the longitudinal and transverse directions can then be adjusted to give the correct EI values (and hence correct stiffness). Further information on this technique is given in O'Brien and Keogh (1999). Great care should be taken to ensure the correct weight of the bridge is used in the model. The weight of road surfacing, footpaths and other additions such as parapets should be estimated as accurately as possible as they may account for 30% or more of the dead weight of the bridge (Keogh and Dempsey 2000).

Once the natural frequencies and mode shapes of the bridge have been calculated, theses are used to calculate the dynamic response of the bridge by the method of modal superposition (Clough & Penzien, 1993), where the response of the bridge is calculated by superimposing the response of the different modes to the applied loading, in order to obtain the displacement-time history of the bridge.

6. Conclusions

This paper describes some of the management tools that have been developed in recent years to assess orthotropic steel deck bridges for fatigue and live load assessment. Such bridges tend to be long span so the ratio of live load to dead load is greater then for other bridge types. Therefore, the fatigue life of these bridges are highly sensitive to traffic loads. Therefore, a B-WIM system for orthotropic decks is described where the actual traffic loads on a particular bridge can be calculated. From, the calculated traffic loads on the bridge, the bridge or bridge detail can be assessed for fatigue damage, using the calculated traffic, influence lines and Miner's law. The effect of dynamics on the fatigue damage can also be calculated if a dynamic orthotropic bridge deck model is used instead of influence lines to calculate the stress-time history of the bridge. Therefore, recommendations for the dynamic modelling of orthotropic decks are also given.

References
- Bignonnet,A., Carracilli, J. & Jacob,B. (1990), 'Comportement en fatigue des ponts metalliques application aux dalles orthotropes en acier' Rapport final, Commission des Comunautes Europeennes, recherche CECA No. 7210 KD/317 (1986-1989) LCPC - IRISD.

- BS5400 : Part 10 (1980) 'Steel, concrete and composite bridges: part 10. Code of practice for fatigue', British Standards Institution.
- Chabrolin, B. (1989), 'Comptage de cycles de fatigue: methode de la goutte d'eau', *Construction Metallique*, No. 4 pp. 73-84.
- Clough,R.W. & Penzien, J. (1993), *Dynamics of Structures*, 2nd Ed., McGraw Hill, Inc, New York.
- Dempsey, A.T., Jacob, B. and Carracilli, J. (1998), 'Orthotropic bridge weigh-in-motion for determining axle and gross vehicle weights' *Pre-Proc. of the 2nd European Conference on weigh-in-motion of road vehicles*, Lisbon, eds. E.J. O'Brien and B. Jacob, pp 435-444.
- Dempsey, A.T., Jacob, B., Carracilli. J., (1999), 'Orthotropic Bridge WIM for determining axle and gross vehicle weights', *Weigh-in-Motion of Road Vehicles: Proceedings of the Final Symposium of the project WAVE (1996-1999)*, ed. B. Jacob, Paris, pp 227-238.
- Eymard, R. & Jacob, B., (1989), 'Un nouveau logiciel : le programme CASTOR pour le Calcul des Actions et des Sollicitations du Traffic dans les Ouvrages Routiers', *Bull. Liaison des LPC*, No. 164, pp 64-77, novembre-decembre.
- ENV 1991-3 (1995), *Eurocode 1: Basis of Design and Actions on Structures*, Part 3: *Traffic loads on road bridges – Assessment of various load models*, final draft August 1994, revised May 1995.
- ENV 1993-2 (1997), Eurocode 3, *Design of steel structures*, part II, *Steel bridges*, October.
- Hambly, E. C. (1991), *Bridge Deck Behaviour*, E&FN Spon, London.
- Heins, C.P. and Firmage, D.A. (1979), 'Design of Modern Steel Highway Bridges', John Wiley and Sons, New York.
- Jacob, D. (1998), 'Application of weigh-in-motion data to fatigue of road bridges', *Pre-proceedings of the 2nd European Conference of Weigh-in-Motion of Road Vehicles*, Lisbon, Portugal, pp 219-229.
- Jacob, B. and Kretz, T., (1996), 'Calibration of bridge fatigue loads under real traffic conditions', *Proceedings of the IABSE Colloqium on Eurocodes*, Delft, pp 479-487.
- Jacob, B. and O'Brien, E.J.,(1996) 'WAVE - A European research project on weigh-in-motion', *NATDAQ 96 - National Traffic Data Acquisition Conference*, Alliance for Transportation Research, Albuquerque, New Mexico.
- Jacob, B. and O'Brien, E.J., (1998), 'European specification on weigh-in-motion of road vehicles (COST 323)', *Pre-proceedings of the 2nd European Conference of Weigh-in-Motion of Road Vehicles*, eds. E. J. O'Brien and B, Jacob, Lisbon, Portugal, pp 171-184.
- Keogh, D. L. and Dempsey, A. T., (2000), 'Numerical Simulation and Modelling of Orthotropic Steel Decks', *Proceedings of Abnormal Loading on Structures, Joint I Struct E / City University Conference, London, April 2000*, in press.
- Kretz, T., and Jacob, B., (1991), 'Convoi de fatigue pour les ponts routes mixtes', *Construction Metallique*, No. 1.
- Moses, F.,(1979) 'Weigh-in-motion system using instrumented bridges', *Transportation Engineering Journal* ASCE, 105, TE3, 233-249.
- O'Brien, E. J. and Keogh, D. L. (1999), *Bridge Deck Analysis*, E&FN Spon, London.
- O'Brien, E. J., Znidaric, A. and Dempsey, A. T., (1999) 'Comparison of two independently developed bridge weigh-in-motion systems', *Heavy Vehicle Systems, A Series of the Int. J. of Vehicle Design*, Vol. 6, Nos 1/4.
- Peters, R.J., (1984) 'AXWAY - A system to obtain vehicle axle weights', *Proceedings 12th Australian Road Research Board Conference*, 12, 1, pp 17-29.
- Rao, S.S.,(1984) *Optimization: Theory and Applications*, 2nd ed., John Wiley & Sons.
- Stanczyk, D. and Jacob, B.(1998), 'European test of weigh-in-motion systems: Continental Motorway Test (CMT) on the A31 motorway', *Pre-proceedings of the 2nd European Conference of Weigh-in-Motion of Road Vehicles*, eds. E.J. O'Brien and B, Jacob, Lisbon, Portugal, pp 389-398.
- Wex, B.P., Gillespie, N.M., and Kinsella, J. (1984), 'Foyle bridge: design and tender in a design and build competition', *Proc. ICE*, Part 1, May 1984, 76, pp 363-383,
- Znidaric, A., Dempsey, A.T., Lavric, I. and Baumgaertner, W., (1999), 'Bridge WIM without axle detectors', *Weigh-in-Motion of Road Vehicles: Proceedings of the Final Symposium of the project WAVE (1996-1999)*, ed. B. Jacob, Paris, pp 101-110.

Analyses of bonded, ring-separated and strengthened masonry arch bridges

Eur Ing S. K. SUMON, BEng (Hons), CEng, MIMechE, Principal Engineer, Infrastructure Transport Research Laboratory, Crowthorne, Berkshire, UK.
Professor C. MELBOURNE, BEng, PhD, CEng, FICE, FIStructE, Department of Civil and Environmental Engineering, University of Salford, Salford, Gtr. Manchester, UK.

ABSTRACT

The European Commission Directive 96/58/EEC, which seeks to harmonise transport policy with regard to heavy traffic has resulted in the introduction, since January 1999, of the 40 tonne lorries on British roads. This has meant that many highway arch bridges require weight restriction, interim measures, strengthening or replacement.

Three arches that had been load tested to failure at the Transport Research Laboratory (TRL) were modelled using finite element (FE) analyses. The first arch modelled was fully bonded, the second ring-separated and the third ring-separated and strengthened. All models were three-ring-brick arches. The development of the two-dimensional FE models and the analyses carried out using the test data to calibrate analytical models are described. A good correlation was obtained between the tested arches and the FE models developed.

1. INTRODUCTION

There are many thousands of masonry arch bridges in the UK and abroad. The earliest in existence are medieval and the amount and weight of traffic they now carry has increased substantially since they were built. It is important that they continue to function because it would be neither practical nor desirable to replace them. The cost would be very large.

For arches that cannot carry modern day traffic loads a number of strengthening methods are available. The methods used vary widely and their relative benefits need to be quantified. Some of the strengthening methods have been in use for many years (Page, 1996). However, the increase in load capacity that they provide has never been experimentally tested or theoretically modelled. This paper describes full-scale testing and finite element (FE) analyses of a fully bonded arch, a ring-separated arch and a ring-separated and strengthened arch (Sumon, 1999). The analyses were carried out using commercially available software (LUSAS). The test data from previous tests was used in the calibration of the FE models. The results were then compared with the tested arches.

2. FULL-SCALE TESTING

A full-scale test frame was constructed to enable a series of 5 m span, 2 m wide, three-ring-brick model arches to be built and load tested to failure. The arches to be constructed and tested differed in two ways from a complete bridge. There were no spandrel walls and no pavement, which had been left out to reduce the number of parameters being studied. The backfill was retained by a steel box, which was designed not to restrain movement of the arch ring. A hydraulic jack was used to apply a load at the centre of a knife-edge (2 m x 0.3 m x 0.3 m); positioned over the three quarter-span (load-line), see Fig. 1. All arches were constructed and tested under controlled environmental conditions.

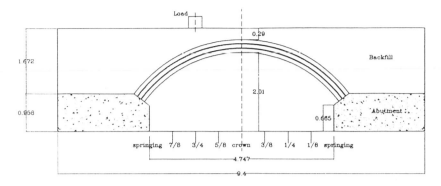

Fig. 1: Typical section through each arch tested and modelled

The salient points from the three arches tested are given below and a summary of test results in given in Table 1.

Fully bonded arch (A1) - this arch consisted of three-rings bonded using traditional mortar. The mortar consisted of cement, lime and sand with a ratio of 1:3:12 to give a low compressive strength.

Ring-separated arch (A2) - this arch was the same as A1 except it was built with a layer of damp sand between the three rings to simulate ring-separation. The purpose of the damp sand was to allow the rings to move independently with minimum resistance in the circumferential direction. Ring-separation is a common defect found in many old masonry arches and much testing has been done in this area (Melbourne et al, 1990, 1995).

Ring-separated and strengthened arch (A3) - after arch A2 had been tested to failure it was restored to its original shape, repaired and then strengthened using 6 mm diameter stainless steel bar reinforcement. The reinforcement was installed in the circumferential, radial and transverse directions by bonding it into slots cut into the soffit of the arch, using a special adhesive (Sumon, 1997).

TESTS TO FAILURE	Load to failure (kN)	Displacement at maximum load applied at the three-quarter span (mm)
A1: Fully bonded	242	7.5
A2: Ring-separated	200	19.4
A3: Ring-separated and strengthened	276	15.7

Table 1: Summary of load tests to failure

3. FINITE ELEMENT (FE) ANALYSES
3.1 MODELS
Three FE models were developed; a bonded model (M1), a ring-separated model (M2) and a ring-separated and strengthened model (M3). In each case, the two-dimensional FE model was represented by an elasto-plastic Mohr-Coulomb constitutive model and split into different components as listed in Table 2 and briefly described, below.

Brick/blocks - the three rings (top, middle and bottom in Fig. 2a) of the arch barrel were split radially into blocks and bonded with the surrounding mortar elements. Each block represents several bricks along the length of each ring.

Fig. 2a: Plot of Brick/block arrangement for arch barrel

Mortar-x1 - the joint in · the circumferential direction represents the material between the Backfill and the top ring. In the experiments, a polythene sheet was partly inserted here to reduce friction.

Mortar-x2 - the joints in the circumferential direction between the three rings (Fig. 2b).

Mortar-y - the joints in between the bricks rings in the radial direction (Fig. 2b).

Fig. 2b: Plot of Mortar-x2 (or Sand) and Mortar-y

Sand - the circumferential mortar joints described under Mortar-x2 were substituted with assumed typical Sand properties to represent ring-separation (Fig. 2b), as it was not possible to test them.

Steel-x - the stainless-steel reinforcement in the circumferential direction was incorporated into the middle of the bottom ring in M3. The reinforcement was placed along the complete length of the bottom ring and anchored at both ends into both concrete abutments (Fig. 2c)

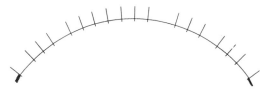

Fig. 2c: Plot of Steel-x and Steel-y

Steel-y - the stainless steel reinforcement in the radial direction (Fig. 2c) was inserted in the radial Mortar-y joints (Fig. 2b) and fully bonded in the matrix. The radial reinforcement (pins) was deliberately put at Mortar-y to give worst case scenario.

Abutments – elastic concrete bases either side of the arch, as shown in Fig. 1.

Backfill - a granular material was used on top of the arch, see Fig. 1.

Knife-edge - steel I-beam positioned at three quarter-span (load-line) directly onto the Backfill (Sumon, 1999).

Except for Steel-x and Steel-y, all components were modelled using 2-d plain strain continuum quadratic and triangle elements (eight and six noded). Steel-x and Steel-y components were modelled using three noded bar elements. The Mortars/Sand joints were represented using the elasto-plastic Mohr-Coulomb 'interface' constitutive material model (LUSAS). This particular material model has been designed to be used at planes of weakness between two discrete bodies. In this case, the Mortar-x2 / Sand were the planes of weakness and the Brick/blocks either side were the discrete bodies. The Backfill was assumed to be a homogeneous material, which can exhibit volumetric plastic strain, non-linear isotropic hardening and drained. This is to be critically discussed in due course. The Backfill was supported at the edges of the arch by springs to allow for horizontal movement. For each model, the loading was applied through the Knife-edge to prevent premature failure in the Backfill and to simulate the experiment. The appropriate loading and boundary conditions were then applied to each model.

COMPONENTS	M1	M2	M3
Brick/blocks	y	y	y
Mortar-x1*	y	y	y
Mortar-x2*	y	-	-
Mortar-y*	y	y	y
Sand*	-	y	y
Steel-x	-	-	y
Steel-y	-	-	y
Abutments	y	y	y
Backfill	y	y	y
Knife-edge	y	y	y

Table 2: Components used in the FE models
* all joints 10 mm thick

3.2 ANALYSES

In the experiments, it was found that there was minimal damage to the abutments, Brick/blocks and Knife-edge so they were kept linear. Non-linear properties were assumed for the Backfill, Mortar-x1, Mortar-x2, Mortar-y, Sand, Steel-x and Steel-y. The 'interface' material model has played an important part in modelling the ring-separation as it allowed replication of "slippage" circumferentially.

Compaction induced stresses in the Backfill were not modelled and the value of the earth pressure coefficient was generated internally in the arch using the strength parameters in Table 3. As in the case of the experimental models the spandrel walls and pavement were not considered. This simplified the model and reduced the number of parameters being analysed. The load was applied to the model in relatively small increments using the Newton-Raphson method. Towards the later stages of the loading it was necessary to reduce the increment size to achieve convergence.

The full range of material properties used is given in Table 3. Most were calculated or estimated using the guidelines provided by Hendry (1990) or CEN (1995) Eurocode 6 in conjunction with the compressive tests carried out on the mortar and bricks. The bricks and mortar were considered to be of low compressive strength and 'poor quality', respectively. The backfill values were obtained from manufacturer's literature except for cohesion Loke (1991), for the other components small differences could be expected between the calculated and actual. The cohesion for mortar and sand was assumed to approach zero but a value of 0.1 N/mm² was used to achieve convergence. Values up to of 10 kN/mm² were tried but they did not effect the outcome significantly, other than slow down the convergence procedure. The density values were measured in-house.

Model	Component	Young's Mod'(E) kN/mm²	Poisson's Ratio (v) -	Density (μ) kg/mm³	Shear Mod'(E) kN/mm²	Compress' Strength(fc) N/mm²	Uniaxial yield strain (εy)	Cohesion N/mm²
M1	Brick/block	7.364	0.20	1.870e-06	3.068	18.41	-	-
	Mortar-x1[α]	0.003	0.10	1.903e-06	0.001	0.50	0.167	0.1
	Mortar-y[α]	0.030	0.15	1.903e-06	0.013	1.28	0.072	0.1
	Mortar-x2[α]	0.030	0.15	1.903e-06	0.013	1.28	0.072	0.1
	Abutments	36.000	0.20	2.400e-06	-	60.00	-	-
	Backfill[β]	0.042	0.30	2.030e-06	0.016	-	0.001	0.1
	Knife-edge	210.00	0.30	7.850e-06	-	-	-	-
M2	Brick/block	7.364	0.20	1.870e-06	3.068	18.41	-	-
	Mortar-x1[α]	0.003	0.10	1.903e-06	0.001	0.50	0.167	0.1
	Mortar-y[α]	0.030	0.15	1.903e-06	0.013	1.87	0.072	0.1
	Sand[α]	0.003	0.10	1.903e-06	0.001	0.50	0.167	0.1
	Abutments	36.00	0.20	2.400e-06	-	60.00	-	-
	Backfill[β]	0.042	0.30	2.030e-06	0.016	-	0.001	0.1
	Knife-edge	210.00	0.30	7.850e-06	-	-	-	-
M3	Brick/block	7.364	0.20	1.870e-06	3.068	18.41	-	-
	Mortar-x1[α]	0.003	0.10	1.903e-06	0.001	0.50	0.167	0.1
	Mortar-y[α]	0.030	0.15	1.903e-06	0.013	1.87	0.072	0.1
	Sand[α]	0.003	0.10	1.903e-06	0.001	0.50	0.167	0.1
	Abutments	36.000	0.20	2.400e-06	-	60.00	-	-
	Backfill[β]	0.042	0.30	2.030e-06	0.016	-	0.001	0.1
	Knife-edge	210.00	0.30	7.850e-06	-	-	-	-
	Steel-x[γ]	205.00	0.30	7.850e-06	78.85	-	0.427	-
	Steel-y[γ]	205.00	0.30	7.850e-06	78.85	-	0.427	-

Table 3: Main material properties used in the FE models

[α] Effective angle of internal friction used = 45°.
[β] Effective angle of internal friction used was 40°.
[γ] Uniaxial yield stress = 480 N/mm². Cross section area = 622 mm² (Steel-x) and 424 mm² (Steel-y).

4. RESULTS AND DISCUSSION

The deformed mesh plot of the arch barrel at maximum load, under the load-line, for M1, is shown in Figs. 3a. Fig. 3b shows the deformed mesh of reinforcement for M3, which was anchored into each of the abutments. Note the staggered effect of the radial pins, which was due to the circumferential movements in the barrel. Fig. 3c gives the combined plot of reinforcement and the barrel. For M2 the deformation was similar to that shown in Figure 3c, but without the reinforcement, since it was also ring-separated. It is clear in M1 (Fig. 3a) that the rings had moved away from each other (caused by the failure in Mortar-x2) and radially Mortar-y joints had opened-up directly under the load-line. Overall in the circumferential direction the bricks and mortar had remained in-line, suggesting the radial mortar joints (Mortar-y) provided the necessary restraint. However, this was not the case in M2 or M3. The main reason for this could be that the ring-separation weakened the ability of the barrel to act in a composite manner and hence increased the movement in the Mortar-y joints. The effect was further amplified in M3 despite the reinforcement (Fig. 3c), but this could be because of the higher load applied (see Table 1).

Fig. 3a: Deformed mesh plot of arch barrel

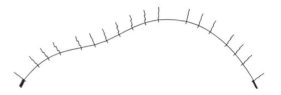

Fig. 3b: Deformed mesh plot of the reinforcement

Fig. 3c: Deformed plot of reinforcement and arch barrel

Displacements

Plot of vertical displacements obtained on top of the arch along its complete length (9.4 m) are given in Fig. 4. For clarification the positions of the three quarter-span (load-line) at 3.345 m and quarter-span at 5.595 m, are shown in-line with Fig. 1. The curves for M1 and M3 gave the expected response, that is as the load was applied at the load-line the Backfill and arch rings immediately below the Knife-edge moved downwards and upwards approximately at the opposite quarter-span (Fig. 4). However, for M2 the upward movement was less because the model's response to load was much more ductile. The arch was simply forced down with little resistance and this prevented much upward movement at the opposite quarter-span.

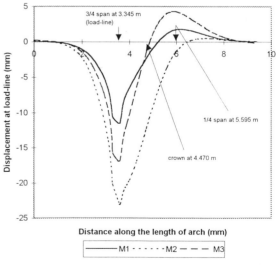

Fig. 4: Plot of displacement at load-line vs length of arch

Fig. 5 shows the load vs displacements obtained from the FE models plotted with the results from the full-scale tests. It can be seen that the theoretical curves were relatively close to the experimental ones, indicating a good relationship between the different pairs. A1 was much stiffer initially than M1, but during the latter stages it showed a similar response. A2 showed gradual failure but highly ductile behaviour. M2 showed a linear response until load approached closer to the maximum load to failure. It was not expected that the curves would correlate closely, but it appears that the stiffness in the M2 was somewhat lower than A2 during the earlier stages of loading. Arch A3 showed an expected response, initially behaving like A2, but as the load was increased the stainless steel reinforcement provided the additional elasticity to give it a higher stiffness; a similar response was achieved in the theoretical model M3.

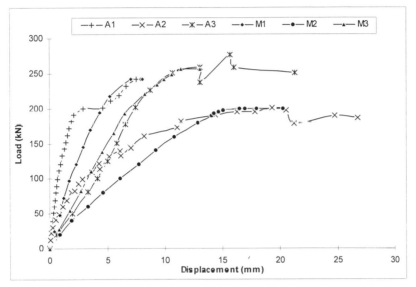

Fig. 5: Plot of load vs displacement from the FE models and full-scale tests

Comparing the response of the reinforced M3 with M1 and M2 it can be seen that the reinforcement worked in two ways. Firstly, the radial pins helped to keep the three rings together to act as a composite or as a fully bonded arch (A1). Secondly, the circumferential reinforcement provided the necessary stiffness to the soffit and delayed it going into tension.

Stresses and strains
In M1 the minimum principal stress (compressive stress) regions were at the top of the barrel and either side of the load-line. In M2 the compressive stresses were concentrated in the individual rings due to the ring-separation. In M3 the reinforcement in the bottom ring spread the load into the middle and top rings, and backfill, but high stress concentration remained in the bottom ring. The concentration was not just at one or two points but tended to run from one end of the ring to the other. In other words, the bottom ring carried most of the load applied and this was only possible because of the reinforcement.

The maximum principal stresses (tensile stresses) in M1 were concentrated at the four hinges that developed as the arch was loaded. M2 showed higher tensile stresses only at the load-line possibly due to the highly ductile failure behaviour between the rings due to lack bond and hence could not sustain any higher tensile stresses. In M3 the tensile zones are spread along the bottom ring, the top ring at the load-line and at quarter-span on the top ring. Here the peak stress was much lower than in M1. But the stress distribution was much more complicated to analyse because of the presence of the reinforcement. Furthermore, the reinforcement (Steel-x) was being pulled at the points where it was anchored into the abutments and subsequently developed reactions there.

The minimum principal strain (compressive strain) in M1 showed that the Backfill below the load-line gave strains above -6000με and most of the radial Mortar-y had failed by crushing. The results of strain from the test were approximately in this range with highest recorded strain of -6000με. The strains monitored were at set positions along the barrel soffit and extrados. M2 and M3 gave strain values that were much higher than recorded in the tests. In the tests, the recorded strains were below -3000με but in the FE analyses these rose to -30000με an increase by a factor of 10. This affect could be the explanation for the highly ductile failure noted in M2 (Fig. 5). However, in M3 the compressive strain was lower in the bottom ring and higher in the Backfill compared to M1. This could be due to the reinforcement and higher compressive strength bonding material used compared to the traditional mortar.

Maximum principal strain (tensile strain) in M1 showed that the Mortar-y failed below the load-line (bottom ring) and between the rings either side of the load-line. Similar findings were noted in M2 but the strain ranges plotted were at a much higher level than found in the tests. The maximum tensile strain recorded for M1 was 6000 με, which was within the test range. However, the strains from A2 and A3 were 3000 με but M2 and M3 gave strain of 20000 με about 6-7 times higher. This high variation could be due to how the FE program calculates the strain across the elements. For example if there is a high local strain the VW strain gauge would have measured the average and would not have picked the peak(s), whereas the FE program did. This could be one of the reason for the big differences. Hence a further investigation would be useful.

The general failure was observed at Mortar-x1, between the top of the barrel and Backfill, where it had failed at key points along the extrados. Further, the material properties to initiate failure here were much lower than for the Sand and the other mortars used. This interface represented the polythene sheet that was partly inserted at the edges to reduce friction. Further, in all the models it was found that below the knife-edge the backfill to the left had failed which potentially permitted the bricks to "fall out" and would eventually lead to collapse. This effect was observed when arch A3 was taken to failure and collapse.

5. CONCLUSIONS
The following conclusions may be drawn:

- Good correlation was obtained between the full-scale tests and the FE models developed.
- The mortar/sand joints used in the FE models played an important role in modelling the test behaviour.
- The FE models have shown that the material properties play an important part, hence input model definition and accuracy is crucial.
- The regions where failure occurred in the models closely followed the experimental arches.
- The ring-separation effect was modelled giving the behaviour expected. The accuracy of this was compared with the displacements, stresses and strains obtained. The displacements gave good correlation, but the strains were generally found higher than those recorded in the tests.
- Strengthening the soffit gave a significant increase to the load carrying capacity. This was achieved by changing the load distribution along the arch rings, instead of load concentrating at the load-line. As a result, the bottom ring carried a significantly higher load in conjunction with the other rings than it would have done if not strengthened.
- The reinforcement would be most beneficial for an arch with weaker or lost mortar joints and an appropriate method of repair would limit further damage.

6. ACKNOWLEDGEMENTS

The authors would like to thank the Bridges, Infrastructure *test-team* for their contribution to this project. Further thanks to LUSAS Software Support Team, FEA Limited - for their useful suggestion during the developments of the finite element models.

7. REFERENCES

BD 21/97 *Assessment of Highway Bridges and Structures*.

BA 16/97 *Assessment of Highway Bridges and Structures*.

CEN, 1995. *Eurocode 6: Design of masonry structures*. ENV 1996-1-1; 1995, CEN, Brussels, Belgium.

HENDRY A. W., 1990. Masonry properties for assessing arch bridges, *TRRL Contractor Report 244*, Department of Transport, Transport and Road Research Laboratory, Crowthorne, UK.

LOKE, 1991. Effects of lateral boundary yielding on large scale unreinforced and reinforced soil walls, *PhD Thesis*, University of Strathclyde.

LUSAS Software versions 12 and 13, 1998. Element Library; Theory Manuals 1 and 2; Command References 1 and 2; Examples; LUSAS User Guide; Modeller User Guide. Forge House, 66 High Street, Kingston upon Thames, Surrey, KT1 1HN, UK.

MELBOURNE, C. and WALKER, P. J., 1990. Load test to collapse on a full-scale model six metre span brick arch bridge. Department of Transport. *TRRL Contractor Report 189*, Transport and Road Research Laboratory, Crowthorne.

MELBOURNE, C. and GILBERT, M., 1995. The behaviour of multi-ring brickwork arch bridges. *The Structural Engineer* / vol. 73/no. 3/7 Feb.

PAGE J., 1996. A guide to repair and strengthening of masonry arch highway bridges. *TRL Report 204*. Transport Research Laboratory, Crowthorne.

SUMON, S. K., 1997. Repair and Strengthening of Damaged Arch with Built-In Ring-Separation. *Proceedings of Seventh International Conference on Structural Faults + Repair-97*, held at Edinburgh, p. 69-75.

SUMON, S. K., 1998. Repair and strengthening of five full scale masonry arch bridges. "Arch Bridges", *Proceedings of the Second International Conference on Arch Bridges* held in Venice, p. 407-415.

SUMON, S. K., 1999. Seedcorn SR68: Modelling masonry arch bridges - final report. *TRL Unpublished Project Report PR/CE/38/99*. Transport Research Laboratory, Crowthorne.

Bridge deck replacement with lightweight exodermic bridge deck panels

Robert A. Bettigole, P. E.
Exodermic Bridge Deck, Inc.
Lakeville, Connecticut, U.S.A.

Summary

An Exodermic bridge deck is a lightweight modular deck system that lends itself to rapid construction. It combines an unfilled steel grid with a 90 mm to 115 mm reinforced concrete slab, making efficient use of the constituent materials, handling spans over 4.4 meters between supports, and weighing as little as 245 kg/m2. It is in use on over 40 bridges in the United States.

By reducing deck dead load, an Exodermic deck permits a bridge to achieve a higher live load rating. An Exodermic deck typically weighs 35% to 50% less than the reinforced concrete deck that would be specified for the same span. Reduced dead load can minimize or eliminate structural strengthening required to achieve a required load rating.

A contract has been let to re-deck the famous Eads Bridge over the Mississippi River in St. Louis. The weight savings possible with the Exodermic design was crucial in the rehabilitation of this bridge for vehicular traffic. In addition, the Exodermic design chosen allows the deck to span continuously over floor beams spaced on 3.66 meter centers, eliminating the need for steel stringers.

Rehabilitating many bridges is greatly complicated by the need to handle traffic while construction work is undertaken. Closing many structures is increasingly untenable, and alternatives must be found. The fast construction possible with the Exodermic bridge deck design has allowed deck replacements on a number of important bridges where time is of the essence.

Precast Exodermic deck panels are being used to re-deck 24 000 square meters of the deck of the 4.9 km long Tappan Zee Bridge over the Hudson River. All work is being done at night, with all seven lanes of the bridge open to traffic each morning by 6 AM. Similar work has been done on a number of other projects.

A revision of the Exodermic design, which dates to the early 1980's, is now in use. The internal shear connection mechanism of the design, where concrete "dowels" engage holes in the upper elements of an unfilled steel grid, has interesting structural applications beyond that of bridge decks alone.

Exodermic Design

The Exodermic design evolved from concrete filled steel grid bridge decks which have been in use in the United States since the early 1930's. In these decks, a fabricated steel grating is filled with concrete, giving a very long lasting deck. When filled flush to the top of the grating with concrete, ride quality suffers, but deck weight is reduced significantly from that of a typical reinforced concrete deck. Some of these original filled grid decks are still in service. Recently, the industry has re-labeled this type of deck as "grid reinforced concrete", which perhaps better describes the design.

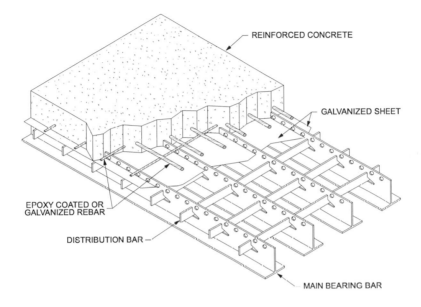

REINFORCED CONCRETE

GALVANIZED SHEET

EPOXY COATED OR
GALVANIZED REBAR

DISTRIBUTION BAR

MAIN BEARING BAR

In an Exodermic deck, the concrete is moved up in relation to the steel grid, providing a number of improvements:

- Much more efficient use is made of the two components

- The concrete component can now be reinforced with a single mat of reinforcing steel, providing a large increase in negative moment capacity.

- The neutral axis shifts close to the welds and punchouts in the grid, reducing live load stress levels, and consequently, enhancing the fatigue resistance of the deck.

Components of the grid are embedded in the concrete, insuring that the concrete and grid will work together in a composite fashion. Thus, in positive bending, the concrete is in compression and the main bars of the grid are in tension. That is, the two materials are used as they are best suited. In negative bending, the reinforcing bars in the concrete handle tensile forces, and the steel grid main bars handle the compressive forces.

The revised Exodermic design will be examined in more detail later in this paper.

Eads Bridge

One of the most historically significant bridges in the United States is the 496 meter Eads Bridge, spanning the Mississippi River in St. Louis, Missouri. The main elements of its three main spans are arch truss structures of cast tubular steel. It was the first bridge in the U.S. to make significant of steel and one of the first bridges in the U.S. to use pneumatic caissons. It was also the first bridge to be built entirely using cantilever construction methods, avoiding the need for false-work and the first bridge designed so that any part could be removed for repair or replacement. Although an innovative structure in many ways, the Eads Bridge was designed and constructed to carry 18th century rail and wagon traffic.

In the 125 years since the Eads Bridge was first opened to traffic, a number of larger bridges were built across the Mississippi in St. Louis, and the Eads Bridge's importance declined. A number of major rehabilitation projects have been done over the years, but in modern times, the bridge had become considerably neglected. In 1990, the maximum allowable truck was restricted to 45.3 kN. By 1991, the upper level roadway was closed to vehicular traffic. In 1993, the lower level railroad was adapted for use by Metro Link, St. Louis's light rail transit system. In the mid 1990's, the upper level roadway and its framing were removed, and consultant Sverdrup Civil, Inc undertook a major design effort.

A key goal of the major rehabilitation of the bridge was to improve the capacity of the bridge sufficiently to allow the resumption of vehicular traffic on the renovated upper level. In order to permit the roadway to be configured for four traffic lanes, dead load had to be reduced significantly in order for the arch trusses to figure for live load. The capacity of the trusses would have been inadequate with conventional steel framing and a reinforced concrete deck. Innovative framing and an Exodermic deck were chosen.

The upper level roadway framing is supported on columns tied into the panel points along the arch trusses, which are on 3.66 meters centers. When a filled grid deck was substituted for the original type of deck (timber) in 1946, floor beams were installed on and between columns. The intermediate floor beams were supported by longitudinal beams between columns, inappropriately putting the latter beams (and the columns) into bending. Conventional steel stringers completed the 1946 framing. In the replacement design now under construction, the columns support new transverse floor beams, which in turn support the deck. The intermediate floor beams have been eliminated, and there are no longitudinal beams supporting the deck; the Exodermic deck is designed for the (up to) 3.66 meter spans. The negative moments in the deck over the floor beams are resisted by appropriate choice of reinforcing in the slab – in this case 19 mm reinforcing bars at 102 mm centers.

The new Exodermic deck and framing weighs approximately 30% less than a conventional reinforced concrete deck and its supporting stringers would have weighed.

Tappan Zee Bridge

The 4900 meter Tappan Zee Bridge is the longest structure on the New York State Thruway, carrying average daily traffic of 130 000 vehicles in seven lanes across the Hudson River twenty miles north of New York City.

As a toll authority, the Thruway is particularly sensitive to the needs of the traveling public, and insists on having all seven lanes open during the morning and evening rush hours. A movable barrier allows lanes to be configured for four lanes in the predominant direction of traffic during such periods.

In replacing over 24 000 sq. meters of deck on this bridge, the Thruway Authority determined that even during this demanding work, all seven lanes of traffic would have to be available for the morning and evening rush hours. Thus, all work had to be done at night. To emphasize the importance of compliance with the daily schedule, the contract provides for up to a $1300 per minute penalty if all seven lanes of the bridge are not reopened to traffic at 6 A.M. each day.

The contractor on this project increased his productivity to nightly replacement of 275 to 315 sq. meters.

The bridge was opened to traffic in 1955, and consists of three different types of span. The main span is a cantilever truss, a total of 736 meters in length. There are 160 "trestle" spans, 15.2 meters in length, now two span continuous, and 20 deck truss spans, averaging 71 meters in length. The 13 deck truss spans east of the main through truss span are the subject of the project described here.

The original reinforced concrete deck of the bridge is only 171 mm thick, with a 51 mm asphaltic concrete overlay. The Thruway Authority's requirement for the deck design was to permit re-decking the bridge working only at night, but there was additional benefit seen in reducing the dead load of the deck to extend the fatigue life of the bridge.

The Exodermic deck chosen weighs 3.5×10^{-3} MPa, or 38% less than the 241 mm reinforced concrete deck that is the standard for new concrete decks in New York. The Exodermic bridge deck being used on the Tappan Zee Bridge is comprised of an unfilled steel grid composite with a 114 mm reinforced concrete slab. A portion of the steel grid extends 25 mm into the bottom of the concrete slab and acts as a shear connector. The Exodermic panels are 190 mm in overall thickness.

Most of the concrete portion of the Exodermic deck was precast, with only closure pours required in the field. The precasting operation provides block-outs in the deck in those areas that will be above stringers. These openings allow headed studs to be welded through the grid to the stringers in the field. A full depth, double female shear key connects panels transversely. In addition, steel angles are welded across the shear keys at the midpoint between stringers.

The concrete mix used in precasting is a newly developed "high performance" mix in which 6% of cement is replaced with microsilica, and 20% is replaced with fly ash. The New York State Department of Transportation developed this mix for its reduced permeability to deicing salts, rather than for its strength. However, 55 to 69 MPa 28 day strengths are being logged by the precaster.

Typical Night's Work

The 25.6 meter wide deck is being replaced in four stages across the bridge. Dowel bar splices between stages provide deck continuity.

In a typical night, work proceeds as follows:

Work Sequence

1. At 8 PM, the Thruway's barrier moving machines begin to shift the centrally located barrier as required. Cones and signs are set up as required by the MPT plan.

2. At 9 PM, two lanes are taken out of service and equipment is mobilized. Saw cutting begins in two separate work zones. Each work zone has a full crew and two cranes. A third crew will uses a single crane to set panels incorporating the new modular joint system, which is replacing the existing finger joints.

3. At 10 PM, a third lane is taken out of service and temporary Jersey barrier and other protective material and equipment is put in place. After traffic protection is in place, the contractor begins to remove sections of the old deck.

4. The tops of the stringers are cleaned by sand blasting and mechanical tools.

5. Haunch forms, prefabricated from galvanized sheet metal, are dropped into place.

6. Exodermic panels are landed and adjusted to surveyed elevation using built-in leveling bolts.

7. Shear studs are welded to the top flange of the stringers through the blockouts in the precast Exodermic panels.

8. Haunch forms are pulled up against the underside of the Exodermic panels and screwed to matching sheet metal forms attached to the panels above the edges of the stringer flanges.

9. For Stages II, III, and IV, dowel bars are threaded into the threaded couplers precast into the panels of the previous stage.

10. Angles are welded across shear keys where specified.

11. Rapid setting concrete is placed full depth over the stringers and at the transverse shear keys using a mobile mixer, consolidated with a pencil vibrator and covered with wet burlap. Typically, the contractor plans to reach this point by 3:30 AM, allowing one hour for the rapid-setting concrete to reach the specified 17 Mpa required before reopening to traffic.

12. Removal of the temporary barriers, cones and signs, and repositioning of the permanent barrier takes approximately one hour.

13. By 6 AM, the bridge is fully reopened to traffic.

Following deck replacement, the deck is being diamond ground for a smooth riding surface, and overlaid with a 9.5 mm epoxy concrete riding surface.

Revised Exodermic Design

As originally constructed, the internal shear connection of Exodermic decks was provided by "tertiary bars', rectangular bars to which short lengths of reinforcing steel or round bar ("vertical studs") were welded at 304 mm intervals. In the revised Exodermic design used on the Tappan Zee Bridge, the tertiary bars were eliminated, and instead, the main bars of the grid were extended 25 mm into the slab. Replacing the vertical studs were 19 mm holes punched in the embedded portion of the main grid bars, and spaced on 51 mm centers.

The concept of using such holes and the resultant concrete "pegs" or "dowels" through them has been explored on a larger scale by others, notably as "perfobond" shear connectors for main members proposed by Leonhardt et al. 1987 and 1990, and Roberts and Heywood 1995. This earlier research demonstrated negligible slip under service load as one of the advantages of this approach.

Although the revised design differed only in the means of shear connection from the original, well tested design which has been in service since 1984, it was felt that additional static and fatigue testing would be appropriate. This was undertaken in 1997 at Clarkson University by Dr. Christopher Higgins.

Testing has shown the importance of aggregate size and confinement to the success of the design. 9.5 mm maximum coarse aggregate is specified. Confinement is provided by 9.5 mm reinforcing steel on 152 mm centers normal to the main bars of the grid.

In the first test (Higgins, 1997), a panel spanning 2.43 meters between supports was loaded to failure. Load was delivered through a steel shoe cushioned by a layer of neoprene, 259 mm × 648 mm sized to simulate an American Association of State Highway and Transportation Officials (AASHTO) HS-25 (118 kN including 30% impact factor) double truck tire footprint (at 0.69 MPa inflation pressure). The deflection and main bar strain were linear to 362 kN loading, indicating full composite behavior to at least that point. The load/deflection curve gradually "softened" to 559 kN, when there was a punching shear failure of the concrete. Far from dramatic, a rectangular area around the load patch dropped approximately 13 mm , and the panel was still carrying 300 kN of load. Concrete used was a standard 24 MPa mix, using 9.5 mm maximum coarse aggregate.

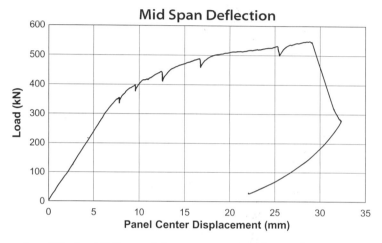

The second test (Higgins, 1998) was a fatigue test, consisting of two million load cycles delivered to a two span continuous panel through two 231 mm 2 579 mm steel loading shoes simulating a 188.5 kN AASHTO HS-20 truck axle including a 30% factor for impact. The tested spans were 2.18 meters. Static tests were conducted at intervals of 250 000 cycles. No significant difference in behavior of the panel was observed from start to finish of the test. For example, strains measured at the bottom of the grid main bars at maximum load didn't change over the course of the test.

After completion of the fatigue test, the panel was cut in half, and a test conducted in which it was sought to fail one half of the deck, with the test set-up being as close as possible to the first such test. Despite the two million fatigue cycles, the panel performed well, demonstrating a linear load-deflection response similar to that of the earlier panel. Due, in part, to the high performance concrete (NYSDOT Class DP, a 34.5 MPa mix incorporating 6% microsilica and 20% fly ash) used in this test, the limits of the test setup were reached before the panel failed.

The applied load was increased gradually to 648 kN , the limit of the load cylinder, and the panel did not reach punching shear as did the first panel tested. Surface cracking did not appear until approximately 535 kN , versus 362 kN in the first test. The crack pattern, when it did appear, was similar.

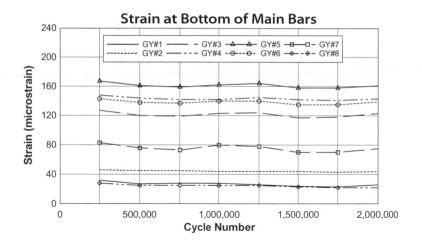

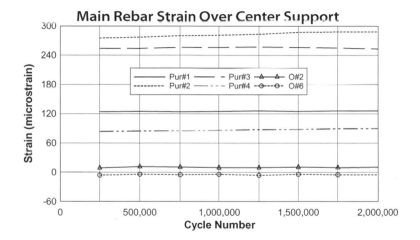

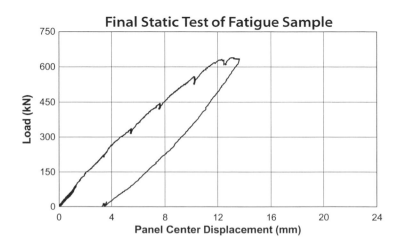

The testing was in accordance with the new American Society for Testing and Materials (ASTM) specification D6275-98, "Standard Practice for Laboratory Testing of Bridge Decks".

The revised design is simpler and less expensive to fabricate, and with the elimination of the tertiary bars, contractors have more working room for installation of shear studs.

Conclusion

The advent of the Exodermic deck design has given designers an important tool in bridge rehabilitation. The reduced dead load of an Exodermic deck can help designers achieve target live load ratings. Exodermic deck panels permit rapid -- often overnight -- deck replacement, while continuing to provide the desirable aspects of reinforced concrete decks - ride quality, including excellent skid resistance, easy and accustomed maintenance, and contractor familiarity.

Bibliography

American Association of State Highway and Transportation Officials (AASHTO). (1998). "AASHTO-LRFD Bridge Design Specifications, Second Edition" AASHTO, Washington, DC, USA

American Association of State Highway and Transportation Officials (AASHTO). (1996). "Standard Specifications for Highway Bridges, 16th Edition," AASHTO Executive Office, Washington, DC, USA

Andrä, Hans-Peter. (1990). "Economical Shear Connectors with High Fatigue Strength," Proceedings of IABSE Symposium, Brussels, pages 167-170.

City of St. Louis (1997). "Eads Bridge Highway Deck Reconstruction," Plans and Specifications. St. Louis, MO, USA.

Exodermic Bridge Deck, Inc. (1999). "Exodermic Bridge Deck Case Study - Tappan Zee Bridge Case Study," Product Literature, EBDI, Lakeville, CT, USA.

Exodermic Bridge Deck, Inc. (1996). "Exodermic Bridge Deck Handbook," Product Literature, EBDI, Lakeville, CT, USA.

Exodermic Bridge Deck, Inc. (1999). "An Introduction to Exodermic Bridge Decks," Product Literature, EBDI, Lakeville, CT, USA.

Higgins, C., and Mitchell, H. (1998). "Fatigue Tests of a Revised Exodermic Bridge Deck Design," Report No. 98-12, Clarkson University, Potsdam, NY, USA.

Higgins, C., and Mitchell, H. (1997). "Tests of a Revised Exodermic Bridge Deck Design," Report No. 97-16, Clarkson University, Potsdam, NY, USA.

Leonhardt, E. F., Andra, W., Andra, H-P., and Harre, W. (1987). "New Improved Shear Connector with High Fatigue Strength for Composite Structures," (in German) Benton-Und Stahlbentonbau, Vol. 12, 325-331.

New York State Thruway Authority (1988). "Tappan Zee Bridge Fact Book", Albany, NY.

New York State Thruway Authority (1997). Plans and specifications for contract TANY 97-32B, Albany, NY.

Roberts, W., and Heywood, R. (1995). "Development and Testing of a New Shear Connector for Steel Concrete Composite Bridges," Fourth International Bridge Engineering Conference, National Academy Press, Washington, D.C., 137-145.

The replacement of Kingsway Canal Viaduct

P.J. CLAPHAM, BSc., C.Eng., M.I.C.E., Principal Engineer (Bridges), Rochdale
Metropolitan Borough Council, Technical Services Department, Rochdale, UK, and
A. COLLINS, BSc., C.Eng., M.I.C.E., Assistant Engineer (Bridges), Rochdale Metropolitan
Borough Council, Technical Services Department, Rochdale, UK.

Abstract
This paper describes the development and design of a scheme to replace an unattractive
reinforced concrete viaduct with an integral bridge and approach embankments supported on
piles with geotextile reinforcement at the base. The reasons for the chosen solution are
described, together with the design of the works. Measures adopted in the design, to facilitate
construction are considered together with the necessary traffic management measures.

Background
Kingsway Canal Viaduct carries the A664 Kingsway over the currently disused Rochdale
Canal, in Rochdale. The road crosses the canal at an angle of about 45°. The bridge was
designed by L.G. Mouchel and Partners using the Hennebique reinforced concrete system and
was built in 1932 (Fig. 1). Peat deposits underlie the site, so that approach viaducts on each
side of the bridge were built to carry the road over the poor ground. The overall length of
these structures was some 150 metres; details are given in Table 1.

Assessment showed that the edge beams could not carry 40 tonne assessment loading and the
parapets and their supports required strengthening. The condition of the reinforced concrete
was generally poor, particularly around the joints. In addition, some elements were subjected
to excessive thermal movements and structural cracking was evident. Extensive concrete
repairs were undertaken in 1982. The environment and appearance of the structure were very
poor and environmental improvements planned.

Options for Refurbishment and Replacement
Various schemes were investigated including:

- Strengthening and repair of the original structures.

- In-filling under the approach viaducts which could then be allowed to deteriorate as self-
  supporting structures and converting the canal bridge into an integral bridge. This was not
  feasible because of the anticipated settlement of the underlying peat.

- Replacing the existing structure with a new single span bridge involving approach
  embankments supported on a basally reinforced load distribution platform utilising the
  original piles.

Replacing the structures was the favoured option, taking into account the whole life cost of
the new bridge compared with refurbishing the original structures. Details of the new
structure are given in Table 2 and Fig. 2.

Investigations

A comprehensive ground investigation was undertaken involving 3 shell and auger boreholes and 23 cone penetrometer tests. Recent filling over an adjacent site had not shown any visible settlements so the site investigation was designed to examine settlements as well as pile capacities. As well as the testing and sampling from the boreholes two pressuremeter tests were undertaken to examine the compressibility of the underlying peat. The investigation determined that significant settlement of the peat deposits was expected which ruled out the option of in-filling under the approach viaducts. The allowable pile loads were principally evaluated from the cone penetrometer tests using the methods in Tomlinson[1].

Figure 1 shows the geotechnical section interpreted from the geotechnical investigation. The following general sequence of strata below the existing structure was determined:
>   Made Ground
>   Peat Deposits
>   Glacial Sand
>   Glacial Till
>   Glacial Sand
>   Glacial Till

This sequence was proven to 15 metres below existing ground level.

Design of the New Bridge

The new bridge is designed as a single-span, integral bridge with a pre-cast, pre-stressed Y-beam deck and in-situ slab (Fig. 3). The main reason for the deterioration of the existing bridge was the poor condition of the reinforced concrete, especially around the joints where water and de-icing salts had penetrated the concrete. The new integral bridge has no movement joints, with the road surfacing being continuous across the structure. Therefore an integral bridge design offered the most cost effective solution for replacing the original bridge as this type of design will stop penetration of the structure by water and de-icing salts and will minimise expensive maintenance work in the future.

The piles supporting the existing structure are to be retained and supplemented by further segmental reinforced concrete piles as used for the design of the approach embankments. By utilising the existing piles the position of the new abutments are fixed in the same position as the existing bridge. Therefore the new bridge will retain the 45° skew of the original bridge.

The outer edge beams, which will carry footway, service bay and parapet loading, were designed as simply-supported beams. The inner edge beam carrying the footway, service bay and carriageway loading and the other inner beams were designed using a space frame analysis to model in-plane forces as well as vertical loads. The sectional properties were calculated using the same principles as those used for a torsionless grillage, using uncracked section properties for bending design. The deck is allowed to expand into the fill material located behind the deck diaphragm wall. This movement is modelled by applying horizontal spring stiffness to each of the idealised beams. The spring stiffness is calculated in accordance with Hambly[2], using recommendations given in accordance with BA 42/96[3] and earth pressure coefficients relevant to the proposed fill material. The bridge deck is allowed to contract horizontally away from the fill material.

The model was designed such that one longitudinal beam of the model represents one longitudinal beam of the proposed structure. The sectional properties for the longitudinal members were calculated assuming that full concrete areas are effective. In calculating the properties of the main beams the values were transformed by the modular ratio to take account of the different strengths of the pre-cast beams and in-situ slab.

The model was prepared using the Superstress space frame analysis computer program. Dead loads were applied to the model directly and live loads were generated for various worst case locations using the H-Load pre-processor program which is linked to Superstress and transfers the loads automatically to the model. Once the basic load cases had been prepared they were formed into load case patterns and ultimately into load case combinations. The model was then run to calculate the forces and moments generated by the combination load cases and these were used in the design to determine the most effective tendon arrangement for the pre-stressed beams.

Design of the Approach Embankments

The new approach embankments are supported using the original piles supplemented by new 200 mm square segmental reinforced concrete piles. The embankment loads being distributed to the piles using geotextiles to form a reinforced soil base to the embankment.

The design of the basally reinforced embankment (Fig. 4) uses the methods in Section 8 of BS8006[4]. In this method the reinforcement is provided at the base of the embankment to support the full weight and distribute it to the piles. A high strength unidirectional reinforcement was adopted for the design with two layers, one acting transversely and the other longitudinally. The BS8006[4] formulae assume that one size of pile and pile cap is used and determine the allowable spacing between the piles based on the allowable pile capacity used with Ultimate Limit State loads. In this case the original piles were already in place, so that the total load on a given area was evaluated and the new piles provided to support the loading in excess of the capacity of the original piles. This gave the maximum number of new piles required in each area. The maximum pile spacing required by the reinforcement was also evaluated and the actual number of new piles was based on the most critical of these two criteria.

The load-carrying capacity of the piles was assessed as part of the design, using the results from geotechnical investigation, principally the cone penetrometer tests. Based on the investigations the areas requiring piled foundations were divided into representative sections and the pile capacities for the original piles and proposed piles were calculated for each section. The capacities assessed for the original piles were based on the pile lengths shown on the original drawings. The record drawings indicate that some piles were lengthened during the contract; unfortunately details of exactly which piles were lengthened are not given. The geotechnical information indicates where piles are likely to have been lengthened, but the design of the new foundations makes no allowance for such increased pile capacities. The proposed new piles are designed to penetrate deeper into the lower layer of sands and gravels.

The design integrates the original and new piles to support the embankments. The design follows BS8006[4] although some of the detailed methods have been adapted to suit the situation. The Authors are not aware of a similar situation where piles of different sizes and capacities have been integrated to support an embankment in this way. However, both the original and proposed piles are precast concrete driven piles transmitting the loads by a combination of skin friction and end bearing.

The final design includes a reinforced earth wall (Fig. 5) to keep the construction within the highway boundary to avoid encroaching on one small area of land. The embankment piled foundation and basal reinforcement support this reinforced earth wall. However, it is hoped that negotiations with the landowner will permit an embankment slope to be constructed which will give the scheme a better appearance.

Design Options Available to the Contractor
In order to ensure that a workable solution is obtained the proposed design uses proprietary piling and reinforcing systems for the embankments. However, the designed solution is not the only option. Therefore the contract allows the Contractor to propose alternatives and provides the necessary design information.

Services
The project has been designed to accommodate the services supported by the original structures. By working with the Statutory Undertakers expensive service diversions have been avoided. The original service bay, along the approach ramps, will be retained whilst the embankments are constructed. The services over the canal will be supported in their original position whilst the new bridge is constructed around them.

Traffic Management
The A664 Kingsway is a busy main road in Rochdale and the bridge is located in an area with housing, three schools and businesses all nearby. In particular, there are a number of new industrial units and an expanding garden centre next to the site. The importance of minimising the effects of this large construction project on road users, local residents and businesses was recognised at an early stage.

There is sufficient open land to the east of the bridge to permit a temporary road to be constructed crossing the canal at low level. The vertical alignment of the temporary road has been selected so that it does not significantly increase ground pressures on the underlying peat. The temporary road will have a speed limit of 20 mph because of the need to access the site to undertake the work, the alignment of the temporary road and the location of the nearby schools. The provision of the temporary road has made this a viable solution. There is a major project to re-open the Rochdale Canal to navigation, but once the canal is navigable then the opportunity to provide a temporary road will be lost.

Conclusions
The replacement of Kingsway Canal Viaduct has been developed by taking whole life costs into account to design a new structure that will be significantly more economic to maintain in the medium to long term than continuing to repair a structure with inherent defects. The use of modern techniques has allowed a solution that retains the existing piled foundations and makes as much use as possible of the original materials within the new scheme. The scheme will also enhance the local environment as part of the regeneration of the area, in contrast to the poor appearance and details of the original structures.

References
1. Tomlinson M.J. (1995) *Foundation Design and Construction*, sixth edition, published by Addison Wesley Longman Ltd.
2. Hambly E.C. (1991) *Bridge Deck Behaviour,* second edition, published by Chapman & Hall.
3. BA 42/96 *The Design of Integral Bridges.* Design Manual for Roads and Bridges, Volume 1, Section 3, Part 12. The Highways Agency, Scottish Office Development Department, Welsh Office, Department of the Environment for Northern Ireland.
4. BS8006:1995 *Code of Practice for Strengthened/Reinforced Soils and Other Fills,* British Standards Institution.

Table 1: Details of the Original Structures

	Canal Bridge	Approach Viaducts
No. independent structures	1	4
Material	Reinforced concrete	Reinforced concrete
Structural type	3-span continuous Spans 7.5 m, 15.0 m & 7.5 m	Continuous, supported on columns forming a 12 ft (3.658 m) grid
Highway details	Carriageway width 40 ft (12.2 m) Footway width 5 ft (1.5 m)	Carriageway width 40 ft (12.2 m) Footway width 5 ft (1.5 m)
Skew	45°	Mainly square with triangular sections adjacent to the canal bridge
Superstructure	Longitudinal main beams with skew beams across the supports and square transverse ribs supporting the deck slab	Longitudinal main beams with square transverse ribs supporting the deck slab
Substructures	Columns each with a bearing supporting a main beam	14" (355 mm) square columns built into the main beams
Foundations	14" (355 mm) square driven piles	14" (355 mm) square driven piles
Bearings	Steel rocker bearings	None, but new bearings added later to the columns adjacent to the canal bridge to improve rotational capacity
Joints	45° skew butt joint with adjacent approach viaduct	Square butt joint between the approach viaducts

Table 2: Details of the New Bridge and Approach Embankments

	Canal Bridge	**Approach Embankments**
No. independent structures	1	2
Material	Reinforced concrete sub-structures and precast pre-stressed concrete superstructure	Granular fill
Structural type	Simply supported single span of 15 m	Basally reinforced earth embankment supported on piled foundations
Highway details	Carriageway width 10.3 m including provision for two 1.5 m cycleways. Footway width 2.5 m	Carriageway width 10.3 m including provision for two 1.5 m cycleways. Footway width 2.5 m
Skew	45°	Regular grid for the piles
Superstructure	Pre-stressed precast concrete beam and slab integral bridge deck	Not applicable
Substructures	Reinforced concrete abutments to incorporate the existing columns and piled foundations	Embankment with basal reinforcement to transfer the loads to the piles
Foundations	Combination of original piles and 200 mm square precast concrete segmental driven piles	Combination of original piles and 200 mm square precast concrete segmental driven piles
Bearings	Elastomeric bearings	Not applicable
Joints	Downstand diaphragm at each end of the bridge deck to retain the embankment fill	Not applicable

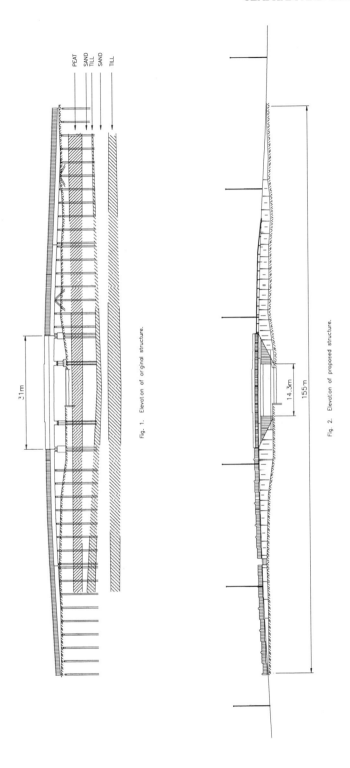

PEAT
SAND
TILL
SAND

TILL

31m

Fig. 1. Elevation of original structure.

14.3m

155m

Fig. 2. Elevation of proposed structure.

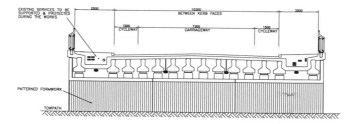

Fig. 3.   Cross—section through proposed bridge.

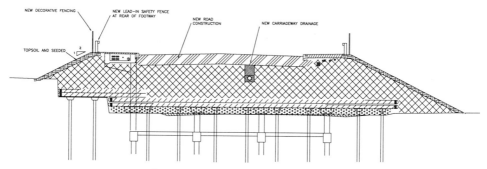

Fig. 4.   Cross—section through North approach.

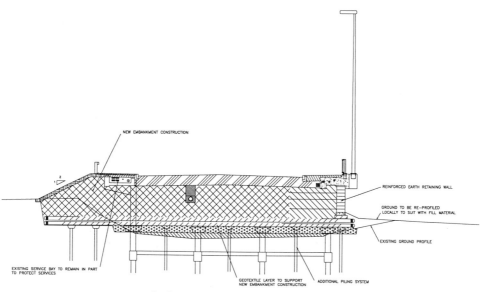

Fig. 5.   Cross—section through South approach.

Replacement of the road deck of the Masnedsund Bridge

M.Sc., Senior Bridge Engineer H. Vagn Jensen, Bridge Maintenance Department
RAMBØLL, Bredevej 2, DK-2830 Virum, Denmark - E-mail: hvj@ramboll.dk

ABSTRACT

The Masnedsund Bridge constructed in 1935 by the English Contractor, Dormann Long Ltd, is approximately 200 metres long, consisting of five 31.5m simply supported normal spans and one 28.4m bascule span. The bridge carries a single track railway to Germany and two local lanes over a sound.

The main structure consists of two longitudinal steel girders with transverse steel girders every 5.25m The steel girders carry two separate concrete troughs - one for the railway and one for the road deck.

Due to a leaking waterproofing membrane on the road deck both the concrete through and also the longitudinal girders suffered from serious damages imposing truck load restrictions which a nearby power station could not live with.

The main functional requirements for the replacement of the superstructure were:
- the railway should be in full service
- the road deck was allowed to be blocked totally during shorter periods only
- the bascule span should be in full service

The new superstructure was designed as orthotropic prefabricated steel decks. Due to the geometrical restrictions the steel troughs were placed perpendicular to the lanes. The unconventional design required particular fatigue analyses.

The new superstructure was prefabricated in Finland and coated and paved under shop conditions. Particular measures were taken during fabrication to keep narrow erection tolerances. The concrete deck was demolished partly by diamond cutting and partly by blasting under controlled conditions.

Introduction
The composite superstructures on the Masnedsund Bridge and on the similar nearby 3km Storstroem Bridge were originally constructed by the English Contractor, Dormann Long Ltd. The main structure consists of two longitudinal steel girders with transverse steel girders every 5.25m on which the concrete trough for the road deck (including a cantilevered sidewalk) was supported by five longitudinal steel girders.

A leaking waterproofing membrane on the concrete road had caused lamination of the concrete deck and serious corrosion of the supporting longitudinal steel girders. These damages imposed load restrictions on truck passages - restrictions which was unacceptable to a nearby power station. Based on a particular inspection the Danish Road Directorate decided that replacement was the only feasible method as repair could not increase the load bearing capacity to the required level.The corrosion problems had primarily been caused due to spreading of de-icing salts on the road deck during the winter seasons. Similar problems have yet not been detected on the rail deck where de-icing salts have not been spread.

An elevation of the main structure and cross sections of the old road deck structure are shown below. The separate concrete deck carrying the rail structure is indicated only by dotted lines.

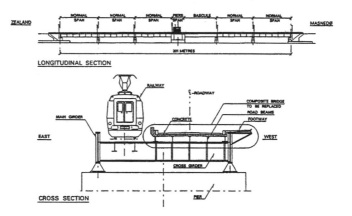

Cross section of the new road deck is shown below.

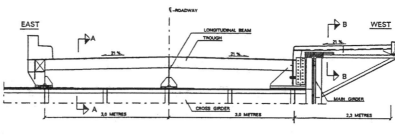

The bridge is administered by two parties. The Danish Road Directorate administers the road deck and the Danish National Railway Agency administers the rail deck and the main bridge structure. The present contract was tendered by The Road Directorate with the contractor Skanska Jensen and RAMBØLL as designer of the project.

The contractor being a subsidiary of Skanska Jensen chose a high degree of subcontracting:
- Erection (incl. fabrication) and paving was subcontracted to two Swedish cousin firms
- shop fabrication was further subcontracted to a north Finnisk manufacturer

Design principles for new orthotropic deck
The height of the new structure had to be the same as the old structure to suit the road levels at the abutments and the intermediate bascule span. The height limitation reduced the possible design principles significantly.

Prefabricated full span concrete spans were too heavy to handle on the shallow waters and in situ concrete superstructure required too long obstruction of the road traffic. For reasons of easy erection a steel deck was chosen for the new superstructure. The distance, however, between the transverse girders did not suit normal orthotropic steel deck designs. A particular concept with transverse troughs spanning 3 metres was developed with one longitudinal girder at each side of the road deck and one at centre line.

The longitudinal girders on the new road deck were, due to the limited vertical load bearing capacity of the existing structure, supported at each transverse girder (5.3m). The longitudinal loads of each span were transferred trough one joint connection between the deck plate and the west main girder. The sidewalk was designed as a cantilevered structure fixed to the west main girder by corbels at each transverse girder. The sidewalk plate was stiffened by "bulb"-profiles. Reference is made to [1] where the particular structural analysis – including fatigue analysis – corresponding to this design is explained.

Fabrication aspects
Surfacing
Surface treatment of the new structure was performed in Finland under shop conditions. The supplier of the paint had given the statement, that although the temperatures of the soffit of the steel during laying the mastic asphalt was estimated to reach approx. 100 C, this high temperature was not expected to damage the actual surface treatment system (epoxy system with thermoplastic top coat (chlorinated rubber)) neither by lime powder deposits nor by reduced gloss. A close visual inspection of the surface treatment after laying the road surfacing, did not indicate any thermal damages.

The road surfacing was a particular Danish type which is only 60mm thick and of the same type mastic asphalt as laid on other steel bridges in Denmark. The advantage of this surfacing is that the surfacing acts as a structural composite material with unique fatigue properties, together with the subjacent orthotropic steel deck. The particular mastic asphalt materials, however, could not be produced or laid to specified quality at the remote Finnish yard. The laying of mastic asphalt was consequently postponed to be performed in Denmark on the barge, shortly before erection.

The expansion joints were the SHW steel/rubber lamella type with only one rubber lamella band enabling the surfacing to be laid before erection. The rubber was inserted after erection.

Measures taken during fabrication to keep narrow final tolerances for erection
Before starting fabrication of the new structure the dimensions of the existing structure should be checked by the contractor. The design was based on dimensions as indicated on the as built drawings of the structure. The possibilities for access, however, were rather poor and limited to platforms along the main girders. Eventually the contractor succeeded to check the longitudinal dimensions by a meter but still with the uncertainty that the spans might be parallelograms and not rectangles.

Before starting fabrication for stage II, re. below table, the contractor tried to eliminate this geometric uncertainty surveying on shore by a distomat but still some lengths could not be checked due to "blind" angles to a few reflectors placed at joints. Luckily the as built documentation on which the design had been based showed very good consistency with the measured dimensions and only a few corrections were implemented for fabrication.

Finally the fabrication tolerances for the new steel structure should be kept within narrow tolerances to match the main steel structure. During erection the new spans were placed to match the joints at the centre transverse girder thus minimising the deviations at the ends.

The trough welds initiated a significant shrinkage of the steel plate. Following measures were taken to keep the narrow tolerances +/-5mm for fixation of the new superstructure:
- Weld shrinkage from one trough was measured on a prefabricated test item
- The troughs were positioned on the deck incorporating the expected weld shrinkage
- The deck plate was produced with moderate surplus length
- At the joints to the transverse girders were the strengthening of the bottom flanges of the three longitudinal girders not fabricated until after all troughs had been welded. The normal flange sections were removed and replaced by prefabricated joint sections.

The longitudinal camber was fabricated utilising the angular weld shrinkage from the transverse trough welds. Similarly the specified transverse camber was fabricated utilising the angular weld shrinkage from the longitudinal girder below the centre line.

On the cross section of the new steel deck involved the splice plate connection (re. the cross section figure) carrying the cantilevered sidewalk several tolerance problems:
- the distance between the gusset plate of the existing and the new structure was specified very narrow, 10mm only
- the splice plates for connecting the two parts, were made from prefabricated hot dip galvanised plates with all holes drilled during fabrication
- at the west horizontal gusset plate had a notch to be introduced to omit clash and to guide easily the new deck horizontally during erection

The narrow tolerance at the above joint implied several clashes during erection and also poor possibilities for repair of surface treatment in the joint area after erection.

Planning serviceability of the bridge during erection works
All erection activities influencing the road, train and navigation traffic should be agreed with relevant authorities and announced 6 weeks in advance. The contract allowed for one week total blocking of road service per span. The period was chosen with consideration to industrial summer holiday and harvest seasons. The bascule span was for commercial reasons not allowed to be out of service for more than one week
The repair works were executed during three summer seasons 1993-1995 re. below table:

stage no/ season	replaced span no.	partly road service	no road service	bascule out of service
I/1993	5 & 6	10 + 4 weeks	2 weeks	-
II/1994	1, 2, 3 & bascule pier deck	15 + 5 weeks	4 weeks	3 days blocked
III/1995	4 (road surfacing at bascule span)	1 week (pedestrians)	1 week	1 week

The road service was during stage I and II reduced to one sidewalk and one road lane during the period with preparatory works and after erection of the new superstructure. During the period with preparatory works following activities were executed:

- Removal of sidewalk and outer part of the road deck above the west main girder
- Surface treatment of upper part of the west main steel girder
- Replacement of rivets by bolts

A ferry service for pedestrians and cyklists was introduced in the period without road service whereas the vehicles were diverted to the nearby Faroe motorway bridge.

After erection, however, the road service was limited for a shorter period until the permanent railings had been erected.

The rail service should during the whole construction period be kept operating without any interruptions. Only a local speed limit during passages was accepted. At the time of replacement most of the south bound international train traffic passed the bridge. After the opening of the Storebælt tunnel in the summer 1997 the number of train passages on the Masnedsundbridge has decreased significantly and the trains carrying goods have been diverted almost totally from the Masnedsund bridge.

All lifting operations were carried out during the night hours with limited train traffic:

stage no	no periods per night	vacant intervals for erection - minutes (week ends)
I	3	36, 36, 60 (80/120)
II	3	25 (99), 44, 104

The crane operations removing the remaining old structures and the erection the new were performed from the "rail side" of the bridge. This side was chosen because the barge transporting the new spans, was too wide to pass by the bascule span. No lifting activities were allowed during train passages. However, fixation of the crane hook and tightening of the crane wires was allowed in a "train free" preceding interval. This way the lifting operations could start as soon as a train had passed.

Only two goods trains were delayed few minutes during stage I. During stage I the erection period had to be extended with one extra night, due to strong winds causing delayed arrival of the crane barge. This extension was given ignoring the normal six weeks announcement period.

Only once during stage II, two lifting operations (removals) were executed within the same train free interval (104 minutes). During stage II also lifting operations were impeded due to strong winds. The time schedule, however, had sufficient slack to succeed in time.

The commercial **navigation service** was not allowed to be obstructed by more than one week per season. The bascule span was out of service for three days in stage II while the old bascule pier deck was replaced. In the following three weeks the bascule was only opened in limited hours for passage of commercial vessels while the railings and pavement on the sidewalk on the bascule span was replaced.

The renewal of the road surfacing on span no 4, the bascule span, was postponed to the summer 1995 in order to reduce the period blocking the operation of the bascule span to less than one week per season. Pedestrians were during stage III allowed to pass using a separate shelter on the sidewalk, where the pavement had been replaced during stage II.

Methods for removal of existing concrete deck
The concrete road deck and the supporting five longitudinal secondary steel girders should be removed and substituted by the new superstructure.

The rivets connecting the longitudinal secondary girders to the transverse girders were removed by flame cutting the heads, and removing the remaining rivets by hammering or jacking. After removal of the rivets a minimum of two temporary bolts were inserted at each joint to assure stability until the elements were removed.

The road deck was cut into segments using diamond cutting.

The first step was to remove the sidewalk part of the deck. One longitudinal cut was made approx. 1.1m east to the sidewalk – just east to the west main girder. The sidewalk lying west to that line was then above each transverse girder (5.3m) cut into smaller segments. These segments were loaded on a truck during the night hours by a smaller crane placed on the remaining concrete road deck.

The concrete deck on the sidewalk was cast on top of the main steel structure at each transverse girder. The concrete was demolished at these joints areas before removal of the adjoining segment. During stage I demolishing was executed by manual jack hammering. During stage II the demolishing was made by controlled blasting operations. Blasting was, for the first time in Denmark, accepted to be used in urban areas nearby rail tracks in full service. The hard manual work for blasting was reduced to drillling a number of minor holes for placing the powder. The blasting was a much quicker method, than the jack hammer and also advantageous from workers health point of view. Also, the top flange of the main girder was not damaged by notches from the chisels impeding the surface treatment.

The next step was preparatory works removing the remaining part of the road deck. The concrete deck was cut at the middle each span into approx. 17m pieces. These cuts and insertion of lifting bars were made during the period without any road service. Shortly before the lifting out operations, the longitudinal girders were by flame cutting cut skew displaced to the transverse girders.

The lifting bars, tensioning bars type, "Betomax 30" were fixed by "dish nuts" which were easy to tighten with a hammer during the short lifting out intervals. A particular lifting gear suitable for the crane and the actual load, was designed by the contractor.

The 17m-segments were removed by a crane vessel and placed nearby on shallow water, to reduce the movements of the barge, during the train free lifting out intervals. During day hours the segments were then placed on shore to be demolished.

Transportation methods
Since the transport was rather long fixation of the spans to the barge, was performed following the principles stated in the Danish off shore standard.

Erection aspects and methods

Removal of the old road deck and erection of the new, was made utilising the largest Danish registered floating crane vessel, Pernille Diver, with a capacity of 125 tonnes on the actual boom. The weight of each span, including road surfacing, was approx. 80 tonnes.

After the removal of the existing concrete deck, the joint areas on the existing structure were prepared by a surface treatment (lead based primer, without need for curing, before the new superstructure was supported on top of the joint area).

In parallel with these surface protection works, a thorough levelling was made at the contact joint areas of the main structure. These levels were compared with the height and the superelevation of the new superstructure as well as with the expected deformations of the main structure when loaded with the new superstructure. The new superstructure was lighter than the old and also the differential deformations of the two main girders varied due to the asymmetric positioning of the road deck. The expected thickness of shim plates were then calculated to adjust the level of the new road deck to the predetermined levels. Prefabricated hot dip galvanised shim plates were placed before erection of the new span. The shim plates were produced in six different thickness' in the interval 1-30 mm in order to keep the number of shims to max. three at each joint.

The new road deck was lifted by a lifting gear designed by RAMBØLL shown belov. The gear utilised the holes for the gussets thus omitting provisional fixtures to the structure. Balance of each span was checked on the barge before beginning the erection.

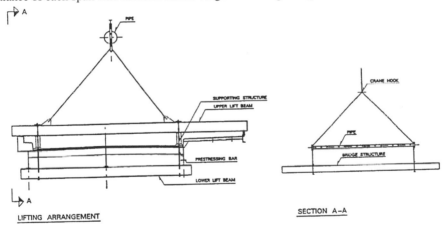

LIFTING ARRANGEMENT   SECTION A-A

After placing the new span, the levels were checked. In case the levels were deviating from the designed final level by more than two millimetres or a slot wider than 0.3 millimetre between the shims was noticed, the heights of the shim plates were adjusted. The erected spans were then fixed horizontally by two bolts at each girder end.

After the final level had been checked once more, holes were drilled, using the existing rivet holes in the supporting structure as templates when drilling holes in the shim plates and the gusset plate of the new structure. This drilling method was quite difficult as it had to be performed "bottom up" thus preventing any cooling liquids to be applied during drilling. While drilling the items were fixed closely adjoined applying temporary brackets. All bolts fixing the new superstructure were locked by centre puncher marks.

At the two vacant support areas of the top flange of the transverse girders, for corrosion reasons, the splice plates were kept closely adjoined by cutting threads and inserting new bolts. During cutting of threads, slots occurred due to insufficient fixation by brackets.

All holes, cut in steel structures during erection, were applied the lead based primer before insertion of bolts. Any slot wider than 0.3mm between the existing splice plates, the shim plates and the new gusset plates, were sealed by an epoxy sealer to prevent any later corrosion.

Function of the new deck
The road traffic load bearing restrictions on the old deck imposing load restrictions on legal truck traffic has been deleted.

Regular inspections since the deck was replaced have confirmed that the new deck fulfil all functional requirements. Only minor problems have been detected at some of the bolts for the railings and gutters, where the nuts have had to be tightened once more and to be locked. Due to vibrations both from train and truck passages, particular attention shall be paid to locking dynamic loaded bolted connections. The smaller the bolts the bigger the problems.

Conclusions
The project verified, that partial replacement of the concrete deck of an old riveted composite superstructure can be performed successfully, both with respect to design and execution. By a careful design, new bolted gusset plate connections to the old steel structure, may be optimised from a performance point of view.

The particular design developed for the bridge may be applicable for a large number of bridges constructed by the same construction method. Limited load bearing capacity of the concrete decks and progressive detoriation of both the concrete decks and subjacent steel structures, necessitate replacement. Repair methods may not be applicable due to time consuming repair methods and insufficient methods to increase the load bearing capacity to normal standard.

Even though this project involved unforeseen co-ordination problems with internordic language interpretation of the Danish contract, and logistical problems due to the remote Finnish manufacturer, the project was successful, meeting the very tight time schedules for erection activities. Sufficient time for the consultant's planning of the design, the contractor's planning of erection activities, open minded co-operation between all parties during execution, and possibility for feed back after stage I, resulted in a successful job, both from the Owner's, the Designer's and Contractor's point of view.

References
[1] Kaern, J. Chr. RAMBØLL & K. Gehrlicher; Danish Road Directorete, Nordic Steel Conference 95, Replacement of the Masnedsund Road Bridge with an orthotropic Steel Deck

Replacement of the Pinet Bridge on the River Tarn

P. MEHUE, Ingénieur Divisionnaire des Travaux Publics de l'Etat. France

INTRODUCTION

In the late twenties a dam was built across the river Tarn as part of the hydroelectric harnessing programme of the valley, which supported a bridge giving access to the village of Pinet, on the south side of the Massif Central.

As many a river in this area, the Tarn is very capricious and can be very peaceful in summer, and wild in winter with extremely sudden and devastating floods drifting a lot of floating debris. Although the dam has experienced a number of important floods over fifty years, a major flood occurred in November 1982 which showed that it presented serious deficiencies with regard to its discharge capacity in such conditions, and led to the behaviour of the as built structure being called into question in case of a bigger flood.

Therefore the decision was immediately made by Electricité de France to increase the discharge capacity of the dam and to carry out a renovation design which mainly consisted in rehabilitating the top of the structure, and subsequently in refurbishing the bridge as well.

DESCRIPTION OF THE ORIGINAL DAM AND BRIDGE

The Pinet dam, built from 1927 to 1929, is a concrete overflow gravity dam, 180 m long at the top and 40 m high, comprising two piers and two side walls delimiting three passes 40 m wide. At the toe of the weir foundation there is a gallery running through the dam, the ventilation of which is supplied by means of vertical air shafts located in the piers and the side walls. Originally each pass was fitted at the top with six radial gates installed on steel and concrete trusses which carried a reinforced concrete bridge adjacent to a gate operating platform. The bridge, 3,60 m wide and 1 m deep, supported a single lane carriageway 2,40 m wide and comprised 18 spans 6,80 m long.

As a matter of fact the 6,50 m clear openings of the gates were insufficient to prevent obstruction by floating debris and particularly by big logs, branches and trees which where blocked under the bridge by the trusses separating the gates. Such an accumulation induced the jamming of varied floating debris on the whole length of the dam, with a significant decrease of its discharge capacity.

RENOVATION DESIGN

The renovation design of the dam consisted in :
1 - lowering and reprofiling the sill of the spillway,
2 - replacing the six radial gates of each pass by a single flap gate 40,26 m long,
in order to improve the discharge capacity without changing the retention water level which was 320 NGF. Hence the new bridge was two be supported only by the piers and the side

walls, which meant three spans approximately 42 m long, with a minimal depth for the structure so as not to be an obstacle to the floating logs and trees.

In the meantime, as this part of the Tarn valley is very pleasant a site, an agreement was found with the Aveyron General Council to meet the development of tourism in this area through a wider bridge supporting a 1 m wide footway and a 4,50 m wide carriageway in order to provide two lanes for light vehicles.

OPTIONS AND STRATEGY
Due to economic reasons :
1 – the immediate downstream power plant had to be operational from the end of November to be beginning of March,
2 – the retention water level was to be kept around 320 NGF in summer since the upstream end of the artificial lake is used for swimming and boating,
which left little time to work for the reprofiling of the spillway which required to lower the water level to 310 NGF. Moreover the dam had to be able to face successfully an occasional severe flood during the rehabilitation works. In such conditions the refurbishment of the dam was planned to extend over three years, one pass being fitted each year with the new flap gates.

As the existing bridge had to be demolished at the outset of the renovation of the dam, prior to the dismantlement of the radial gates, and could not be replaced before the completion of all works on the three passes, the road to Pinet was practically to be closed for four years. Since there are few bridges over this portion of the Tarn, so long a closure would have had a dramatic effect on the local economy and traffic pattern and therefore could not actually be faced.

Thus the construction of a temporary bridge was obviously required to give access to Pinet and the neighbouring villages throughout the dam renovation, but, due to the depth of water, difficulties arose for setting up piers immediately upstream the dam. The option of a floating bridge was considered as well, however it would have been an obstacle to floating debris and could have been swept off by a sudden flood.

After consulting the Bridge Division of the Service d'Etudes Techniques des Routes et Autoroutes it came out that it was possible :
1 – to build temporarily the new bridge along the existing one, on provisional cantilever beams set on the piers and the side walls and extending upstream above the water,
2 – after completion of the spillway renovation on the three passes, to shift the new bridge downstream from the provisional cantilever beams to the piers and the side walls to be laid on its definitive bearings,
which solved the problems regarding both traffic and construction.

According to these conditions, the choice was made of steel spans comprising and orthotropic deck in order to minimize the imbalance effect on the cantilever beams and to increase as much as possible the vertical clearance above the retention water level.

BRIDGE DESIGN

The bridge design carried out by Electricité de France with the technical assistance of the SETRA Steel Bridges Department comprised three independent elements 43,42 m long, with a span of 42,70 m in the temporary situation reduced to 41,80 m in the definitive position, and four provisional steel cantilever beams 12,30 m long. All of them where fully welded.

Each span consisted of two main plate girders 4,51 m apart and 1,40 m deep, which meant a slenderness ratio of 1/30. The 14 mm thick deck plate was supported by continuous longitudinal trapezoidal ribs 0,60 m spaced and 0,95 m deep floobeams 3,80 m spaced. The ribs were made of 750 x 6 mm folded plates, with bevelled edges prior to the welding so as to ensure high quality performance. The skid resistant surfacing of the carriageway and the footway was 10 mm thick. The weight was approximately 105 t.(Figure 1).

Each cantilever beams consisted of a rectangular box girder 0,80 m wide and 1,20 m deep supported by steel bearings 3,60 m apart. For the 6,80 m long overhanging the top flange was 1,20 m wide. The beam was anchored in the dam at its rear end by means of prestressed cables fixed through a vertical air shaft on a temporary frame installed in the toe gallery.

In addition to the standard live loads both spans and cantilever beams were designed to carry a 60 t crane vehicle which was to be used for handling operations onto the spillway.

CONSTRUCTION

The construction of the bridge started in late 1986 with the cantilever beams, which were entirely shop manufactured so that no welding had to be made on site, then were erected in early 1987 (Figure 2).

The spans were fabricated by halves, longitudinally and transversely, in order to minimize transport difficulties on the narrow roads giving access to the dam.

Next spring and summer, the elements of the spans were successively assembled end to end on the existing bridge, then shifted upstream onto the cantilever beams prior to be assembled side to side. A provisional wooden surfacing was fixed onto the carriageway so as to protect the deck plate from engines operating during the spillway refurbishment. Loading tests were carried out in September 1987 then the bridge was immediately opened to traffic in this temporary position(Figure 3).

In November 1990, after completion of the renovation of the spillway and the installation of the new gates on the three passes, the spans were shifted downstream from the cantilever beams onto the refurbished piers and side walls (Figure 4 & figure 5).

In spring 1991 the provisional beams were dismantled and the wooden deck was removed to give place to a thin prefabricated surfacing (Figure 6).

CONCLUSION

The replacement of the Pinet bridge gives an interesting example of the facilities provided by steel works and orthotropic decks for the reconstruction of short spans in particular conditions.

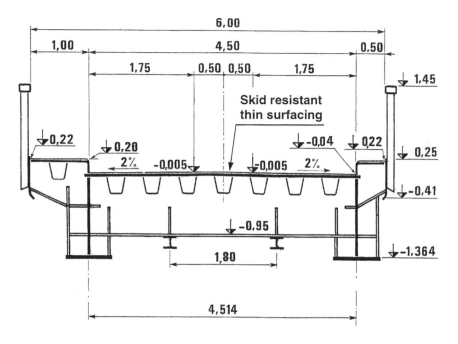

Figure 1 - Cross section of the new bridge

Figure 2 - Provisional cantilever beams

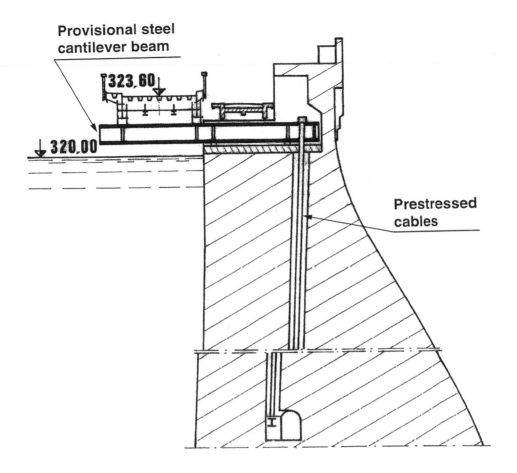

Figure 3 - Temporary position of the bridge

REFERENCE

J.M. DEVERNAY & M. GUERINET. L'évacuateur de crue du barrage de Pinet.
CHANTIERS DE France. Fevrier 1990.

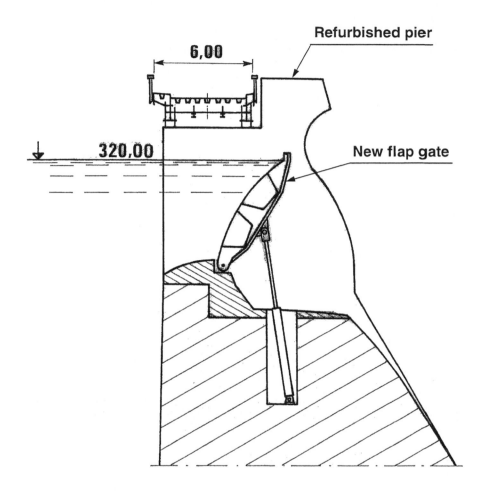

Refurbished pier

6,00

320,00

New flap gate

Figure 4 - Definitive position of the bridge

Figure 5 - Erecting the bridge on the provisional steel cantilever beams

Figure 6 - Shifting the spans from the provisional steel cantilever beams onto the refurbished concrete piers and side walls of the dam

Lifting, strengthening, widening of a motorway bridge

H. POMMER, Head of the bridge department
Waagner Biro Binder AG, Vienna, Austria

HISTORY

The Prater-bridge across the Danube in Vienna was built in 1967-1969. The bridge was designed as a three span (81m+216m+81m) continuous two box girder steel bridge with three lanes in each direction.

Due to the higher water level after the construction of a hydraulic power station a few km downstream the bridge had to be lifted in 1997 by 1.8m to ensure the minimum distance for ships between the steel structure and the water.

At the time the bridge was built 70,000 cars per day were expected to cross it. Today more than 150,000 cars use the bridge every day. Two additional lanes became necessary for the increasing traffic. The bridge only had two narrow footpaths. In summertime many pedestrians and cyclists cross the bridge to get to a newly-built recreation area.

From 1996 to 1998 the bridge was lifted and got two new traffic lanes and two new foot- and bicycle paths.

First the pathways were removed and an additional traffic lane was installed on each side of the bridge. Now the bridge has eight lanes. While raising the bridge, four lanes had to be able to carry the traffic, so the bridge was divided into two halves. One after the other, both halves of the bridge were lifted at the abutments by 4.5m. This was the phase for strengthening because the bridge has the lowest stresses. Additional deck and bottom plates were welded. Then the supporting pillars were raised by 1.8m and the abutments were lowered back.

After widening, lifting and strengthening the two bridges were reconnected and on each side a suspended path- and bicycle lane was added. All together more than 2,000 tons of steel were added to the existing structure of 8,000 tons.

ACCIDENT DURING THE INSTALLATION

This bridge was built in 1969 from both sides. While fitting the bridge construction with the last missing parts, an extraordinary drop in temperature caused a buckling of the webs and bottom plates near of the zones where the lowest bending moments are expected in the final position of a three span girder. The central part,where welding was almost finished, was strong enough to carry the shifted bending moments.

The bridge could be repaired and was been finalised with a plastic deformation of 50 cm in the middle of the large span. Finally the cross section of S2 had additional load capacity caused by the shifting of bending moments to the repaired strengthened cross section of S1. This circumstance was useful during the recent strengthening works.

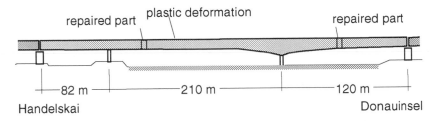

Figure 1. View of the Prater bridge in 1995

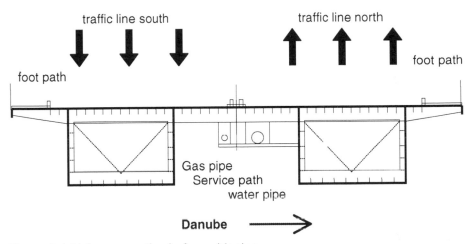

Figure 2. initial cross section before widening

WIDENING AT THE UPSTREAM SIDE

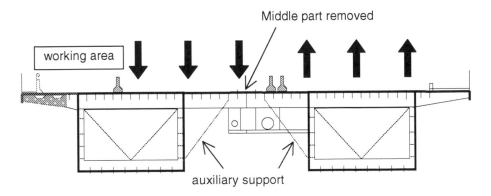

Figure 3. Widening at the upstream side

The same works were done on the downstream side. During this time the traffic lanes was directed to the widened upstream side.

WIDENING AT THE DOWNSTREAM SIDE

new cantilever
with emergency exit

new cantilever
with security path

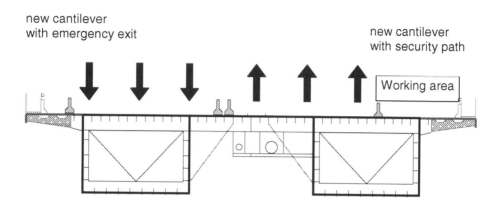

Working area

Figure 4. Widening at the downstream water side

STRATEGY FOR STRENGTHENING THE BRIDGE

We studied various combinations of lifting heights at the supports to get minimum stresses and minimum strengthening work and thus to make the total work efficient and economical. Every different lifting heights or strengthening methods caused changes in different areas – plates, stiffeners, cross girders - which have to be checked.

LONGITUDINAL DISCONNECTION OF THE STEEL BRIDGE

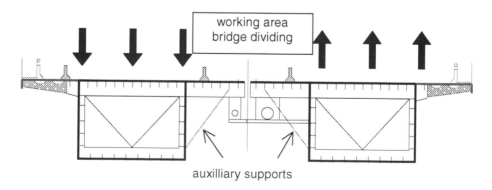

working area
bridge dividing

auxilliary supports

Figure 5. Disconnection of the bridge, cross section

First auxiliary supports which carried the cross girders of the orthotropic deck had to be installed.
After the two box girders have been separated, one side with four lanes could still carry the traffic.

LIFTING AND STRENGTHENING OF THE UPSTREAM BRIDGE

Before lifting the bridge many stiffeners and cross girders in the bottom plates had to be strengthened against buckling caused by the additional loads
All possible dead load, like the road surface had been removed. The edge of the steel construction at the abutment S/F had to be lifted by 4.5m and the new bearings of the support S2 had to be lifted by 1.8m. Through this method the lowest stresses were in the top and bottom plates. This was the correct time to weld additional bottom and deck plates to strengthen the steel structure for carrying the raised dead load. The dead load was increased by the additional walk paths and the additional material for widening and strengthening.

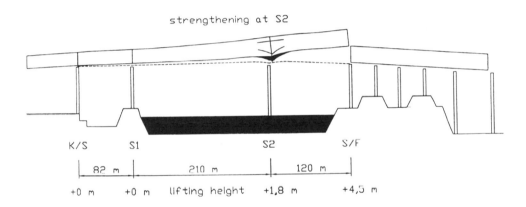

Figure 6. Lifting at the abutment S/F

After strengthening all plates the steel edge at the abutment S/F was lowered to the final lifting height of 1.4m and the other bridge end at the abutment K/S was lifted by 0,8m. The bearings at S1 were renewed and lifted to the final position. The strengthening of the bottom and deck plates could start.

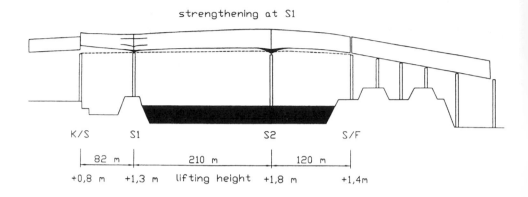

Figure 7. Lifting at the abutment K/S

After finalising the strengthening the bridge was lowered in its final position.

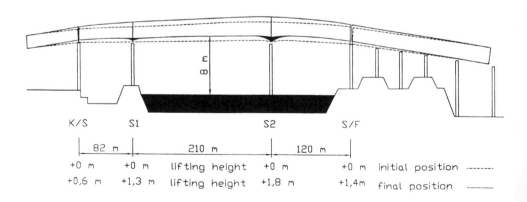

Figure 8. New position of the upper part, view

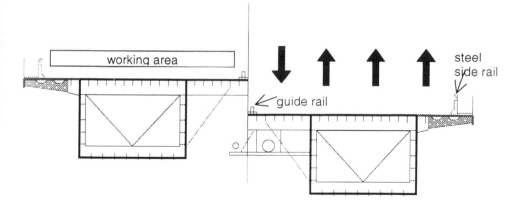

Figure 9. New position of the of the upper part, cross section

LIFTING OF THE DOWNSTREAM PART
In the same sequence the down stream part was executed. After the middle part was lifted the water level could be raised by about 1.8m which is necessary for the economical work in the hydro electric power station Freudenau.

RECONNECTING OF BOTH LIFTED AND STRENGHTENED BOX GIRDERS
For the proper use of the raised bridge it was necessary to connect all cross girders and the orthotropic steel deck. After finalising the road surface the bridge was ready to carry 8 lanes of traffic without any obstacles.

INSTALLATION OF FOOT AND BICYCLE PATHS
After the bridge was strengthened and in its new position the installation works of the suspended walk way could be started.

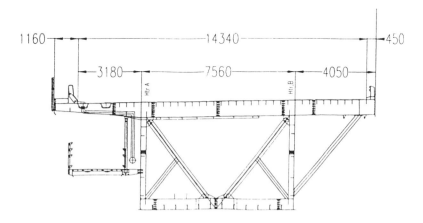

Figure 10. New position of the total bridge with foot- and bicycle path underneath

CONCLUSION

It took two years to successfully complete this project. About 2,000 tons of strengthening materials and 400 tons of additional footpath construction were installed.

The total dead load of the bridge is about 10,000 tons, which was lifted 1.8m.

The total cost for widening, strengthening, lifting and additional foot paths for the steel bridge was 15 million pounds.

The Prater bridge in Vienna is a good example for the flexibility of a steel bridge, for what can be done, if the requirements change during its lifetime.

Tamar Bridge strengthening and widening – a case study

RICHARD J. FISH, Cornwall County Council, Truro, UK,
JOLYON A. GILL, Hyder Special Structures, London, UK and
PETER J. LADD, Cleveland Bridge, Darlington, UK

INTRODUCTION
The Tamar Bridge, carrying the A38 Trunk Road over the River Tamar between the county of Cornwall and the city of Plymouth on Britain's south west peninsula, was the first of the country's modern suspension bridges. Opened in 1961, the 374 metre (1100 feet) main span was then the longest in the land and the structure remains the fourth longest suspension bridge behind Humber, Forth and Severn.

The bridge was originally built through the initiative of the two public local authorities of Cornwall and Plymouth. With no central government support, the councils promoted a private Act of Parliament to secure the necessary consents, including the approval to borrow the capital needed to commission designers (Mott, Hay and Anderson) and to appoint contractors (The Cleveland Bridge and Engineering Co. Ltd) to construct the bridge.

A similar sequence of events was to be closely followed some 40 years later when the need for major strengthening works was identified and the project described in detail in this paper had to be implemented.

ASSESSMENT
In 1991, the UK Government introduced a major programme of bridge assessment and strengthening which was part of an initiative to harmonise heavy vehicle weights throughout the European Union. Increased axle and gross vehicle weights were to be introduced on 1 January 1999, and this became the target date for the completion of all strengthening work arising from the assessment programme. Bridge owners, mostly local councils, were allocated grants to undertake the structural assessments and such funding was provided for the Tamar Bridge in 1994. Commissions for the assessment and the independent assessment check were issued to Mott MacDonald and Hyder Special Structures, respectively.

The assessment indicated that the major structural components of anchorages, towers and suspension system were adequate for the projected live loading (although in the last of these, the cables and hangers, were almost at full capacity). The outcome was not as reassuring for the road deck and stiffening trusses which were found to be significantly sub-standard, even for existing design live loading. Furthermore, the original 150mm (6 inch) concrete deck slab had not stood the test of time and was showing signs of deterioration.

FEASIBILITY STUDY

As soon as these shortcomings had been identified a number of interim measures were introduced to try to ensure that the critical elements were not overstressed and, where appropriate, enhanced inspection regimes and structural monitoring were put in place.

In 1995, the Joint Committee approved a proposal to commission a feasibility study into strengthening options with the absolute requirement that the bridge should remain open to traffic throughout the works. The brief for the study was further extended to encompass solutions for existing traffic management problems for *all* users of the bridge: namely, to provide improved facilities for pedestrians and cyclists and to offer a way of achieving priority for public transport.

As two consultants now had an intimate knowledge of the bridge, including its structural inadequacies, a fee competition between the two parties was instigated for the appointment for the next phase of the project. This was the first procurement of professional engineering services for the project to be based on both quality and price criteria - a pattern that was to be followed throughout. Hyder Special Structures secured this commission and embarked on a thorough analysis of several series of options, each with the above criteria considered in detail.

The eventual preferred and recommended solution was the project described in this paper. With the conceptual design now in place and an estimate for the works prepared, an application for grant funding was made to central government for the essential strengthening works. It was at this point that the Government position was made clear: no funding was to be made available and that the project should be totally funded from existing reserves and toll revenue. Furthermore, complex rules restricting public sector borrowing had to be followed which meant that at no time could the undertaking borrow money or work within a deficit budget. This now became the fundamental criterion for the procurement of the works - certainty of financial outcome.

THE CONCEPT

The elements of the bridge failing the assessment were the truss chords and the concrete deck. It was considered that the only long term solution for the latter was to replace it with a lightweight steel orthotropic deck. Similar construction could be used for new cantilever decks necessary to provide the limited increase in capacity for local traffic, public transport, cyclists and pedestrians. To undertake this work whilst keeping the bridge open to traffic, the new cantilever steel orthotropic deck panels would be welded to the truss top chords: this gives the chords the extra strength required during construction as well as greater resistance for life. Bottom chord strengthening is also required.

This solution removes approximately 4000 tonnes of composite deck, carrying 3 lanes of traffic, and replaces it with the same weight of steel deck able to carry 5 lanes. The suspension system, although working near capacity, could probably take the increased dead load but could not take the maximum calculated live load. More importantly, during construction, the cantilevers have to be added to the bridge before the heavy concrete deck is removed.

To accommodate the increased loading, additional stay cables are to be added to the suspension system. Another pair of horizontal cables is attached to the underside of each truss bottom chord, masked by the chord strengthening, to ensure that the stay cables do not dictate an unwanted change of bridge articulation.

The work also requires major strengthening of the approach and anchorage spans and widening of the approach roads through the construction of retaining walls.

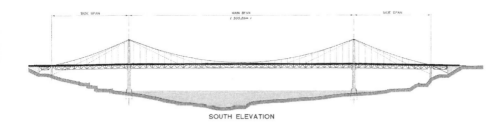

SOUTH ELEVATION

Tamar Bridge - Elevation showing existing and proposed cable systems

PROCUREMENT STRATEGY

The provisional works estimate from the feasibility study was £22m at 1996 prices. With all other fees and expenses, some advanced works, utility diversions, and the fabrication and installation of maintenance gantries and runway beams, the total project estimate was between £29m and £31m.

With a preferred start on site on or about 1 January 1999 (to coincide with the new heavy vehicle loading), it was clear that sufficient capital for the project would not be available for the start of the works but, by keeping the bridge open to traffic throughout the construction period, a revenue stream could be guaranteed. Hence, it was to be cash flow that became the crucial financial consideration and from this point of view, a longer construction period was desirable. In order to meet this need, it was decided to opt for a partnering arrangement, with shared objectives and incentives, but further supplemented by a desire for innovation and value engineering to help reduce costs throughout the life of the project.

Another innovative part of the procurement strategy was to appoint a contractor well in advance of the construction period. This would allow him the opportunity to work alongside the consultant and ensure that the detailed design developed in a way that would exploit his fabrication and erection expertise.

THE LEGAL POSITION

As stated above, the bridge had originally been built through the powers granted by the 1957 Tamar Bridge Act. Following Local Government reorganisation in 1974, another Act was eventually granted Royal Assent to become the 1979 Tamar Bridge Act. This mostly clarified the roles of the constituent authorities but also removed some of the original Act's powers.

When this project, therefore, was first being considered there were no legal powers to widen the existing bridge.

This was quickly rectified in November 1996 when a Private Bill was deposited in Parliament. After a lengthy process, including careful consideration by a House of Lords Select Committee, the 1998 Tamar Bridge Act received Royal Assent in July 1998.

The new Act also gave outline planning permission but detailed consent had to be given by the two planning authorities, one on each side of the river. It was only after this had been achieved, in December 1998, that a contract for the works could be entered into.

APPOINTMENT OF CONSULTANTS

In accordance with EU requirements, a new appointment of a consultant was needed for the remainder of the project and it was agreed that, again, a quality and price system was to be used. As a result of this competition, Hyder Special Structures were appointed in September 1996 as design consultants for the duration of the project and as retained consultants for the five years after completion of the works.

Unsuccessful tenderers were given a second opportunity to secure work on the project, as independent checking consultants, as required by UK technical approval recommendations. This commission was awarded to Mouchel International Consultants in March 1998.

DESIGN

The design work was subdivided into phases that fitted with the philosophy and timing chosen for procurement of the contractor:-

Phase 1: preliminary design between Hyder Special Structures and Cornwall County Council/Plymouth City Council to develop the Feasibility Study to give a good cost estimate.

Phase 2A: definitive design that provided enough detail to produce tender documentation in order to procure a contractor. Key elements of the project were designed in detail.

Phase 2B: design development that was carried out together with the successful contractor, Cleveland Bridge. This was where the benefits of the partnering approach became more apparent once the full team of Client, Consultant and Contractor were assembled. The aim of this phase of design was to produce the Target Price.

Phase 3: detailed design with all parties in the Tamar Bridge Project Team taking part but with overall responsibility remaining with Hyder Special Structures. Completion of the detailed design was phased to construction to suit the detailed programme for the works.

It is clear that the resultant design achieved by this process has produced a scheme that is not only <u>cheaper</u> than that possible with a traditional client design/contractor build contract, it has also been produced <u>quicker</u> than a traditional design and build contract.

The design achieved is as innovative as the contract process adopted.

A lightweight steel deck will replace the original concrete and simultaneously provide the dual function of strengthening the trusses and widening the bridge without compromising the performance of the cable supporting structure.
Additionally, in order to be able to carry out this strengthening and widening whilst the bridge continues to carry traffic, new cables will be added to the existing cable structure. They have the appearance of stay cables, but are actually a simple form of suspension system. These cables will also perform a dual function. Initially they carry the additional weight of cantilevers, added before the concrete deck is removed. Finally, when the work is completed, they help carry the new European 40t loading which is significantly heavier than the traffic loading for which the existing cable structure was designed 40 years ago.

It is believed that no suspension bridge has been strengthened in this fashion whilst continuing to carry traffic.

Existing Cross Section

Proposed Cross Section

CONTRACTUAL CONSIDERATIONS
Although the concept of partnering had been accepted as the preferred way to achieve the desired objectives (listed below) it was essential that a contractual form was determined to act as the framework within which the project could be managed and constructed.

- Certainty of financial outcome
- A start on site on or near 1 January 1999
- A teamwork approach
- A fair return for all parties
- Scope for innovation
- Risk transfer
- Total safety for the travelling public
- Minimal traffic disruption

A relatively new form of contract was used: the New Engineering Contract (recently renamed the Engineering and Construction Contract) had evolved from a number of studies into the UK construction industry and with a desire to improve the traditional methods of working.

An option within this contract allowed for a Target Price to be submitted but for the works to be paid against actual costs incurred. Any surplus between Target Price and actual cost may be shared by using a predetermined formula. Similarly, any overspend is distributed through a similar arrangement.

RISK
An early decision was taken to use risk analysis techniques to give a complete understanding of the potential risks and their consequences. The first risk schedule was compiled in 1996 and has been updated throughout the life of the project. By using a Monte Carlo analysis, a financial contingency has been derived which has been used as the project contingency within the total project estimate.

The principle of the analysis also allowed decisions to be taken on which party had control of any risk and where the ownership of a risk should lie. This was to be a major consideration in the tendering process for selecting a contractor.

APPOINTMENT OF CONTRACTOR
The project was advertised in the European Journal and expressions of interest sought. The long list was carefully scrutinised to ensure that prospective tenderers had adequate technical and project management experience and were of a suitably strong financial standing. Another requirement was that the tenderers should either have their own in-house steel fabrication facility or have a steel fabricator as part of a joint venture bid. In this way it was hoped to avoid a large percentage of the works being the subject of a major sub-contract which would be difficult to control within the partnering approach.

Tenders were invited in December 1997 from a short list of 6 firms or consortia, although one withdrew during the tender period. Again, bids were invited on a quality and price basis, the latter relating to the new steel orthotropic deck which, purely for the purpose of tendering, had been designed, detailed and billed. As well as a total price for this element, tenderers also had to give unit rates which were eventually to be used in the actual costs calculated during the contract. All tenderers were asked to give presentations and attend interviews before the final decisions were taken on the quality elements of the bids.

In March 1998, Cleveland Bridge of Darlington was appointed as preferred contractor, but initially only as an engineering advisor. At this time the legal position was still uncertain and no firm commitment could be given that the project would proceed. A contract was formally entered into in March 1999.

TARGET PRICE
One of the first tasks that Cleveland Bridge had to concentrate on was the preparation of the Target Price to be used in the eventual contract. This required a detailed appraisal of the existing preliminary design and the preparation of a comprehensive scope of the works. A detailed programme was also drawn up to ensure that time related costs were included. All this work was carried out in close co-operation with the engineering teams from Hyder and Cornwall, reflecting the spirit of partnership that was also being nurtured within the unified project team. A figure of £23.528m was eventually agreed and accepted.

PARTNERING

Although this project represents a major technical challenge and day to day problems are inevitable, the only way to ensure success is to establish a spirit of mutual trust and co-operation and to work at creating strong professional relationships and a team culture.

With each of the three parties committed to these objectives from the outset, it was essential to develop the ideas into something workable. An initial partnering workshop was held in April 1998, followed by others in October 1998 and March 1999. An independent facilitator was used, with no engineering background, to explore the needs of the project and to arrive at a number of commitments that everyone was able to sign up to. These are embodied in a charter:

• The project being completed to specified quality, time and target price.	• Minimising disruption to the public
• A fair return for all parties.	• Working in a spirit of honesty, openness and trust and co-operation
• Setting a partnering model for future projects.	• The overall partnering aims being more important than the aims of the individual parties
• Providing high quality and sufficient resources to unified project team.	• A happy and fulfilling working environment
• Achieving high environmental standards	• Always striving for improvements
• A commercially open book environment	• Respecting the needs of others
• Safety being of paramount concern	

The workshops also produced a draft Issue Resolution Process which was later to be formalised and included within the Contract.

PROGRAMME

The overall project programme was created in1997 setting down five clearly defined stages on the basis of contracting in a partnership. This is summarised below.

Activity	1997	1998	1999	2000	2001
1 Preliminary Design	▬▬				
2A Definitive Design		▬▬			
2B Design Development		▬			
3 Detailed Design			▬▬		
4 Construction				▬▬▬▬	

Simplified Programme

Under the partnering arrangements the period of 9 months between tender submission and acceptance of the target price was used by the parties to develop the design into a more clearly defined scope and accurate price for the project thus providing a more secure out turn cost.

FABRICATION

The fabricated steelwork comprises the following categories:

Deck Panels	3 624
Truss Strengthening	372
Approach Spans	153
Miscellaneous	90
	4 239 tonnes

The deck panels are similar to those on other major bridges. Cleveland Bridge have been fabricating such orthotropic decks since 1963. During this period the process has been developed to reflect the advances in understanding this complex construction form. The fabrication costs have been accurately determined and as the deck panels represent 85 per cent of the fabricated work a large portion of the target fabrication price should be secure.

The deck panels are produced on a semi-automated deck panel machine with twin 16 metre beds allowing one panel to be assembled whilst another is pre-cambered and welded. The main plates are prepared by flame stripping and bevelling but the trough stiffening material is milled and not burnt. This process eliminates stress induced distortion and ensures a straight edge that provides a trough to deck plate gap not greater than 0.12mm. This maximum gap will allow 80 per cent penetration of the weld without arc blow through and provides a better fatigue resistance by a smoother transfer in the root area of the weld.

ERECTION

The limitation of loads on the structure essentially determined the sequence of working. The bridge carries a significant volume of traffic so any erection scheme must ensure the safety of bridge users and limit their disruption as much as possible. A decision was made not to use cranes on the bridge and to keep the three lanes of traffic open for most of the time. The bridge footpaths were closed and pedestrians and cyclists provided with a shuttle bus service across the bridge.

The erection of the new cantilever decks, the removal and replacement of the main deck is achieved by lifting the components from a runway beam supported by davits. The deck sections are delivered to the work front by a special vehicle with steering on all wheels. The new cantilevers are constructed first starting from the four anchorages and working towards the middle of the bridge whilst traffic remains running on the existing three lanes.

Two existing adjacent traffic lanes are then closed and traffic diverted to the two new lanes on the cantilevers. This allows one half of the existing concrete deck to be removed and replaced with the new steel orthotropic panels with work again proceeding from the anchorages to the middle of the bridge. This process is then repeated for the remaining half of the main deck.

VALUE ENGINEERING
In order to ensure that the out-turn cost of the project remains within (or below) the budget, a continuous programme of value engineering has been adopted. The drive to achieve positive value engineering is further enhanced by the financial advantages to all parties through the bonus/penalty clauses in the contract.

STATUS
Work officially started on site in March 1999, just 2 months after the preferred date determined in 1995. A formal start of works ceremony was held on 26 April 1999, 38 years to the day of the official opening of the bridge by Her Majesty Queen Elizabeth the Queen Mother in 1961. With a 31 month contract period, completion is expected in November 2001.

At the time of submitting this paper (October 1999), utility diversions are complete and the widening of the approaches, including retaining wall construction is well advanced. Truss strengthening has started and the approach span strengthening completed.

The site office has a Visitors Centre where models of the modified bridge and other material are displayed. A web site (www.tamarbridge.org.uk) has been established to help keep both the general public and the profession informed.

CONCLUSION
Since the structural shortcomings were identified in 1994, considerable effort has been given to finding the right solution for the Tamar Bridge. This is a technically challenging project: a major suspension bridge is being strengthened, widened, having its deck replaced and having a supplementary cable system fitted, all with the bridge continuing to stay open and allowing three lanes of traffic at peak times to be available throughout the works. This can only be achieved by a thorough understanding of the existing bridge and an innovative approach to the structural design solutions and by a highly experienced and competent team to complete the works.

Tamar Bridge

Introducing the PML aluminium pedestrian bridge system

UNIV.-PROF. DR.-ING. DR.-ING. HABIL. DIMITRIS KOSTEAS,
UNIV. DIPL.-ING. MENNO MEYER-STERNBERG
Section for Light Metal Structures and Fatigue, Institute for Structures, Technische
Universitaet Muenchen, Arcis St 21, D-80333 Munich, Germany

INTRODUCTION

Standardised structural elements based on specifically designed aluminium extruded sections
are used for a flexible pedestrian bridge system, composed of prefabricated welded walking
platform elements bolted together forming the lower chord and simple hollow shapes for
diagonals and verticals, also joined by bolting to the upper chord. Free span lengths between
15 and 45 m may be constructed, the bridge width is variable. Various materials to cover
sides or ceiling, an anti-skid or noise dampening walking surface are available. A stair and/or
ramp bridge access system is provided. Options include single-sided, clamped-in-place,
permanent or temporary (for repair purposes) applications on existing structures. Further
possibilities as a folding bridge or heavier elements for vehicle crossings are being
considered.

Development of the system and analytical or experimental studies concerning its structural
efficiency and integrity are being carried out by the Light Metal Structures section at the
Tech. Univ. of Munich. Significant points of the design procedure (initially following the
German DIN 4113 and the European code ENV 1999) will be outlined and compared to the
recently developed Aluminium Design Manual of the Aluminium Association, following
initial interest and discussions with DOT authorities in the United States. Special attention is
given to vibration behaviour of the bridge with analytical and experimental investigations
providing characteristic values in practice - field measurement results and comparison with
other materials are also evaluated.

Erection is very simple and rapid with light-weight elements. Significant advantage: a
maintenance-free structure. Life-cycle-cost concepts and parameters are discussed.

THE CONCEPT

A new bridge system for pedestrians and bicyclists with extruded aluminium shapes in the alloy 6082 T6 (AlMgSi1) has been developed and patented by the PML-LOGIS-Bridge Systems Company in Singen, Germany. The walking platform is composed of shop-welded aluminium hollow extrusions with respective end parts joined to one another by a bar, a bolted in-situ connection, forming the lower or tension chord of the structure. The bar holds the special Y-shaped extrusion together with the platform and the adjoining diagonals. Their connections are also bolted to the Y-shape and to the upper chord or handrail extrusion, Fig. 1 and Fig. 3. The height of the unit varies with the number of basic elements welded together to form the platform unit, thus enabling various structural dimensions and respective loading capacity or free span lengths. The latter are typically in the range of 15 to 27 m. The bridge width may vary freely, as the basic elements come in any desired length, usually between 1.2 to 2.6 m. The bridge may be constructed as a „trough" cross section or closed frame section, providing greater stiffness and bearing capacity. Stair or ramp platforms are also provided as a uniform access system to the bridge.

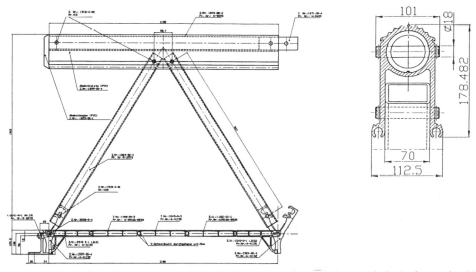

Fig. 1: The basic unit composed of the hollow extrusions of the welded platform, the Y-shaped connector, the diagonals, and the handrail tubular extrusion with side flaps.

The new structural shape of the so-called „tower-profile", a tubular cross section of 200 mm diameter, allowed the development of the second generation of bridge types, Fig. 2. This forms the lower chord of the trusses on each side, joined by tubular cross bars and the same shop-welded basic platform units. The upper chord or hand rail is composed either of the same „tower profile" - in the case of the heavier bridge type, with higher load carrying capacity - or of the former lighter tubular extrusion with flaps - in the case of the lighter version. The bridge cross section is again either an open U-shape or a closed frame. The special extrusions on the tubular „tower profile" enable an easy bolted assembly of the parts, and in this case also verticals are possible, which by means of attached stiffeners provide a higher stiffness to the whole structure. This second generation bridge type allows for spans up to 45 m. The „tower profile" got its name from the fact that these elements were initially developed as columns for a tower-like staircase structure - the latter combined with the bridge offers further possibilities of interesting solutions and architectural forms.

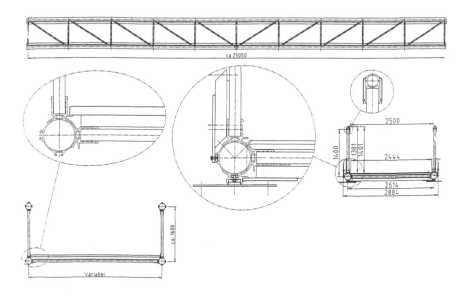

Fig. 2: The second generation bridge truss

Fig. 3: Bridge of type I over a river

DESIGN

The bridges have been designed in general for a load of 500 kg/m² or roughly 103 lbs/sqft according to the German loading specifications and design specifications DIN 4113. The design has been checked according to the latest European specifications ENV 1999 - May 1998, too. AASHTO Specifications and the corresponding provisions of the new Aluminium Design Manual by The Aluminium Association, Washington, D.C., have also been accounted for. The author being a member of the corresponding German, European and US technical committees has carried comparative studies at the Technical University of Munich involving these codes. The AASHTO Specifications ask for somewhat lower live load capacity, and there are certain differences in the allowable stresses of welded joints between the codes.

In some cases seismic design provisions have to be accounted for. This affects the design as far as piers constructed by the „tower profiles" are used. Concerning the vibration behavior of the bridge see below.

ASSEMBLY, ERECTION AND SERVICE PERFORMANCE

The aluminium lightweight system allows easy handling - most structural parts weigh less than 50 kg, a typical full unit weighs around 90 kg - and a rapid assembly. Two technicians assemble in two hours a bridge of 15 m. A light crane lifts and positions the whole bridge in place in a very short time. This can be a significant advantage for crossings with a high volume of traffic. The bridge can be a permanent structure in the case of crossings over roads, rail tracks, rivers or as a connecting skywalk between buildings or in water resource plants. As a temporary structure it serves at building sites, Fig. 4, in case of emergencies or at expositions, sports events, amusement grounds, etc. The bridge elements may be used over and over again, they are corrosion and wear resistant and their light weight is of advantage. In comparison to traditional structures in steel or timber they perform more favourably as they are practically maintenance-free and offer a long service life. Assembly, erection or transport cost is minimal, even from one continent to the other, as the standardised elements fit in great numbers in a transport pallet. Through serialisation a 50% lower manufacturing cost is reached in comparison to traditional structures.

The excellent environmental behaviour of the alloy 6082 (AlMgSi) has been proven in decades of successful applications in buildings and transportation structures, and is also covered by the respective provisions of the Eurocode.

Side coverings may be chosen in a wide variety of materials and shapes. Composite materials, acrylic glass, perforated sheets, steel cables, aluminium plates, fabric, timber panels, etc., as well as a variety of surface finishings and colours fulfil all aesthetic or technical/safety needs. A polyurethane anti-skid and wear-resistant surface layer can be applied to the walking platforms.

Fig. 4: A temporary crossing at a construction site

VIBRATION BEHAVIOR

Of special concern has always been the vibration behaviour of lightweight bridges, not only in aluminium. The Eurocode ENV 1999 for aluminium structures demands that

- structures where people walk regularly the lowest natural frequency of the structure should not be lower than 3 cycles per second (or Hertz), and

- structures where people jump or move in a rhythmical manner the lowest frequency should not be lower than 5 Hertz.

Of course the respective deflections are of concern, too, but these may be relaxed where justified by high damping values.

In a series of analytical and experimental or field studies we have begun to study the behaviour of the various PML bridge systems. As a first step the bridge of type II in Ettlingen was chosen, Fig. 2, with a span of 20 m and a width of 2,644 m, the bridge weight being approximately 3.740 kg. Several analytical estimations of the bridge frequency were undertaken with the following results:

a) according to M. Gerold: „Dynamically loaded Timber Structures", Bautechnik 1998, p. 518 the frequency was 6.412 Hz estimated from the deflection of the bridge of 0.76 cm under dead weight

b) according to the frequency equation in Grundmann/Knittel: Structural Dynamics, Publication of the Institute of Structures, 1983 the value was 7,79 Hz under dead weight, and 4,69 Hz or 3,66 Hz for a pedestrian equally distributed live load of 2.5 kN or 5.0 kN respectively

c) a similar result of a frequency value of approximately 7 Hz was reached applying the British Standard 5400

Initial field measurements support the above values, Fig. 5, 6 and 7. Lastly the variation of the bridge frequency was estimated for varying span lengths over the range of 20 to 60 m. The width of the bridge was kept constant with 2.644 m, the deflection also constant at $^1/_{500}$ of the span length. The moment of inertia was varied to ensure respective bending moment capacity, rising from 1.069×10^4 cm^4 at the 20 m span to 28.882×10^4 cm^4 for the 60 m span, the total mass of 187 kg per meter was kept constant. The frequency value declined from 7.85 Hz at 20 m to 4.53 Hz at 60 m.

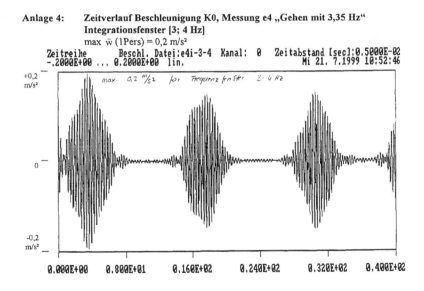

Anlage 4:   **Zeitverlauf Beschleunigung K0, Messung e4 „Gehen mit 3,35 Hz"**
Integrationsfenster [3; 4 Hz]
max ẅ (1Pers) = 0,2 m/s²

Fig. 5: Measurement of acceleration for case „walking/running at 3.35 Hz" with spectrum between extremes of ± 0.2 m/sec² for a frequency window [3; 4 Hz]

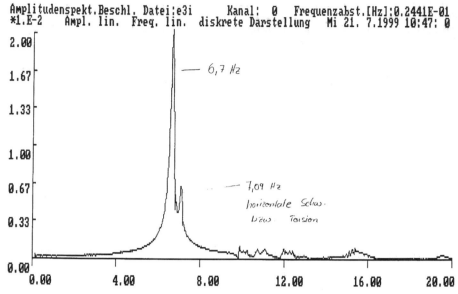

Fig. 6: Frequency peak at 6.70 Hz and 7.09 Hz respectively for horizontal vibration (torsion)

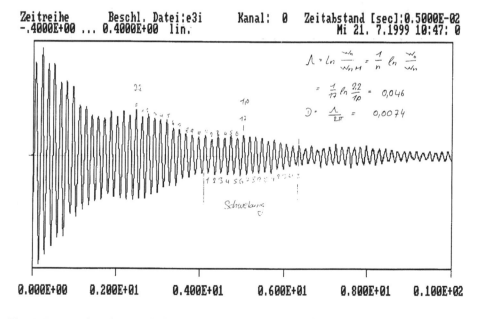

Fig. 7: Dampening characteristic - respective value estimated at 0.0074

FURTHER DEVELOPMENTS

The field vibration measurements, accompanied by further analytical estimations, will be continued for more bridge types, bridge spans and lengths.

A series of experimental investigations of the serviceability level behaviour of the bridge and its ultimate carrying capacity, with special attention to the various joint configurations, their tolerances and strength, their contribution to the vibration dampening mechanism, are planned to start within the next months.

Other options of the bridge include single-sided, clamped-in-place solutions, as permanent or temporary (repair purposes) applications on existing bridges or buildings. The development of a folding bridge is also planned for experimental tests.

REFERENCES

[1] DIN 4113 "Aluminiumkonstruktionen unter vorwiegend ruhender Belastung" 05/80, German national code for the design of aluminium structures.
[2] ENV 1999-1-1 "Design of Aluminium Structures" 1997.
[3] AASHTO Bridge Design Specifications, Washington, D.C., 1994
[4] M. Gerold: „Dynamically loaded Timber Structures", Bautechnik 1998, p. 518.
[5] Grundmann/Knittel: Structural Dynamics, Publication of the Institute of Structures, 1983.
[6] Vibration Measurement in the field, 22.06.1999, unpublished report of Technische Universität München

Songsu Bridge, widening of an existing multi-span Truss Bridge

MJ King, JD Howells, PA Sanders
High-Point Rendel, 61 Southwark Street, London Tel. +44 20 7928 8999

INTRODUCTION

As traffic densities and vehicle weights rise many key road bridges in the most heavily congested cities are found to be in need of widening and strengthening. The technical challenges for designers and contractors working on these schemes include maintaining traffic flow through construction, building onto existing structures that may already be heavily loaded and planning of erection sequences to achieve the required state of stress and profile in the widened structure. This paper describes the widening of Songsu Bridge, an example of where these criteria are major factors influencing the design solution.

BACKGROUND

Songsu Bridge was originally constructed during the mid 1970's providing a 4 lane crossing running approximately North-South. In October 1994, one of the suspended trusses of the main bridge collapsed with 32 fatalities. The cause of the collapse was failure of the fabricated hanger at the point of the welded connection between the thicker eye plates at each end of the hangers and the thinner plates forming the 'H' section hanger. Inspection of the collapsed structure identified other poor details and defective welds.

In April 1995 the bridge owners, Seoul Metropolitan Government (SMG) awarded a rehabilitation contract to Hyundai Engineering & Construction (Hyundai). SMG appointed High-Point Rendel (H-PR) in association with local consultant Yooshin as Supervisor for the rehabilitation. H-PR was required to report on the 're-usability' of the existing steelwork and concluded that the repairs necessary to welded joints throughout the main truss chords were so extensive as to be uneconomical and H-PR recommended complete reconstruction of the superstructure steelwork. H-PR advised on modifications to substantially improve the design and detailing including the introduction of an orthotropic. In June 1997 the reconstructed bridge was re-opened to traffic.

THE NEED FOR WIDENING

At the time the bridge collapsed it was carrying total traffic flows of around 110,000 vehicles per day. Loss of the crossing caused major congestion on adjacent routes. Although it was realised at the time that there would be a need for increased capacity in the future, the most urgent requirement was to re-establish the crossing to at least it's previous capacity. During 1997 H-PR were asked to advise SMG on options for widening and were able to confirm in principle the viability of integrally widening the newly rebuilt crossing. In September 1997, as soon as the reconstruction contract was completed, SMG announced the procurement of the expansion of the bridge and the provision of interchanges by Turnkey competition. H-PR joined with Hyundai and

Chun-Il to prepare a submission. In May 1998 the submission from Hyundai, H-PR and Chun-Il was assessed as winning the competition with both the highest mark and largest margin ever recorded in S. Korea. Detailed design was completed at the end of 1998.

DESIGN

Figures 1-3 show plans, elevations and cross sections through the existing and widened bridge.

An assessment of the existing bridge before widening indicated that the truss sections were heavily utilised. A key objective of the widening was therefore to devise an arrangement which ensured that minimal additional load was applied to the existing trusses in the widened bridge. An erection sequence was developed which controlled (and effectively minimised) additional dead load transfer to the existing bridge and redistributed permanent loads from the existing trusses to the new additional trusses. This redistribution is achieved by placing temporary ballast on the new truss prior to a structural connection being formed to the existing truss. After a structural connection between the existing and new trusses is formed the temporary ballast is removed thus relieving the load from the existing trusses. The temporary ballast varies between 1.4 and 2.1 tonnes/metre. The erection sequence is explained in some detail below.

Figure 4 provides details of the new transverse members. Key connections to the existing truss are made via site welded connections details of which are shown in Figures 5A & 5B. A main feature of the transverse arrangement is the provision of articulation in the temporary condition. At every node point the upper and lower transverse members will be connected to the new and existing truss during erection by a single M48 diameter bolt effectively forming a joint hinged in the vertical plane. This configuration stabilises the new truss during erection whilst permitting relative vertical displacement between the new and existing trusses until the new diagonal bracing members are installed.

Total load effects in the new truss are greater than in the existing truss because of the load transferred from the existing trusses during widening. Hence the new truss chord and web sections are somewhat heavier than the existing truss sections. Two standard truss sections are used corresponding to the dual three lane section (650mm overall width) and dual four lane sections (750mm overall width). Plate thicknesses vary between 20mm and 70mm along the truss as required by the design. Typical member make up is shown in Figure 6.

The widened orthotropic deck is similar to the existing deck. It is welded to the prepared edge of the existing deck to form a continuous surface. Although the design does not assume composite action between the deck and the trusses, shear plates are provided at regular intervals to transfer longitudinal forces due to braking etc.

The ramp slip roads were aligned to permit the most favourable arrangement with respect to both highway alignment criteria and structural configurations. Several options were investigated for the ramp structural configuration including simply supported trusses and plate girder spans supported off the main truss. However, whilst continuity with the main truss resulted in an extremely low span to depth ratio at the connection to the main truss it provided an aesthetically acceptable solution, improved details and obviated the need for movement joints.

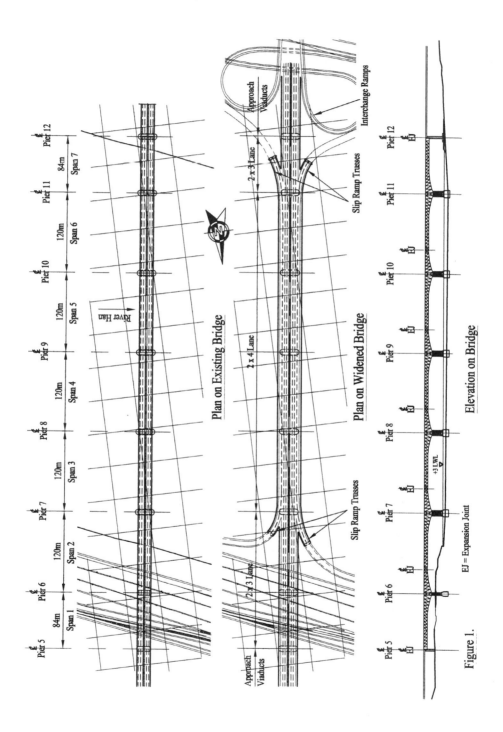

Plan on Existing Bridge

Plan on Widened Bridge

Elevation on Bridge

Figure 1.

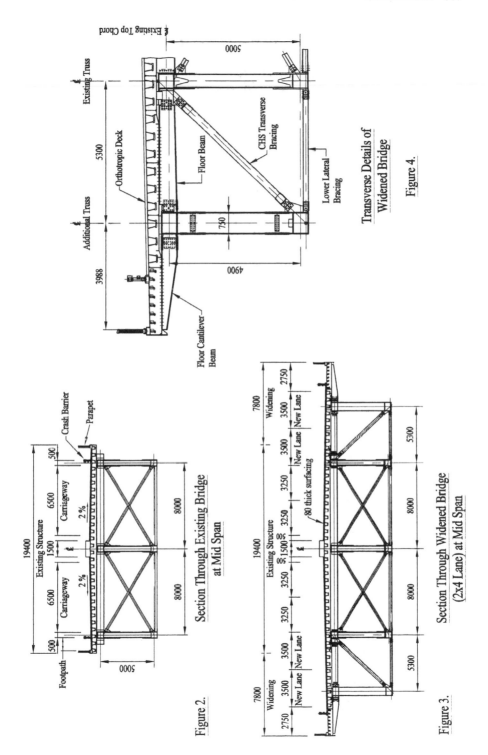

Figure 2.

Section Through Existing Bridge
at Mid Span

Figure 3.

Section Through Widened Bridge
(2x4 Lane) at Mid Span

Transverse Details of
Widened Bridge

Figure 4.

The foundations had been enlarged as part of the reconstruction, thus providing sufficient additional foundation capacity to carry the loads imposed by the widened bridge. However, widening of the piers above water level will be necessary. This is to be achieved by extending the existing piers in prestressed concrete construction. The prestressing arrangement adopted gives a positive connection between new and existing substructure.

The design was undertaken in accordance with the Standard for Korean Highway Bridges (which is similar to AASHTO) and supplemented by international standards. Generally all reinforced concrete elements were designed using factored load techniques whereas steel elements were designed on a working stress approach. Highway live loading was DB24 loading and fatigue loading was equivalent to Class I of AASHTO. In addition load effects arising from stream flow, ship impact, wind and earthquake loading were also considered. A peak ground acceleration of 0.14g was used to produce seismic response spectrum.

CONSTRUCTION ASPECTS

Consideration of the construction method and the sequence of operations required to convert the existing structure into the required new configuration is an essential part of the design process for bridge widening schemes. This is particularly so when, as in this case, some redistribution of dead load from the existing to the new additional structure is required.

The construction methods and sequence of working must be arranged to satisfy the following criteria:
i.   The required distribution of loading in the final modified structure must be achieved.
ii.   At all stages during the widening work, stress levels in both the existing and new parts of the structure must be kept within defined design limits.
iii.   Disturbance to traffic using the existing structure must be kept to the minimum practicable.
iv.   The required transverse profile across the widened deck must be achieved by introduction of camber in the new trusses.

In the case of the Songsu Bridge, the construction methods are largely dictated by the requirement to transfer some of the existing dead load to the new additional main trusses, with this being achieved as described previously. This load transfer operation causes significant vertical deflections to occur between the existing and new structural members during construction. As a result, the new steelwork can not be rigidly connected to the existing structure immediately after its erection, and provision must be made for the rotations which will occur during the construction sequence in the end connections of the new deck cross-beams and lower transverse bracing members. This is achieved by incorporating a temporary hinge pin for the initial connections between the new truss and the existing structure.

The new truss steelwork is of welded construction, but with all site joints made with high strength bolts. The new orthotropic deck is of all welded construction. Where the additional transverse steelwork is connected to the existing structure, the only practicable means of achieving this was found to be the use of site welded brackets and stiffeners, see Figures 5A and 5B. However, the number of these has been kept to a minimum, and the details and weld

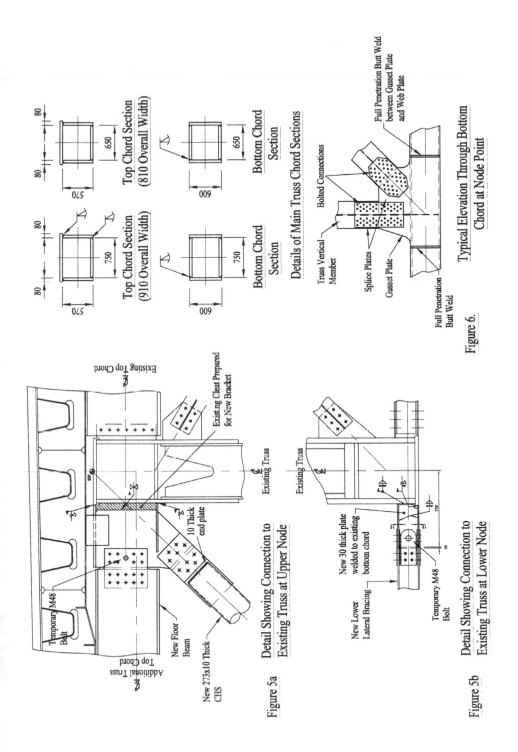

Top Chord Section
(810 Overall Width)

Bottom Chord
Section

Top Chord Section
(910 Overall Width)

Bottom Chord
Section

Details of Main Truss Chord Sections

Full Penetration Butt Weld
between Gusset Plate
and Web Plate

Bolted Connections

Truss Vertical
Member

Splice Plates

Gusset Plate

Full Penetration
Butt Weld

**Typical Elevation Through Bottom
Chord at Node Point**

Figure 6.

Existing Top Chord

Existing Cleat Prepared
for New Bracket

Existing Truss

10 Thick
end plate

Temporary M48
Bolt

New Floor
Beam

New 273x10 Thick
CHS

Additional Truss
Top Chord

**Detail Showing Connection to
Existing Truss at Upper Node**

Figure 5a

Existing Truss

New 30 thick plate
welded to existing
bottom chord

New Lower
Lateral Bracing

Temporary M48
Bolt

**Detail Showing Connection to
Existing Truss at Lower Node**

Figure 5b

arrangement made as simple as possible consistent with achieving the structural requirements of the connection.

As the method of erection of the additional steelwork and the sequence of its connection to the existing bridge determines the load distribution in the completed structure, a detailed description and analysis of the effects of these have formed an essential part of the design process. Construction of the widening works will commence at the north end of the structure (Pier 5) and will proceed progressively across the river until the south end pier (Pier 12) is reached. Traffic volumes on the existing structure are currently very high, and the closure of one lane, except for very brief periods, would clearly cause unacceptable levels of delay and disruption. The widening scheme has therefore been designed to enable all principal construction activities to be carried out using floating plant, so that the bridge can remain fully open to traffic throughout the works except for the closure of the existing footpaths. The construction work will proceed generally in the following sequence:

i.   The existing footpaths will be closed, and temporary traffic barriers erected at the edge of the existing slow lanes

ii.   The existing parapets, drainage channels, and other deck edge furniture will be removed.

iii. The deck edge member will be removed.

iv.   Brackets and stiffeners required for the connection of the new steelwork will be welded onto the existing outer main trusses.

v.   The additional main truss steelwork will be pre-assembled into blocks typically 9.6 metres long and with weights ranging up to 60 tonnes. These blocks will be brought to the bridge site by barge and lifted into position by a barge mounted crawler crane. Erection of the truss steelwork will be carried out as balanced cantilevers from each pier, with these then being linked by the erection of a central 48m long section. Erection of the deck and upper and lower cross-beams will proceed in parallel with truss erection, with their connections made but loosely bolted allowing free rotation in the vertical plane at their connection to the trusses. Figure 7 shows the cross section of the structure after completion of additional truss sections.

vi.   The new bottom plan diagonal bracing and the deck cantilever girders will be erected. The deck cantilever joints will be fully bolted, but those of the bottom plan bracing will only be loosely connected at this stage.

vii. The new orthotropic deck plates will be erected and temporarily tack bolted down to the new cross-girders.

viii. The joints between the deck plates will be welded, working in a predetermined sequence to minimise any misalignment and distortion resulting from weld shrinkage effects. At this stage, the joint between the new and existing orthotropic deck will remain unwelded.

ix.   With erection of the new steelwork substantially complete, the operation to transfer a proportion of the existing structure dead load to the new trusses can be carried out. This will commence by placing temporary ballast loading directly over the centre-line of the new trusses. The effect of this will be to deflect these downwards sufficiently to correctly align the joints in the transverse members connecting them to the existing

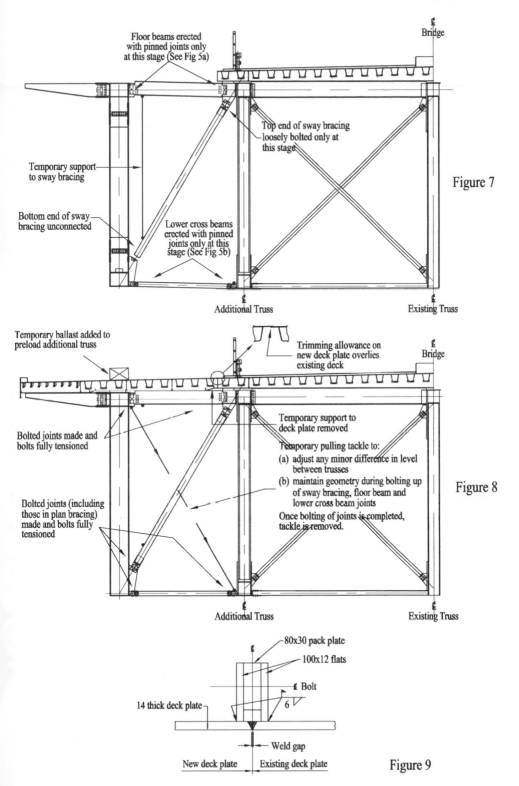

Floor beams erected
with pinned joints only
at this stage (See Fig 5a)

Top end of sway bracing
loosely bolted only at
this stage

℄
Bridge

Temporary support
to sway bracing

Bottom end of sway
bracing unconnected

Lower cross beams
erected with pinned
joints only at this
stage (See Fig 5b)

Figure 7

℄
Additional Truss

℄
Existing Truss

Temporary ballast added to
preload additional truss

Trimming allowance on
new deck plate overlies
existing deck

℄
Bridge

Temporary support to
deck plate removed

Bolted joints made and
bolts fully tensioned

Temporary pulling tackle to:
(a) adjust any minor difference in level
between trusses
(b) maintain geometry during bolting up
of sway bracing, floor beam and
lower cross beam joints
Once bolting of joints is completed,
tackle is removed.

Bolted joints (including
those in plan bracing)
made and bolts fully
tensioned

Figure 8

℄
Additional Truss

℄
Existing Truss

80x30 pack plate

℄

100x12 flats

℄ Bolt

14 thick deck plate

6

New deck plate

Weld gap

Existing deck plate

Figure 9

bridge. The cross-section of the bridges at this stage of construction will be as shown in Figure 8.

x.   As the stiffness of the existing structure is not precisely known, some variation of the ballast loading, together with the use of temporary pulling tackle may be necessary to exactly align the joints. As this will slightly affect the redistribution of dead loading, allowance has been made in the design for the anticipated variations in stiffness due to bolt slippage and residual welding stresses.

xi.   The transverse joints between the new and existing steelwork will be fully bolted, following which the temporary ballast will be removed, causing the load transfer to occur.

xii. The final longitudinal deck plate weld between the new and existing orthotropic deck plates will then be made. As the existing deck is subject to vibration and stresses from traffic using the bridge, before welding is commenced, temporary connections will first be installed at frequent intervals (Figure 9) to rigidly clamp the joint and prevent any relative movement during welding.

xiii. The widening of the structure will be completed by the erection of the connecting ramp trusses and the concreting of their composite deck.

xiv. Finally the widened carriageway will be surfaced, and new parapets, lighting columns etc will be installed.

CONCLUSION

The design of the Songsu Bridge widening has proved a significant technical challenge given the various constraints. The solution developed by High-Point Rendel permits widening of the existing bridge whilst it remains open to traffic by the addition of new parallel trusses and an extension to the existing deck plate. The design employs a complex erection sequence to achieve the required state of stress and profile in the widened structure. This requires a full understanding of the existing bridge and how it can be modified to achieve widening. The techniques used in the widening of the Songsu Bridge could be employed for the widening of other bridges in similar situations. High-Point Rendel continue to provide input during the construction stage.

ACKNOWLEDGEMENTS

The authors acknowledge both the Seoul Metropolitan Government (SMG) and Hyundai Engineering & Construction Ltd. for permission to publish this paper. Also to Chun-Il Engineering Consultants Ltd., particularly Mr. Park, Young-Han who provided invaluable assistance with communication and translation in the preparation of our design deliverables.

Widening and strengthening of Kingston Bridge, London

J.H.W. COUNSELL (SYMONDS GROUP LTD)
P. A. NOSSITER (M.J. GLEESON GROUP PLC)
England.

ABSTRACT: This paper describes the planning, design and construction of a scheme to widen and strengthen a historic multi-span-masonry arch bridge across the River Thames. The history of a river crossing at this point is discussed and the features that make Edward Lapidge's structure of 1828 a part of the English national heritage are described. In order to deal with modern day traffic levels it has been necessary to develop a scheme to widen the structure and strengthen the existing arches. This has had to be carried out in a manner that maintains the bridge's attractive appearance and recognises all local environmental conditions.

1. INTRODUCTION AND HISTORY

Figure 1

Kingston Bridge, London, UK, Figure 1, is a strategic crossing point over the River Thames in South West London. It carries a total of approximately 50,000 vehicles per day with some 2000 vehicles per hour crossing in each direction at peak times.

There has been a crossing of the Thames at Kingston since Roman times. It is believed that the first bridge of timber construction was built in AD43 following the Roman invasion of England. Over the centuries it was rebuilt and repaired on several occasions.

The wooden bridge was eventually replaced in 1828 by a masonry arched bridge, Figure 2, which was widened on the south side in 1914. Spanning the river are five elliptical arches with Portland stone façades with a bold cornice and balustraded parapet. Semi-circular cut waters carry flat panelled piers surmounted by balcony projections that form recesses in the balustrading. Above the springings the 1914 bridge is separated from the 1828 bridge by a small gap which can only be seen from the river. The façade of the 1914 bridge is a replica of the 1828 original using similar Portland stone and ornamental features.

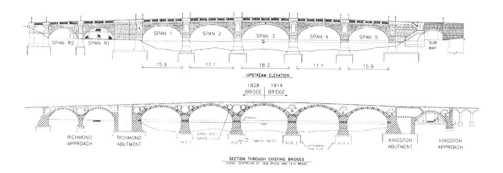

Figure 2

The nearest alternative road crossing points on the Thames are at Richmond, 7 km downstream, and Hampton, 4 km upstream.

The bridge is part of the national heritage and in 1951 was granted listed status. This places extra control on any work that is carried out to maintain the structure and also any alterations.

2. SITE CONSTRAINTS

The bridge is an integral part of the road network serving Kingston town centre, the wider area of south west London and beyond. It is therefore essential that any work on the bridge does not disrupt the flow of traffic and if unavoidable should be carried out only during off peak times.

The river at the bridge is non-tidal with flow being controlled at Teddington weir 4 km downstream. The west bank of the river lies within the Thames flood plain so ground levels in this area cannot be increased because of the risk of flooding elsewhere during high flow conditions. Although no longer used by any significant commercial traffic, the river is a valuable amenity used by pleasure craft particularly in the summer season. Navigation must be maintained whilst carrying out any maintenance work to the bridge.

The areas surrounding the bridge are environmentally sensitive with parkland on the west side and the historic market town of Kingston on the east. Restaurants and cafes overlook the river on the Kingston side. Both sides have riverside footpaths and are within designated conservation areas which recognises their historical and visual importance with the bridge acting as a focal point.

The bridge carries numerous services including electricity and telecommunication cables. A majority of these are located in the south side footway.

3. INSPECTION AND LOAD ASSESSMENT

Kingston Bridge has been the responsibility of The Royal Borough of Kingston Upon Thames since 1986. In September 1993, consulting engineers Travers Morgan were commissioned to inspect the condition of the bridge and carry out a load assessment.

Prior to carrying out the load assessment detailed inspections were undertaken to examine the bridge's condition. The inspections revealed numerous defects including:

1. Displaced and cracked brickwork in the hidden voids over the piers. (These internal voids are shown in Figure 3).

Figure 3
Internal Voids

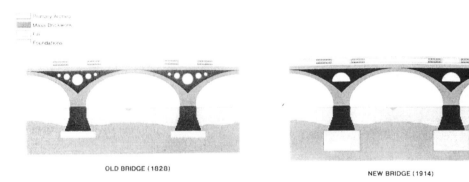

OLD BRIDGE (1828)

NEW BRIDGE (1914)

2. Lack of effective drainage and waterproofing causing penetration of water through the bridge fabric.

3. Open joints, loss of mortar and minor cracking to the brickwork arches.

As a result of the inspections, exceptionally heavy vehicles (over 38 tonnes) were prohibited from crossing the bridge and a programme of regular inspections instigated to monitor its condition.

The subsequent load assessment identified a number of elements of the bridge with a reduced load bearing capacity as shown in Table 1 below.

Table 1: Assessed Load Carrying Capacity

Element	Assessed Load Carrying Capacity (tonnes)	
	1828 Bridge	1914 Bridge
Pier Voids	3	40
River Arches	3	3
Richmond Approach Arch	3	3
Kingston Approach Arch	10	13

The imposition of a permanent 3 tonne weight limit was rejected due to the severe disruption it would cause to traffic, local businesses and the community as a whole. Access to Kingston town centre by public transport would also be severely restricted if buses were banned from crossing the bridge. It was therefore decided to carry out investigations to identify methods of strengthening the bridge. Whilst the investigations were being undertaken the bridge would continue to be monitored.

4. STRENGTHENING

An established technique for strengthening arches is to construct new external arches to support the existing. Alternatively, the strengthening could be contained within the fabric of the bridge which upon completion would be hidden from view. In deciding which method to adopt for Kingston Bridge priority was given to maintaining the appearance of the bridge and hence the external method was rejected.

A comparative study was undertaken to assess the relative technical merits of five different strengthening techniques. As part of this process close consultations took place with English Heritage who, whilst not objecting to the principle of strengthening, requested as little interference as possible with the existing internal fabric of the bridge. The study concluded that, on balance, a concrete composite saddle was the preferred option as it preserved a majority of the internal fabric and in particular the hidden voids over the piers.

The strengthening scheme would require work to be carried out from the carriageway surface with a reduced number of traffic lanes and for traffic to be diverted around the working areas. Studies identified that two traffic lanes were the maximum that could be kept open during the works which would mean a lane reduction in each direction. To assess the implications of this on the area-wide road network a detailed traffic analysis was carried out. This showed substantial impact over the whole area and the option was rejected. Investigations were then undertaken to identify possible means of maintaining four lanes of traffic.

5. TRAFFIC SOLUTIONS

Studies were undertaken to examine temporary bridge and widening options. This exercise involved widespread consultations with affected parties including the public, businesses, local authorities, and English Heritage. An environmental impact study was also carried out to assess and compare their likely impact on the area.

A permanent widening scheme was eventually chosen due to the reduced environmental impact and also the opportunity to introduce permanent facilities for buses and cyclists.

6. DETAILS OF WIDENING

The widening is on the upstream side of the 1914 bridge and adds 6.6 m to the current width of 17.5 m. A section through the widened bridge is shown in Figure 4.

The river spans match the existing arrangement of 5 elliptical arches symmetrical about the centre line of the central arch.

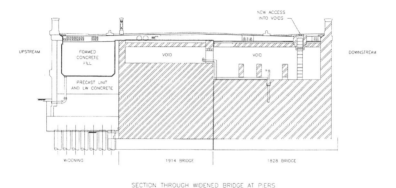

SECTION THROUGH WIDENED BRIDGE AT PIERS

Figure 4

The new piers comprise reinforced concrete stems founded on a reinforced concrete pile cap with bored cast in-situ concrete piles. The piers are faced with natural Portland stone.

Precast concrete arch units are used for constructing the spans of the widened bridge to minimise the disruption to river traffic during construction. Lightweight materials have also been specified to minimise the loading on the foundations and hence reduce potential settlement.

The precast units forming the spans of the widened bridge are made continuous over the crown using coupled reinforcing bars with in-situ concrete. These precast units are faced with masonry and brick to match the existing arch profile and finishes. An in-situ lightweight reinforced concrete saddle is cast on top of the precast units to form a composite section. Starter bars project from the extrados of the precast units to provide continuity with the saddle.

The haunches of the arch are filled with a low density foamed concrete fill overlain with half a metre of higher density foamed concrete below the carriageway surfacing. Reinforced concrete spandrel walls are cast in-situ on top of the arches and faced with new Portland stone to match the existing 1914 elevation.

7. DESIGN ANALYSIS

7.1 Widened Bridge

The spans of the widened bridge were analysed using a LUSAS 2-dimensional finite element model. Each span was idealised as being restrained at each end by the piers. To accommodate shrinkage and thermal effects movement joints are incorporated at the end of each arch.

The transverse stiffness of the arch was ignored in the analysis. A plane stress analysis of unit width was therefore carried out using plane stress elements. In addition the stiffening effects of the spandrel walls were not taken into account in the design of the composite reinforced concrete arch barrel. A local transverse analysis was carried out to determine the forces in the spandrel walls.

End support conditions were modelled to simulate the degree of fixity provided during the construction of the bridge and also in the permanent situation. The analysis therefore considered the three pinned, two pinned and fixed situations.

7.2 Strengthening

The superstructure was modelled globally as a five span masonry arch using the program ARCHIE. This analysis was used to investigate the out of balance forces in the piers and hence to check their stability during the construction phases and when strengthening is completed.

The analysis of the lightweight reinforced concrete saddle and brick arch composite behaviour was based on a non-linear analysis of the central river span. A plane stress two dimensional finite element model of unit width was developed using LUSAS. The material model for the existing brick arches assumed no tensile stresses. The reinforced concrete saddle was modelled using linear elastic properties based on uncracked section behaviour.

The behaviour of the finite element model was benchmarked against a single span ARCHIE model by comparing the order and position of hinge formation under a point load positioned close to midspan.

8. RIVER WORKS

Cast in-situ bored piles were designed on the basis of mobilising skin friction within the London clay substrata. Experience of this material in the London area is extensive and it is generally uniform and consistent. The 600 mm diameter concrete piles were reinforced to accommodate the possible lateral loading on the pile group arising from differential live loads on adjacent arch spans.

Piles were sleeved through a new 2 m thick base of mass concrete installed at the bottom of the cofferdams. This base acts as a seal to the clay formation and prevents softening of the upper zones of the clay . It also helps resist the effects of heave following excavation of the cofferdam. Once widening works are complete this mass concrete also provides resistance to scour effects.

9. NEW ARCHES

The restrictions on temporary works in the river and the potential for disturbance to navigation were the factors that primarily led to the choice of precast reinforced concrete for the main structural members of the new arches. In addition there is the advantage of factory conditions for manufacture which was utilised in the design by incorporating the brickwork facing to the soffit to match the existing bridges as an integral part of the precast unit.

Twelve units are cast for each span, each single unit being a half span and contained within dimensions which can be carried on UK roads without special escort arrangements. Figure 5 shows details of the units.

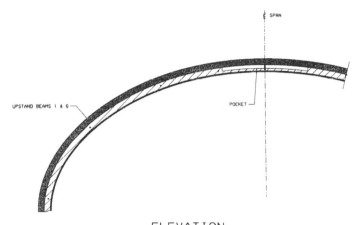

ELEVATION

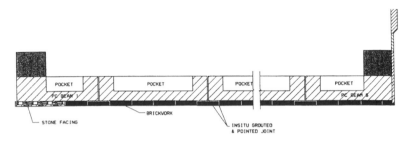

CROSS SECTION

Figure 5
Precast Arch Units

10. CONSTRUCTION.

The contract to widen and strength the bridge was awarded to M.J. Gleeson Group Plc in October 97 with a construction period of 32 months under the NEC 2nd Edition, Engineering & Construction Contract.

10.1 Access & Stone Removal

The area of the construction site is extremely restricted by the River Thames, by the close proximity of the protected tree area owned by Hampton Court Palace on the Richmond side and by numerous commercial properties on the Kingston side. Access to the site is only available via a narrow temporary access across a tree protection area, to the riverbank immediately south of the bridge. All equipment and materials have been delivered to site via this access.

The planning consent for the bridge involved the re-use of as much of the original Portland stone façade as possible from the southern face. To this end the embrasure and cutwater stone together with the cornice, bottles, plinths and copings were surveyed, plotted and numbered. After careful removal these were taken to store off site ready for re use in exactly the same configuration but on the newly widened part of the bridge at a later stage.

10.2 Cofferdam Construction To Piers

Cofferdams could not be completely closed and had to be sealed onto the existing piers by the installation of bulkheads cut to the profile of the brick and stone piers. 9m long (20W) sheet piles were driven to provide a 4m penetration into the London clay. Then excavation and placing of concrete to seal the base was undertaken underwater to prevent the need for a bottom frame. After installation of the top frame the cofferdam was pumped out to expose the existing base. This revealed a steel caisson which had been used for the construction of the foundation to the 1914 structure as shown in figure 6.

Figure 6

This caisson was not indicated on the original as-built drawings and proved a major problem due to the extensive use of steel plate in its construction. The piles had to be sleeved through this caisson. Numerous methods were attempted to break through the steel and concrete construction to provide a 650mm-dia hole to allow the piles clearance from the existing base. These methods included thermic lancing, high pressure water jetting, diamond stitch drilling, percussive drilling and rock drilling with bursting compound. The most successful method was the traditional compressor and breaker together with financially motivated operatives. The lack of space within the cofferdams precluded the use of an auger rig and the piles were installed by tripod rig restricting output to one pile per rig per day.

The pier construction above the piles was heavily reinforced concrete with new 200 mm thick Portland stone cladding to the sides. The original cutwater stone was returned to the front of each pier. Portland stone Roach bed was used below the splash zone and Whitbed above.

10.3 Pre-Cast Units

The design called for the arches to be pre-cast concrete with brick slips and Portland stone bands to match the existing arches. The shape of the southern face of existing structure had to be accurately determined by undertaking a detailed laser survey. In addition, the stonework subcontractor, Universal Stone, was given the task of drawing and plotting every stone in the voussoir face. These plots were used for the stone manufacture. To ensure that an accurate representation was achieved these two surveys were overlaid and a tracing made of the true shape of the arch with the exact location of each voussoir stone. This tracing was then supplied to the pre-cast manufacturer Macrete (Northern Ireland) and used to manufacture the timber moulds for the pre-cast arches. When the moulds were made Macrete sent timber templates to site and these were erected on the bridge face to check the shape prior to casting. This proved to be a time consuming, but worthwhile exercise as it produced extremely accurate results in the shape and alignment of the final units.

Because of the dimensional constraints which required matching the existing arch geometry the units were manufactured to much tighter tolerances than normally required for precast beam construction. In addition to the shape of the units great care had to be exercised over the width of each unit. The horizontal alignment of the new voussoir was not necessarily parallel with the existing voussoir face. Consequently the first unit to be installed against the existing face was tapered to accommodate any mis-alignment and to produce a uniform joint. The pre-cast units were match-cast against the preceding unit to ensure extremely close fitting of the units when installed. Figure 7 shows one of the 80 pre-cast units being lifted into position by the 50 tonne crawler crane mounted on a unifloat pontoon. The design required a 20mm gap between the units at the crown and that they be supported until the concrete placed in the pocket (see figure 5) had reached the required strength. This support was achieved by the use of Mabey Towers and tie rods.

Figure 7

Once the Pre-cast units had been installed the heavily reinforced saddle slab could be constructed. To reduce the weight on the pre-cast arches the saddle slabs and spandrel walls are cast with 40N Lytag concrete with a max oven dry density of $1800Kg/m^3$ and max fresh wet density of $2000Kg/m^3$. The shape of the units and the requirement to balance loading over the arches called for complex curved shutter as shown in figure 8.

Figure 8

A waterproofing membrane was spray applied to the saddle slabs and spandrel walls and the remaining void filled with foam concrete with a density of $700Kg/m^3$ to within 0.5m of the road construction and topped off with foam concrete of $1400Kg/m^3$ density. A 46-way service duct was cast into the foam concrete and, in some spans within the reinforced saddle slab. These enabled the diversion of services from the existing structure to allow access for the strengthening works in the later phases. With the road construction laid the traffic could be diverted and work could start on the strengthening of the existing bridges.

10.4 Strengthening Work

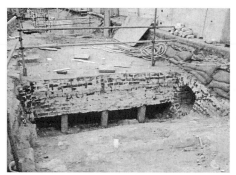

Figure 9
1828 Bridge showing voids

The listed building consent was granted subject to English Heritage approval of the contractor's detailed method statement for the part demolition of the voids in the 1828 structure and the installation of the saddle slab. In figure 9 the existing small void can be seen on the right of the photograph. At this location the void has been removed completely and supports have been installed. Only the brickwork to enable the saddle slab to be constructed has been removed from the intermediate void.

The reinforced slab is bonded to the extrados of the brick arch by epoxy grouted stainless steel hoops drilled 300mm into the 600mm brickwork. In total there are 6500 hoops which act in shear to transfer load into the brickwork arch and down into the piers. Lightweight construction has again been used to reduce the load on the existing foundations. The saddle slabs are cast with Lytag concrete and foam concrete is used to backfill to the underside of the road construction.

10.5 Stone Facing

As much stone as possible was recovered from the existing structure but not the original voussoir and spandrel wall stones which form an integral structural part of the bridge. As a consequence in excess of 2300 individual items were manufactured for the new structure from Portland stone and installed to exactly replicate the original façade.

11. CONCLUSION

The solution to strengthen Kingston bridge has had to take into account a wide range of issues beyond the pure engineering. By permanently widening the bridge a balance is believed to have been achieved between the need to minimise disruption to users of the bridge and to preserve the character of the bridge and its surroundings. Construction has had to recognise these issues and methods have been adopted which preserve most of the fabric of the historic structure. In addition, the development of the scheme has required close liaison with all affected parties including commercial interests, government bodies, local authorities and the public. The client, The Royal Borough of Kingston upon Thames, has striven to ensure that all these parties have been closely involved and kept informed of progress both during design and construction.

A14 Huntingdon Railway Viaduct

A. WAKEMAN

Thorburn Colquhoun, Bedford, UK

JOHN St LEGER

W. S. Atkins, Cambridge, UK

Description of Viaduct

Huntingdon Railway Viaduct is a six span structure carrying the A14 Trunk Road over the East Coast Mainline Railway and Huntingdon Station, to the west of Huntingdon. A bridge carrying the heavily trafficked Brampton Road into Huntingdon over the railway at the same location also lies beneath the viaduct.

The main span of the viaduct consists of post tensioned concrete cantilevers with a drop in span of pre-cast concrete box beams, supported on half joints at either end. The overall length of the main span is 64 metres. The approach spans, which are 32 metres long, consist of pre-cast beams. The piers and abutments are of conventional reinforced concrete construction and the wing walls are formed of concrete faced reinforced earth.

Photo 1

The viaduct was constructed in 1975 as part of the A604 (now A14) Huntingdon Bypass Scheme. The design was a contractor's alternative to the conforming steel structure.

Photo 2

The cantilever spans, which support the main drop in span, were formed adjacent to the railway and slid into place during possessions of the track. General views of the structure are shown in Photographs 1 and 2.

Bridge Management 4, Thomas Telford, London, 2000.

Inspection and Assessment prior to 1996

Prior to 1996 the structure had been assessed as part of the Roads for Prosperity Widening Programme. A number of areas of non-compliance were reported including insufficient bursting reinforcement beneath the bearings in the tops of the piers. This was confirmed by vertical cracking observed around the tops of the piers. The half joints also gave cause for concern because their strength relied on some of the deck's main post-tensioning tendons, which were anchored in the half joints. The condition of these tendons could not be ascertained but cracking had been observed on the outer faces of the half joints.

As a result of these findings vehicles crossing the bridge was restricted to those complying with the 40Tonne Construction and Use Regulations.

A special post-tensioned investigation of the structure was carried out in 1994. Intrusive investigations had been undertaken but their number and location were severely restricted by the lack of readily available access to the tendons. The viaduct, although of cellular construction, contained no man access holes into the deck voids. The limited number of holes drilled into the ducts indicated voiding of the protective grout and some corrosion of the tendons. A permanent access port into one of the exposed ducts was installed to enable future monitoring of the condition of the tendons.

Cracking of the concrete at the half joints was also noted at this time. This gave considerable cause for concern because of the location of the suspended span above the East Coast Mainline, Huntingdon Station and Brampton Road.

The approaches to the viaduct abutments are supported by concrete faced reinforced earth retaining walls. These are among the first structures of this type to be constructed on the UK trunk road network. An investigation of the condition of the reinforcement that ties the pre-cast concrete facing panels into the granular fill that forms the approach embankment showed that the top layers were badly corroded. The top section of the wall was strengthened in 1994 by installing stainless steel anchors in holes drilled through the concrete facing panels and into the backfill. A GRP cover protects the exposed heads of these anchors.

Management of the Structure from 1996

In 1996 Thorburn Colquhoun took over management of the structure from the Local Authority as part of their new responsibilities for Trunk Road Super Agency No. 8. In these duties they are assisted by W S Atkins, who previously worked for the Local Authority.

An immediate priority was to develop a maintenance and refurbishment strategy for the viaduct. All previous reports into the structure were reviewed and a strategy for implementation of remedial schemes discussed and agreed with the Highway's Agency. It was clear from the start that the available windows for work on the topside of the bridge would be severely restricted by the volume of traffic using the viaduct and the proximity of other highway maintenance schemes requiring traffic management measures on the carriageway of the A14.

Works identified included the following:
- Strengthening of the pier tops to resist the bearing splitting forces
- Provision of access into the deck voids to allow further investigation of the post-tensioning tendons
- Investigation of the post-tensioning tendons
- Sensitivity analysis on the post-tensioning system to determine what loss of pre-stress could be tolerated
- Replacement of worn out expansion joints
- Replacement of worn out deck waterproofing membrane
- Identification and installation of monitoring system to warn of failures in the post-tensioning system.

Pier Top Strengthening

Strengthening the pier tops was identified as a first priority. Steel collars were designed to clamp round each pier and the concrete put into triaxial compression by means of flat jacks installed between the collar and the concrete. The space between the steel collar and the pier was then filled with non-shrink grout to provide permanent containment of the concrete. The flat jacks are filled with hydraulic oil and could be re-inflated if the need arose.

Photo 3

The A14 trunk road is extremely heavily trafficked with 65,000 vehicles per day passing over the viaduct. For operational reasons, lane or carriageway closures are generally only permitted at night, placing severe restrictions on any work affecting the top surface of the structure. The pier strengthening required night time lane closures during jacking operations.

Removable GRP fascia panels to improve the aesthetics of the strengthened piers covered the new steelwork. This work was completed during the first half of 1998 (See Photograph 3).

X Ray Investigation Trials

The main span of the viaduct is formed of post tensioned back and cantilever spans supporting a pre-cast prestressed beam suspended span on half joints. The post tensioning tendons lie within the walls of the cellular deck section and follow a sinusoidal profile between the end of the back span and the half joints. Since no man access into the deck cells was provided in the original design, the tendon ducts could only be examined at the high point of their profile and from the outside edge of the deck. Previous examination at these locations had identified areas of voiding in some ducts and a means of examining the other tendons, preferably without breaking into the ducts, was sought.

Trials had previously been undertaken in similar circumstances on other structures using X Ray equipment to photograph ducts and pre-stress tendons inside the concrete. The results had been encouraging but the images produced were indistinct. One possible cause of this was thought to be under exposure due to time constraints on site, another the lack of a stable platform for the X Ray equipment which had typically been mounted on an underbridge unit.

The overall thickness of the deck on the viaduct is 1.5metres. However the combined concrete thickness of the top and bottom flanges is only 600mm. Offsite trials by Materials Measurements Ltd using Betatron X Ray equipment indicated that it would be possible to obtain a successful image if the X Ray source was located on a stable platform beneath the deck and the film on top of the deck surfacing. A trial was undertaken during the pier-strengthening contract in order to take advantage of the traffic management arrangements.

The Betatron equipment was supported on a scissors lift under the deck for stability and the film placed on the road surface beneath sandbags for protection and to keep it still and pressed against the deck. Considerable liaison with HSE was undertaken to establish the size of exclusion zones both below and on top of the deck. The Zone beneath was 25 metres. The A14 above could not be closed entirely and thus the pictures were taken at night using rolling blocks to ensure that no traffic was on the deck being investigated whilst the film was exposed.

A particular feature of the equipment used enabled the X Ray generator to be immediately shut down and made safe, thus allowing emergency vehicles to pass through the works area at short notice.

An exposure time of approximately 15 minutes was required to produce a satisfactory image and, because the film and source were kept perfectly still during the process, this was usually achieved using an exposure counter to accurately control a series of shorter exposure times for each shot.

The results of the trial very good, bearing in mind that the X Rays were taken through 1500mm of deck and the tendons are located in the top flange. The picture definition was better than anticipated and no voids were detected in the top flange ducts examined. However most of the tendons in the deck are in the internal webs of the cellular box construction and this method could not be used to radiograph them. It was therefore decided that the only way to examine the bulk of the tendons was to provide access holes into the deck voids and to carry out a more extensive X Ray from inside the deck.

Access to the Deck Voids

The scaffolding erected during the pier strengthening was also used to provide access to form enlarged drain holes into some of the viaduct box cells. This allowed photographs of the interior to be taken using a modified hand held camera. The pictures revealed considerable amounts of construction debris, including timber shuttering. This, coupled with a need to gain access to a larger selection of tendons to examine by X Ray methods, led to preparation of designs to form man access holes into the deck voids.

A trial installation of four new access holes was undertaken to prove the viability of the proposed method. X Ray was used to locate the existing reinforcement and the stress release method was used to determine the stress in the bottom flange of the cantilevers adjacent to the position of the proposed holes. Vibrating wire strain gauges were placed on the concrete adjacent to the new hole, as shown in Photograph 4, and the concrete removed using water jets in order to reduce possible micro-cracking in the surrounding concrete. The reinforcement in the hole was also strain gauged and, once the readings had

Photo 4

stabilised and were within limits predicted by a sophisticated finite element model of the bottom flange, the reinforcement was cut. A steel collar was then inserted into the hole, welded to the projecting deck reinforcement and the space between collar and concrete filled with a proprietary mortar.

Following these trials a further seventeen access holes were constructed allowing more extensive investigation of the condition of the tendons to be undertaken.

Investigation from Inside the Deck Cells

Access holes were provided into 58% of the deck voids. Visual inspection identified large amounts of debris left inside the voids during construction of the viaduct. In addition a number of areas of severely honeycombed and voided concrete were found. A more extensive X Ray survey was then carried out from inside the cells with the Betatron equipment mounted inside the voids on scaffold rails (see Photograph 5). Approximately 53% of the available ducts were X Ray'd.

Photo 5

The results of the survey indicated some degree of voiding in 75% of the ducts examined, of which 18% were considered significant. Some of the voiding in the ducts coincided with the areas of voided cover concrete and the opportunity was taken to open the duct and confirm the findings of the radiographs for those locations. Photograph 6 shows the type of the voiding found within the deck.

Photo 6

Acoustic Monitoring

Concern over the condition of the post tensioning cables and the critical location of the viaduct prompted a review of available methods of monitoring the structure for evidence of strand fracture. An acoustic monitoring system, developed by Pure Technologies of Canada, was identified as the most appropriate system currently available. The Transport Research Laboratory (TRL) had been conducting trials on the system at their research centre at Crowthorne and had been impressed by the results in comparison with other methods examined by them.

The system comprises sensors placed on the viaduct soffit in a predetermined pattern. The sensors respond to sound events on the structure and transfer the information to a data-logger installed in a secure cabin adjacent to the viaduct. Software within the system filters out all events that have a significantly different sound signature from that of a breaking strand. Events, which fall within a pre-set number of parameters, are reported to Pure Technologies office in Calgary, Canada via the Internet where a detailed analysis of the results is undertaken.

The Highways Agency ordered a trial of the system in a working location, and acoustic sensors have been installed to monitor one of the viaduct's cantilever and back-spans. The system was commissioned in June 1998. Photograph 7 shows a typical sensor on the deck soffit and the connecting cable running back to the data-logger.

In acoustic terms, Huntingdon Viaduct has been a particularly noisy structure during most of the period whilst the monitoring system has been installed. The road joints are failing and have generated significant noise levels and a series of contracts have been undertaken to strengthen the pier tops and install access holes into the deck voids. The acoustic monitoring system has had to be finely tuned to exclude most noises outside the wire break spectrum in order to limit the amount of data being transferred to Canada for analysis.

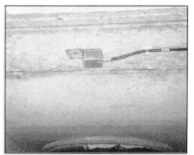

Photo 7

Once installed, firing a Schmidt Hammer onto the concrete to check if the event was recorded, tested the systems response. A Schmidt Hammer on concrete gives a readily identifiable signature within the range monitored for wire breaks. More accurate simulations of a wire break were also caused by the impact of a spring-loaded ball bearing onto the surface of the concrete. These trials were conducted both at pre-selected locations on the viaduct and blind.

In order to establish the response of the system to strand breakage whilst the viaduct is in service TRL, in association with Pure Technologies, developed and undertook simulated strand breaks within one of the deck voids. The trial consisted of a short length of prestressing strand installed in a steel frame attached to the inner face of one of the deck voids. Individual wires that make up the strand were partly cut through and the strand tensioned by means of a hand pumped hydraulic jack. The acoustic monitoring system recorded wire break events when each of the wires fractured.

It is intended to monitor the structure for two years after which the need for continued monitoring will be reviewed. The programme is approximately half way through and to date no fractures have been recorded.

Sensitivity Analysis

A finite element model of the structure has been developed to predict the sensitivity of the structure to deterioration of the post tensioning system. Work continues to calibrate this model against the actual stresses measured during the access hole cutting exercise. This model will be used in conjunction with the acoustic monitoring to ensure continued safety of the structure.

Replacement of Joints and Waterproofing

Replacement of joints and waterproofing was carried out in the summer of 1999. A thin wearing course surfacing was also laid to reduce road traffic noise. The opportunity was taken at this time to investigate the amount of chloride ingress into the deck.

When completed, the work on the viaduct will ensure continued and improved service from this critical structure on the heavily trafficked A14 M1/A1 Link Road.

Acknowledgements

The Highways Agency for kind permission to publish this paper.

Saving a 75 year old Free State Bridge

APC Oosthuizen, Pr.Eng. C.Eng. MICE, F(SA)ICE,AIStrucE.
AOA Consulting Engineers, Sandton, Rep. South Africa

REHABILITATION SUMMARY

The 75 year old unstable bridge across the Zand River has been saved by re-engineering to serve the Free State community for a further estimated 75 years. The design and construction techniques include a massive new prestressed concrete foundation with robust piles, hydro-demolition of concrete and the increase of the deck vertical clearance. The all inclusive cost was BStg 300 000 which is less than 50% of the replacement value. This project was completed within budget, on time and to high quality workmanship standards.

The regional community greatly needs this vital road bridge which is the link to schools, shops and farming activities. On closure for repair works, despite prior negotiations, the community annexed the bridge with arms. The project was then halted for 3 months while a temporary causeway was constructed to avoid a 45 km detour.

Since the bridge was built in 1921 large floods undermined the caisson foundation in the river causing tilting in the upstream direction. When the resulting kink in the road surface reached over 300mm, stabilisation of the foundation to prevent instability became vital. Further works required were a new pier head to house the straightened deck, modification of the front portals in structural steel for modern vertical clearance (to avoid recurring damage), a new concrete road surface and new expansion joints.

To strengthen the bridge against future floods new and innovative design and construction techniques employed were:

- hydro-demolition of concrete to ensure that the remaining concrete is crack-free and homogeneous;
- prestressing to eliminate drying shrinkage and to guarantee load transfer;
- internal water cooling to guarantee homogeneous concrete;
- structural steel replacement portals to ensure horizontal stability; and
- robust piling methods to penetrate boulders and hard rock.

FOUNDATION STRENGTHENING

The two classic structural steel spans were shipped out from the United Kingdom. The durability of the works has been quite remarkable and should be recognised as a monument of sound design practice. Apart from the tilting of the pier foundation there was no other deterioration. The primary reason for such outstanding durability characteristics is that the initial design stresses were kept low.

The tilting caisson (8,6 x 3,35 x 5m deep) is equivalent to two large double storey office buildings on top of which the four storey high pier (12,2m) stands, making the total constructed height in the river equal to six storeys.

The forces to arrest the further tilting of this substantial structure and to withstand the flood water and debris up to its full height, ie. 7 storeys (21m depth), demand substantial structural members. Accordingly, 6 x 1,05m dia. piles founded about 5,0m through the boulder layer onto sound bedrock about 10m below riverbed level, were required. In order to minimise obstruction forces 3 piles were grouped upstream as well as downstream for the existing caisson.

These piles, which are used for the most severe conditions worldwide, were socketed 0,3m into very hard rock (R5) and then fixed into the pile cap which surrounds the existing caisson.

In order to guarantee the load transfer between pile cap and caisson, the pile cap is prestressed onto the existing caisson. The interface between the existing caisson and the new pile cap is at the heart of the long term durability of the project. For this reason two new construction techniques were employed, namely hydro-demolition of the existing surfaces and prestressing of the new pile cap onto the caisson. The advantages of these two techniques are described as follows:

HYDRO-DEMOLITION OF EXISTING CONCRETE

In order to ensure sound aggregate interlock at the interface, the existing smoothly shuttered concrete surface had to be made rough. The traditional method of employing pneumatic percussion tools results in micro cracks in the in-situ concrete which stays behind. Apart from the structural jeopardy of the remaining concrete, the presence of micro cracks allows the ingress of water which could freeze during winter periods with resultant further cracking and eventual collapse of the structural integrity. In addition, the ingress of water encourages corrosion of the reinforcement which inhibits durability.

In contrast, hydro-demolition eats away and converts the concrete under attack to its original constituents of sand and stone without the introduction of micro-cracking., The body of concrete thus remains homogeneous and sound with a roughened surface to ensure aggregate interlock.

PRESTRESSING OF THE NEW PILE CAP

In order to avoid the development of a drying shrinkage crack between the new massive pile cap (16,0m x 5,0m x 3,8m deep) and the existing caisson, the introduction of prestressing was employed. The prestressing is only located within the new construction but in such a curvilinear shape that the new concrete is stressed onto the existing caisson.

The massive size of the footing, which in all probability is the largest river foundation ever constructed in the RSA, demanded continuous internal cooling of the concrete. The internal concrete temperatures were controlled to within 10 C of the external surface temperature.

NEW STRUCTURAL STEEL PORTAL FRAMES

The existing vertical clearance of the deck for vehicular traffic was 4.3m. Modern day standards require 5.1m. Accordingly, the existing structural steel portals were continually requiring maintenance.

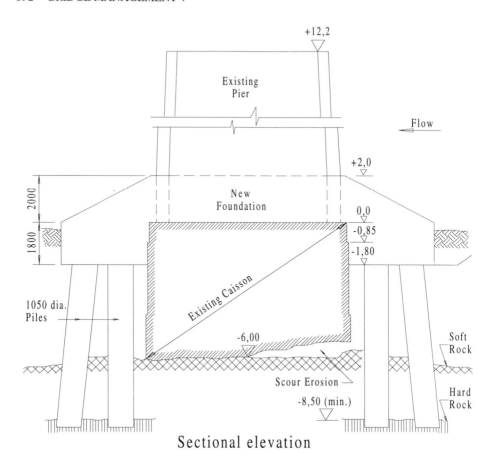

Sectional elevation

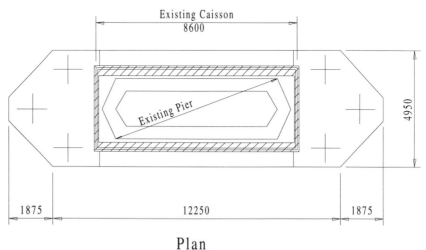

Plan

Figure 1. Existing caisson and new piled foundation

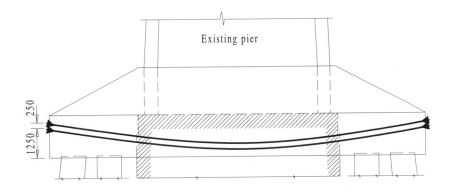

Existing pier

250

1250

Elevation

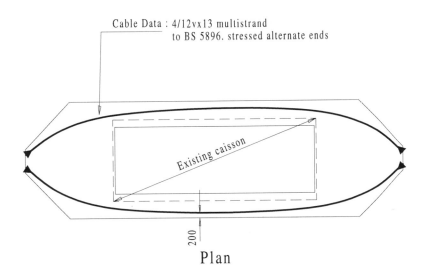

Cable Data : 4/12vx13 multistrand
to BS 5896. stressed alternate ends

Existing caisson

200

Plan

Figure 2. Prestressing of new foundation

Various schemes have been employed for the raising of the vertical clearance to these truss spans. In this instance, the new portal action to the truss ends has been extended from the top of the compression chord but at the same inclination of the sloping top chords. Thereafter, the upper portal members have been increased in section so as to eliminate the need for the conventional braced upper truss. The internal portal members have similarly been extended by short vertical and horizontal portal members. Finally, the beam in plan action was reinstated with horizontal in-plane bracing.

CONCLUSION

The rehabilitation scheme set out to save the 75 year old bridge and to enable it to serve the community for another 75 years. As for the initial design this concept demands that the design stresses are kept low and that only basic materials with a known long life history be employed, namely concrete and steel.

The rehabilitation elements of the scheme are visual and obvious. The new strengthened foundation, the enlarged pier head, the straightened concrete roadway and painted structural steel all blend together to form a new and much needed community facility.

ACKNOWLEDGEMENTS

Permission from the Chief Directorate – Roads and Transport, Department of Public Works, Free State Provincial Government to publish the paper is acknowledged and greatly appreciated.

The combination of the valuable inputs by all the various key personnel from the Department and the Contractor at all levels, including the unskilled labourers from the local community resulted in a stimulating force which delivered the much valued rehabilitated structure. This is even more relevant as the Chief Executive of the Contractor has since tragically passed away.

Horkstow Suspension Bridge

H. Ambarchian, AMIStructE, Posford Duvivier Consulting Engineers

SCOPE OF THIS PAPER

The aim of this paper is twofold;
First it will introduce the Ancholme bridge structures and briefly discuss the Environment Agency's responsibility for their operational maintenance and secondly, it will present the case study on Horkstow Suspension Bridge from the initial inspection and assessment stage, through the feasibility study and the subsequent refurbishment and strengthening stages of the site work.

INTRODUCTION

The Environment Agency Anglian Region is responsible for the operational maintenance of the Seven River Ancholme bridges under the River Ancholme Act 1767. The bridges are Horkstow Suspension Bridge, Saxby Bridge, Broughton Bridge, Castlethorpe Bridge, Cadney Bridge, Hibaldstow Bridge and Snitterby Bridge. The 1767 Act states that " ...the said owner of the river channel (New River Ancholme) shall build proper bridges over the waterways and furthermore forever after support and maintain them for the safe passage of the farm land owner/occupier with cattle and/or carriages ". The validity of the 1767 Act was recently interpreted by the Agency's legal department, which came to the conclusion that "the grant of right of way for carriages will give right of way to most modern vehicles that could physically and conveniently use the access ". This meant that the Agency as owners of these bridges were not only responsible for the upkeep of the structures but also responsible for ensuring that the structures are of sufficient strength and carrying capacity to maintain a safe passage for the general public.

An earlier inspection and assessment of the structures carried out in 1993 under a different commission had reported that the bridges were in need of refurbishment and that the timber decking of the bridges, being Douglas Fir softwood, had an assessed live load capacity of 0 tonnes. From the Agency's point of view maintaining the status quo would have led to closure of the bridges and amounted to a non-conformity with the 1767 Act. See Diagram 1.

FEASIBILITY STUDY

In order to honour their obligations and conform with the requirements of the 1767 Act, the Environment Agency appointed Posford Duvivier in 1995 to undertake a feasibility study of the River Ancholme bridges and their surrounding areas of the river channel.

The study was given two main themes;

To explore possible strengthening schemes in order to increase the potential Live Load Capacity of the bridges without compromising their aesthetics and with due consideration for their listed status, and

To identify key potential environmental impacts associated with the proposed strengthening schemes both during the construction phase and the operational phase for future usage.

The options considered for each bridge comprised the following theme;

" Do - Nothing " : allowing the structures to progressively deteriorate.

" Maintain Only " : maintaining the existing structures in good repair and extending their longevity , but retaining their capacities unaltered at 0 tonnes .

"Strengthening " :with emphasis on keeping any structural changes to a minimum in order to retain the aesthetics of the listed status.

Subsequent to the submission of the Feasibility Study Report, the Agency commissioned Posford Duvivier to produce an Engineer's Report for the preferred strengthening options. MAFF Grant Aid was awarded in early 1996, which then prompted the preparation of tender drawings and documentation for invitation to tender, and for submission to Consultative and Local Authorities.

Following the tendering procedure, C.Spencer Limited was appointed as the Principal Contractor for the works. The site work commenced in the middle of July 1996.

HORKSTOW SUSPENSION BRIDGE
General Information
Horkstow Suspension Bridge is a listed Grade II* structure which is one grade of archaeological status higher than the other bridges in the group. It therefore represents an important building of more than special interest (approximately 4% of the listed buildings in this country). The bridge is a single span structure crossing the River Ancholme to the south of South Ferriby. From the information available there has been a bridge at this location for well over 200 years, although earlier bridges were most probably built in timber.

Sir John Rennie designed the structure circa 1830 for the then owners of the waterway, The Commissioners of the Ancholme Drainage and Navigation. The ironworks of the suspension system may well have come from the Rotherham Ironworks of Joshua Walker and Co. who were also involved with John Rennie Senior and his son in the casting of the ironworks for Southwark Bridge completed in 1819. The masonry work was probably carried out by Jolliffe and Banks, a large contracting firm who were similarly connected with the Rennies on the Waterloo, Southwark and London Bridges.

Photo1. Horkstow Suspension Bridge

ELEVATION
(EXISTING)

PLAN
(EXISTING)

Suspended Span	*42.6m (c/c of masonry towers)*
Back Stay	20.0m (from CL of masonry tower to anchorage end)
Sag of the Chains	3.0m
Height of Masonry Tower	6.5m (From ground level to top of tower parapet)
Deck Width at Entry	3.5m (Through archway in masonry)
Deck Width Between Rails	4.3m

Description of the Structure

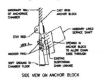

SIDE VIEW ON ANCHOR BLOCK

Diagram 3. Suspension details

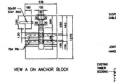

VIEW A ON ANCHOR BLOCK

The structure comprises two pairs of wrought iron suspension chains stretched between masonry towers on each side supporting the bridge deck via suspension rods. The chains are made up of pairs of links (not dissimilar to a bicycle chain), each being approximately 2.4m long, joining to the neighbouring links via 25mm thick connecting plates. On each side of the suspended deck, there are 33 suspension rods hung from the middle of the connecting plates. The suspension rods are then attached to U-shaped straps at the deck level, which house the transverse spanning timber crossbeams.

The suspension chain system supporting the bridge deck pass over rollers housed in circular holes in cast iron casings above each masonry tower, then follow an inclined path down the backstays towards the cast iron anchorage blocks located at each corner of the bridge, 2.5m below ground level. Each chain is passed round a 75mm diameter wrought iron pin at its anchorage point.

The anchorages are accessible via inspection chambers located at each corner of the bridge. The inspection chambers are inter connected through an arched passage (see Diagrams 2 and 3).

Assessment of the Bridge

The Assessed Live Load capacity of the structure was determined using BD 21/93 considering HA UDL & KEL Live Loads. It was generally believed that although no traffic count had ever been carried out at the bridge it was fundamentally recognised that the bridge was neither subjected to heavy load traffic nor was it heavily trafficked.

A number of options were considered for the strengthening of the bridge which included replacing the existing deck timber sections with a Greenheart Tropical Hardwood species. This option was preferred and approved by all involved. It not only strengthened the structure to 5t ALL capacity, just below that of the suspension chain system (chain & hanger 7.5t ALL), but more importantly it maintained the physical appearance of the listed structure. In view of the light and infrequent traffic over the bridge, measures to strengthen the ironwork members were considered unnecessary.

SITE WIDE ISSUES
Hardwood Timber Supply

Photo 2. Timber approved for use

Photo 3. Timber rejected for use

One of the first problems to overcome was the procurement of the proposed hardwood material, which under the strict requirement of the EA had been specified to be procured from an environmentally sensitive source subject to approval by EA.

Having approved the supply source, the next step was to vet the quality of the delivered timber. This was jointly carried out by Posford Duvivier and C.Spencer Limited in a manner designed to ensure a dimensional as well as a physical uniformity.

The General Public

The bridge was officially closed to all pedestrian and vehicular traffic during the full period of the site work. However due to the Public Right of Way over the bridge notices had to be erected at the site well before the works began in order to inform the general public of the relevant dates for commencement and completion of the works.

Environmental Requirements During Site Works

The landscape surrounding the bridge is generally flat, arable, open farmland. An old brickwork site on the western side of the bridge is designated as a site of Nature Conservation Interest, containing flora and grassland species uncommon to the Ancholme valley.

The river is used by the local people and visitors who come for its tranquillity throughout the year. There are local angling, canoeing, rowing and boat clubs, which make extensive use of the river and the Public Right of Way along the west bank.

During the site works notices were issued to the various interested parties in order to make them aware of the length of the bridge closure. Discussions were also held with the same in order to avert clashes with annual events and to reduce any disruption caused to a minimum.

Photo 5. Construction on half of the deck

River Ancholme is a navigable watercourse. Therefore only half the bridge at any one time could be scaffolded for access. The other half was left free for river traffic. The Agency has stringent navigation requirements in pursuant to various Navigation Acts. These were strictly exercised and adhered to during the works.

From liaison with the EA maintenance engineers it was possible to establish that the existing paint which had been applied some 30 years earlier, was likely to be lead based. Therefore in order to avoid cross-contamination of the surrounding land and the water course, the site blasting and subsequent painting of the existing ironwork was carried out entirely in an enclosed environment, all in accordance with the Contractors' Health & Safety Management Plan.

GENERAL NOTES ON SPECIFICATIONS

Ironworks

The ironworks were cleaned by sand blasting in accordance with BS 7079, first quality and to a visual standard of SA 2½ , washed down and dried in accordance with the Agency's standard specification. No subsequent painting was carried out above 85% relative humidity or when the air temperature was below 5 degrees centigrade. The paint system comprised three layers to give a total dft of 325 microns.

Masonry

The tower and entry chamber masonry was wet blasted to BS 6270:Part 1:1982, using a low-pressure air/water abrasive system (with a weak mixture of hydro flouric acid). A nozzle pressure of 20 psi was initially used on a trial patch which gave unsatisfactory results. The pressure was subsequently increased to 30 psi with an improved outcome.

INSPECTION OF THE ROLLERS / SPINDLES, AND THE SLEEVED BACKSTAY CHAINS

Rollers / Spindles

During the works a more thorough inspection was carried out of the hidden elements of the bridge structure. The Contractors scaffolding was used to gain access to the spindle/roller assembly located on top of the masonry towers. Air jetting was employed to remove debris in the form of birds' nests and weathered masonry built up around the roller/spindle assembly over the years.

The inspection of the linkage assembly above each tower showed that the rollers were corroded superficially although, more significantly, they appeared seized. The rollers appeared to have lost their spherical form and become somewhat oval within their housing. There was no evidence of cracking on the surrounding masonry to indicate structural distress caused by the seizure.

Photo 4.Each suspension chain passes over 4No rollers in cast iron casing

It was decided that no remedial work was necessary at this stage for either the masonry towers or the rollers although monitoring would be carried out during the normal maintenance period.

Sleeved Backstay Chains

Another area of concern was the sleeved length of the backstay chains. These were inspected using fibre optics and the findings for each chain were recorded on a videotape for future reference. The first significant point to note was the complete lack of a protective finish on the sleeved chains. The loss of section through corrosion became more extensive with increase in depth. Near the mid length of the sleeved chain the loss of section was estimated to be in excess of 2mm on each face of the chain components compounded with lamination effects which is a characteristic feature of wrought iron material. Considering the corroded condition of the chains and the structural integrity of the bridge it was recommended that the defective sections be replaced.

Photo 5. Corroded sleeved backstay chains

RE - ASSESSMENT OF THE SUSPENSION SYSTEM

In view of the new findings, and with the knowledge that the replacement of the corroded chain sections would require building consent application, the structure was re-assessed incorporating the section losses in order to determine if the bridge should remain closed whilst the consent application was under review.

For the assessment of the suspension chains a computer model was utilized. The section losses due to corrosion were determined at 0.035 mm per year per exposed face using the criteria given in the BSC piling handbook. The results showed that under the existing conditions, the bridge suspension chains were capable only of carrying approximately 3kN per hanger position, and with 33 hangers per side, this amounted to a maximum crowd loading at any given time of approximately 200 people.

The Immediate Future

Considering the remote location of the structure the above crowd loading on the bridge was deemed unlikely. Therefore it was proposed and subsequently accepted that, whilst the Consent Application for replacement of the sleeved backstay chains was under review, the bridge remained closed to vehicular traffic but reopened for pedestrian usage.

In the mean time the work on replacing the existing softwood timber with Greenheart hardwood and the refurbishment of the bridge was completed in January 1997 and the " Defects Correction Period " commenced.

At this stage it was envisaged that building consent application for the replacement of the corroded chain sections would be realised within the Defects Correction Period. However, in the event, the consultation period took longer than anticipated and consent was not granted until the following year.

Proposed Replacement of the Tie Rods

The proposed strengthening of the backstays entailed the replacement of each of the existing sleeved wrought iron chain sections with a high tensile 50mm diameter Macalloy bar equipped with a purpose made end piece to fit around the pin at the anchorage block and a turn buckle/coupler connection piece at the top end for linkage with the backstay chains, all fabricated in Grade 43 steel.

The inclusion of a turnbuckle allowed final adjustment to be made on site when taking up slack. The tie rods were protected against the corrosive elements by a wrapping of protective tape and smearing with petroleum jelly. One sample of each connection piece was fabricated and tested to the required strength before fabrication approval was given.

Following the completion of preliminary design work, drawings were produced and submitted to local and consultative authorities for Building Consent Application. After a lengthy period of consultation Building Consent was finally granted in late February 1998.

Tenders were invited in April 98 and C Spencer Ltd. were subsequently appointed Principal Contractors.

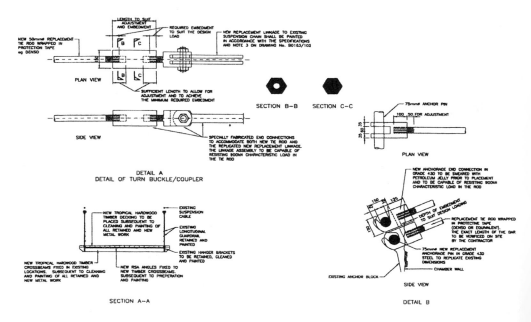

Diagram 4 End connection details of replacement sleeved suspension tie rods

Temporary Support System

A preliminary design for the proposed temporary support system (TSS) was required to be submitted at the Tender stage. In view of the age of the structure the TSS was specified to have the facility and the mechanism to take up load from and return load to the structure in a smooth manner, with little or no deviation from the inclination of the backstay chain from the saddle towards the anchorages.

The TSS proposed by C Spencer Ltd. successfully implemented the specified requirements of the tender and was therefore approved for site operation. It consisted of a temporary anchor pile driven into the ground at each end of the bridge, A claw arrangement which gripped the suspension chain linkage at the

Photo 7. Anchor point at foot of tower

top of the backstay, One tension wire which connected the claw to a bracket positioned at the foot of the tower masonry approximately 300 mm below ground level in order to maintain the line of thrust as close as possible to that on the backstay with little or no deviation, and another tensioning wire attaching the claw arrangement to the temporary anchor piles fitted with a load cell and jack.

Photo 6. Temporary Support System tension wire

Site Works

The site works to replace the corroded sleeved backstay sections began in late October 1998. Sheet piles were driven at each end of the bridge and the TSS tension wires were set up on the first of the eight chains to be replaced. For reasons of safety work was carried out on only one chain at a time. A jacking system combined with a load cell was lined up at a convenient height within the

Photo 9. Tie rods being replaced

tension wire stretched between the " claw " and the temporary anchor pile. Before commencing the replacement work, a theodelite was set up pointed on a marking placed on the top chain linkage adjacent to the rollers in order to record the chain at rest position. This position was monitored and maintained during each replacement. The tension wire was then tensioned to a pre-determined force and monitored at all times using the load cell.

Photo 8. Men carrying tie rod replacement

After a cautious slow start on the first two chains, confidence was gained in the system and the following chains were successfully replaced over a period of eight weeks.

CONCLUSION

The above refurbishment and strengthening works on Horkstow Suspension Bridge commissioned by the Environment Agency, Designed by Posford Duvivier and executed by C Spencer Ltd have extended the operational life of the structure by another 30 years. This has been achieved with neither compromising the existing aesthetics of the bridge or any long-term damage to the environment of the Ancholme corridor.

Restoration of the Ouse Valley Viaduct

MARK HUBAND B.Eng. C.Eng. M.I.C.E.
Railtrack, London, UK

Ouse Valley Viaduct is located on the London to Brighton mainline in the parish of
Balcombe, just north of Haywards Heath in West Sussex . It carries the two track railway
across a valley and the River Ouse. The viaduct has 37 circular arches, is 492 yards in length
and 92 feet high. It is known for its pierced piers, ornate limestone parapets and eight
pavilions . The viaduct was designed by the London and Brighton Railway's Engineer, John
Rastrick, who was one of the three judges at the Rainhill trials on the Liverpool and
Manchester Railway. The ornamental parapets and pavilions are credited to the architect
David Mocatta. Work on the viaduct started in 1839 and it opened for traffic in 1841 with one
line in use and was completed with its parapets and pavilions in 1842. The limestone for the
parapets and pavilions was shipped over from Caen in Normandy to Newhaven and then
brought up the River Ouse in barges to a wharf in the vicinity of the viaduct.

Railtrack has recently completed a £6.5 million restoration of the viaduct, which started on
site in March 1996 and finished in September 1999.

The condition of the limestone parapets and pavilions was the cause of concern for a number
of years and this issue pre-dates Railtrack. There is correspondence on file going back to
1956 regarding this with quotes being sought. However, the cost of undertaking the work was
always a major stumbling block and the high cost of the various schemes proposed resulted in
the work being shelved.

Bridge Management 4, Thomas Telford, London, 2000.

The parapet balusters had weathered severely in places. This, combined with a deterioration in the stone fixings, was leading to a lack of stability of the parapets. There were areas such as the refuges, where the weathering of the stone varied. Certain refuges were in extremely good condition and others had lost all their detail. Fortunately, this was not causing a structural problem to the parapets. Some areas of the parapet - notably the coping stones - had weathered extremely well. A number of the cornice stones had broken off and fallen down to the field below.

The eight pavilions were in a similar state of disrepair with some stones in near perfect condition and others very badly spalled. The pavilions had been propped internally with scaffolding since the early eighties to avoid any further collapses of the roof stones.

When British Railways looked at the problem in the early nineteen eighties, the preferred option was to dismantle the parapets and pavilions and rebuild them in reconstituted stone. Not surprisingly this did not receive a warm response from English Heritage. The scheme was not implemented due to financial constraints.
When Railtrack came into existence in 1994, Ouse Valley Viaduct was back on the agenda. There was an acceptance within Railtrack that something had to be done. An initial meeting

was held between Railtrack and English Heritage to discuss the proposed restoration. English Heritage were extremely keen to see the viaduct restored and restored with real stone. It was suggested that English Heritage would contribute to the funding, if real stone was used. Consequently, it was decided to rebuild the parapet using as much of the original stone as possible and replacing any decayed stone with new.

A pilot scheme was developed and undertaken in 1996 over the first eight arches length of the viaduct and comprised of the following:-

Vegetation clearance - both on and below the viaduct.
Careful dismantling and rebuilding of the parapets, utilising existing and new stone.
Replacement of the missing cornice oversail blocks
Installation of stainless steel reinforcement bars in all cornice blocks
New lead flashing to cornice oversail
Brickwork repairs to arches and piers
Removal of heavy calcite deposits from brickwork
Clearance of downpipes in piers
Installation of access holes/lintels in jack arch chambers

The pilot scheme enabled any difficulties in undertaking the restoration work to be resolved for the remainder of the project. A particular success was the erection of a chain-link fence between the parapet and the track. This gave complete segregation between the operational

railway and the works and produced a safe working environment. It was also more cost-effective than look-out protection. It protected men from walking out into the railway and was also high enough to deflect any projectiles thrown from passing trains. It is worth noting that none of the parapet or pavilion restoration work resulted in buses replacing any trains - the safety screen was erected at night with one line remaining open to traffic. Undertaking works in a railway environment often imposes constraints, such as night working, which can make a relatively simple job difficult and increase costs. The safety screen effectively turned the site into a green field environment and meant that work could proceed without interruption from the operational railway.

Once the safety screens were in position, lifting beams were erected, which enabled the parapets to be dismantled carefully. The stone was then taken down to ground level via a powered hoist, where it was either set aside for re-use or discarded.

The largest stone handled was just under two tonne in weight. Where the cornice oversail blocks had broken off, the remaining stone was prepared and a new stone lifted into position and set on a 150 mm ledge. Cores were made through the new and remaining stone and stainless steel anchors were then inserted to tie the new cornice in. Stainless steel anchors were also installed into all the remaining cornices to avoid future falls.

As the pilot scheme progressed we were able to utilise the expertise of the stonemasons. An important lesson that we learnt was the value of using the knowledge of a stonemason (whether a contractor or a consultant) at an early stage. This enabled a cost-effective solution to be identified. More of the existing Caen stone was salvaged than originally proposed. This was partly the result of stone repairs; the quality of some of these repairs was outstanding.

The replacement stone originally proposed was a Cadeby stone. This appears to have been matched to a rogue example from the parapet and was not a good match to the original Caen stone. Instead, we used a French limestone, Richemont blanc, from a quarry near Bordeaux. This was a close match in colour and texture to the original Caen stone. The mortar used for the joints in the stonework was a 4:5:2:1 (silver sand: Portland dust: slate lime putty: white cement). This was chosen to ensure that the mortar was softer than the actual stone.

During the pilot scheme, the opportunity was taken to survey the London End pavilions and working with the main and stonemasonry sub-contractor a methodology was developed for their restoration. In the past there was never any clear idea how this work would be undertaken alongside a mainline railway. There were thoughts of utilising rail cranes; this would have been both expensive and very disruptive to a critical part of the rail network. In fact the methodology utilised for the parapet works was developed for the pavilions with an overhead gantry beam, which could traverse on beams. Combined with a high safety screen, this proved to be a very effective system and the fact that this work has been undertaken in a railway environment without planned service disruption is an achievement. The pavilion restoration was undertaken under stage two of the works (1997-99). The pavilions were dismantled down to the base level and rebuilt using new and salvaged stone. This work revealed the most magnificent cast iron fish-bellied lintel beams, which were blast-cleaned, repainted and reused.

During the pilot scheme, the opportunity was taken to undertake a detailed survey of the remaining parapets from a cherry-picker to enable the remaining works to be scoped up.

The scheme obviously involved a considerable amount of scaffolding. The cost of the scaffold for the whole scheme was £1.2 million. This provided access throughout the section of the structure being repaired during each stage. Powered personnel/goods hoists were provided on both sides of the scaffold, on each section and at the pavilions. The stone storage on the scaffold required careful management to ensure that it was not overloaded.

Whilst there had been talk over the years of undertaking repairs to the stonework, little thought had been given to the brickwork. It was always intended to undertake brickwork repairs to the arches and piers during the pilot scheme. However, when the scaffold was erected and the heavy calcite deposits removed, it became apparent that the quantity of brickwork repairs would be greater than originally envisaged. In fact at one stage of the project in 1997 - the first year of the second phase - the brickwork repairs became the lead item. The condition of the piers worsened as the height of the piers increased with the profile of the valley.

There were a considerable number of repairs undertaken in the 1890's, where the piers were reskinned in a strong engineering brick with very strong mortar cement amongst the weak mortar and brick of the original structure. This mistake is made often and results in the repair attracting a greater load and then failing. Furthermore, the headers were not bonded in at every level and snapped headers were used instead. Not surprisingly, these repairs had debonded in many cases, resulting in drummy brickwork. It is interesting to note that a very poor repair was undertaken at this time.

There were a number of fractures at the quoins of the piers, often where repairs had previously be carried out. It is likely that this is due to the extreme fibre stresses being present here with respect to sway movement. Pier 14 was particularly bad and the inside face of the pierced pier had to be rebuilt, requiring temporary works to support the pier arch.
The bricks used in the repairs were specially (hand)made to a variety of sizes present in the structure. The repair mortar used was a 1:1:6 (cement:lime:sand) mix.

The skills required by bricklayers working on structural repairs should never be underestimated. On major brick repair schemes, problems are often encountered finding sufficient skilled tradesman. Bricklayers, used to straightforward new build, are likely to experience difficulty in undertaking repair works to tapered piers or may be unable to turn arches.

Many of these skills are dying out, so nurturing some of them on site was a positive aspect of the project. Contractors, engineers and clients should never underestimate the amount of management effort and supervision required to ensure that these repairs are executed efficiently.

The main arches had various fractures, none of which were surprising for a structure of this age. These were probably caused by freeze-thaw action due to the failure of the drainage system. Where repairs were undertaken, the arches were returned using the traditional method of laggings and irons. Irons, which are made to suit the radius of the arch, are secured to the brickwork by doweling threaded bar into the arch rings. These give support to timber laggings, which are placed as the arch is re-turned. These laggings then hold the newly placed bricks in position, whilst the arch is being rebuilt.

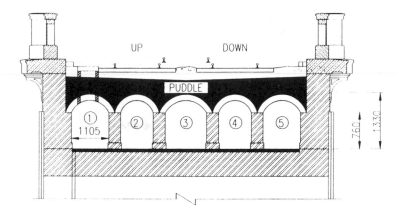

There are five jack arches above each pier underneath the track. These form part of the drainage system. However, over the years, the chambers had become blocked with sludge. Part of the project involved making openings in the skin walls and the installation of lintels. This enabled the chambers to be cleared out and inspected. This work was particularly difficult and involved cutting through 500 mm of brickwork with hydro-demolition. This work was undertaken under a series of full possessions planned for track renewals works. This meant that we had to allow engineer's trains to pass across the structure during the possession; this necessitated temporary propping of the holes at certain times during the progress of this work.

There are drainage downpipes in the piers formed by 100 mm holes in the brickwork. These too had become blocked and were cleared by rodding up from the bottom, finding the first blockage, breaking in and clearing it and rodding up to the next.

With the experience gained from the pilot scheme, the remainder of the restoration works were let as a three year contract with a shut down over the winter periods. The work was undertaken over a three-year period to suit external heritage funding as well as reducing the call on a limited supply of skilled resources. No work was planned over the winter months due to problems of working at height with strong winds and the problems of undertaking masonry in cold weather.

Overall, the restoration of Ouse Valley Viaduct will have cost £6.5 million including a £500k contribution from English Heritage and a £550k contribution from The Railway Heritage Trust.

During the construction phase, the combination of the main contractor's, Osborne Rail's, expertise with the specialist knowledge of their sub-contractor, Cathedral Works Organisation (both part of the Geoffrey Osborne Group) worked very well and was a major contribution to the success of this project.

The sympathetic restoration of the Ouse Valley Viaduct is a project, of which Railtrack is justifiably proud. It has preserved one of this Country's finest viaducts. The re-use of much of the original Caen stone, with new stone where required, has not only maintained the heritage of the structure but proved to be a cost-effective solution.

Managing sub-standard bridges

G. COLE
Environment, Surrey County Council, Epsom, UK

Abstract

The assessment programme has been underway for some time and the first phase is now substantially complete. Until recently, bids for assessments managed by local highway authorities had been met in full through the annual TPP submissions. Unfortunately, bids for strengthening schemes have only partially been met. The result of this situation is that there are a growing number of Sub-standard bridges which need to be maintained at an adequate level of safety but where there is little prospect of a strengthening scheme being promoted.

The publication of Highways Agency Advice Note BA79 meant that, for the first time, guidance on the management of Sub-standard bridges was available to the bridge manager. This paper describes the experiences of a U.K. shire county in the use of this document. The effectiveness of various types of interim measure, including traffic management and weight restrictions, are discussed. The production of detailed monitoring procedures are considered and evaluated. The role that further detailed assessments and testing can play in reducing the number of strengthening schemes is discussed.

The paper concludes that it is possible to adequately manage the stock of Sub-standard bridges without having a major impact on the travelling public. However, adequate levels of funding need to be maintained in order to avoid unacceptable levels of safety on the highway network.

1 Introduction

The assessment programme has been underway for some time and the first phase is now substantially complete. Surrey County Council (SCC), in common with many other bridge owners, has been left with a large number of bridges which have failed the assessment process but there is little prospect these being strengthened in the near future. The assessment code, BD21 [1], requires that in the event of an assessment failure one of a series of formal interim measures is imposed on the bridge. Many bridges are unsuitable for propping and weight restrictions or lane closures would often cause significant disruption to traffic flows. The diversion of traffic away from a restricted bridge may create additional problems for safety on the diversion routes. Consequently, many authorities had been operating monitoring schemes with the intention of allowing the bridge to remain open to unrestricted traffic. The publication of BA79 [2] provided, for the first time, a rational method for considering the whole concept of Sub-standard bridges and provided a framework for instigating a robust monitoring regime in approved cases. The definitions referred to in this paper can be found in the Advice Note [2].

2 Surrey Bridgestock and Assessment Position

SCC started its assessment programme in 1986 [3]. The position at September 1999 is shown in Table 1. It can be seen that out of a total of 702, 81 County owned bridges had failed their assessment and were considered to be Sub-standard. There were a further 21 bridges owned by Railtrack which were also classed as Sub-standard. The task of the bridge manager was to:

- ensure the safety of the travelling public
- minimise delays to the public
- maintain the integrity of the highway network
- prioritise the use of limited resources

Table 1 - Assessment Status (September 1999)

	Private	Railtrack	Thames Water Utilities	Surrey County Council	TOTAL
38T to BD21/84				46	46
Not yet Assessed	3	14	3	49	69
Passed 40T to BD21/97	3	47	5	290	345
Passed (Provisional)		18		1	19
Failed 40T		12	2	81	95
Failed 40T (Forecast)		9			9
Strengthened				119	119
TOTALS	6	100	10	586	702

Until recently, bids for assessments managed by local highway authorities had been met in full by Transport Supplementary Grant through the annual TPP submissions. Unfortunately, bids for strengthening schemes have only partially been met. The relation between total bids and funding over the last few years at SCC is shown in fig. 1. The bid figure was reduced after 1996 to reflect the lower level of grant and does not imply a reduction in demand. This situation had the effect of extending the strengthening programme and hence increasing the need to consider the use of interim measures. More recently, assessment bids have not been met in full and this has increased pressures on the ability to implement interim measures.

Figure 1

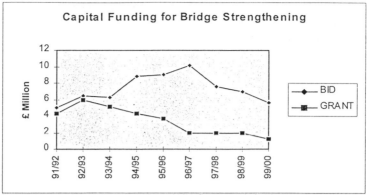

Note : 99/00 figures exclude Thames Crossing

3   The Advice Note

The Advice Note [2] is divided into 5 main chapters:

• Assessment
• 'Immediate Risk Structures'
• Interim Measures During Assessment
• Interim Measures on Completion of Assessment
• Prioritisation for Strengthening

This document makes more references to the role of the Technical Approval Authority (TAA) than any other Highways Agency Standard or Advice Note. It also leaves a number of areas open for interpretation. It was, therefore, necessary for SCC to consider its role as TAA and bridge manager and produce its own interpretation of the Advice Note. Some of this interpretation is discussed further below.

One of the problems with assessment failures is the need to explain to members of the public or elected members why it is necessary to impose weight restrictions or other measures in cases where there are no obvious signs of distress in the bridge concerned. Laboratory testing suggests that a typical ductile reinforced concrete slab will deflect to approximately half of its overall depth before failure occurs, giving ample warning of the event. The engineering judgement of many bridge owners suggested that some theoretical assessment failures could be safely monitored without the need to impose restrictions. The publication of the Advice Note [2] meant that, for the first time, guidance on the management of Sub-standard bridges was available to the bridge manager. The use of robust procedures would ensure the maintenance of adequate levels of safety.

4   The Assessment Process

This section of the Advice Note contains a flow chart which illustrates the processes involved. SCC simplified this chart (fig. 2) for presentation to, and endorsement by, elected members. It was particularly important that members were aware of the steps being taken to ensure the safety of the highway network.

Figure 2   BA79/98 The Management of Sub-standard Highway Structures

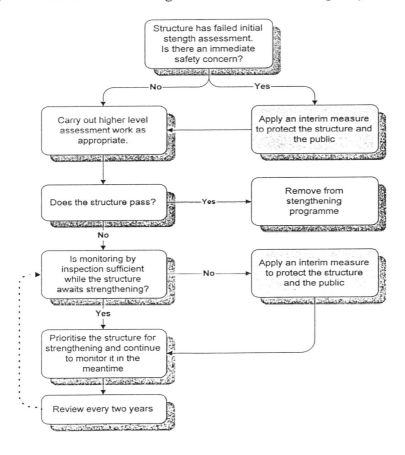

Appendix B to the Advice Note introduces the concept of five levels of assessment. SCC have used all of these levels with the exception of level 5, reliability analysis. In general, a whole life costing approach to the Sub-standard bridge problem shows that it is more cost effective to refine the assessment method to obtain a 'pass' than to continue a prolonged monitoring exercise. This approach is also encouraged by the Advice Note. However, there are clearly some circumstances where further assessment is unlikely to be worthwhile and formal interim measures would be necessary to safeguard the bridge.

A key aspect of the Advice Note is the requirement to keep adequate records throughout the assessment process with details of the decisions taken at each stage. This is done by the use of a proforma (known as Form E1) throughout all of the processes shown on the flow chart (fig. 2). The use of this proforma is recommended for all Sub-standard structures in Surrey, irrespective of ownership. The Form E1 is kept on the bridge file at each stage or level of the assessment process. Where the structure is considered to be monitoring appropriate then a monitoring procedure document (see below) is appended to the Form E1. All of these procedures are carried out by the assessment engineer. Some of the detailed requirements in Surrey are as follows:

- the level of assessment completed is shown on the assessment certificate
- amendments to Approval in Principle (AIP) forms are required for each level of assessment proposed by way of an addendum to the original AIP
- Sub-standard bridges must be restricted to Construction and Use (C&U) loading if no lower limit is imposed (see below).

One objective of BA79 is to limit the disruptive effect of weight restrictions by the use of monitoring on appropriate structures. The use of abnormal indivisible loads (AIL) on the highway network is increasing and the selection of suitable routes is not made any easier by the large number of Sub-standard bridges that exist. In particular circumstances, this problem can be overcome by a special assessment for a specific vehicle. The existence of one C&U restricted bridge on a motorway link can create significant problems for the alternative AIL route on the parallel county road.

5 Immediate Risk Structures

These structures are where 'an immediate and unacceptable risk to public safety is identified' [2]. This diagnosis needs to be confirmed by the TAA. The Advice Note recommends that 'factors such as the nature of the structural weakness, any corresponding signs of distress, the recent load history of the structure and the level of assessment completed should be taken into account'.

A technique for determining whether or not a bridge is at immediate risk has been produced [4]. A sensitivity check is carried out which involves reducing partial factors of dead and live load to the minimum values which would still give an acceptable short term risk level. If the assessment live load rating is still less than 40 tonnes after applying the reduced partial safety factors then formal interim measures should be introduced without delay.

6 Interim Measures During Assessment

As noted above, the assessment process can involve several stages of work and be extended over a considerable period of time. It is, therefore, recommended that interim measures are implemented whilst the assessment process is completed unless the structure is considered to be low risk. The use of this procedure might require, for example, the erection of weight restriction signs and their subsequent removal without the members of the public being aware of any apparent change in the condition of the bridge. This type of situation has occurred in Surrey in order to maintain the safety of the public but extensive explanations were required for elected members and the local Member of Parliament.

The definition of 'low risk' requires that there is likely to be a considerable reserve of strength inherent in the structure. It should be remembered that at this stage of the process it will not have been possible to have completed the full range of options open to the assessment engineer. Therefore, the definition can only be subjective. SCC have adopted the policy of using the other two definitions i.e. 'Where consequences of failure are low or where the live load capacity factor is greater than 0.7'.

7   Interim Measures on Completion of Assessment

The purpose of these measures is to reduce the risks of a failure to acceptable levels until strengthening is carried out. The financial situation described above means that these measures are likely to be in place for some time. The major change from previous documents is the introduction of the definition 'Monitoring-appropriate Structures'. Previously some bridge owners had resorted to monitoring in order to avoid the imposition of weight restrictions or traffic management measures where these were likely to cause 'excessive disruption to traffic or incur disproportionate costs' [2]. In order to qualify as 'Monitoring-appropriate', structures need to satisfy all of the following criteria (except in the case of small spans generally less than 5 metres):

- no significant signs of critical distress
- hidden problems unlikely to be present
- predictable and gradual failure mode
- monitoring should be meaningful and effective

The use of monitoring is treated as a departure from Standard. Fortunately, further guidance is contained in Appendices C and D of the Advice Note.

The assessment engineer may be able to offer alternative Formal Interim Measures. For example, a bridge may be assessed at 7.5t with three lanes of traffic but at 17t with only two lanes of traffic. The lower Assessment Live Loading (ALL) will handicap buses and most commercial vehicles but permit unrestricted use by private cars. Conversely, the higher ALL is suitable for most buses and a larger number of commercial vehicles but the reduction in traffic lanes may cause queuing of private cars at peak times. The bridge manager will need to discuss the options with the highway manager and record the outcome on Form E1. The lower the level of posted weight restriction, the more likely it is that it will be ignored.

The Trading Standards Department monitored 4 weight restricted bridges throughout Surrey on behalf of Environment. Each bridge was visited twice a week and observed for two hours on each occasion. A total of 139 contraventions were recorded and 78 prosecutions have resulted. The average fine is £132 which makes interesting comparison with the potential cost of damage to a weak bridge. The average figures conceal the fact that one of the bridges had 39 prosecutions and 26 cautions. If the sample rate is multiplied by 12, say, then the total number of contraventions at this site is likely to be nearly 800 in a single year. The Trading Standards Department recommended that SCC be encouraged to strengthen structures and develop better warning signs and physical barriers.

Very low levels of weight restriction (3t) allow the weight restriction to be accompanied by a width restriction. This generally ensures enforcement of the weight limits but the physical restriction may need to be substantial to prevent all but the most determined of drivers.

Therefore, SCC prefers to use traffic management measures and propping rather than weight restrictions. These techniques also provide a more positive method of load reduction on the structure. However, they may require a more politically sensitive approach as well as being more expensive to implement.

Weight restrictions are imposed by the use of traffic regulation orders. These are normally only valid for eighteen months unless extensions are granted by the Secretary of State. Failing this they either have to be removed or made permanent. Objections may be raised against a permanent order which could result in a public inquiry with the obvious programming and resource implications.

8   Prioritisation for Strengthening

SCC has included its stock of Sub-standard bridges in the draft five year Local Transport Plan (LTP) which was published in June 1999. The following weighting criteria, which are consistent with the overall objectives of the LTP, are used in the prioritisation process:

Safety (61%)
- assessment rating
- condition
- risk of sudden failure
- consequences of sudden failure
- adequacy of parapets and alignment
- suitability of diversion routes

Economy (21%)
- network importance/road classification
- usage by 40 tonne vehicles (or equivalent)
- level of inconvenience of restrictions to business

Access (5%)
- extent to which bridge caters for pedestrians/cyclists

Integration (5%)
- usage by buses/providing access to transport facilities

Cost/benefit (5%)
- cost of scheme relative to network benefit

Environment (3%)
- bridge unsightly/perceived to be unsafe

It is important to note that, despite the importance of road hierarchy, safety is the primary consideration in the prioritisation process. As a result, whilst Primary and Principal routes predominate in the early years of the LTP programme, some bridges on unclassified routes also have high priority ratings. This can be due to economic factors where, for example, a bridge provides sole access to an industrial premises or due to particularly advanced state of deterioration. The relative weightings of these parameters will be reviewed as experience is gained with the implementation of the LTP.

9   Monitoring

The following points taken from the Advice Note have been incorporated into the SCC procedural guide:

- monitoring is not required for Low Risk structures
- monitoring is not permitted for an Immediate Risk structure
- a monitoring procedure document is required

A monitoring inspection has a completed standard BE11 form as well as a Monitor Inspection Update Sheet which is copied to the TAA immediately following inspection. The monitoring procedure document or specification is treated as an Appendix to the Form E1. The document should be clearly written and define the objectives, frequency and period of the monitoring. The live load factor 'C' is included as well as the road surface/traffic flow matrix defined in BD21/97 [1]. The potential failure modes should be described together with limits on deflections and crack widths. Finally, emergency procedures should be in place in the event that the monitoring inspection reveals that certain trigger values have been met.

The importance of well documented monitoring reports can not be understated. There are several methods of monitoring which are beyond the scope of this paper. However, the chosen method should be robust and repeatable.

The Advice Note recognises that monitoring in itself does not prevent damage occurring and the longer the process is continued the worse the situation will become. It is necessary to review the status of the structure every two years to see whether or not it is possible to continue with monitoring or if Formal Interim Measures are necessary if it has not been possible to strengthen. The costs of monitoring the average stock of Sub-standard bridges to the appropriate standard contained in the Advice Note is significant and should not be overlooked.

10 Continuous Assessment Process

It has been acknowledged by the Highways Agency that the assessment of the bridge stock is likely to become a 'continuous' process rather than single, discrete exercises like Bridgeguard. This change of emphasis will enable bridge maintenance strategies to become safety related and based on whole life performance techniques. However, an assessment result is only strictly valid for the day of the site inspection and knowledge of the rate of deterioration must be obtained by further inspection or testing. The derivation of a safety led approach, which will improve targeting of funds, has been defined (Flint and Das, 1996). The further development of such techniques, together with more refined assessment methods, should enable the burden of Sub-standard bridges to be reduced.

11 Conclusions

Experience has shown that it is possible to adequately manage the stock of Sub-standard bridges. Monitoring methods when adopted, should be robust, well-defined and well recorded. However, adequate levels of funding need to be maintained in order to continue monitoring, refine assessments, promote research and progressively strengthen Sub-standard bridges. The alternative is to accept reduced levels of safety on the highway network.

12 Acknowledgements

The author wishes to thank the many colleagues in the County Surveyors Society and Concrete Bridge Development Group for the discussions which have contributed to this paper. The permission of Callum Findlay, Head of Engineering at Surrey County Council, to publish this paper is gratefully acknowledged.

13 References

1. Highways Agency. *The Assessment of Highways Bridges and Structures.* Highways Agency, London, 1997, Departmental Standard BD21/97.

2. Highways Agency. *The Management of Sub-Standard Highway Structures.* Highway Agency, London, 1998, Departmental Advice Note BA79/98.

3. Palmer, J., and Cogswell, G. *Management of the Bridge Stock of a UK County for the 1990s.* First International Conference on Bridge Management, Guildford, UK, 1990.

4. Cronshaw, J. *Hampshire County Council Internal Guidance Note AG2/98,* 1998.

5. Flint, A.R. *Whole life performance-based assessment rules: development programme and issues.*
    Safety of Bridges (ed. P.Das), Thomas Telford, 1997.

Management of bridges in Germany

DR.-ING. J. KRIEGER, DR.-ING. P. HAARDT
Federal Highway Research Institute, Bergisch Gladbach, Germany

INTRODUCTION

The federal road network of Germany contains a total of 35.272 bridges with a bridge deck area of 24.79 million m^2 (as of 31.12.1998), 157 tunnels with a total length of 115,8 km, as well as a large number of other highway structures like retaining walls, noise reduction walls and bridges for traffic signs. The maintenance programs to be prepared for this purpose not only require a high budget, but also influence the traffic infrastructure and, thus, the economy and society as a whole. Due to growing volumes of traffic and higher weights of trucks, bridges in roads are subjected to increasing loads which implies that maintenance costs will be rising in the future. Considering the fact that financial resources become continuously tighter, the maintenance costs have to be spent in a way to obtain the greatest possible benefits. This task will in the future be supported by the application of a Management System (Bridge Management System, BMS).

The competencies and tasks related to road construction are specified in the constitution of the Federal Republic of Germany. In accordance with the German basic law, the Federal Government is the owner of the federal road network. The states administer these roads under their own responsibility as authorities commissioned by the Federal Government.

Firstly, the planned Management System is to provide the Federal Ministry with an overview of the current situation of the structures at the network level, and allow it to come to statements concerning future maintenance costs as well as strategies for realising long-term objectives and fulfilling basic conditions for maintenance routines. Secondly, the states and authorities are to be supplied with recommendations for performing improvements at the object level in compliance with strategies, long-term objectives, basic conditions and budgetary restrictions.

Some of the above mentioned topics have already been realised. Other sub-modules are currently being prepared. However, a large portion of the essential issues are still in the planning phase. In the following the existing regulations and procedures as well as those which are going to be developed are presented.

EXISTING RULES, REGULATIONS AND PROCEDURES

Database

An important task of administrators is the observation and inspection of the structural inventory. To ensure a constant supply of actual data concerning existing structures, the structural data are registered, stored and evaluated by the administrative authorities of the states with the help of electronic data processing equipment. In Germany these data are acquired and stored in accordance with ASB (Road Database Instructions) [1]. The SIB-Structures program system (Road

Information Database) [2] was conceived and realised simultaneously to the ASB; this system is intended for the registration, storage and evaluation of the structural data. In addition SIB-structures is also used for registration of data concerning inspection results and damage, maintenance measures as well as maintenance costs. SIB-Structures serves as one of the cornerstones of a planned Management System.

To improve the information situation at the federal level, the Federal Ministry of Transport, Building and Housing is presently developing a database titled BISStra (Federal Road Information System). State-level data in accordance with the ASB can be included into analyses as part of BIS-Stra. This ensures the availability of data for the federal authorities. BISStra will also include a linkage to traffic- and accident related data.

Inspection
DIN 1076 "Engineering Structures in Connection with Roads; Observation and Inspection" [3] regulates the technical observation, inspection and testing of the stability and traffic-safety of bridges and other engineering structures in connection with roads. Inspections are performed by an experienced civil engineers who record damages and faults directly at the structure supported by the SIB-structures program system.

The RI-EBW-PRÜF "Guideline for Standardised Registration, Processing and Analysis of the Results of Inspection in Accordance with DIN 1076" [4] contains rules for a simple and standardised logging of the results of inspection. The standardised procedure allows structural conditions to be rated in accordance with various criteria and the inspection results to be linked with the construction-related data in accordance with ASB. For condition assessment every single damage is evaluated separately in terms of its effect on the stability, traffic safety and durability of the structure with the help of SIB-structures. The damage evaluation is used as a basis for automatically calculating the condition grade for the entire structure. The extent of damage and the number of individual occurrences of damage is also considered in this process. RI-EBW-PRÜF contains detailed definitions of assessment criteria and sample catalogues for a standard evaluation of damage.

Procedures currently used in the frame of maintenance planning
The findings obtained from the bridge inspections and additional object-related analyses are used to prepare the maintenance concepts at the construction agencies. The preparation of performance specifications, announcements/allocations as well as the processing and documentation of projects are also performed here. Annual construction programs are prepared by the agencies and co-ordinated with the responsible higher-level authorities. Priorisation is performed on the basis of the existing severity of damage, operational and traffic-related circumstances and available financial resources. Cost estimates prepared by the agencies generally result in the requirement reports for the Federal Ministry.

To exercise its influential rights, a number of controlling techniques are implemented by the Federal Ministry. Fig. 1 shows different reports submitted by the states, as well as the possibilities which the federal authorities have of influencing maintenance planning. For controlling purposes, the states annually supply the Federal Ministry with average structural condition grades. The states have to report a medium-term requirement program for maintenance and an annual program plan for the following year. These reports provide the federal authorities with information concerning major upcoming repair measures and the corresponding resource requirements.

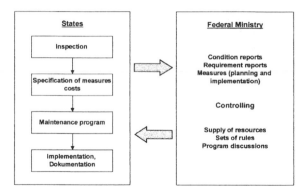

Fig. 1: Tasks of the federal and state authorities in the frame of maintenance planning

Being responsible for strategic tasks, the federal government requires techniques which allow it to include specifications, i.e. objectives, strategies and boundary conditions in the planning process and obtain the information required for this purpose. This information comprises current conditions, the benefits of invested resources as well as the maintenance strategies currently targeted by the individual states. Furthermore, information on costs is needed: costs given requirement-oriented maintenance, costs given alternative strategic objectives, as well as costs given modified boundary conditions. A matter of decisive importance to the federal authorities is the use of standardised techniques of prioritisation, program creation and determination of requirements by the states. Only information acquired by the states on a homogeneous platform using standardised techniques can allow the federal authorities to formulate and implement strategies, objectives and boundary conditions. In future the planning tasks of the state agencies as well as of the federal authorities will be supported by a comprehensive management system.

ESTIMATION OF FUTURE MAINTENANCE COSTS
One of the most important modules of a BMS is the possibility to carry out analyses with the aim of estimating future maintenance costs. This task has to be done by local agencies to estimate repair and replacement costs on object level or for a small local network. These analyses use in most cases data which are available on object level such as information about the structural condition from the bridge inventory as well as information from bridge inspections. With these data and additional information added by maintenance experts measures can be planned. For these measures costs can be estimated. Together with additional costs e. g. for traffic deviation the total costs for repair and replacement can be estimated.

On network level this kind of approach is not useful because of the necessity to use expert knowledge for the deviation of measures on the basis of information from bridge inspections. Therefore methods which use aggregated data are more suited to generate information about the future network level costs for repair and replacement. For the use in process of the BMS at federal level two different methods were developed and applied using the available data.

One of the methods uses information about the age distribution of the bridge stock, rates for repair and replacement together with accompanying costs. Similar approaches have been used for the prognosis of the maintenance costs 1992 [5] or by the Highways Agency [6].

Fig. 2 shows the age distribution of the bridges on Federal Roads in Germany. It becomes obvious that a large proportion of the bridge stock was built during the years 1965 to 1984. Most of these structures are prestressed concrete bridges.

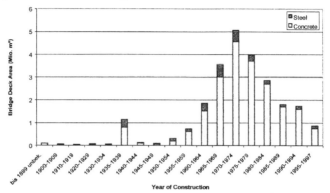

Fig. 2: Age distribution of the German bridge stock on Federal roads

Fig. 3 shows the assumed distribution for replacement of bridges. These distributions were derived from bridge inspection data and additional information of maintenance experts for various construction materials of the bridge superstructure (concrete, prestressed concrete, steel, composite). These distributions suppose that maintenance activities are carried out in regular intervals.

In Fig. 4 the repair rates for bridges are shown. For the example presented here constant maintenance intervals of 20 years were used. This distributions were developed using bridge inventory and bridge inspection data. Medium life time intervals for the most important bridge parts were calculated (expansion joints, pavement and sealing, concrete repair, corrosion protection).

Additional input data for the estimation of the maintenance budget are costs for repair and replacement of bridges. These information were collected during the progrosis 1992 [5]. For the replacement of concrete and prestressed concrete bridges values of 3.000 DM/m^2 and 4.800 DM/m^2 for steel bridges were assumed. Repair costs were estimated at 650 DM/m^2 for concrete and prestressed concrete bridges and 700 DM/m^2 for steel bridges.

The first step of the analysis is the superposition of the age distribution with the distributions for repair and for replacement. The result of this process are bridge deck areas for repair and replacement as a function of time. Costs for repair and replacement are gained by multiplying these areas with the appropriate repair and replacement costs.

Fig. 5 shows the estimated costs for repair and replacement for the years 1999 to 2020. According to this calculation model 800 to 1.200 Mio. DM/year will be needed from 1999 to 2013 and about 1.100 Mio. DM/year during the years 2014 to 2020. The result of this simple approach are costs for repair and replacement over a certain time period under the assumption that all maintenance activities are carried out in time. As the maintenance costs spent during the last years amount about 600 Mio. DM/year it becomes obvious that the assumptions of the above presented approach are not fully satisfied.

Therefore a second approach using bridge rating data, which is capable of analysing different financial scenarios was developed. Input data for this approach are condition rating data, deterioration functions for bridges and costs for repair and replacement of bridges.

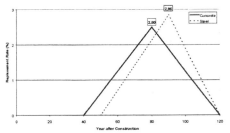

Fig. 3: Rates for the replacement of bridges

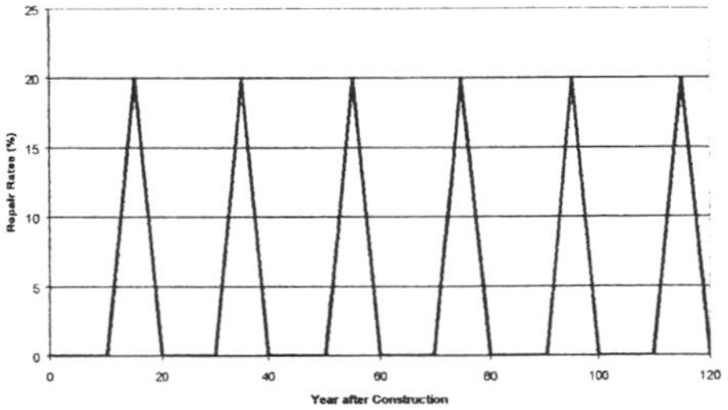

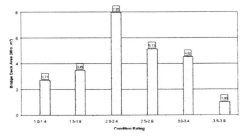

Fig. 4: Rates for repair of bridges

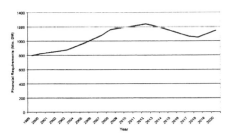

Fig. 5: Estimated costs for repair and
replacement area for of bridges
on Federal roads

Fig. 6: Condition Rating and bridge deck
bridges on Federal Roads
(State 31.12.1998)

Fig. 6 shows the condition rating data for bridges on Federal roads (State: 31.12.1998), whereas a rating of 1.0 – 1.4 means that these bridges are in a very good condition and bridges with a rating of 3.5– 3.9 are in a very poor condition.

By analysing inspection data from the state "Thüringen" [7] a deterioration function for bridges was set up. The superposition of this deterioration function with the distribution of condition rating (Fig. 6) yields in condition distributions as a function of time. Fig.7 shows the development of the condition rating for the years 2000, 2005, 2010, 2015 and 2020. This scenario supposes that no money for repair and replacement of bridges is spent. It becomes obvious that the condition of the bridge stock decreases strongly from 1998 to 2020. On the basis of this case (expanses for repair and replacement = 0) other szenarios can now be analysed. As a parameter for the condition the value of a bridge stock can be used. The value of the bridge stock can be defined by the replacement value minus the costs for repair and replacement which are necessary to bring a bridge back to a very good condition (rating 1.0 – 1.4). The development of the value of the bridge over the years 1999 to 2020 is shown in Fig. 8 for different szenarios. The figure shows that for keeping the value of the bridge stock on the level of 1998 about 800 to 1.200 Mio. DM/year have to be spent for repair and replacement. This corresponds to the results of the first approach. Additionally the influence of different strategies for repair and replacement on the condition of the bridge stock can be shown.

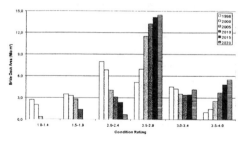

Fig. 7: Development of the condition of bridges   Fig. 8: Actual value of the bridge stock for
different scenarios

DEVELOPMENT OF A MANAGEMENT SYSTEM FOR THE MAINTENANCE OF STRUCTURES

The planning task at state level comprises development and comparison of the various alternative measures, resulting in a priorisation of variants, as well as budget planning, i.e. a comparison of all projects scheduled in the planning period, resulting in a maintenance program. Taking into account co-operation between the federal and state authorities, the planning task should be viewed as a two-stage process. Whereas the planning process supplies a program draft for the planning period as a result - although this result still needs to be tuned in the controlling process - the realisation process is intended to ensure final program formation and execution. In contrast, the controlling process at federal level involves control with the aim of positively influencing the quality of the decision-making processes in terms of the attainment of the specified objectives [8].

The first step of the planning process consists of the registration and evaluation of damages and conditions at the object level in accordance with DIN 1076 and RI-EBW-PRÜF. This makes it possible to determine the requirement for intervention and followed by a complete evaluation of the object. The ideal times of intervention are determined on the basis of response forecasts. If a concurrence of several projects is established, it is necessary to carry out a priorisation at the network level in order to identify priorities. The type of measure is drafted and used to prepare a forecast of the changes in condition to be achieved thereby. Alternative actions are to be developed and evaluated. With the help of cost analyses, variants are prioritised and a selection is made at the object level. The resulting draft concepts are used to draft the program for the planning period and determine financial requirements at the network level. Whereas the database and inspection procedure still exists, the following sub-modules will be developed in future:

- Evaluation and selection of measures
- Overall evaluation and priorisation
- Program draft and determination of financial requirements

The controlling process of higher-level administrative units is based on an registration of detailed, relevant information which originates from a comprehensive database and also results from the preceding planning process. More specifically, the supplied data are used as a basis for performing comparisons of actual and planned values in the form of balances, evaluating and analysing condition attributes, forecasting condition developments and expected costs, as well as performing evaluations of special issues. The analyses and evaluations are followed by discussions of the program drafts, supply of resources, updating of sets of technical rules as well as direct interventions in the planning process.

The formation of the final program forms an essential step in the realisation process. In addition to the above-mentioned input parameters, implementability at the agencies, possible grouping of measures and other boundary conditions need to be taken into account. Interfaces to other Management Systems also exist. The result, again, is a prioritisation of measures, but taking into account the specified restrictions this time. The measures are to be announced, allocated, implemented and documented. Balance sheets are to be repaired. The results are to be submitted to the planning and controlling processes. A sub-modulus which supports the program creation by taking restrictions into account will be developed as a first step. The realisation modulus will be extended by sub-modules for project preparation, administration of measures and documentation, balancing and preparation of statistics.

Fig. 9: Linkage of planning, controlling and realisation process

A linkage of the processes is of significance for the implementation of the maintenance strategy at the federal level; in this case, optimization of maintenance planning is ultimately achieved in iteration cycles (Fig. 9).

A phase plan for the development of the Management System in combination with the formulation of sub-projects for completing the system was prepared. Based on this plan, the realization of a complete BMS on the federal and state levels will be completed by the year 2005 [8].

SUMMARY
The task of those responsible for the maintenance of structures is to implement the complex decision making process based on information of the best possible quality. A Management System can contribute significantly towards achieving this goal. In the Federal Republic of Germany, too, it is desirable to use a system which allows an integrated treatment of all the sub-aspects and provides a basis for making decisions concerning the use of maintenance resources. At present, decisions related to maintenance planning are based mainly on subjective estimates. In contrast, an economi-

cal deployment of resources requires transparency, objective decision –making processes, integrated structures and controlling instruments.

The Federal Ministry of Transport, Building and Housing is aiming for a comprehensive Management System taking into account state-specific and federal requirements as well as existing regulations in the frame of maintenance planning. The system is subdivided into three processes, each comprising special tools for valuation, analysing and planning. Some of the sub-modules have already been realised, e. g. the database at state level, the inspection procedure in combination with the procedures of damage and condition assessment and the estimation of future maintenance costs at federal level. Other sub-modules are currently being prepared, e.g. tools for evaluation and selection of measures. However, a large portion of the essential issues are still in the planning phase, e. g. program creation by taking restrictions into account. The realisation of the complete Management System will be finished by the year 2005.

REFERENCES

[1] ASB, (1998), Anweisung Straßeninformationsbank, Teilsystem Bauwerksdaten, Verkehrsblatt Verlag, Dortmund.

[2] Programmsystem SIB-Bauwerke, (1998), DV-Programm zur Erfassung, Speicherung und Auswertung von Bauwerksdaten, WPM-Ingenieure, Neunkirchen.

[3] DIN 1076, (1983), Ingenieurbauwerke im Zuge von Straßen und Wegen, Überwachung und Prüfung, Beuth Verlag, Berlin.

[4] RI-EBW-PRÜF, (1998), Richtlinie zur einheitlichen Erfassung, Bewertung, Aufzeichnung und Auswertung von Ergebnissen der Bauwerksprüfungen nach DIN 1076, Verkehrsblatt Verlag, Dortmund.

[5] Erhaltung der Brücken und anderen Ingenieurbauwerke der Bundesfernstraßen (West), Prognose des Finanzbedarfs, (1992), Der Bundesminister für Verkehr, Abteilung Straßenbau, Aufgestellt: Bund/Länder Fachausschuß Brücken- und Ingenieurbau, not published.

[6] Das, P. C., (1998), New Developments in Bridge Management Methodology, Structural Engineering International, pp. 299 – 302, 04/1998.

[7] König, (1999), Verhaltensfunktionen für Brücken, Auswertung für das Land Thüringen, 1999, not published.

[8] Haardt, P., (1998), Konzeption eines Managementsystems zur Erhaltung von Brücken- und Ingenieur-bauwerken, Schriftenreihe der Bundesanstalt für Straßenwesen, **B25**, Verlag für neue Wissenschaft, Bremerhaven.

Probabilistic-based assessment of Swedish Road Bridges

I. ENEVOLDSEN, M.Sc. Ph.D., RAMBOLL, Denmark
Bredevej 2, DK-2830 Virum, Denmark. ibe@ramboll.dk, www.ramboll.dk, phone +4545986674
S. PUP, M.Sc. Swedish National Road Administration, Sweden
Röda vägen 1, S-78187 Borlänge Sweden, stefan.pup@vv.se, www.vv.se, phone +4624375663

ABSTRACT
A revised decision process of the Swedish National Road Administration for administration and assessment of the safety of bridges with probability-based approached as standard when bridges need a higher classification, is presented. The legal justification for application of probabilistic-based assessment is given, followed by two examples showing the possibilities including the potential future large cost savings, due to avoided or reduced strengthening and rehabilitation projects.

INTRODUCTION
The situation in Sweden is very traditional regarding management of bridge stocks: Increasing demand on load carrying capacity, combined with low budgets for rehabilitation and strengthening of older bridges. As a consequence the Swedish National Road Administration has, during the last years, assessed approximately 9000 bridges for administration of the bridge safety and for adaptation of the EC traffic load. The results of the assessments are classification of the individual bridges, which are then used in a bridge management system for permitting special heavy transports with a weight of 60 – 500 t to pass the bridges. The assessments are performed according to a deterministic code developed specifically for the assessment of the existing bridges /9/. This is a well functioning approach, that has proven its efficiency. In some cases, however, the code seems to be too general or conservative as some of the bridges only obtain a relatively low classification, even though the bridges show no evidence of overload. Based on this observation the Swedish National Road Administration in 1998, with RAMBOLL as consultant, instead of expensive strengthening projects, initiated a project using probabilistic-based assessments on a number of Swedish road bridges that obtained an insufficient classification level according to the deterministic approach.

The basic idea in the probabilistic-based assessment approach is, that instead of using the general code which necessarily must generalise in order to cover all types of bridges, an individual bridge approach is applied. In the individual bridge approach, the assessment or bridge safety is determined, applying the FORM β-method, which directly takes the bridge specific uncertainties and local traffic situation into account consistently. The uncertainties arise from physical uncertainties in the use and identification of materials, from the load and truck models, from statistical sources, and finally from the simplification in the structural evaluation models. Based on the FORM β-method one could say that a bridge specific code is established with bridge specific partial safety factors. The result is in general a classification permitting considerably higher traffic load following the principle, that a bridge does not necessarily have to fulfil the specific requirements of the general deterministic code, as long as the overall level of safety defined by the code is satisfied. As a consequence the individual approach cuts or reduces the strengthening or rehabilitation cost without compromising the level of safety.

The drawback is of cause, that it is more expensive to perform the analysis in the individual approach compared to the general approach. The difference, however is marginal compared with the potential cost savings.

The results are so promising, that the Swedish National Road Administration intend to revise the decision process for upgrading existing bridges, from the traditional one shown in Figure 1 (left) to the decision process shown in Figure 1 (right), where an intermediate phase considering probabilistic assessment techniques is introduced before decision on strengthening are taken.

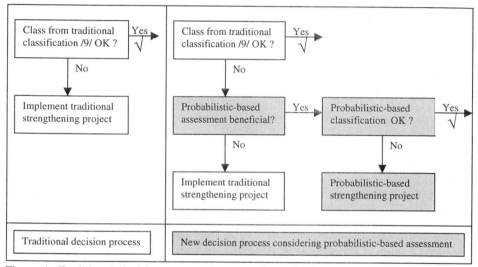

Figure 1. Traditional decision process for upgrading of existing bridges (left) and revised decision process including consideration of the probabilistic-based approach (right).

The conclusion from Figure 1 is, that probabilistic-based classification should be considered as a possibility (especially for bridges intended for heavy special transports), before an expensive strengthening project is initiated. Probabilistic-based assessment will in many cases be fruitful, especially if the bridge has problems with limit states which are believed to be conservative modelled according to /9/. If, however, the probabilistic-based assessment is not sufficient for obtaining the required classification, the strengthening project can be probabilistic-based, and under normal circumstances less costly, than a strengthening project based on traditional assessment methods according to /9/.

In the following section the process of interpretation of Swedish codes for legal application of probabilistic-based bridge assessment, followed by a section on uncertainty and traffic load modelling, are given. Following, the experience with assessment of two of the Swedish bridges is presented as examples. Finally, a conclusion is given.

LEGAL JUSTIFICATION AND PROCEDURES
It is obvious, that it is a fundamental requirement for the application of probabilistic-based assessment, that the authorities in the countries where the methods are used, establish the legal basis. In Sweden it is stated in the code for safety of structures /6/ that probabilistic-based approaches may be applied with reference to ISO /3/ for definition of the reliability index β, and NKB /1/ for specification of safety levels. NKB /1/ is the background documentation for the codes in the Nordic countries (Sweden, Denmark, Finland, Iceland and Norway), and describes

how a probabilistic-based assessment can be performed in order to fulfil the required safety level, including requirements for modelling of uncertainties. The safety requirement in the Ultimate Limit State (ULS) is specified with reference to failure types and consequences as requirements for the formal yearly probability of failure p_f, which trough the standard normal distribution function Φ is related directly to the reliability index β by the relation $\beta = -\Phi^{-1}(p_f)$. In Table 1 the requirements in NKB /1/ for the ULS is seen

Failure consequence (Safety class)	Failure type I, Ductile failure with remaining capacity	Failure type II, Ductile failure without remaining capacity	Failure type III, Brittle failure
Less Serious (Low safety class)	$p_f \leq 10^{-3}$ $\beta \geq 3.09$	$p_f \leq 10^{-4}$ $\beta \geq 3.71$	$p_f \leq 10^{-5}$ $\beta \geq 4.26$
Serious (Normal safety class)	$p_f \leq 10^{-4}$ $\beta \geq 3.71$	$p_f \leq 10^{-5}$ $\beta \geq 4.26$	$p_f \leq 10^{-6}$ $\beta \geq 4.75$
Very Serious (High safety class)	$p_f \leq 10^{-5}$ $\beta \geq 4.26$	$p_f \leq 10^{-6}$ $\beta \geq 4.75$	$p_f \leq 10^{-7}$ $\beta \geq 5.20$

Table 1. Safety requirements in the Ultimate Limit State specified as the formal yearly probability of failure p_f and the corresponding reliability index β, NKB /1/

From Table 1. it is seen, that if it is possible to calculate the formal yearly probability of failure p_f or the corresponding reliability index β, it is possible to determine whether the requirements for the structural safety is fulfilled or not. It can therefore be concluded, that the legal justification for application of probabilistic-based approaches in Sweden is present. Furthermore, NKB /1/ specify the principles for modelling of uncertainties, including model uncertainties, which makes NKB /1/ superior to other specifications for probabilistic based safety assessment. Requirements and principles for probabilistic-based analyses may also be found in Eurocode 1 /2/ and ISO /3/.

The reliability index β is obtained by applying (First Order Reliability Methods). In FORM, please see /4/ or /5/, the basic uncertain variables are modelled as stochastic variables $\mathbf{X}$ and transformed into the space of uncorrelated standardised normal distributed variables $\mathbf{U}$: $\mathbf{U} = \mathbf{T}(\mathbf{X})$. The Limit State of the considered failure mode is similarly transformed $g_u(\mathbf{u}) = g_x(\mathbf{T}^{-1}(x))$. In the U-space the limit state $g_u(\mathbf{u})$ is linearised in the so-called β-point, u^*, i.e. in the point closest to the origin with $g_u(\mathbf{u}) = 0$ which corresponds to the most likely failure point. This β-point can in general not be obtained from hand-calculations, but from the solution of the non-linear optimisation problem: $\min_u \beta = |u|$, s.t. $g(u) = 0$. Hereafter, the probability of failure can be obtained from: $p_f \approx \Phi(-\beta) = \Phi(-|u^*|)$, see /4/ or /5/ for a closer description.

The overall procedure in probabilistic-based assessment of bridges can be stated as: 1) *Pre-evaluation* including evaluation according to the deterministic code with identification of the critical limit states, 2) *Modelling and programming of critical limit states*, 3) *Traffic load modelling* depending on the limit state and actual bridge load situation, 4) *Modelling of stochastic variables* related to load and resistance according to the bridge specific information concerning materials, loads and mechanical modelling, 5) *Calculation of the reliability index for each limit state* applying a standard software package, 6) *Comparison of the safety level with the requirement for the structural safety* and 7) *Post evaluation* including sensitivity analysis and comparison with the deterministic evaluation.

MODELLING OF UNCERTAINTIES AND TRAFFIC LOAD

The basic resistance variables for e. g. strength, are often modelled as Log-normal distributed stochastic variables for which the statistical parameters (mean and standard deviation) are obtained from the project information, including information from as-built information (drawings, specifications of materials etc.) and later material testing. The contributions to the uncertainty arise from 1) physical uncertainty, 2) statistical uncertainty and 3) model uncertainties. For the fundamental definition of uncertainties, see /1/, /4/ or /5/. The loads due to permanent actions are in general modelled by normal distributions with added contributions from model uncertainties, /1/. Variable loads are modelled applying extreme distributions (e.g. Gumbel distributions) obtained with a 1-year reference period. Further, model uncertainties are included.

The dominating load in assessment of existing bridges are in most cases the traffic load, which is divided into populations of: 1): Ordinary traffic load with main contribution from the largest loaded trucks (weight 40 – 60 t) and 2): Heavy transports with special permission for passing of the bridge (typical weight 50 – 500 t). Based on these populations, the load combinations are made. Several situations can be relevant. However, the basic cases for a two-lane bridge with a relatively small influence length (< ~ 40 m) in general are: 1) Appearance and meeting of ordinary trucks with ordinary trucks and 2) Appearance and meeting of special heavy transports with ordinary trucks. For both cases the extreme distribution function of the load effects can be obtained from so-called thinned Poisson processes, see /7/ and /8/. If only a two-lane bridge is considered the extreme distribution F_{max} of the considered load effect q can be obtained from:

$$F_{max}(q) = \exp(-(v_1 - v_{12})T(1 - F_1(q)))\exp(-(v_2 - v_{12})T(1 - F_2(q)))\exp(-v_{12}T(1 - F_{12}(q)))$$

where v_1 and v_2 are the intensities of the considered traffic in lane 1 and 2, respectively. v_{12} is the intensity of meetings. T is the considered reference period for the extreme distribution (One traffic year). Further, the distributions for the load effect in lane 1, $F_1(q)$, lane 2, $F_2(q)$, and the distribution of load effects due to simultaneously traffic load in both lanes, $F_{12}(q)$, must be determined. These three distributions do in general include modelling of a) the number, configuration and weight of the trucks, b) the longitudinal and transverse appearance in the bridge lanes, c) the dynamic amplification of the static truck load, d) the mechanical models for the relation between traffic load and traffic load effects and e) the relative importance for the reliability of load in the actual lanes.

From this it is seen, that the modelling of the resistance variables, based on information from the specific bridge and the traffic load effects based on the traffic situation on the actual bridge in the individual approach with probabilistic-based assessment, avoid the unnecessary conservatism present in many deterministic codes.

EXPERIENCE WITH PROBABILISTIC-BASED ASSESSMENT OF BRIDGES
Before a probabilistic-based assessment can be performed it is very important to adapt the probabilistic-based approach to the code format, regulations and administrative procedures for administration of special heavy transports in Sweden. I.e. it is a requirement for the probabi-listic-based assessment that the results can be interpreted and applied as the results from a traditional deterministic assessment. This was done in a study in which the background in the Swedish rules for the resistance and load modelling was investigated. Based on this study the probabilistic-based modelling was formulated including load combinations and modelling of the heavy special transports with bogie configuration B according to the code /9/. The modelling was applied in the classification of the actual bridges and therefore also used in the two following examples of probabilistic-based classifications:

Bridge C295 over Sävja stream, Motorway E4 Stockholm-Uppsala

Bridge C295 is a traditional 4-span Motorway Bridge from 1971 constructed as two 103-m long independent box-girder bridges in post-tensioned concrete, see Figure 2. The individual bridges are supported at 3 centrally located circular columns and 3 supports at each abutment.

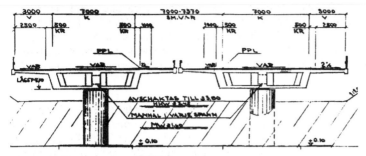

Figure 2. Cross section of C295.

The pre-evaluation of the bridge previously performed as a classification according to /9/ identified ULS of torsion in the cross section close to the abutment, as the critical limit state with B_{max} = 115 kN (~ truck weight 38 t /9/). All other limit states could be classified with B_{max} > 240 kN (~ truck weight > 80 t /9/), i.e. it was concluded that the probabilistic-based assessment should focus at the critical limit state in torsion. However, before the probabilistic-based assessment was initiated, the capacity model was revised according to Elfgren & Grzybowski /10/ resulting in B_{max} = 180 kN which, however, still was insufficient. Therefore the critical limit state was formulated and implemented as a FORTRAN subroutine for the reliability analysis. The limit state included influence functions from 3D FEM grillage analysis of cross section moments in bending and torsion transferring any traffic load to load effects in the critical cross section.

Based on this limit state modelling the safety requirement can be found to $\beta \geq 4.26$, see /1/.

The traffic load situation in the general code /9/, is a heavy truck with a load corresponding to 0.8 B_{max} overtaking a heavy special transport with a load corresponding to B_{max}. The vehicles are running in the worst case lanes with the special heavy vehicle driving in the emergency lane as shown in Figure 3 (left). This is, however, seen from a probabilistic point of view a situation with a probability of occurrence ~ 0, i.e. more likely load cases should be considered. With a influence length in torsion of approximately 70 m the case of a heavy special transport followed by an ordinary truck is more likely, i.e. a convoy of two trucks is considered in stead of the conservative generalised case in /9/. In Figure 3 the two traffic load situations are illustrated. Based on this and local traffic counting, the load model was made up according the description in the previous section.

The uncertain variables were modelled as stochastic variables according to the Swedish regulations applied in 1971 and present codes. The uncertain variables are: a) Yield stresses of the longitudinal and transverse reinforcement, b) loss coefficient of cable forces, c) cross sectional moments from bending and torsion for both dead load and superimposed dead load, d) *B* of the heavy special transport corresponding to a standard configuration in /9/ and the total weight of the special vehicles, e) weight of the identified heaviest group of the ordinary trucks (In Sweden trucks with 7 axles are dominating in both number and load), f) two independent variables for transverse load location of the special heavy transport and the following ordinary truck, respectively, g) dynamic amplifications and finally h) model uncertainties of the load model. Model uncertainties are included in the above modelling in a), b) and c). Further a number of deterministic quantities are specified in the modelling.

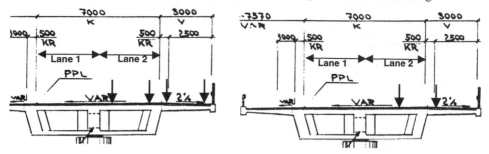

Figure 3. Illustration of the worst case loading with two heavy vehicles side by side from the general code /9/ (left) and the more likely extreme case from the performed probabilistic-based traffic load modelling of two trucks in convoy (right)

Based on this limit state, a calibration with the expected value of B_{max} corresponding to $\beta = 4.26$ as result was performed. From this result of the probabilistic-based assessment was obtained as $B_{max} = 543$ kN which is a drastically increase compared to the initial 180 kN. The post-evaluation of the analysis showed that the results are not more sensitive to the model parameters than acceptable. Further, the bridge specific partial safety factors in Table 2 were obtained. It is seen that the partial safety factors are increased at the traffic load but decreased at the strength variables. Furthermore, the more realistically traffic load modelling is of major importance for the results.

Stochastic Variable	Partial safety factors	
	Deterministic /9/	Probabilistic-Based
Yield stress, longitudinal reinforcement	1.32	1.07
Yield stress, transverse reinforcement	1.32	0.96
Weight, Special heavy transport	1.3	1.61
Death weight	1.0	1.00
Loss coefficient, cable force	1.0	1.05
Superimposed death weight	1.2	1.00
Weight, ordinary transport	-	1.76

Table 2 Comparison of partial safety factors from /9/ and the probabilistic-based assessment.

However, the other limit states from the deterministic assessment resulted in classifications $B_{max} > 240$, i.e. the obtained classification of the bridge from the performed probabilistic analysis can be set to $B_{max} > 240$ for all limit state and thereby for the entire bridge. From this it is seen that the probabilistic-based assessment including a more realistic modelling of the traffic load situations and capacity modelling according to the specific bridge conditions avoided a possible costly strengthening project.

Bridge E129 over Motala stream, Road 1142 Kimstads-Finsprång

The E129 Bridge is a simply supported post-tensioned concrete bridge with a 49.4 m span constructed in 1962. The bridge is basically two beams supporting a slab which carries a traditional two-lane main-road, see Figure 4. The pre-evaluation of this bridge according to /9/ identified the serviceability limit state SLS as the critical limit state due to a replacement of the edge beams which therefore are not carrying dead weight but only traffic load. The determined classification was $B_{max} = 170$ kN for the critical limit state whereas all other limit states obtained a classification $B_{max} > 215$ kN which is sufficient for this bridge. The SLS can formally be formulated as:

$$g = f_{ct} - \sigma_D(\eta_1, \eta_2, F)$$

whith f_{ct} : Tensile strength of concrete, σ_D : Tension in cross section, η_1 : Loss coefficient, phase A → B, η_2 : Loss coefficient, phase C → D and F: Applied load. Where B and C corresponds to the time just before and just after the replacement of the edge beams. It is seen that the limit state is based on a formulation in which track must be kept on all time varying parameters in the large post-tensioned concrete cross-section.

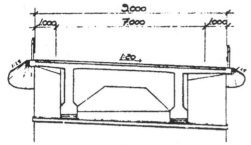

Figure 4 Cross-section of E129.

It is slightly more complicated to determine the requirement for the safety in the SLS because more individual judgement must be applied. NKB /1/ gives a guideline as $1.3 \le \beta \le 2.3$ which corresponds to the requirement in BBK 94 /11/. ISO2394-1998 and other references divide the SLS into Irreversible SLS (e.g. permanent unacceptable large deformations, permanent cracks or locale damages) and Reversible SLS (e.g. large vibrations causing discomfort, temporary large deformations or temporary cracks). In JJCS /12/ another approach is taken stating, that the requirement should be $\beta = (1.3; 1.7; 2.3)$ for (large; moderate; low) repair costs, respectively. In this case with a SLS requiring that the normal stress in tension must be not higher than the concrete tension strength capacity, it is clear that the limit state is reversible. Further, it is clear that the repair costs are high. Therefore, it is concluded that the $\beta \ge 1.3$ is a suitable requirement.

The traffic modelling corresponds to the modelling described in the above section on traffic load with meetings between heavy special transports and ordinary loaded 7-axle trucks on the bridge. The limit state is dependent on a relatively large number of uncertain variables modelled as stochastic variables. The most important are cross sectional forces due to the cable forces, which are corrected for the influence of creep and shrinkage in the phases before and after the replacement of the edge beams. The stochastic modelling of these concrete parameters are performed according to CEB-FIB /13/. Further, the dead weight and the superimposed dead weight are modelled as stochastic variables dependent on the considered phase. The concrete tension strength is modelled as a stochastic variable based on a number of split test performed with test specimens from the bridge.

Based on this modelling the classification is obtained as $B_{\max} = 233$ kN, i.e. a classification higher than the deterministic class of 170 kN and higher than 215 kN corresponding to the deterministic classification of the other limit states. (It should be mentioned that it is controlled that the FORM approximation is trustworthy in this case with a small reliability index).

It can therefore be concluded, that the probabilistic-based classification of E129 as result gives, that a costly strengthening project can be avoided and that the bridge can be classified with $B_{\max} = 215$ kN. This is mainly due to a carefully modelling of the phases including uncertainties in the model, a bridge specific traffic load modelling, a material modelling based on actual bridge material tests and finally a bridge specific determination of the relevant safety requirement in the actual SLS.

CONCLUSIONS

The conclusion of this paper can be outlined as:

- The general approach for assessment of existing bridges applying standard general codes is fast and efficient, but can be costly in the case of problems with the load carrying capacity due to expensive strengthening or rehabilitation projects.
- The individual approach applying probabilistic-based methods for assessment of existing bridges is based on the concept, that a bridge does not necessarily have to fulfil the specific requirements of a general code, as long as the overall level of safety defined by the code is satisfied.
- The individual approach is able to cut or reduce the strengthening or rehabilitation cost without compromising the level of safety.
- The background legal justification for the individual approach applying probabilistic methods is present in background documents for the codes in Sweden.
- The methodologies for application of probabilistic-based assessment are available and have been proven to work in practice.
- The practical experience from Sweden has shown that potential strengthening projects can be avoided.
- The new decision route for upgrading of existing bridges introducing consideration of the individual approach applying probabilistic-based assessment, can save millions of SEK in the future.

REFERENCES

/1/ Nordic Committee for Building Structures (NKB) "Recommendation for Loading and Safety Regulations for Structural Design" NKB report no. 35, 1978 & NKB report no. 55, 1987.

/2/ Eurocode ENV 1991-1: 1995 "Basis of design and actions on structures – Part 1: Basis of Design", September 1994.

/3/ ISO 2394-1998 "General principles on reliability for structures".

/4/ H. O. Madsen, S. Krenk & N. C. Lind "Methods of Structural Safety" Prentice-Hall, 1986.

/5/ O. Ditlevsen & H. O. Madsen "Structural Reliability Methods" John Wiley, 1996.

/6/ Boverkets konstruktionsregler, BFS 1993:58 med ändringar t.o.m. BFS 1998:39, November 1998. (In Swedish) "Code of practice for the safety and loads for the design of structures"

/7/ O. Ditlevsen & H. O. Madsen "Stochastic Vehicle-Queue-Load Model for Large Bridges" Journal of Engineering Mechanics. Vol 120, No. 9, pp 1829 - 1847, 1994.

/8/ O. Ditlevsen "Traffic Loads on Large Bridges Modelled as White-Noise Fields" Journal of Engineering Mechanics. Vol 120, No. 4, pp 681 - 694, 1994

/9/ Swedish Road Administration (Vägverket) "Allmän teknisk beskrivning för Klassnings-beräkning av vägbroar", 1998. (In Swedish) "Guideline for classification of the load carrying capacity of bridges".

/10/ Elfgren, L. & M. Grzybowski "Uppsprickning av betongkonstruktioner på grund av vridning"

/11/ Boverket BBK 95 "Betongkonstruktioner, Band 1 – Konstruktion", 1995. (In Swedish) "Concrete Structures, Code of practice".

/12/ Joint Committee of Structural Safety JCSS "Assessment of existing structures" Annex C, May 1999.

/13/ CEB-FIB Model Code 1990, May 1993.

Probabilistic-based assessment of a large steel bascule bridge and a concrete slab bridge

I. ENEVOLDSEN, M.Sc. Ph.D., RAMBOLL, Denmark
Bredevej 2, DK-2830 Virum, Denmark. ibe@ramboll.dk, www.ramboll.dk, phone +4545986674
C. T. von SCHOLTEN, M.Sc. & **J. LAURIDSEN**, M.Sc., Danish Road Directorate
Niels Juels Gade 13, DK-1059 Cph. K, Denmark, cts@vd.dk & jl@vd.dk, www.vd.dk, phone +4533933338

ABSTRACT

An individual bridge approach applying probabilistic-based bridge assessment is presented as a part of the standard methodology for administration of bridge safety of the Danish Road Directorate. The basic steps of the approach are presented and exemplified in two practical cases. Assessment of the load carrying capacity of a large steel bascule bridge and a concrete slab bridge. The latter demonstrates the benefits of combining probabilistic-based analysis with plastic limit state analysis. The approaches have already saved large costs from avoided strengthening projects.

INTRODUCTION

In the past years large effort has been put into the assessment of the load carrying capacity of the bridges managed by the Danish Road Directorate. The assessments are mostly dealing with the passing of heavy vehicles, but for bridges sensitive to other loads as ship impact or wind, these actions are considered too. This paper deals with the passage of heavy vehicles only, but the methods are applicable to other types of loads as well. A heavy vehicle is defined as a vehicle that needs special permission of the police with a weight larger than 48 t.

A Classification system for bridges and vehicles has been developed for administration of heavy vehicles on bridges. The system is based on the idea that the assessment of bridges regarding heavy vehicles should be carried out only once. Once the bridges are classified it is easy for the police and bridge administrators to decide if a specific heavy vehicle can pass the bridge. The classification system /10/ is based on a set of standard vehicles representing vehicles with a total weight ranging from 20 t to 200 t. In a regular deterministic assessment of the bridge class a relatively conservative traffic load combination is applied comprising the standard vehicle pointing out the class parallel with a standard vehicle of 50 t. The bridge class is equal to the weight in tons of the heaviest standard vehicle in the load combination. Furthermore, bridge classes are determined for passages of heavy vehicles with imposed restrictions such as speed limits and exclusion of other traffic.

A relatively detailed so-called blue road net has been established comprising roads with no bridges having a class less than 100. As the blue road net includes all motorways and many other major roads it ties together the whole country. The police uses the map of the blue road net when preparing the permissions and the haulage contractors when planing the transports.

It is the aim of the Danish Road Directory that all state roads and as many other main roads as possible are included in the blue road net. As about 98 % of all heavy vehicles are classified below 100 t, and hence needs no special investigation when passing on blue roads, an easy and efficient administration for the police and bridge administrators is hereby established together with the provision of satisfactory service to the industry.

The bridges are in general assessed applying deterministic methods with elastic or plastic limit state analysis according to the Danish guideline for classification of bridges /10/. However, if the obtained bridge class is insufficient (<class 100) probabilistic-based approaches are always considered, before an expensive strengthening or rehabilitation project is implemented. The probabilistic-based approach is often combined with advanced response models which raises the bridge class considerable, and in many cases these analysis results in a satisfactory bridges class minimising or avoiding strengthening projects. Therefore these methods has proven to be very beneficial for the bridge managers with large cost savings as results.

In the following section the approach for probabilistic-based classification will be described followed by two sections demonstrating the efficiency of the approach by two cases finished in 1997 and 1999, respectively. Finally a conclusion is given.

APPROACH FOR PROBABILISTIC-BASED ASSESSMENT OF BRIDGES
The basic idea in the probabilistic-based assessment approach is that in stead of using a <u>general approach</u> with a general code that necessarily must generalise in order to cover all types of bridges an <u>individual bridge</u> approach is used including application of the FORM β-method which directly takes the bridge specific uncertainties and local traffic situation into account, consistently. The results are in general a classification permitting considerably higher traffic load following the principle that a bridge does not necessarily have to fulfil the specific requirements of the general deterministic code, as long as the overall level of safety defined by the code is satisfied. As consequence the individual approach cut or reduces the strengthening or rehabilitation cost without compromising the level of safety. The basic steps applying the probabilistic based approach   are:

1) *Pre-evaluation*: In order to focus the probabilistic-based analysis the bridge is evaluated according to the relevant deterministic code. Thereby, the critical limit states are identified.
2) *Modelling of identified critical limit states* with implementation as computer code in subroutines for the analyses.
3) *Traffic load modelling* with a set-up depending on the actual load situation at the actual bridge and type of the examined limit state.
4) *Modelling of uncertain variables as stochastic variables*: The uncertain variables related to load, resistance and modelling are modelled as stochastic variables with corresponding statistical distributions including parameters according to the specific project material and level of knowledge concerning materials, loads and mechanical modelling.
5) *Calculation of the reliability index for each limit state* applying a standard software package which automatically performs the necessary transformations and solve the relevant optimisation problems when subroutines with the limit states and modelling of the stochastic variables are provided. For further explanation, see /4/ or /5/.
6) *Evaluation of the safety level* with comparison of the obtained reliability indices with the requirement for the structural safety in e.g. /1/. (/1/ contains the background documentation for the required safety levels in the Nordic Countries)
7) *Post evaluation*, which normally include a comprehensive sensitivity analysis and a comparison with the results from the deterministic analysis.

EXAMPLE 1 - LOAD CARRYING CAPACITY OF THE VILSUND BRIDGE

The Vilsund Bridge, see Figure 1, is a 381 m long steel bridge from 1939 consisting of 5 ordinary 67.8 m spans and a 34 m bascule span. The two lane road bridge has a width of 8.6 m with a concrete slab deck that is supported by a steel girder system with cross girders for every 5.58 m.

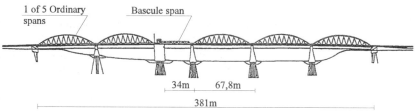

Figure 1. The Vilsund steel bridge with 5 ordinary spans and a bascule span.

The Danish Road Directorate decided that it would be desirable to allow trucks with a total weight of 100 metric tons to pass the Vilsund Bridge instead of a maximum truck weight of 50 metric tons. (Trucks with a weight above 50 metric tons must take a 150 km long detour if the Vilsund Bridge can not be passed). RAMBOLL was assigned the re-classification job including technical analysis and if necessary preparation of strengthening projects.

Analyses of the bridge were performed according to general approach for classification of existing bridges in Denmark /10/ with Class 50 as result which corresponds to a situation where two Class 50 trucks with an approximate weight of 50 metric tons, see Figure 2, are passing each other on the bridge in the most critical situation for the considered structural element. In the general load model the truck loads shown in Figure 2 are increased with a dynamic amplification factor of 1.25, further a distributed load of 2.5 kN/m² is applied 10 m before and 10 m after the trucks. The loads in the desired Class 100 are obtained similar as in Class 50 except that a Class 100 truck with a weight of approximately 100 metric tons is passing a Class 50 truck in stead of two Class 50 trucks passing each other as in Class 50.

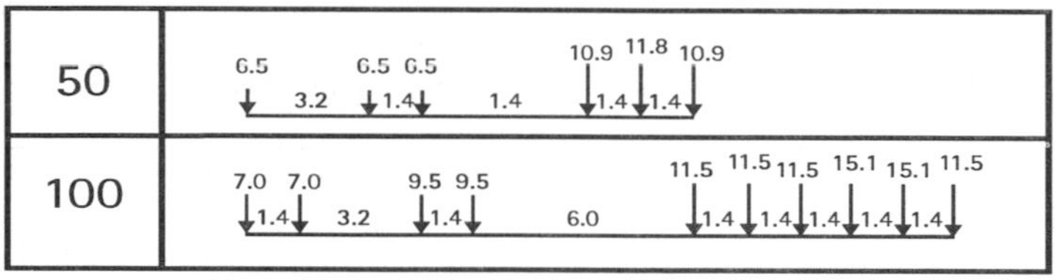

50	6.5 6.5 6.5 10.9 11.8 10.9
	↓ 3.2 ↓1.4↓ 1.4 ↓1.4↓1.4↓
100	7.0 7.0 9.5 9.5 11.5 11.5 11.5 15.1 15.1 11.5
	↓1.4↓ 3.2 ↓1.4↓ 6.0 ↓1.4↓1.4↓1.4↓1.4↓1.4↓

Figure 2 Characteristic axle-loads (t) and axle spacing (m) Class 50 and Class 100 trucks /10/.

The results of the analyses showed that the critical structural members are the main cross girders supporting the concrete deck and some of the steel truss members in the ordinary spans. A rough estimate of cost for strengthening the Vilsund Bridge to Class 100 was found to 20 mill DKK.

Safety Assessment by the Individual Probabilistic-Based Approach

In stead of performing the strengthening project, RAMBOLL suggested to perform a bridge specific reliability analysis of the critical elements in order to reduce or eliminate the costs of the strengthening project. The reliability evaluations were split into probabilistic-based evaluation of a) the main cross girders and b) of the critical steel truss members in the main structure.

Probabilistic-Based Reliability Evaluation of Main Cross Girders

The objective of the probabilistic-based reliability evaluation of the main cross girders is to decide whether the girders could be assigned a higher Class than 50 based on the bridge specific conditions. The probabilistic-based assessment is based on FORM, /4/ or /5/. The failure of the main cross girders is assigned the failure consequence class denoted "very serious" and failure type II including ductile failure without any extra load carrying capacity according to /1/ from which it is possible to obtain the requirement for the reliability index as $\beta \geq 4.75$.

The reliability evaluation of the main cross girder is based on the meeting event of two heavy trucks. The yearly probability of failure will then depend not only on the weight of the trucks but also on the yearly number of trucks passing the bridge. It is obvious that the chance that one heavy truck meets another heavy truck is lower for a bridge with a relatively low heavy truck frequency and vice versa. The largest load effect within a given time interval is in the reliability evaluation modelled as a so-called thinned Poisson load-pulse process /7/. The reliability model includes a model for determination of the cross-sectional forces in a gross girder and a modelling of the uncertain variables as stochastic variables based on the available information concerning the bridge from material characteristics and as-built data from the thirties and later. This information is obtained from the maintenance program of the bridge and traffic counting on the bridge. The variables modelled as stochastic are: The weight of the trucks with the related total weight of the two meeting trucks, the yield stress, the dynamic amplification factors, the position of the trucks, the dead load of the bridge deck and a number of model uncertainties.

Based on traffic counting on the bridge and distributions of the weight of the ordinary trucks, the investigations of the meeting situations are divided into 4 cases: 1) Meetings of all trucks with a total weight larger than 2 tons in both lanes, 2) Exclusively meetings of only the articulated lorries (tractor with trailer) in both lanes, 3) Meetings of all trucks with a total weight larger than 2 tons in one lane and a nominal number of heavy trucks corresponding to Class 100 in the other lane and 4) Exclusively meetings of articulated lorries in one lane and a nominal number of heavy trucks corresponding to Class 100 in the other lane. The results of the reliability analyses of the critical cross section of the main cross girder are shown in Table 1 for the 4 cases:

Case	1	2	3	4
β	5.66	5.77	4.75	4.75
No.	-	-	180	300

Table 1 Results of the reliability analyses of the main cross girders

In Table 1 it is seen that between 180 and 300 Class 100 trucks yearly can pass the bridge without compromising the required reliability level. This shall be compared with an expected yearly number below 12. The results of the sensitivity analysis show that the results are in-sensitive to large changes in the ordinary traffic volume and the reliability analysis is dominated by the uncertainties related to the yield stress of the main cross girder and the load distribution model.

The results show that the main cross girders can be classified in Class 100, i.e. an 15 million DKK strengthening project is avoided without compromising the safety of the bridge deck.

Reliability Evaluation of Critical Steel Truss Members in the Main Structure

Encouraged by the above results it was decided to proceed with a probabilistic-based reliability evaluation of the critical truss members in the ordinary span. A deterministic analysis performed according to the general approach /10/ showed that the problem was a two-fold. As it can be seen in Figure 3 the critical members are the first 3 beam trusses in the head of the span. This interferes with another problem in the structure. In Figure 3 is shown a bracing, which was requested removed because the clearance profile is to small for high trucks. However, it was decided to perform a probabilistic-based evaluation with the purpose of evaluating if it is possible to remove the bracing or strengthen the structure in the area connected to the bracing and afterwards classify the beams in Class 100.

The failure of the members is assigned failure consequence class "very serious" and failure type I for ductile failure with extra load carrying capacity (elastic limit state) resulting in the safety requirement $\beta \geq 4.26$ corresponding to a formal yearly probability of failure $P_f \leq 10^{-5}$ /1/.

The fact that the main spans of the Vilsund Bridge are 67.8 m long makes the probabilistic-based reliability evaluations more complicated because when the load effects are evaluated in a beam truss member in the ordinary span it becomes necessary to examine the traffic mix of trucks and cars over the entire 67.8 m long span. Further, the bascule makes the probabilistic-based reliability evaluations even more complicated because the queue release after closure of the bascule must be modelled individually. However, the considered load situations are: 1) Maximal yearly load effect from freely moving traffic, 2) Maximal yearly load effect from queue release after closure of the bascule and 3) Maximal yearly load effect from a number of Class 100 trucks which meets the load effect from the instantaneous traffic (Here defined as the maximum distribution for a 8 hour interval). The instantaneous traffic includes the probability of the meeting of a queue release after closure of the bascule.

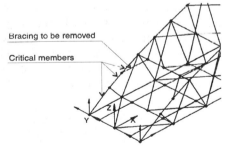

Figure 3 Critical members and bracing to be removed.

Further, the wind load was introduced in the probabilistic modelling applying a relatively crude model because it turned out that the stress level in the critical members due to wind load was significantly smaller than the stress levels due to traffic load.

The theory behind the used probabilistic-based reliability models can be found in /7/ & /8/. The main idea is that based on the influence functions statistical distributions of the load effects in the structural member can be found based on the traffic intensity and mix between cars and heavy vehicles. The model takes among others the following parameters as input: a) daily number of vehicles, b) yearly number of Class 100 trucks, c) daily distribution of traffic, d) The traffic speed, e) distribution of the weight of the traffic and f) probability of queue release. Further, the relative part of the time the traffic is congested in a queue release after closure of the bascule and other parameters for description of the traffic are included in the models.

After the statistical distribution of the load effect has been found it is possible to perform a more ordinary probabilistic-based reliability evaluation. The modelled stochastic variables are: Stresses due to traffic loads, stress due to dead load, dynamic amplification factors, yield stress and a number of model uncertainties. The result of the reliability analysis can be seen in Table 2 for the critical truss element. Further, it turned out that the reliability in the situation with removed bracing is slightly higher in the traffic load situations (The reliability is lower in the extreme wind situation but this is not critical for the structure).

Load situation	1	2	3
β	5.86	4.65	4.54

Table 2 Results of the reliability analyses of the critical steel truss members

It is seen that the reliability of the individual cases is higher than the required 4.26 with the conservative number of 100 yearly class 100 trucks. The 3 situations should in principle be combined in a series system that, however, can not reduce the reliability level below the requirement. The results of the analyses do further show that the load effects due the traffic load are reduced by 35 % compared to the worst case model in the general approach /10/. The sensitivity analysis show that the results in general are relatively insensitive to large model changes.

Hereby, the Vilsund Bridge is classified as a Class 100 bridge even with removed bracing. The use of bridge specific probabilistic-based methods have in this case saved approximately 20 million DKK (Excluding road users inconvenience costs).

Example 2 – Probabilistic and Plasticity Based Assessment of a concrete slab bridge
In this example plastic response analysis and probabilistic-based safety analysis are combined in RAMBOLL's program called PROCON. The idea behind PROCON is to apply the combination of advanced non-linear mechanical models and reliability models of concrete bridges. It is hereby the intention to use models which both are closer to the actual structural behaviour in the failure situation and to the safety of the bridge. Hence, a safety evaluation closer to reality and not as conservative as in the usual general approach is obtained. PROCON contains both a response module for plasticity based limit state analysis of concrete structures and a probabilistic module for more advanced safety evaluations. The mechanical response models in PROCON are based on lower bound stress or force equilibrium finite element formulations as formulated in /12/ and /13/. The application of a finite element formulations makes the calculations more operational compared to the traditional upper bound kinematic mechanism based formulations with yield lines or plastic hinges /14/.

In this example a relatively small simply supported concrete slab bridge is considered. The bridge from 1942 is constructed for under passing a 4 lane motorway for a small factory railroad has a span of 4.43 m, a with of 24 m and a screw angle of 25.6°, see Figure 4.

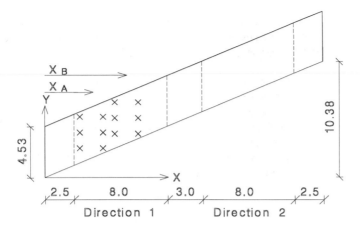

Figure 4 Concrete slab bridge at 4-lane motorway.
The intention for a bridge at motorways in Denmark is that they minimum should be classified in class 100. The deterministic classification according to /10/ is based on a worst case situation which for this bridge means that the 3 heaviest axles of two vehicles are placed at the critical location at the bridge corresponding to X_A = 4.0 m and X_B = 7.0 m, see Figure 4. Both vehicles are multiplied by a dynamic amplification factor of 1.25 (98 % quintile) and the partial safety factors are 1.3 at the class 100 truck and 1.0 at the class 50 truck. The partial safety factor at the characteristical 10 % quintile concrete strength of 37.5 MPa is 1.90. The partial safety factor at the characteristical 0.1 % quintile steel strength of 235 MPa is 1.5. The thickness of the slab is 0.4 m. The slab is only reinforced in the bottom with reinforcement areas in the x and y directions as a_x = 319 mm²/m and a_y = 1510 mm²/m .

The result of an elastic analysis with 110 slab-elements with 9 nodal degree of freedom is that only 29 % of the load from the class 100 truck can be carried together with the death load and the class 50 truck based on a strip-method.

The analysis with PROCON using the finite equilibrium element lower bound method is performed with 48 triangular elements. The result of the analysis similar to the elastic case is that 79% of the load from the class 100 truck can be carried. This is a drastically increase in the load carrying capacity and show the efficiency of a redistribution of the forces in a plastic ultimate limit state analysis. It must in this connection be mentioned that the serviceability limit state not is critical in this case.

In the probabilistic-based individual approach the plastic failure of the slab is assigned the failure consequence class "very serious" failure type II including ductile failure without any extra load carrying capacity resulting in the requirement $\beta \geq 4.75$ according to /1/.

The reliability evaluation is based on the passing event of two heavy trucks in which the heavy class 100 trucks with a speed of 40 km/h is overtaken by the faster (80 km/h) ordinary trucks in one of the directions of the 2 lane motorway. The yearly probability of failure will then depend not only on the weight of the trucks but also on the yearly number of trucks at the bridge. The largest load effect within a given time interval is in the reliability evaluation modelled as a thinned Poisson load-pulse process according to the description in /7/.

The reliability modelling of the uncertain variables as stochastic variables is based on the available information concerning the bridge from material characteristics and as-built data from the forties and later. This information is obtained from the maintenance program and traffic counting on the bridge. The variables modelled as stochastic are: The weight of the trucks with the related total weight of the two meeting trucks, the yield stress, the dynamic amplification factors, the position of the trucks, the dead loads of the bridge and a number of model uncertainties.

The result of the probabilistic based safety analysis is that 116 % of the load from the class 100 truck can be carried It is hereby seen that the use of plastic analysis in combination with probabilistic analysis makes a strengthening project redundant and that the cost saving is considerably.

Examining the results in details gives that the partial safety factors from the probabilistic analysis are with the partial safety factors from /10/ in ( ): Yield stress $\gamma_s = 1.12$ (1.5), concrete strength $\gamma_c = 0.85$ (1.9), dead load $\gamma_d = 1.05$ (1.0), class 100 truck $\gamma_A = 1.69$ (1.3), class 50 truck $\gamma_B = 1.52$ (1.0). Further the dynamic amplification factors turn out to be $S_A = 1.03$ (1.25) and $S_B = 1.06$ (1.25) which is due to the fact that the dynamic amplification factors are modelled as stochastic variables uncorrelated to the stochastic variables modelling the weight of the trucks in contradiction to the principles in /10/.

The sensitivity analysis shows that the results are relatively insensitive to large changes in the ordinary traffic volume and that the reliability analysis is dominated by the uncertainties related to the yield stress and the load model.

The overall conclusion of the probabilistic and plasticity based safety assessment of the bridge in this example is that the bridge is classified as class 100 whereas a traditional analysis would require a costly strengthening or replacement project.

CONCLUSIONS

An individual approach applying probabilistic-based methods alone or in combination with advanced plastic limit state analysis for assessment of existing bridges will in many cases be able to cut or reduce the strengthening or rehabilitation cost without compromising the overall level of safety defined by the relevant national code.

The methodologies for use of probabilistic-based assessment are available and have been proven to work in practice as e.g. in the two examples presented in this paper. Therefore, the approach is adopted as a standard approach in the bridge administration of the Danish Road Directorate.

It is estimated that the Danish Road Directorate already has saved more than EUR 13.5 million applying probabilistic-based approaches for classification and establishment of management plans of bridges. Based on this it is believed that the probabilistic-based approaches will save millions of EUR for the Danish Road Directorate in the future.

REFERENCES

/1/ Nordic Committee for Building Structures (NKB) "Recommendation for Loading and Safety Regulations for Structural Design" NKB report no. 35, 1978 & NKB report no. 55, 1987.

/2/ Eurocode ENV 1991-1: 1995 "Basis of design and actions on structures – Part 1: Basis of Design", September 1994.

/3/ ISO 2394-1998 "General principles on reliability for structures", 1998.

/4/ H. O. Madsen, S. Krenk & N. C. Lind "Methods of Structural Safety" Prentice-Hall, 1986.

/5/ O. Ditlevsen & H. O. Madsen "Structural Reliability Methods" John Wiley, 1996.

/7/ O. Ditlevsen & H. O. Madsen "Stochastic Vehicle-Queue-Load Model for Large Bridges" Journal of Engineering Mechanics. Vol 120, No. 9, pp 1829 - 1847, 1994.

/8/ O. Ditlevsen "Traffic Loads on Large Bridges Modelled as White-Noise Fields" Journal of Engineering Mechanics. Vol 120, No. 4, pp 681 - 694, 1994

/10/ Danish Road Directorate (Vejdirektoratet) "Rules for Determination of the Load Carrying Capacity of Existing Bridges, Danish Road Directorate, 1996. (Written in Danish: Beregningsregler for beregning af eksisterende broers bæreevne, April 1996).

/12/ Damkilde L. & O. Høyer "An Efficient Implementation of Limit State Calculations Based on Lower-Bound Solutions. Computers & Structures Vol. 49, No 6, pp. 953-962, 1993.

/13/ Krenk S., L. Damkilde & O. Høyer "Limit analysis and Optimal Design of Plates with Tri-angular Equilibrium Elements" Engineering Mech. Pap. No.16, Aalborg University 1993.

/14/ Nielsen, M.P. "Limit Analysis and Concrete Plasticity" Printice-Hall, 1984.

Proposed LoBEG prioritisation system for bridge strengthening, maintenance and upgrading works

NAVIL SHETTY & MIKE CHUBB   WS Atkins Consultants Ltd., Epsom, UK
DAVID YEOELL   Westminster City Council, London, UK
RICHARD McFARLANE   Royal Borough of Kingston, London, UK

ABSTRACT

The paper presents a pragmatic methodology for prioritising bridge works which is comprehensive and yet easy to use and implement. The system prioritises some 30 different work types which are classified into: (a) strengthening and re-instatement, (b) structural preventive maintenance, (c) non-structural maintenance, and (d) improvement and upgrade works. The system was developed for the London Bridges Engineering Group (LoBEG) with inputs from bridge engineers, traffic planners and bridge managers. The priority score for a bid is evaluated on a scale of 1 to 100 using five main criteria with relative weights: Risk of failure (40%), Cost of works (20%), Durability (15%), Social impact (15%), Cost of interim safety measures (10%). The weights have been derived using formal methods of multi-criteria decision analysis based on responses from bridge managers during structured interviews. Detailed algorithms have been developed for the evaluation of the above criteria and presented in an easy to use tabular form and require data that is readily available to engineers.

INTRODUCTION

Bridge Authorities world wide are faced with the difficult task of allocating scarce resources to various bridge works. As the available funding generally falls short of the needs, there is a need to prioritise bridge works with a view to maximise the benefits from the money spent.

London has a complex transport system with a high concentration of bridges of varying age, size, material of construction and strategic importance. Within Greater London there are 33 Boroughs which manage the highways network within their own areas. The work relating to highway bridge assessment, strengthening and maintenance is currently co-ordinated by the London Bridges Engineering Group (LoBEG). A package bid, called the "London Package" is submitted in October of each year by LoBEG to the Government Office for London (GOL) for funding of all bridge related works under the Transport Supplementary Grant (TSG) scheme. Once funding has been obtained, which is generally lower than what is required to complete all proposed works, the available funding is allocated by LoBEG to the individual Boroughs depending on the priority of the bridge works proposed in that bid year.

Over recent years attention has been focused on the assessment and strengthening programme which attracted all funding. Structural maintenance has received little funding as its importance relative to strengthening could not be demonstrated. This situation is common to many Local Authorities in the UK. There is therefore a need to determine priorities of different types of strengthening, maintenance and upgrade works on a consistent basis.

The paper presents a pragmatic methodology for prioritising bridge works which is comprehensive and yet easy to use and implement. The methodology can be used manually but can also be computerised with limited effort. The system prioritises over 30 different types of bridge works covering strengthening, maintenance and upgrade across the London area. This is an extension of the previous strengthening prioritisation system which has been successfully used by LoBEG for the last three years.

PRIORITISATION FRAMEWORK

Objectives of the Prioritisation System

The objectives of the prioritisation system are:

- to provide a consistent comparison of different bridges
- to provide a consistent basis for comparing strengthening, maintenance, improvement and upgrade works
- to provide a "fair" basis for the allocation of funding between the London Boroughs, and
- to maximise the benefits from the available resources

In order to be practicable, the prioritisation system should be simple, robust and auditable, and should account for the differences in commercial, industrial and traffic characteristics between the Boroughs. At the same time it has to cater to a wide range of structural and non-structural maintenance and upgrade works.

Method of Approach

A number of discussions were held with bridge engineers and managers in order to develop a comprehensive list of works which the system has to cover and to classify the work types into a limited number of categories. Examples of previous bids from the Boroughs were also used in this process. Criteria which the bridge managers regard as relevant in deciding priorities for each category of work were also noted. This gave a large number of factors or attributes which would influence the ranking of work types and of different bridges.

Next, the different factors were carefully analysed to produce an "Attribute Hierarchy Model" which expresses "Bridge Priority Score" as a function of five main criteria, each of which is a function of a number of sub-criteria down to the level of some basic parameters for which bridge engineers can easily provide values for each bridge. As discussed later, structured interviews were carried out with bridge engineers, bridge managers and traffic planners to determine the relative weights for the main criteria and various sub-criteria.

Detailed algorithms were developed to evaluate scores for each of the five main criteria as a function of the sub-criteria and parameters input for a bridge. Scoring sheets were developed to facilitate the scoring process. A comprehensive pro-forma was developed for bridge engineers to supply the necessary engineering, traffic and cost data on each bridge where works are proposed.

A master spreadsheet then combines the scores for different bridges and ranks them in terms of the Bridge Priority Score. Funding is allocated to bridges in the order of their priority until the available budget is fully utilised. The main features of the prioritisation system are summarised in the following.   The system will be shortly put to trial before full implementation.

Categorisation of Works

The various work types are grouped into: (1) Mandatory, (2) Routine and (3) Prioritised works. Funding for Mandatory works (such as assessment, inspection and monitoring, interim safety measures, etc.) is "ring-fenced" from out of the annual funding from the Government. On the other hand, for Routine works (such as cleaning of drains, maintenance of services, lighting, etc.) a blanket allocation is made the funding for which comes mainly from the Boroughs' own budget. All other works are subject to prioritisation which are classified into four categories as shown in Table 1.

Work Category	Description
Strengthening and Reinstatement	
SS-1	Strengthening of substandard bridges failing 40 tonne A.L.L. criteria
SS-2	Strengthening/protection of piers vulnerable to impact loads
SS-3	Parapet, crash barrier and safety fence strengthening/replacement
SS-4	Provision/repair/improvement of scour protection system
SS-5	Repair of damage/deterioration on deck elements
SS-6	Repair of damage to substructure due to impact/deterioration/scour etc.
SS-7	Embankment repair/strengthening
Structural Maintenance - Preventive	
SP-1	Renewal/repair of drainage system
SP-2	Repair/replace expansion joints (not linked to carriageway resurfacing)
SP-3	Replace/install deck waterproofing (not linked to carriageway resurfacing)
SP-4	Corrosion protection of metal members, e.g. painting
SP-5	Minor structural concrete repairs
SP-6	Installation/maintenance of cathodic protection of concrete members
SP-7	Silane impregnation of concrete members
SP-8	Desalination/Re-alkalisation
SP-9	Re-pointing brickwork/masonry
SP-10	Maintaining river training works and scour protection
SP-11	Replace bearings
SP-12	Maintaining BD21 road roughness category
SP-13	Signing for inadequate headroom
Non-Structural - Preventive	
NP-1	Renewal of mechanical and electrical equipment
NP-2	Renewal of lighting on the structure
NP-3	Cladding repairs
Improvement Works	
IW-1	Traffic management arrangements for the bridge
IW-2	Improving sight distance to prevent accidents and impact on structure
IW-3	Provision/widening of footpaths or cycle lanes on a bridge
IW-4	Handrailing
IW-5	Health and safety work
IW-6	Increasing headroom
IW-7	Strengthening for HB loading if a bridge passes 40 tonne A.L.L.

Table 1: Work types considered for prioritisation

Bid Preparation

A bid for funding typically combines all works proposed on a single bridge. Where the works are of a minor nature costing small amounts, a bid can combine works on different bridges. Works on different bridges on a route can also be combined into a bid to minimise traffic delay costs during works. A priority score is then worked out for the whole of the bid.

Pro-forma for Data Collection

A comprehensive pro-forma is used for the Borough engineers to bid for funds and to supply all the data necessary for prioritising the bids in an electronic form. The pro-forma is structured into different sections depending on the type of data, for example Scheme, Structure, Assessment, Maintenance, Social Impact, and Traffic. The sections are linked to the work categories to ensure that only the minimum necessary data need to be provided depending on the categories of work contained in a bid.

SCORING OF BRIDGES

Criteria for Scoring

The priority score for a bid is evaluated on a scale of 1 to 100 using five main criteria as shown in Figure 1 below which also gives the relative weights for the criteria.

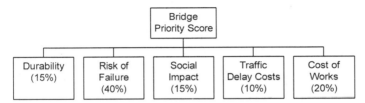

Figure 1: Attribute hierarchy for Bid Priority

Each criteria has a number of sub-criteria and factors, and the relative weights between the main criteria and several sub-criteria have been derived using the Analytical Hierarchy Process (AHP) method, [Ref.1], based on responses from bridge managers. The responses were obtained through structured interviews in which a number of hypothetical bridge decision scenarios involving choices between pairs of criteria at a time were presented to bridge managers. The final weights for the criteria were established by aggregating all responses and deriving a consensus matrix of preferences.

Bid Priority Score

The priority of a bid is evaluated in terms of the five main criteria mentioned above and by aggregating the scores for the different works proposed in the bid as

$$\text{Bid Priority Score} = 0.40 \times \text{Risk Score} + 0.20 \times \text{Cost of Works Score} +$$

$$0.15 \times \text{Durability Score} + 0.15 \times \text{Social Impact Score} +$$

$$0.10 \times \text{Cost of Interim Measures Score}$$

The Risk and other scores are evaluated on a scale of 0 to 100. The scores for each criteria are calculated by aggregating the scores for all work categories included in a bid. The weighting can be varied by other bridge authorities to reflect their policy and local factors.

ALGORITHMS FOR SCORING

Algorithms have been developed for the evaluation of the five main criteria given in Figure 1 which use the data supplied in the pro-forma by engineers.

Risk Score

The algorithm for risk covers structural hazards such as bridge failure due to traffic loading, parapet impact, pier collision, deck impact, scour, and traffic accidents due to certain deficient features of a bridge such as inadequate sight distance, narrow footway, etc. The procedures for evaluating risk and other criteria are presented in an easy to use tabular form and require data that is readily available to engineers. The factors considered in evaluating the risk score are shown in Figure 2.

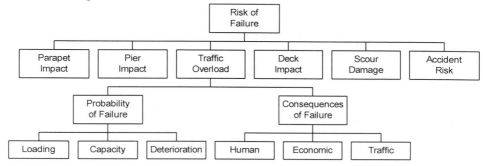

Figure 2: Attribute hierarchy for risk score

The risk due to each of the above hazards is evaluated in terms of the probability and consequences of the event. The Risk Score due to each hazard is then obtained from the risk matrix shown in Table 2. This matrix takes account of the fact that public react differently to a catastrophic accident with multiple casualties and many accidents with minor consequences.

Table 2: Risk score as a function of probability and consequences

Probability of Failure	Consequences of Failure					
	$\geq £\,25M$	20M	15M	10M	5M	$\leq £\,1M$
$\geq 1\times10^{-2}$	100	80	60	40	20	10
1×10^{-3}	80	64	48	32	16	8
1×10^{-4}	60	48	36	24	12	6
1×10^{-5}	40	32	24	16	8	4
1×10^{-6}	20	16	12	8	4	2
$\leq 1\times10^{-7}$	10	8	6	4	2	1

In view of the limitations on the length of the paper, only the evaluation of traffic loading risk is summarised here.

Probability of Failure

The probability of failure is initially calculated from the assessed capacity for the critical bridge element as below. This relationship has been established based on detailed structural reliability analysis calculations carried out for representative bridges, see [Ref.2].

	Assessed Capacity to BD21/97 [Ref.3]						
	≥40 t	38 t	25 t	17 t	7.5 t	3 t	0 t
Carriageway	10^{-6}	10^{-5}	10^{-4}	10^{-3}	10^{-2}	10^{-1}	1
Footway	10^{-9}	10^{-8}	10^{-7}	10^{-6}	10^{-5}	10^{-4}	10^{-3}

The probability of failure from the above is then multiplied by the following modification factors. The default modification factor is unity.

Level of Assessment
If the assessment has not been carried out to the highest feasible Level in accordance with BA70, [Ref.4], and if it is believed that the assessed capacity can be improved by higher Levels of assessment then the following modification factors are used.

Expected Improvement	None	Some	Significant
Modification Factor	1	10^{-1}	10^{-2}

Redundancy & Mode of Failure
The following modification factors are used depending on the type of element failure (brittle or ductile) and whether element failure causes only local damage or leads to collapse.

	Brittle	Ductile
Element failure leads to local damage	10^{-1}	10^{-2}
Element failure leads to bridge collapse	10	1

Signs of Distress
If the inspection of the structure has revealed signs of distress due to overloading of the element considered of the type that could be expected for the failure mode, then the a modification factor of 100 is used. The factor for no signs of distress is unity.

Deterioration
If the element considered is deteriorating and is likely to deteriorate further in the future, the following modification factors based on the Extent (on a scale of A~D) and Severity (on a scale of 1~4) codes from inspection (Highways Agency BE11 forms) are used.

Severity - Extent	1	2-B	2-C	2-D	3-A	3-B	3-C	3-D	4
Modification Factor	1	5	10	20	20	50	100	200	1000

Consequences of Failure
The consequences of a bridge failure include: (a) fatalities and injury, (b) repair/re-building costs, and (c) traffic delay/diversion costs while the bridge is re-built. There are other indirect costs such as cost of public enquiry/litigation, cost of re-assessments nationally of similar bridges, adverse public opinion and loss of morale, etc. These are difficult to evaluate but are seen to be closely related to the number of fatalities and injuries, and hence a constant factor of 4 on casualty costs is used to account for the indirect costs.

All the consequences are evaluated in monetary terms as explained below and then combined together to obtain the total cost of failure as

Consequences of Failure = Casualty Costs×4 + Re-building Costs + 0.5 × Traffic Costs

The traffic costs are factored by 0.5 in line with the lower weight given for the "Cost of Interim Measures" compared to the "Direct Cost of Works" as shown in Figure 1.

Casualty Costs
Fatalities/injuries due to a bridge failure can occur both on the bridge and on the facility underneath (highway/railway) the bridge. These are evaluated separately and <u>added</u> together to obtain the total casualty costs.

The number of fatalities and injuries on the bridge are evaluated as a function of the number lanes, span, road speed and proportion of different vehicle types in the flow. Fatalities and injuries below the bridge are determined depending on the total length of the failed spans, the traffic flow on the facility underneath (highway, navigable waterway or a railway line).

Re-building Costs
Re-building costs are obtained from the pro-forma and include direct cost of works, design & supervision costs and traffic management costs. If a temporary bridge is erected at the site while the bridge is being re-built, its cost and the cost of any other interim measures, e.g. propping, are also included.

Traffic Costs
The traffic delay and diversion costs can be obtained from the pro-forma. The effect of measures such as a temporary bridge in reducing the traffic disruption is taken into account in arriving at the traffic costs.

Durability Score
The durability score for an element is determined as

Durability Score = Element Importance × (Severity Score + Extent Score)

The Severity and Extent Scores are determined based on the "Severity" and "Extent" codes from inspections. The importance of an element is determined in terms of the effect the non-functioning of the element would have on the deterioration of remaining elements of the bridge. For example, expansion joint and waterproofing membrane are given an importance rating of 5, silane coating a rating of 3 while most other elements are given a rating of 2.

Social Impact Score
If the proposed works are not carried out on the bridge and if weight restrictions are imposed, it can cause undue impact on the local community. These effects are in addition to the traffic delay and diversion costs accounted for separately as "Cost of Interim Measures". This impact on the community can often lead to adverse public opinion and "bad press" for the bridge authority. The following effects are considered to be important and are collectively termed as "Social Impact":

- Inconvenience to the community due to, for example, buses being diverted away from their usual stops, HGV's being routed through residential roads, increased risk if emergency vehicles have to take a longer route, etc. This effect is very difficult to quantify but is seen to be closely related to the route category and its importance.
- Economic impact on the local business and industry

- Effect on the aesthetics of the bridge if maintenance work is not carried out.

The factors considered in evaluating the Social Impact Score are shown in below.

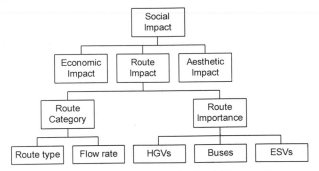

Figure 3: Attribute hierarchy for Social Impact Score

The Social Impact Score is evaluated by adding the scores for the above three effects with appropriate weights as below.

Social Impact Score = 0.5×Route Score + 0.3×Economic Score + 0.2×Aesthetics Score

Cost of Interim Measures Score

Pending full strengthening, one or more interim safety measures such as weight restriction, lane closure, propping, monitoring, etc. will be implemented on a substandard bridge. The cost of these are calculated to obtain the total cost of interim measures.

The traffic diversion costs due to weight restrictions are based on no. of vehicles diverted, road speed, vehicle composition and diversion length. It is assumed that the weight restrictions will be in place for 1 year and hence the traffic diversion costs are evaluated for this period. Simple look-up tables have been prepared to allow engineers to calculate the traffic costs quickly. The cost of traffic delays during maintenance/repair works is ignored as it is small in comparison with the diversion costs due to interim measures.

Cost of Works Score

The total cost for all works on a bridge are first calculated assuming the works would be carried out in the proposed year. The cost includes direct cost, supervision, traffic management and access costs. The different elements of the costs are obtained from the pro-forma prepared by engineers.

The savings in future costs as a result of carrying out the works in the Bid Year are evaluated in Net Present Value (NPV) terms considering the likely increase in costs over the next 5 years if the works were to be deferred. The savings are deducted from the cost of works and the net cost is then converted to "Cost of Works Score" which decreases as the cost increases.

SUMMARY AND CONCLUSIONS

Bridge Authorities world wide are faced with the difficult task of allocating scarce resources to various bridge works. As the available funding generally falls short of the needs, there is a need to prioritise bridge works with a view to maximise the benefits from the money spent.

The paper presents a logical and rational methodology for prioritising bridge works which is comprehensive and yet easy to use and implement. The system was developed for London Bridges Engineering Group (LoBEG) with practical inputs from bridge engineers, traffic planners and bridge managers.   The system prioritises some 30 different work types which are classified into: (a) strengthening and re-instatement, (b) structural preventive maintenance, (c) non-structural maintenance, and (d) improvement and upgrade works.   A comprehensive pro-forma is provided for the Boroughs to bid for funds and to supply all the data necessary for prioritising the bids in an electronic form. The priority score for a bid is evaluated on a scale of 1 to 100 using five main criteria with relative weights: Risk of failure (40%), Cost of works (20%), Durability (15%), Social impact (15%), Cost of interim safety measures (10%). The weights have been derived using formal methods of multi-criteria decision analysis based on responses from bridge managers during structured interviews. Detailed algorithms have been developed for the evaluation of the above criteria and presented in an easy to use tabular form and require data that is readily available to engineers.

The proposed system is very practicable and provides a rational basis for allocating funds to competing bridge needs. It also provides a consistent basis for comparing strengthening, maintenance, improvement and upgrade works. This should provide a strong justification for funding maintenance works which are often neglected in the face of other demands on funds.

ACKNOWLEDGEMENTS

The authors would like to acknowledge the financial support from LoBEG and the contribution of colleagues from WS Atkins, City of Westminster, Royal Borough of Kingston and other London Boroughs in carrying out the work reported in this paper.

REFERENCES

1. Saaty, T. L. (1980): The Analytic Hierarchy Process Method. John Wiley.
2. Shetty, N.K., Chubb, M.S. and Halden, D. (1997): "A risk-based assessment procedure for substandard bridges", in *Safety of Bridges*, (ed) Parag Das, Thomas Telford, London.
3. Highways Agency (1997): BD21 - Assessment of Short Span Bridges, Design Manual for Roads and Bridges, Vol.2. HMSO, London.
4. Highways Agency (1998): BA 70: Management of Substandard Bridges, HMSO.

Management of bridge maintenance – a local authority perspective

H. BROOMAN & N. WOOTTON, Surrey County Council, Epsom, UK

KEYWORDS: Bridge management, local authority.

INTRODUCTION

Surrey County is located in the south east of England and has one of the highest population densities of any county within the UK. As a consequence, the motorways and primary route network carries some of the highest traffic flows in the country. This in turn puts pressure on the entire county road network, including the unclassified roads. All classes of road carry over twice the national average traffic flows (Surrey County Council 1998)[1].

Road	5 Day Average 24hr Flow	Direction
M25	194634	Junction 13 - 14
A322	32478	Swift Lane to M3 Junc 3
A331	27192	Lynchford Rd - Aldershot Rd
A316	39922	County Boundary - Park Rd

Surrey County Council's Infrastructure Group is responsible for the operation and maintenance of over 2,000 structures on the county roads and rights of way network. The current cost of replacing these structures is in the region of £450 million. With approximately 750 structures in Surrey built between 1880 to 1920, we are rapidly approaching the end of the 120 year design life for a considerable percentage.

WHAT IS A BRIDGE MANAGEMENT SYSTEM

Various definitions of Bridge Management Systems have been given.
The Highways Agency state that: " the overall purpose .....is to enable the structures management process to fulfill the following objectives:-

1)   To provide relevant information for answering the following questions:-
   a) Why are maintenance activities necessary?
   b) What are the likely consequences if the works are not carried out?
   c) Are the funds being used effectively?
2)   To obtain better value for money through whole life costing of work options.
3)   To make safety the primary objective while minimizing disruption to functional use.
4)   To take into account long term sustainability in respect of future logistical and funding requirements based on strategic forecasts of maintenance needs." (Das, P 1998)[2]

The American Association of State Highway and Transportation Official's (AASHTO) suggest that a bridge management system include four basic components: data storage, cost and deterioration models, optimization models for analysis and updating functions.

The Danish Road Department have stated that; "Bridge Management involves a number of activities, such as the collection of inventory data, inspections, assessment, management of special transports, allocation of funds, all with the purpose of ensuring traffic safety and maintaining the bridge stock in the desired condition at the lowest possible cost. The purpose of a Bridge Management System is to assist in the management of these activities." (Lauridsen, J and Lassen, B 1998)[3]

Of all the definitions available, this one lies closest to the philosophy of bridge maintenance and Bridge Management Systems developed and used by Surrey County Council. Most systems will have an inventory list, some form of inspection regime, maintenance prioritisation and cost analysis/control function. These modules are common to DANBRO, PONTIS, SIHA, HiSMiS and COSMOS.

PREVIOUS MANAGEMENT SYSTEMS
Originally, paper records were used to list the inventory. Maintenance was 'the Cinderella of bridge works' (Palmer, J and Cogswell, G, 1990)[4], done on an ad-hoc basis when and if the finance was available.

The first computer system implemented at Surrey was a mainframe based system called STREG (STructures REGister). BRIDGIT was developed by Surrey County Council in conjunction with Howard Humphreys as an inventory and inspection system and replaced STREG in 1990. This system should not be confused with BRIDGIT that was developed in the United States about the same time as PONTIS. Data was entered into BRIDGIT from the original paper STREG Cards.

BRIDGIT was a major step forward in that it allowed inspections to be carried out on site using handheld data capture devices (DCDs), but this system was never fully utilised due to the vulnerability and unfriendly nature of the DCDs. BRIDGIT was run on a mainframe system across the County network which hindered it's long term viability with the increase in desktop power.

BRIDGIT's ability to deal with the day to day requirements of the inspection/maintenance process declined as new and improved database systems appeared on the market. In 1993 the decision was made to take the data from BRIDGIT and load it into a Microsoft Access database. Attempts to write an in-house system using Access were hampered by its inability to handle both the number of structures on the register and large amounts of historical information. With the amount of data being held increasing each year it was eventually decided to buy a pre-written bridge management system in 1995.

The system currently used for bridge maintenance in Surrey is COSMOS. COmputerised System for the Management Of Structures is the interface between the user and an Oracle database system. It was written by the Babtie Group and first saw service with Berkshire County Council. COSMOS came on line at Surrey in 1996.

From the dates shown above it can be seen that for two or three years Surrey was not operating a formal bridge inventory system. In many ways Surrey had reverted to using only the minimal computer systems with an inventory list and inspection programme. Maintenance had always been undertaken at the inspectors discretion, based on their knowledge of the area and the structures within it. Although this system worked, it was not the most efficient means of spending the maintenance budget.

DEVELOPING THE SYSTEM
A good Bridge Management System should:-

- be able to store up to date relevant information on every structure and more importantly, it should be able to retrive this information quickly and with minimum fuss.
- be able to prioritise maintenance based on established and understood principals.
  Obscure lists of numbers that cannot be checked or understood are no use.
- be flexible enough to ensure that changes in working practice or requirements can be incorporated without a major rebuild or large amounts of downtime.
- allow the job to be done without a major shake up in established procedures and work practice. Although there is no such thing as a seemless integration of new systems, the rollout for the new system should be subtle enough not to scare off the end users.

While the original COSMOS is a fully working Bridge Inventory and Management System, Surrey have used several of the core data-tables to further develop our own systems in line with the above requirements. These include a site instruction and cost monitoring module, inspection generation module, anti graffiti and pump maintenance patrol modules and a full featured Intranet system for the dissemination of up to date structural information.

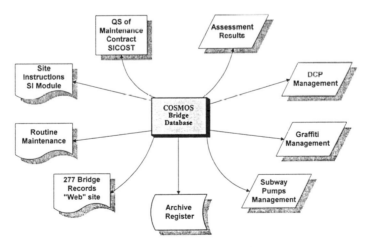

Maintenance Subsystems outside of COSMOS

INSPECTIONS

At the start of the financial year a full inspection programme is produced for the whole bridge stock. The system prints out paper BE11 inspection forms and risk assessment forms with all the relevant structure details filled in. Paper BE11's are used because we have not found a suitably reliable, user friendly, tough and cheap system that can withstand the rigours of the field while remaining useable. An inspection is carried out and the paperwork is brought back to the office where it is passed to a data entry clerk. The inspection is entered into the BMS and the paper copy is filed.

MAINTENANCE

Maintenance has historically been undertaken at the inspectors discretion, based on their knowledge of the area and the structures within it. This has not necessarily been a very efficient use of the maintenance budget, as an overall view of the County stock was not being taken.

The BMS system now runs a Maintenance Priority Factor (MPF) algorithm that produces a list of elements on structures County wide that require maintenance. This list has a highest priority of 3.7 and a lowest of 100. MPF's are based on element, location, the link that is carried and to some extent the importance of that link in relation to its peers.

The trouble with this list is that it doesn't really do anything. It can be printed out, but that is only the beginning of the maintenance process. Site instructions have to be written, the term maintenance contractor mobilised and when the works are completed the financial paperwork sorted out with the correct finance codes and the job billed to the correct client. Time is spent sorting these issues out, delays in updating the budget can occur and then suddenly you have no budget left.

The new system takes the best of both worlds. COSMOS calculates the MPF's and the data is transferred to a Site Instruction module where the list can be searched, modified and a complete set of site instructions produced. These SI's have all the correct financial coding, all the relevant structural information and extracts from the inspection relating to the works required. The engineer only has to add any minor comments or instructions of his own before the works can be given to the maintenance contractor.

The system currently gives priority to all structural elements with a MPF of less than 20. Any other works to the structure (up to a MPF value of 40) are included on the SI as a form of preventative maintenance. This range can be increased or decreased based on other issues or we can exclude certain structure types or even elements. This allows the flexibility to undertake maintenance within budgetary constraints.

However, the human element is not ignored. The engineer sees all the inspections and can edit the SI's that are inappropriate or wrong. This can be due to the sheer scale of the works required - replacing the bearings on a bridge over a dual carriageway or human error - inexperienced staff mis-coding the severity of a problem on a rights of way bridge.

Maintenance Subsystems

In addition to the main SI Module there are two sub systems that run specialised maintenance contracts outside of our biennial contract. These are an anti-graffiti patrol of subways and other structures and a contract for the maintenance of subway pumping systems.

Under these contracts the structures are visited at predefined intervals to undertake specific work: removal of graffiti or electrical and mechanical servicing of the pumps. Should emergencies occur; offensive graffiti or a flooded subway, the system will generate a request for action that can be faxed across to the contractor for resolution within 24 hours.

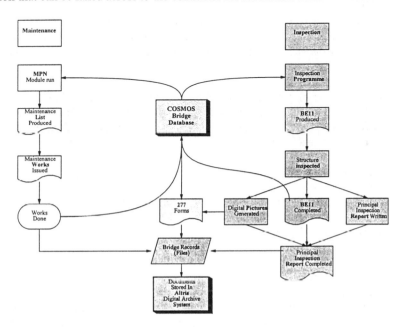

Inspection & Maintenance Cycle

COST CONTROL

Although the existing system does not have any facility for Cost Benefit Analysis or maintenance priority based on costs, it does allow for on-going cost monitoring. Costs associated with Site Instructions are entered into the SICOST module which allows us to see the money spent to date, the break down of costs by inspector, job type, or several other filters.

This module is part of the Bridge Management System, but is run by our Quantity Surveyors.

NEW TECHNOLOGY IN BRIDGE MANAGEMENT

New technology has been fully embraced by Surrey County Council as part of its Bridge Management philosophy. Some of the key components of this are detailed below.

Endoscope Unit

As part of the investigation of post tensioned structures that was instigated several years ago by the Highways Agency, Surrey purchased an endoscope unit. This is a remote controlled camera on a 3m long fibre optic link that can be steered by the operator. The camera is only 10mm wide and can be inserted into joints, weepholes or cracks to allow investigation of previously hidden details. Images can be stored on to video tape and digital stills produced for inclusion into reports or other documents.

Endoscope Unit

Although the Highways Agency programme has now finished, the unit is still used on a regular basis by Surrey staff and hired out to third parties for investigations. The unit is self contained and includes a remote viewing monitor allowing the engineer to sit in a warm car and watch the inspection while the inspector gets cold and wet on an underbridge unit. The system includes a boroscope which is designed specifically for investigating cavities in buildings.

Digital Cameras

Surrey were spending over £4000 per year on developing film from inspections. In addition to the developing costs, staff were involved in scanning the pictures into digital files for inclusion into reports. As a large numbers of these photographs would never make it in to the final report, a large proportion of the budget was being wasted solely on developing.

With the introduction of digital cameras, we have cut our photographic outlay by over 80%. Digital images can be downloaded from the camera into the computer, enhanced (lightened etc.) and then inserted directly into the document. The only outlay is for batteries, and by using rechargeables this cost becomes almost nothing. An interesting side effect of these digital images was the development of a photographic database of all the structures in Surrey. This system has formed the core of the new Intranet.

The use of digital cameras on site has also allowed us to produce CDROMs that contain before and after images of major schemes. One CD takes up considerable less room that photo albums and costs on average only £10 to produce.

CD ROM Technology
By combining the ease of use of a digital camera and a powerful word processor, high quality, visually useful reports can be produced. The finished reports, complete with photographs, maps etc. etc. can be stored on CD. Information from these reports can be copied into other documents or reused at the next principal inspection. One CD holds 650Mb of data. With an average of 15Mb per inspection report, over 40 reports can be stored on one CD.

By using CD's we can send a copy to our Client, keep a working copy in a central library for reference, keep a copy in a fire safe in the office and still send a copy off site for backup storage for a cost of about £40.

Intranet and Internet Technology
Taking best aspects of the technologies discussed above, digital images, storage of Principal Inspections as electronic media, geographic map systems and forge them together with Internet technologies, you can create a system for disseminating information across an organisation for very little capital outlay.

Most offices already have a large computer network throughout the building for day to day working. This same network is all that is needed as a basis for the Intranet. The Structures Intranet allows staff to search for any structure by number, name, owner or location. It will display photographs of the structure from a database of digital images, assessment information and produce a single A4 sized summary sheet that is incorporated into every new report relating to that structure. It can link to a digital map base for Statutory Undertakers and other geographic information.

The system links into the rest of the County Council Intranet (S-Net) allowing access to on-line documentation, discussion forums, staff notice boards and other relevant areas.

ADVANTAGES & DISADVANTAGES
Using the system we have developed over the past two years it has been possible to show an improvement in the overall condition of the county bridge stock. While this improvement is small, it has been possible to demonstrate it.

The weakest link in the whole inspection/maintenance/management cycle is the entry of BE11 inspection forms back into COSMOS. Any delays or inaccuracies at this stage can cause a problem in the maintenance part of the cycle. In practice it has been the delay in entering the inspection that has caused the most problems.

Inappropriate coding of defects on the BE11 form can also cause problems. These errors are not corrected at data input stage, but as the new Site Instructions are produced. The engineer will look at each SI at time of issue and thin out those that are inappropriate. Incorrect fault codes can be altered in COSMOS and will come in to affect on the next maintenance cycle.

THE FUTURE
Workstyle
Surrey County Council are in the process of implementing a radical change in the way that staff work. Staff are being encouraged to change their working habits, to work hours that are more suitable for their lifestyle and family arrangements. Offices around the County will be equipped in such a way that any member of staff can go into any office and work. Staff can work at home with their line managers approval or in the workstyle office in the nearest town.

County Chief Executive, Paul Coen, said: "This organisation is totally committed to improving services for our customers. Surrey Workstyle is an innovative and ambitious, long-term enterprise, which helps us to achieve our aim. The flexibility to work from any number of Workstyle offices across the county will bring greater opportunities to match our working day more closely with our own lifestyle."

3,400 staff across some 74 offices will see changes in their working environment and practices.

The proposals include:
- people will not be so tied to one desk or office.
- different Council services sharing the same buildings and resources, or being located with partner organisations.
- one headquarters and four area offices.
- up to 26 local offices.
- reduction, approaching 50%, of our current office accommodation.
- standardised workstations and information & communications technology (ICT): teleworking, video-conferencing and hot-desking will become commonplace.
- drop-in facilities at Workstyle offices so staff can work from various locations.

Workstyle is designed to improve the quality of service provided to Surrey residents while lowering the costs to the Council. Working in partnership with several large companies, Surrey plan to use the Workstyle ethos to reduce traffic congestion on the county roads at rush hour by up to 20%.

The Bridge Management Team is already extremely mobile. By equipping staff with equipment that not only fully embraces this philosophy, but actually extends it to embrace our own working practices, we can dramatically reduce the costs associated with the inspection process.

The laptop system now in use allows the inspector to use all of the software available in the office in the field. This includes standard word processor and spreadsheet packages, geographic information systems for locating structures, inspection software and access to e-mail facilities which allows contact with the rest of the team for retrieval of information.

The results from inspections can be mailed electronically back to the main office and loaded directly into COSMOS without the need for physical input by a member of staff. It will allow them to edit previous Principal Inspections on site, reducing the time spent on creating new documents from scratch.

CONCLUSIONS

We did not envisage creating a full intranet system from scratch when we started to run COSMOS. It allows structural information to be shared across our organisation almost instantly, but it has taken time to set up.

Without some form of BMS, there is no realistic way in which the overall condition of any bridge stock can be maintained or improved. The 'value for money' ethic now being extolled by central government has put a strain on our ability to deal with routine and emergency maintenance. With an increasing demand on our already limited resources, the need for the next century is not on building new infrastructure but on maintaining our existing systems. As engineers we are faced with the challenge of meeting the greater expectations of the public with less resources at our disposal.

A Bridge Management System should not take the decisions about the bridge stock away from the engineer. The problems associated with structures on the network are unique to each and as such, there cannot be a definitive rule for all structures. No computer is *currently* capable of dealing with the level of intuitive decisions that this would require. Given enough information, a computer could make general cost comparisons based on a simple set of predetermined values and rules, but we lack a large enough dataset to establish this information.

It remains the job of the engineer to assess the information that he or she is presented with. It follows that information is the key to everything. Without up to date information, even the best system in the world is rendered useless. This information does not need to be specific cost evaluations, it can simply be a list of structures that require maintenance based on an established priority algorithm.

We use the BMS to manage the stock by controlling the flow of information through the organisation; we do not let the BMS manage the stock for us. The decisions are still made by experienced engineers but using the information that the system produces. The skill in developing and managing a Bridge Management System lies not in complicated programming or procedures, but in careful control of the information. With the right information at the right time it becomes possible to ensure that money is being spent when and where it is needed. Surely at its most basic level, this is all that a BMS is designed to do.

REFERENCES

1.   Surrey County Council: *Transport Policies & Programme 1998-1999*
2.   Das, P: *Development of a comprehensive structures management methodology for the Highways Agency*. The Management of Highway Structures Conference 1998
3.   Lauridsen, J and Lassen, B: *The Danish Bridge Management System (DANBRO)*. The Management of Highway Structures Conference 1998
4.   Palmer, J and Cogswell, G: *Management of the Bridge Stock of a UK County for the 1990's*. First International Conference on Bridge Management 1990.

ACKNOWLEDGEMENT

The permission of Mr.Callum Findlay, Head of Engineering, Surrey County Council Environment, to publish this paper is gratefully acknowledged.

Managing bridge management

C.R. HUTCHINSON, J.W. BOYD and J. KIRKPATRICK,
DoE (NI) Roads Service, Northern Ireland.

SYNOPSIS

In Northern Ireland the Roads Service, an Agency within the Department of the Environment, has sole responsibility for the management of all public road bridges. With a total of approximately 6392, there is a considerable volume of information to be managed. The effective management of these bridges depends on the proper management of the relevant information.

In 1998 Roads Service developed in-house a Microsoft Access based Bridge Management System. This paper traces the development of the system, from its inception to the current state of successful operation. The thought processes behind the development, highlighting hardware and software issues that had to be overcome to produce an integrated system are discussed. This paper will hopefully guide others who are considering developing their own systems.

INTRODUCTION

In 1973, local Government reorganisation in Northern Ireland saw the dissolution of County Councils with responsibility for the various functions of those Councils being passing to Government Departments. Roads Service, now an Agency within the Department of the Environment, became the sole road authority in Northern Ireland, with Headquarters in Belfast and six regional offices, called Divisions, located across the province.

As part of its responsibilities Roads Service took on board over 6392 bridges on both trunk and non-trunk roads. That bridge stock is comprised of 64% masonry structures, 23% concrete structures, with the remainder being of steel or composite construction.
In 1990 the Roads Service acquired a new bridge management system, which combined a records database for trunk road bridges with a compatible on site inspection module. It was a significant step forward in terms of how much information could be held on any one bridge, and was considered to provide best value at that time.

However during the early 1990s advances in technology led to a culture where information became a valuable commodity and where ease of access to more information presented opportunities for improving the delivery of services. For example budgets could be more appropriately targeted, estimates more accurately predicted and results more closely monitored. The increasing access to more friendly Graphical User Interfaces running on much faster computers led to ever increasing expectations in the availability and processing of information. This was no different for engineers in Roads Service involved in the management of bridges.

Given what had become standard by the mid nineties, it was inevitable that users of our bridge management system became increasingly frustrated with a system which was proving to be both unfriendly and slow to use. Subsequently use of the system virtually ceased with users reverting to simple spreadsheets and in some cases paper systems, to store and retrieve basic information. Obviously the time had come to review the bridge management system.

THE SEARCH FOR A BETTER SYSTEM

In compliance with Government initiatives Roads Service, in seeking best value, had established Continuous Improvement Project procedures, whereby all perceived improvements were formally assessed.

In March 1998, a Continuous Improvement Project was initiated and an in-house team comprised of system users was tasked with making a recommendation on how best to address bridge management within Roads Service.

In order to assess the extent of the perceived problems with the existing system, the team sent a questionnaire to all users and potential users. The response from that survey identified the following four main issues: -

- The system was cumbersome to use and interrogate
- Access to and response from the system was slow
- Few users had confidence in the accuracy of output
- 80% of users did not use the system as their primary source of bridge information. Most used some form of paper system.

The overall summary (Fig. 1) shows that 64% held negative views about the system with only 12% holding positive views. These views applied equally to the main database and the on-site inspection module.

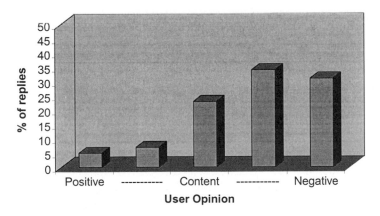

Overall feeling of users about the existing bridge management system.

Fig. 1

We investigated five basic options as to how we could progress, namely: -
- Do Nothing
- Update the existing system
- Purchase an Alternative 'Off the Shelf' Package
- Third party Bespoke System
- In-house bespoke system

The team drew up the following shopping list of essential and/or desirable criteria against which each option would be assessed and compared: -
- Meet the particular operational requirements of Roads Service users
- Easy to amend and update
- User friendly and easy to interrogate
- Windows based
- Readily available System support
- GIS compatible
- Support the storage of drawings and photographs
- Require minimum staff training

Consideration also had to be taken of the cost implication of each option, including the long-term operation and maintenance.

Do Nothing
Given the background to the project, that easy access to a significant amount of information was necessary to manage our bridges, the "do nothing" option was discarded immediately

Update the existing system
Assessment of an upgrade to the existing system against our shopping list showed that it would provide only short-term minimal improvement and was also discarded.

Purchase an Alternative 'Off the Shelf' Package
"Off the shelf" packages, which varied in functionality, scored well in many of the attributes, but suffered in that they had not been developed specifically for Roads Service. Tailoring these systems would have been possible but at further, and significant, cost.

Third party Bespoke System
With a third party developed bespoke system the user is not the system developer, thus the result is an interpretation by the developer of the end users brief. Given the previous experiences within the team and the fact that procurement would necessarily have been through competitive tendering, it was felt that there was significant potential for cost overruns and for functionality to be less than satisfactory - particularly if the developer was not fully familiar with the management of bridges. Bespoke systems are expensive, as there is only one client and a very limited market across which the developer might spread the cost. Additionally Roads Service was preparing to go through a radical restructuring with the result that further amendments and refinements would be required shortly after implementation.

In-house bespoke system
The in-house development of a bespoke system would not always be a viable option given the need to have expertise in both bridge management and database development. However we were fortunate to have a member of the team with expertise in both which meant that this option scored highly against all eight criteria. We also had the advantage of knowing the skill levels of the end users and could therefore steer the development with their specific capabilities in mind.

SYSTEM DEFINITION AND CHOICE OF DBMS (Database Management System)
The environment necessary for development of the system was dictated by the list of essential/desirable criteria. Inexpensive to develop, easy to amend and update, windows based and system support readily available all pointed to a mainstream database development tool. Using such a tool would have the added benefit that others with similar development expertise, either within or outside Roads Service, could make amendments if necessary.

The selected DBMS would have to support the future integration with GIS as this was seen to offer substantial benefits for the management of bridges.

While there were various suitable database development products, such as Foxpro and Microsoft SQL Server, Microsoft Access was our preferred option as it had been installed as standard on all users PC's. This option also eliminated the need for any additional licensing costs and meant that the various levels of security ranging from data input only through to full administration could be provided.

Windows NT was the selected Operating System as it is capable of coping with the large number of files associated with the storage of drawings and photographs. The project team thus concluded that a new Bridge Management System developed in-house on Windows NT based Microsoft Access offered the best value. This recommendation was presented to the Roads Service Quality Council and approval was subsequently given to develop the system.

REQUIREMENTS ANALYSIS AND NETWORK ISSUES
The first step was to determine how many fields of information should be held for each bridge. Reaching the right balance was central to the success or otherwise of the whole project. With too little information it would be ineffective and of little use, whereas too many fields would introduce complexities comparable with the previous system.

The agreed way forward was to prepare a draft list of required fields, produce a mock up of the interface for the viewing and adding of bridge information, and present this to the users for comment and consideration. The mock up did not have any functionality menus, its sole purpose being to demonstrate to users the style of the application and the fields that were proposed.
From that presentation a final list of over 100 data fields was agreed. The team was encouraged by the fact that all users who saw the presentation were impressed with the interface and saw the potential of the new system, even at this very early stage.

By this time the restructuring of the Roads Service into Client and Consultant Provider and from 6 Divisions to 4 was about to take place. This raised the question of access for the various users.

While any one Division would not require access to information in another Division, Headquarters would require access to some of the information in all Divisions.

As all Roads Service offices are linked through a Wide Area Network, (WAN), one option was to copy the set up used for the previous system where the database was held on a central server at Headquarters, and accessed through the WAN by the various Divisional users. However the large volumes of network traffic being generated throughout the Roads Service caused access to be slow and erratic. That set up also suffered from a lack of ownership.

It was a requirement of the system that photographs and drawings should be easily accessed. Whilst the availability, at reasonable cost, of large capacity hard discs meant that the storage of photographs and drawings was an economically viable option, the problems with WAN performance dictated that these should be held locally rather than centrally.

Thus we chose the option whereby each Division would hold their own data on a local machine, with Headquarters up loading data as necessary. With the up load being an infrequent operation, the net benefit in overall user time would be significant.

THE DEVELOPMENT OF THE DATABASE
Having decided upon a DBMS and prepared a user requirements/design specification it was time to develop the application. The development of the Database Management system was broken down into three essential elements: -
- Data Input
- Data Query and Viewing
- Reporting

Data Input
A major feature, which was considered to be essential to user friendliness, was that all information on a particular bridge would be available on one screen. Generally database systems are developed to display of the order of 10 fields of data per screen. This keeps the volume of data to a level where the speed of transfer over a WAN, between the server and the user PC, does not result in an unacceptable delay to the user. However this requires a cascade of screens to view all the fields.

In our case, with over 100 fields to be displayed, the use of one screen was not a viable option. While it would be theoretically possible to design such a screen, in practice it would be impossible to read. Additionally even though we were proposing to use stand-alone workstations, with the possibility of LAN connections, the time taken to display such a screen would still be too slow to be acceptable to the user. Thus having already decided not to use the conventional method of data display we had to develop an alternative method. The solution adopted was to provide access to all the fields on one screen by using a series of tab controls to display groups of associated fields. The top of the screen displayed the basic details of a bridge with the lower part of the screen consisting of a series of overlapping tab pages, each holding specific details (Fig.2).

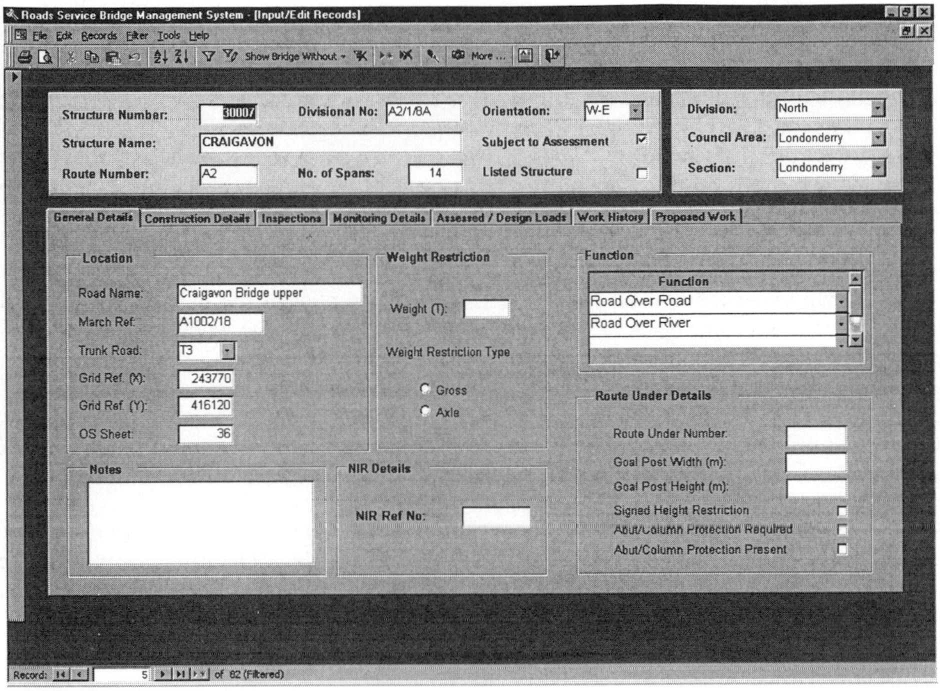

Figure 2

This gives the user the impression that all the data is being loaded when the form is initially opened, whereas the data on each of the tabs is not loaded until the tab is selected. As such, if the user does not select a tab the associated data is not loaded - thus saving time.

Producing this single interface screen was laborious, but it was essential that it should be visually acceptable, yet remain simple to operate, if the system was to meet the needs of the end user. This same screen format was used for data input, viewing and querying.

Data Query
Development of the query interface would prove to be the most complex and challenging stage of the system development. A query involving data in a single table is straightforward. If there are two or more tables queries have to contain the correct join syntax.

Queries contain many logic statements such as AND and OR. The relationship between even these two simple logic statements can become complex leading to confusion and often-incorrect database querying. Thus, unlike other applications on commercially available databases, we designed a query interface that removed, as far as possible, the users need to set logic statements. The search interface screen looks similar to that of the input form. The search criterion is simply selected from or typed into the relevant field boxes and the search carried out by selecting the search button. A limited number of key words and logic statement options are available to allow for more complex searches if the user so desires.

The query module was developed to read the user entered search criteria and build an SQL (Structured Query Language) statement taking account of the tables containing the fields being queried and thus the joins required. Those records, ie bridges in this database, meeting the search criteria are known as the found set. The first record in a found set is displayed on the view screen and the user is able to page through the remainder that are displayed in turn.

Reporting
The user does not generally wish to go through each record in a found set, but rather requires a report displaying particular details of each record. For example the user might wish to visit all concrete bridges having a span equal to 10m. The report however would possibly only need to list the name and grid reference of each bridge.

Effective report building requires a sound knowledge of the reporting tool and the data structure within a database. Since it was unlikely that users would have that level of knowledge and expertise, a wide range of "standard" reports were built, covering the most common combinations of fields against which reporting would be required. The user simply selects the report that most closely matches the output requirements. If necessary new reports can be quickly generated and added to each Divisional system.

Having developed the main modules they were interconnected by a series of user friendly menus to provide easy navigation between each one.

ADDITIONAL FUNCTIONALITY
At this stage a number of additional features were developed to enhance the system.
Frequently requested reports such as bridge condition priority listings, inspection programmes and the like have been set up as standard queries, thus providing quick results with limited input required from the user.

As GIS was not being made available immediately, a function whereby the system pulls up all bridges having a particular overall condition factor and within a user defined radius of a given bridge, has been developed. This feature aids programming of work.

A major feature of the system is the ability to quickly view digital photographs, drawings and other documents related to. This has been found to greatly improve the understanding and technical appreciation of the data presented.

Electronic on-site capture of inspection data offered a further opportunity to enhance the effectiveness of the system by offering time savings, minimisation of errors and consistency of approach. The previous on-site data capture unit failed for the same reasons as the main database in that it was cumbersome and slow, both in terms of entering inspections and downloading to the main system. Any new system would have to avoid these downfalls. A considerable amount of time was put into sourcing a suitable device that would facilitate fast data entry whilst being able to withstand the rough conditions encountered during bridge inspection. Due to memory limitations, data capture devices generally require specialist, and thus expensive programming. However the hardware selected uses a Windows 95 Operating System, allowing development by conventional programming tools such as Microsoft Access and a pen-pointing device that radically speeds up the input of data.

IMPLEMENTATION

Having developed the application, testing and debugging followed. Data was entered and all the functions of the system were tested repeatedly over a period of 3 months in one Division Division. This revealed a number of bugs that were duly fixed before the application was released to other users. To ease the future debugging, the system logs all errors that occur and the circumstances under which they occur. This allows the system administrator to view the error message displayed to the user when the error occurred.

Prior to release of the system to the Divisions, training was carried out to encourage full and effective use. Electronic and hard copy manuals were produced.

CONCLUSION

The Roads Service Bridge Management System (RSBMS) as it is now known is in use throughout the Roads Service in Northern Ireland and has rapidly become the primary source of bridge information.

While the in-house development of this database has been successful it required a systematic approach and considerable in-house expertise and resource. The benefits however are equally significant in that, having been tailored to meet the needs and capabilities of the users, the system has been adopted by them. This is seen to augur well for its long-term success and consequently the effective management of our bridges. User feedback about the performance and functionality is encouraged in order to identify possible improvements or enhancements.

Like most things in life, the work environment is a dynamic and ever changing place. What was acceptable or the norm quickly becomes dated or obsolete. As reflections of the work environment, database applications almost by definition are subject to change or improvement. A well designed and implemented application should cope with change and modification without the threat of replacement or early obsolescence.

ACKNOWLEDGEMENTS

The authors would like to thank Mr V C Crawford, Director of Engineering of the DoE (NI) Roads Service for permission to publish this paper, Mr R F S McCandless, Head of Roads Service Consultancy and Mr J Irvine Roads Service Head Quarters for their encouragement and advice, and all other members of Roads Service staff who gave us their support.

Views expressed are those of the authors, and not necessarily those of Roads Service.